Environmental Toxicology and Human Health

Environmental Toxicology and Human Health

Editors

Esref Demir
Sam Kacew

MDPI

Basel • Beijing • Wuhan • Barcelona • Belgrade • Novi Sad • Cluj • Manchester

Editors
Esref Demir
Department of Medical Services and Techniques, Vocational School of Health Services, Antalya Bilim University
Antalya, Turkey

Sam Kacew
R. Samuel McLaughlin Centre for Population Health Risk Assessment, Institute of Population Health, University of Ottawa
Ottawa, Canada

Editorial Office
MDPI
St. Alban-Anlage 66
4052 Basel, Switzerland

This is a reprint of articles from the Topic published online in the open access journals *International Journal of Molecular Sciences* (ISSN 1422-0067), *International Journal of Environmental Research and Public Health* (ISSN 1660-4601), *Cells* (ISSN 2073-4409), and *Toxics* (ISSN 2305-6304) (available at: https://www.mdpi.com/topics/EnvironmentalToxicologyHumanHealth).

For citation purposes, cite each article independently as indicated on the article page online and as indicated below:

Lastname, A.A.; Lastname, B.B. Article Title. *Journal Name* **Year**, *Volume Number*, Page Range.

ISBN 978-3-0365-9983-0 (Hbk)
ISBN 978-3-0365-9984-7 (PDF)
doi.org/10.3390/books978-3-0365-9984-7

Contents

International Journal of
Molecular Sciences

MDPI

Editorial

Environmental Toxicology and Human Health

Esref Demir [1,2,3,*] and Sam Kacew [4,*]

1 Department of Medical Services and Techniques, Medical Laboratory Techniques Programme, Vocational School of Health Services, Antalya Bilim University, 07190 Antalya, Turkey
2 F.M. Kirby Neurobiology Center, Boston Children's Hospital, 300 Longwood Avenue, Boston, MA 02115, USA
3 Department of Neurobiology, Harvard Medical School, Boston, MA 02115, USA
4 Institute of Population Health, R. Samuel McLaughlin Centre for Population Health Risk Assessment, University of Ottawa, Ottawa, ON K1N 6N5, Canada
* Correspondence: esref.demir@antalya.edu.tr (E.D.); sam.kacew@uottawa.ca (S.K.)

Humans and animals may be exposed on a continuous daily basis to a mixture of environmental contaminants that may act on several organ systems through differing mechanisms [1] resulting in adverse consequences. Environmental contamination now constitutes a major global issue with adverse effects on health of the ecosystem and food security. Globally, air pollution alone produces millions of premature deaths annually, predominantly associated with from lung cancer, chronic obstructive pulmonary disease (COPD), asthma, stroke, heart failure, and respiratory infections, according to the World Health Organization (WHO) [2]. It is noteworthy that 99% of humanity breathes air containing contaminants above recommended levels.

In order to mitigate contamination and diminish e our burden of pollutant -related diseases, we need to devise target-specific strategies to prevent or decrease exposure. To that end, risk assessment attributed to exposure to synthetic or and naturally occurring contaminants is necessary and; thus evidence obtained from toxicity studies appears to be of critical importance. Comprehensive efforts need to be undertaken to search for possible underlying mechanisms of action for each pollutant to establish toxic potential and safe limits through both in vitro and in vivo animal testing approaches. This issue focused on environmental pollutants including heavy metals, pesticides, nanoparticles, micro-nanoplastics, indoor air pollutants, pharmaceuticals, and industrial toxicants with effects on human health, risk assessment, and relationship between various diseases and environmental pollutants. Human exposure to environmental pollutants may initiate adverse effects including neurotoxicity, carcinogenicity, infertility, and metabolic disorders. Therefore, research into possible mechanisms of action for environmental contaminants is of critical importance for the well-being of humans and animals [3].

Over the last couple of decades, novel in vitro and in vivo methods and techniques were developed in the scientific discipline genotoxicology, enabling investigators to quantify genotoxicity attributed to exposure to certain compounds [4,5]. Acute or chronic exposure to environmental contaminants is known to be associated with several adverse health conditions, including cancer, impaired immune and reproductive function, as well as imbalanced gastrointestinal microbiota, which regulates a range of host metabolic and immune processes. The aims of this topic are to present a comprehensive overview of different studies carried out with in vivo and in vitro model organisms and the potential risk of environmental pollutants exposure to human health. In this Topic, 20 original articles, 6 reviews and 1 communication were collected, as presented in Table 1 with a particular focus on alcohol-based hand sanitizers, polycyclic aromatic hydrocarbons, monochromatic light pollution, paraben as an endocrine disruptors, heavy metal pollution attributed to antimony and arsenic of mines in the soil, water, and sediments, groundwater with high fluoride, virus transmission from heating, ventilation, and air conditioning systems of urban subways, chronic home radon exposure, organotin compounds, heavy metal

Citation: Demir, E.; Kacew, S. Environmental Toxicology and Human Health. *Int. J. Mol. Sci.* **2024**, *25*, 555. https://doi.org/10.3390/ijms25010555

Received: 27 December 2023
Accepted: 27 December 2023
Published: 31 December 2023

pollutants including mercury, lead, cadmium, polypropylene microplastics, ventral body wall defects in chick embryos, microcystin-LR as an aquatic toxin, N-nitroso compounds, methylmercury as a global pollutant, triazine herbicides, persistent organic pollutants, bisphenol A and trace metals, autophagy, nano- and micro-sized polystyrene particles, tributyltin as an environmental contaminant, polybrominated diphenyl ethers, and per- and polyfluoralkyl substances. Most of the examined compounds originated from natural sources, whereas some semi-synthetic derivatives were also identified and discussed. The most recent findings on the effects of compounds and their constituents in treating various toxic outcomes and genotoxicity are discussed. These studies summarize our current knowledge based upon previous in vitro and in vivo research that scrutinized the influence of several environmental contaminants on various mammalian and non-target model organisms at several genetic, cellular, and molecular levels, as well as potential mechanisms underlying toxicity.

Table 1. Original articles and reviews collected in the six journals participating in the Topic using different in vitro and in vivo model systems.

Title	Author	Journal	Year	DOI
Evaluation of the Safety and Efficacy of Hand Sanitizer Products Marketed to Children Available during the COVID-19 Pandemic	[6]	*IJERPH*	2022	https://doi.org/10.3390/ijerph192114424 (accessed on 1 February 2022)
Health Risk Assessment of Dermal Exposure to Polycyclic Aromatic Hydrocarbons from the Use of Infant Diapers	[7]	*IJERPH*	2022	https://doi.org/10.3390/ijerph192214760 (accessed on 1 February 2022)
Monochromatic Light Pollution Exacerbates High-Fat Diet-Induced Adipocytic Hypertrophy in Mice	[8]	*Cells*	2022	https://doi.org/10.3390/cells11233808 (accessed on 1 February 2022)
Impact of Paraben Exposure on Adiposity-Related Measures: An Updated Literature Review of Population-Based Studies	[9]	*IJERPH*	2022	https://doi.org/10.3390/ijerph192316268 (accessed on 1 February 2022)
Leaching Mechanism and Health Risk Assessment of As and Sb in Tailings of Typical Antimony Mines: A Case Study in Yunnan and Guizhou Province, Southwest China	[10]	*Toxics*	2022	https://doi.org/10.3390/toxics10120777 (accessed on 1 February 2022)
Relationship of Fluoride Concentration to Well Depth in an Alluvial Aquifer in a Semiarid Area	[11]	*Environments*	2022	https://doi.org/10.3390/environments9120155 (accessed on 1 February 2022)
Reducing Virus Transmission from Heating, Ventilation, and Air Conditioning Systems of Urban Subways	[12]	*Toxics*	2022	https://doi.org/10.3390/toxics10120796 (accessed on 1 February 2022)
Chronic Home Radon Exposure Is Associated with Higher Inflammatory Biomarker Concentrations in Children and Adolescents	[13]	*IJERPH*	2023	https://doi.org/10.3390/ijerph20010246 (accessed on 1 February 2022)
Organotin Antifouling Compounds and Sex-Steroid Nuclear Receptor Perturbation: Some Structural Insights	[14]	*Toxics*	2023	https://doi.org/10.3390/toxics11010025 (accessed on 1 February 2022)
Health Risk Assessment for Human Exposure to Heavy Metals via Food Consumption in Inhabitants of Middle Basin of the Atrato River in the Colombian Pacific	[15]	*IJERPH*	2023	https://doi.org/10.3390/ijerph20010435 (accessed on 1 February 2022)
Exposure to Polypropylene Microplastics via Oral Ingestion Induces Colonic Apoptosis and Intestinal Barrier Damage through Oxidative Stress and Inflammation in Mice	[16]	*Toxics*	2023	https://doi.org/10.3390/toxics11020127 (accessed on 1 February 2022)
Y-27632 Impairs Angiogenesis on Extra-Embryonic Vasculature in Post-Gastrulation Chick Embryos	[17]	*Toxics*	2023	https://doi.org/10.3390/toxics11020134 (accessed on 1 February 2022)
Downregulation of LncRNA GCLC-1 Promotes Microcystin-LR-Induced Malignant Transformation of Human Liver Cells by Regulating GCLC Expression	[18]	*Toxics*	2023	https://doi.org/10.3390/toxics11020162 (accessed on 1 February 2022)
Association of Dietary Nitrate, Nitrite, and N-Nitroso Compounds Intake and Gastrointestinal Cancers: A Systematic Review and Meta-Analysis	[19]	*Toxics*	2023	https://doi.org/10.3390/toxics11020190 (accessed on 1 February 2022)
Subchronic Low-Dose Methylmercury Exposure Accelerated Cerebral Telomere Shortening in Relevant with Declined Urinary aMT6s Level in Rats	[20]	*Toxics*	2023	https://doi.org/10.3390/toxics11020191 (accessed on 1 February 2022)

Table 1. *Cont.*

Title	Author	Journal	Year	DOI
Triazine Herbicides Risk Management Strategies on Environmental and Human Health Aspects Using In-Silico Methods	[21]	*IJMS*	2023	https://doi.org/10.3390/ijms24065691 (accessed on 1 February 2022)
Development of an Improved Sulfur-Oxidizing Bacteria-Based Ecotoxicity Test for Simple and Rapid On-Site Application	[22]	*Toxics*	2023	https://doi.org/10.3390/toxics11040352 (accessed on 1 February 2022)
A Realistic Mixture of Persistent Organic Pollutants Affects Zebrafish Development, Behavior, and Specifically Eye Formation by Inhibiting the Condensin I Complex	[23]	*Toxics*	2023	https://doi.org/10.3390/toxics11040357 (accessed on 1 February 2022)
Protective Effects of Selenium Nanoparticles against Bisphenol A-Induced Toxicity in Porcine Intestinal Epithelial Cells	[24]	*IJMS*	2023	https://doi.org/10.3390/ijms24087242 (accessed on 1 February 2022)
Drosophila as a Robust Model System for Assessing Autophagy: A Review	[25]	*Toxics*	2023	https://doi.org/10.3390/toxics11080682 (accessed on 1 February 2022)
Uptake of Breathable Nano- and Micro-Sized Polystyrene Particles: Comparison of Virgin and Oxidised nPS/mPS in Human Alveolar Cells	[26]	*Toxics*	2023	https://doi.org/10.3390/toxics11080686 (accessed on 1 February 2022)
Environmental Health and Toxicology: Immunomodulation Promoted by Endocrine-Disrupting Chemical Tributyltin	[27]	*Toxics*	2023	https://doi.org/10.3390/toxics11080696 (accessed on 1 February 2022)
Toxic Effects and Mechanisms of Polybrominated Diphenyl Ethers	[28]	*IJMS*	2023	https://doi.org/10.3390/ijms241713487 (accessed on 1 February 2022)
Maternal Serum Concentrations of Per- and Polyfluoroalkyl Substances in Early Pregnancy and Small for Gestational Age in Southern Sweden	[29]	*Toxics*	2023	https://doi.org/10.3390/toxics11090750 (accessed on 1 February 2022)
Transfer of Bisphenol A and Trace Metals from Plastic Packaging to Mineral Water in Ouagadougou, Burkina Faso	[30]	*IJERPH*	2023	https://doi.org/10.3390/ijerph20206908 (accessed on 1 February 2022)
Environmental Endocrinology: Parabens Hazardous Effects on Hypothalamic–Pituitary–Thyroid Axis	[31]	*IJMS*	2023	https://doi.org/10.3390/ijms242015246 (accessed on 1 February 2022)
Mixture Effects of Bisphenol A and Its Structural Analogs on Estrogen Receptor Transcriptional Activation	[32]	*Toxics*	2023	https://doi.org/10.3390/toxics11120986 (accessed on 1 February 2022)

Table 1 schematically illustrates the content of this Topic, with all the contributions published in the five participating journals.

Author Contributions: Writing, review and editing, E.D. and S.K. All authors have read and agreed to the published version of the manuscript.

Conflicts of Interest: The authors declare no conflict of interest.

References

1. Demir, E.; Turna Demir, F. *Drosophila*: A promising model for evaluating the toxicity of environmental pollutants. *Karaelmas Sci. Eng. J.* **2022**, *12*, 101–118.
2. WHO Global Health Statistics. WHO Resources on Sound Management of Pesticides. 2015. Available online: https://www.who.int/neglected_diseases/vector_ecology/pesticide-management/en/ (accessed on 1 February 2022).
3. Bianchi, J.; Mantovani, M.S.; Marin-Morales, M.A. Analysis of the genotoxic potential of low concentrations of malathion on the *Allium cepa* cells and rat hepatoma tissue culture. *J. Environ. Sci.* **2015**, *36*, 102–111. [CrossRef] [PubMed]
4. Salk, J.J.; Kennedy, S.R. Next-generation genotoxicology: Using modern sequencing technologies to assess somatic mutagenesis and cancer risk. *Environ. Mol. Mutagen.* **2020**, *61*, 135–151. [CrossRef] [PubMed]
5. IARC (International Agency for Research on Cancer). IARC Monographs on the Evaluation of Carcinogenic Risks to Humans. Chemical Agents and Related Occupations. 2012, Volume 100, pp. 1–599. Available online: https://publications.iarc.fr/Book-And-Report-Series/Iarc-Monographs-On-The-Identification-Of-Carcinogenic-Hazards-ToHumans/Chemical-Agents-And-Related-Occupations-2012 (accessed on 1 February 2022).
6. Gloekler, L.E.; de Gandiaga, E.J.; Binczewski, N.R.; Steimel, K.G.; Massarsky, A.; Kozal, J.; Vincent, M.; Zisook, R.; LaGuardia, M.J.; Dotson, S.; et al. Evaluation of the Safety and Efficacy of Hand Sanitizer Products Marketed to Children Available during the COVID-19 Pandemic. *Int. J. Environ. Res. Public Health* **2022**, *19*, 14424. [CrossRef] [PubMed]

7. Bernard, A.; Dudler, V. Health Risk Assessment of Dermal Exposure to Polycyclic Aromatic Hydrocarbons from the Use of Infant Diapers. *Int. J. Environ. Res. Public Health* **2022**, *19*, 14760. [CrossRef] [PubMed]
8. Guan, Q.; Li, Y.; Wang, Z.; Cao, J.; Dong, Y.; Ren, F.; Chen, Y. Monochromatic Light Pollution Exacerbates High-Fat Diet-Induced Adipocytic Hypertrophy in Mice. *Cells* **2022**, *11*, 3808. [CrossRef] [PubMed]
9. Xu, X.; Wu, H.; Terry, P.D.; Zhao, L.; Chen, J. Impact of Paraben Exposure on Adiposity-Related Measures: An Updated Literature Review of Population-Based Studies. *Int. J. Environ. Res. Public Health* **2022**, *19*, 16268. [CrossRef]
10. Bai, Z.; He, Y.; Han, Z.; Wu, F. Leaching Mechanism and Health Risk Assessment of As and Sb in Tailings of Typical Antimony Mines: A Case Study in Yunnan and Guizhou Province, Southwest China. *Toxics* **2022**, *10*, 777. [CrossRef]
11. Espino-Valdés, M.S.; Rodríguez-Lozano, D.F.; Gutiérrez, M.; Silva-Hidalgo, H.; Pinales-Munguía, A. Relationship of Fluoride Concentration to Well Depth in an Alluvial Aquifer in a Semiarid Area. *Environments* **2022**, *9*, 155. [CrossRef]
12. Nazari, A.; Hong, J.; Taghizadeh-Hesary, F.; Taghizadeh-Hesary, F. Reducing Virus Transmission from Heating, Ventilation, and Air Conditioning Systems of Urban Subways. *Toxics* **2022**, *10*, 796. [CrossRef]
13. Taylor, B.K.; Smith, O.V.; Miller, G.E. Chronic Home Radon Exposure Is Associated with Higher Inflammatory Biomarker Concentrations in Children and Adolescents. *Int. J. Environ. Res. Public Health* **2023**, *20*, 246. [CrossRef] [PubMed]
14. Beg, M.A.; Beg, M.A.; Zargar, U.R.; Sheikh, I.A.; Bajouh, O.S.; Abuzenadah, A.M.; Rehan, M. Organotin Antifouling Compounds and Sex-Steroid Nuclear Receptor Perturbation: Some Structural Insights. *Toxics* **2023**, *11*, 25. [CrossRef] [PubMed]
15. Caicedo-Rivas, G.; Salas-Moreno, M.; Marrugo-Negrete, J. Health Risk Assessment for Human Exposure to Heavy Metals via Food Consumption in Inhabitants of Middle Basin of the Atrato River in the Colombian Pacific. *Int. J. Environ. Res. Public Health* **2023**, *20*, 435. [CrossRef] [PubMed]
16. Jia, R.; Han, J.; Liu, X.; Li, K.; Lai, W.; Bian, L.; Yan, J.; Xi, Z. Exposure to Polypropylene Microplastics via Oral Ingestion Induces Colonic Apoptosis and Intestinal Barrier Damage through Oxidative Stress and Inflammation in Mice. *Toxics* **2023**, *11*, 127. [CrossRef] [PubMed]
17. Duess, J.W.; Gosemann, J.-H.; Kaskova Gheorghescu, A.; Puri, P.; Thompson, J. Y-27632 Impairs Angiogenesis on Extra-Embryonic Vasculature in Post-Gastrulation Chick Embryos. *Toxics* **2023**, *11*, 134. [CrossRef] [PubMed]
18. Huang, X.; Su, Z.; Li, J.; He, J.; Zhao, N.; Nie, L.; Guan, B.; Huang, Q.; Zhao, H.; Lu, G.-D.; et al. Downregulation of LncRNA GCLC-1 Promotes Microcystin-LR-Induced Malignant Transformation of Human Liver Cells by Regulating GCLC Expression. *Toxics* **2023**, *11*, 162. [CrossRef] [PubMed]
19. Seyyedsalehi, M.S.; Mohebbi, E.; Tourang, F.; Sasanfar, B.; Boffetta, P.; Zendehdel, K. Association of Dietary Nitrate, Nitrite, and N-Nitroso Compounds Intake and Gastrointestinal Cancers: A Systematic Review and Meta-Analysis. *Toxics* **2023**, *11*, 190. [CrossRef]
20. Wu, X.; Li, P.; Tao, J.; Chen, X.; Zhang, A. Subchronic Low-Dose Methylmercury Exposure Accelerated Cerebral Telomere Shortening in Relevant with Declined Urinary aMT6s Level in Rats. *Toxics* **2023**, *11*, 191. [CrossRef]
21. Yao, T.; Sun, P.; Zhao, W. Triazine Herbicides Risk Management Strategies on Environmental and Human Health Aspects Using In-Silico Methods. *Int. J. Mol. Sci.* **2023**, *24*, 5691. [CrossRef]
22. Eom, H. Development of an Improved Sulfur-Oxidizing Bacteria-Based Ecotoxicity Test for Simple and Rapid On-Site Application. *Toxics* **2023**, *11*, 352. [CrossRef]
23. Guerrero-Limón, G.; Nivelle, R.; Bich-Ngoc, N.; Duy-Thanh, D.; Muller, M. A Realistic Mixture of Persistent Organic Pollutants Affects Zebrafish Development, Behavior, and Specifically Eye Formation by Inhibiting the Condensin I Complex. *Toxics* **2023**, *11*, 357. [CrossRef] [PubMed]
24. Pan, Z.; Huang, J.; Hu, T.; Zhang, Y.; Zhang, L.; Zhang, J.; Cui, D.; Li, L.; Wang, J.; Wu, Q. Protective Effects of Selenium Nanoparticles against Bisphenol A-Induced Toxicity in Porcine Intestinal Epithelial Cells. *Int. J. Mol. Sci.* **2023**, *24*, 7242. [CrossRef] [PubMed]
25. Demir, E.; Kacew, S. *Drosophila* as a Robust Model System for Assessing Autophagy: A Review. *Toxics* **2023**, *11*, 682. [CrossRef] [PubMed]
26. Laganà, A.; Visalli, G.; Facciolà, A.; Celesti, C.; Iannazzo, D.; Di Pietro, A. Uptake of Breathable Nano- and Micro-Sized Polystyrene Particles: Comparison of Virgin and Oxidised nPS/mPS in Human Alveolar Cells. *Toxics* **2023**, *11*, 686. [CrossRef] [PubMed]
27. da Silva, R.C.; Teixeira, M.P.; de Paiva, L.S.; Miranda-Alves, L. Environmental Health and Toxicology: Immunomodulation Promoted by Endocrine-Disrupting Chemical Tributyltin. *Toxics* **2023**, *11*, 696. [CrossRef] [PubMed]
28. Xue, J.; Xiao, Q.; Zhang, M.; Li, D.; Wang, X. Toxic Effects and Mechanisms of Polybrominated Diphenyl Ethers. *Int. J. Mol. Sci.* **2023**, *24*, 13487. [CrossRef] [PubMed]
29. Malm, E.; Vilhelmsson, A.; Högfeldt, H.; Deshayes, I.; Källén, K.; Hansson, S.R.; Lindh, C.H.; Rylander, L. Maternal Serum Concentrations of Per- and Polyfluoroalkyl Substances in Early Pregnancy and Small for Gestational Age in Southern Sweden. *Toxics* **2023**, *11*, 750. [CrossRef]
30. Sawadogo, B.; Konaté, F.O.; Konaté, Y.; Traoré, O.; Sossou, S.K.; Sawadogo, E.; Sourabié Ouattara, P.B.; Karambiri, H. Transfer of Bisphenol A and Trace Metals from Plastic Packaging to Mineral Water in Ouagadougou, Burkina Faso. *Int. J. Environ. Res. Public Health* **2023**, *20*, 6908. [CrossRef]

31. Azeredo, D.B.C.; de Sousa Anselmo, D.; Soares, P.; Graceli, J.B.; Magliano, D.C.; Miranda-Alves, L. Environmental Endocrinology: Parabens Hazardous Effects on Hypothalamic–Pituitary–Thyroid Axis. *Int. J. Mol. Sci.* **2023**, *24*, 15246. [CrossRef]
32. Lee, H.; Park, J.; Park, K. Mixture Effects of Bisphenol A and Its Structural Analogs on Estrogen Receptor Transcriptional Activation. *Toxics* **2023**, *11*, 986. [CrossRef]

International Journal of
Environmental Research and Public Health

MDPI

Article

Evaluation of the Safety and Efficacy of Hand Sanitizer Products Marketed to Children Available during the COVID-19 Pandemic

Lauren E. Gloekler [1,*], Elise J. de Gandiaga [1], Natalie R. Binczewski [1], Katie G. Steimel [1], Andrey Massarsky [1], Jordan Kozal [2], Melissa Vincent [3], Rachel Zisook [2], Mark J. LaGuardia [4], Scott Dotson [5] and Shannon Gaffney [2]

1 Stantec (ChemRisk), Aliso Viejo, CA 92656, USA
2 Stantec (ChemRisk), San Francisco, CA 94104, USA
3 Stantec (ChemRisk), Cincinnati, OH 45242, USA
4 Virginia Institute of Marine Science, William & Mary, Gloucester Point, VA 23062, USA
5 Insight Exposure and Risk Sciences Group, Cincinnati, OH 45249, USA
* Correspondence: lauren.gloekler@stantec.com

Abstract: Hand sanitizer use in the United States (U.S.) increased after the SARS-CoV-2 outbreak. The U.S. Food and Drug Administration (FDA) released temporary manufacturer guidance, changing impurity level limits for alcohol-based hand sanitizers (ABHSs). Since the guidance took effect, the FDA has recommended against using these hand sanitizers due to concerns over safety, efficacy, and/or risk of incidental ingestion. To address current gaps in exposure characterization, this study describes a survey of ABHSs marketed to children available in the U.S., as defined by several inclusion criteria. A subset of ABHSs (n = 31) were evaluated for ethanol and organic impurities using a modified FDA method. Products with detectable impurity levels were compared to the FDA's established interim limits. Seven children's products had impurity levels exceeding the FDA's recommended interim limits, including benzene (up to 9.14 ppm), acetaldehyde (up to 134.12 ppm), and acetal (up to 75.60 ppm). The total measured alcohol content ranged from 52% to 98% in all hand sanitizers tested, ranging from 39% below, and up to 31% above, the labeled concentration. Future studies should confirm impurity contamination sources. A risk assessment could determine whether dermal application or incidental ingestion of impurity-containing hand sanitizers pose any consumer risk.

Keywords: hand sanitizer; consumer products; children; impurities; ethanol; COVID-19

Citation: Gloekler, L.E.; de Gandiaga, E.J.; Binczewski, N.R.; Steimel, K.G.; Massarsky, A.; Kozal, J.; Vincent, M.; Zisook, R.; LaGuardia, M.J.; Dotson, S.; et al. Evaluation of the Safety and Efficacy of Hand Sanitizer Products Marketed to Children Available during the COVID-19 Pandemic. *Int. J. Environ. Res. Public Health* **2022**, *19*, 14424. https://doi.org/10.3390/ijerph192114424

Academic Editor: Atin Adhikari

Received: 23 September 2022
Accepted: 25 October 2022
Published: 3 November 2022

Publisher's Note: MDPI stays neutral with regard to jurisdictional claims in published maps and institutional affiliations.

1. Introduction

In 2019, the emergence of the new coronavirus SARS-CoV-2 resulted in increased sales and use of alcohol-based hand sanitizer (ABHS) because of concern of virus spread from contaminated hand surfaces. Increased demand for ABHSs from the general public and healthcare workers, however, led to difficulties accessing raw materials used in manufacturing, such as ethanol [1]. Hand sanitizers are classified as over-the-counter (OTC) drugs, and companies that manufacture these products are regulated by the U.S. Food and Drug Administration (FDA) [2]. To alleviate ABHS supply chain issues, the FDA allowed additional manufacturing entities that were not currently regulated as drug manufacturers to produce hand sanitizers, as long as the Secretary of Health and Human Services (HHS) declared a public health emergency [3]. To outline current manufacturing requirements for hand sanitizer production, the FDA promulgated several guidance documents [2–4] (Figure 1). These guidance documents provided temporary policies for preparing and manufacturing ABHS, as well as guidance for testing ethanol for contamination [5].

Per the FDA guidelines, hand sanitizers should be manufactured with no less than 94.9% by volume (v/v) ethanol or isopropyl alcohol, in addition to glycerin, hydrogen peroxide, and sterile water; the final formulation should have at least 80% ethanol (v/v) or 75% isopropyl alcohol (v/v) in an aqueous solution [3,6]. The FDA temporarily specified that fuel or technical grade ethanol were deemed acceptable for use in hand sanitizers. Regardless of the source, however, the ethanol should meet the following guidelines: (a) produced using fermentation and distillation processes "typically used for consumable goods" and contains no additives or additional chemicals; (b) meets United States Pharmacopoeia (USP) or Food Chemical Codex (FCC) grade requirements (including testing for impurities); and (c) screened for additional impurities not specified by USP and FCC requirements [3] (p. 7) [4,7]. Both the FDA temporary guidance and the World Health Organization (WHO) guidance state that hand sanitizers should not contain other active or inactive ingredients, such as additives, gelling agents, or fragrances to improve the smell or taste of the product, specifically "due to the risk of accidental ingestion in children" [4,8].

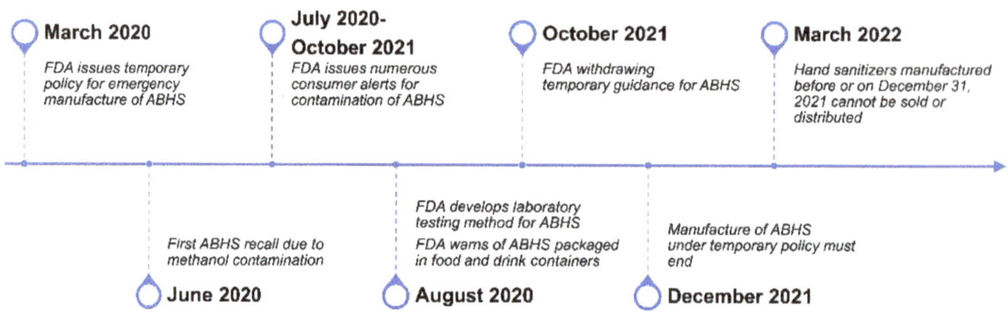

Figure 1. Timeline on Alcohol-Based Hand Sanitizer (ABHS)-Related U.S. Food and Drug Administration (FDA) Actions and Policies.

Impurities may potentially be introduced to hand sanitizer formulations through utilization of ethanol produced from fuel or technical grade ethanol via fermentation and distillation, or via the manufacturing environment (e.g., equipment and containers) [3]. Technical grade ethanol can be produced from a variety of feedstock, such as corn, sugar cane, sugar beet, wheat, and barley, and can contain byproducts and impurities that can potentially produce unpleasant odors or flavors and/or cause adverse human health effects [9]. Interim limits for specific impurities in ethanol have been provided for several compounds, and if the concentration of these in ethanol exceeds a sum of 300 ppm, then they must also meet interim limits for additional impurities [3]. In August 2020, FDA developed analytical procedures in support of its guidance, providing a methodology for manufacturers to help assure hand sanitizer products contain accurately labeled ingredients, and a screen for potentially harmful levels of impurities. The method provides guidance for detecting the impurities quantitatively or by using a "limit test approach", in which an impurity is either greater or less than the corresponding peak [10]. Impurities in alcohol products, including methanol, acetates, aldehydes, butanols, amyl alcohols, propanols, and pentanols, are generated during grain fermentation. These compounds can form azeotropes that are co-distilled with the ethanol fraction, making purifying ethanol difficult [3,11].

Throughout the pandemic, the FDA tested and issued alerts for several hand sanitizer products from domestic and international manufacturers and identified products with unacceptable levels of methanol, 1-propanol, benzene, acetaldehyde, and acetal, and "concerningly low levels of ethyl alcohol or isopropyl alcohol" [12]. As of 10 August 2021, 255 hand sanitizer products were listed on the do-not-use list, the majority of which were manufactured in Mexico (n = 201) [12]. Other countries with products recalled or listed were China (n = 22), Guatemala (n = 6), Turkey (n = 3), Poland (n = 3), Korea (n = 1), and the U.S. (n = 13). Additionally, alerts were issued for hand sanitizers packaged in containers that resembled food pouches, water bottles, juice bottles, vodka bottles, or beer cans, as well as those containing food flavors, as these products may lead to incidental hand sanitizer ingestion, especially in children [12]. These containers were also labeled with children's characters or movie references, including Barbie, Hot Wheels, Paw Patrol, Minions, JoJo Siwa, and Trolls World Tour [12].

Some hand sanitizers have particular characteristics (e.g., colorful or food-like packaging, cartoon characters, or candy and fruit scents) that may make them more attractive to children [13]. Children represent a susceptible group who may be at an increased risk of negative impacts from hand sanitizer impurities. Hand sanitizers are intended to be applied to hand surfaces; dermal contact with potential impurities is therefore likely the most common consumer exposure route. Intentional (as a result of substance abuse or mental health issues) and unintentional (accidental) ingestion of alcohol-based hand sanitizer in adults and children, however, does occur [14–16]. Increased exposure to hand sanitizers, including ingestion, was reported in children 12 years old and younger during the COVID-19 pandemic [2,17].

While inhalation of compounds present in hand sanitizers is possible, inhalation would not be expected to represent a primary exposure route relative to dermal and ingestion exposures, based on the National Poison Data System (NPDS) and the FDA Adverse Event Reporting System (FAERS) database call frequency and type [16,18–20]. Furthermore, between January 2020 and July 2021, the American Association of Poison Control Centers (AAPCC) [17] received thousands of accidental exposure reports of children age 12 and younger. Certain impurities may pose acute human health risks at sufficient doses. Methanol exposure at sufficient levels, for example, may result in metabolic acidosis, headaches, dizziness, the inability to coordinate muscle movement (ataxia), and, in extreme cases, blindness, coma, seizures, and death via ingestion or dermal routes [21,22]. Further, ingesting impurities at certain levels, including benzene, acetal, acetone, ethyl acetate, 2-butanol, isobutanol, 1-butanol, isoamyl alcohol, and amyl alcohol, may result in nausea, vomiting, diarrhea, staggering gait, drowsiness, central nervous system (CNS) depression, loss of consciousness, altered mental status, and/or abdominal pain. Dermal exposure to these same impurities at sufficient levels may result in irritation, redness, burning sensation, blistering, inflammation, and/or dry skin [22–31]. Additionally, benzene and acetaldehyde specifically are listed as "carcinogenic to humans" and "possibly carcinogenic to humans", respectively, by the International Agency for Research on Cancer (IARC) [32,33]. Whether the levels potentially present in hand sanitizers and their frequency of use would pose a risk for cancer to consumers, however, remains unknown.

This study's primary objectives were to: (a) perform a survey of hand sanitizers marketed to children available for purchase in the U.S. during the COVID-19 global pandemic, including information on the prevalence of certain inclusion criteria that may increase attractiveness of these products to children; (b) quantify the alcohol (ethanol) percentage in the final formulation; and (c) determine if any impurities, such as methanol and other organic impurities, were present in a selection of hand sanitizers marketed to children using a modified version of FDA analytical methods.

2. Materials and Methods

2.1. Survey of Hand Sanitizers Marketed to Children

A survey of hand sanitizers marketed to children available for purchase in the U.S. during the COVID-19 pandemic was performed between January and April 2021. Qualitative data were collected for hand sanitizers marketed to children available for purchase in 'brick and mortar' retail stores located in California, Colorado, and Hawaii, or available online for purchase from retailers' websites or solely online marketplaces. Store types targeted for the market survey, either online or in-person, fell into distinct store categories, including grocery stores, full-line discount stores, pharmacy/drug stores, specialty stores, and department stores. In some cases, retailers fit into more than one category, having both 'brick and mortar' locations and functioning as an online retailer. Store type categories and definitions are outlined in Table 1.

Table 1. Store Type Categories and Definitions.

Store Type	Definition	
Type 1	Brick and mortar	Store with physical location that is visited in person
	Online retailer	Store with 'online' presence that is visited from any device with internet access
Type 2	Grocery store	Store that sells primarily food for consumption
	Full line discount store	Store that sells various types of products, including groceries
	Pharmacy/drug store	Store that sells prescription drugs, OTC drugs, as well as medical supplies and groceries
	Distillery	A place or establishment that distills liquor
	Specialty store	Store that carries a deep assortment of brands, styles, or models within a relatively narrow category of goods
	Department store	Retail establishment offering a wide range of consumer goods in different areas of the store; each area ('department') specializes in a product category

Hand sanitizers marketed to children were defined as having one or more of the following characteristics: (a) labeling or packaging with cartoon characters, children's television shows, movies, toy images or references, or other pop-culture references well-known by children; (b) bright, colorful, or glittery packaging and/or liquid/gel; and/or (c) scented gel or liquid, especially a food or candy scent. Descriptive information was collected for each product, including: (a) store of purchase; (b) manufacturer name; (c) country of manufacture; (d) lot number; (e) active ingredient type and percentage; (f) scent; and (g) whether the product fit inclusion criteria.

2.2. *Quantification of Impurities in a Selection of Hand Sanitizers Marketed to Children*

A subset of 31 hand sanitizers marketed to children were purchased from 'brick and mortar' or online retailers and analyzed for methanol and other organic impurities listed in the FDA guidance (Table 2). One bottle of ABHS from an established and well-known manufacturer was included as a control comparison, resulting in a total of 32 samples analyzed for impurities. The control product was colorless, odorless, and did not have any packaging characteristics (e.g., colorful packaging, scents, etc.) as defined in the market survey methods. The test products were selected based on the following criteria: (a) all contained alcohol [either denatured alcohol, unspecified alcohol, or ethanol (ethyl alcohol)]; and (b) represented a variety of different manufacturers or distributors in order to characterize the maximum number of entities. Two of each hand sanitizer product were purchased at the same store, and one set was shipped to the laboratory for analysis. The other set was retained in case reanalysis was needed.

Table 2. List of FDA Interim Limits and Limits of Detection/Limits of Quantification (LODs/LOQs) for Impurities Assessed in Hand Sanitizers Marketed to Children.

Chemical Name	CAS Number	Interim Limit Listed in FDA Guidance (ppm) [a,b]	LOD/LOQ (μg/g) [c]
1-Propanol	71-23-8	NMT 1000	68.3
Methanol	67-56-1	NMT 630	67.2
Benzene	71-43-2	NMT 2	3.74
Acetaldehyde	75-07-0	NMT 50	30.0
Acetal	105-57-7	NMT 50	5.30
Acetone	67-64-1	NMT 4400	67.2

Table 2. *Cont.*

Chemical Name	CAS Number	Interim Limit Listed in FDA Guidance (ppm) [a,b]	LOD/LOQ (µg/g) [c]
Ethyl Acetate	141-78-6	NMT 2200	76.6
2-Butanol	78-92-2	NMT 6200	68.7
Isobutanol	78-83-1	NMT 21,700	68.3
1-Butanol	71-36-3	NMT 1000	68.9
Isoamyl alcohol	123-51-3	NMT 4100	68.8
Amyl alcohol	71-41-0	NMT 4100	68.9

[a] NMT = Not more than. [b] Methanol, benzene, and acetaldehyde have specific individual limits. The sum of all other impurities should not exceed 300 ppm. If the sum of all other impurities is >300 ppm, then each impurity in this category must meet the individual limit specified above. [c] Based on sample volume of 0.1 mL, density 0.850 g/mL.

2.3. Sample Analysis

The FDA developed an analytical method to assess the quality of finished hand sanitizer products entitled "Direct Injection Gas Chromatography Mass Spectrometry (GC/MS) Method for the Detection of Listed Impurities in Hand Sanitizers" in August 2020 [10]. This procedure was intended to evaluate products formulated with either ethanol or isopropyl alcohol (also called isopropanol or 2-propanol) as the labeled active ingredient and 12 potential impurities (Table S1). This method was modified to include mass spectrum matching for each analyte of interest to those contained within the NIST/EPA/NIH Mass Spectral Library [34]. Each compound of interest was also quantified using a five-point calibration curve produced by the analysis of analytical standards for each analyte. These changes greatly improved identifying and quantifying compounds over the proposed method. Quality Assurance and Quality Control (QA/QC) requirements were also modified to adhere to Good Laboratory Practices (GLP). GLP regulations are intended to assure data quality and data integrity. Analytical equipment, procedures, QA/QC, and reporting requirements are listed below.

All samples were analyzed on an Agilent 5975C, Quadrupole GC/MS equipped with an Agilent 7890A Gas Chromatography and Gerstel MultiPurpose MPS2 AutoSampler with an Agilent DB-624 Capillary GC Column, 30 m × 0.25 mm × 1.4 µm. The reference standards (purity, >95%) were purchased from Sigma Aldrich, St. Louis, MI, USA, and the diluent acetonitrile was LCMS/HPLC grade, 99.9% (purchased from J.T. Baker in Phillipsburg, NJ, USA). To determine the hand sanitizer sample's density, a 200 µL sample was weighed with an analytical balance to the nearest tenth of a milligram (±0.1 mg). Reporting units were g mL^{-1}. To measure ethanol content, samples were prepared by transferring 100 µL via a micropipette to a 10 mL volumetric flask containing approximately 8 mL of acetonitrile. Samples were diluted to volume with acetonitrile and mixed. Next, 100 µL of the acetonitrile mixture was transferred to a 2 mL GC vial containing 1 mL of the internal standard (400 ng, toluene) in acetonitrile. To measure impurities, each sample was prepared by transferring 100 µL via a micropipette to a 2 mL GC vial containing 1 mL of the internal standard (400 ng, toluene) in acetonitrile. Peak assignment for each analyte is based on matching the chromatographic retention time to the reference standard (Figure S1). Peak identification is determined by mass spectral matching to the reference mass spectrum contained in NIST's 2011 library (NIST 2021). To quantify active ingredients and impurities, a five-point calibration curve ($R^2 > 0.985$) was used at concentrations bracketing the compound of interest. The curves for each compound were produced by plotting the area of each Quantitation Ion (Table S1) for the internal standard (400 ng, toluene) and reference standards. Reporting units were µg mL^{-1} (ppm), µg g^{-1} (ppm), and % by weight. Quantitation software (MSD ChemStation E.02.00.493, Agilent Technologies, Inc., Santa Clara, CA, USA) was used to determine the LOQ for each compound. The limit of quantification (LOQ) for each impurity is shown in Table 2. The hand sanitizer active ingredients and impurities, chromatogram peak assignments, LOQ, and calibration range are summarized in Table S2.

A blank and a duplicate sample were analyzed with each batch of 20 or fewer hand sanitizer samples. For samples that were measured at concentrations above the FDA interim limit, a duplicate analysis was performed to verify the results. Additionally, some products were chosen at random and a duplicate analysis was performed as an additional QC step. A total of 12 samples were tested in duplicate to confirm impurity concentrations and alcohol content. The target relative % deference (RPD) for the duplicate analysis was <30%.

2.4. Data Analysis

Contaminant concentrations reported as non-detect (ND) during sample analysis were calculated as one half the reported LOQ (i.e., LOQ/2) when graphing data and comparing concentrations to FDA interim limits.

Statistical analyses were conducted using SigmaPlot (SPW 14.0; Systat Software, Inc., San Jose, CA, USA). Briefly, normality was tested using Shapiro–Wilk test. For data that were normally distributed (parametric), Student's *t*-test (two groups) or One-Way ANOVA (three groups) were used to deduce statistical differences. For data that were not normally distributed (non-parametric), Mann–Whitney Rank Sum test (two groups) or Kruskal–Wallis or One-Way ANOVA on Ranks (three groups) were used to deduce statistical differences. Additionally, paired *t*-test (parametric) or Wilcoxon Signed Rank test (non-parametric) were used to assess statistical differences between original testing and upon repeated testing (or duplicate). Specifically, statistical differences between impurity concentrations in U.S. and non-U.S. hand sanitizers, gel and liquid products, active ingredient type (ethyl alcohol, denatured alcohol, or 'unspecified alcohol'), scented or non-scented products, as well as between colored and non-colored hand sanitizers were examined (Figures S2–S7). In all cases, p-value < 0.05 was considered significant. The statistical tests for all comparisons are provided in Table S2.

3. Results

3.1. Survey of Hand Sanitizers Marketed to Children Available during the COVID-19 Pandemic

A total of 139 hand sanitizers potentially marketed to children were identified in the survey, 74 of which were available from brick and mortar stores, and 65 of which were available from online retailers. The majority of these products (n = 120, 87%) were available in full line discount stores and grocery stores. The hand sanitizers marketed to children were manufactured in the U.S. (n = 51, 37%), China (n = 85, 61%), Turkey (n = 1, 0.72%), or South Korea (n = 2, 1.44%). The hand sanitizers contained a variety of active ingredients, including ethyl alcohol (ethanol), 'denatured alcohol', 'unspecified alcohol', chloroxylenol, or benzalkonium chloride. Ethyl alcohol was the most common active ingredient, accounting for 64% (n = 89), and the labeled percentage of active ingredient ranged between 62 to 80%. Gel was the most common hand sanitizer type, accounting for 78% (n = 108) of the products, and the remainder were spray (n = 19), liquid (n = 8), or foam (n = 4). Almost half the hand sanitizers (n = 68, 49%) had labeling or packaging featuring cartoon characters, children's television shows, movies, toy references, or other well-known pop-culture references that children would likely recognize (e.g., Disney characters). Most of the products (n = 105, 75%) were scented, and the rest were unscented (n = 33, 25%). Over 50 different scents were identified (e.g., Berry; Bubble Gum; and Strawberry Pound Cake). Three of the samples had packaging reminiscent of food packaging, such as a baby food type pouch with a screw top or honey bear shaped bottle. Several of the hand sanitizers had explicit instructions on the label regarding their use on or by children. These instructions read: (a) "Not for children under 3 years. If swallowed, get medical help or contact a Poison Control Center right away"; (b) "Keep out of reach of children"; or (c) "For children under 8, use under adult supervision".

3.2. Selection of Hand Sanitizers Marketed to Children Purchased for Impurity Analysis and Ethanol Content

Table 3 provides a summary of the subset of hand sanitizers marketed to children sampled ($n = 31$), including information such as an assigned manufacturer unique identifier, active ingredient type and percentage, scent, color, and packaging information. The products selected for chemical analysis had similar attributes as those noted for all surveyed hand sanitizers. All the products analyzed in this study had at least one of the inclusion criteria that made the product potentially attractive to children, such as overt children's packaging with cartoon characters, food or candy scents, and/or colored gel.

The key attributes of the hand sanitizer subset selected for sampling are summarized in Figure 2. The selected hand sanitizers were manufactured in China ($n = 25$, 83%), U.S. ($n = 4$, 13%), or South Korea ($n = 2$, 6%), and represented 21 separate manufacturers. The products were labeled as containing at least one of these active ingredients: (1) ethyl alcohol ($n = 20$, 65%); (2) denatured alcohol ($n = 8$, 26%); or (3) unspecified alcohol ($n = 3$, 9%). Denatured alcohol and alcohol, however, are other names for ethyl alcohol. The percentage of labeled active ingredient ranged from 62 to 80%. The products were purchased at a wide variety of store types and locations, including at full line discount stores ($n = 18$), department stores ($n = 2$), grocery stores ($n = 3$), an online marketplace ($n = 2$), pharmacies/drug stores ($n = 2$), or specialty stores ($n = 4$). The majority of the products were in gel form ($n = 28$, 90%). The products tested primarily had a fruit scent or fruit in combination with another scent (e.g., cosmic cherry; vanilla bean + coconut + sugared musk) (n= 15, 48%), followed by a dessert or candy scent (e.g., cotton candy; creamy cappuccino) ($n = 5$, 16%), and the remaining samples had floral or herbal scents, or were labeled as unscented or unspecified ($n = 11$, 35%). Most products had a colored gel or liquid ($n = 20$, 64%). Only three of the products had no discernable packaging that would appeal to children specifically, whereas the rest of the products had cartoon characters or children's TV characters on the packaging or holders attached to the bottles themselves, food or candy images, or was a well-known children's brand.

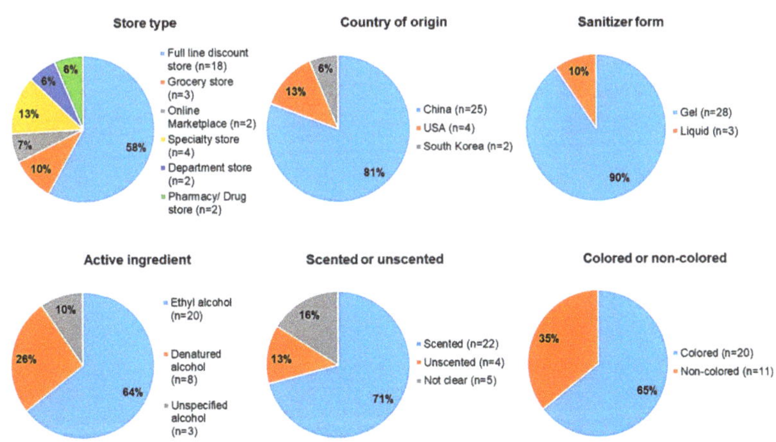

Figure 2. Key attributes of Hand Sanitizers Marketed to Children Selected for Testing.

Table 3. Selection of Hand Sanitizers Marketed to Children Tested for Impurities.

Sample Ref. No.	Manufacturer Identification Code	Country of Manufacture	Active Ingredient	Labeled Active Ingredient (%)	Store Type	Labeled Scent	Scent Category	Sanitizer Form	Gel/Liquid Color	Packaging Category
CCR-1	Manufacturer A	China	Ethyl Alcohol	70	Full line discount store	Unscented	No scent	Gel	Non-colored	Cartoon Character
CCR-3	Manufacturer B	China	Ethyl Alcohol	62	Full line discount store	Choco Orange	Fruit/Candy	Gel	Orange	Candy Image
CCR-5	Manufacturer C	China	Ethyl Alcohol	63	Full line discount store	Cotton Candy	Candy	Gel	Purple	Cartoon Character; Candy Image
CCR-6	Manufacturer C	China	Ethyl Alcohol	63	Full line discount store	Berry	Fruit	Gel	Pink	Cartoon Character; Food Image
CCR-7	Manufacturer D	China	Denatured alcohol	70	Full line discount store	Cotton Candy	Candy	Gel	Pink	Cartoon Character; Glitter
CCR-10	Manufacturer E	China	Ethyl Alcohol	70	Full line discount store	Ocean Breeze	Un-categorized	Gel	Non-colored	Cartoon Character
CCR-11	Manufacturer F	China	Ethyl Alcohol	75	Full line discount store	Strawberry	Fruit	Gel	Pink	Food Image
CCR-15	Manufacturer F	China	Ethyl Alcohol	75	Full line discount store	Vanilla Sugar Cookie	Dessert	Gel	Green	Food Image
CCR-19	Manufacturer G	China	Ethyl Alcohol	75	Full line discount store	Unscented	No scent	Gel	Non-colored	Cartoon Character (outer cover)
CCR-21	Manufacturer H	China	Unspecified Alcohol	65	Full line discount store	Cotton Candy	Candy	Gel	Pink	Cartoon Character (outer cover); Candy Image
CCR-22	Manufacturer I	China	Denatured alcohol	75	Full line discount store	Citrus	Fruit	Gel	Non-colored	Cartoon Character
CCR-23	Manufacturer J	China	Denatured alcohol	62	Pharmacy/Drug Store	Berry Lollipop	Fruit/Candy	Gel	Pink	Candy Image
CCR-28	Manufacturer J	China	Denatured alcohol	62	Pharmacy/Drug Store	Fruit Punch	Fruit	Gel	Pink	Children's TV Character
CCR-32	Manufacturer D	China	Denatured alcohol	70	Grocery Store	Grape	Fruit	Gel	Purple	Cartoon Character
CCR-41	Manufacturer D	China	Denatured alcohol	70	Grocery Store	Raspberry	Fruit	Liquid	Pink	Cartoon Character

Table 3. *Cont.*

Sample Ref. No.	Manufacturer Identification Code	Country of Manufacture	Active Ingredient	Labeled Active Ingredient (%)	Store Type	Labeled Scent	Scent Category	Sanitizer Form	Gel/Liquid Color	Packaging Category
CCR-47	Manufacturer D	China	Denatured alcohol	70	Grocery Store	Strawberry	Fruit	Liquid	Pink	Children's TV Character
CCR-55	Manufacturer K	China	Ethyl Alcohol	68	Full line discount store	Unscented	No scent	Liquid	Non-colored	Cartoon Character (outer packaging)
CCR-59	Manufacturer L	USA	Ethyl Alcohol	80	Full line discount store	Lavender	Floral	Gel	Non-colored	Honey bear shaped packaging
CCR-63	Manufacturer M	USA	Ethyl Alcohol	68	Full line discount store	Strawberry Pound Cake	Dessert	Gel	Non-colored with blue beads	None
CCR-117	Manufacturer N	USA	Ethyl Alcohol	75	Online retailer	Unscented	No scent	Gel	Blue	Kid's Brand
CCR-120	Manufacturer O	China	Ethyl Alcohol	63	Online retailer	Unscented	No scent	Gel	Non-colored	Cartoon Character
CCR-122	Manufacturer H	China	Denatured alcohol	66	Pharmacy/Drug Store	Melon	Fruit	Gel	Pink	Food Image
CCR-128	Manufacturer K	China	Ethyl Alcohol	68	Specialty Store	Unscented	No scent	Gel	Non-colored	Cartoon Character (outer packaging)
CCR-129	Manufacturer P	South Korea	Ethyl Alcohol	70	Department Store	Candy Apple	Fruit/Candy	Gel	Non-colored	Cartoon Character; Candy Image; Pouch packaging
CCR-133	Manufacturer B	China	Unspecified Alcohol	62	Full line discount store	Berries & Cream	Dessert	Gel	Pink	Food Image
CCR-134 (Control)	Manufacturer Q	USA	Ethyl Alcohol	70	Full line discount store	Unscented	No scent	Gel	Non-colored	None
CCR-150	Manufacturer P	South Korea	Ethyl Alcohol	70	Department Store	Mixed Berry	Fruit	Gel	Non-colored	Pouch packaging
CCR-151	Manufacturer R	China	Ethyl Alcohol	75	Pharmacy/Drug Store	Watermelon	Fruit	Gel	Pink	Food Image
CCR-152	Manufacturer S	China	Unspecified Alcohol	67	Specialty Store	Unscented	No scent	Gel	Orange	Holographic
CCR-153	Manufacturer K	China	Ethyl Alcohol	68	Specialty Store	Unscented	No scent	Gel	Turquoise	Cartoon Character
CCR-154	Manufacturer T	China	Ethyl Alcohol	63	Pharmacy/Drug Store	Unscented	No scent	Gel	Non-colored	Cartoon Character
CCR-155	Manufacturer U	USA	Ethyl Alcohol	62	Specialty Store	Cupcake	Dessert	Gel	Purple	None

3.3. Impurities and Ethanol Content Measured in a Selection of Hand Sanitizers Marketed to Children

Table 4 lists the concentrations of organic impurities and percentage of active ingredients (%) in the children's hand sanitizer analyzed in this study. No impurities were detected above LOQ for the control sample. Several compounds, such as 1-butanol, 1-pentanol, 2-butanol, acetone, and isoamyl alcohol, were not detected at levels above the LOQ in all children's samples tested. The remaining impurities were measured in some of the products at the following concentrations: 1-propanol (n = 4, BLOQ-651 μg/g); acetal (n = 13, BLOQ-76 μg/g); acetaldehyde (n = 4, BLOQ-134 μg/g); benzene (n = 3, BLOQ-9.14 μg/g); ethyl acetate (n = 2, BLOQ-363 μg/g); and methanol (n = 9, BLOQ-400 μg/g). Of all the impurities tested, the most commonly measured were acetal (42% of samples) and methanol (29% of samples). Of the 31 children's hand sanitizers, seven had concentrations of impurities above the FDA interim limits. Specifically, four bottles (CCR-11, CCR-47, CCR-59, and CCR-129) had levels of acetal higher than the FDA interim limit of 50 ppm (μg/g). With respect to benzene, all three bottles (CCR-5, CCR-129, CCR-153) containing measured concentrations were in exceedance of the FDA interim limit of 2 ppm (μg/g). Further, three bottles (CCR-11, CCR-41, CCR-59) had levels of acetaldehyde higher than the FDA interim limit of 50 ppm (μg/g). Figure 3 details a comparison between impurity concentrations measured in hand sanitizers and FDA interim limits.

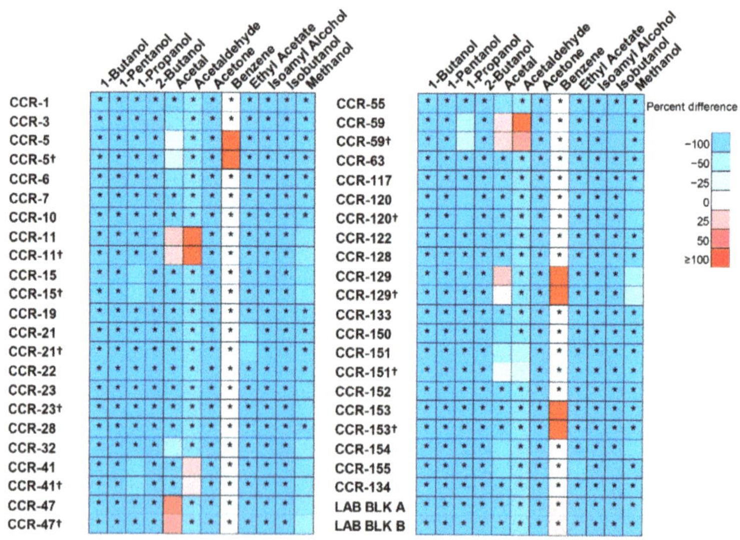

Figure 3. Impurity concentrations (expressed as percent difference relative to FDA specifications). Duplicate samples noted with † (expressed as percent difference relative to FDA specifications). Cells denoted with asterisks (*) refer to instances when impurity concentrations were equivalent to one half LOQ.

Table 4. Concentrations of Impurities Measured in Selected Hand Sanitizers Marketed to Children.

	Actives Measured Concentration (%)		Impurities Measured Concentration (μg/g)											
Sample Ref. No.	Ethanol	Isopropanol	1-Butanol	1-Pentanol	1-Propanol	2-Butanol	Acetal	Acetaldehyde	Acetone	Benzene	Ethyl acetate	Isoamyl alcohol	Isobutanol	Methanol
CCR-1	61	-	BLOQ	BLOQ	BLOQ	BLOQ	BLOQ	BLOQ	BLOQ	BLOQ	BLOQ	BLOQ	BLOQ	BLOQ
CCR-3	57	-	BLOQ	BLOQ	BLOQ	BLOQ	**13.9**	BLOQ	BLOQ	BLOQ	BLOQ	BLOQ	BLOQ	BLOQ
CCR-5	55	-	BLOQ	BLOQ	BLOQ	BLOQ	**41.8**	BLOQ	BLOQ	**9.14**	BLOQ	BLOQ	BLOQ	BLOQ
CCR-5*	55	-	BLOQ	BLOQ	BLOQ	BLOQ	**37.0**	BLOQ	BLOQ	**4.85**	BLOQ	BLOQ	BLOQ	BLOQ
CCR-6	80	-	BLOQ	BLOQ	BLOQ	BLOQ	**11.7**	BLOQ	BLOQ	BLOQ	BLOQ	BLOQ	BLOQ	BLOQ
CCR-7	66	-	BLOQ	BLOQ	BLOQ	BLOQ	BLOQ	BLOQ	BLOQ	BLOQ	BLOQ	BLOQ	BLOQ	BLOQ
CCR-10	56	-	BLOQ	BLOQ	BLOQ	BLOQ	BLOQ	BLOQ	BLOQ	BLOQ	BLOQ	BLOQ	BLOQ	BLOQ
CCR-11	88	-	BLOQ	BLOQ	BLOQ	BLOQ	**61.8**	**134**	BLOQ	BLOQ	BLOQ	BLOQ	BLOQ	**134**
CCR-11 *	88	-	BLOQ	BLOQ	BLOQ	BLOQ	**59.6**	**133**	BLOQ	BLOQ	BLOQ	BLOQ	BLOQ	**187**
CCR-15	69	-	BLOQ	BLOQ	**189**	BLOQ	BLOQ	BLOQ	BLOQ	BLOQ	BLOQ	BLOQ	BLOQ	**109**
CCR-15 *	69	-	BLOQ	BLOQ	**192**	BLOQ	BLOQ	BLOQ	BLOQ	BLOQ	BLOQ	BLOQ	BLOQ	**156**
CCR-19	86	-	BLOQ	BLOQ	BLOQ	BLOQ	BLOQ	BLOQ	BLOQ	BLOQ	BLOQ	BLOQ	BLOQ	BLOQ
CCR-21	80	-	BLOQ	BLOQ	BLOQ	BLOQ	BLOQ	BLOQ	BLOQ	BLOQ	**340**	BLOQ	BLOQ	BLOQ
CCR-21 *	80	-	BLOQ	BLOQ	BLOQ	BLOQ	BLOQ	BLOQ	BLOQ	BLOQ	**363**	BLOQ	BLOQ	BLOQ
CCR-22	61	-	BLOQ	BLOQ	BLOQ	BLOQ	BLOQ	BLOQ	BLOQ	BLOQ	BLOQ	BLOQ	BLOQ	BLOQ
CCR-23	63	-	BLOQ	BLOQ	BLOQ	BLOQ	BLOQ	BLOQ	BLOQ	BLOQ	BLOQ	BLOQ	BLOQ	**55.3**
CCR-23 *	63	-	BLOQ	BLOQ	BLOQ	BLOQ	BLOQ	BLOQ	BLOQ	BLOQ	BLOQ	BLOQ	BLOQ	**71.5**
CCR-28	59	-	BLOQ	BLOQ	BLOQ	BLOQ	BLOQ	BLOQ	BLOQ	BLOQ	BLOQ	BLOQ	BLOQ	BLOQ
CCR-32	98	-	BLOQ	BLOQ	BLOQ	BLOQ	**28**	BLOQ	BLOQ	BLOQ	BLOQ	BLOQ	BLOQ	**150**
CCR-41	52	-	BLOQ	BLOQ	**166**	BLOQ	BLOQ	**60**	BLOQ	BLOQ	BLOQ	BLOQ	BLOQ	**107**
CCR-41 *	52	-	BLOQ	BLOQ	**206**	BLOQ	BLOQ	**56**	BLOQ	BLOQ	BLOQ	BLOQ	BLOQ	**138**
CCR-47	69	-	BLOQ	BLOQ	BLOQ	BLOQ	**76**	BLOQ	BLOQ	BLOQ	BLOQ	BLOQ	BLOQ	**174**
CCR-47 *	69	-	BLOQ	BLOQ	BLOQ	BLOQ	**68**	BLOQ	BLOQ	BLOQ	BLOQ	BLOQ	BLOQ	**273**
CCR-55	84	-	BLOQ	BLOQ	BLOQ	BLOQ	**8**	BLOQ	BLOQ	BLOQ	BLOQ	BLOQ	BLOQ	BLOQ
CCR-59	55	-	BLOQ	BLOQ	**563**	BLOQ	**59**	**104**	BLOQ	BLOQ	BLOQ	BLOQ	**175**	BLOQ
CCR-59 *	55	-	BLOQ	BLOQ	**651**	BLOQ	**59**	**69**	BLOQ	BLOQ	BLOQ	BLOQ	**221**	BLOQ
CCR-63	69	3	BLOQ	BLOQ	BLOQ	BLOQ	BLOQ	BLOQ	BLOQ	BLOQ	BLOQ	BLOQ	BLOQ	BLOQ
CCR-117	75	-	BLOQ	BLOQ	BLOQ	BLOQ	BLOQ	BLOQ	BLOQ	BLOQ	BLOQ	BLOQ	BLOQ	BLOQ
CCR-120	66	-	BLOQ	BLOQ	**50.1**	BLOQ	BLOQ	BLOQ	BLOQ	BLOQ	BLOQ	BLOQ	BLOQ	**37.8**
CCR-120 *	66	-	BLOQ	BLOQ	**41.4**	BLOQ	BLOQ	BLOQ	BLOQ	BLOQ	BLOQ	BLOQ	BLOQ	**66.6**
CCR-122	67	-	BLOQ	BLOQ	BLOQ	BLOQ	BLOQ	BLOQ	BLOQ	BLOQ	BLOQ	BLOQ	BLOQ	BLOQ
CCR-128	64	-	BLOQ	BLOQ	BLOQ	BLOQ	BLOQ	BLOQ	BLOQ	BLOQ	BLOQ	BLOQ	BLOQ	BLOQ
CCR-129	74	2	BLOQ	BLOQ	BLOQ	BLOQ	**63**	BLOQ	BLOQ	**6.94**	BLOQ	BLOQ	BLOQ	**328**
CCR-129 *	74	2	BLOQ	BLOQ	BLOQ	BLOQ	**54.4**	BLOQ	BLOQ	**4.36**	BLOQ	BLOQ	BLOQ	**400**
CCR-133	55	-	BLOQ	BLOQ	BLOQ	BLOQ	BLOQ	BLOQ	BLOQ	BLOQ	BLOQ	BLOQ	BLOQ	BLOQ
CCR-150	75	-	BLOQ	BLOQ	BLOQ	BLOQ	**9.16**	BLOQ	BLOQ	BLOQ	BLOQ	BLOQ	BLOQ	BLOQ
CCR-151	80	-	BLOQ	BLOQ	BLOQ	BLOQ	**25.4**	**29.6**	BLOQ	BLOQ	BLOQ	BLOQ	BLOQ	BLOQ
CCR-151 *	80	-	BLOQ	BLOQ	BLOQ	BLOQ	**43.3**	**37.5**	BLOQ	BLOQ	BLOQ	BLOQ	BLOQ	BLOQ
CCR-152	63	-	BLOQ	BLOQ	BLOQ	BLOQ	BLOQ	BLOQ	BLOQ	BLOQ	BLOQ	BLOQ	BLOQ	BLOQ
CCR-153	65	-	BLOQ	BLOQ	BLOQ	BLOQ	BLOQ	BLOQ	BLOQ	**4.67**	BLOQ	BLOQ	BLOQ	BLOQ
CCR-153 *	65	-	BLOQ	BLOQ	BLOQ	BLOQ	BLOQ	BLOQ	BLOQ	**7.91**	BLOQ	BLOQ	BLOQ	BLOQ
CCR-154	83	-	BLOQ	BLOQ	BLOQ	BLOQ	**11.6**	BLOQ	BLOQ	BLOQ	BLOQ	BLOQ	BLOQ	**88**
CCR-155	62	-	BLOQ	BLOQ	BLOQ	BLOQ	**9.3**	BLOQ	BLOQ	BLOQ	**235**	BLOQ	**119**	BLOQ
CCR-134 (Control)	74	3	BLOQ	BLOQ	BLOQ	BLOQ	BLOQ	BLOQ	BLOQ	BLOQ	BLOQ	BLOQ	BLOQ	BLOQ

Table 4. *Cont.*

	Actives Measured Concentration (%)		Impurities Measured Concentration (μg/g)											
Sample Ref. No.	**Ethanol**	**Isopropanol**	**1-Butanol**	**1-Pentanol**	**1-Propanol**	**2-Butanol**	**Acetal**	**Acetaldehyde**	**Acetone**	**Benzene**	**Ethyl acetate**	**Isoamyl alcohol**	**Isobutanol**	**Methanol**
Lab Blank A	-	-	BLOQ	BLOQ	BLOQ	BLOQ	BLOQ	BLOQ	BLOQ	BLOQ	BLOQ	BLOQ	BLOQ	BLOQ
Lab Blank B	-	-	BLOQ	BLOQ	BLOQ	BLOQ	BLOQ	BLOQ	BLOQ	BLOQ	BLOQ	BLOQ	BLOQ	BLOQ
Average (STD)	68.5% (± 11%)	0.2% (± 0.7%)	34.7 (± 1.009)	34.7 (± 1.01)	75.8 (± 127.4)	34.6 (± 1.006)	18.7 (± 23.848)	26.9 (± 29.837)	33.8 (± 0.984)	2.5 (± 1.697)	57.8 (± 72.241)	34.6 (± 1.007)	44.0 (± 37.175)	78.8 (± 84.541)
Median	66%	0.0%	34.5	34.5	34.6	34.4	2.8	15.2	33.6	1.9	38.8	34.4	34.2	34.5
Range	52–98%	0–3.3%	32.8–37	32.9–37.1	32.6–651	32.8–36.9	2.5–75.6	14.3–134.1	32.1–36.1	1.8–9.1	36.6–363	32.8–37	32.6–221	32.1–400

Footnote: BLOQ: below limit of quantification; (-): not measured; (*): duplicative sample; descriptive statistics (avg, STD, median, range) did not include sample CCR-134, or Lab Blank A and B. LOQ for each analyte was calculated using sample volume 0.1 mL, and density of each sample, which ranged between 0.8115–1.0 g mL^{-1}.

To better understand differences between impurity concentrations measured in the children's products and various characteristics, several statistical comparisons were made for U.S. and non-U.S. products, clear and non-clear liquid, and original and duplicate samples (Figures S2–S7). Samples collected from hand sanitizers manufactured in the U.S. had higher 1-propanol and isobutanol concentrations when compared to non-U.S. brand hand sanitizers. Additionally, acetal and methanol concentrations were higher for non-U.S. brands when compared to U.S. brands. However, no impurity concentrations were significantly different when comparing U.S. to non-U.S, except for methanol, which was detected in concentrations significantly higher in non-U.S. manufactured hand sanitizers. Further, samples collected from hand sanitizers with non-colored liquid reported higher 1-propanol, isobutanol, and methanol concentrations when compared to hand sanitizers with colored liquid. Additionally, acetal, benzene, and ethyl acetate concentrations were higher for colored liquid hand sanitizers compared to non-colored hand sanitizers. All impurity concentrations, however, were not significantly different when comparing non-colored and colored hand sanitizer liquid. With the exception of methanol, all impurities measured were not statistically different between duplicate samples. For sample CCR-120, methanol had significantly higher concentrations in the duplicate compared to the original dataset. No statistically significant differences in other analyte concentrations were seen in any of the 12 hand sanitizer bottles that were tested in duplicate (Figure S7).

3.4. Ethanol Content Measured in a Selection of Hand Sanitizers Marketed to Children

The measured alcohol content ranged from 52% to 98% in all tested hand sanitizers marketed to children. The alcohol percentage reported on the bottle labels, however, ranged from 62% to 80%. Figure S8 shows a comparison between labeled and measured alcohol concentrations for each product. Overall, when comparing the measured and labeled alcohol ranges, the measured concentrations in the bottles ranged from 31% lower than the labeled concentration to up to 28% higher than the labeled concentration, depending on the product. Almost one third (n = 10) of the bottles tested had alcohol concentrations lower than the required 60% needed for germicidal properties, as recommended by FDA and CDC [3,22].

4. Discussion

The survey identified numerous products potentially attractive to children available for purchase in the U.S. The majority of the products contained fragrances and coloring additives that might increase their appeal, despite FDA recommending against these additives because of exposure risk in children. Further, we found that products continued to be packaged in containers resembling food or drink containers, despite FDA alerts to similar products because of their possible accidental ingestion risk by children [12]. Hence, understanding the possible health risks associated with incidental ingestion and dermal absorption – especially by children – of hand sanitizer impurities is needed not only to effectively characterize potential future health risks, but also to help successfully manage them. Tse et al. (2021) reported that, based on the toxicology of common impurities found in technical-grade ethanol, the majority pose a low risk during normal use [35]. However, acetaldehyde and benzene, in particular, may pose a risk especially to susceptible populations like children, and potential risk from exposure to several impurities is still uncertain. Further, certain formulation additives (such as water or gelling agents) may increase the risk of dermal exposure and should be studied further [35]. In a concurrent study by Kozal et al. presented at the Society of Toxicology (SOT), a screening level risk assessment was performed to estimate systemic exposure doses (SEDs) that may result from repeated dermal exposure, as well as incidental ingestion of hand sanitizers, and were compared to health-based guidance values and toxicity thresholds [36].

Numerous product recalls and alerts were issued by FDA during the COVID-19 pandemic due to methanol contamination. The full list of recalled products and associated warning letters detailing the methanol concentrations are available on FDA website [12]. As an example, methanol concentrations in hand sanitizers manufactured by 4E Global SAPI de CV were reported to range from 65% to 74% methanol v/v [37]. Although several alerts were reported for contamination with benzene, acetaldehyde, or acetal, detailed warning letters specifying concentrations of these contaminants were not available on FDA website. In one alert, a product was found to contain 6 ppm benzene [38]. Therefore, it is unclear if the concentrations reported in our study are similar to the concentrations found by FDA, as this information is not publicly available. One of the products (CCR-129) with levels of acetal that exceeded FDA interim limits was listed on the 'FDA's Do-Not-Use List' in April 2022 due to the product being manufactured "at the same facility that produced benzene contaminated product" [12]. Additionally, numerous products from the same distributor as several products tested in this study were also included on the FDA list because they were manufactured at the same facilities in which a product was found to have methanol and benzene contamination. Another product tested in this study was recommended for recall, and was added to an import alert in March 2022.

Several other studies have reported impurity concentrations measured in ABHSs. In an unpublished study, 260 unique batches of hand sanitizer were analyzed and 44 batches (17%) contained benzene at 0.1 ppm or above, while 21 batches (8%) contained benzene at 2 ppm or above [39]. The highest benzene concentration detected was 16.1 ppm, over eight times the FDA interim limit of 2 ppm [39]. Of the 21 batches that contained >2 ppm benzene, one batch reported 8680 ppm methanol and 147 ppm acetaldehyde, which are fourteen and three times the FDA interim limit, respectively. Another batch contained 709 ppm methanol [39]. In another study, 42 liquid and gel hand sanitizers were analyzed for nine different impurities, and 11 of the samples were non-compliant with interim Health Canada guidelines [11]. The authors noted that these samples primarily contained acetaldehyde at levels above the Canadian interim guidance of NMT 75 ppm total acetaldehyde and acetal [40]. In a study conducted in the U.S., 51 samples from bulk refillable hand sanitizer dispensers at community settings, such as restaurants, malls, and fitness centers, across the U.S., and 40 samples from a single school district in South Carolina were analyzed for methanol, benzene, acetaldehyde, and acetal [41]. Only one sample from the school district was positive for acetal (511 ppm) at levels above the FDA interim limit, while 35.29% (18/51), 33.33% (17/51), and 5.88% (3/51) of the community-acquired samples had acetal, acetaldehyde, and methanol concentrations, respectively, above the FDA's interim limits. Benzene was not detected in any of the 91 samples [41]. Further, in a study conducted in Malaysia, of the 121 samples purchased from retail locations, 7.4% contained methanol above the LOD of 4.4% (v/v), while 18.8% of the 265 samples collected from freely deployed public dispensers contained methanol above the LOD [42]. A recent study by Pal et al. (2022) tested 200 hand sanitizers for concentrations of benzene, toluene, and styrene [43]. Similar to our study, the authors found that a selection of products ($n = 10$) exceeded FDA interim limits for benzene (>2 ppm). It was also noted that products also contained toluene (25%) and styrene (32%); however, these compounds have no FDA limits. Similar to our study, the hand sanitizers purchased were primarily manufactured in the U.S. or China; however, they also identified additional countries (such as India, Mexico, and United Arab Emirates) that were not represented in our study. Therefore, it is possible that there are additional impurities present in hand sanitizers that were not quantified in our study, but their presence may differ based on the country of manufacture. Further, the authors performed an exposure assessment and determined that the benzene exposures would "increase the EPA's benchmark for the *de minimus* cancer risk" in children, teenagers, and adults [43].

Certain ABHSs identified in this study may possibly be less effective at preventing SARS-CoV-2 transmission from hand surfaces due to lower ethanol content than current regulatory standards; however, more information would be needed to verify that hypothesis. Some studies have reported on ethanol content in hand sanitizer products. For example, Tse et al. (2021) [11] reported ethanol concentrations between 63% and 90% v/v, with one sample having ethanol content below 60%. Another study reported ethanol concentrations between 16.21% and 87.33% v/v, with a total of 33.62% of the samples collected (39/116) containing under 60% ethanol [41].

The current study was not without limitations. One limitation included the GC/MS analysis sensitivity for detecting benzene. The GC/MS analytical method utilized in this study resulted in an LOQ of 0.44 µg/mL for benzene. The determined LOQ met the FDA's concentration ranges reported in its published method entitled "Direct Injection Gas Chromatography Mass Spectrometry (GC/MS) Method for the Detection of Listed Impurities in Hand Sanitizers" (0.044–2.19 µg/mL). In terms of sensitivity, however, the determined LOQ was one order of magnitude larger than the lowest range that FDA reported. Overall, the determined LOQ met method specifications; however, when converting two reported benzene non-detects from one hand sanitizer marketed to children (CCR-120 and CCR-120 duplicate) utilizing the measured sample density (0.9145 g/mL) to a µg/g value in order to compare to the FDA interim limit, the calculated concentration appeared to be above the FDA interim limit. All other benzene non-detects, when converted to µg/g utilizing hand sanitizer densities measured between 0.8345–0.8865 g/mL, were below the FDA interim limit. An additional study limitation was that our dataset provides impurity levels in the final formulation, which contain other ingredients such as fragrances, dyes, water, and other compounds, and not in the raw ethanol. The concentrations therefore may be higher in the raw ethanol than reported in this study, and additional exceedances cannot be ruled out. Further, the impurities possibly could have come from ingredients other than the raw ethanol.

5. Conclusions

Given increased ABHS consumer use during the COVID-19 pandemic and the potential ingestion risk by children, characterization of the availability of ABHSs that may be attractive to children proved vital. Further, due to increased reports of incidental ingestion and potential contamination of these products with impurities and ineffective ethanol levels, identification and quantification of potential health hazards to the public, especially to children, who represent a unique at-risk population, was performed.

As this study shows, numerous ABHSs with labeling or packaging characteristics that may increase their attractiveness to children are available for purchase in the U.S. Some products were packaged in containers resembling food, which is concerning because of the possibility of accidental ingestion. In our impurity analysis of a subset of available ABHSs marketed to children, the majority of the products did contain measurable concentrations of one or more organic impurities at levels above the respective LOQs. However, only some of the products exceeded FDA interim limits for several organic impurities, such as acetaldehyde, benzene, and acetal. All products were analyzed in their final formulation; however, the FDA interim limits were developed to address the acceptable impurity levels in the raw ethanol used to manufacture these products. Whether the impurity levels seen in this study would be higher in the raw ethanol remains unclear. While we observed exceedances, it is unclear if the impurities are present at concentrations that would possibly pose a health risk to consumers—especially children—should dermal absorption or accidental ingestion occur. Given the potential for exposure to children, however, and the reported accidental poisoning increase in children following the start of the COVID-19 pandemic, this issue is concerning, and should be further evaluated.

Future studies on the potential source(s) of these impurities in ethanol and/or other ingredients used in hand sanitizer products are thus needed. Understanding if potential contamination was a direct result of new manufacturers and improper manufacturing procedures during the COVID-19 pandemic would be helpful, as would knowing if the presence of such impurities will linger even after the FDA interim limits are no longer applicable and the ethanol shortage has diminished. Additional assessments should be made to determine whether the criteria examined in this study, such as packaging, gel coloring, and scent, encourage ABHS attraction, use, and potential ingestion. We can reasonably assume that future events may once again increase ABHS demand, and result in supply shortages.

Supplementary Materials: The following supporting information can be downloaded at: https://www.mdpi.com/article/10.3390/ijerph192114424/s1, Table S1: Hand sanitizer active ingredients and impurities, chromatogram peak assignments, LOQ and calibration range; Table S2: Statistical tests for comparing impurity concentrations in U.S. and non-U.S. products; Figure S1: Total Ion Chromatogram (TIC) of Hand Sanitizer Active Ingredients and Impurities; Figure S2: Impurity concentrations in U.S. and non-U.S. manufactured products; Figure S3: Impurity concentrations in gel and liquid products; Figure S4: Impurity concentrations in products by type of active ingredient; Figure S5: Impurity concentrations in scented or non-scented products; Figure S6: Impurity concentrations in non-colored and colored products; Figure S7: Impurity concentrations in original and duplicate samples; Figure S8: Labeled vs. measured alcohol concentrations in hand sanitizers marketed to children (expressed as a ratio fraction).

Author Contributions: Conceptualization, L.E.G., E.J.d.G., N.R.B., J.K., M.V., R.Z., S.D. and S.G.; Formal Analysis, L.E.G., E.J.d.G., A.M. and N.R.B.; Investigation, L.E.G., E.J.d.G., N.R.B., A.M., J.K. and K.G.S.; Methodology, L.E.G., E.J.d.G., J.K., N.R.B., M.J.L., M.V., R.Z., S.D. and S.G.; Project Administration, L.E.G.; Resources, M.J.L., Visualization, L.E.G., E.J.d.G., N.R.B., A.M. and M.J.L.; Writing—Original Draft Preparation, L.E.G., E.J.d.G., N.R.B., A.M. and K.G.S.; Writing—Review and Editing, L.E.G., E.J.d.G., N.R.B., A.M., K.G.S., J.K., M.J.L., R.Z., M.V., S.D. and S.G. All authors have read and agreed to the published version of the manuscript.

Funding: This research received no external funding.

Institutional Review Board Statement: Not applicable.

Informed Consent Statement: Not applicable.

Data Availability Statement: Not applicable.

Acknowledgments: The authors would like to thank Carrie Kahn (Stantec's ChemRisk group) for providing assistance with gathering literature, and copy editing of the manuscript.

Conflicts of Interest: Ten of the authors [L.E.G., E.J.d.G., N.R.B., K.G.S., A.M., J.K., M.V., R.Z., S.D. and S.G.] are currently or were formerly employed by ChemRisk (Stantec), a consulting firm that provides scientific advice to the government, corporations, law firms, and various scientific and professional organizations. One of the authors [M.J.L.] is a marine scientist senior at the Virginia Institute of Marine Science at the College of William and Mary. One of the authors [M.V.] has provided consulting services for a manufacturer of hand sanitizers. Neither M.V. nor any other authors have received any funding from hand sanitizer manufacturers in support of this manuscript.

References

1. RFA. Essential Energy: 2021 Ethanol Industry Outlook. 2021. Available online: https://d35t1syewk4d42.cloudfront.net/file/314/RFA_Outlook_2021_fin_low.pdf (accessed on 1 May 2021).
2. FDA. Q&A for Consumers: Hand Sanitizers and COVID-19. Content Current as of 17 May 2022. Available online: https://www.fda.gov/drugs/information-drug-class/qa-consumers-hand-sanitizers-and-covid-19 (accessed on 29 July 2022).
3. FDA. Temporary Policy for Preparation of Certain Alcohol-Based Hand Sanitizer Products During the Public Health Emergency (COVID-19): Guidance for Industry. March 2020. Available online: https://collections.nlm.nih.gov/catalog/nlm:nlmuid-9918227352906676-pdf (accessed on 10 February 2021).
4. FDA. Temporary Policy for Manufacture of Alcohol for Incorporation into Alcohol-Based Hand Sanitizer Products during the Public Health Emergency (COVID-19): Guidance for Industry. March 2020. Available online: https://www.regulations.gov/document/FDA-2020-D-1106-0038 (accessed on 10 February 2021).
5. FDA. Policy for Testing of Alcohol (Ethanol) and Isopropyl Alcohol for Methanol, Including during the Public Health Emergency (COVID-19): Guidance for Industry. January 2021. Available online: https://www.fda.gov/regulatory-information/search-fda-guidance-documents/policy-testing-alcohol-ethanol-and-isopropyl-alcohol-methanol-including-during-public-health (accessed on 22 September 2022).
6. FDA. Policy for Temporary Compounding of Certain Alcohol-Based Hand Sanitizer Products during the Public Health and Human Services Food and Drug Administration Center of Drug Evaluation and Research (CDER). March 2020. Available online: https://doh.sd.gov/boards/pharmacy/assets/Temporary-Policy-Compounding-Hand-Sanitizer-Guidance.pdf (accessed on 7 August 2020).

7. FDA. Hand Sanitizers: COVID-19. Content Current as of 10 January 2022. Available online: https://www.fda.gov/drugs/coronavirus-covid-19-drugs/hand-sanitizers-covid-19 (accessed on 29 July 2022).
8. WHO. Guide to Local Production: WHO-Recommended Handrub Formulations. Available online: https://www.who.int/publications/i/item/WHO-IER-PSP-2010.5 (accessed on 1 April 2010).
9. Onuki, S.; Koziel, J.A.; Jenks, W.S.; Cai, L.; Grewell, D.; van Leeuwen, J.H. Taking ethanol quality beyond fuel grade: A review. *J. Inst. Brew.* **2016**, *122*, 588–598. [CrossRef]
10. FDA. Direct Injection Gas Chromatography Mass Spectrometry (GC-MS) Method for the Detection of Listed Impurities in Hand Sanitizers. 24 August 2020. Available online: https://www.fda.gov/media/141501/download (accessed on 29 July 2022).
11. Tse, T.J.; Nelson, F.B.; Reaney, M.J.T. Analyses of commercially available alcohol-based hand rubs formulated with compliant and non-compliant ethanol. *Int J. Environ. Res. Pub. Health* **2021**, *18*, 3766. [CrossRef] [PubMed]
12. FDA. FDA Updates on Hand Sanitizers Consumers Should Use. Content Current as of 27 July 2022. Available online: https://www.fda.gov/drugs/drug-safety-and-availability/fda-updates-hand-sanitizers-consumers-should-not-use (accessed on 29 July 2022).
13. FDA. Safely Using Hand Sanitizer 3 November 2021. Available online: https://www.fda.gov/consumers/consumer-updates/safely-using-hand-sanitizer (accessed on 29 July 2022).
14. Gormley, N.J.; Bronstein, A.C.; Rasimas, J.J.; Pao, M.; Wratney, A.T.; Sun, J.; Austin, H.A.; Suffredini, A.F. The rising incidence of intentional ingestion of ethanol-containing hand sanitizers. *Crit. Care Med.* **2012**, *40*, 290–294. [CrossRef] [PubMed]
15. Mahmood, A.; Eqan, M.; Pervez, S.; Alghamdi, H.A.; Tabinda, A.B.; Yasar, A.; Brindhadevi, K.; Pugazhendhi, A. COVID-19 and frequent use of hand sanitizers; human health and environmental hazards by exposure pathways. *Sci. Total Environ.* **2020**, *742*, 140561. [CrossRef] [PubMed]
16. Santos, C.; Kieszak, S.; Wang, A.; Law, R.; Schier, J.; Wolkin, A. Reported adverse health effects in children from ingestion of alcohol-based hand sanitizers-United States, 2011–2014. *MMWR* **2017**, *66*, 223–226. [CrossRef] [PubMed]
17. AAPCC. Hand Sanitizer. Available online: https://aapcc.org/track/hand-sanitizer (accessed on 29 January 2022).
18. Arndt, T.; Schrofel, S.; Gussregen, B.; Stemmerich, K. Inhalation but not transdermal resorption of hand sanitizer ethanol causes positive ethyl glucuronide findings in urine. *Sci. Int.* **2014**, *237*, 126–130. [CrossRef] [PubMed]
19. Brewer, C.; Streel, E. Is alcohol in hand sanitizers absorbed through the skin or lungs? Implications for disulfiram treatment. *Alcohol Alcohol.* **2020**, *55*, 354–356. [CrossRef] [PubMed]
20. FDA. FDA Drug Safety Communication. 16 June 2021. Available online: https://www.fda.gov/media/150127/download (accessed on 29 July 2022).
21. Chan, A.P.L.; Chan, T.Y.K. Methanol as an unlisted ingredient in supposedly alcohol-based hand rub can pose serious health risk. *Int. J. Environ. Res. Public Health* **2018**, *15*, 1440. [CrossRef] [PubMed]
22. NIOSH. Emergency Response Safety and Health Database: Methanol: Systemic Agent. CAS No.: 67-56-1. Page Last Reviewed 12 May 2011. Available online: https://www.cdc.gov/niosh/ershdb/emergencyresponsecard_29750029.html (accessed on 22 July 2022).
23. ILO. Acetal. CAS No.: 105-57-7. International Chemical Safety Card (ICSC) 1070. November 2003. Available online: https://inchem.org/documents/icsc/icsc/eics1070.htm (accessed on 20 July 2022).
24. ILO. Acetaldehyde. CAS No.: 75-07-0. International Chemical Safety Card (ICSC) 0009. November 2003. Available online: https://inchem.org/documents/icsc/icsc/eics0009.htm (accessed on 20 July 2022).
25. ILO. Isobutanol. CAS No.: 78-83-1. International Chemical Safety Card (ICSC) 0113. April 2005. Available online: https://www.ilo.org/dyn/icsc/showcard.display?p_lang=en&p_card_id=0113&p_version=2 (accessed on 20 July 2022).
26. ILO. 1-Butanol. CAS No.: 71-36-3. International Chemical Safety Card (ICSC) 0111. April 2005. Available online: https://www.ilo.org/dyn/icsc/showcard.display?p_lang=en&p_card_id=0111&p_version=2 (accessed on 20 July 2022).
27. ILO. Acetone. CAS No: 67-64-1. International Chemical Safety Card (ICSC) 0087. April 2009. Available online: https://inchem.org/documents/icsc/icsc/eics0087.htm (accessed on 20 July 2022).
28. ILO. 1-Pentanol. CAS No.: 71-41-0. International Chemical Safety Card (ICSC) 0535. April 2009. Available online: https://www.ilo.org/dyn/icsc/showcard.display?p_lang=en&p_card_id=0535&p_version=2 (accessed on 20 July 2022).
29. ILO. Ethyl Acetate. CAS No.: 141-78-6. International Chemical Safety Card (ICSC) 0367. April 2014. Available online: https://inchem.org/documents/icsc/icsc/eics0367.htm (accessed on 20 July 2022).
30. ILO. Isoamyl Alcohol. CAS No.: 123-51-3. International Chemical Safety Card (ICSC) 0798. June 2015. Available online: https://www.ilo.org/dyn/icsc/showcard.display?p_lang=en&p_card_id=0798&p_version=2 (accessed on 20 July 2022).
31. ILO. 2-Butanol. CAS No.: 78-92-2. International Chemical Safety Card (ICSC) 0112. May 2018. Available online: https://www.ilo.org/dyn/icsc/showcard.display?p_lang=en&p_card_id=0112&p_version=2 (accessed on 20 July 2022).
32. IARC. International Agency for Research on Cancer (IARC)-Summaries & Evaluations: Acetaldehyde (Group 2B). CAS No.: 75-07-0. Section 5: Summary of Data Reported and Evaluation. Last Updated 12 April 1999. Available online: https://inchem.org/documents/iarc/vol71/005-acetaldehyde.html (accessed on 22 July 2022).
33. IARC. *IARC Monographs on the Evaluation of Carcinogenic Risks to Humans, Vol. 120: Benzene*; International Agency for Research on Cancer (IARC); World Health Organization (WHO): Lyon, France, 2018.
34. Bedner, M.; Murray, J.A.; Urbas, A.A.; MacCrehan, W.A.; Wilson, W.B. *A Comparison of Measurement Methods for Alcohol-Based Hand Sanitizers*; National Institute of Standards and Technology: Gaithersburg, MD, USA, 2021; pp. 1–56. [CrossRef]

35. Tse, T.J.; Purdy, S.K.; Shen, J.; Nelson, F.B.; Mustafa, R.; Wiens, D.J.; Reaney, M.J.T. Toxicology of alcohol-based hand rubs formulated with technical-grade ethanol. *Toxicol. Rep.* **2021**, *8*, 785–792. [CrossRef] [PubMed]
36. Kozal, J.S.; Vincent, M.J.; Gloekler, L.E.; de Gandiaga, E.J.; Massarsky, A.; Zisook, R.E.; Binczewski, N.R.; Gibbs, K.E.; Gaffney, S.H.; Dotson, G.S. Risk Assessment of Organic Impurities Detected in Hand Sanitizers Marketed to Children during the COVID-19 Pandemic. In Proceedings of the Society of Toxicology 61st Annual Meeting & ToxExpo, San Diego, CA, USA, 27–31 March 2022. Abstract no. 3341.
37. FDA. Warning Letter 320-21-04 to J. Gonzalez Olvera, General Director, 4E Global, S.A.P.I. de C.V. MARCS-CMS 608940. 23 October 2020. Current as of 3 November 2020. Available online: https://www.fda.gov/inspections-compliance-enforcement-and-criminal-investigations/warning-letters/4e-global-sapi-de-cv-608940-10232020 (accessed on 2 August 2022).
38. FDA. Warning Letter 320-22-16 to M.S. Eun, President, Mirfeel Korea Co., Ltd. MARCS-CMS 627401. 22 June 2022. Current as of 28 June 2022. Available online: https://www.fda.gov/inspections-compliance-enforcement-and-criminal-investigations/warning-letters/mirfeel-korea-co-ltd-627401-06222022 (accessed on 2 August 2022).
39. Light, D.; Kucera, K.; Wu, Q. Letter to FDA Re: Valisure Citizen Petition on Hand Sanitizer Products Containing Benzene Contamination and Other Significant Issues. 24 March 2021. Available online: https://assets-global.website-files.com/6215052733f8bb8fea016220/626fee4c78a2e4769ba2b7fe_FDA-2021-P-0338-0001_attachment_1.pdf (accessed on 28 March 2021).
40. Health Canada. Public Advisory: Recall of Certain Hand Sanitizers That May Pose Health Risks (Part 1—17 June 2020 to 24 March 2021). 20 May 2021. Available online: https://recalls-rappels.canada.ca/en/alert-recall/recall-certain-hand-sanitizers-may-pose-health-risks-part-1-june-17-2020-march-24-2021 (accessed on 29 July 2022).
41. Manuel, C.S.; Yeomans, D.J.; Williams, J.A.; Fricker, C.; Kucera, K.; Light, D.; Arbogast, J.W. Presence of unsafe chemical impurities, accelerated evaporation of alcohol, and lack of key labeling requirements are risks and concerns for some alcohol-based hand sanitizers and dispenser practices during the COVID-19 pandemic. *PLoS ONE* **2022**, *17*, e0265519. [CrossRef] [PubMed]
42. Ng, J.K.; Tay, F.H.; Wray, P.S.; Mohd Saberi, S.S.; Ken Ting, K.K.; Khor, S.M.; Chan, K.L.A. Inexpensive Portable Infrared Device to Detect and Quantify Alcohols in Hand Sanitizers for Public Health and Safety. *Anal. Chem.* **2021**, *93*, 15015–15023. [CrossRef] [PubMed]
43. Pal, V.K.; Lee, S.; Naidu, M.; Lee, C.; Kannan, K. Occurrence of and dermal exposure to benzene, toluene and styrene found in hand sanitizers from the United States. *Environ. Int.* **2022**, *167*, 107449. [CrossRef] [PubMed]

International Journal of
Environmental Research and Public Health

MDPI

Article

Health Risk Assessment of Dermal Exposure to Polycyclic Aromatic Hydrocarbons from the Use of Infant Diapers

Alfred Bernard [1,*] and Vincent Dudler [2]

1 Institut de Recherche Expérimentale et Clinique (IREC), Université catholique de Louvain, 74 Avenue Hippocrate, 1200 Bruxelles, Belgium
2 Federal Food Safety and Veterinary Office, Schwarzenburgstrasse 155, 3003 Bern, Switzerland
* Correspondence: alfred.bernard@uclouvain.be

Abstract: In September 2021, the European Chemicals Agency evaluated a dossier for restricting polycyclic aromatic hydrocarbons (PAHs) in infant diapers and concluded that risks were not demonstrated, because of inconclusive exposure data. To fill this gap, we measured the 16 priority PAHs of the U.S. Environmental Protection Agency in the diaper core of four brands and in the sheets and fastening tapes of six brands of commercially available diapers. Health risks were conservatively assessed by assuming that dermally absorbed PAHs can cause both local (skin cancer) and systemic critical effects (neurobehavioral changes). Total concentrations of PAHs in the diaper core and top sheet, the only significant contributors to skin exposure, averaged 26.5 µg/kg and 66.6 µg/kg, respectively. Excess skin cancer risks and hazard quotients for neurobehavioral effects calculated with the daily dose of total PAHs from the combined diaper core and top sheet averaged 1.44×10^{-7} and 1.19×10^{-2}, respectively. The median daily doses of total PAHs and of its benzo[a]pyrene-equivalent from breast milk estimated worldwide are 171 and 30 times greater than that from the combined diaper core and top sheet, respectively. Altogether, these findings indicate that trace levels of PAHs found in infant diapers are unlikely to pose health risks.

Keywords: diaper; polycyclic aromatic hydrocarbons; skin cancer; neurobehavioral changes

Citation: Bernard, A.; Dudler, V. Health Risk Assessment of Dermal Exposure to Polycyclic Aromatic Hydrocarbons from the Use of Infant Diapers. *Int. J. Environ. Res. Public Health* **2022**, *19*, 14760. https://doi.org/10.3390/ijerph192214760

Academic Editors: Esref Demir and Sam Kacew

Received: 11 October 2022
Accepted: 8 November 2022
Published: 10 November 2022

1. Introduction

Polycyclic aromatic hydrocarbons (PAHs) are ubiquitous pollutants occurring as complex mixtures throughout the environment. They are formed during the incomplete combustion of fossil fuels and organic matter such as wood or tobacco. They also occur in food as a result of environmental pollution or of some food cooking methods such as grilling and roasting. For nonsmokers, the major sources of exposure are food and, to some extent, polluted air [1,2].

PAHs are organic compounds consisting of multiple fused aromatic rings. The physicochemical properties of PAHs that determine their potential toxicity greatly vary with the number of rings. Two-ring PAHs and, to a lesser extent, three- and four-ring PAHs can partly dissolve in water and are sufficiently volatile to be released in air. By contrast, PAHs with five or more rings have a very low volatility and water solubility. These large PAHs, therefore, are tightly adsorbed onto the surface of solid materials, which reduces their biological accessibility. The strong hydrophobicity of these large PAHs also reduces their solubilization into aqueous liquids while facilitating their transfer across biological membranes and barriers [1,2].

Mixtures of PAHs have long been recognized as potent human carcinogens [3]. In humans, as with animals, the sites of tumors induced by PAHs largely depend on the route of exposure. By inhalation, benzo[a]pyrene (BaP), one of the most potent PAHs, induces only respiratory tract tumors in both humans and rodents, whereas, administered orally in animals, it causes gastro-intestinal tumors. By the dermal route, PAHs caused squamous cell carcinoma of the skin in humans with high occupational exposure [3]. Lifetime carcinogenicity bioassays in several strains of mice have demonstrated that dermally applied BaP induces only skin tumors, which, thus, should be considered as the critical effect of PAHs by the dermal route [1–3]. This conclusion is strengthened by the fact that dermal slope factors are more than one order of magnitude greater than oral slope factors. The formation rate of DNA adducts is also orders of magnitude greater in the skin of mice exposed to BaP compared to internal organs [4].

Animal studies have also shown that PAHs can cause various systemic effects including developmental, reproductive, and immunological effects [1]. Among these effects, the critical one occurring at the lowest exposure level is the altered neurobehavior observed in rats following gavage or inhalation of BaP during early life. Epidemiological studies among nonsmoking pregnant women mainly exposed to PAHs mixtures via food have reported associations between biomarkers of BaP exposure (benzo[a]pyrene diol epoxide-DNA adducts) and adverse birth outcomes, neurobehavioral effects, and decreased fertility [1].

Recently, public concern arose over exposure to PAHs through the use of disposable infant diapers. A report published in 2019 by the French Agency for Food, Environmental, and Occupational Health and Safety (ANSES), largely echoed in the media, suggested that adverse systemic effects of PAHs, in particular, cancers and neurobehavioral changes, cannot be excluded from the long-term [5]. Therefore, the French agency concluded the need of regulatory actions in order to ensure the safety of diapers. In October 2020, ANSES submitted to the European Chemical Agency (ECHA) a dossier for restricting in infant diapers levels of several hazardous contaminants including PAHs [6]. In September 2021, however, both the ECHA's Committee for Risk Assessment (RAC) and the Committee for Socio-Economic Analysis (SEAC) considered that the evidence was insufficient to conclude that some chemicals found in diapers may pose a risk to babies [7]. In particular, RAC found that data on the concentrations of some of these substances in diapers were inconclusive. This was especially the case for PAHs as the risk assessment conducted by ANSES relied on the limits of quantification of a poorly sensitive analytical method. In the absence of accurate measurements, RAC was not in a position to completely exclude risks from PAHs in diapers. To fill this gap, we measured PAHs in commercially available infant diapers with an adequate analytical method. We also estimated health risks through a very conservative approach assuming that dermally absorbed PAHs can cause both local (skin cancer) and systemic critical effects (neurobehavioral changes).

2. Materials and Methods

2.1. Composition of Disposable Diapers

Disposable infant diapers consist of several layers of materials with different functional properties. The most important layer in weight and volume is the diaper core, which is a superabsorbent polymer typically made of sodium polyacrylate granules. In contact with urine, this superabsorbent polymer forms a gel-like material that absorbs up to 30 times its weight in liquid. The top sheet is the inner porous layer of the diaper in direct contact with the baby's skin. This layer is designed for a rapid transfer of the urine to the diaper core. The back sheet is the water-proof outer layer of the diaper that prevents urine from leaking out of the diaper. Diapers also contain various additional features such as fastening tapes to ensure a good fit [8–10].

2.2. Measurement of PAHs

This health risk assessment is based on analyses of PAHs performed by the Swiss Federal Institute of Metrology (METAS), which is the reference laboratory in Switzerland for the determination of PAHs in food. The laboratory analyzed the 16 priority PAHs of the U.S. Environmental Protection Agency (U.S. EPA) by gas chromatography–mass spectrometry (GC–MS) using a method adapted from the European standard EN 16619:2015. PAHs were extracted from diapers with n-hexane by liquid-pressure extraction (LPE). For each analysis, about 2 g of material was mixed in a LPE cell with glass beads and 200 μL of a standard solution (mixture of deuterated PAHs). The extraction was conducted in two cycles under the following conditions: pressure 1500 psi, temperature 100 °C, heating time 5 min, static time 10 min, and rinsing volume 10 mL. The extract was concentrated to 5 mL with a rotary evaporator and the analysis was performed by gas chromatography-mass spectrometry (GC-MS). PAHs were measured in the back and top sheets and in the fastening tapes of six international brands of diapers purchased in commercial stores in Switzerland. PAHs were also measured in the diaper core of four of these brands. The diaper core and top sheet had a mean weight of 13 and 4 g, respectively. Results are reported as the mean of three measurements performed on mixtures of three to five randomly selected diapers. The recovery of the PAHs content in the samples was determined from the deuterated labeled PAHs according to the EN 16619:2015 standard, and was always higher than 90%. The limit of quantification (LOQ) is sample-dependent and was measured specifically for each PAH/sample pair. For the 16 substances considered, the range of LOQs and the average were 0.01–2.18 μg/kg and 0.51 μg/kg, respectively. For PAHs that were not quantifiable, we used the LOQ value divided by two to calculate the total PAHs concentrations. In the assessment of skin cancer risks, we calculated the BaP potency equivalent concentrations of individual and total PAHs using the same potency equivalent factors (PEF) as in the ANSES report and our previous study [5,11,12].

2.3. Health Risk Assessment

Because the diaper core is not in direct contact with the skin, ingredients in diaper core require urine as an aqueous carrier to reach the skin. The amount of liquid resurfacing back to the skin is, however, strongly limited by the potent absorbency of polyacrylate that allows the achievement of a very low rewetting fraction. In the study of Dey et al. [9], the proportion of urine returning to the top sheet under pressure was estimated on average at 0.46% with a range of 0.32–0.66%. As in our previous study [12], we conservatively used a rewet factor of 1%. By contrast, chemicals in the top sheet may be directly transferred to the skin. For ingredients in the top sheet intended to be applied onto the skin (i.e., top-sheet lotions), Odio et al. [13] estimated that less than 7% is actually transferred to the skin. For ingredients not intended for skin care, their absorption is strongly limited by the fact these ingredients are integrated within the polymeric matrix resin of the sheet. We nevertheless adopted this transfer percentage of 7% as a conservative upper bound estimate of the skin transfer of PAHs from the top sheet.

As shown previously [12], potential health risks related to infant diaper wearing are expectedly highest in infants aged 0–6 months when the number of diapers used per day and the surface/body weight ratio are highest. We, therefore, focused our risk assessment on this period of age by adopting the same conservative exposure parameters as previously [12] (7.98 diapers per day, body weight of 3.9 kg). The daily dose of PAHs from the diaper core or the top sheet for an infant aged 0–6 months was calculated using the following equations:

$$\text{Diaper core}: \ DD = \frac{C \times W \times N \times R \times A}{SDAF} \tag{1}$$

$$\text{Top sheet}: \ DD = \frac{C \times W \times N \times T \times A}{SDAF} \tag{2}$$

where DD is the daily dose (μg/day); C is the concentration of the chemical in the diaper (μg/kg); W is the weight of the diaper layer (kg); N is the number of diapers used per day; T is the fraction transferred to the skin (7%); R is the rewet factor (1%); A is the fraction absorbed by the skin conservatively set at 100%; SDAF is the solvent-dependent adjustment factor incorporated to adjust for the difference in bioaccessibility of PAHs from the diaper core or top sheet when the extraction is performed with a biological fluid at 37 °C or with n-hexane at 100 °C. As there is a close inverse relationship between log Ko/w and water solubility [14] and only a small deviation of 0.4 between log Ko/w and the log Kn-hexane/water of alkyl-aromatic molecules [15], we adopted a SDAF of 100. This adjustment factor of 100 can be considered conservative as it is 20 times lower than the

Kow value of the most water-soluble PAH congener (naphthalene) among the 16 priority U.S. EPA PAHs. For the top sheet in direct contact with the skin, we used a lower SDAF to account for the fact that transfer to the skin does not require solubilization in urine. PAHs from the top sheet can indeed directly dissolve in sweat or sebum that is present in variable proportions at the surface of the skin. We adopted a default SDAF of 10 to take into account that the transfer factor of 7% was derived for a product intended for skin care, which presumably should be more easily extracted with n-hexane than PAHs embedded in the top sheet matrix.

We also estimated the daily dose of PAHs from the diaper core on the basis of the limits of solubility in water of individual PAHs using the following equation:

$$DD = C \times R \times V \tag{3}$$

where DD is the daily dose (μg/day); C is the limit of water solubility (μg/L); R is the rewet factor of 1%; V is the daily urine output estimated at 500 mL for a 0–6-months-old infant. The limits of water solubility of individual PAHs were obtained from PubChem https://pubchem.ncbi.nlm.nih.gov (accessed on 1 February 2022).

Excess skin cancer risks of PAHs from the diaper core or the top sheet were calculated by using the following equation:

$$ECR = \frac{DD \times PEF \times CSF \times T}{S \times 70} \tag{4}$$

where ECR is the excess skin cancer risk; DD is the daily dose (μg per day); PEF is the BaP relative potency equivalent factor; CSF is the skin cancer slope factor of 3.5 ($\mu g/cm^2$ per day)$^{-1}$ developed by Knafla et al. [4]; T is the duration of exposure (0.5 years); 70 is the duration of lifetime conventionally set at 70 years; S is the skin surface area in contact with the diaper (234 cm^2) [16,17]. Unlike our previous study [12], we decided not to incorporate an age-dependent adjustment factor (ADAF) in agreement with the recent opinion of the ECHA risk assessment committee, concluding that there is no need to apply an ADAF for PAHs in addition to the high to low dose extrapolation [18].

Although there are no experimental or epidemiological data to support an extrapolation from ingestion or inhalation to the dermal route of exposure for PAHs, ANSES assumed that PAHs from diapers can cause behavioral changes as observed in rats following oral or inhalation exposure during early life. In case this assumption proves to be correct, we also estimated the risks of neurobehavioral changes by calculating the hazard quotient with the following equation:

$$HQ = \frac{DD}{TDI} \tag{5}$$

where HQ is the hazard quotient; DD is the daily dose and TDI is the tolerable daily intake, both expressed in μg/kg body weight and per day, respectively. We assumed a body weight of 3.9 kg for a 0–6-months-old infant and we used as TDI the U.S. EPA reference dose (RfD) of 0.3 μg/kg body weight and per day as used by ANSES [5] and our previous study [12].

3. Results

3.1. Diaper Core

All investigated PAHs congeners were found in the core of the four tested diaper brands and, depending on the brand, 38% to 81% of them were present in quantifiable amounts (see Supplementary materials, Table S1). Of these, there were the two most potent PAHs, BaP and dibenzo[a,h]anthracene (DBA) that were quantified in three and two brands, respectively. The concentrations of BaP and DBA, however, did not exceed 1 μg/kg at the exception of one brand that contained 1.4 μg/kg of DBA. Of note, the concentrations and patterns of PAHs congeners were very consistent across the four brands, being all dominated by naphthalene, which, on average, contributed to 52% of the total PAHs concentrations. Table 1 shows the mean concentrations of PAHs in the diaper core of the four brands and the estimated ECR. The total PAHs concentration in the diaper core averaged 26.5 μg/kg and 1.26 μg/kg when expressed as BaP-equivalent. The ECRs for BaP and DBA were lower than 10^{-9}, while the ECR for the total PAHs concentration remained below 10^{-8}. As shown in Table 2, the hazard quotient for neurobehavioral effects of total PAHs in the diaper core averaged 2.35×10^{-4}.

Table 1. Excess skin cancer risks of polycyclic aromatic hydrocarbons (PAHs) in the diaper core.

PAHs	Concentration (μg/kg) *	Dose from Diaper Core (μg/day)	PEF	Dose from Diaper Core (μg BaP eqv./day)	Dose from Diaper Core (μg BaP eqv./cm²/day)	Excess Skin Cancer Risk
Benzo[a]pyrene	0.55	5.65×10^{-6}	1	5.65×10^{-6}	2.42×10^{-8}	6.04×10^{-10}
Dibenzo[a,h]anthracene	0.48	4.98×10^{-6}	1	4.98×10^{-6}	2.13×10^{-8}	5.32×10^{-10}
Benzo[g,h,i]perylene	0.76	7.87×10^{-6}	0.01	7.87×10^{-8}	3.37×10^{-10}	8.41×10^{-12}
Naphtalene	13.7	1.42×10^{-4}	0.001	1.42×10^{-7}	6.07×10^{-10}	1.52×10^{-11}
Anthracene	0.81	8.44×10^{-6}	0.01	8.44×10^{-8}	3.61×10^{-10}	9.02×10^{-12}
Benzo[a]anthracene	0.54	5.56×10^{-6}	0.1	5.56×10^{-7}	2.38×10^{-9}	5.94×10^{-11}
Indeno[1,2,3-cd]pyrene	0.61	6.29×10^{-6}	0.1	6.29×10^{-7}	2.69×10^{-9}	6.72×10^{-11}
Chrysene	0.43	4.49×10^{-6}	0.01	4.49×10^{-8}	1.92×10^{-10}	4.80×10^{-12}
Benzo[b]fluoranthene	0.31	3.23×10^{-6}	0.1	3.23×10^{-7}	1.38×10^{-9}	3.45×10^{-11}
Benzo[k]fluoranthene	0.51	5.24×10^{-6}	0.1	5.24×10^{-7}	2.24×10^{-9}	5.60×10^{-11}
Acenaphthene	0.78	8.09×10^{-6}	0.001	8.09×10^{-9}	3.46×10^{-11}	8.65×10^{-13}
Acenaphthylene	0.35	3.63×10^{-6}	0.001	3.63×10^{-9}	1.55×10^{-11}	3.88×10^{-13}
Phenanthrene	3.58	3.71×10^{-5}	0.001	3.71×10^{-8}	1.59×10^{-10}	3.97×10^{-12}
Fluoranthene	1	1.04×10^{-5}	0.001	1.04×10^{-8}	4.43×10^{-11}	1.11×10^{-12}
Fluorene	1.37	1.42×10^{-5}	0.001	1.42×10^{-8}	6.07×10^{-11}	1.52×10^{-12}
Pyrene	0.76	7.88×10^{-6}	0.001	7.88×10^{-9}	3.37×10^{-11}	8.42×10^{-13}
Total PAHs	26.5	1.94×10^{-4}		1.31×10^{-5}	5.60×10^{-8}	1.40×10^{-9}

* Average of the concentrations measured in the core of four brands of diaper.

Table 2. Assessment of neurobehavioral risks of polycyclic aromatic hydrocarbons (PAHs) in the diaper core.

PAHs	Concentration (μg/kg) *	Dose from Diaper Core (μg/kg/day)	Hazard Quotient
Benzo[a]pyrene	0.545	1.45×10^{-6}	4.83×10^{-6}
Dibenzo[a,h]anthracene	0.48	1.28×10^{-6}	4.26×10^{-6}
Benzo[g,h,i]perylene	0.76	2.02×10^{-6}	6.73×10^{-6}
Naphtalene	13.7	3.64×10^{-5}	1.21×10^{-4}
Anthracene	0.81	2.17×10^{-6}	7.22×10^{-6}
Benzo(a)anthracene	0.54	1.43×10^{-6}	4.75×10^{-6}
Indeno(1,2,3-cd)pyrene	0.61	1.61×10^{-6}	5.37×10^{-6}
Chrysene	0.43	1.15×10^{-6}	3.84×10^{-6}
Benzo[b]fluoranthene	0.31	8.27×10^{-7}	2.76×10^{-6}
Benzo[k]fluoranthene	0.51	1.34×10^{-6}	4.48×10^{-6}
Acenaphthene	0.78	2.07×10^{-6}	6.92×10^{-6}
Acenaphtylene	0.35	9.31×10^{-7}	3.10×10^{-6}
Phenanthrene	3.58	9.52×10^{-6}	3.17×10^{-5}
Fluoranthene	1.00	2.66×10^{-6}	8.87×10^{-6}
Fluorene	1.37	3.64×10^{-6}	1.21×10^{-5}
Pyrene	0.76	2.02×10^{-6}	6.74×10^{-6}
Total PAHs	26.5	7.06×10^{-5}	2.35×10^{-4}

* Average of the concentrations measured in the core of four brands of diaper.

The importance of water solubility in reducing the skin transfer and cancer risk of PAHs from the diaper core clearly emerges from Table 3 showing ECRs calculated with the water solubility limits of individual PAHs. For the most potent PAHs (PEF $\geq$ 0.1) with water solubility limits lower than 10 μg/L, ECRs were all below 10^{-6}. The water solubility limits of the less potent PAHs (PEF 0.01 and 0.001) were much higher, ranging from 135 up to 31,000 μg/L in the case of naphthalene. The ECR of total PAHs reached 2.52×10^{-5} and 8.52×10^{-6} after exclusion of naphthalene.

Table 3. Excess skin cancer risks of polycyclic aromatic hydrocarbons (PAHs) in the diaper core calculated with their respective limit of solubility in water.

PAHs	Number of Rings	Water Solubility (µg/L)	Dose from Diaper Core (µg/day) *	PEF	Dose from Diaper Core (µg BaP eqv./day)	Dose from Diaper Core (µg BaP eqv./cm²/day)	Excess Skin Cancer Risk
Benzo[a]pyrene	5	1.62	8.10×10^{-3}	1	8.10×10^{-3}	3.46×10^{-5}	8.65×10^{-7}
Dibenzo[a,h]anthracene	5	1.66	6.60×10^{-3}	1	6.60×10^{-3}	3.55×10^{-5}	8.87×10^{-7}
Benzo[g,h,i]perylene	6	0.26	1.30×10^{-3}	0.01	1.30×10^{-5}	5.56×10^{-8}	1.39×10^{-9}
Naphthalene	2	31,000	$1.55 \times 10^{+2}$	0.001	1.55×10^{-1}	6.62×10^{-4}	1.66×10^{-5}
Anthracene	3	43.4	2.17×10^{-1}	0.01	2.17×10^{-3}	9.27×10^{-6}	2.32×10^{-7}
Benzo[a]anthracene	4	9.4	4.70×10^{-2}	0.1	4.70×10^{-3}	2.01×10^{-5}	5.02×10^{-7}
Indeno[1,2,3-cd]pyrene	6	0.19	9.50×10^{-4}	0.1	9.50×10^{-5}	4.06×10^{-7}	1.01×10^{-8}
Chrysene	4	2.0	1.00×10^{-2}	0.01	1.00×10^{-4}	4.27×10^{-7}	1.07×10^{-8}
Benzo[b]fluoranthene	5	1.5	7.50×10^{-3}	0.1	7.50×10^{-4}	8.33×10^{-5}	8.01×10^{-8}
Benzo[k]fluoranthene	5	0.8	4.00×10^{-3}	0.1	4.00×10^{-4}	8.40×10^{-5}	4.27×10^{-8}
Acenaphthene	3	3900	1.95×10^{1}	0.001	1.95×10^{1}	8.33×10^{-5}	2.08×10^{-6}
Acenaphthylene	3	3930	1.97×10^{1}	0.001	1.97×10^{1}	8.40×10^{-5}	2.10×10^{-6}
Phenanthrene	3	1300	6.50	0.001	6.50×10^{-3}	2.78×10^{-5}	6.94×10^{-7}
Fluoranthene	4	260	1.30	0.001	1.30×10^{-3}	5.56×10^{-6}	1.39×10^{-7}
Fluorene	3	1690	8.45	0.001	8.45×10^{-3}	3.61×10^{-5}	9.03×10^{-7}
Pyrene	4	135	6.75	0.001	6.75×10^{-3}	2.88×10^{-6}	7.21×10^{-8}
Total PAHs			2.11×10^{2}		2.36×10^{-1}	1.01×10^{-3}	2.52×10^{-5}
Total PAHs without naphthalene			5.64×10^{1}		8.07×10^{-2}	3.45×10^{-4}	8.62×10^{-6}

* The daily dose is calculated by assuming a daily urine output of 500 mL for a 0–6-months-old infant.

3.2. Top Sheet

In contrast to the diaper core, concentrations and patterns of PAHs congeners in the top sheet widely varied between the six tested brands (see Supplementary materials, Table S2). Depending on the brand, 31% to 50% of PAHs congeners could be quantified. However, only one brand contained a quantifiable amount of the two most potent PAHs congeners (BaP). Table 4 shows the mean concentrations of PAHs in the top sheet of the six tested brands and the associated ECR. The total PAHs concentration averaged 66.6 µg/kg and 3.51 µg/kg when expressed as BaP-equivalent. ECRs for BaP and DBA were lower than 10^{-8} and the ECR for total PAHs remained below 10^{-7}. As shown in Table 5, the hazard quotient for neurobehavioral effects of total PAHs in the top sheet was estimated, on average, at 1.27×10^{-2}.

3.3. Diaper Core Plus Top Sheet

Because of the great heterogeneity of top sheet PAHs concentrations between the six diaper brands, we refined our risk assessment by calculating the ECR and hazard coefficients of total PAHs from the combined core and top sheet of the four diaper brands. As shown in Table 6, the ECR ranged from 5.1×10^{-10} to 3.1×10^{-9} for PAHs in the diaper core and from 5.2×10^{-8} to 2.0×10^{-7} for PAHs in the top sheet. The ECR for PAHs from the combined diaper core and top sheet averaged 1.44×10^{-7} (range, 2.0×10^{-7} to 8.60×10^{-7}), while the hazard quotient for neurobehavioral changes averaged 1.19×10^{-2} (range, 9.6×10^{-3} to 1.41×10^{-2}).

3.4. Back Sheet and Fastening Tapes

Diaper materials used on the outside of the diaper chassis (e.g., back sheet and fastening tapes) have a very limited direct contact with the skin. Even though there would be some contact with the skin, the contribution of fastening tapes and the back sheet would be totally insignificant when one considers the very low concentrations of PAHs and the patterns of congeners (Table S2 of Supplementary materials). The two most potent PAHs congeners, BaP and DBA, were mostly undetectable or present in unquantifiable concentrations with the exception of BaP, which was quantified in one brand of fastening tape.

Regarding other congeners, only 27% and 29% of them could be quantified in the back sheet and fastening tapes, respectively.

Table 4. Excess skin cancer risks of polycyclic aromatic hydrocarbons (PAHs) in the top sheet of diapers.

PAHs	Concentration (μg/kg) *	Dose from Top Sheet (μg/day)	PEF	Dose from Top Sheet (μg BaP eqv./day)	Dose from Top Sheet (μg BaP eqv./cm²/day)	Excess Skin Cancer Risk
Benzo[a]pyrene	1.6	3.58×10^{-4}	1	3.58×10^{-4}	1.53×10^{-6}	3.82×10^{-8}
Dibenzo[a,h]anthracene	0.73	1.63×10^{-4}	1	1.63×10^{-4}	6.97×10^{-7}	1.74×10^{-8}
Benzo[g,h,i]perylene	1.33	2.97×10^{-4}	0.01	2.97×10^{-6}	1.27×10^{-8}	3.17×10^{-10}
Naphtalene	5.19	1.15×10^{-3}	0.001	1.15×10^{-6}	4.96×10^{-9}	1.24×10^{-10}
Anthracene	5.61	1.25×10^{-3}	0.01	1.25×10^{-5}	5.36×10^{-8}	1.34×10^{-9}
Benzo(a)anthracene	6.74	1.51×10^{-3}	0.1	1.51×10^{-4}	6.44×10^{-7}	1.61×10^{-8}
Indeno(1,2,3-cd)pyrene	0.43	9.61×10^{-5}	0.1	9.61×10^{-6}	4.11×10^{-8}	1.03×10^{-9}
Chrysene	3.09	6.90×10^{-4}	0.01	6.90×10^{-6}	2.95×10^{-8}	7.38×10^{-10}
Benzo[b]fluoranthene	1.66	3.71×10^{-4}	0.1	3.71×10^{-5}	1.59×10^{-7}	3.96×10^{-9}
Benzo[k]fluoranthene	1.54	3.44×10^{-4}	0.1	3.44×10^{-5}	1.47×10^{-7}	3.68×10^{-9}
Acenaphthene	8.48	1.89×10^{-3}	0.001	1.89×10^{-6}	8.10×10^{-9}	2.02×10^{-10}
Acenaphtylene	8.16	1.82×10^{-3}	0.001	1.82×10^{-6}	7.79×10^{-9}	1.95×10^{-10}
Phenanthrene	2.37	5.30×10^{-4}	0.001	5.30×10^{-7}	2.26×10^{-9}	5.66×10^{-11}
Fluoranthene	2.04	4.56×10^{-4}	0.001	4.56×10^{-7}	1.95×10^{-9}	4.87×10^{-11}
Fluorene	14.2	3.17×10^{-3}	0.001	3.17×10^{-6}	1.36×10^{-8}	3.30×10^{-10}
Pyrene	3.4	7.60×10^{-4}	0.001	7.60×10^{-7}	3.25×10^{-9}	8.12×10^{-11}
Total PAHs	66.6	1.49×10^{-2}		7.85×10^{-4}	3.35×10^{-6}	8.38×10^{-8}

* Average of the concentrations measured in the top sheet of six brands of diaper.

Table 5. Assessment of neurobehavioral risks of polycyclic aromatic hydrocarbons (PAHs) in the top sheet of diapers.

PAHs	Concentration (μg/kg) *	Dose from Top Sheet (μg/kg/day)	Hazard Quotient
Benzo[a]pyrene	1.6	9.17×10^{-5}	3.06×10^{-4}
Dibenzo[a,h]anthracene	0.73	4.18×10^{-5}	1.39×10^{-4}
Benzo[g,h,i]perylene	1.33	7.62×10^{-5}	2.54×10^{-4}
Naphtalene	5.19	2.97×10^{-4}	9.91×10^{-4}
Anthracene	5.61	3.21×10^{-4}	1.07×10^{-3}
Benzo(a)anthracene	6.74	3.86×10^{-4}	1.29×10^{-3}
Indeno(1,2,3-cd)pyrene	0.43	2.46×10^{-5}	8.21×10^{-5}
Chrysene	3.09	1.77×10^{-4}	5.90×10^{-4}
Benzo[b]fluoranthene	1.66	9.51×10^{-5}	3.17×10^{-4}
Benzo[k]fluoranthene	1.54	8.82×10^{-5}	2.94×10^{-4}
Acenaphthene	8.48	4.86×10^{-4}	1.62×10^{-3}
Acenaphtylene	8.16	4.68×10^{-4}	1.56×10^{-3}
Phenanthrene	2.37	1.36×10^{-4}	4.53×10^{-4}
Fluoranthene	2.04	1.17×10^{-4}	3.90×10^{-4}
Fluorene	14.2	8.14×10^{-4}	2.71×10^{-3}
Pyrene	3.40	1.95×10^{-4}	6.49×10^{-4}
Total PAHs	66.6	1.49×10^{-2}	1.27×10^{-2}

* Average of the concentrations measured in the top sheet of six brands of diaper.

3.5. Comparison of the Daily Dose of PAHs from Diaper Core or Top Sheet with That from Breast Milk

Table 7 compares the daily dose of total PAHs and of its BaP-equivalent from the combined diaper core and top sheet with that from breast milk of non-smoking women reported worldwide. The daily dose of total PAHs varied over two orders of magnitude between countries and over more than three orders of magnitude when expressed as BaP-equivalent. These wide variations in the BaP-equivalent daily dose are largely due to BaP and DBA whose concentrations in human milk were undetectable in some studies, while in other studies, they could be quantified with concentrations up to 4.36 μg/kg [19]. The median daily dose of total PAHs and of its BaP-equivalent from breast milk estimated worldwide are 171 and 30 times greater than that from the combined diaper core and top sheet, respectively.

Table 6. Excess skin cancer and neurobehavioral risks of polycyclic aromatic hydrocarbons (PAHs) in the core and the top sheet of four brands of diaper, separately or combined.

Brands	Part of Diaper	Total PAHs Concentration (μg/kg)	Dose from Diaper (μg BaP eqv./cm^2/day)	Excess Skin Cancer Risk	Dose from Diaper (μg/kg/day)	Hazard Quotient
A	Top sheet	57.2	8.1×10^{-6}	2.0×10^{-7}	3.28×10^{-3}	1.09×10^{-2}
	Core	17.8	3.0×10^{-8}	7.6×10^{-10}	4.74×10^{-6}	1.58×10^{-5}
	Top sheet + core		8.0×10^{-6}	2.0×10^{-7}	3.29×10^{-3}	1.10×10^{-2}
B	Top sheet	65.6	2.1×10^{-6}	5.2×10^{-8}	$3,74 \times 10^{-3}$	1.25×10^{-2}
	Core	34.9	4.7×10^{-8}	1.2×10^{-9}	9.28×10^{-5}	3.09×10^{-4}
	Top sheet + core		2.1×10^{-6}	1.7×10^{-7}	3.84×10^{-3}	1.28×10^{-2}
C	Top sheet	72.3	4.5×10^{-6}	1.1×10^{-7}	4.13×10^{-3}	1.36×10^{-2}
	Core	30.2	1.3×10^{-7}	3.1×10^{-9}	8.03×10^{-5}	2.68×10^{-4}
	Top sheet + core		4.7×10^{-6}	1.2×10^{-7}	4.21×10^{-3}	1.40×10^{-2}
D	Top sheet	49.4	3.4×10^{-6}	8.5×10^{-8}	2.82×10^{-3}	9.40×10^{-3}
	Core	22.5	2.0×10^{-8}	5.1×10^{-10}	6.00×10^{-5}	2.00×10^{-4}
	Top sheet + core		3.4×10^{-6}	8.6×10^{-8}	2.88×10^{-3}	9.60×10^{-3}
All	Top sheet + core		3.79×10^{-6}	1.44×10^{-7}	3.56×10^{-3}	1.19×10^{-2}

Table 7. Comparison of the BaP-equivalent daily dose of PAHs from breast milk of non-smoking women with that from diapers.

Country	Authors	Dose of PAHs from Breast Milk *		Breast Milk/Diaper Dose Ratio	
		μg/kg/Day	μg BaP eqv./kg/Day	μg/kg/Day	μg BaP eqv./kg/Day
Colombia	Torres-Moreno et al., 2022 [20]	0.62	6.13×10^{-3}	224	29
Portugal	Oliveira et al., 2020 [21]	1.28	6.48×10^{-2}	462	304
USA	Acharya et al., 2019 [22]	0.52	1.86×10^{-2}	188	87
Ghana	Asamoha et al., 2019 [23]	0.43	6.14×10^{-3}	155	29
Italy	Santonicola et al., 2017 [19]	16.1	1.00	5812	4695
Czech Republic	Pulkrabova et al., 2016 [24]	0.093	9.53×10^{-4}	34	4.5
Turkey	Cok et al., 2012 [25]	0.30	2.18×10^{-3}	108	10
USA	Kim et al., 2008 [26]	0.084	2.90×10^{-4}	30	1.3
Italy	Zanieri et al., 2007 [27]	2.58	1.16×10^{-2}	931	54
Japan	Kishikawa et al., 2003 [28]	0.11	6.75×10^{-3}	40	32

* Daily dose from breast milk was calculated for an infant of 5 kg of body weight fed daily with 700 mL of maternal milk containing 25 g/L of lipids. The daily dose from the diaper is the daily dose of PAHs from combined diaper core and top sheet.

4. Discussion

To accurately assess exposure to PAHs from infant diapers, we measured the 16 priority U.S. EPA PAHs in the different layers of commercially available infant diapers in Switzerland. In the diaper core and top sheet, which are the only significant contributors to skin exposure, the proportion of quantifiable PAHs congeners varied depending on the brand between 38% and 81%, and 31% to 50%, respectively. BaP and DBA could be quantified in the diaper core of, respectively, 2 and 3 brands out of the four tested. In the top sheet, only BaP could be quantified in one brand out of the six tested. Of interest, concentrations and patterns of PAHs congeners in the diaper core were remarkably consistent across the brands with a predominance of naphthalene. This consistency suggests a common source of contamination, which is probably linked to ingredients used in the manufacture of the diaper. A possible source of contamination might be the glue used to fix the core as naphthalene is a common contaminant of glues [29]. By contrast, the marked variations in the concentrations and patterns of PAHs congeners in the diaper top sheet suggest the existence of brand-specific sources of contamination that may come from the ingredients, the factory environment, or the widespread environmental pollution by PAHs.

The present study confirms as anticipated [12] that ANSES has largely overestimated cancer risks from PAHs in infant diapers. Our ECR estimates for total PAHs concentration were lower than 10^{-8} for the diaper core and 10^{-7} for the top sheet. These ECRs are more than four orders of magnitude lower than those calculated by ANSES, which, for BaP and DBA alone, exceeded 10^{-3}. Interestingly, our study shows that in the case of the diaper core, such a high exceedance is impossible to achieve for a mere physical reason, which is the very poor water solubility of HAPs and especially of BaP and DBA. The ECRs of BaP and DBA in the diaper core calculated at their respective limit of water solubility were indeed in the range of 10^{-6}. Furthermore, because of the conservative assumptions made in our study, we probably overestimated the skin exposure to PAHs from the diaper core. Not only did we assume a dermal absorption of 100% but we also conservatively used a SDAF value of 100, which is more than one order of magnitude lower than the lowest Kow value of the 16 priority U.S. EPA PAHs. By incorporating a SDAF value of 100, we assume that the bioavailability of diaper core PAHs is at least 100 times lower when the extraction is performed with urine at 37 °C than with n-hexane at 100 °C. This assumption unavoidably leads to an overestimation of the exposure and health risks of large (≥4 rings) strongly lipophilic PAHs, which have much higher Kow values (log Kow between 5.61 and 6.84) and include the most potent carcinogens (PEF ≥ 0.1). For the top sheet in direct contact with the skin, the transport of PAHs to the skin does not require solubilization in urine.

They can be directly absorbed across the skin after solubilization in sweat or sebum present in variable proportions at the surface of the skin. For top sheet PAHs, we adopted a SDAF of 10 to take into account that the 7% transfer factor was established for substances intended to be delivered to the skin and also that PAHs embedded in the top sheet are presumably more efficiently extracted with n-hexane at 100 °C than with an aqueous solvent at 37 °C. With a skin absorption assumed to be 100%, this SDAF results in a bioavailability of 10%, which can be regarded conservative in regard to the experimental data in the literature. In a human ex vivo skin model, Bourgart et al. [30], for instance, estimated the dermal absorption of unchanged BaP at less than 5% and that of 3-hydroxybenzo[a]pyrene, the metabolite presumably responsible for the neurobehavioral effects of BaP, at less than 0.1% [31]. Moreover, these estimates were made with BaP dissolved in acetone, i.e., under conditions that are known to facilitate the skin absorption of BaP [1,3,4]. Recently, Luo et al. [14] estimated at less than 6% the dermal availability of BaP adsorbed onto indoor dust, which is presumably much more bioavailable than PAHs embedded in the matrix of the diaper top sheet.

Our estimates of skin cancer are in accordance with epidemiological or case report studies that provide no evidence of dermal carcinogenicity of PAHs in infant diapers. In the hypothesis that PAHs from diapers would cause skin cancer, there is no doubt that the critical skin site would be the highly permeable scrotum. It is difficult to believe that such a cancer risk could have passed undetected after more than five decades of widespread use of infant diapers [32]. Squamous cell carcinoma (SCC) is the type of scrotal malignancy that has been reported after high occupational exposure to PAHs. Now, with the early recognition of this hazard and implementation of preventive measures, SCC has become a very rare cancer with a steady incidence through the 20th century. Of note also, the median range of age at SCC diagnosis is 52–57 years, which makes it unlikely that SCC could be initiated during infancy, even though the median SCC latency is close to 30 years [33].

The risk assessment conducted by ANSES was based on both an exposure route and species extrapolation, assuming that dermally absorbed PAH_S can cause digestive tract tumors and neurobehavioral changes as evidenced in animals following oral and/or inhalation early exposure. The first assumption about digestive tract cancers is strongly challenged by experiments in rodents showing that dermally applied BaP causes only skin tumors. The second assumption cannot be formally refuted as developmental effects of PAHs have not been investigated in animals by the dermal route. However, if one assumes, as with ANSES, that PAHs absorbed by the skin can cause neurotoxic effects, our findings clearly show that these risks are totally unlikely as the estimated absorbed doses are about two orders of magnitude lower than the U.S. EPA reference dose. This conclusion is indirectly supported by comparing the PAHs daily dose from diapers with that from breast milk, which, in some countries, can be one to three orders of magnitude higher. There is no epidemiological evidence whatsoever associating breastfeeding with increased risks of cancer or neurotoxic effects. On the contrary, breastfeeding is well recognized as protective against a number of diseases or disorders including cancers (e.g., leukemia) and as beneficial to the child's neurodevelopment, improving the IQ and reducing the risk of behavioral disorders [34,35].

Our study presents some limitations. The first is the lack of experimental data about the bioavailability of PAHs in the diaper core and top sheet. Because extraction was performed with n-hexane, we had to make some conservative assumptions to adjust for the much lower solubility of PAHs in biological fluids compared to n-hexane. One might argue that it would have been more relevant to perform the extraction with artificial urine, the carrier transporting PAHs from the diaper core to the skin. The issue is that given the already very low concentrations found after n-hexane extraction, it is very likely that with an aqueous solvent, most PAHs would have been, if not undetectable, unquantifiable. In addition, extraction with artificial urine would have led to an underestimation of the risks of PAHs in the top sheet, which can be directly transported to the skin without solubilization in an aqueous carrier. Another limitation is the lack of neurotoxicity data for individual PAHs as well as following exposure to PAHs by the dermal route. We conservatively assumed that the 16 U.S. EPA PAHs can cause neurobehavioral changes by the dermal route with the same neurotoxic potency as BaP, which could result in a risk overestimation. Nevertheless, for both the carcinogenic and neurotoxic effects of PAHs, our evaluation has inherent uncertainties due to the extrapolation between species and the possibility of synergistic interactions between PAHs congeners. However, given the very conservative assumptions adopted in our study, we think that these limitations and uncertainties should not change our conclusions.

5. Conclusions

We measured the 16 priority U.S. EPA PAHs in the different layers of commercially available infant diapers in Switzerland. In the diaper core and top sheet, the main sources of skin exposure, the proportion of quantifiable PAHs in the different brands varied between 38% and 81%, and 31% to 50%, respectively. The concentrations and patterns of PAHs congeners in the diaper core were remarkably consistent across the brands, which suggests a common source of contamination probably linked to an ingredient used in the manufacture of the diaper. By contrast, both the concentrations and patterns of PAHs congeners greatly varied between brands, which points to different sources of contamination linked to the ingredients and/or the environment. Excess skin cancer risks and hazard quotients for neurobehavioral effects of total PAHs from the combined diaper core and top sheet averaged 1.44×10^{-7} and 1.19×10^{-2}, respectively. The median daily dose of total PAHs and of its BaP-equivalent from breast milk of non-smoking women estimated worldwide are 171 and 30 times greater than that from the combined diaper core and top sheet, respectively. Altogether, these findings indicate that trace levels of PAHs found in infant diapers are unlikely to pose health risks to babies.

Supplementary Materials: The following supporting information can be downloaded at: https://www.mdpi.com/article/10.3390/ijerph192214760/s1, Table S1: concentration of polycyclic aromatic carbons (PAHs) in the diaper core of infant diapers from four brands; Table S2: concentrations of polycyclic aromatic hydrocarbons (PAHs) in the top sheet (A), back sheet, (C) and fastening tapes (F) of infant diapers from six brands.

Author Contributions: Conceptualization, A.B.; methodology, A.B. and V.D.; formal analysis, A.B.; investigation, V.D.; resources, V.D.; writing—original draft preparation, A.B.; writing—review and editing, A.B. and V.D.; funding acquisition, A.B. and V.D. All authors have read and agreed to the published version of the manuscript.

Funding: This study was supported by the International Disposables and Nonwovens Association (EDANA).

Institutional Review Board Statement: Not applicable.

Informed Consent Statement: Not applicable.

Data Availability Statement: Concentrations of PAHs in the different parts of diapers are provided in the Supplementary Material.

Acknowledgments: Alfred Bernard is Honorary Research Director of the National Fund for Scientific Research (Belgium).

Conflicts of Interest: The authors declare no conflict of interest. The funder had no role in the design of the study; in the collection, analyses, or interpretation of data; in the writing of the manuscript, or in the decision to publish the results.

References

1. U.S. EPA (U.S. Environmental Protection Agency). Toxicological Review of Benzo[a]pyrene. 2017; pp. 1–234. Available online: https://iris.epa.gov/static/pdfs/0136tr.pdf (accessed on 1 February 2022).
2. ADSTR. Toxicological Profile for Polycyclic Aromatic Hydrocarbons. U.S. Department of Health and Human Services Public Health Service. Agency for Toxic Substances and Disease Registry; 1995. Available online: https://www.atsdr.cdc.gov/toxprofiles/tp69.pdf (accessed on 1 February 2022).
3. IARC (International Agency for Research on Cancer). IARC Monographs on the Evaluation of Carcinogenic Risks to Humans. Chemical Agents and Related Occupations. 2012. 100F. pp. 1–599. Available online: https://publications.iarc.fr/Book-And-Report-Series/Iarc-Monographs-On-The-Identification-Of-Carcinogenic-Hazards-To-Humans/Chemical-Agents-And-Related-Occupations-2012 (accessed on 1 February 2022).
4. Knafla, A.; Petrovic, S.; Richardson, M.; Campbell, J.; Rowat, C. Development and application of a skin cancer slope factor for exposures to benzo[a]pyrene in soil. *Regul. Toxicol. Pharmacol.* **2011**, *59*, 101–110. [CrossRef] [PubMed]
5. ANSES. Agence Nationale de Sécurité Sanitaire de L'alimentation, L'environnement et du Travail, France. Sécurité des Couches Pour Bébé. Avis Révisé. 2019, pp. 1–206. Available online: https://www.anses.fr/fr/content/avis-revise-et-rapport-de-lanses-relatif-a-la-securite-des-couches-pour-bebe (accessed on 1 February 2022).
6. ANSES. French Agency for Food, Occupational and Environmental Health and Safety, Dossier Submitter. Annex xv Restriction Report with Its Annex. Hazardous Chemicals in Single-Use Baby Diapers. 2020, pp. 1–85; (annex pp. 1–198). Available online: https://echa.europa.eu/documents/10162/99f020fd-e8ae-1b66-4fe6-0ec40789db8a (accessed on 1 February 2022).
7. ECHA. European Chemicals Agency. Committee for Risk Assessment (RAC) Committee for Socio-Economic Analysis (SEAC) Opinion on an Annex XV Dossier Proposing Restrictions on Substances in Single-Use Baby Diapers. Available online: https://echa.europa.eu/documents/10162/58e146ec-f113-09f3-1d7e-6e03b2061928 (accessed on 1 February 2022).
8. Rai, P.; Lee, B.M.; Liu, T.Y.; Yuhui, Q.; Krause, E.; Marsman, D.S.; Felter, S. Safety evaluation of disposable baby diapers using principles of quantitative risk assessment. *J. Toxicol. Environ. Health, A.* **2009**, *72*, 1262–1271. [CrossRef] [PubMed]
9. Dey, S.; Purdon, M.; Kirsch, T.; Helbich, H.M.; Kerr, K.; Li, L.; Zhou, S. Exposure factor considerations for safety evaluation of modern disposable diaper. *Regul. Toxicol. Pharmacol.* **2016**, *81*, 183–193. [CrossRef]
10. Bae, J.; Kwon, H.; Kim, J. Safety evaluation of absorbent hygiene pads: A review and assessment framework and test methods. *Sustainability* **2018**, *10*, 4146. [CrossRef]
11. INERIS (Institut National de l'Environnement Industriel et des Risques). Hydrocarbures Aromatiques Polycycliques (HAPs). Évaluation de la Relation Dose-Réponse Pour Des Effets Cancérigènes: Approche Substance Par Substance (Facteurs D'équivalence Toxique-FET) et Approche Par Mélanges. Évaluation de la Relation Dose-Réponse Pour des Effets non Cancérigènes: Valeurs Toxicologiques de Référence (VTR). 2003. Available online: https://www.ineris.fr/fr/hydrocarbures-aromatiques-polycycliques-haps (accessed on 1 February 2022).
12. Bernard, A. Dermal exposure to hazardous chemicals in baby diapers: A re-evaluation of the quantitative health risk assessment conducted by the French Agency for Food, Environmental and Occupational Health and Safety (ANSES). *Int. J. Environ. Res. Public Health* **2022**, *19*, 4159. [CrossRef]
13. Odio, M.R.; O'Connor, R.J.; Sarbaugh, F.; Baldwin, S. Continuous topical administration of a petrolatum formulation by a novel disposable diaper. 2. Effect on skin condition. *Dermatology* **2000**, *200*, 238–243. [CrossRef] [PubMed]

14. Luo, K.; Zeng, D.; Kang, Y.; Lin, X.; Sun, N.; Li, C.; Zhu, M.; Chen, Z.; Man, Y.B.; Li, H. Dermal bioaccessibility and absorption of polycyclic aromatic hydrocarbons (PAHs) in indoor dust and its implication in risk assessment. *Environ. Pollut.* **2020**, *264*, 114829. [CrossRef]
15. Garrido, N.M.; Economou, I.G.; Queimada, A.J.; Jorge, M.; Macedo, E.A. Prediction of the n-hexane/water and 1-octanol/water partition coefficients for environmentally relevant compounds using molecular simulation. *AIChE J.* **2012**, *58*, 1929–1938. [CrossRef]
16. Boniol, M.; Verriest, J.P.; Pedeux, R.; Doré, J.F. Proportion of skin surface area of children and young adults from 2 to 18 years old. *J. Investig. Dermatol.* **2008**, *128*, 461–464. [CrossRef]
17. U.S. EPA (U.S. Environmental Protection Agency). *Exposure Factors Handbook. Edition (Final)*; EPA/600/R-09/052F; U.S. Environmental Protection Agency: Washington, DC, USA, 2011. Available online: https://cfpub.epa.gov/ncea/risk/recordisplay (accessed on 1 February 2022).
18. ECHA. Minutes of the 56th Meeting of the Committee for Risk Assessment (RAC-56). 19 March 2021. Available online: https://echa.europa.eu/documents/10162/32022442/rac56_minutes_en.pdf/5618cedf-c4b2-becf-968d-bf7a5a1493c8 (accessed on 1 February 2022).
19. Santonicola, S.; De Felice, A.; Cobellis, L.; Passariello, N.; Peluso, A.; Murru, N.; Ferrante, M.C.; Mercogliano, R. Comparative study on the occurrence of polycyclic aromatic hydrocarbons in breast milk and infant formula and risk assessment. *Chemosphere* **2017**, *175*, 383–390. [CrossRef] [PubMed]
20. Torres-Moreno, C.; Puente-DelaCruz, L.; Codling, G.; Villa, A.L.; Cobo, M.; Klanova, J.; Johnson-Restrepo, B. Polycyclic aromatic hydrocarbons (PAHs) in human breast milk from Colombia: Spatial occurrence, sources and probabilistic risk assessment. *Environ. Res.* **2022**, *204*, 111981. [CrossRef] [PubMed]
21. Oliveira, M.; Duarte, S.; Delerue-Matos, C.; Angelina Pena, A.; Morais, S. Exposure of nursing mothers to polycyclic aromatic hydrocarbons: Levels of un-metabolized and metabolized compounds in breast milk, major sources of exposure and infants' health risks. *Environ. Pollut.* **2020**, *266*, 115243. [CrossRef] [PubMed]
22. Acharya, N.; Gautam, B.; Subbiah, S.; Rogge, M.M.; Anderson, T.A.; Gao, W. Polycyclic aromatic hydrocarbons in breast milk of obese vs normal women: Infant exposure and risk assessment. *Sci. Total Environ.* **2019**, *668*, 658–667. [CrossRef] [PubMed]
23. Asamoah, A.; Fini, M.N.; Essumang, D.K.; Muff, J.; Søgaard, E.G. PAHs contamination levels in the breast milk of Ghanaian women from an e-waste recycling site and a residential area. *Sci. Total Environ.* **2019**, *666*, 347–354. [CrossRef] [PubMed]
24. Pulkrabova, J.; Stupak, M.; Svarcova, A.; Rossner, P.; Rossnerova, A.; Ambroz, A.; Sram, R.; Hajslova, J. Relationship between atmospheric pollution in the residential area and concentrations of polycyclic aromatic hydrocarbons (PAHs) in human breast milk. *Sci. Total Environ.* **2016**, *562*, 640–647. [CrossRef]
25. Çok, I.; Mazmanci, B.; Mazmanci, M.A.; Turgut, C.; Henkelmann, B.; Schramm, K.W. Analysis of human milk to assess exposure to PAHs, PCBs and organochlorine pesticides in the vicinity Mediterranean city Mersin, Turkey. *Environ. Int.* **2012**, *40*, 63–69. [CrossRef]
26. Kim, S.R.; Halden, R.U.; Buckley, T.J. Polycyclic Aromatic Hydrocarbons in Human Milk of Nonsmoking U.S. Women. *Environ. Sci. Technol.* **2008**, *42*, 2663–2667. [CrossRef]
27. Zanieri, L.; Galvan, P.; Checchini, L.; Cincinelli, A.; Lepri, L.; Donzelli, G.P.; Del Bubba, M. Polycyclic aromatic hydrocarbons (PAHs) in human milk from Italian women: Influence of cigarette smoking and residential area. *Chemosphere* **2007**, *67*, 1265–1274. [CrossRef]
28. Kishikawa, N.; Wada, M.; Kuroda, N.; Akiyama, S.; Nakashima, K. Determination of polycyclic aromatic hydrocarbons in milk samples by high-performance liquid chromatography with fluorescence detection. *J. Chromatogr. B. Analyt. Technol. Biomed. Life Sci.* **2003**, *789*, 257–264. [CrossRef]
29. Danish Environmental Protection Agency. Survey of Naphthalene (CAS 91-20-3). *Part of the LOUS-Review Environmental Project No. N°1721*. 2015. Available online: https://www2.mst.dk/Udgiv/publications/2015/06/978-87-93352-34-6.pdf (accessed on 1 February 2022).
30. Bourgart, E.; Barbeau, D.; Marques, M.; Von Koschembahr, A.; Béal, D.; Persoons, R.; Leccia, M.T.; Douki, T.; Maitre, A. A realistic human skin model to study benzo[a]pyrene cutaneous absorption in order to determine the most relevant biomarker for carcinogenic exposure. *Arch. Toxicol.* **1999**, *93*, 81–93. [CrossRef]
31. Wang, D.; Rietdijk, M.M.; Kamelia, L.; Boogaard, P.J.; Rietjens, I.M.C.M. Predicting the in vivo developmental toxicity of benzo[a]pyrene (BaP) in rats by an in vitro–in silico approach. *Arch. Toxicol.* **2021**, *95*, 3323–3340. [CrossRef] [PubMed]
32. Dyer, D. Seven Decades of Disposable Diapers. A Record of Continuous Innovation and Expanding Benefit. The Wintrop Group, Inc. on Behalf of EDANA. Available online: http://www.edana.org/docs/default-source/absorbent-hygiene-products/edana---seven-decades-of-diapers.pdf?sfvrsn=3e24d (accessed on 1 February 2022).
33. Vyas, R.; Zargar, H.; Di Trolio, R.; Di Lorenzo, G.; Autorino, R. Squamous cell carcinoma of the scrotum: A look beyond the chimneystacks. *World J. Clin. Cases* **2014**, *16*, 654–660. [CrossRef] [PubMed]
34. Meek, J.Y.; Noble, L. Section on Breastfeeding. Policy statement: Breastfeeding and the use of human milk. *Pediatrics* **2022**, *150*, e2022057988. [CrossRef] [PubMed]
35. Binns, C.; Lee, M.; Low, W.Y. The long-term public health benefits of breastfeeding. *Asia Pac. J. Public Health* **2016**, *28*, 7–14. [CrossRef] [PubMed]

cells

Article

Monochromatic Light Pollution Exacerbates High-Fat Diet-Induced Adipocytic Hypertrophy in Mice

Qingyun Guan [1], Yixuan Li [2], Zixu Wang [1], Jing Cao [1], Yulan Dong [1], Fazheng Ren [2] and Yaoxing Chen [1,2,*]

[1] Neurobiology Laboratory, College of Veterinary Medicine, China Agricultural University, Haidian, Beijing 100193, China

[2] Key Laboratory of Precision Nutrition and Food Quality, Key Laboratory of Functional Dairy, Ministry of Education, Beijing Laboratory of Food Quality and Safety, Department of Nutrition and Health, China Agricultural University, Beijing 100083, China

[*] Correspondence: yxchen@cau.edu.cn; Tel.: +86-10-62733778; Fax: +86-10-62733199

Abstract: Light pollution worldwide promotes the progression of obesity, which is widely considered a consequence of circadian rhythm disruptions. However, the role of environmental light wavelength in mammalian obesity is not fully understood. Herein, mice fed a normal chow diet (NCD) or a high-fat diet (HFD) were exposed to daytime white (WL), blue (BL), green (GL), and red light (RL) for 8 weeks. Compared with WL and RL, BL significantly increased weight gain and white adipose tissue (WAT) weight, and it disrupted glucose homeostasis in mice fed with HFD but not NCD. The analysis of WAT found that BL significantly aggravated HFD-induced WAT hypertrophy, with a decrease in IL-10 and an increase in NLRP3, p-P65, p-IκB, TLR4, Cd36, Chrebp, Srebp-1c, Fasn, and Cpt1β relative to WL or RL. More interestingly, BL upregulated the expression of circadian clocks in the WAT, including Clock, Bmal1, Per1, Cry1, Cry2, Rorα, Rev-erbα, and Rev-erbβ compared with WL or RL. However, most of the changes had no statistical difference between BL and GL. Mechanistically, BL significantly increased plasma corticosterone (CORT) levels and glucocorticoid receptors in the WAT, which may account for the changes in circadian clocks. Further, in vitro study confirmed that CORT treatment did promote the expression of circadian clocks in 3T3-L1 cells, accompanied by an increase in Chrebp, Cd36, Hsp90, P23, NLRP3, and p-P65. Thus, daily BL, rather than RL exposure-induced CORT elevation, may drive changes in the WAT circadian clocks, ultimately exacerbating lipid dysmetabolism and adipocytic hypertrophy in the HFD-fed mice.

Keywords: monochromatic light pollution; adipose hypertrophy; high-fat diet; circadian clocks; corticosterone

Citation: Guan, Q.; Li, Y.; Wang, Z.; Cao, J.; Dong, Y.; Ren, F.; Chen, Y. Monochromatic Light Pollution Exacerbates High-Fat Diet-Induced Adipocytic Hypertrophy in Mice. *Cells* **2022**, *11*, 3808. https://doi.org/10.3390/cells11233808

Academic Editors: Esref Demir and Sam Kacew

Received: 6 November 2022
Accepted: 25 November 2022
Published: 28 November 2022

1. Introduction

With the acceleration of urbanization, the massive introduction of artificial light at night (LAN) has blurred the normal time division of day and night, making light pollution a potential global health risk factor [1,2]. Meanwhile, growing exposure to light from electronic products has also exacerbated the progress of light pollution. In modern society, people are exposed to intelligent device screens, such as mobile phones and computers, for a long time during the daytime, which emits blue light (BL) that is harmful to health. The unreasonable or irregular use of artificial light may be responsible for bringing terrible environmental disasters to the health of humans and animals.

Interestingly, the influence of light pollution on metabolic homeostasis has raised increasing attention in recent years [3]. Irregular light exposure has been proven to be a hazardous factor for obesity [4–6], as it promotes hepatic steatosis [7,8], increases fasting blood glucose (FBG), and induces insulin resistance (IR) [9,10]. The effects of ambient light exposure on metabolic homeostasis depend on light period, intensity, and wavelengths [11]. As reported, the extent to which dim LAN (dLAN) and constant light (LL) exposure dysregulate metabolisms differs [12]. Exposure to LAN has an intensity-dependent acute detrimental effect on glucose metabolism [13]. However, studies on the link between wavelength and metabolism are limited. The existing findings have not fully revealed the role of light wavelength, barely diving into the physiological mechanisms involved in the light wavelength's impact at the lipid metabolism level.

There are currently more than 1.9 billion adults worldwide who are overweight or obese. These staggering numbers continue to climb, posing inestimable health and economic burdens [14]. The increased white adipose tissue (WAT) in obesity is a highly active but dysfunctional metabolic organ, which is one of the main stages of complex mechanisms that underlie the pathogenesis of obesity, namely low-grade inflammation [15]. On the contrary, thermogenic brown adipose tissue (BAT) contributes to weight loss due to its abundant mitochondria. Nevertheless, BAT dysfunction leads to obesity, a process linked to the "whitening" of the tissue [16]. Moreover, abnormal changes in genes involved in lipid synthesis, transport, and oxidation can affect lipid accumulation in adipose and lead to WAT hypertrophy or BAT whitening [17].

Notably, obesity exacerbated by light pollution is now commonly linked to the disruption of the circadian clock. In the core cycle of the circadian clock, the main components include Circadian Locomotor Output Cycles Kaput (CLOCK), Brain and Muscle Arnt-like Protein 1 (BMAL1), Cryptochrome (CRY1, CRY2), and Period (PER1, PER2, PER3). Additional feedback loops include nuclear receptors retinoic acid-related orphan receptors (RORs), REV-ERBs., etc., to maintain the robustness of the clock system [18]. The circadian system is important for energy metabolism, which is synchronized mainly through the standard light-dark (LD) cycle [19]. On the contrary, irregular light is transmitted to the suprachiasmatic nucleus (SCN) via intrinsic photosensitive retinal ganglion cells (ipRGCs), disturbing the central circadian rhythm and driving changes in peripheral circadian clocks, which is likely to be mediated by hormone signals such as melatonin and corticosterone (CORT) [18]. In the adipose tissue, biological clock genes in the adipose tissue regulate the proliferation and differentiation of adipocytes, lipid metabolism, and endocrine [20,21].

Although some knowledge has been gained about the relationship between light wavelength and obesity, the effect and mechanism on the adipose tissue remain unknown, especially for diurnal light exposure. Therefore, this study aims to explore the effect and potential pathways of long-term exposure to diurnal monochromatic lights on obesity in mice from the perspective of the adipose tissue, thus providing a brand-new theoretical reference for the prevention or treatment of obesity.

2. Materials and Methods

2.1. Animal Treatment and Light Exposure

The experiment is based on the "Guidelines for the Care and Use of Laboratory Animals" issued by the Animal Welfare Committee of the Agricultural Research Organization, China Agricultural University (Approval No.AW18079102-1-2). Six-week-old male C57BL/6 mice (Charles River Co., Ltd., Beijing, China) were raised under optimum conditions (at a temperature of 21 ± 1 °C, relative humidity of 50 ± 10%, 14-h:10-h LD cycle) and obtained food and water freely. The mice were firstly housed under white light (WL, 400–700 nm) and a normal chow diet (NCD, 4% of energy from lipid, Charles River Co. Ltd., Beijing, China) for 1-week adaption. Then, the mice were randomly divided into (1) mice exposed to WL fed on NCD (WN) or on a high-fat diet (HFD, 45% of energy from lipid, Beijing HFK Bioscience Co., Ltd., Beijing, China) (WF), (2) mice exposed to BL (peak at 444 nm) fed on NCD (BN) or HFD (BF), (3) mice exposed to green light (GL, peak at 528 nm) fed on NCD (GN) or HFD (GF), and (4) mice exposed to red light (RL, peak at 624 nm) fed on NCD (RN) or HFD (RF). The lights were powered by a light-emitting diode system (Zhongshan Junsheng Lighting Technology Co., Ltd. Zhongshan, China) with a constant intensity of 150 lx under 14-h:10-h LD cycle (lights on from 7:00 AM to 9:00 PM). The lights' parameters are shown in Table 1. During the experiment, animal body weight, food intake, and water drinking were recorded. After 8 weeks, the plasma and adipose tissues were harvested for the following experiments.

Table 1. Light parameters.

Items	Light Exposure			
	WL	BL	GL	RL
Light wavelength (nm)	400–700	Peak at 444	Peak at 528	Peak at 624
Light intensity (lux)	150	150	150	150
Light: dark cycle (h)	14:10	14:10	14:10	14:10

2.2. Glucose and Insulin Tolerance Test

After 12 h overnight, mice were intraperitoneally injected with glucose (1 g/kg body weight, Sigma, St. Louis, MO, USA) for a glucose tolerance test (GTT). Blood samples from the tip of the tail were measured at 0, 15, 30, 60, 90, and 120 min after glucose injection by a portable glucose monitor (Aike, Lingrui, Hangzhou, China). The area under the curve (AUC) was calculated by Graphpad prism (version 9.4, GraphPad Software Inc., San Diego, CA, USA). For the insulin tolerance test (ITT), mice were fasted overnight for 6 h to inject insulin (0.75 IU/kg, Novolin R, Novo Nordisk, Denmark), and their blood glucose was measured as described.

2.3. Commercial Kits Detection

The concentration of CORT, melatonin, and noradrenalin in plasma samples and the concentrations of inflammatory factors (IL-6 and IL-10) in the WAT were detected by a competitive enzyme-linked immunosorbent assay (Uscn Life Science, Inc., Wuhan, China). Commercial kits (Jiancheng Institute of Biotechnology, Nanjing, China) were used to determine the plasma levels of total cholesterol (TC) and triglyceride (TG). All the tests were carried out according to the manufacturer's instructions.

2.4. Histology Staining

The harvested epididymis (Epi)-WAT and BAT were maintained in 4% formalin for 48 h, followed by a series of alcohol dehydrations, and finally embedded in paraffin. Epi-WAT and BAT (5 μm) were sectioned using a semiautomatic rotary microtome, stained with Hematoxylin and Eosindye (H&E), and observed microscopically (BX51, Olympus, Tokyo, Japan). The density of adipocytes in WAT was measured as the ratio of the number of cells to the total area of the section field of view. Measurement of mean adipocyte area was found using automated Image-pro plus software (Media Cybernetics, Inc., Rockville, MD, USA). The WAT was examined from six animals in each group.

2.5. Quantitative Real Time (RT)-PCR Analysis

We used a previously described method [22]. In short, total mRNA was first extracted from the Epi-WAT, BAT, and cell, followed by reverse transcription to obtain cDNA. Then, RT-PCR experiments were conducted via designed gene primers and demonstrated the expression level of each gene with glyceraldehyde-3-phosphate dehydrogenase (Gapdh) as an internal control. The experiments were repeated in triplicate. The primers used are shown in Table 2.

Table 2. Sequences of primers used for RT-PCR.

Gene	Primer Sequence (5′ to 3′)		Product Size	Accession
Cd36	F: GTGCAAAACCCAGATGACGT	R: TCCAACAGACAGTGAAGGCT	180	NM_001159558.1
Chrebp	F: GTGTGTGGTTTCGTGACCC	R: CACTTGTGGTATTCGCGCATC	128	NM_001359237.1
Srebp-1c	F: ATCGCAAACAAGCTGACCTG	R: AGATCCAGGTTTGAGGTGGG	115	NM_001388389.1
Fasn	F: TCCTGGAACGAGAACACGATCT	R: GAGACGTGTCACTCCTGGACTTG	138	NM_007988.3
Cpt1β	F: GGCACCTCTTCTGCCTTTAC	R: TTTGGGTCAAACATGCAGAT	136	NM_009948.2
Clock	F: ATGGTGTTTACCGTAAGCTGTAG	R: CTCGCGTTACCAGGAAGCAT	197	XM_011249402.3
Bmal1	F: CAGAGCCGGAGCAGGAAAAATAGGT	R: CAGGGGGAGGCGTACTTGTGATGT	128	NM_001374642.1
Per1	F: CGGATTGTCTATATTTCGGAGCA	R: TGGGCAGTCGAGATGGTGTA	142	NM_001159367.2
Per2	F: GAAAGCTGTCACCACCATAGAA	R: AACTCGCACTTCCTTTTCAGG	186	NM_011066.3
Cry1	F: CACTGGTTCCGAAAGGGACTC	R: CTGAAGCAAAAATCGCCACCT	153	NM_007771.3
Cry2	F: CACTGGTTCCGCAAAGGACTA	R: CCACGGGTCGAGGATGTAGA	102	NM_009963.4

Table 2. *Cont.*

Gene	Primer Sequence (5′ to 3′)		Product Size	Accession
Rorα	F: TCCAAATCCCACCTGGAAAC	R: GGAAGGTCTGCCACGTTATCTG	70	NM_001289916.1
Rve-erbα	F: TACATTGGCTCTAGTGGCTCC	R: CAGTAGGTGATGGTGGGAAGTA	127	NM_145434.4
Rve-erbβ	F: GGAAACACTCATCCGTGCACTA	R: ATCGAAGATCTGGCAACTTTAGAA	101	NM_001145425.2
Ucp1	F: TAAGCCGGCTGAGATCTTGT	R: GGCCTCTACGACTCAGTCCA	84	NM_009463.3
Ucp3	F: CTGCACCGCCAGATGAGTTT	R: ATCATGGCTTGAAATCGGACC	191	NM_009464.3
Pgc-1α	F: TATGGAGTGACATAGAGTGTGCT	R: CCACTTCAATCCACCCAGAAAG	134	NM_008904.3
Hsp70	F: CGGTGCCCGCCTACTTC	R: TCCTTCTTGTGCTTCCTCTTGA	322	NM_005346.6
Hsp90	F: ACGAGGAAGAGAGAAGAAGAAAATGG	R: GCAGGGTGAAGACACAAGCC	131	NM_001271971.2
P23	F: ATGCGTTTGGAGAAGGACAGA	R: CAGGGATGAAGTGATGGTGAG	210	NM_001289785.1
Gapdh	F: CCGAGAATGGGAAGCTTGTC	R: TTCTCGTGGTTCACACCCATC	232	NM_001289726.1

F = forward primer; R = reverse primer.

2.6. Western Blot Analysis

Epi-WAT and cell proteins were extracted and performed western blot detection according to the previously described method [22]. The specific antibodies and related concentrations were as follows: NOD-like receptor thermal protein domain associated protein 3 (NLRP3, 1:1000; CST), p-P65 (1:1000; CST), p-IκB (1:1000; CST), toll-like receptor 4 (TLR4, 1:1000; Santa Cruz, Dallas, TX, USA), circadian locomotor output cycles kaput (CLOCK, 1:1000; Abcam, Cambridge, UK), brain and muscle arnt-like protein 1 (BMAL1, 1:1000; Santa Cruz, Dallas, TX, USA), glucocorticoid receptor (GR, 1:2000; Proteintech, Rosemont, IL, USA), and β-actin (1:8000; Proteintech, Rosemont, IL, USA) overnight at 4 °C. The membranes were washed with Tris-buffered saline Tween and incubated with horseradish peroxidase-conjugated goat anti-mouse/rabbit antibody (1:8000; CoWin Biotech Co., Inc., Cambridge, MA, USA). The target band values were normalized to those of β-actin. The results were based on three independent experiments.

2.7. Cell Culture and Treatment

3T3-L1 cell lines were maintained in Dulbecco's modified Eagle's medium supplemented with 10% fetal bovine serum and 4 mM l-glutamine + 100 U/mL penicillin + 100 μg/mL streptomycin + 4500 mg/L glucose, and cultured at 37 °C in a 95% air/5% CO_2 humidity environment. Cells were plated at a density of 8×10^3 cells/well in 96-well plates for 3-(4,5-dimethylthiazol-2-yl)-2,5-diphenyltetrazolium bromide (MTT) assay to observe the effect of CORT on cell's viability. Briefly, cells were seeded and cultured in 96-well plates for 6 h. Subsequently, we replaced the complete medium with the basal medium and continued to culture for 12 h. Then, cells were treated with various concentrations (1 μM, 5 μM, 10 μM, 20 μM, and 30 μM) of CORT (MedChemExpress, Prinbceton, NJ, USA) for 24 h. Next, MTT solution (5 mg/mL in PBS, Sigma, St. Louis, MO, USA) was added and incubated for 4 h. Afterwards, the supernatant was removed and the crystals were dissolved with 150 μL of DMSO (Sigma, St. Louis, MO, USA). The absorbance at 490 nm was measured. The 3T3-L1 cells were then treated with 1 μM (L-CORT) and 10 μM (H-CORT) CORT respectively for the subsequent RT-PCR and WB assay.

2.8. Statistical Analysis

Data are expressed as the mean ± standard error and were analyzed using Graphpad Prism (version 9.4, GraphPad Software Inc., San Diego, CA, USA). One-way ANOVA was used to statistically analyze differences between groups. $p < 0.05$ was considered statistically significant.

3. Results

3.1. Effects of Monochromatic Light Exposure on Metabolic Disorders in Mice

To determine the effect of different wavelengths of light exposure on body weight and obesity, 7-week-old mice fed with NCD or HFD were respectively raised under WL, BL, GL, and RL exposure for 8 weeks (Figure 1A). As shown in Figure 1B, there were no significant changes in weight gain between light colors under NCD feeding ($p > 0.05$). However, under HFD feeding, BL increased weight gain by 14.2% compared with WL ($p < 0.05$), while having an insignificant change with GL ($p > 0.05$). Consistent with the changes in body weight gain, light colors did not change the weights of the heart (Figure 1C), liver (Figure 1D), spleen (Figure 1E), kidney (Figure 1F), Epi-WAT (Figure 1G), inguinal subcutaneous adipose tissue (Ing-SAT, Figure 1H), and BAT (Figure 1I) under NCD feeding. However, in the HFD-fed groups, BL increased the weights of most of these organs. Besides, plasma total cholesterol (TC) was significantly elevated by BL compared to WL and RL under both NCD and HFD ($p < 0.05$, Figure 1J), while there was no statistical difference in triglycerides (TG, $p > 0.05$, Figure 1K). To examine the effect on glucose homeostasis, we measured the level of fasting blood glucose (FBG) and found that BL exposure accentuated HFD-induced increases in FBG compared with WL and RL, by 31.3% and 114.6% ($p < 0.01$, Figure 1L), respectively. Results of the GTT and ITT (Figure 1M–O) and AUC values further showed that BL sharpened insulin resistance and insulin resistance impaired by HFD compared with WL and RL ($p < 0.01$).

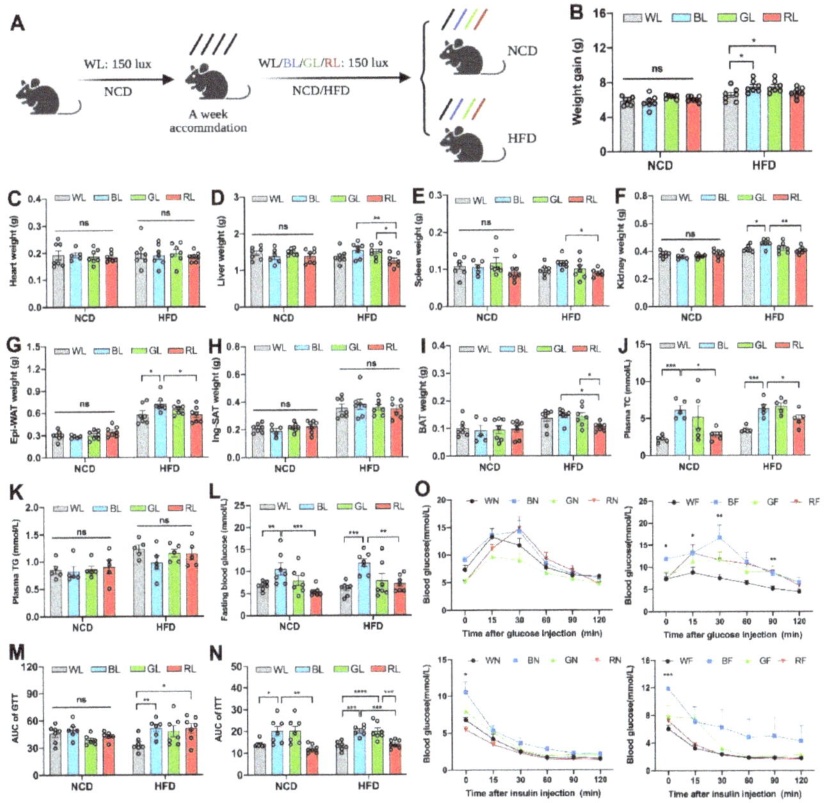

Figure 1. Influences of monochromatic light exposure on metabolic disorders in mice. (**A**) Schematic diagram of animal experiments. (**B**) Body weight gain (n = 7). (**C–I**) Weights of the heart, liver, spleen, kidney, Epi-WAT, Ing-WAT, and BAT (n = 7). (**J**,**K**) Plasma TC and TG concentrations (n = 5). (**L**) Fasting blood glucose level (n = 7). (**M–O**) GTT and ITT curve and relevant AUC (n = 7). The circles represent the number of samples. The results are presented as the means ± SEM. * $p < 0.05$, ** $p < 0.01$, *** $p < 0.001$, **** $p < 0.0001$.

3.2. *Effects of Monochromatic Light Exposure on WAT Hypertrophy and Inflammation in Mice*

As obesity and the increased Epi-WAT weight were previously observed, we further measured the size of adipocytes gained from the Epi-WAT sections via H&E staining (Figure 2A). Consistently, there was also no significant change in the adipocyte density

($p > 0.05$, Figure 2B) and mean adipocyte size ($p > 0.05$, Figure 2C) between light colors under NCD feeding. Nevertheless, BL reduced the adipocyte density ($p < 0.001$) and enlarged the mean adipocyte size ($p < 0.001$) compared with WL, GL, and RL under HFD. Based on the insignificant differences in NCD, we focused on HFD feeding in the following study. As shown in Figure 2D–I, BL significantly inhibited the levels of IL-10 in the Epi-WAT as compared with RL ($p < 0.05$), though no changes were exerted in the level of IL-6 ($p > 0.05$). In parallel, BL consistently increased relative protein levels of NLRP3 ($p < 0.01$, Figure 2F), p-P65 ($p < 0.01$, Figure 2G), p-IκB ($p < 0.05$, Figure 2H), and TLR4 ($p < 0.05$, Figure 2I) relative to WL and RL. Likewise, there was no statistical difference between BL and GL at the level of inflammation ($p > 0.05$). Next, we further examined the expression of genes related to lipid metabolism in the Epi-WAT (Figure 2J–N), including those involved in fatty acid transporters (Cd36), de novo lipogenesis (carbohydrate response element binding protein, Chrebp; sterol regulatory element binding protein-1C, Srebp-1c; fatty acid synthase, Fasn), and fatty acid (FA) β-oxidation (carnitine palmitoyl transferase 1β, Cpt1β). Compared with WL or RL, BL significantly increased the levels of Cd36, Chrebp, Srebp-1c, Fasn, and Cpt1β ($p < 0.05$), while GL did not ($p > 0.05$).

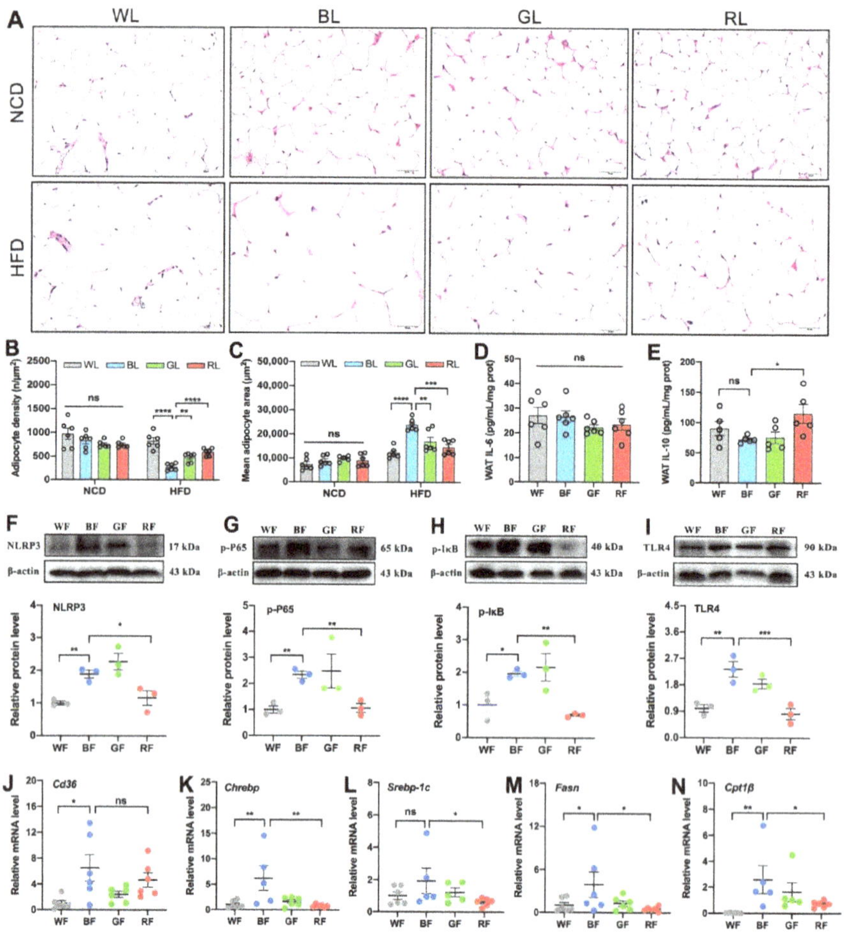

Figure 2. Influences of monochromatic light exposure on WAT hypertrophy and inflammation in mice. (**A**) H&E staining of the Epi-WAT (scale: 50 μm). (**B**) Adipocyte density (n = 6). (**C**) Mean

adipocyte size ($n = 6$). (**D**,**E**) Concentrations of IL-6 and IL-10 in the Epi-WAT ($n = 6$). (**F–I**) Relative protein expression levels of NLRP3, p-P65, p-IκB, and TLR4 in the Epi-WAT ($n = 3$). (**J–N**) Relative mRNA expression levels of Cd36, Chrebp, Srebp-1c, Fasn, and Cpt1β ($n = 6$). The circles represent the number of samples. The results are presented as the means ± SEM. * $p < 0.05$, ** $p < 0.01$, *** $p < 0.001$, **** $p < 0.0001$.

3.3. Effects of Monochromatic Light Exposure on the Expression of Circadian Clock in the WAT

Since the metabolic abnormalities affected by light pollution may be controlled by biological clocks, we then examined the expression of circadian-related molecules in the WAT, including transcripts encoding Clock, Bmal1, Per1, Per2, Cry1, Cry2, Rorα, Rev-erbα, and Rev-erbβ. Compared with WL or RL, BL consistently increased the mRNA expression levels of Clock ($p < 0.001$, Figure 3A), Bmal1 ($p < 0.01$, Figure 3B), Per1 ($p < 0.01$, Figure 3C), Cry1 ($p < 0.05$, Figure 3E), Cry2 ($p < 0.05$, Figure 3F), Rorα ($p < 0.05$, Figure 3G), Rev-erbα ($p < 0.05$, Figure 3H), and Rev-erbβ ($p < 0.05$, Figure 3I), except for Per2 ($p > 0.05$, Figure 3D). In parallel, BL exposure elevated the expression levels of CLOCK and BMAL1 compared to WL or RL at the protein level ($p < 0.05$, Figure 3J,K).

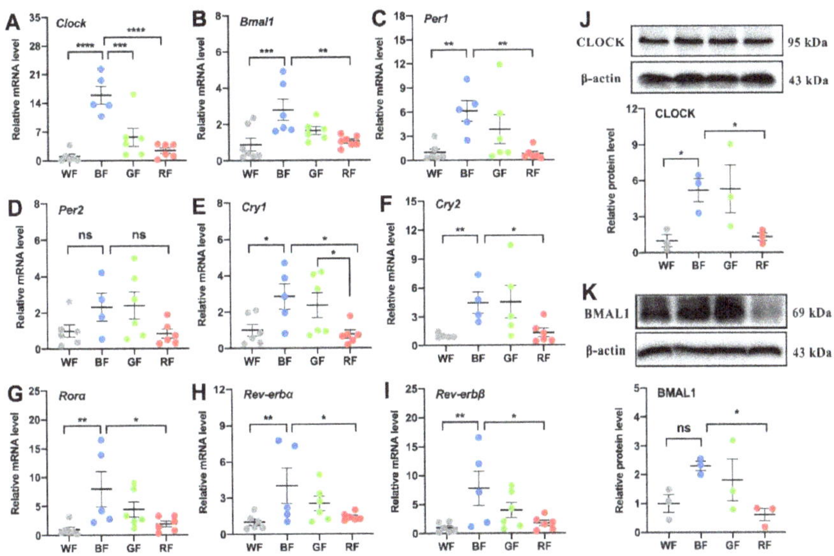

Figure 3. Influences of monochromatic light exposure on the expression of circadian genes in the WAT. (**A–I**) Relative mRNA expression levels of Clock, Bmal1, Per1, Per2, Cry1, Cry2, Rorα, Rev-erbα, and Rev-erbβ in the Epi-WAT ($n = 6$). (**J**,**K**) Relative protein expression levels of CLOCK and BMAL1 in the Epi-WAT ($n = 3$). The results are presented as the means ± SEM. * $p < 0.05$, ** $p < 0.01$, *** $p < 0.001$, **** $p < 0.0001$.

3.4. Effects of Monochromatic Light Exposure on BAT Whitening and Expression Levels of Circadian Genes in the BAT

The proper function of BAT is critical to the fight against obesity. As shown in Figure 4A, HFD-fed mice exhibited markedly enhanced lipid accumulation (i.e., "whitening") under WL, BL, and GL, except for RL exposure. Then we examined the expression of genes related to lipid metabolism in the BAT. Compared to WL, BL significantly increased the mRNA expression of uncoupling protein (Ucp) 1 ($p < 0.05$, Figure 4B) and Cd36 ($p < 0.01$, Figure 4E), while RL significantly increased the expression of Ucp1 ($p < 0.05$). There were no statistical differences in the expression of Ucp3 ($p > 0.05$, Figure 4C) and Cpt1β ($p > 0.05$, Figure 4G) among the groups. Furthermore, consistent with the result in the WAT, BL increased the expression of clock genes in the BAT compared with WL or RL, including

Clock ($p < 0.01$, Figure 4H), Bmal1 ($p < 0.05$, Figure 4I), Per1 ($p < 0.01$, Figure 4J), Rorα ($p < 0.05$, Figure 4N), Rev-erbα ($p < 0.01$, Figure 4O), and Rev-erbβ ($p < 0.05$, Figure 4P), except Per2 ($p > 0.05$, Figure 4K), Cry1 ($p > 0.05$, Figure 4L), Cry2 ($p > 0.05$, Figure 4M).

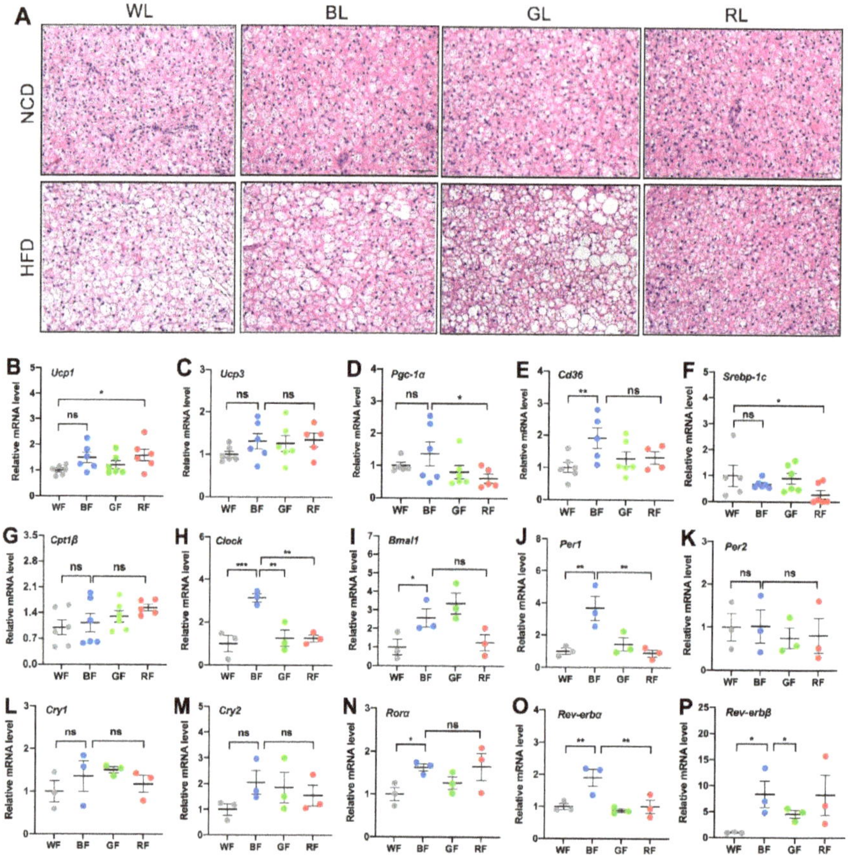

Figure 4. Influences of monochromatic light exposure on BAT whitening and expression of circadian genes in BAT. (**A**) H&E staining of BAT (scale: 50 μm). (**B–G**) Relative mRNA expression levels of genes involved in lipid metabolism in the BAT, including Ucp1, Ucp3, Pgc-1α, Cd36, Srebp-1c, Cpt1β ($n = 6$). (**H–P**) Relative mRNA expression levels of Clock, Bmal1, Per1, Per2, Cry1, Cry2, Rorα, Rev-erbα, and Rev-erbβ in the BAT ($n = 3$). The results are presented as the means ± SEM. * $p < 0.05$, ** $p < 0.01$, *** $p < 0.001$.

3.5. The Role of CORT in Interference with the Adipose Circadian Clock

Finally, to investigate the mechanism by which light exposure influenced the adipose circadian clock, we focused on the hormonal pathways of output and feedback of the central circadian system. As shown in Figure 5A, the concentration of plasma CORT in the HFD-fed group was significantly higher by 21.3% and 27.6% in BL compared with WL and RL ($p < 0.05$), respectively. However, the concentrations of plasma melatonin ($p > 0.05$, Figure 5B) and noradrenaline ($p > 0.05$, Figure 5C) were not statistically significant among light colors under HFD feeding. Then, we investigated whether different wavelengths affect the expression level of GR in the WAT. Consistently, the expression of GR in BL was 65.9% more than that of RL ($p < 0.05$, Figure 5D). We then examined whether GR transport was also affected. Compared with RL, BL significantly increased the expression of Hsp90 mRNA ($p < 0.05$, Figure 5E), while there was no statistically significant difference in the expressions of Hsp70 ($p > 0.05$, Figure 5F) and P23 ($p > 0.05$, Figure 5G) mRNA. Therefore, BL increased plasma CORT levels and enhanced GR synthesis and transport in the HFD-fed mice.

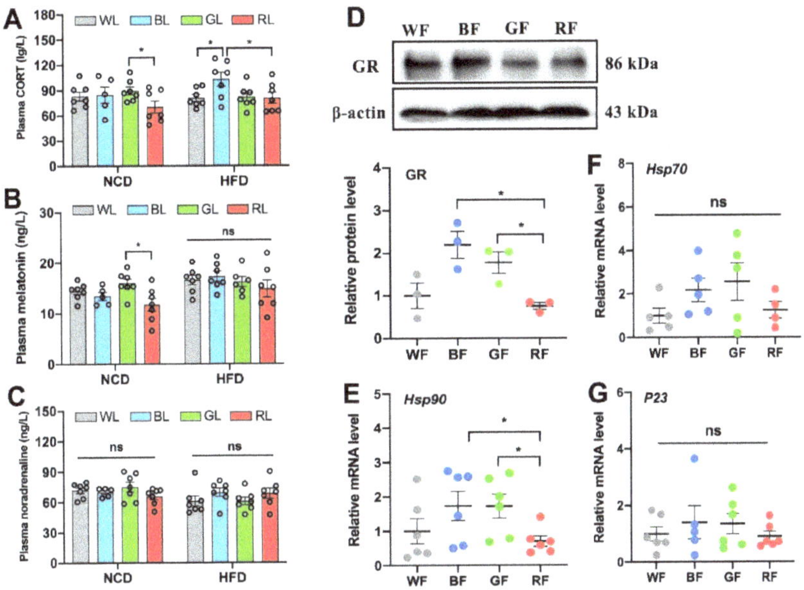

Figure 5. Influences of monochromatic light exposure on CORT concentrations. (**A–C**) Plasma concentrations of CORT, melatonin, and noradrenaline ($n = 7$). (**D**) Relative protein expression level of GR in the Epi-WAT ($n = 3$). (**E–G**) Relative mRNA expression levels of Hsp90, Hsp70, and P23 in the Epi-WAT ($n = 6$). The circles represent the number of samples. The results are presented as the means ± SEM. * $p < 0.05$.

As CORT is one of the output and feedback signals of the central circadian system that may drive changes in the peripheral biological clock, we hypothesized that adipose circadian clocks may be affected by the elevated CORT. To test our hypothesis, we added CORT to treat 3T3-L1 cell lines. As shown in Figure 6A, CORT promoted cell death in a dose-dependent manner ($p < 0.001$). Compared with the control group, both L-CORT and H-CORT increased the mRNA expression levels of Clock (Figure 6B), Per1 (Figure 6D), and Per2 (Figure 6E). Moreover, CORT increased the mRNA expression levels of Cry1 (Figure 6F) and Cry2 (Figure 6G). Besides, CORT treatment significantly increased the mRNA expression levels of Chrebp (Figure 6K), Cd36 (Figure 6M), Hsp90 (Figure 6O), and P23 (Figure 6P), as well as increased levels of NLRP3 and p-P65 proteins ($p < 0.05$, Figure 6Q–S).

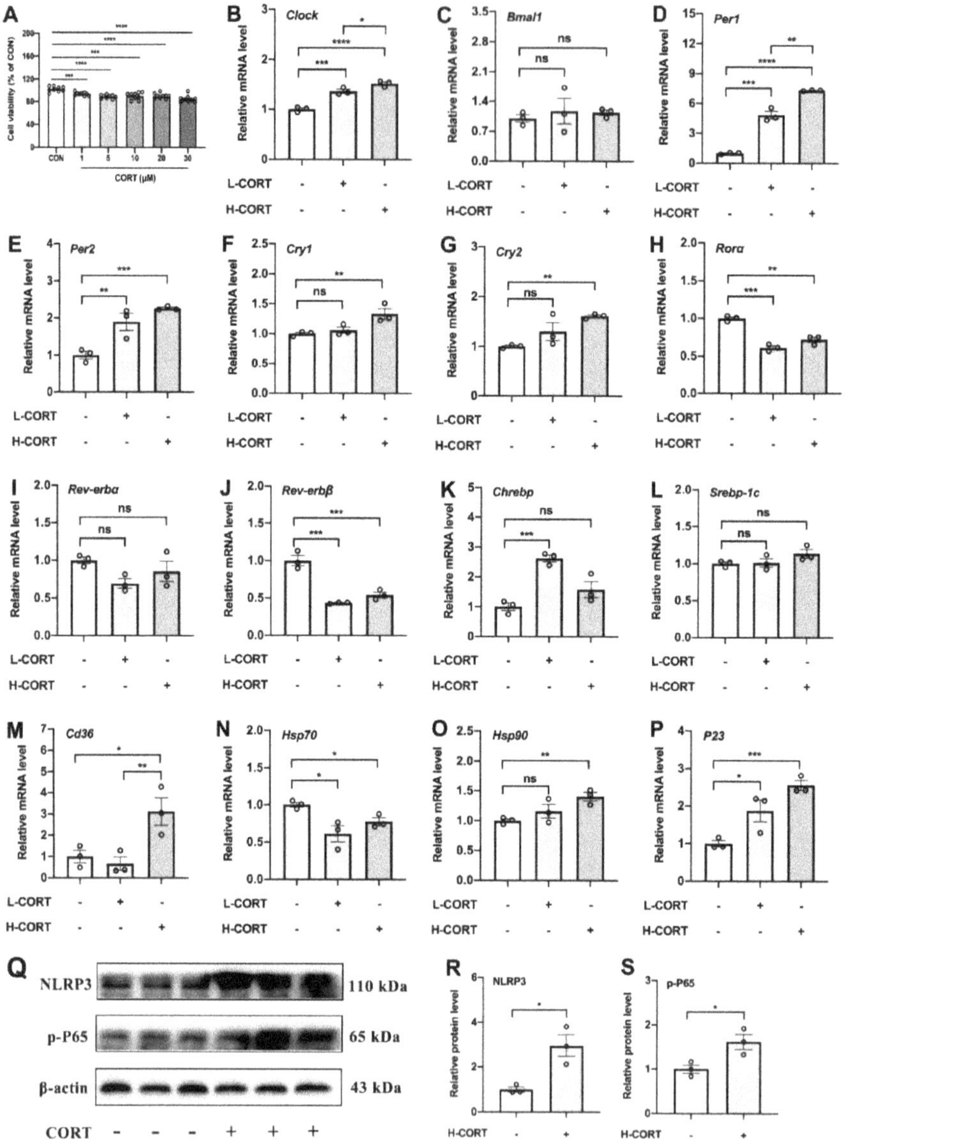

Figure 6. The role of CORT in interference with the adipose circadian rhythms. (**A**) Cell viability (% of CON). (**B–P**) Relative mRNA expression levels of Clock, Bmal1, Per1, Per2, Cry1, Cry2, Rorα, Rev-erbα, Rev-erbβ, Chrebp, Srebp-1c, Cd36, Hsp70, Hsp90, and P23 in 3T3-L1 cells treated by CORT (n = 3). (**Q–S**) Relative protein levels of NLRP3 and p-P65 (n = 3). The circles represent the number of samples. The results are presented as the means ± SEM. * $p < 0.05$, ** $p < 0.01$, *** $p < 0.001$, **** $p < 0.0001$.

4. Discussion

The wide application of artificial lighting technology and electronic products inevitably exposes organisms to light pollution. The effects of light pollution on metabolic diseases such as obesity [23], impaired glucose tolerance [24], and non-alcoholic fatty liver disease [25] are partially understood, but most of this knowledge is based on nighttime WL exposure. There is a growing force to uncover the role and ways in which light wavelengths in the development of obesity, especially daytime exposure.

Previous studies suggested that BL exposure at night disrupted glucose metabolism [24,26], and continuous BL exposure aggravated HFD-induced obesity in mice [22]. Besides, daytime GL exposure promoted the development of hepatic steatosis and pancreatic dysfunction [27]. However, there is no direct evidence of the effect of light colors on adipose tissues, which displayed the most immediate role in obesity. Herein, we constructed a mouse model exposed to different diurnal light wavelengths under an NCD or HFD for 8 weeks to investigate the influence of the diurnal spectrum on adipose tissues. Our results showed that under the NCD feeding, there were no statistical differences in body weight gain, adipose tissue weight, and WAT expansion under different light colors. Nevertheless, these changes could be exaggerated in the case of HFD feeding. Compared with WL or RL, BL exposure increased body weight gain, WAT's weight, and disrupted glucose homeostasis. Further analysis of WAT found that BL significantly promoted WAT hypertrophy and inflammatory response in the HFD-fed mice, accompanied by a reduction in IL-10 and an increase in the expression of NLRP3, p-P65, p-IκB, and TLR4 in BL relative to WL and RL. Adipocytic hypertrophy is related to the abnormal increase in lipid accumulation. Lipid metabolism in the body involves a series of processes. Activation of transcription factors such as Chrebp and Srebp-1c contribute to FA synthesis, which is transported by FA transporters such as Cd36 [28]. Our results showed that BL increased the mRNA level of Cd36, Chrebp, and Cpt1β in the WAT, which may further promote lipid accumulation and hypertrophy in the adipocyte. In parallel, BL promoted BAT whitening induced by HFD compared with RL. However, there was no significant difference between BL and GL. Altogether, BL may play a more critical role in adipocytic hypertrophy than WL or RL in HFD-fed mice, and nutritional signaling may be involved in the role of synergistic light wavelengths in mammalian lipid dysmetabolism.

To a large extent, metabolic disorders induced by irregular light have to do with circadian clock disturbance. A normal circadian system is essential for maintaining lipid metabolic homeostasis [29]. For instance, Clock mutant mice have a diminished daily feeding rhythm and promote obesity [30]. Deletion of Bmal1 may induce higher or lower body fat [31,32]. Normalization of the circadian clock in adipose tissue controls processes such as lipogenesis and lipolysis, as many key enzymes involved in lipolysis and lipogenesis are directly regulated by the circadian clocks [33]. In our study, BL exposure significantly increased the expression levels of clock genes in the WAT compared with WL or RL, including Clock, Bmal1, Per1, Cry1, Cry2, Rorα, Rer-erbα, and Rer-erbβ. The same was true in BAT, i.e., BL increased the expression of Clock, Per1, and Rer-erbα compared with RL. Therefore, we hypothesized that changes in clock genes in the WAT may be related to abnormalities in genes involved in lipid metabolism. In addition, differences among light wavelengths are most likely due to the short wavelength, perceived as BL, being the strongest synchronizer of the circadian system, which synchronizes most biological and psychological rhythms internally [34]. However, the circadian system is less sensitive to RL because RL cannot activate melanopsin-containing retinal ganglion cells that project to the SCN [18,35]. Hence, BL instead of RL exposure interfered with the adipose circadian rhythms in the HFD-fed mice, which, to some extent, may support the above results.

As thus, light color-induced disruptions in the adipose circadian clock caught further attention. Hormonal signals, including melatonin and CORT, are one of the key pathways that SCN controlled the peripheral clock [18,36]. Hence, we inferred that light-induced changes in the adipose circadian clock were driven by altered central circadian clock output hormonal signals. Through measuring plasma melatonin and CORT levels, it was found that BL significantly increased plasma CORT compared with WL and RL, while there were no significant effects on melatonin. In addition, the BL exposure group was accompanied by an increase in GR synthesis and transport in the WAT. Therefore, CORT may participate in the process of driving the biological clock changes of the WAT. To further verify the direct effect of CORT on circadian clocks, we treated 3T3-L1 cells with CORT. Both low and high concentrations of CORT increased the overall expression levels of the circadian clock in cells, including Clock, Per1, and Per2, and high CORT concentration increased the mRNA expression levels of Cry1 and Cry2. CORT treatment also increased the mRNA levels of Chrebp, Cd36, Hsp90, and P23 mRNA, and protein levels of NLRP3 and p-P65. Consistent in vivo and in vitro results suggest that CORT does play a vital role in disrupting the circadian clocks of adipose tissues, which may further contribute to lipid metabolic disorder and inflammatory response. In conclusion, in our model, long-term BL exposure was likely to exacerbate adipose hypertrophy induced by HFD in mice, which may be related to the change of adipose clock induced by CORT (Figure 7).

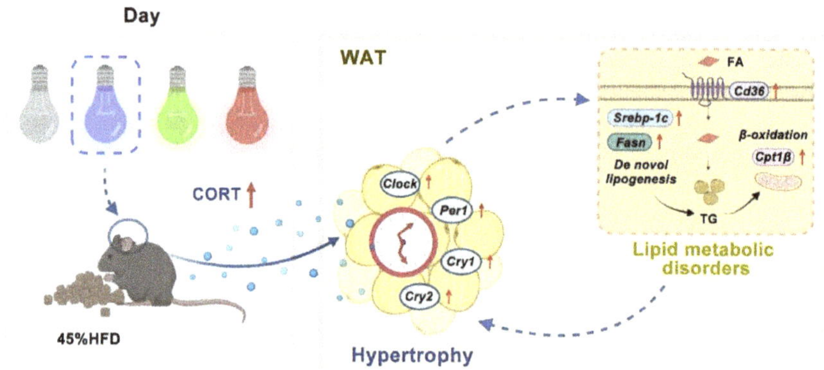

Figure 7. A schematic of our proposed model. Effects of long daytime monochromatic light exposure on adipose hypertrophy in mice fed with HFD. BL exposure increases the level of plasma CORT, which disturbs the circadian clocks in the WAT. Changes in circadian clocks may further regulate the process of lipid metabolism and ultimately lead to adipose hypertrophy. CORT, corticosterone; FA, fatty acid; HFD, high-fat diet; TG, triglyceride; WAT, white adipose tissue.

Notably, there are still some limitations within this study. Firstly, the nocturnal rodents used here may not fully mimic the effects of daytime spectral effects in humans, as nocturnal rodents and diurnal humans may have different physiological responses to light exposure. Secondly, the changes in the peripheral clocks we examined were only at one point in time rather than at various Zeitgeber times. Furthermore, we focused on the role of CORT in interfering with the adipose clock, yet whether the same is true in BL exposure has not been fully established. Little is known about how diurnal monochromatic light affects CORT, which is another shortcoming of our study.

5. Conclusions

In general, our results highlight the critical role of BL exposure in shaping obesity in HFD-fed mice. Long-term diurnal BL exposure exacerbated adipocyte hypertrophy, most likely due to its changes in the adipose circadian clock affected by the increased CORT. The role of monochromatic light pollution on adipose tissue opens a new avenue for interventions against ambient monochromatic light pollution and obesity.

Author Contributions: Conceptualization, methodology, formal analysis, writing the original draft, Visualization, Y.C. and Q.G.; resources, supervision, project administration, Y.L. and Z.W.; funding acquisition, methodology, supervision, J.C., Y.D. and F.R.; funding acquisition, Y.C. All authors have read and agreed to the published version of the manuscript.

Funding: This research was funded by "Chinese National Natural Science Foundation, grant number 32172801 and 31873000" and "Beijing Natural Science Foundation, grant number 6222019".

Institutional Review Board Statement: The study was conducted in accordance with the Declaration of Helsinki and approved by the Animal Welfare and Ethics Committee of China Agricultural University (No.AW18079102-1-2).

Informed Consent Statement: Not applicable.

Data Availability Statement: Data are contained within the article.

Acknowledgments: The authors would like to thank all members of the neurobiology laboratory.

Conflicts of Interest: The authors declare no conflict of interest.

References

1. Gaston, K.J.; Visser, M.E.; Hölker, F. The biological impacts of artificial light at night: The research challenge. *Philos. Trans. R. Soc. Lond. B Biol. Sci.* **2015**, *370*, 20140133. [CrossRef] [PubMed]
2. Batra, T.; Malik, I.; Kumar, V. Illuminated night alters behaviour and negatively affects physiology and metabolism in diurnal zebra finches. *Environ. Pollut.* **2019**, *254*, 112916. [CrossRef] [PubMed]
3. Guan, Q.; Wang, Z.; Cao, J.; Dong, Y.; Chen, Y. The role of light pollution in mammalian metabolic homeostasis and its potential interventions: A critical review. *Environ. Pollut.* **2022**, *312*, 120045. [CrossRef]
4. Zhang, D.; Jones, R.R.; Powell-Wiley, T.M.; Jia, P.; James, P.; Xiao, Q. A large prospective investigation of outdoor light at night and obesity in the NIH-AARP Diet and Health Study. *Environ. Health* **2020**, *19*, 74. [CrossRef]
5. Esaki, Y.; Obayashi, K.; Saeki, K.; Fujita, K.; Iwata, N.; Kitajima, T. Bedroom light exposure at night and obesity in individuals with bipolar disorder: A cross-sectional analysis of the APPLE cohort. *Physiol. Behav.* **2021**, *230*, 113281. [CrossRef]
6. Lin, L.Z.; Zeng, X.W.; Deb, B.; Tabet, M.; Xu, S.L.; Wu, Q.Z.; Zhou, Y.; Ma, H.M.; Chen, D.H.; Chen, G.B.; et al. Outdoor light at night, overweight, and obesity in school-aged children and adolescents. *Environ. Pollut.* **2022**, *305*, 119306. [CrossRef] [PubMed]
7. Borck, P.; Batista, T.; Vettorazzi, J.; Soares, G.; Lubaczeuski, C.; Guan, D.; Boschero, A.; Vieira, E.; Lazar, M.; Carneiro, E. Nighttime light exposure enhances Rev-erbα-targeting microRNAs and contributes to hepatic steatosis. *Metab. Clin. Exp.* **2018**, *85*, 250–258. [CrossRef] [PubMed]
8. Yue, F.; Xia, K.; Wei, L.; Xing, L.; Wu, S.; Shi, Y.; Lam, S.M.; Shui, G.; Xiang, X.; Russell, R.; et al. Effects of constant light exposure on sphingolipidomics and progression of NASH in high-fat-fed rats. *J. Gastroenterol. Hepatol.* **2020**, *35*, 1978–1989. [CrossRef] [PubMed]
9. Borck, P.C.; Rickli, S.; Vettorazzi, J.F.; Batista, T.M.; Boschero, A.C.; Vieira, E.; Carneiro, E.M. Effect of nighttime light exposure on glucose metabolism in protein-restricted mice. *J. Endocrinol.* **2021**, *252*, 143–154. [CrossRef]
10. Fan, X.; Chen, D.; Wang, Y.; Tan, Y.; Zhao, H.; Zeng, J.; Li, Y.; Guo, X.; Qiu, H.; Gu, Y. Light intensity alters the effects of light-induced circadian disruption on glucose and lipid metabolism in mice. *Am. J. Physiol. Endocrinol. Metab.* **2022**, *322*, E1–E9. [CrossRef]
11. Bourgin, P.; Hubbard, J. Alerting or somnogenic light: Pick your color. *PLoS Biol.* **2016**, *14*, e2000111. [CrossRef] [PubMed]
12. Rumanova, V.S.; Okuliarova, M.; Zeman, M. Differential effects of constant light and dim light at night on the circadian control of metabolism and behavior. *Int. J. Mol. Sci.* **2020**, *21*, 5478. [CrossRef] [PubMed]
13. Opperhuizen, A.L.; Stenvers, D.J.; Jansen, R.D.; Foppen, E.; Fliers, E.; Kalsbeek, A. Light at night acutely impairs glucose tolerance in a time-, intensity- and wavelength-dependent manner in rats. *Diabetologia* **2017**, *60*, 1333–1343. [CrossRef]
14. Kwaifa, I.K.; Bahari, H.; Yong, Y.K.; Noor, S.M. Endothelial dysfunction in obesity-induced inflammation: Molecular mechanisms and clinical implications. *Biomolecules* **2020**, *10*, 291. [CrossRef] [PubMed]
15. Kawai, T.; Autieri, M.V.; Scalia, R. Adipose tissue inflammation and metabolic dysfunction in obesity. *Am. J. Physiol. Cell. Physiol.* **2021**, *320*, C375–C391. [CrossRef]
16. Shimizu, I.; Walsh, K. The Whitening of brown fat and its implications for weight management in obesity. *Curr. Obes. Rep.* **2015**, *4*, 224–229. [CrossRef]
17. Xu, P.; Wang, J.; Hong, F.; Wang, S.; Jin, X.; Xue, T.; Jia, L.; Zhai, Y. Melatonin prevents obesity through modulation of gut microbiota in mice. *J. Pineal Res.* **2017**, *62*, e12399. [CrossRef] [PubMed]
18. Fonken, L.; Nelson, R. The effects of light at night on circadian clocks and metabolism. *Endocr. Rev.* **2014**, *35*, 648–670. [CrossRef]
19. Potter, G.D.; Skene, D.J.; Arendt, J.; Cade, J.E.; Grant, P.J.; Hardie, L.J. Circadian rhythm and sleep disruption: Causes, metabolic consequences, and countermeasures. *Endocr. Rev.* **2016**, *37*, 584–608. [CrossRef] [PubMed]
20. Kiehn, J.T.; Tsang, A.H.; Heyde, I.; Leinweber, B.; Kolbe, I.; Leliavski, A.; Oster, H. Circadian rhythms in adipose tissue physiology. *Compr. Physiol.* **2017**, *7*, 383–427. [CrossRef]
21. Froy, O.; Garaulet, M. The Circadian clock in white and brown adipose tissue: Mechanistic, endocrine, and clinical aspects. *Endocr. Rev.* **2018**, *39*, 261–273. [CrossRef]
22. Guan, Q.; Wang, Z.; Cao, J.; Dong, Y.; Chen, Y. Monochromatic blue light not green light exposure is associated with continuous light-induced hepatic steatosis in high fat diet fed-mice via oxidative stress. *Ecotoxicol. Environ. Saf.* **2022**, *239*, 113625. [CrossRef] [PubMed]
23. Koo, Y.S.; Song, J.Y.; Joo, E.Y.; Lee, H.J.; Lee, E.; Lee, S.K.; Jung, K.Y. Outdoor artificial light at night, obesity, and sleep health: Cross-sectional analysis in the KoGES study. *Chronobiol. Int.* **2016**, *33*, 301–314. [CrossRef] [PubMed]

24. Masís-Vargas, A.; Hicks, D.; Kalsbeek, A.; Mendoza, J. Blue light at night acutely impairs glucose tolerance and increases sugar intake in the diurnal rodent Arvicanthis ansorgei in a sex-dependent manner. *Physiol. Rep.* **2019**, *7*, e14257. [CrossRef] [PubMed]
25. Hong, F.; Pan, S.; Xu, P.; Xue, T.; Wang, J.; Guo, Y.; Jia, L.; Qiao, X.; Li, L.; Zhai, Y. Melatonin orchestrates lipid homeostasis through the hepatointestinal circadian clock and microbiota during constant light exposure. *Cells* **2020**, *9*, 489. [CrossRef] [PubMed]
26. Nagai, N.; Ayaki, M.; Yanagawa, T.; Hattori, A.; Negishi, K.; Mori, T.; Nakamura, T.J.; Tsubota, K. Suppression of Blue Light at Night Ameliorates Metabolic Abnormalities by Controlling Circadian Rhythms. *Investig. Ophthalmol. Vis. Sci.* **2019**, *60*, 3786–3793. [CrossRef] [PubMed]
27. Zhang, S.; Xu, M.; Shen, Z.; Shang, C.; Zhang, W.; Chen, S.; Liu, C. Green light exposure aggravates high-fat diet feeding-induced hepatic steatosis and pancreatic dysfunction in male mice. *Ecotoxicol. Environ. Saf.* **2021**, *225*, 112802. [CrossRef] [PubMed]
28. Canbay, A.; Bechmann, L.; Gerken, G. Lipid metabolism in the liver. *Z. Für Gastroenterol.* **2007**, *45*, 35–41. [CrossRef]
29. Gooley, J.J. Circadian regulation of lipid metabolism. *Proc. Nutr. Soc.* **2016**, *75*, 440–450. [CrossRef]
30. Turek, F.W.; Joshu, C.; Kohsaka, A.; Lin, E.; Ivanova, G.; McDearmon, E.; Laposky, A.; Losee-Olson, S.; Easton, A.; Jensen, D.R.; et al. Obesity and metabolic syndrome in circadian Clock mutant mice. *Science* **2005**, *308*, 1043–1045. [CrossRef]
31. Shimba, S.; Ogawa, T.; Hitosugi, S.; Ichihashi, Y.; Nakadaira, Y.; Kobayashi, M.; Tezuka, M.; Kosuge, Y.; Ishige, K.; Ito, Y.; et al. Deficient of a clock gene, brain and muscle Arnt-like protein-1 (BMAL1), induces dyslipidemia and ectopic fat formation. *PLoS ONE* **2011**, *6*, e25231. [CrossRef] [PubMed]
32. Kennaway, D.J.; Varcoe, T.J.; Voultsios, A.; Boden, M.J. Global loss of bmal1 expression alters adipose tissue hormones, gene expression and glucose metabolism. *PLoS ONE* **2013**, *8*, e65255. [CrossRef]
33. Ribas-Latre, A.; Eckel-Mahan, K. Nutrients and the circadian clock: A partnership controlling adipose tissue function and health. *Nutrients* **2022**, *14*, 2084. [CrossRef]
34. Wahl, S.; Engelhardt, M.; Schaupp, P.; Lappe, C.; Ivanov, I.V. The inner clock-Blue light sets the human rhythm. *J. Biophotonics* **2019**, *12*, e201900102. [CrossRef] [PubMed]
35. Martin, J.S.; Laberge, L.; Sasseville, A.; Bérubé, M.; Alain, S.; Lavoie, J.; Houle, J.; Hébert, M. Timely use of in-car dim blue light and blue blockers in the morning does not improve circadian adaptation of fast rotating shift workers. *Chronobiol. Int.* **2021**, *38*, 705–719. [CrossRef]
36. Sage, D.; Ganem, J.; Guillaumond, F.; Laforge-Anglade, G.; François-Bellan, A.M.; Bosler, O.; Becquet, D. Influence of the corticosterone rhythm on photic entrainment of locomotor activity in rats. *J. Biol. Rhythm.* **2004**, *19*, 144–156. [CrossRef] [PubMed]

International Journal of
Environmental Research and Public Health

MDPI

Review

Impact of Paraben Exposure on Adiposity-Related Measures: An Updated Literature Review of Population-Based Studies

Xinyun Xu [1], Haoying Wu [1], Paul D. Terry [2], Ling Zhao [1,*] and Jiangang Chen [3,*]

1 Department of Nutrition, The University of Tennessee, Knoxville, TN 37996, USA
2 Department of Medicine, Graduate School of Medicine, The University of Tennessee, Knoxville, TN 37920, USA
3 Department of Public Health, The University of Tennessee, Knoxville, TN 37996, USA
* Correspondence: lzhao7@utk.edu (L.Z.); jchen38@utk.edu (J.C.)

Abstract: Parabens are alkyl esters of p-hydroxybenzoic acid that are commonly used in pharmaceutical and cosmetic products. Humans are exposed to parabens when they use these products and through diet. There are growing concerns that paraben exposure can adversely impact human health. The endocrine-disrupting and obesogenic properties of parabens have been observed in animal studies and in vitro, prompting the increase in population-based studies of paraben exposure and adiposity-related endpoints. In this review, we summarize epidemiological studies published between 2017 and 2022 that examined paraben exposure in utero, between birth and adolescence, and in adulthood, in relation to adiposity-related measures. Overall, these studies provide some evidence that suggests that paraben exposure, especially during critical development windows, is associated with adiposity-related measures. However, we have noted several limitations in these studies, including the predominance of cross-sectional studies, inconsistent sample collection procedures, and small sample sizes, which should be addressed in future studies.

Keywords: parabens; human exposure; adiposity-related measures (ARM); population-based studies; endocrine disruptors

Citation: Xu, X.; Wu, H.; Terry, P.D.; Zhao, L.; Chen, J. Impact of Paraben Exposure on Adiposity-Related Measures: An Updated Literature Review of Population-Based Studies. *Int. J. Environ. Res. Public Health* **2022**, *19*, 16268. https://doi.org/10.3390/ijerph192316268

Academic Editors: Esref Demir and Sam Kacew

Received: 4 November 2022
Accepted: 22 November 2022
Published: 5 December 2022

Publisher's Note: MDPI stays neutral with regard to jurisdictional claims in published maps and institutional affiliations.

1. Introduction

Parabens are a series of alkyl esters of *p*-hydroxybenzoic acid. They are low-cost, broad-spectrum antimicrobial, and antiseptic preservatives with greater effectiveness against Gram-positive bacteria than Gram-negative bacteria. Therefore, parabens are widely used in cosmetics, personal-care products, and pharmaceuticals to suppress microbial growth and extend product shelf life [1–3]. Additional paraben exposure is linked to indoor air and dust, foodstuffs, and tap water [4]. Parabens have been detected in human serum, umbilical cord blood, urine, milk, and placental tissues, indicating systemic and early exposure to parabens in humans [5–11]. There is an appreciable correlation between the urinary paraben concentrations of pregnant women and their newborn infants, indicating the transfer of the compound from the mother to the fetus [12,13].

Parabens are endocrine disruptors with demonstrated estrogenic [14–16] and antiandrogenic effects [17,18]. Parabens also activate the thyroid hormone receptor [19], peroxisome proliferator-activated receptors (PPARs) [20,21], and glucocorticoid receptor (GR) [22,23]. Activation of PPARs and GR correlates with their adipogenic effects. Consistently, parabens have been shown to promote adipogenesis in cell models of mice and human origins [23]. Moreover, post-weaning exposure to methylparaben increases fat pad mass, whereas post-weaning exposure to either methylparaben or butylparaben suppresses serum markers of bone formation in chow-fed female C57BL/6J mice [24]. Furthermore, both parabens, butylparaben, in particular, modulate mesenchymal stem cell fates by promoting the adipocyte lineage at the expense of osteocytes and chondrocytes [20]. These in vitro and animal studies suggest the obesogenic potentials of parabens.

In this review, we summarized population-based studies published in the past five years to assess the association of paraben exposure with the risks of adiposity-related measures. Studies were obtained from searches of PubMed, using search terms such as "paraben," "obesity," and "BMI." Studies were grouped according to the life stage of the study participants during exposure from in utero, birth through to adolescence, and adulthood. Studies that only focused on paraben biomonitoring using various biological matrices were not included in this review. In vivo and in vitro studies of parabens' actions, the safety of parabens in cosmetics, and the role of parabens in reproductive systems and breast cancer have been reviewed elsewhere [25–27]; therefore, they are not in the scope of the current review.

2. Results and Discussion

2.1. Paraben Exposure in Utero and Adiposity-Related Measures

From 2017 to 2022, 13 studies explored the association between prenatal or in-utero paraben exposure and adiposity-related measures in children (Table 1). The majority (11 out of 13) of the studies were prospective in design. Biological matrices were collected at various gestational ages, with urine as the predominantly measured biological matrix. A few studies measured paraben concentrations using placentae or amniotic fluid [28–31]. The majority of these studies showed that parabens could cross the placental barrier [28–31]. Early life exposure to parabens is associated with increases in the gestational weight gain (GWG) rate [32], changes in height, head, hip, or arm circumference [28,30,33,34], z-scores of bodyweight [35], BMI or BMI z-scores [31,36,37], and overweight status in early childhood [38], although the direction of associations depends on the type of parabens and gender (Table 1). However, results from one large-scale (n = 1015 mother–child pairs) and one small-scale study (n = 99 mother–newborn pairs) did not show any clear association. In the larger study [39], most (>76%) of the children had already reached puberty at the time the adiposity-related measures were evaluated at age 11 [39]. Because the onset of puberty has an impact on body lean and fat mass [40], the dynamic and substantial increase in sex hormones in the participants could mask the impact of weak exogenous chemical exposures on body weight and fat mass. In the smaller study [29], only 20% of the amniotic fluid samples had detectable levels of methylparaben (MeP), and all other parabens were detected in less than 2% of the samples, reducing the statistical power of the analysis.

Table 1. Association of paraben exposure in utero and adiposity-related measures (ARM).

Study/Country	Sample Size	Year of Recruitment	Type of Study	Time of Sample Collection	Matrices	ARM	Major Shortcomings	Major Findings
Güil-Oumrait et al. 2022, Spain [39] [a]	1015 mother–child pairs (500 males; 515 females)	2003–2008	Prospective	First and third trimesters	Urine	BMI; BMI z-score	High temporal variability in paraben concentrations due to the lack of serial urine collections over the pregnancy; health effects were assessed at asingle time point (11 years).	No association was identified between maternal urinary concentrations of parabens and adiposity measures of the child at 11 years of age.

Table 1. *Cont.*

Study/Country	Sample Size	Year of Recruitment	Type of Study	Time of Sample Collection	Matrices	ARM	Major Shortcomings	Major Findings
Golestanzadeh et al. 2022, Iran [28] [b]	128 pregnant women and 142 newborns	2019–2021	Cross-sectional	During C-section	Amniotic fluid	Newborn weight, head, chest, hip, and arm circumference	Only women who underwent C-sections were enrolled (selection bias); small sample size; only one spot sample was collected; cross-sectional nature of the study design; no maternal caloric intake was considered in the analysis.	BuP concentrations in the amniotic fluids were positively associated with weight, hip, and arm circumference but negatively correlated with height, head, and chest circumference of the newborns. MeP concentrations in the amniotic fluids were negatively associated with head circumference, chest, hip, and arm circumference but positively associated with the height of the newborns; EtP concentrations in the amniotic fluids were negatively associated with arm circumference but positively associated with the height of the newborns; PrP concentrations in the amniotic fluids were negatively associated with arm circumference of the newborns.
Reimann et al. 2021, Belgium [31] [c]	218 mother–child pairs, (112 males and 106 females)	2014–2017	Prospective	During delivery	Placenta	BMI z-score	No information was available for the child's breastfeeding status, caloric intake, or postnatal paraben exposure; only one spot sample was collected.	Placental EtP concentrations were negatively associated with children's BMI z-scores.
Hojsager et al. 2021, Demark [36] [d]	312 mother–child pairs	2010 and 2012	Prospective	Early third trimester	Urine	Fat mass, body mass; BMI z-score	The women enrolled in the study were older and more often nulliparous compared to the background population (selection bias); no information was available about weight gain during pregnancy; only one spot sample was collected; childhood exposure to parabens was not assessed; childhood caloric intake was not considered in the analysis.	Maternal urinary BuP concentrations were positively associated with total body fat percentage and android fat percentage in boys.
Karzi et al. 2021, Greece [29]	99 mother–newborn pairs (41 males; 44 females)	Unknown	Prospective	Second trimester	Urine/amniotic fluid	Birth weight, length, and head circumference	Low prevalence of parabens detected in amniotic fluid; small sample size; only one spot sample was collected; maternal caloric intake was not considered in the analysis.	No association was identified between maternal parabens in either urine or amniotic fluid and adiposity-related measures in the newborns.

Table 1. *Cont.*

Study/Country	Sample Size	Year of Recruitment	Type of Study	Time of Sample Collection	Matrices	ARM	Major Shortcomings	Major Findings
Berger et al. 2021, United States [38] [e]	309 mother–child pairs	Since 1999	Prospective	First and second trimesters	Urine	BMI, BMI z-score; overweight/obese status	Lack of ability to assess sex-specific associations; childhood exposure to parabens was not assessed.	Maternal urinary PrP concentrations were positively associated with BMI z-scores and overweight/obesity status of the child at 5 years of age.
Vrijens et al. 2020, Belgium [30]	142 mother–child pairs (74 males; 66 females)	2014–2016	Cross-sectional	During delivery	Placenta	Birth weight, length, and head circumference	Cross-sectional nature of the study design; small sample size; only one spot sample was collected; maternal caloric intake was not considered in the analysis.	Placental total parabens were negatively associated with birth weight and head circumference in girls; EtP concentrations were negatively associated with head circumference in girls.
Wen, et al. 2020, China [32]	613 pregnant women	2014–2015	Prospective	During each trimester	Urine	Gestational weight gain (GWG) and GWG rate (kg/week)	Neither the frequency of use of personal care products during pregnancy nor maternal caloric intake was considered in the analysis.	First-trimester MeP, EtP, PrP, and total parabens levels in the women's urine were positively associated with an increase in the GWG rate of the first-trimester, and these associations were stronger than those of the second or third trimesters.
Jamal et al. 2020, Iran [33]	189 mother–child pairs (66 males; 92 females)	2016	Prospective	First trimester	Urine	Birth weight, length, and head circumference	Small sample size; only one spot sample was collected; maternal caloric intake was not considered in the analysis.	Maternal urinary BuP concentrations were positively associated with the birth weight of boys; maternal urinary PrP concentrations were negatively associated with the birth length of girls; maternal urinary MeP and BuP concentrations were positively associated with the head circumference in girls.
Leppert, et al. 2020, German [37] [f]	626 mother–child pairs, (108 males; 115 females)	2006–2008	Prospective	Third trimester	Urine	BMI	No information was available for postnatal paraben exposure; only one spot sample was collected; childhood caloric intake was not considered in the analysis.	Maternal urinary BuP concentrations were positively associated with the overweight status of the children within the first eight years of life, with a stronger trend observed in girls.
Chang et al. 2019, Taiwan [34]	199 mother–child pairs (99 males; 100 females)	2014–2015	Prospective	Third trimester	Urine	Birth weight, body length, head, and thoracic circumference; Ponderal Index	Small sample size; highly educated pregnant women were enrolled in the study (selection bias); only one spot sample was collected; maternal caloric intake was not considered in the analysis.	Maternal urinary MeP concentrations were positively associated with the head circumference and Ponderal Index in boys; maternal urinary MeP concentrations were negatively associated with the birth weight, length, head circumference, and thoracic circumference in girls.

Table 1. *Cont.*

Study/Country	Sample Size	Year of Recruitment	Type of Study	Time of Sample Collection	Matrices	ARM	Major Shortcomings	Major Findings
Wu, et al. 2019, China [35] [g]	850 mother–child pairs (446 males; 404 females)	2014–2015	Prospective	During each trimester	Urine	Z-scores for weight and height	High temporal variability in paraben concentrations throughout pregnancy; postnatal paraben exposure was assessed.	Maternal urinary EtP concentrations were negatively associated with the weight z-scores of the child at birth; third-trimester urinary EtP concentrations were negatively associated with the weight z-scores at birth, 1 and 2 years in boys.
Wu, et al. 2017, China [41]	1016 mother–child pairs (527 male; 489 females)	2012–2014	Prospective	Within three days before delivery	Urine	Birth length and weight	Only one spot sample was collected; maternal caloric intake was not considered in the analysis.	Maternal urinary MeP concentrations were positively associated with the birth length in boys.

Footnotes: [a] BMI information was collected at 11 years of age; [b] number of male and female newborns was not provided; [c] a subset of 63 newborns was followed for 29 months to monitor BMI z-scores; [d] number of male and female newborns was not provided; adiposity-related measures were assessed at 7 years of age; [e] number of male and female newborns was not provided; adiposity-related measures were assessed at 5 years of age; [f] newborns were followed from 1 to 8 years of age; [g] adiposity-related measures were assessed at birth, 6 months, 1, and 2 years of age.

Overall, these 13 studies suffer from methodological issues that hamper the interpretation of the results. Most of the studies only collected single-spot samples. One major weakness of all the studies is the lack of information regarding whether urine samples were collected consistently in a defined manner or at a defined time point (for example, first-morning fasting urine), which would help ensure measurement validity and reliability. The variability in the timing of urine collection could have affected the exposure estimation for chemicals such as parabens with short half-lives [42]. Furthermore, with single-spot samples, the varied timing of exposure assessment in relation to gestational age may introduce exposure bias, in that the urinary paraben concentrations measured during the later gestational stage may not reflect the level of paraben exposure during the earlier pregnancy [43]. The standardization of exposure assessments should be, therefore, an important consideration in future studies. In addition, the concentrations of parabens in a sample depend on the source of paraben exposure and the time when the last exposure occurred before the sample collection [44]. When exposure to different parabens varies over time, repeated measures of exposure will reduce bias in exposure estimation. Finally, in all the studies that carried out postnatal follow-ups with the children, postnatal paraben exposure was not measured for these children; therefore, it is unknown how postnatal exposure of parabens might have contributed to the trajectories of adiposity measures in these studies.

Although the overall evidence supports an association between paraben exposure in utero and the risks of subsequent adiposity-related measures, we cannot make any conclusions regarding potential effect modifications by factors such as gender, which have shown conflicting results. Hu et al. reported that the adipogenic potency of parabens depends on the length of the linear alkyl chain [23]. Longer chain parabens are also more estrogenic compared to their shorter counterparts [45]. Estrogens play a critical role in the development of obesity [46,47] and parabens interfere with the estrogen metabolism pathway [48,49]. In vitro maternal butylparaben (BuP) exposure increases the weight of female offspring, but not male offspring mice [37], which may support the observed gender-specific effects in some of the studies, although the latter has not been consistent (Table 1) [30,33–37,41]. Both estrogen and androgen are critical in regulating adiposity and metabolism. The concentrations of endocrine disruptors in local tissues and organs may determine the unbalanced steroid hormone action [50]. Therefore, the exact molecular mechanisms responsible for the gender-specific effects of prenatal paraben exposure on several adiposity-related measures warrant further investigation.

It is also possible the different results from these studies are due to the heterogeneity in paraben exposure profiles in different populations [32]. Propylparaben (PrP) was the most relevant chemical in the mixture in Berger's study conducted in the United States [38], while in the Spanish Environment and Childhood project (INMA), PrP minimally contributed to the mixture [39]. Paraben content can also vary substantially among different food products [51]. Furthermore, among the early-life adiposity-related measures (for example, gestational weight gain (GWG), GWG rates, length at birth, head, arm, hip, chest circumference, BMI, BMI z-score and percentage of fat), the most reliable predictor(s) for subsequent later-life health outcomes is yet to be determined [36,52–55].

The current literature suggests that prenatal exposure to parabens interferes with the programming of endocrine signaling pathways that could lead to the observed changes in adiposity in children. Therefore, as a precaution, pregnant women should consider reducing their use of cosmetic products with high paraben contents, using alternative products that do not contain parabens, and possibly reducing paraben exposure from dietary sources [51]. Future studies on prenatal paraben exposure and adiposity-related outcomes in children may need to quantify dietary paraben exposure and frequency of cosmetic product applications or the type of cosmetic products used before and during pregnancy in mothers, to improve the accuracy of exposure estimation. The collection of data from matrices, in addition to urine, should also be considered. For instance, as the largest fetal organ, the placenta may better reflect fetal paraben exposure [56]. Caution is also needed when interpreting data with a low detection rate of certain parabens in the biological samples used. In some studies, the concentrations of certain parabens in most samples were lower than the limit of detection (LOD); therefore, dichotomization (e.g., paraben levels were treated as a binary variable, either below LOD or above LOD) was used for data analysis [33,35,36], leading to the loss of information and statistical power [57]. Dichotomization may also increase the chance of false-positive findings [58]. In addition, the threshold effects and dose-dependent associations cannot be examined when dichotomization is applied [36]. A few studies used mixed-pollutant models to account for the complex exposure patterns of common endocrine disruptors while controlling for potential co-pollutant confounding [38,39]. However, models such as Bayesian hierarchical models (BMH) and Bayesian kernel machine regression (BKMR) cannot function with missing data and/or are sensitive to outliers [38,39,59], leading to reduced sample sizes. Future studies should continue to develop statistical models that more accurately and completely capture human chemical exposures.

2.2. *Paraben Exposure during Adolescence and Adiposity-Related Measures*

Six studies have explored the association between postnatal paraben exposure and adiposity-related measures in children of various ages (all under 20 years of age) (Table 2). Postnatally, pharmacological, personal-care products and dietary intake are the major sources of paraben exposure [60]. The majority (5 out of 6) of the studies were cross-sectional. Therefore, causal inference is limited, and the causal direction between these factors cannot be established. A one-time urine sample was the predominant matrix, although one study used daily intake of parabens from food sources to assess adiposity-related measures [41,51]. Results from these studies are mixed, showing positive [41,51,61], inverse [62,63], and null [64,65] associations. Further complicating the interpretations, one study found an association only in boys [61], whereas another found an association only in girls [51]. Thus, these studies do not clearly support any association between postnatal adolescent paraben exposure and adiposity-related measures. A major challenge with studies on adolescent exposure is the dynamic changes in sex hormones and the circadian rhythms of other hormones during the adolescent years, which are likely to affect paraben exposure. Standardized sample collection may help to minimize such confounding effects. In addition, these studies relied on a single-spot urine sample, which is another potential source of bias. Furthermore, most postnatal exposure studies did not collect dietary information [51,63]. If foods containing the most parabens are also the most calorific, then dietary confounding factors would present considerable threats to causal inference regarding paraben exposure [66].

Overall, the literature on paraben exposure and adiposity-related measures in adolescents remains sparse and inconsistent. As mentioned above, both methodologic and adolescent-specific sources of error hamper the interpretation of this literature. Prospective studies with standardized and repeated sample collection are needed. Studies should also capture dietary information to accurately estimate paraben exposure and identify confounding dietary factors, such as total energy intake.

Table 2. Association of paraben exposure during adolescence and adiposity-related measures (ARM).

Study/Country	Sample Size (Male/Female)	Year of Recruitment	Type of Study	Matrices	Time of ARM Assessment (Years of Age)	ARM	Major Shortcomings	Major Findings
Monteagudo et al. 2021, Spain [51]	585 (313/272)	2017–2018	Cross-sectional	Food	12–16	BMI	Only the dietary sources of paraben exposure were assessed; cross-sectional nature of the study design; paraben concentrations were not measured in biological samples; caloric intake was not considered in the analysis.	High total parabens and MeP daily intake were associated with high BMI in girls but not boys.
Feizabadi et al. 2020, Iran [62]	100 (50/50)	unknown	Cross-sectional	Urine	12–20	Weight, BMI; waist circumference	Small sample size; cross-sectional nature of the study design; only one spot sample was collected; caloric intake was not considered in the analysis.	Inverse associations were identified between urinary concentrations of paraben (MeP and EtP) and BMI.
Kim et al. 2020, Canada [64]	1418 (695/723)	2014–2015	Cross-sectional	Urine	3–17	Height, weight, BMI; waist circumference	Cross-sectional nature of the study design; only one spot sample was collected.	No association was identified between urinary paraben concentrations and BMI z-score or waist circumference.
Quirós-Alcalá et al. 2018, United States [63]	1324 (684/640)	2007–2012	Cross-sectional	Urine	6–19	BMI z-score; waist circumference	Cross-sectional nature of study design; only one spot sample was collected.	Negative associations were identified between urinary MeP, PrP, and total parabens concentrations and the prevalence odds ratios of being obese vs. normal weight; negative associations were identified between urinary MeP, PrP, and total parabens concentra-tions and waist circumference. The associations were stronger in girls.
Deierlein et al. 2017, United States [65] [a]	1017 (0/1017)	2004–2007	Prospective	Urine	12.8–18.4	Weight, height, waist circumference, BMI; percent body fat	Only females were enrolled in the study; only one spot sample was collected; caloric intake was not considered in the analysis.	No association was identified between baseline total paraben concentrations in urine and girls' adiposity-related measures.

Table 2. *Cont.*

Study/Country	Sample Size (Male/Female)	Year of Recruitment	Type of Study	Matrices	Time of ARM Assessment (Years of Age)	ARM	Major Shortcomings	Major Findings
Guo et al. 2017, China [61]	436 (221/215)	2012–2013	Cross-sectional	Urine	3	Weight z-score, height z-score, weight for height z-score; BMI z-score	Cross-sectional nature of study design; only one spot sample was collected; caloric intake was not considered in the analysis.	Urinary EtP concentrations were positively associated with weight z-scores and height z-scores. Total paraben concentrations were associated with anthropometric measures only in boys.

Footnotes: [a] Girls were enrolled at 6–8 years of age. Adiposity-related measures were assessed yearly or biannually until they were 15.6 years of age on average (range: 12.8–18.4 years of age); weight, height, and waist circumference were recorded at baseline. BMI, waist circumference, and percentage of fat were measured at each visit.

2.3. Paraben Exposure during Adulthood and Adiposity-Related Measures

As in the case of adolescents, adults are also primarily exposed to parabens through pharmacological and personal-care products (PPCPs) and dietary sources [51]. Twelve studies in the past five years have investigated the association between the concentration of parabens in urine and adiposity-related measures in adults (Table 3). Overall, seven studies suggest an inverse association between paraben exposure and adiposity-related measures (Table 3). Urinary concentration of parabens has been shown to increase with the frequency of the use of personal-care products, specifically "leave-on products" or products applied over a wider skin surface [67]. Urinary concentration of parabens also varies with age, which may reflect age-related changes in lifestyle, food exposure, and endocrine disruptor metabolism rates [60]. As urine is the most common biological matrix in these studies, a major limitation is the lack of consistency in urine sample collection. In some studies, participants provided first-morning/early-morning urine samples [62,68,69], whereas, in other studies, the timing of urine collection was different, even within the same study [65,67,70]. In some studies, fasting morning samples were collected from at least a portion of the participants [7,67], while in other studies, no such information was provided [29,62,64,68,71–73]. Seasonal variability in EDS concentrations in urine, including parabens, is well documented [69,70]; however, seasonal timing of exposure was not considered in most of these studies.

BMI and/or waist circumference (predictors of abdominal adiposity) were measured in all twelve studies. Yet, the diet was not considered in most studies, which is an important source of paraben exposure and a confounder when adiposity-related measures are considered. The majority (10 of 12) of these studies were cross-sectional (Table 3); therefore, the results cannot be used to determine the causal relationship. Regarding prospective studies, one study had a small sample size (n = 73), with a single spot-urine sample collected nine years prior to the assessment of the adiposity-related measures [73]. Therefore, it is unknown whether the level of paraben exposure during this 9-year period remained constant or not. In another study, women in one group (the metabolic group) were approximately three years older than the non-metabolic group, but the participants' menopausal status was not examined [73]. The menopausal transition period in aging women is strongly associated with weight gain [74].

One-third of the studies (4 of 12) did not find an association between parabens and BMI (Table 3), including the only study that considered dietary sources of parabens [7]. Moreover, these four studies were statistically underpowered, especially for sub-analysis by gender or other potential effect-modifying factors. Among the remaining eight studies, all but one [72] demonstrated an inverse relationship between paraben exposure and BMI, or that higher baseline paraben exposure was associated with reduced weight loss in response to a calorie-restriction intervention [75]. Interestingly, one of the largest studies, conducted in Korea with 3782 participants [72], showed that urinary ethylparaben (EtP) concentrations were positively associated with BMI, whereas a study conducted in the US using NHANES showed an inverse association [63]. The reasons for the inconsistency are unknown. However, the relatively short half-lives of parabens, varying uses of personal-care products by the two different populations [72], differences in the time frame during which the participants were recruited (2007–2014 in NHANES vs. 2015–2017 in Korean study) [63,72], the timing of sample collection, and unknown methodological issues or bias may have affected the estimation of urinary paraben concentrations, therefore influencing the strength and/or direction of the associations. For example, no fasting was required for those participants with afternoon or evening appointments to provide urine in the NHANES cohort (2007–2008) [76].

Table 3. Association of paraben exposure during adulthood and adiposity-related measures (ARM).

Study/Country	Sample Size (Male/Female)	Year of Recruitment	Type of Study	Matrices	Time of ARM Assessment (Years of Age)	ARM	Major Shortcomings	Major Findings
Jala et al. 2022, India [71]	52 (0/52)	2020	Cross-sectional	Urine	18–31	BMI; waist-to-hip ratio (WHR)	Small sample size; cross-sectional nature of the study design; only females were enrolled in the study (selection bias); only one spot sample was collected; caloric intake was not considered in the analysis.	No association was identified between urinary paraben concentrations and BMI or waist-to-hip ratio.
Vindenes et al. 2021, Norway [67]	496 (258/238)	2014–2015	Cross-sectional	Urine	18.1 to 47.5	BMI	Cross-sectional nature of the study design; only one spot sample was collected; caloric intake was not considered in the analysis.	Urinary concentrations of MeP and EtP were negatively associated with BMI.
van der Meer et al. 2021, the Netherlands [75] [a]	218 (70/148)	2008–2010	Prospective intervention	Urine	Average 52	BMI, waist circumference, body fat percentage	The study did not have a control group.	Higher baseline urinary paraben exposures were associated with reduced weight loss in a calorie-restriction intervention.
Lee et al. 2021, Korea [72]	3782 (1648/2134)	2015–2017	Cross-sectional	Urine	19–86	BMI	Cross-sectional nature of the study design; only one spot sample was collected; caloric intake was not considered in the analysis.	Urinary EtP concentrations were positively associated with BMI.
Zamora et al. 2021, Mexico [73]	73(0/73)	2008	Prospective	Urine	46.6 ± 6.3	BMI; waist circumference	Small sample size; potential confounding effect caused by the menopausal status of the participants; only females were enrolled (selection bias); only one spot sample was collected; paraben exposure was assessed 9 years prior to the measurement of adiposity-related outcomes; caloric intake was not considered in the analysis.	No association was identified between urinary paraben concentrations and BMI or abdominal obesity.

Table 3. *Cont.*

Study/Country	Sample Size (Male/Female)	Year of Recruitment	Type of Study	Matrices	Time of ARM Assessment (Years of Age)	ARM	Major Shortcomings	Major Findings
Karzi et al. 2021, Greece [29] [b]	99 (0/99)	Unknown	Cross-sectional	Urine	18.0 to 44.0	BMI	Small sample size; cross-sectional nature of the study design; only females were enrolled (selection bias); only one spot sample was collected; caloric intake was not considered in the analysis.	No association was identified between urinary paraben concentrations and BMI.
Kim et al. 2020, Canada [64]	1137 (568/569)	2014–2015	Cross-sectional	Urine	46.2 ± 0.3	BMI; waist circumference	Cross-sectional nature of the study design; only one spot sample was collected.	Negative associations were identified between urinary MeP and total parabens concentrations and obesity and waist circumference in women.
Kiani Feizabadi et al. 2020, Iran [68]	178 (75/103)	Unknown	Cross-sectional	Urine	43.7 ± 11.8	BMI	Small sample size; cross-sectional nature of the study design; only one spot sample was collected; caloric intake was not considered in the analysis.	Negative associations were identified between urinary MeP concentrations and BMI.
Hajizadeh et al. 2020, Iran [69] [c]	95 (0/95)	2018	Cross-sectional	Urine	34.2 ± 8.2	BMI; waist circumference	Small sample size; cross-sectional nature of the study design; only females were enrolled (selection bias); only one spot sample was collected; caloric intake was not considered in the analysis.	Negative associations were identified between urinary EtP concentrations and BMI.
Bethea et al. 2020, United States [70] [d]	766 (0/766)	2010–2012	Cross-sectional	Urine	23–34	BMI	Participants were recruited from a single urban area (selection bias); cross-sectional nature of the study design; only one spot sample was collected; no dietary information was available.	Urinary concentrations of MeP and BuP were negatively associated with morbid obesity (BMI $\geq 35\,kg/m^2$) compared to BMI < 25.
Yu et al. 2019, China [7]	562 (550/12)	2013–2015	Cross-sectional	Urine	22–59	BMI	Cross-sectional nature of the study design; only one spot sample was collected; only 12 women were enrolled (unbalanced study design); caloric intake was not considered in the analysis.	No correlations were identified between urinary concentrations of parabens (MP, EtP, and PrP) and BMI.
Quirós-Alcalá et al. 2017, United States [63]	4730 (2306/2424)	2007–2014	Cross-sectional	Urine	49.6 ± 17.4	BMI; waist circumference	Only one spot sample was collected; cross-sectional nature of the study design.	Urinary MeP concentrations were negatively associated with prevalence odds ratios for obesity and adiposity measures; Stronger associations were observed in females.

Footnotes: [a] All participants with a BMI > = 27 kg/m^2 were enrolled; [b,c] only pregnant women were enrolled in these studies; [d] a study of a vulnerable population from Detroit, Michigan metropolitan area.

The association between urinary paraben concentrations and adiposity-related measures may also depend on how urinary paraben concentrations were adjusted. Urinary creatinine (UC), specific urinary gravity (SG), and covariate-adjusted standardization (CAS) are methods of adjustment for urine dilution, with UC and SG being the most common. The choice of the method should not be arbitrary; for example, in one study [72], a positive association between EtP and BMI was revealed when CAS was used to adjust paraben concentrations in urine but was not observed when UC or SG was used. The concentration of urinary creatinine is influenced by age, race/ethnicity, gender, and muscle mass. Future studies should determine if the use of the CAS adjustment method is valid across populations or race-/ethnicity-dependent. Therefore, the choice of urine normalization method should be validated and standardized.

3. Expert Opinion and Future Directions

Early-life determination of adult health theory suggests that there is a vulnerable biological window during which exposure to sufficient doses of endocrine disruptors is associated with an increased risk of adverse health outcomes [77]. Results of human studies from multiple countries indicate that parabens can cross the placental barrier [28–31]. Fetal exposure to parabens is associated with altered gestational weight gain [32], height, head, hip, or arm circumference [28,30,33,34,41], z-scores of weight and length [35], BMI z-scores [31,38], and overweight status in early childhood [38]. One of the most significant challenges in the interpretation of this literature is that few studies were initially designed to examine parabens. For instance, the environmental exposure components were often added at a later research stage, resulting in only one single-spot urine sample, or urine samples were missing for many study participants. Thus, estimates of paraben concentrations from many of these studies may not be representative of exposure over the targeted time periods. Several recent longitudinal cohort studies did not have an adequate sample size to detect small effects or allow stratified analyses based on gender, race/ethnicity, or pre-pregnancy BMI status [29,71,73]. Thereby, there is a great need to carry out large, long-term cohort studies with repeated measurements of chemical exposures across a broad range of developmental periods to elucidate their effects on childhood and adult health outcomes.

The absorption and metabolism of parabens depend on the length of the carbon chain, which varies across the types and brands of consumer products. The composition of paraben profiles in human samples is often population- and/or region-specific, influenced by the local environment, how the specific pharmaceutical and personal-care products are used, as well as by varying dietary cultures/lifestyles [51,68,72]. The existing literature suggests several other ways that future investigations may advance the current knowledge. For example, BMI or BMI-z scores may not be an appropriate outcome variable when measuring body development in newborns and preadolescents [53,54]. Instead, whole-body dual X-ray absorptiometry [36] might be more effective in estimating body composition in these age groups.

As in adolescence, pregnancy is a biological stage characterized by dynamic changes in circulating hormones, glucose, proteins, and kidney function, which could affect the osmolarity of urine samples [78,79]. Urinary analyte concentrations are also susceptible to variations by the time of the day [80] and the season when the samples are collected [69,70], inherited inter-individual differences in toxicokinetics, and physiological characteristics of the biomonitoring matrix [13]. Therefore, in addition to spot urine samples, these sources of variation need to be addressed when designing future studies. Urinary creatinine remains the most widely used method for adjusting urine dilution. Research from Lee's group, however, showed that different methods of urine dilution adjustment might change the magnitude, and even the direction, of the associations between paraben exposure and metabolic-syndrome-related components [72]. Measurement of urinary analytes over 24 h is currently the most definitive method to quantify endocrine disruptor exposure. However, prolonged urine collection is inconvenient and often inaccurate due to frequent collection errors [81]. Alternatively, sampling biometrics, such as hair, may help to improve exposure assessment due to the accumulation of endocrine disruptors during hair growth. Hair is a relatively easy sample to obtain and analyze [82] and can, therefore, be sampled frequently, but the validity and repeatability of this method require further study.

Humans are exposed to multiple environmental pollutants daily. However, many published human studies reviewed here used single-pollutant models to assess paraben exposure and the associated health impacts. Bayesian mixture pollutant models might be a better analytical approach to account for complex exposure patterns and potential co-pollutant synergy [38,39]. Furthermore, many studies reported in this review measured exposure to parabens several years ago [7,32,33,36,39,63,64,67,70,72,73], which may not accurately reflect the types or levels of parabens prevalent in humans today. Paraben content in consumer products changes over time, partially due to the rise of public awareness and tightening regulatory guidelines in personal-care products and food [83,84]. Age, race, location, and gender-specific associations between paraben exposure and adiposity-related measures should be priorities of future studies due to the likelihood of higher endocrine disruptor exposure in vulnerable populations.

Finally, increasing evidence also highlights the importance of epigenetics as a functional modifier of the genome and a key determinant of disease risk [85]. For example, it has been suggested that the interaction of leptin receptor polymorphism and dietary intake of parabens may increase BMI [31]. Leptin receptors, together with leptin secreted from adipocytes, serve as a signal of satiety in the central nervous system, which controls food intake and subsequent energy expenditure. The inclusion of an epigenetic and/or genetic polymorphism component in future population studies may help to elucidate the interaction between paraben exposure and the key genes underlying obesity predisposition and outcomes [31,86,87].

Author Contributions: J.C. and L.Z. developed the main conceptual ideas and participated in manuscript writing; X.X. drafted the manuscript and created the tables; P.D.T. provided critical input on the outline of the manuscript and participated in writing; H.W. provided critical feedback and participated in manuscript writing. All authors have read and agreed to the published version of the manuscript.

Funding: The work was partially supported by the National Institutes of Health (1R15DK132728-01) to Ling Zhao and the University of Tennessee Open Publishing Support Fund to Jiangang Chen.

Institutional Review Board Statement: Not applicable.

Informed Consent Statement: Not applicable.

Data Availability Statement: Not applicable.

Conflicts of Interest: The authors declare no conflict of interest.

References

1. Matwiejczuk, N.; Galicka, A.; Brzoska, M.M. Review of the safety of application of cosmetic products containing parabens. *J. Appl. Toxicol.* **2020**, *40*, 176–210. [CrossRef] [PubMed]
2. Tade, R.S.; More, M.P.; Chatap, V.; Deshmukh, P.; Patil, P. Safety and toxicity assessment of parabens in pharmaceutical and food products. *Inventi Rapid Pharm. Pract.* **2018**, *3*, 1–9.
3. Review, C.I. Final report on the safety assessment of methylparaben, ethylparaben, propylparaben, and butylparaben. *J. Am. Coll. Toxicol.* **1984**, *3*, 147–209.
4. Bledzka, D.; Gromadzinska, J.; Wasowicz, W. Parabens. From environmental studies to human health. *Environ. Int.* **2014**, *67*, 27–42. [CrossRef] [PubMed]
5. Fisher, M.; MacPherson, S.; Braun, J.M.; Hauser, R.; Walker, M.; Feeley, M.; Mallick, R.; Berube, R.; Arbuckle, T.E. Paraben Concentrations in Maternal Urine and Breast Milk and Its Association with Personal Care Product Use. *Environ. Sci. Technol.* **2017**, *51*, 4009–4017. [CrossRef] [PubMed]
6. Frederiksen, H.; Jorgensen, N.; Andersson, A.M. Parabens in urine, serum and seminal plasma from healthy Danish men determined by liquid chromatography-tandem mass spectrometry (LC-MS/MS). *J. Expo. Sci. Environ. Epidemiol.* **2011**, *21*, 262–271. [CrossRef]
7. Yu, Y.; Li, W.; Lu, S.; Wu, S.; Wang, F.; Tse, L.A.; Kang, L.; Ma, S. Urinary parabens in adults from South China: Implications for human exposure and health risks. *Ecotoxicol. Environ. Saf.* **2019**, *182*, 109419. [CrossRef]

8. Park, N.Y.; Cho, Y.H.; Choi, K.; Lee, E.H.; Kim, Y.J.; Kim, J.H.; Kho, Y. Parabens in breast milk and possible sources of exposure among lactating women in Korea. *Environ. Pollut.* **2019**, *255*, 113142. [CrossRef]
9. Pycke, B.F.; Geer, L.A.; Dalloul, M.; Abulafia, O.; Halden, R.U. Maternal and fetal exposure to parabens in a multiethnic urban U.S. population. *Environ. Int.* **2015**, *84*, 193–200. [CrossRef]
10. Jimenez-Diaz, I.; Vela-Soria, F.; Zafra-Gomez, A.; Navalon, A.; Ballesteros, O.; Navea, N.; Fernandez, M.F.; Olea, N.; Vilchez, J.L. A new liquid chromatography-tandem mass spectrometry method for determination of parabens in human placental tissue samples. *Talanta* **2011**, *84*, 702–709. [CrossRef]
11. Philippat, C.; Wolff, M.S.; Calafat, A.M.; Ye, X.; Bausell, R.; Meadows, M.; Stone, J.; Slama, R.; Engel, S.M. Prenatal exposure to environmental phenols: Concentrations in amniotic fluid and variability in urinary concentrations during pregnancy. *Environ. Health Perspect.* **2013**, *121*, 1225–1231. [CrossRef] [PubMed]
12. Kang, S.; Kim, S.; Park, J.; Kim, H.J.; Lee, J.; Choi, G.; Choi, S.; Kim, S.; Kim, S.Y.; Moon, H.B.; et al. Urinary paraben concentrations among pregnant women and their matching newborn infants of Korea, and the association with oxidative stress biomarkers. *Sci. Total Environ.* **2013**, *461*, 214–221. [CrossRef] [PubMed]
13. Brauner, E.V.; Uldbjerg, C.S.; Lim, Y.H.; Gregersen, L.S.; Krause, M.; Frederiksen, H.; Andersson, A.M. Presence of parabens, phenols and phthalates in paired maternal serum, urine and amniotic fluid. *Environ. Int.* **2022**, *158*, 106987. [CrossRef] [PubMed]
14. Vo, T.T.; Jeung, E.B. An evaluation of estrogenic activity of parabens using uterine calbindin-d9k gene in an immature rat model. *Toxicol. Sci.* **2009**, *112*, 68–77. [CrossRef]
15. Sun, L.; Yu, T.; Guo, J.; Zhang, Z.; Hu, Y.; Xiao, X.; Sun, Y.; Xiao, H.; Li, J.; Zhu, D.; et al. The estrogenicity of methylparaben and ethylparaben at doses close to the acceptable daily intake in immature Sprague-Dawley rats. *Sci. Rep.* **2016**, *6*, 25173. [CrossRef]
16. Okubo, T.; Yokoyama, Y.; Kano, K.; Kano, I. ER-dependent estrogenic activity of parabens assessed by proliferation of human breast cancer MCF-7 cells and expression of ERalpha and PR. *Food Chem. Toxicol.* **2001**, *39*, 1225–1232. [CrossRef]
17. Chen, J.; Ahn, K.C.; Gee, N.A.; Gee, S.J.; Hammock, B.D.; Lasley, B.L. Antiandrogenic properties of parabens and other phenolic containing small molecules in personal care products. *Toxicol. Appl. Pharm.* **2007**, *221*, 278–284. [CrossRef]
18. Ozdemir, E.; Barlas, N.; Cetinkaya, M.A. Assessing the antiandrogenic properties of propyl paraben using the Hershberger bioassay. *Toxicol. Res.* **2018**, *7*, 235–243. [CrossRef]
19. Taxvig, C.; Vinggaard, A.M.; Hass, U.; Axelstad, M.; Boberg, J.; Hansen, P.R.; Frederiksen, H.; Nellemann, C. Do parabens have the ability to interfere with steroidogenesis? *Toxicol. Sci.* **2008**, *106*, 206–213. [CrossRef]
20. Hu, P.; Overby, H.; Heal, E.; Wang, S.; Chen, J.; Shen, C.L.; Zhao, L. Methylparaben and butylparaben alter multipotent mesenchymal stem cell fates towards adipocyte lineage. *Toxicol. Appl. Pharm.* **2017**, *329*, 48–57. [CrossRef]
21. Fujino, C.; Watanabe, Y.; Sanoh, S.; Hattori, S.; Nakajima, H.; Uramaru, N.; Kojima, H.; Yoshinari, K.; Ohta, S.; Kitamura, S. Comparative study of the effect of 17 parabens on PXR-, CAR- and PPARalpha-mediated transcriptional activation. *Food Chem. Toxicol.* **2019**, *133*, 110792. [CrossRef] [PubMed]
22. Kolsek, K.; Gobec, M.; Mlinaric Rascan, I.; Sollner Dolenc, M. Screening of bisphenol A, triclosan and paraben analogues as modulators of the glucocorticoid and androgen receptor activities. *Toxicol. Vitr.* **2015**, *29*, 8–15. [CrossRef] [PubMed]
23. Hu, P.; Chen, X.; Whitener, R.J.; Boder, E.T.; Jones, J.O.; Porollo, A.; Chen, J.; Zhao, L. Effects of parabens on adipocyte differentiation. *Toxicol. Sci.* **2013**, *131*, 56–70. [CrossRef]
24. Hu, P.; Kennedy, R.C.; Chen, X.; Zhang, J.; Shen, C.L.; Chen, J.; Zhao, L. Differential effects on adiposity and serum marker of bone formation by post-weaning exposure to methylparaben and butylparaben. *Environ. Sci. Pollut. Res. Int.* **2016**, *23*, 21957–21968. [CrossRef] [PubMed]
25. Giulivo, M.; Lopez de Alda, M.; Capri, E.; Barcelo, D. Human exposure to endocrine disrupting compounds: Their role in reproductive systems, metabolic syndrome and breast cancer. A review. *Environ. Res.* **2016**, *151*, 251–264. [CrossRef] [PubMed]
26. Cherian, P.; Zhu, J.; Bergfeld, W.F.; Belsito, D.V.; Hill, R.A.; Klaassen, C.D.; Liebler, D.C.; Marks, J.G., Jr.; Shank, R.C.; Slaga, T.J.; et al. Amended Safety Assessment of Parabens as Used in Cosmetics. *Int. J. Toxicol.* **2020**, *39*, 5S–97S. [CrossRef]
27. Hager, E.; Chen, J.; Zhao, L. Minireview: Parabens Exposure and Breast Cancer. *Int. J. Environ. Res. Public Health* **2022**, *19*, 1873. [CrossRef]
28. Golestanzadeh, M.; Ebrahimpour, K.; Daniali, S.S.; Zarean, E.; Yazdi, M.; Basirat, Z.; Goodarzi-Khoigani, M.; Kelishadi, R. Association between parabens concentrations in human amniotic fluid and the offspring birth size: A Sub-study of the PERSIAN birth cohort. *Environ. Res.* **2022**, *212*, 113502. [CrossRef]
29. Karzi, V.; Tzatzarakis, M.N.; Hatzidaki, E.; Katsikantami, I.; Alegakis, A.; Vakonaki, E.; Kalogeraki, A.; Kouvidi, E.; Xezonaki, P.; Sifakis, S.; et al. Determination of prenatal exposure to parabens and triclosan and estimation of maternal and fetal burden. *Toxicol. Rep.* **2021**, *8*, 808–815. [CrossRef]
30. Vrijens, K.; Van Overmeire, I.; De Cremer, K.; Neven, K.Y.; Carollo, R.M.; Vleminckx, C.; Van Loco, J.; Nawrot, T.S. Weight and head circumference at birth in function of placental paraben load in Belgium: An ENVIRONAGE birth cohort study. *Environ. Health* **2020**, *19*, 83. [CrossRef]
31. Reimann, B.; Vrijens, K.; Roels, H.A.; Wang, C.; Cosemans, C.; Van Overmeire, I.; Nawrot, T.S.; Plusquin, M. In utero exposure to parabens and early childhood BMI z-scores—Associations between placental ethyl paraben, longitudinal BMI trajectories and cord blood metabolic biomarkers. *Environ. Int.* **2021**, *157*, 106845. [CrossRef]

32. Wen, Q.; Zhou, Y.; Wang, Y.; Li, J.; Zhao, H.; Liao, J.; Liu, H.; Li, Y.; Cai, Z.; Xia, W. Association between urinary paraben concentrations and gestational weight gain during pregnancy. *J. Expo. Sci. Environ. Epidemiol.* **2020**, *30*, 845–855. [CrossRef] [PubMed]
33. Jamal, A.; Rastkari, N.; Dehghaniathar, R.; Nodehi, R.N.; Nasseri, S.; Kashani, H.; Shamsipour, M.; Yunesian, M. Prenatal urinary concentrations of environmental phenols and birth outcomes in the mother-infant pairs of Tehran Environment and Neurodevelopmental Disorders (TEND) cohort study. *Environ. Res.* **2020**, *184*, 109331. [CrossRef] [PubMed]
34. Chang, C.H.; Wang, P.W.; Liang, H.W.; Huang, Y.F.; Huang, L.W.; Chen, H.C.; Pan, W.C.; Lin, M.H.; Yang, W.; Mao, I.F.; et al. The sex-specific association between maternal paraben exposure and size at birth. *Int. J. Hyg. Environ. Health* **2019**, *222*, 955–964. [CrossRef] [PubMed]
35. Wu, C.; Xia, W.; Li, Y.; Li, J.; Zhang, B.; Zheng, T.; Zhou, A.; Zhao, H.; Huo, W.; Hu, J.; et al. Repeated Measurements of Paraben Exposure during Pregnancy in Relation to Fetal and Early Childhood Growth. *Environ. Sci. Technol.* **2019**, *53*, 422–433. [CrossRef]
36. Hojsager, F.D.; Kyhl, H.B.; Frederiksen, H.; Juul, A.; Andersson, A.M.; Andersen, M.S.; Grontved, A.; Jensen, T.K. Prenatal Exposure to Butyl Paraben Is Associated with Fat Percentage in 7-Year-Old Boys. *J. Clin. Endocrinol. Metab.* **2021**, *106*, e2633–e2638. [CrossRef]
37. Leppert, B.; Strunz, S.; Seiwert, B.; Schlittenbauer, L.; Schlichting, R.; Pfeiffer, C.; Roder, S.; Bauer, M.; Borte, M.; Stangl, G.I.; et al. Maternal paraben exposure triggers childhood overweight development. *Nat. Commun.* **2020**, *11*, 561. [CrossRef]
38. Berger, K.; Hyland, C.; Ames, J.L.; Mora, A.M.; Huen, K.; Eskenazi, B.; Holland, N.; Harley, K.G. Prenatal Exposure to Mixtures of Phthalates, Parabens, and Other Phenols and Obesity in Five-Year-Olds in the CHAMACOS Cohort. *Int. J. Environ. Res. Public Health* **2021**, *18*, 1796. [CrossRef]
39. Guil-Oumrait, N.; Cano-Sancho, G.; Montazeri, P.; Stratakis, N.; Warembourg, C.; Lopez-Espinosa, M.J.; Vioque, J.; Santa-Marina, L.; Jimeno-Romero, A.; Ventura, R.; et al. Prenatal exposure to mixtures of phthalates and phenols and body mass index and blood pressure in Spanish preadolescents. *Environ. Int.* **2022**, *169*, 107527. [CrossRef]
40. Baker, E.R. Body weight and the initiation of puberty. *Clin. Obs. Gynecol.* **1985**, *28*, 573–579. [CrossRef] [PubMed]
41. Wu, C.; Huo, W.; Li, Y.; Zhang, B.; Wan, Y.; Zheng, T.; Zhou, A.; Chen, Z.; Qian, M.; Zhu, Y.; et al. Maternal urinary paraben levels and offspring size at birth from a Chinese birth cohort. *Chemosphere* **2017**, *172*, 29–36. [CrossRef] [PubMed]
42. Sandanger, T.M.; Huber, S.; Moe, M.K.; Braathen, T.; Leknes, H.; Lund, E. Plasma concentrations of parabens in postmenopausal women and self-reported use of personal care products: The NOWAC postgenome study. *J. Expo. Sci. Environ. Epidemiol.* **2011**, *21*, 595–600. [CrossRef] [PubMed]
43. Jain, R.B. Impact of Pregnancy on the Levels of Parabens and Bisphenol A: Data from NHANES 2005–2010. *J. Chem.* **2016**, *2016*, 1529071. [CrossRef]
44. Kim, S.; Lee, S.; Shin, C.; Lee, J.; Kim, S.; Lee, A.; Park, J.; Kho, Y.; Moos, R.K.; Koch, H.M.; et al. Urinary parabens and triclosan concentrations and associated exposure characteristics in a Korean population-A comparison between night-time and first-morning urine. *Int. J. Hyg. Environ. Health* **2018**, *221*, 632–641. [CrossRef] [PubMed]
45. Terasaki, M.; Kamata, R.; Shiraishi, F.; Makino, M. Evaluation of estrogenic activity of parabens and their chlorinated derivatives by using the yeast two-hybrid assay and the enzyme-linked immunosorbent assay. *Environ. Toxicol. Chem.* **2009**, *28*, 204–208. [CrossRef]
46. Leeners, B.; Geary, N.; Tobler, P.N.; Asarian, L. Ovarian hormones and obesity. *Hum. Reprod. Update* **2017**, *23*, 300–321. [CrossRef]
47. Grantham, J.P.; Henneberg, M. The estrogen hypothesis of obesity. *PLoS ONE* **2014**, *9*, e99776. [CrossRef]
48. Engeli, R.T.; Rohrer, S.R.; Vuorinen, A.; Herdlinger, S.; Kaserer, T.; Leugger, S.; Schuster, D.; Odermatt, A. Interference of Paraben Compounds with Estrogen Metabolism by Inhibition of 17beta-Hydroxysteroid Dehydrogenases. *Int. J. Mol. Sci.* **2017**, *18*, 2007. [CrossRef]
49. Boberg, J.; Taxvig, C.; Christiansen, S.; Hass, U. Possible endocrine disrupting effects of parabens and their metabolites. *Reprod. Toxicol.* **2010**, *30*, 301–312. [CrossRef]
50. Konings, G.; Brentjens, L.; Delvoux, B.; Linnanen, T.; Cornel, K.; Koskimies, P.; Bongers, M.; Kruitwagen, R.; Xanthoulea, S.; Romano, A. Intracrine Regulation of Estrogen and Other Sex Steroid Levels in Endometrium and Non-gynecological Tissues; Pathology, Physiology, and Drug Discovery. *Front. Pharm.* **2018**, *9*, 940. [CrossRef]
51. Monteagudo, C.; Robles-Aguilera, V.; Salcedo-Bellido, I.; Galvez-Ontiveros, Y.; Samaniego-Sanchez, C.; Aguilera, M.; Zafra-Gomez, A.; Burgos, M.A.M.; Rivas, A. Dietary exposure to parabens and body mass index in an adolescent Spanish population. *Environ. Res.* **2021**, *201*, 111548. [CrossRef] [PubMed]
52. Roberge, J.B.; Harnois-Leblanc, S.; McNealis, V.; van Hulst, A.; Barnett, T.A.; Kakinami, L.; Paradis, G.; Henderson, M. Body Mass Index Z Score vs Weight-for-Length Z Score in Infancy and Cardiometabolic Outcomes at Age 8–10 Years. *J. Pediatr.* **2021**, *238*, 208–214.e2. [CrossRef] [PubMed]
53. Vanderwall, C.; Eickhoff, J.; Randall Clark, R.; Carrel, A.L. BMI z-score in obese children is a poor predictor of adiposity changes over time. *BMC Pediatr.* **2018**, *18*, 187. [CrossRef]
54. Vanderwall, C.; Randall Clark, R.; Eickhoff, J.; Carrel, A.L. BMI is a poor predictor of adiposity in young overweight and obese children. *BMC Pediatr.* **2017**, *17*, 135. [CrossRef] [PubMed]
55. Walter, J.R.; Perng, W.; Kleinman, K.P.; Rifas-Shiman, S.L.; Rich-Edwards, J.W.; Oken, E. Associations of trimester-specific gestational weight gain with maternal adiposity and systolic blood pressure at 3 and 7 years postpartum. *Am. J. Obs. Gynecol.* **2015**, *212*, 499.e1–499.e12. [CrossRef]

56. Burton, G.J.; Jauniaux, E. Development of the Human Placenta and Fetal Heart: Synergic or Independent? *Front. Physiol.* **2018**, *9*, 373. [CrossRef]
57. Altman, D.G.; Royston, P. The cost of dichotomising continuous variables. *BMJ* **2006**, *332*, 1080. [CrossRef]
58. Austin, P.C.; Brunner, L.J. Inflation of the type I error rate when a continuous confounding variable is categorized in logistic regression analyses. *Stat. Med.* **2004**, *23*, 1159–1178. [CrossRef]
59. Bottegal, A.Y.A.G.; Hjalmarsson, H.; Pillonetto, G. Outlier robust system identification: A Bayesian kernel-based approach. *IFAC Proc. Vol.* **2014**, *47*, 6. [CrossRef]
60. Kang, H.S.; Kyung, M.S.; Ko, A.; Park, J.H.; Hwang, M.S.; Kwon, J.E.; Suh, J.H.; Lee, H.S.; Moon, G.I.; Hong, J.H.; et al. Urinary concentrations of parabens and their association with demographic factors: A population-based cross-sectional study. *Environ. Res.* **2016**, *146*, 245–251. [CrossRef]
61. Guo, J.; Wu, C.; Lu, D.; Jiang, S.; Liang, W.; Chang, X.; Xu, H.; Wang, G.; Zhou, Z. Urinary paraben concentrations and their associations with anthropometric measures of children aged 3 years. *Environ. Pollut.* **2017**, *222*, 307–314. [CrossRef] [PubMed]
62. Kiani Feizabadi, G.; Hajizadeh, Y.; Feizi, A.; Ebrahimpour, K. Urinary Concentrations of Parabens in a Population of Iranian Adolescent and Their Association with Sociodemographic Indicators. *Arch. Environ. Contam. Toxicol.* **2020**, *79*, 195–207. [CrossRef] [PubMed]
63. Quiros-Alcala, L.; Buckley, J.P.; Boyle, M. Parabens and measures of adiposity among adults and children from the U.S. general population: NHANES 2007-2014. *Int. J. Hyg. Environ. Health* **2018**, *221*, 652–660. [CrossRef] [PubMed]
64. Kim, J.; Chevrier, J. Exposure to parabens and prevalence of obesity and metabolic syndrome: An analysis of the Canadian Health Measures Survey. *Sci. Total Environ.* **2020**, *713*, 135116. [CrossRef] [PubMed]
65. Deierlein, A.L.; Wolff, M.S.; Pajak, A.; Pinney, S.M.; Windham, G.C.; Galvez, M.P.; Rybak, M.; Calafat, A.M.; Kushi, L.H.; Biro, F.M.; et al. Phenol Concentrations During Childhood and Subsequent Measures of Adiposity Among Young Girls. *Am. J. Epidemiol.* **2017**, *186*, 581–592. [CrossRef]
66. Yeo, G.S.H. Is calorie labelling on menus the solution to obesity? *Nat. Rev. Endocrinol.* **2022**, *18*, 453–454. [CrossRef]
67. Vindenes, H.K.; Svanes, C.; Lygre, S.H.L.; Real, F.G.; Ringel-Kulka, T.; Bertelsen, R.J. Exposure to environmental phenols and parabens, and relation to body mass index, eczema and respiratory outcomes in the Norwegian RHINESSA study. *Environ. Health* **2021**, *20*, 81. [CrossRef]
68. Kiani Feizabadi, G.; Hajizadeh, Y.; Feizi, A.; Ebrahimpour, K. Urinary concentrations of parabens amongst Iranian adults and their associations with socio-demographic factors. *J. Environ. Health Sci. Eng.* **2020**, *18*, 1227–1238. [CrossRef]
69. Hajizadeh, Y.; Kiani Feizabadi, G.; Ebrahimpour, K.; Shoshtari-Yeganeh, B.; Fadaei, S.; Darvishmotevalli, M.; Karimi, H. Urinary paraben concentrations and their implications for human exposure in Iranian pregnant women. *Environ. Sci. Pollut. Res. Int.* **2020**, *27*, 14723–14734. [CrossRef]
70. Bethea, T.N.; Wesselink, A.K.; Weuve, J.; McClean, M.D.; Hauser, R.; Williams, P.L.; Ye, X.; Calafat, A.M.; Baird, D.D.; Wise, L.A. Correlates of exposure to phenols, parabens, and triclocarban in the Study of Environment, Lifestyle and Fibroids. *J. Expo. Sci. Environ. Epidemiol.* **2020**, *30*, 117–136. [CrossRef]
71. Jala, A.; Varghese, B.; Dutta, R.; Adela, R.; Borkar, R.M. Levels of parabens and bisphenols in personal care products and urinary concentrations in Indian young adult women: Implications for human exposure and health risk assessment. *Chemosphere* **2022**, *297*, 134028. [CrossRef] [PubMed]
72. Lee, I.; Park, Y.J.; Kim, M.J.; Kim, S.; Choi, S.; Park, J.; Cho, Y.H.; Hong, S.; Yoo, J.; Park, H.; et al. Associations of urinary concentrations of phthalate metabolites, bisphenol A, and parabens with obesity and diabetes mellitus in a Korean adult population: Korean National Environmental Health Survey (KoNEHS) 2015–2017. *Environ. Int.* **2021**, *146*, 106227. [CrossRef] [PubMed]
73. Zamora, A.N.; Jansen, E.C.; Tamayo-Ortiz, M.; Goodrich, J.M.; Sanchez, B.N.; Watkins, D.J.; Tamayo-Orozco, J.A.; Tellez-Rojo, M.M.; Mercado-Garcia, A.; Baylin, A.; et al. Exposure to Phenols, Phthalates, and Parabens and Development of Metabolic Syndrome Among Mexican Women in Midlife. *Front. Public Health* **2021**, *9*, 620769. [CrossRef] [PubMed]
74. Kodoth, V.; Scaccia, S.; Aggarwal, B. Adverse Changes in Body Composition During the Menopausal Transition and Relation to Cardiovascular Risk: A Contemporary Review. *Womens Health Rep.* **2022**, *3*, 573–581. [CrossRef] [PubMed]
75. Van der Meer, T.P.; Thio, C.H.L.; Van Faassen, M.; Van Beek, A.P.; Snieder, H.; van Berkum, F.N.R.; Kema, I.P.; Makris, K.C.; Wolffenbuttel, B.H.R.; van Vliet-Ostaptchouk, J.V. Endocrine disrupting chemicals during diet-induced weight loss—A post-hoc analysis of the LOWER study. *Environ. Res.* **2021**, *192*, 110262. [CrossRef]
76. Lakind; J.S.; Levesque, J.; Dumas, P.; Bryan, S.; Clarke, J.; Naiman, D.Q. Comparing United States and Canadian population exposures from National Biomonitoring Surveys: Bisphenol A intake as a case study. *J. Expo. Sci. Environ. Epidemiol.* **2012**, *22*, 219–226. [CrossRef]
77. Braun, J.M. Early-life exposure to EDCs: Role in childhood obesity and neurodevelopment. *Nat. Rev. Endocrinol.* **2017**, *13*, 161–173. [CrossRef]
78. Teasdale, S.; Morton, A. Changes in biochemical tests in pregnancy and their clinical significance. *Obs. Med.* **2018**, *11*, 160–170. [CrossRef]
79. Cheung, K.L.; Lafayette, R.A. Renal physiology of pregnancy. *Adv. Chronic Kidney Dis.* **2013**, *20*, 209–214. [CrossRef]

80. Nassan, F.L.; Coull, B.A.; Gaskins, A.J.; Williams, M.A.; Skakkebaek, N.E.; Ford, J.B.; Ye, X.; Calafat, A.M.; Braun, J.M.; Hauser, R. Personal Care Product Use in Men and Urinary Concentrations of Select Phthalate Metabolites and Parabens: Results from the Environment and Reproductive Health (EARTH) Study. *Environ. Health Perspect.* **2017**, *125*, 087012. [CrossRef]
81. Mann, S.J.; Gerber, L.M. Addressing the problem of inaccuracy of measured 24-hour urine collections due to incomplete collection. *J. Clin. Hypertens.* **2019**, *21*, 1626–1634. [CrossRef] [PubMed]
82. Robin, J.; Binson, G.; Albouy, M.; Sauvaget, A.; Pierre-Eugene, P.; Migeot, V.; Dupuis, A.; Venisse, N. Analytical method for the biomonitoring of bisphenols and parabens by liquid chromatography coupled to tandem mass spectrometry in human hair. *Ecotoxicol. Environ. Saf.* **2022**, *243*, 113986. [CrossRef] [PubMed]
83. Sasseville, D.; Alfalah, M.; Lacroix, J.P. Parabenoia" Debunked, or "Who's Afraid of Parabens? *Dermatitis* **2015**, *26*, 254–259. [CrossRef] [PubMed]
84. Snodin, D. Regulatory risk assessments: Is there a need to reduce uncertainty and enhance robustness? Update on propylparaben in relation to its EU regulatory status. *Hum. Exp. Toxicol* **2017**, *36*, 1007–1014. [CrossRef]
85. Alavian-Ghavanini, A.; Ruegg, J. Understanding Epigenetic Effects of Endocrine Disrupting Chemicals: From Mechanisms to Novel Test Methods. *Basic Clin. Pharm. Toxicol.* **2018**, *122*, 38–45. [CrossRef]
86. Lu, X.; Fraszczyk, E.; Van der Meer, T.P.; Van Faassen, M.; Bloks, V.W.; Kema, I.P.; Van Beek, A.P.; Li, S.; Franke, L.; Westra, H.J.; et al. An epigenome-wide association study identifies multiple DNA methylation markers of exposure to endocrine disruptors. *Environ. Int.* **2020**, *144*, 106016. [CrossRef]
87. Wauman, J.; Zabeau, L.; Tavernier, J. The Leptin Receptor Complex: Heavier Than Expected? *Front. Endocrinol.* **2017**, *8*, 30. [CrossRef]

toxics

MDPI

Article

Leaching Mechanism and Health Risk Assessment of As and Sb in Tailings of Typical Antimony Mines: A Case Study in Yunnan and Guizhou Province, Southwest China

Ziyou Bai [1], Yinping He [2], Zhiwei Han [1,3,*] and Fuzhong Wu [2]

1 Resource and Environmental Engineering College, Guizhou University, Guiyang 550025, China
2 School of Materials and Metallurgy, Guizhou University, Guiyang 550025, China
3 Key Laboratory of Karst Georesources and Environment, Ministry of Education, Guiyang 550025, China
* Correspondence: zwhan@gzu.edu.cn

Abstract: The weathering and leaching of mining tailings have released large amounts of antimony (Sb) and arsenic (As), causing serious pollution in the surrounding soil, water, and sediments. To understand the leaching characteristics of Sb and As in mining tailings, Zuoxiguo and Qinglong mining tailings were collected for analysis. The average content of Sb in Zuoxiguo and Qinglong tailings was 5902.77 mg/kg and 1426.43 mg/kg, respectively, while that of As was 412.53 mg/kg and 405.26 mg/kg, respectively, which exceeded the local background value. Furthermore, the concentrations of Sb in the leachate of Zuoxiguo and Qinglong increased with time; the average Sb concentration in the leachate of Zuoxiguo and Qinglong was 1470.48 μg/L and 70.20 μg/L, respectively, while that of the As concentration was 31.20 μg/L and 6.45 μg/L, respectively. This suggests that the concentrations of Sb and As in the leachate of Zuoxiguo are both higher than those in the leachate of Qinglong and that the pH of the leachate of Zuoxiguo and Qinglong significantly changed within the first day under different initial pH conditions, and tended to be between 6 and 8, after one day. The results of the average health risk index showed that As in the leachate from Zuoxiguo and Qinglong for children was 5.67×10^{-4} and 9.13×10^{-5}, respectively, and 4.43×10^{-4} and 7.16×10^{-5}, respectively, for adults. As in the leachate from Zuoxiguo poses serious carcinogenic risks for residents, and in the study area, As poses a serious threat to human health. Therefore, the local government must manage As in these areas.

Keywords: mining tailings; leachate; chemical Sb and As species; scanning electron microscope; health risk assessment

Citation: Bai, Z.; He, Y.; Han, Z.; Wu, F. Leaching Mechanism and Health Risk Assessment of As and Sb in Tailings of Typical Antimony Mines: A Case Study in Yunnan and Guizhou Province, Southwest China. *Toxics* **2022**, *10*, 777. https://doi.org/10.3390/toxics10120777

Academic Editors: Esref Demir and Sam Kacew

Received: 5 November 2022
Accepted: 8 December 2022
Published: 12 December 2022

1. Introduction

Mining tailings are one of the most important contributors to heavy metal pollution [1]. Indeed, mining activities increase the levels of heavy metals in agricultural soils, surface water, groundwater, and plants in surrounding areas [2–4] and pose significant health risks to residents and miners [5,6]. Mineral resources dominate China's energy consumption [7]; however, despite China's development has gained economic benefits, heavy metal pollution has increased [8]. Heavy metals in mining tailings can be transferred through weathering and rainfall to the surrounding soils and water bodies [9]. Liquids are a key medium for the migration of heavy metals from tailings [10], which can be leached by rainfall and subsequently migrate with surface runoff or leachate before being absorbed by the mechanistic components of soils and sediments [11]. Heavy metal pollution from mining activities, tailing weathering, and leaching is a significant challenge in southwest China [12]. The reported levels of antimony (Sb) and arsenic (As) in soil, water bodies, and vegetation around the Xikuangshan mine, an active antimony mine located in southwest China, exceed those of local background levels [13–15], which can adversely affect the surrounding environment and residents.

Sb is a heavy metal with a density of >5 g/cm^3 [16], while As is categorized as a heavy metal [17]. As and Sb pollution in the natural environment is persistent, non-degradable, and irreversible [18], which has become a major concern [19,20], especially in antimony mining areas, where As and Sb pose a significant health risk to residents and animals [21]. Sb and As contents in hair samples from residents near antimony mine were observably higher compared with those in non-mining areas, suggesting that the Sb and As released from Qinglong mining areas adversely affect human health [22]. Sb has attracted significant attention due to its toxicity and environmental pollution [23]. However, along with As, data on Sb environmental and health risks are limited compared with other heavy metals such as Pb, Zn, Cd, and Cu [24].

The study of the leaching characteristics of heavy metals in tailings can provide insights into their impact on groundwater and soil [25,26]. Previous studies have identified the risks associated with the leaching of heavy metals in tailings on the soil environment [27]. Moreover, the release of heavy metals in tailings is not only influenced by the total concentration of heavy metals and the pH of tailings but also the chemical species of heavy metals [27,28]. In addition, the leaching of heavy metals under different environmental conditions can be simulated [29]; hence, their threat to the ecology in the area and surrounding populations can be assessed.

Therefore, in this study, Zuoxiguo and Qinglong antimony mines were selected in southwest China as the study areas to: (1) investigate the total content of As and Sb in the tailings and tailings components; (2) study the leaching characteristics of As and Sb in the tailings based on the concentration of As and Sb in the leachate and the chemical species of As and Sb in the tailings under different initial pH conditions; and (3) derive the health risks following human consumption of As and Sb in the leachate and provide suggestions to policy makers to mitigate such risks.

2. Materials and Methods

2.1. Study Area

Qinglong antimony mine (approximately 25°33′—26°11′ N, 105°01′—105°25′ E), located in Dachang Town, Qinglong County, Guizhou Province, has an average annual temperature of 17 °C and precipitation of approximately 1500 mm, as the area hosts a subtropical monsoon climate. Zuoxiguo antimony mine (approximately 23°22′—23°41′ N, 103°31′—103°57′ E), located in Zuoxiguo Village, Beige Town, Kaiyuan City, Yunnan Province, hosts a subtropical plateau monsoon climate, with distinct dry and wet seasons. The area has an annual average temperature of 24.2 °C and rainfall of 1450 mm. The mining history of these two typical Sb mines is extensive [18], posing a serious heavy metal pollution challenge. Qinglong and Zuoxiguo antimony mines have terminated production for many years [30,31].

2.2. Sampling and Sample Preparation

According to the different landforms of the two antimony mines, the sampling layout in the two tailings were also different. After removing the sundries horizons from the sampling sites, samples were drilled via the method of mixed sampling. The landform of the Zuoxiguo antimony mine is an open-type, and heavy metals could easily migrate along the direction of water flow; therefore, six 30 cm samples were collected from high to low using a stainless-steel shovel in each sample square via the double diagonal five-point mixing method. In contrast, the Qinglong antimony mine is surrounded by mountains, while the middle is a basin where heavy metals are more likely to migrate and diffuse downwards and are less likely to migrate over long distances. A total of one tailing sample was collected (Figure 1). Each sample was marked with a marker pen, and the location of every sampling site was recorded using the global positioning system (GPS). All samples were freeze-dried for 5–7 days to prevent oxidation. Once dried, part of each sample was ground with an agate mortar by hand, and the powdered samples were sieved through a <10-mesh sieve for sand property determination and <200-mesh sieve for element analysis. Tailing sample pieces (0.05 g) were accurately weighed and dissolved in a 3 mL acid mixture with a 2:1 volume ratio of concentrated HNO_3 and concentrated HF to analyse the total content of heavy metals. Sb and As concentrations in the extracts were analysed using an atomic fluorescence spectrometer (AFS-8510, China).

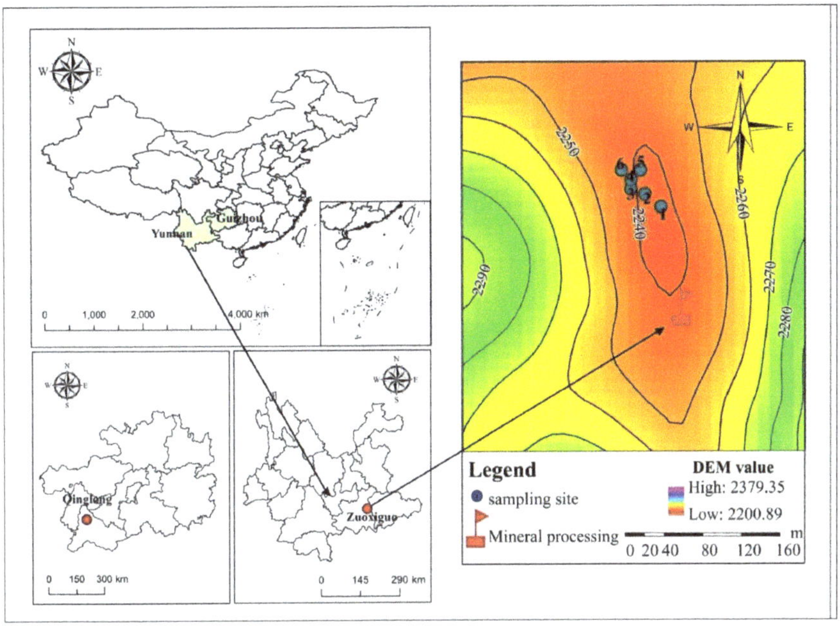

Figure 1. Location of the two typical antimony mines.

2.3. Sequential Extraction and Leaching Test

Chemical species of tailings were extracted using improved sequential extraction (BCR). Briefly, the acid-soluble fraction (F1) was extracted using 40 mL 0.11 mol·L^{-1} CH_3COOH; then, the reducible fraction (F2) was extracted with 40.00 mL 0.50 mol·L^{-1} hydroxylamine hydrochloride ($NH_2OH{\cdot}HCl$), while the oxidizable fraction (F3) was digested with 10.00 mL 8.80 mol·L^{-1} $NH_2OH{\cdot}HCl$ and extracted with 50.00 mL 1.00 mol·L^{-1} ammonium acetate (CH_3COONH_4). Sb and As concentrations in F1, F2, and F3 were measured using an atomic fluorescence spectrometer (AFS-8510, China). The residual fraction (F4) was calculated as the difference between the concentration of the sum of the three fractions (F1 + F2 + F3) and the total Sb and As concentrations [32].

The pH-static leaching experiments were performed for 9 days. Tailing samples (2 $\pm$ 0.01 g) were placed in a 50 mL polyethylene bottle, and 40 mL of prepared extraction was added. Separate bottles were used for each sampling to ensure identical ratios. The pH values of 3, 5, 6.87, and 9 were to account for acidic, neutral, and basic conditions encountered in the mining areas. Sulphuric acid and sodium hydroxide were added to adjust the pH. The reactor was placed in a gas bath thermostatic shaker at 120 r/min. The leachate was sampled after days 1, 3, 5, 7, and 9 and filtered through a 0.45 μm membrane. A PHS-3c pH meter was used to continuously measure the tailing supernatant three times, and the average value was the leachate used for the analysis of Sb and As concentrations.

On the basis of the previously established standard for determining heavy metals in soil (HJ680-2013), a total of 6 samples to be analysed were inserted using a blank sample, in accordance with the 10% standard. During analysis, procedural gaps, parallel experiments, and soil samples (obtained as per national standard GSS-4) were considered for quality control. The calibration curves of Sb and As standard solutions showed determination coefficients exceeding 0.999 with recovery rates between 80.0 and 105.0%. Ultrapure water was used as the experimental water, while all reagents utilized were guaranteed reagents. The container was soaked in 10% HNO_3 solution for more than 24 h, rinsed with ultra-pure water, and dried prior to use. All experimental equipment was dipped in nitric acid (12%) for more than 24 h and washed with deionized water at least three times. Arcmap10.6, Origin 2021, and Microsoft Excel 2017 were used for data processing and graphics drawing.

2.4. Health Risk Assessment

The human health risk assessment was divided into carcinogenic and non-carcinogenic assessments. As is a carcinogen via the drinking route, while Sb is a non-carcinogen. In this study, health risk was considered to the population through the drinking water route when leachate containing As and Sb contaminates groundwater. Therefore, the carcinogenic risk of As and the non-carcinogenic risk of Sb were analysed separately.

2.4.1. Carcinogenic Risk Model

In general, even a very small number of carcinogenic risk substances can have negative affect on human health. The expression of the evaluation model is as follows:

$$R_c = (D_i \times SF_i)/70 \tag{1}$$

where R_c is the per capita carcinogenic risk generated by chemical carcinogens through the surface of water sources, a^{-1}; D_i is the average daily exposure dose per unit of toxic substances through the surface of drinking water sources, mg·(kg·d)^1; SF_i is the carcinogenic coefficient of chemical carcinogens ingested through the surface of drinking water sources, mg·(kg·d)$^{-1}$ [33].

2.4.2. Non-Carcinogenic Risk Model

Non-carcinogens are only hazardous to human health if they exceed a threshold value. The non-carcinogenic risk is described by the commonly used risk index (HI) evaluated as follows:

$$R_n = \frac{D_i}{RfD \times 70} \times 10^{-6} \tag{2}$$

where R_n is the per capita annual health risk from non-carcinogenic substances exposed through drinking water sources, a^{-1}; D_i is the average daily exposure dose per body of toxic substances exposed through water sources, mg·(kg·d)$^{-1}$; *RfD* is the carcinogenic coefficient of non-carcinogenic substances ingested through drinking water sources, mg·(kg·d)$^{-1}$; 70 is the average life expectancy, in years [34].

The average daily exposure dose *Di* by the drinking water route is calculated separately for adults and children using the following formula:

$$Adult: D_i = 2.2C_i/64.3 \tag{3}$$

$$Child: D_i = (1.0C_i)/22.9 \tag{4}$$

where C_i is the concentration of the chemical carcinogen, mg/L; 2.2 is the average daily water intake of adults, L/d; 64.3 is the average body weight of adults, in kg; 1.0 is the average daily water intake of children, L/d; 22.9 is the average body weight of children, in kg [34].

The classification system is based on the International Agency for Research on Carcinogenesis (IARC) and the World Health Organization's (WHO) comprehensive evaluation of the carcinogenicity of chemical substances, combined with the US EPA recommended values. Among the two toxic elements measured in this study, the carcinogenic intensity factor SF for As was 15 mg·(kg·d)$^{-1}$ and the drinking water exposure reference dose *RfD* for Sb was 0.004 mg·(kg·d)$^{-1}$. The health risk in the water environment was classified into five levels: $R_c/R_n < 1.0 \times 10^{-6}$ a^{-1}, low risk; 1.0×10^{-6} $a^{-1} \leq R_c/R_n < 1.0 \times 10^{-5}$ a^{-1}, considerable; $1.0 \times 10^{-5}\ a^{-1} \leq R_c/R_n < 5.0 \times 10^{-5}\ a^{-1}$, medium; $5.0 \times 10^{-5}\ a^{-1} \leq R_c/R_n < 1.0 \times 10^{-4}\ a^{-1}$, high risk; $R_c/R_n > 1.0 \times 10^{-4}\ a^{-1}$, serious risk [35].

3. Results and Discussion

3.1. Mineral Composition and Toxic Elements Content in Tailings

The peak XRD analysis pattern intensities primarily represent the crystallinity of different phases; hence, the contents of different components in the samples cannot bean-alyzed. The XRD patterns of the tailings showed different diffraction patterns (Figure 2). The compositions of the Zuoxiguo antimony mining tailings were SiO_2, $CaCO_3$, FeS_2, and Sb_2S_3, a typical sulphide mining tailing. Conversely, the fractions of the Qinglong antimony mine tailings were SiO_2, $CaCO_3$, $CaSiO_3$, and $CaAl_2Si_4O_{12}$-$2H_2O$. The primary component of the tailings of both antimony mines was SiO_2. Compared with the Zuoxiguo antimony mine, the tailings of the Qinglong antimony mine contained less Sb (Table 1); therefore, no significant peaks in the Sb phase were detected. Yet, the oxides of secondary Sb, such as cubic Sb_2O_3 and rhombic Sb_2O_3, were the most important weathering of Sb_2S_3 products. The tailing sand contained SiO_2 and $CaCO_3$, which are associated with silicification and carbonatization in wall rock alteration, respectively.

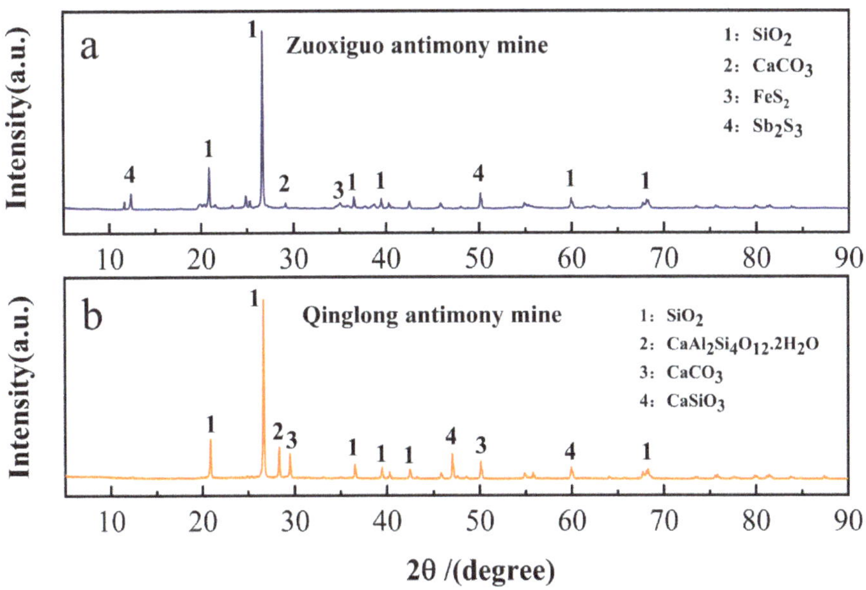

Figure 2. XRD pattern of two mining tailings. ((**a**): XRD analyse for Zuoxiguo mining tailings; (**b**): XRD analyse for Qionglong mining tailings).

The tailings of Zuoxiguo and Qinglong antimony mines contained large amounts of Sb and As (Table 1), with the contents of Sb and As in the Qinglong antimony mine being 637 and 20 times higher than the local soil background values, respectively. The average content of Sb and As in the Zuoxiguo mine was 771 and 27 times higher than the local soil background values, respectively. Notably, the pH of the Qinglong antimony mine tailings was much higher than that of the Zuoxiguo tailings, which is neutral to weakly alkaline. Meanwhile, the pH of the Zuoxiguo antimony mining tailings was acidic, which is more conducive to the migration and transformation of heavy metals under acidic conditions [36]. This may explain the significantly higher Sb in the Zuoxiguo antimony mining tailings than Qinglong antimony mining tailings. In addition, combined with the XRD in Figure 2, pyrite was identified in the tailings of the Zuoxiguo antimony mine. Under natural conditions, the pyrite in the tailings undergoes a redox reaction ($2FeS_2 + 7O_2 + 2H_2O \rightarrow 2Fe_2 + 4SO_4^{2-} + 4H^+$), thereby releasing a large number of hydrogen ions into the tailings [37], which lower the pH. Heavy metals associated with pyrite are released into the environment along with heavy metal carbonates [38], which may be another reason for the high Sb and As content in the tailings.

Table 1. Total heavy metals in the Zuoxiguo and Qinglong mining tailings (mg/kg).

Item	Sb	As	pH
QL	1426.43	405.26	7.38
Background value of soil in Guizhou	2.24	20.00	n.d
Z1	13,689.70	541.20	3.73
Z2	9663.30	771.50	3.05
Z3	4450.90	218.00	2.99
Z4	3004.42	600.50	3.62
Z5	2428.35	206.00	3.49
Z6	2179.92	138.00	4.26
Background value of soil in Yunnan	7.64	14.90	n.d

n.d means that there is no pH background value in Guizhou and Yunnan.

3.2. Leaching Characteristics of Different Antimony Mining Tailings

Leaching Characteristics of Sb and As in Tailings, and pH Changes of Leachate during Leaching

Figure 3 shows the total Sb released from each mining tailing, which, although different, was increased with time, especially in Zuoxiguo mine tailings. The Sb release of the Zuoxiguo mine tailings was larger than the Qinglong mine tailings.

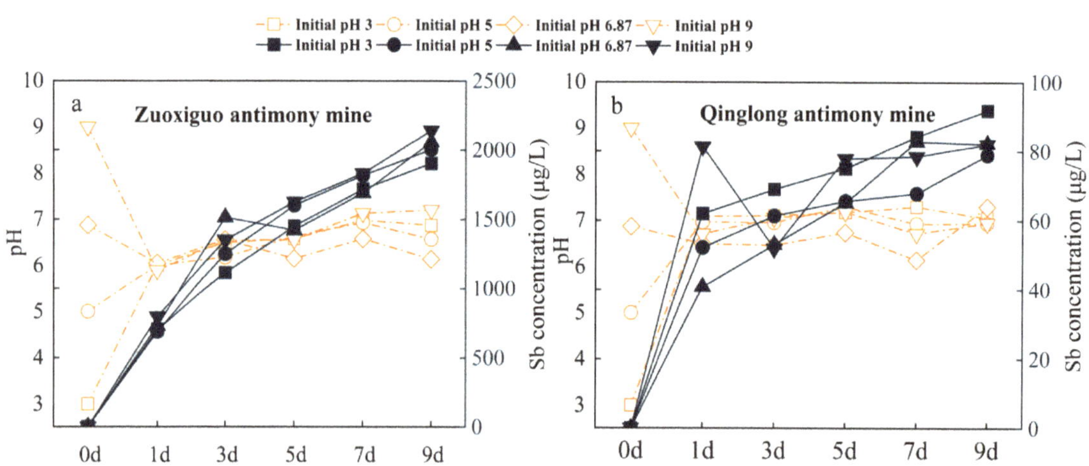

Figure 3. Leached concentration changes of Sb in different mines and pH changes of the leached solution (corresponding dashed labels). ((**a**): Sb's leaching characteristics and pH changes in Zuoxiguo-tailings leachate; (**b**): Sb's leaching characteristics and pH changes in Qinglong-tailings leachate).

Contrarily, the pH showed different trends, changing significantly within the first day regardless of the initial pH, and remained between 6 and 8, after one day, for days 2–9. Regardless of the pH at the initial condition, the release of Sb in the tailings of Zuoxiguo and Qinglong antimony mines was the highest within the first day, with an average release rate of 37%. This indicates that the artificially given pH value within the first day helped the Sb leaching process. Notably, the alkaline conditions were more favourable to Sb release than acidic conditions during the leaching process of the Zuoxiguo antimony mining tailings [39]. The antimony mining tailings consist of pyroxene, which can dissolve in water to form the hydroxide $Sb(OH)_3$ [40], the expression of which is:

$$Sb_2S_3 + 6H_2O \Longleftrightarrow 2Sb_2(OH)_3 + 3H_2S$$

However, $Sb(OH)_3$ behaves more like an acid than a hydroxide; therefore, it is often written as H_3SbO_3 (antimonous acid) and can dissociate to form an anion as follows:

$$H_3SbO_3 \Longleftrightarrow H_2SbO_3^- + H^+$$

As the pH increases, more H_3SbO_3 is converted to the anionic form; $H_2SbO_3{}^-$ is very soluble in water [41,42]. Sb in the tailings of the Qinglong antimony mine was more readily released at an initial pH of 3. The trend of release under other pH conditions was consistent with that of Zuoxiguo. However, the Sb release in the tailings of Zuoxiguo was higher than that of Qinglong regardless of the initial pH conditions since the Sb content of the tailings of Zuoxiguo antimony mine was higher than that of Qinglong. The presence of FeS_2 helped to facilitate Sb release and increased th4 Sb_2S_3 oxidation rate; therefore, in systems with more FeS_2 input, Sb accumulation is increased [43]. Herein, the interaction between FeS_2 and Sb_2S_3 increased Sb dissolution from Sb_2S_3. FeS_2 has a higher resting potential than Sb_2S_3, 0.66 V vs. 0.12 V, respectively [44], which may stimulate Sb_2S_3 to release large amounts of Sb. Therefore, the Sb concentration in the Zuoxiguo leachate was higher than Qinglong.

Conversely, the As release differed between the Zuoxiguo mining tailings and that of Qinglong mining tailings (Figure 4). The release of As from the Zuoxiguo mine tailings was much higher, while As concentration in the Zuoxiguo leachate increased with time before day 3, then decreased. Likewise, the pH showed different trends, varying significantly within the first day regardless of the initial conditions and remained between 6 and 8, after one day, for days 2–9.

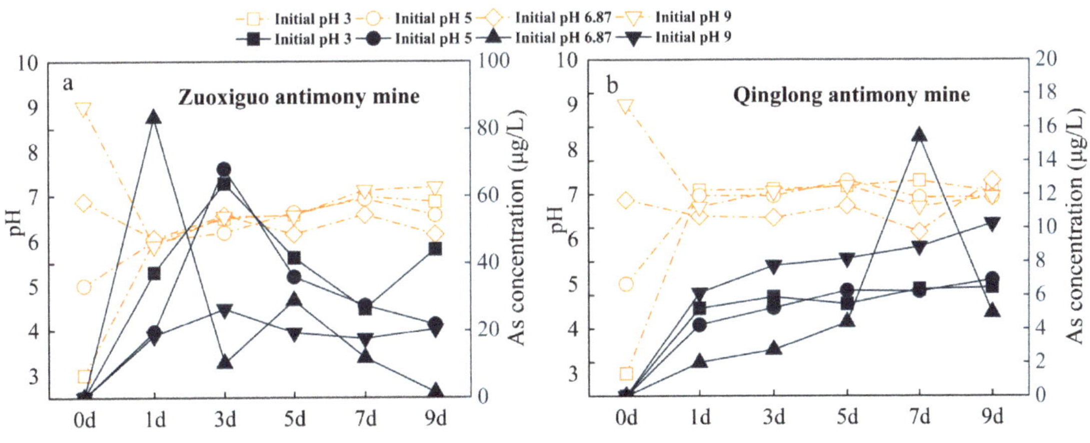

Figure 4. Leached concentration changes of As in Zuoxiguo and Qinglong mines and pH changes of the leached solution (corresponding dashed labels). ((**a**): As's leaching characteristics and pH changes in Zuoxiguo-tailings leachate; (**b**): As's leaching characteristics and pH changes in Qinglong-tailings leachate).

This suggests that the $CaCO_3$ and other related components in the samples provided sufficient acid neutralization and buffering capacity. The release of As in the leachates from Zuoxiguo and Qinglong was significantly lower than that of Sb, possibly due to the source of As being toxic sand, which decomposed more slowly than pyroxene. This may be attributed to the formation of oxidation edges on the toxic sand particles consisting of Fe, As, S, Sb, and Ca that slowed down dissolution, which was not observed on the pyroxene [45]. In addition, the S content in the tailings of the Zuoxiguo antimony mine ranged from 3.065 g/kg to 14.718 g/kg, while the sulphur and sulphate in the tailings supported the leaching of As. As is chemically similar to S; hence, it can replace elemental S in sulphides, resulting in sulphides with high levels of As, such as FeAsS [46]. Reduction zones, such as rivers, sediments, and mines, can form authigenic FeS_2 in which some dissolved As is bound during the formation of these FeS_2 minerals; during transformation, As is bound to FeS_2 and released during FeS_2 dissolution [47]. Compared with other initial pH values, the release of As in the leachate from Zuoxiguo and Qinglong varied when the initial pH was 6.87; the release of As in the leachate from Zuoxiguo began to decrease within 1–3 days, while the release of As in the leachate from Qinglong sharply increased within 5–7 days. This indicates that the leaching variation of As is greater under neutral conditions, and the elemental activity is higher and easily influenced by other factors [48].

3.3. *Influence of Chemical Sb and As Species in Tailings on Leaching Mechanism*

The improved BCR method was used to determine the Sb and As chemical species in the tailings of the Zuoxiguo and Qinglong antimony mines. The chemical species of Sb and As in the Qinglong mining tailings were predominantly F4 and F2, with overall proportions in the following order: F4 > F2 > F3 > F1. Qun et al. (2022) have reported that too high of a residue fraction indicated high contents of elements in the lattice, such as silicate, primary minerals, and secondary minerals. However, when the reducible was too high, elements bound with Fe and Mn hydrated oxides showed high content and activity, which are easily hydrolysed under pH < 7 with a certain bioavailability [49]. Figure 5 shows that F1 and F2 of Sb and As values in the Zuoxiguo mining tailings were higher than those of the Qinglong mining tailings, indicating that Sb in the Zuoxiguo mining tailings had stronger mobility and bioavailability.

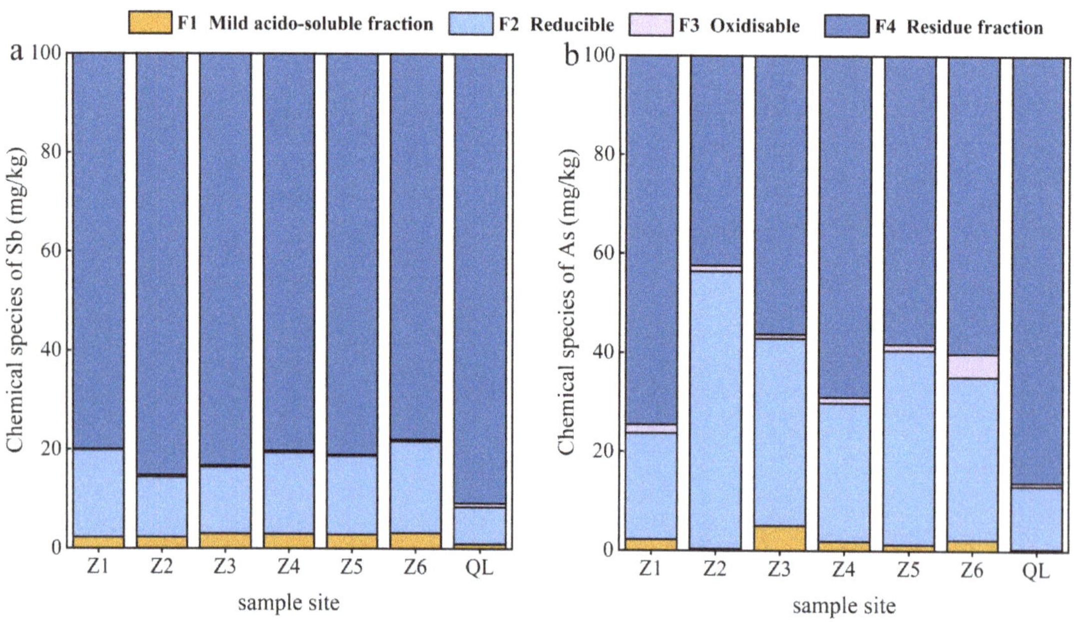

Figure 5. As and Sb chemical proportions at different points in tailings of Zuoxiguo and Qinglong.((**a**): Proportions of chemical Sb species in tailings of Zuoxiguo; (**b**): Proportions of chemical As species in tailings of Qinglong).

Hence, the release of Sb and As from the tailing sand in the leaching experiment of Zuoxiguo was higher than that of Sb and As from the tailing sand of Qinglong. Meanwhile, the F4 of Sb and As in the Qinglong mining tailings was higher than that of As in the Zuoxiguo mining tailings. The tailing sand of Qinglong had a higher binding degree with Sb and As elements, thereby hindering migration and transformation; hence, the release of Sb and As in the tailing sand of the Qinglong leaching experiment was significantly lower than that of Zuoxiguo.

3.4. Surface Changing of Tailings during Leaching

The surface shape of the tailings from the scanning electron microscope (SEM) is shown in Figure 6. The roughness of the tailing surfaces increased post leaching. The originally smooth surface of the tailings was eroded, while small holes appeared on the surface after 9 days of the alkaline initial leaching condition, and raised layers appeared on the surface after 9 days of the acid initial leaching condition (Figure 6). This illustrates that the dissolution of the tailings began with the tailing–water interface, while different initial conditions led to different tailings changes. However, it is possible that the reaction mechanism at the tailing–water interface is different, and so is the production under different initial pH conditions during leaching.

Figure 6. Tailing surface morphology changes before and after leaching: (**a**) Zuoxiguo antimony mine and (**b**) Qinglong antimony mine. Scale bar: 10 μm and 20 μm.

3.5. Health Risk Assessment of Leachate

During rainfall, the tailings' leachate flows into the river from high to low along the terrain in the study area, which has exacerbated the negative health effects of Sb and As; therefore, a health risk assessment of Sb and As in the leachate was conducted. The highest leaching amount was used to calculate the health risk index under different pH conditions (Table 2). The non-carcinogenic and carcinogenic risks of Sb and As in the leachate at different pH were higher for children than adults. Children were at higher risk from As, while the health risk index of As in the leaching solution of Zuoxiguo was higher than that of Qinglong; hence, contaminated drinking water must be more strictly managed. In Table 2, the carcinogen element As in the leachate of the Zuoxiguo antimony mining tailings for the adult and children health risk index was higher than 1×10^{-4}, a serious risk level, while the health risk index of Sb was less than 1.0×10^{-6}, a low-risk level. The health risk index of carcinogenic As in the tailing leachate of the Qinglong antimony mine was greater than 1×10^{-4} for both adults and children at pH 6.87, a serious risk level. At pH 3, the health risk index for adults ranged from 1.0×10^{-5} to 5.0×10^{-5}, a medium-risk level, while at pH 5 and 9, the health risk index for adults ranged from 5.0×10^{-5} to 1.0×10^{-4} at pH 5 and 9, a high-risk level. The health risk index for children was between 5.0×10^{-5} to 1.0×10^{-4} at pH 3, 5, and 9, a high-risk level; As had the highest health risk index for adults and children at pH 6.87, consistent with Zuoxiguo. On the contrary, the health risk indices of Sb were less than 1.0×10^{-6}, a low-risk level. Therefore, the non-carcinogenic risk of Sb in the leachate is low-risk for both adults and children, i.e., within the acceptable range, while the carcinogenic risk of As exceeds a certain value, where As in the leachate of Zuoxiguo is a serious risk for both adults and children, which may be associated with the different toxicity of different heavy metals [49,50]. Therefore, it is necessary to introduce health risk assessment into water quality monitoring and evaluation around the study area and develop and implement appropriate contaminant control strategies.

Table 2. Annual health risks caused by Sb and As in the leachate of Zuoxiguo and Qinglong (a^{-1}).

Item		pH	Sb	As
			Rn	Rc
Zuoxiguo	Adult	3.00	2.32×10^{-7}	4.67×10^{-4}
		5.00	2.45×10^{-7}	5.00×10^{-4}
		6.87	2.51×10^{-7}	6.11×10^{-4}
		9.00	2.62×10^{-7}	1.94×10^{-4}
	Child	3.00	2.97×10^{-7}	6.00×10^{-4}
		5.00	3.12×10^{-7}	6.38×10^{-4}
		6.87	3.21×10^{-7}	7.80×10^{-4}
		9.00	3.34×10^{-7}	2.48×10^{-4}
Qinglong	Adult	3.00	1.12×10^{-8}	4.74×10^{-5}
		5.00	9.64×10^{-9}	5.07×10^{-5}
		6.87	1.01×10^{-8}	1.13×10^{-4}
		9.00	1.00×10^{-8}	7.52×10^{-5}
	Child	3.00	1.43×10^{-8}	6.05×10^{-5}
		5.00	1.23×10^{-8}	6.47×10^{-5}
		6.87	1.29×10^{-8}	1.44×10^{-4}
		9.00	1.28×10^{-8}	9.59×10^{-5}

4. Conclusions

In this study, Zuoxiguo and Qinglong antimony mining tailings were collected and analysed. The total contents of As and Sb in Zuoxiguo and Qinglong tailings, as well as the leaching mechanism of As and Sb, were determined, while the health risk evaluation of As and Sb in the leachate was conducted. The results showed that: (1) SiO_2 was the main component of the Zuoxiguo and Qinglong antimony mining tailings; notably, Sb_2S_3 and FeS_2 were identified in the tailings of the Zuoxiguo antimony mine. The average content of Sb in the Zuoxiguo and Qinglong tailings was 5902.77 mg/kg and 1426.43 mg/kg, respectively, while that of As was 412.53 mg/kg and 405.26 mg/kg, respectively, which were higher than the background values. (2) The average Sb concentration in the leachate of Zuoxiguo and Qinglong was 1470.48 μg/L and 70.20 μg/L, respectively, while that of As concentration in the leachate of Zuoxiguo and Qinglong was 31.20 μg/L and 6.45 μg/L, respectively. The release of Sb in the leachate of the Zuoxiguo and Qinglong tailings increased with time; the alkaline conditions contributed to the release of Sb. The release of As in the leachate of Zuoxiguo was first increased, then decreased, while the release of As in the leachate of Qinglong showed an increasing trend, in which the changes in As release in the leachate of Zuoxiguo and Qinglong at pH 6.87 differed from those of As release under other pH conditions. (3) The health risk index showed that the average carcinogenic risk of As in the Zuoxiguo and Qinglong leachates for children was 5.67×10^{-4} and 9.13×10^{-5}, respectively, and 4.43×10^{-4} and 7.16×10^{-5} for adults, respectively, suggesting that As has a significantly higher risk for children compared with adults. Moreover, the carcinogenic risk index of As in the leachate of Zuoxiguo was higher than that of Qinglong. In contrast, the average non-carcinogenic risk index of Sb in the leachate of Zuoxiguo and Qinglong for children was 3.16×10^{-7} and 1.31×10^{-8}, respectively, and 2.48×10^{-7} and 1.02×10^{-8} for adults, respectively, which were below the minimum limitation and within the acceptable range.

Author Contributions: Conceptualization, Z.B. and Y.H.; methodology, Z.B. and Y.H.; software, Z.B.; validation, Z.B., Y.H. and Z.H.; formal analysis, Z.B.; investigation, Y.H.; resources, Z.B.; data curation, Z.B. and Y.H.; writing—original draft preparation, Z.B.; writing—review and editing, Z.B.; visualization, Z.B.; supervision, Z.H. and F.W.; project administration, Z.H.; funding acquisition, Z.H. All authors have read and agreed to the published version of the manuscript.

Funding: This study was financially supported by the National Key R&D Program of China (2018YFC1801701), the National Natural Science Foundation of China (U1612442), and Guizhou Talent Base Project (RCJD2018-21).

Institutional Review Board Statement: Not applicable.

Informed Consent Statement: Not applicable.

Data Availability Statement: Reasonable data requests can be obtained by email at bzy823396@163.com.

Conflicts of Interest: The authors declare no competing interest.

References

1. Puyang, X.; Deng, R.; Wang, C.; Huang, J.; Song, G. Leaching characteristics of heavy metals from tailings under simulated rainfall in dexing copper mine (in Chinese). *Environ. Eng.* **2019**.
2. Liu, L.; Lu, Y.; Shan, Y.; Mi Jimin Zhng, Z.; Ni, F.; Zhang, J.; Shao, W. Pollution characteristics of soil heavy metals around two typical copper mining and beneficiation enterprises in Northwest China. *Environ. Monit. Assess.* **2022**, *194*, 788. [CrossRef] [PubMed]
3. Yu, K.; Li, J.; Li, H.; Chen, K.; Lv, B.; Zhao, L. Statistical characteristics of heavy metals content in groundwater and their interrelationships in a certain antimony mine area. *J. Groundw. Sci. Eng.* **2016**, *4*.
4. Zhan, T.; Huang, Y.; He, T.B.; Teng, Y.; Shi, W.; Hou, C.; Lou, Y.; Zhao, Q. Heavy metal pollution and health risk in major agricultural products in Tongren mercury mine area, Guizhou. *J. Saf. Environ.* **2017**, *17*, 1524–1529.
5. Xie, Q.; Ren, B. Pollution and risk assessment of heavy metals in rivers in the antimony capital of Xikuangshan. *Sci. Rep.* **2022**, *12*, 14393. [CrossRef]
6. Peng, J.; Zhang, S.; Han, Y.; Bate Ke, H.; Chen, Y. Soil heavy metal pollution of industrial legacies in China and health risk assessment. *Sci. Total Environ.* **2021**, *816*, 151632. [CrossRef]
7. Yan, K.; Wang, H.; Lan, Z.; Zhou, J.; Fu, H.; Wu, L.; Xu, J. Heavy metal pollution in the soil of contaminated sites in China: Research status and pollution assessment over the past two decades. *J. Clean. Prod.* **2022**, *373*, 133780. [CrossRef]
8. Xu, W.; Yi, J.; Cheng, J. The Heterogeneity of High-Quality Economic Development in China's Mining Cities: A Meta Frontier Function. *Int. J. Environ. Res. Public Health* **2022**, *19*, 6374. [CrossRef]
9. Liu, A.; Ma, Y.; Gunawardena, J.; Egodawatta, P.; Ayoko, G.; Goonetolleke, A. Heavy metals transport pathways: The importance of atmospheric pollution contributing to stormwater pollution. *Ecotoxicol. Environ. Saf.* **2018**, *164*, 696–703. [CrossRef]
10. Qiao, P.; Wang, S.; Zhao, Q.; Wei, Y.; Lei, M.; Yang, J.; Zhang, Z. Process, influencing factors, and simulation of the lateral transport of heavy metals in surface runoff in a mining area driven by rainfall: A review. *Sci. Total Environ.* **2022**, *857*, 159119. [CrossRef]
11. Zhou, L.; Zhao, X.; Meng, Y.; Fei, Y.; Teng, M.; Song, F.; Wu, F. Identification priority source of soil heavy metals pollution based on source-specific ecological and human health risk analysis in a typical smelting and mining region of South China. *Ecotoxicol. Environ. Saf.* **2022**, *242*, 113864. [CrossRef] [PubMed]
12. Tang, X.; Wu, X.; Xia, P.; Lin, T.; Huang, X.; Zhang, Z.; Zhang, J. Health risk assessment of heavy metals in soils and screening of accumulating plants around the Wanshan mercury mine in Northeast Guizhou Province, China. *Environ. Sci. Pollut. Res.* **2021**, *28*, 48837–48850. [CrossRef] [PubMed]
13. Wang, X.; He, M.; Xie, J.; Xi, J.; Lu, X. Heavy metal pollution of the world largest antimony mine-affected agricultural soils in Hunan province (China). *J. Soils Sediments* **2010**, *10*, 827–837. [CrossRef]
14. Sheng, L.; Hao, C.; Gaun, S.; Huang, Z. Spatial distribution, geochemical behaviors and risk assessment of antimony in rivers around the antimony mine of Xikuangshan, Hunan Province, China. *Water Sci. Technol.* **2022**, *85*, 1141–1154. [CrossRef] [PubMed]
15. Guo, D.; Fan, Z.; Lu, S.; Ma, Y.; Nie, X.; Tong, F.; Peng, X. Changes in rhizosphere bacterial communities during remediation of heavy metal-accumulating plants around the Xikuangshan mine in southern China. *Sci. Rep.* **2019**, *9*, 1947. [CrossRef] [PubMed]
16. Wang, Y.; Liu, X.; Yan, J.; Ye, S. Selective extraction of arsenic and antimony from gold-bearing sludge using two-stage alkaline leaching. *Resour. Conserv. Recycl.* **2021**, *167*, 105388. [CrossRef]
17. Dousova, B.; Buzek, F.; Lhotka, M.; Krejcova, S.; Boubinava, R. Leaching effect on arsenic mobility in agricultural soils. *J. Hazard. Mater.* **2016**, *307*, 231–239. [CrossRef]
18. He, Y.; Han, Z.; Wu, F.; Xiong, J.; Gu, S.; Wu, P. Spatial Distribution and Environmental Risk of Arsenic and Antimony in Soil Around an Antimony Smelter of Qinglong County. *Bull. Environ. Contam. Toxicol.* **2021**, *107*, 1043–1052. [CrossRef]
19. He, M.; Wang, X.; Wu, F.; Fu, Z. Antimony pollution in China. Science of the total environment. *Sci. Total Environ.* **2012**, *421*, 41–50. [CrossRef]

20. McArthur, J.; Ravenscroft, P.; Safiulla, S. Arsenic in groundwater: Testing pollution mechanisms for sedimentary aquifers in Bangladesh. *Water Resour. Res.* **2001**, *37*, 109–117. [CrossRef]
21. Liu, B.; Wu, F.; Li, X.; Fu, Z.; Deng, Q.; Mo, C.; Zhu, J.; Zhu, Y.; Liao, H. Arsenic, antimony and bismuth in human hair from potentially exposed individuals in the vicinity of antimony mines in Southwest China. *Microchem. J.* **2011**, *97*, 20–24. [CrossRef]
22. Ye, L.; Qiu, S.; Li, X.; Jiang, Y.; Jing, C. Antimony exposure and speciation in human biomarkers near an active mining area in Hunan, China. *Sci. Total Environ.* **2018**, *640*, 1–8. [CrossRef] [PubMed]
23. Zhang, Y.; Ding, C.; Gong, D.; Deng, Y.; Huang, Y.; Zheng, J.; Xiong, S.; Tang, R.; Wang, Y.; Su, L. A review of the environmental chemical behavior, detection and treatment of antimony. *Environ. Technol. Innov.* **2021**, *24*, 102026. [CrossRef]
24. Li, Z.; Ma, Z.; Kuijp Tsering, J.; Yuan, Z.; Huang, L. A review of soil heavy metal pollution from mines in China: Pollution and health risk assessment. *Sci. Total Environ.* **2014**, *468*, 843–853. [CrossRef]
25. Zhang, X.; Yang, H.; Cui, Z. Evaluation and analysis of soil migration and distribution characteristics of heavy metals in iron tailings. *J. Clean. Prod.* **2018**, *172*, 475–480. [CrossRef]
26. Luo, C.; Routh, J.; Dario, M.; Sarkar, S.; Wei, L.; Luo, D.; Liu, Y. Distribution and mobilization of heavy metals at an acid mine drainage affected region in South China, a post-remediation study. *Sci. Total Environ.* **2020**, *724*, 138122. [CrossRef]
27. Sun, R.; Gao, Y.; Yang, Y. Leaching of heavy metals from lead-zinc mine tailings and the subsequent migration and transformation characteristics in paddy soil. *Chemosphere* **2022**, *291*, 132792. [CrossRef]
28. Jiang, L.; Sun, H.; Peng, T.; Ding, W.; Liu, B.; Liu, Q. Comprehensive evaluation of environmental availability, pollution level and leaching heavy metals behavior in non-ferrous metal tailings. *J. Environ. Manag.* **2021**, *290*, 112639. [CrossRef]
29. Guo, J.; Yuan, C.; Zhao, Z.; He, Q.; Zhou, H.; Wen, M. Soil washing by biodegradable GLDA and PASP: Effects on metals removal efficiency, distribution, leachability, bioaccessibility, environmental risk and soil properties. *Process Saf. Environ. Prot.* **2022**, *158*, 172–180. [CrossRef]
30. Zhang, M.; Zhou, J.; Song, C.; Yuan, L. Discovery and Study on Exploitation and Utilization of Shazi Large-scale Anatase Deposit in Qinglong County of Guizhou Province. *Min. Res. Dev.* **2015**. Available online: http://en.cnki.com.cn/Article_en/CJFDTotal-KYYK201501002.htm (accessed on 7 December 2022).
31. Jian, P.; Rui, H.; Liang, Q.; Guo, J. REE geochemistry of fluorite from the Qinglong antimony deposit and its geological implications. *Chin. J. Geol.* **2002**, *37*, 277–287.
32. Rauret, G.; Lopez, S.; Sahuquillo, A.; Rubio, R.; Davidson, C.; Ure, A. Quevauviller Ph. Improvement of the BCR three step sequential extraction procedure prior to the certification of new sediment and soil reference materials. *J. Environ. Monit.* **1999**, *1*, 57–61. [CrossRef] [PubMed]
33. Negi, P.; Mor, S.; Ravindra, K. Impact of landfill leachate on the groundwater quality in three cities of North India and health risk assessment. *Environ. Dev. Sustain.* **2020**, *22*, 1455–1474. [CrossRef]
34. Wu, F.; Fu, Z.; Liu, B.; Mo, C.; Chen, B.; Corns, W.; Liao, H. Health risk associated with dietary co-exposure to high levels of antimony and arsenic in the world's largest antimony mine area. *Sci. Total Environ.* **2011**, *409*, 3344–3351. [CrossRef]
35. Mishra, H.; Karmakar, S.; Kumar, R.; Kadambala, P. A long-term comparative assessment of human health risk to leachate-contaminated groundwater from heavy metal with different liner systems. *Environ. Sci. Pollut. Res.* **2018**, *25*, 2911–2923. [CrossRef] [PubMed]
36. Naz, M.; Dai, Z.; Hussain, S.; Tariq, M.; Danish, S.; Khan, I.; Qi, S.; Du, D. The soil pH and heavy metals revealed their impact on soil microbial community. *J. Environ. Manag.* **2022**, *321*, 115770. [CrossRef]
37. Wang, Y.; Kim, S.; Marcano, M.; Cao, Y.; Becher, U. Effect of EDTA Complexation on the Kinetics and Thermodynamics of Uranium Redox Reactions Catalyzed by Pyrite: A Combined Electrochemical and Quantum-Mechanical Approach. *ACS Earth Space Chem.* **2022**, *6*, 830–846. [CrossRef]
38. Qian, G.; Fan, R.; Short, M.; Schumann, R.; Li, J.; St, C. The effects of galvanic interactions with pyrite on the generation of acid and metalliferous drainage. *Environ. Sci. Technol.* **2018**, *52*, 5349–5357. [CrossRef]
39. Hu, X.; He, M.; Li, S.; Guo, X. The leaching characteristics and changes in the leached layer of antimony-bearing ores from China. *J. Geochem. Explor.* **2017**, *176*, 76–84. [CrossRef]
40. Hu, X.; Guo, X.; He, M.; Li, S. pH-dependent release characteristics of antimony and arsenic from typical antimony-bearing ores. *J. Environ. Sci.* **2016**, *44*, 171–179. [CrossRef]
41. Guo, X.; Wang, K.; He, M.; Liu, Z.; Yang, H.; Li, S. Antimony smelting process generating solid wastes and dust: Characterization and leaching behaviors. *J. Environ. Sci.* **2014**, *26*, 1549–1556. [CrossRef] [PubMed]
42. Long, J.; Tan, D.; Zhou, Y.; Zhou, D.; Luo, Y.; Bin, Y.; Wang, Z.; Wang, J.; Lei, M. The leaching of antimony and arsenic by simulated acid rain in three soil types from the world's largest antimony mine area. *Environ. Geochem. Health* **2022**, *44*, 4253–4268. [CrossRef]
43. Adelman, J.; Beauchemin, S.; Hendershot, W.H.; Kwong, Y.T.J. Change in the oxidation rate of stibnite as affected by pyrite in an oxygenated flow-through system. *Geochem.: Explor. Environ. Anal.* **2012**, *12*, 227–239. [CrossRef]
44. Abraitis, P.; Pattrick, R.; Vaughan, D. Variations in the compositional, textural and electrical properties of natural pyrite: A review. *Int. J. Miner. Process.* **2004**, *74*, 41–59. [CrossRef]
45. Radková, A.B.; Jamieson, H.E.; Campbell, K.M. Antimony mobility during the early stages of stibnite weathering in tailings at the Beaver Brook Sb deposit, Newfoundland. *Appl. Geochem.* **2020**, *115*, 104528. [CrossRef]
46. Ma, C.; Zhang, G.; Chen, J.; Wang, Q.; Liu, F. Transfer of FeS-bound arsenic into pyrite during the transformation of amorphous FeS to pyrite. *Appl. Geochem.* **2020**, *119*, 104645. [CrossRef]

47. Choppala, G.; Bush, R.; Moon, E.; Ward, N.; Wang, Z.; Bolan, N.; Suullivan, L. Oxidative transformation of iron monosulfides and pyrite in estuarine sediments: Implications for trace metals mobilisation. *J. Environ. Manag.* **2017**, *186*, 158–166. [CrossRef]
48. Hwang, S.; Her, Y.; Jun, S.; Song, J.; Lee, G.; Kang, M. Characteristics of Arsenic Leached from Sediments: Agricultural Implications of Abandoned Mines. *Appl. Sci.* **2019**, *9*, 4628. [CrossRef]
49. Paikaray, S.; Peiffer, S. Arsenic fractionation and mobilization in agricultural soils of NE Punjab, India. *Appl. Geochem.* **2022**, *139*, 105255. [CrossRef]
50. Chaudhary, R.; Nain, P.; Kumar, A. Temporal variation of leachate pollution index of Indian landfill sites and associated human health risk. *Environ. Sci. Pollut. Res.* **2021**, *28*, 28391–28406. [CrossRef]

environments MDPI

Communication

Relationship of Fluoride Concentration to Well Depth in an Alluvial Aquifer in a Semiarid Area

María Socorro Espino-Valdés [1], Daniel F. Rodríguez-Lozano [1], Mélida Gutiérrez [2,*], Humberto Silva-Hidalgo [1] and Adán Pinales-Munguía [1]

[1] Facultad de Ingeniería, Circuito Universitario, Campus II, Universidad Autónoma de Chihuahua, Chihuahua 31124, Mexico
[2] Department of Geography, Geology and Planning, Missouri State University, Springfield, MO 65897, USA
* Correspondence: mgutierrez@missouristate.edu; Tel.: +1-417-836-5967

Abstract: Groundwater of northern Mexico contains high concentrations of geogenic fluoride (F^-), a contaminant known to affect human health. The origin of F^- in groundwater in this region has been related to the weathering of rhyolite and other volcanic rocks present in the alluvium. However, the relationship of F^- concentration to water depth has not been established. F^- concentrations, pH, and total dissolved solids (TSD) were determined for 18 wells within the Meoqui-Delicias aquifer in 2021. The F concentrations varied between 0.62 mg L^{-1} and 4.84 mg L^{-1}, and 61% of the wells exceeded the 1.5 mg L^{-1} guideline. F^- concentrations did not correlate to TDS but correlated to well depth ($r = -0.52$, $p < 0.05$). Because of the less-than-strong correlation coefficient value obtained, a diagram of F^- concentrations vs. well depth was constructed. The diagram showed a distinct enrichment of F^- in shallow wells, suggesting that groundwater residence time and evaporation may be important factors in explaining the F^- content within the aquifer. This pattern was confirmed after plotting 2003 and 2006 data for the same wells. These findings are important to better understand the distribution of F^- in neighboring alluvial aquifers as well as in alluvial aquifers elsewhere.

Keywords: basin-fill aquifer; evaporation; fluoride; Chihuahua; groundwater withdrawal

Citation: Espino-Valdés, M.S.; Rodríguez-Lozano, D.F.; Gutiérrez, M.; Silva-Hidalgo, H.; Pinales-Munguía, A. Relationship of Fluoride Concentration to Well Depth in an Alluvial Aquifer in a Semiarid Area. *Environments* **2022**, *9*, 155. https://doi.org/10.3390/environments9120155

Academic Editor: Simeone Chianese

Received: 28 September 2022
Accepted: 10 December 2022
Published: 16 December 2022

1. Introduction

Groundwater with high fluoride (F^-) concentrations occurs in regions of more than 25 countries in the world [1,2]. Recent estimates establish that 180 million people are potentially affected by F^-, most of them in Asia and Africa [2]. Although the F^- sources are believed to be geogenic for the most part, there are notable hydrologic, climatic, and geologic differences among the affected regions. Arid and semiarid areas are prone to groundwater with high F^- concentrations due to prevailing conditions that favor the dissolution of F^-, such as high pH and alkalinity, warm temperatures, well depth, mean annual precipitation, aquifer lithology, and long residence times [2,3]. Within a particular aquifer, there can also be variations in the F^- concentration. For example, the F^- concentration may increase in the deeper parts of a large aquifer [3], but in other cases, it may increase near the discharge area because of a longer residence time, and, since the water is by then found in the shallow part of the aquifer, evaporation may also play a role [4,5]. Hence, geogenic high F^- groundwaters have been classified into three major types [6]: high F^- in shallow groundwater, high F in deep groundwater, and high F^- in geothermal water. Studies that narrow down the affected areas and report their F spatial distribution and the factors responsible for their content have increased in recent years [6–9].

Ingestion of groundwater containing F^- is the most common pathway of exposure leading to health problems [10], including a condition known as fluorosis [1,11]. The standard guideline for F^- concentration in drinking water set by the World Health Organization and by many countries is 1.5 mg L^{-1} F^-; however, a limit of 1.0 mg L^{-1} F^- is advisable in places where people drink more water [1]. Ingesting water above 1.5 mg L^{-1} F causes teeth discoloration (dental fluorosis), whereas chronic ingestion of higher concentrations (e.g., 4.0 mg L^{-1} F^-) has more serious consequences to human health, affecting the bones (skeletal fluorosis) and vital organs, as well as developing neurotoxic and metabolic effects [10–12]. The health predicament complicates when arsenic (As) is found co-occurring with F, as is the case in many regions worldwide [12–24].

The guideline in Mexico for drinking water is 1.5 mg L^{-1} F^- [25], but in the future, it will be 1.0 mg L^{-1} F^-, a change that will take a few years to be fully implemented [14]. High concentrations of F^- in the groundwater of northern Mexico are common, where concentrations up to 28 mg L^{-1} F have been observed [10]. Among the most affected regions in Mexico with high F^- concentrations are the states of Chihuahua, Durango, and Zacatecas [14]. In the state of Chihuahua, a median of 1.4 mg L^{-1} F^- with 45% of sampled wells exceeding the 1.5 mg L^{-1} F^- guideline was reported for samples collected in 2017–2019 [10]. Another study, based on 445 groundwater samples from rural communities of southeast Chihuahua, found F^- concentrations varying between 0.05 and 11.8 mg L^{-1} F^-, and 37.2% of these samples exceeded the 1.5 mg L^{-1} F^- guideline [15]. Most of the F^- studies report the presence of both F^- and As [3,16–18,24].

In the alluvial aquifers of northern and central Mexico, the origin of F has been reported as geogenic, associated with the weathering of silicate-rich rocks such as rhyolite and ignimbrite [9,13,15,24]. These studies also report that the distribution of F^- concentrations varies greatly with location [13–15]. Therefore, factors that might explain this variability are constantly sought, among them the depth of the well, groundwater extraction, and total dissolved solids (TDS). Well-known factors responsible for high F^- concentrations include aridity, alkalinity, and the presence of silicate-rich rocks [2,3]. Less explored factors include well depth, residence time, and the input of anthropogenic contaminants [10,16]. The objectives of this study were to determine the relationship between F concentration and well depth in an alluvial aquifer in northern Mexico and to infer about the variation in this pattern in space and time.

2. Materials and Methods

2.1. Description of the Study Area

The study area comprises the Meoqui-Delicias aquifer, an overexploited alluvial aquifer located in the central part of the state of Chihuahua, Mexico. This aquifer underlies a region of irrigated agriculture and dairy farm operations known as Distrito de Riego 005. The aquifer occupies a surface area of 4830 km^2 and has an irregular geometry, with a maximum thickness of 500 m and an average thickness of 300 m [26]. The aquifer is recharged primarily in the many arroyos and the alluvial fans that form at the base of hills that rise on its western part, but a significant recharge likely occurs at the fields, which are irrigated with surface water [26]. The local discharge areas are the Rio Conchos and Rio San Pedro, and the regional groundwater flows in a northerly direction. Although this is a primarily unconfined aquifer, under clay lenses, it operates as a confined aquifer.

The climate is semiarid, with an average annual precipitation of 284 mm. Most of the precipitation occurs during the monsoon season (July to September). The aquifer provides drinking water to several communities (total pop. ≈200,000) and contributes 17% of the water used to irrigate crops [27]. However, this amount may vary depending on the amount of surface water available, e.g., during dry periods, when more groundwater is extracted.

2.2. Sampling and Analyses

The groundwater was sampled from 18 wells in 2021 according to standard procedure [25]. The location of these wells, natural groundwater flow direction, and the prevalent trajectory of monsoon rains are shown in Figure 1. The wells included here were selected to match those wells for which F^- concentrations were reported in 2003 and 2006 [26]. The major ion concentration of groundwater in these wells sampled in 2006 is shown in Figure 2 (Tables S1 and S2 in Supplementary Material).

Temperature, pH, electrical conductivity, and TDS were measured in the field using a HANNA HI9828 multiparameter probe. Groundwater samples were kept cool during their transport to the laboratory, where they were analyzed for F^- concentration using a selective ion electrode, according to the standard method [28]. All reagents were of analytical quality. The F^- electrode and multiparameter probe were calibrated daily. The probe was calibrated using pH 4, 7, and 10 calibrating solutions and a 1413 µS conductivity calibrating solution, whereas the F^- electrode was calibrated using 1.0 mg/L F^- and 10 mg/L F^- standard solutions. Replicates were determined at least every 10 samples.

The locations of the sampled wells were plotted using ArcMap with WGS 1984 coordinate System and a Transverse Mercator Projection, and the map was constructed at a 1:380,000 scale. A Spearman correlation was utilized to determine the correlation between F^- concentrations and TDS, as well as between F^- concentration and well depth.

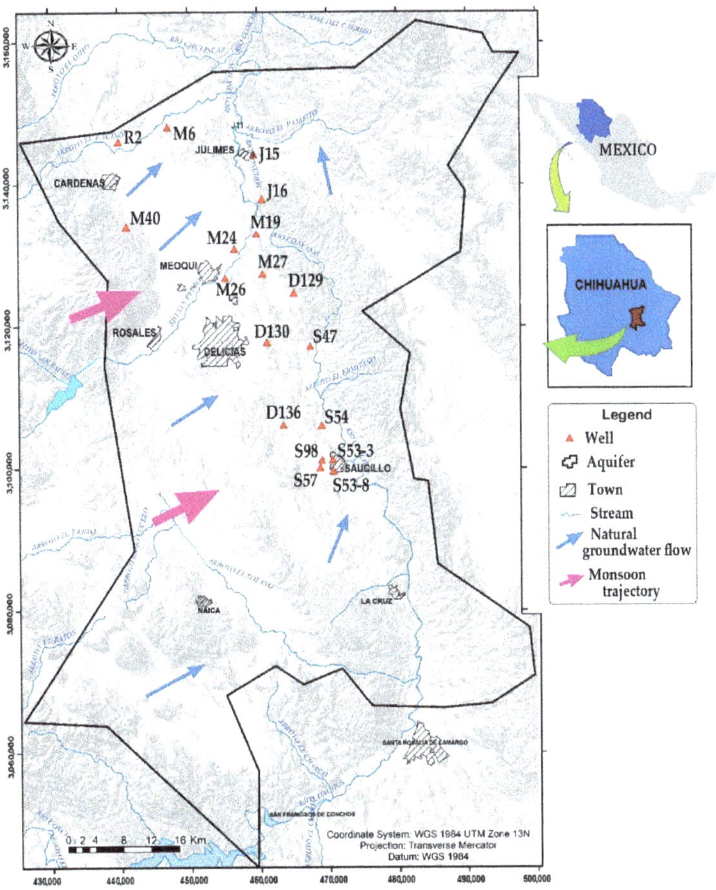

Figure 1. Meoqui-Delicias aquifer and location of the sampled wells. Samples are labeled using the first letter of the nearest community (D for Delicias, M for Meoqui, R for Rosales, J for Julimes, and S for Saucillo) followed by a number.

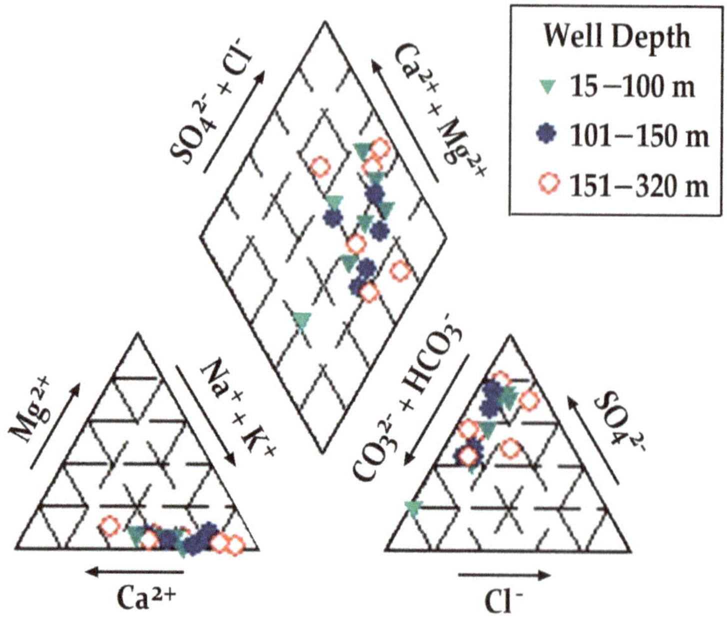

Figure 2. Piper diagram showing major ion concentration of the sampled wells.

3. Results

The results are listed in Table 1. The median of the 2021 F^- concentration values is 1.58 mg L^{-1} F^-, and 61% of the samples exceeded the 1.5 mg L^{-1} F^- guideline. This result would imply health problems in the form of dental fluorosis. However, this problem has not spread through the population because of small inverse osmosis filters fitted to many of the wells in the city of Delicias and in most of the rural communities [27]. The water quality reflected in Figure 2 shows Ca and Na are major cations that vary over a broad range of values and shows a similar behavior for anions SO_4 and HCO_3. According to Figure 2, concentration variations seem to be independent of well depth. This behavior is likely the result of the heterogeneity of the alluvial fill and solutes leaked down after the intensive agricultural practices taking place on the surface (increase in TDS and soil salinization) [26,27,29].

Table 1. Fluoride (F^-) concentrations, pH, and TDS in groundwater.

Well	Location	Depth m	F 2003 [1] mg L^{-1}	F 2006 [1] mg L^{-1}	F 2021 mg L^{-1}	pH 2021	TDS 2021 mg L^{-1}
D139	La Merced	70	1.70	1.67	1.56	7.70	794
D130	Santa Fe	320	0.67	0.82	0.94	7.30	2059
D136	Est. Armendáriz	200	2.89	2.50	2.61	7.19	863
J15	Julimes	79	3.84	4.13	4.84	7.04	1193
J16	Ex-hacienda H.	15	2.93	3.28	3.57	7.43	890
M6	Potrero del Llano	181	3.38	3.16	3.39	7.91	953
M19	El Torreón	60	1.86	2.58	3.12	7.98	701
M24	Las Puentes	36	2.46	2.40	2.85	7.43	539
M26	Fco. Portillo	150	1.09	0.98	0.91	7.54	355
M27	Nuevo Loreto	150	1.65	1.53	1.67	7.12	675
M40	Est. Consuelo	150	1.95	1.77	1.85	7.03	398
R2	Barranco Blanco	152	1.54	1.50	1.60	7.38	587

Table 1. *Cont.*

Well	Location	Depth m	F 2003 [1] mg L^{-1}	F 2006 [1] mg L^{-1}	F 2021 mg L^{-1}	pH 2021	TDS 2021 mg L^{-1}
S47	Orranteño	90	1.63	1.30	1.31	7.03	735
S54	Gomeño	60	1.55	1.30	1.52	7.31	786
S56-3	Saucillo P3	185	0.20	0.70	0.75	6.74	834
S56-8	Saucillo P8	250	0.20	0.20	0.62	7.23	843
S57	Vicente Guerrero	180	1.02	1.00	0.89	7.58	456
S98	Est. Saucillo	137	1.40	2.40	1.32	7.22	668

([1]) F concentrations for 2003 and 2006 reported in [26].

A concentration map of each of F^- and TDS allows for a better visualization of their concentration patterns. The spatial distributions of F^- concentration and TDS for 2021 data are shown in Figure 3. TDS values are lower in wells within or near the natural recharge area (wells R2, M40, M24, and M26) and increase as they approach the discharge areas. The pronounced increase in TDS in two wells, one in the northern part of the aquifer and one in the center of the aquifer, may be due to infiltration of irrigation return flows and other wastes containing large amounts of dissolved salts, e.g., dairy farm effluents [29]. F^- concentrations followed a different pattern than TDS as was confirmed by their low correlation. F^- concentrations increased at discharge areas, and their overall distribution pattern is rather irregular, as reported in other studies in northern Mexico [14].

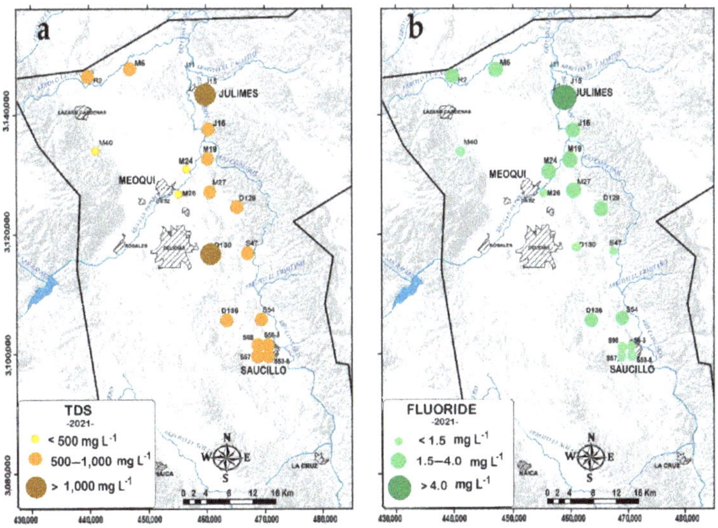

Figure 3. (**a**) Total dissolved solids (TDS) and (**b**) F^- concentrations in the sampled wells.

The Spearman correlation coefficient between F^- concentrations and TDS was low (r = 0.24), and p, the probability of r being significant at 95% ($\alpha = 0.05$, two tailed), was not significant at $p < 0.05$. A visual comparison between the plotted concentrations (Figure 3) shows a roughly similar pattern, except for wells M24 and D130, whose difference in concentrations was high enough to lower the correlation coefficient below the 95% confidence threshold. In contrast, the correlation coefficient between F^- concentration and well depth was −0.52 and was significant at $p < 0.05$. As a way to validate the relationship between well depth and F^- concentrations, and since the correlation was moderate and not a strong one, the 2021 data were plotted (Figure 4). The diagram clearly shows decreasing F^- concentrations with increasing well depth, which means that the F^- concentration is highest in shallower wells and, for the most part, coincides with discharge areas.

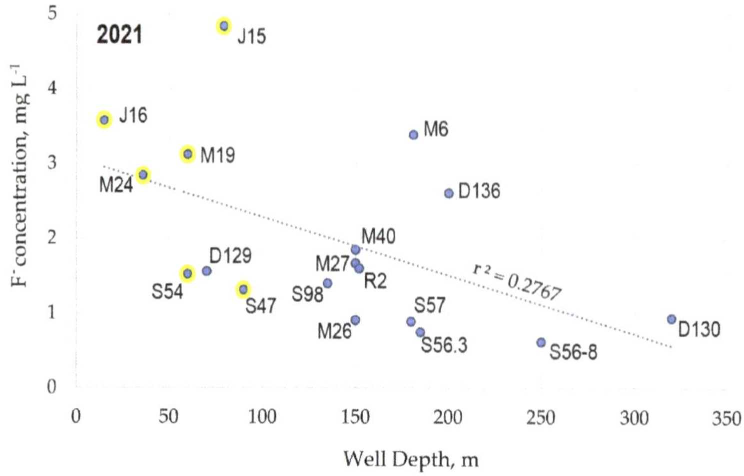

Figure 4. F⁻ concentration vs. well depth for 2021 data. Points with a yellow rim are located at a discharge area.

4. Discussion

Although TDS is reportedly associated with high F⁻ concentrations [3], this relationship was not validated in the Meoqui-Delicias aquifer. The low correlation between F⁻ concentrations and TDS is likely the result of human activities taking place at the surface and the infiltration of both excess surface water used to irrigate crops and domestic and dairy farm wastes [29]. However, F⁻ concentrations correlated, albeit weakly and inversely, to water depth. Few studies report the relationship between F⁻ concentration and well depth; however, high F⁻ concentration in the shallow part of overexploited alluvial aquifers have been reported in the western United States [30] and in northern China [6,7,31].

The moderate correlation (correlation coefficient −0.52 for 2021 data) between F⁻ concentrations and well depth may be a reflection of the heterogeneity of the alluvium and change in groundwater flow direction near some wells, as extensive extractions of groundwater are common in overexploited aquifers such as the Meoqui-Delicias aquifer [27]. To validate the F⁻–well depth association, data from 2003 and 2006 [26] were plotted. The graphs, shown in Figure 5, indicate a behavior that is similar to the one observed for 2021. Therefore, the higher F⁻ concentrations being associated with shallow wells was not a one-time occurrence but rather a confirmed pattern.

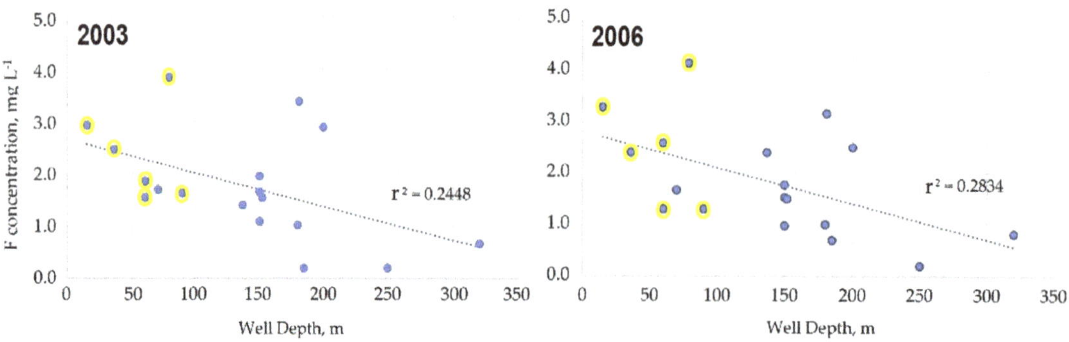

Figure 5. F concentration vs. well depth for 2003 and 2006 data. Points with a yellow rim are located at a discharge area.

The above results help develop a conceptual model for F^- concentration in groundwater, as follows: Rain infiltrates through the arroyos and alluvial fans at the base of hills and starts flowing as groundwater toward the discharge areas. As groundwater comes in contact with the silicate-rich rock fragments in the alluvial material, it picks up F^- that is naturally released via chemical weathering, incorporating it into the aquifer through the vertical flow of F-rich water. As groundwater approaches the central part of the aquifer, it mixes with irrigation drainage water, and the direction of flow changes according to the new potentiometric levels created by groundwater withdrawals of some wells, resulting in an increase in its residence time and its F^- concentration. Once the groundwater reaches the shallow depth near the discharge areas, evaporation further heightens the F^- concentration.

From the public health point of view, 67%, 61%, and 61% of wells surpassed the 1.5 mg L^{-1} F^- guideline in 2003, 2006, and 2021, respectively, and only one well (J15) had a concentration above 4 mg L^{-1} F^-, although this occurred consistently in all sampled years. The results obtained here agree with the F^- concentration behavior reported for other arid areas worldwide contaminated with geogenic F^- [3–5]. Based on this information, an effective strategy to mitigate the problem would be to direct remediation actions to wells where the highest F^- concentrations are to be expected, including wells near groundwater discharge areas as well as those with a historically high F^- content.

5. Conclusions

The results of the 2021 study show F^- concentrations varying between 0.62 and 4.84 mg L^{-1} F^-. Once these concentrations were plotted on a map, F^- concentrations were lower in recharge areas and higher in discharge areas, which highlights the groundwater residence time as a controlling process to the F content in groundwater. The shallow groundwater depth in discharge areas will also allow a further increase in F^- concentration through evaporation. TDS in this area has a large anthropogenic component, i.e., irrigation return flows and dairy farm effluents, which affected the correlation to F^- concentration. Since the Spearman correlation between F^- concentrations and well depth was only moderate, the data were then plotted into F^- concentrations vs. well depth, and the resulted diagram showed F concentration decreasing with increasing depth. These results were confirmed after comparing this graph to the graphs of 2003 and 2006 data for the same wells. The moderate correlation coefficients and low regression values obtained for the relation between F^- concentration and well depth could be due to the heterogeneity of the alluvium and the changes in local water flow direction generated by irregular and often large groundwater withdrawals in some wells used to irrigate crops, especially during dryer years. In sum, groundwater residence time and evaporation, in this order, seemed to be additional factors significantly affecting F^- concentration besides the well-known factors of alluvium mineral composition and alkalinity.

Supplementary Materials: The following supporting information can be downloaded at: https://www.mdpi.com/article/10.3390/environments9120155/s1, Table S1: Laboratory analytical procedures and Mexican technical standards describing analytical methodologies and quality controls for the analytical determination of arsenic, fluoride, pH and physicochemical parameters of water.; Table S2: Water quality of major ions in groundwater samples collected in 2006 [26].

Author Contributions: Conceptualization, M.S.E.-V. and H.S.-H.; methodology, M.S.E.-V.; software, M.S.E.-V.; validation, M.S.E.-V., A.P.-M. and H.S.-H.; formal analysis, M.S.E.-V.; investigation, M.S.E.-V. and M.G.; resources, A.P.-M.; data curation, H.S.-H.; writing—original draft preparation, M.S.E.-V. and D.F.R.-L.; writing—review and editing, M.G.; visualization, D.F.R.-L.; supervision, M.S.E.-V.; project administration, A.P.-M. and H.S.-H.; funding acquisition, M.S.E.-V. All authors have read and agreed to the published version of the manuscript.

Funding: This research received no external funding.

Institutional Review Board Statement: Not applicable.

Informed Consent Statement: Not applicable.

Data Availability Statement: Data and methodology are reported in detail in Rodríguez-Lozano D.F. Evolución espacial y temporal de la presencia de fluoruros en el acuífero Meoqui-Delicias, Chihuahua. M.S. Thesis. Universidad Autónoma de Chihuahua, Facultad de Ingenieria, Mexico. 2022.

Acknowledgments: The authors are grateful for the constructive comments and insightful reviews from three anonymous reviewers. We also thank Nayeli Villalobos and Alejandro Hernández for helping with driving to the study area and taking the sample and field measurements and the farmers who kindly allowed access to their well to take the samples and measurements.

Conflicts of Interest: The authors declare no conflict of interest.

References

1. Kimambo, V.; Bhattacharya, P.; Mtalo, F.; Mtamba, J.; Ahmad, A. Fluoride occurrence in groundwater systems at global scale and status of defluoridation—State of the art. *Groundw. Sustain. Dev.* **2019**, *9*, 100223. [CrossRef]
2. Podgorsky, J.; Berg, M. Global analysis and prediction of fluoride in groundwater. *Nat. Commun.* **2022**, *13*, 4232. [CrossRef] [PubMed]
3. McMahon, P.B.; Brown, C.J.; Johnson, T.D.; Belitz, K.; Lindsey, B.D. Fluoride occurrence in United States groundwater. *Sci. Total Environ.* **2020**, *732*, 139217. [CrossRef] [PubMed]
4. Guo, Q.; Wang, Y.; Ma, T.; Ma, R. Geochemical processes controlling the elevated fluoride concentrations in ground waters of the Taiyuan Basin, Northern China. *J. Geochem. Explor.* **2007**, *93*, 1–12. [CrossRef]
5. Jia, H.; Qian, H.; Qu, W.; Zheng, L.; Feng, W.; Ren, W. Fluoride occurrence and human health risk in drinking water wells from southern edge of Chinese Loess Plateau. *Int. J. Environ. Res. Public Health* **2019**, *16*, 1683. [CrossRef] [PubMed]
6. Wang, Y.; Zheng, C.; Ma, R. Review: Safe and sustainable groundwater supply in China. *Hydrogeol. J.* **2018**, *26*, 1301–1324. [CrossRef]
7. Li, J.; Wang, Y.; Zhu, C.; Xue, X.; Qian, K.; Xie, X.; Wang, Y. Hydrogeochemical processes controlling the mobilization and enrichment of fluoride in groundwater of the North China Plain. *Sci. Total Environ.* **2020**, *730*, 138877. [CrossRef]
8. Su, H.; Kang, W.; Kang, N.; Liu, J.; Li, Z. Hydrogeochemistry and health hazards of fluoride-enriched groundwater in the Tarim Basin, China. *Environ. Res.* **2021**, *200*, 111476. [CrossRef]
9. Gutiérrez, M.; Alarcón-Herrera, M.T. Fluoruro en aguas subterráneas de la región centro-norte de México y su posible origen. *Rev. Int. Contam. Ambient.* **2022**, *38*, 389–397. [CrossRef]
10. Kumar, S.; Singh, R.; Venkatesh, A.S.; Udayabhanu, G.; Sahoo, P.R. Medical Geological assessment of fluoride contaminated groundwater in parts of Indo-Gangetic Alluvial plains. *Sci. Rep.* **2019**, *9*, 16243. [CrossRef]
11. Edmunds, W.M.; Smedley, P.L.; Selinus, O. Fluoride in natural waters. In *Essentials of Medical Geology*, 2nd ed.; Alloway, B., Centeno, J.A., Finkelman, R.B., Fuge, R., Lindh, U., Smedley, P.L., Eds.; Springer: New York, NY, USA, 2013; pp. 311–336.
12. Jiménez-Córdova, M.I.; Sánchez-Peña, L.C.; Barrera-Hernández, A.; González-Horta, C.; Barbier, O.; Del Razo, L.M. Fluoride is associated with altered metabolism of arsenic in an adult Mexican population. *Sci. Total Environ.* **2019**, *684*, 621–628. [CrossRef] [PubMed]
13. Alarcón-Herrera, M.T.; Martin-Alarcon, D.A.; Gutiérrez, M.; Reynoso-Cuevas, L.; Martín-Domínguez, A.; Olmos-Márquez, M.A.; Bundschuh, J. Co-occurrence, possible origin, and health-risk assessment of arsenic and fluoride in drinking water sources in Mexico: Geographical data visualization. *Sci. Total Environ.* **2020**, *698*, 134168. [CrossRef] [PubMed]
14. González-Horta, C.; Ballinas-Casarrubias, L.; Sánchez-Ramírez, B.; Ishida, M.C.; Barrera-Hernández, A.; Gutiérrez-Torres, D.; Zacarias, O.L.; Saunders, R.J.; Drobná, Z.; Mendez, M.A.; et al. A Concurrent Exposure to Arsenic and Fluoride from Drinking Water in Chihuahua, Mexico. *Int. J. Environ. Res. Public Health* **2015**, *12*, 4587–4601. [CrossRef] [PubMed]
15. Navarro, O.; Gonzalez, J.; Júnez-Ferreira, H.E.; Bautista, C.-F.; Cardona, A. Correlation of arsenic and fluoride in the groundwater for human consumption in a semiarid region of Mexico. *Procedia Eng.* **2016**, *186*, 333–340. [CrossRef]
16. Armienta, M.A.; Segovia, N. Arsenic and fluoride in the groundwater of Mexico. *Environ. Geochem. Health* **2008**, *30*, 345–353. [CrossRef] [PubMed]
17. Fernández-Macías, J.C.; Ochoa-Martínez, A.C.; Orta-García, S.T.; Varela-Silva, J.A.; Pérez-Maldonado, I.N. Probabilistic human health risk assessment associated with fluoride and arsenic co-occurrence in drinking water from the metropolitan area of San Luis Potosí, Mexico. *Environ. Monit. Assessm.* **2020**, *192*, 712. [CrossRef]
18. Li, C.C.; Gao, X.B.; Zhang, X.; Wang, Y.X.; Howard, K. Groundwater fluoride and arsenic mobilization in a typical deep aquifer system within a semi-arid basin. *J. Hydrol.* **2022**, *609*, 127767. [CrossRef]
19. He, X.; Li, P.; Ji, Y.; Wang, Y.; Su, Z.; Elumalai, V. Groundwater arsenic and fluoride and associated arsenicosis and fluorosis in China: Occurrence, distribution and management. *Exp. Health* **2020**, *12*, 355e368. [CrossRef]
20. Maity, J.P.; Vithanage, M.; Kumar, M.; Ghosh, A.; Mohan, D.; Ahmad, A.; Bhattacharya, P. Seven 21st century challenges of arsenic-fluoride contamination and remediation. *Groundw. Sustain. Dev.* **2021**, *12*, 100538. [CrossRef]

21. Cinti, D.; Vaselli, O.; Poncia, P.P.; Brusca, L.; Grassa, F.; Procesi, M.; Tassi, F. Anomalous concentrations of arsenic, fluoride and radón in volcanic sedimentary aquifers from central Italy: Quality indexes for management of the water resource. *Environ. Pollut.* **2019**, *253*, 525–537. [CrossRef]
22. Kumar, M.; Goswami, R.; Patel, A.K.; Srivastava, M.; Das, N. Scenario, perspectives, and mechanism of arsenic and fluoride co-occurrence in the groundwater: A review. *Chemosphere* **2020**, *249*, 126126. [CrossRef] [PubMed]
23. Ortega-Guerrero, M.A. Presencia, distribución, hidrogeoquimica y origen de arsénico, fluoruro y otros elementos traza disueltos en agua subterránea, a escala de cuenca hidrológica tributaria de Lerma-Chapala, Mexico. *Rev. Mex. Cienc. Geol.* **2009**, *26*, 143–161.
24. Reyes-Gómez, V.M.; Alarcón-Herrera, M.T.; Gutiérrez, M.; Núñez-López, D. Fluoride and arsenic in an alluvial aquifer system in Chihuahua, Mexico: Contaminant levels, potential sources, and co-occurrence. *Water Air Soil Pollut.* **2013**, *224*, 1433. [CrossRef]
25. *NOM014-SSA1-1993*; Norma Oficial Mexicana. Procedimientos Sanitarios Para el Muestreo de Agua Para uso y Consumo Humano en Sistemas de Abastecimiento de Agua Públicos y Privados. Secretaría de Salud: Mexico City, Mexico, 1993. Available online: https://dof.gob.mx/nota_detalle.php?codigo=4801645&fecha=12/11/1993#gsc.tab=0 (accessed on 9 December 2022).
26. Rodríguez-Lozano, D.F. Evolución Espacial y Temporal de la Presencia de Fluoruros en el Acuífero Meoqui-Delicias, Chihuahua. Master's Thesis, Facultad de Ingenieria, Universidad Autónoma de Chihuahua, Chihuahua, Mexico, 2022.
27. Gutiérrez, M.; Calleros-Rincón, E.Y.; Espino-Valdés, M.S.; Alarcón-Herrera, M.T. Role of nitrogen in assessing the sustainability of irrigated areas: Case study of northern Mexico. *Water Air Soil Pollut.* **2021**, *232*, 148. [CrossRef]
28. *NMX-AA-077-SCFI-2001*; Determination of Fluoride in natural Wastewaters Treated Wastewaters-Test Method. Secretaría de Economía: Mexico City, Mexico, 2001. Available online: https://www.gob.mx/cms/uploads/attachment/file/166793/NMX-AA-077-SCFI-2001.pdf (accessed on 9 December 2022).
29. Lucero, B.A.R.; Gutiérrez, M.; Magaña, J.E.M.; Salcido, F.M.; Fierro, W.M. Salt Content of Dairy Farm Effluents as an Indicator of Salinization Risk to Soils. *Soil Syst.* **2018**, *2*, 61. [CrossRef]
30. Rosecrans, C.; Belitz, K.; Ransom, K.M.; Stackelberg, P.E.; McMahon, P.B. Predicting regional fluoride concentrations at public and domestic supply depths in basin-fill aquifers of the western United States using a random forest model. *Sci. Total Environ.* **2022**, *806*, 150960. [CrossRef]
31. Li, D.; Gao, X.; Wang, Y.; Luo, W. Diverse mechanisms drive fluoride enrichment in groundwater in two neighboring sites in northern China. *Environ. Pollut.* **2018**, *237*, 430–441. [CrossRef]

toxics

MDPI

Article

Reducing Virus Transmission from Heating, Ventilation, and Air Conditioning Systems of Urban Subways

Ata Nazari [1,*], Jiarong Hong [2,*], Farzad Taghizadeh-Hesary [3] and Farhad Taghizadeh-Hesary [4,5]

1 Department of Mechanical Engineering, University of Tabriz, Tabriz 51666-16471, Iran
2 Mechanical Engineering & Saint Anthony Falls Laboratory, University of Minnesota, Minneapolis, MN 55455, USA
3 ENT and Head and Neck Research Center and Department, The Five Sense Health Institute, School of Medicine, Iran University of Medical Sciences, Tehran 14535, Iran
4 TOKAI Research Institute for Environment and Sustainability (TRIES), Tokai University, Hiratsuka-shi 259-1292, Kanagawa-ken, Japan
5 School of Global Studies, Tokai University, Hiratsuka-shi 259-1292, Kanagawa-ken, Japan
* Correspondence: ata.nazari.u@gmail.com (A.N.); jhong@umn.edu (J.H.)

Abstract: Aerosols carrying the virus inside enclosed spaces is an important mode of transmission for severe acute respiratory syndrome coronavirus 2 (SARS-CoV-2), as supported by growing evidence. Urban subways are one of the most frequented enclosed spaces. The subway is a utilitarian and low-cost transit system in modern society. However, studies are yet to demonstrate patterns of viral transmission in subway heating, ventilation, and air conditioning (HVAC) systems. To fill this gap, we performed a computational investigation of the airflow (and associated aerosol transmission) in an urban subway cabin equipped with an HVAC system. We employed a transport equation for aerosol concentration, which was added to the basic buoyant solver to resolve the aerosol transmission inside the subway cabin. This was achieved by considering the thermal, turbulent, and induced ventilation flow effects. Using the probability of encountering aerosols on sampling surfaces crossing the passenger breathing zones, we detected the highest infection risk zones inside the urban subway under different settings. We proposed a novel HVAC system that can impede aerosol spread, both vertically and horizontally, inside the cabin. In the conventional model, the maximum probability of encountering aerosols from the breathing of infected individuals near the fresh-air ducts was equal to 51.2%. This decreased to 3.5% in the proposed HVAC model. Overall, using the proposed HVAC system for urban subways led to a decrease in the mean value of the probability of encountering the aerosol by approximately 84% compared with that of the conventional system.

Keywords: indoor air quality; air circulation; aerosol; *OpenFOAM*; SARS-CoV-2; urban subway; ventilation

Citation: Nazari, A.; Hong, J.; Taghizadeh-Hesary, F.; Taghizadeh-Hesary, F. Reducing Virus Transmission from Heating, Ventilation, and Air Conditioning Systems of Urban Subways. *Toxics* **2022**, *10*, 796. https://doi.org/10.3390/toxics10120796

Academic Editors: Esref Demir and Sam Kacew

Received: 17 November 2022
Accepted: 15 December 2022
Published: 17 December 2022

1. Introduction

Since December 2019, the novel coronavirus disease (COVID-19) has emerged as a significant global concern. It has led to 6,299,323 deaths worldwide as of 6 June 2022 and has evoked an urgent response from all disciplines to control the crisis [1–5]. Airborne transmission through droplet nuclei (aerosols) is an important mode of transmission of SARS-CoV-2 [6]. Airborne transmission depends on the distance traveled by the aerosol, congregation density of people, and ambient conditions, such as wind speed, temperature, relative humidity, particle size, and particle shape. Respiratory droplets and aerosols can be exhaled during coughing, sneezing, talking, or breathing [7–27]. To date, studies have evaluated SARS-CoV-2 spread through heating, ventilation, and air conditioning (HVAC) systems in various practical settings, such as underground car parks [28], restaurants [29], restrooms [30–32], elevators [33], public spaces [34], cleanrooms [35], urban buses [36], classrooms [37–41], cafeterias [42], airplanes [43], dental clinics [44,45], escalators [46], orchestral wind instrument performances [47], and other confined spaces [48,49]. Many studies have been conducted to investigate airborne transmission and develop appropriate mitigation strategies [50–52]. However, investigation of airborne transmission in subways is lacking. Subways are potential hotspots for airborne transmission, considering the high passenger density. Contact tracing has revealed cases where the infection transmission likely occurred during the subway transition [53].

Subway HVAC systems have been utilized to ventilate cabins for thermal comfort [53–56]. The conventional ventilation system of an urban subway comprises supply and exhaust ducts spanning the cabin ceiling and a mechanism for moving pollutants, such as breath aerosols toward the exhaust and recirculated-air ducts. Air is recycled through high-efficiency particulate air (HEPA) filters at the recirculation ducts. Air is constantly sucked up through vents, cooled, and filtered before being pushed back through supply air ducts [55]. Chang et al. demonstrated that such a subway HVAC system exhibits poor thermal comfort and energy-utilization efficiency and proposed a novel hybrid ventilation model to solve these problems [54]. Subway HVAC systems based on conventional exhaust ducts can spread respiratory droplets in the airflow direction and increase the risk of viral transmission. The number of passengers in the subway car and the exhaust duct configuration affect the risk of viral transmission. Given the characteristics of respiratory aerosol transmission on the HVAC streamline pattern, the virus can spread within an urban subway when a COVID-19 patient releases respiratory aerosols near the supply duct [53,55]. Urban subway ventilation redesign and passenger behavior modifications have rarely been studied directly in this setting.

Nazari et al. proposed antiviral adaptations for ventilation systems using jet fans inside underground car parks, focusing on the spread of respiratory droplets and aerosols and recommended several learning points to reduce the risk of viral transmission in this setting [28]. Liu et al. evaluated the effects of indoor airflow and associated aerosol transport on the risk of airborne viral transmission in a restaurant setting [29]. They identified that aerosols returning from HVAC systems, owing to low-efficiency filtration, can expose individuals near the outlet ducts to the possibility of infection. Dbouk and Drikakis [33] studied the impact of air-ventilation systems on airborne viral transmission in elevators and confined spaces. They concluded that the position of the inlets and outlets significantly influences the flow circulation and aerosol dispersion inside an elevator and indicated that using a single pair of inlet and outlet reduces droplet dispersion. Zhang et al. investigated the effects of wearing a face mask, opening windows and doors, and using an HVAC system on the aerosol spread inside an urban bus, and found that by opening doors and windows, the concentration of aerosols was reduced by approximately 50% [36]. Foster and Kinzel [41] estimated the risk of SARS-CoV-2 transmission in a classroom setting using computational fluid dynamics (CFD) and the Wells–Riley model. They showed that a well-designed HVAC system, face mask usage, reduction in the exposure time, and number of occupants are more important parameters than physical distancing. Wu et al. numerically studied virus dispersion from sneezing inside a cafeteria and stated that maintaining a safe distance in small indoor spaces (such as cafeterias) does not offer sufficient protection for activities without wearing face masks [42].

This study aimed to (i) provide a deep understanding of the routes of viral transmission in an urban subway setting, (ii) delineate the flaws of conventional subway HVAC systems and measures that enhance the transmissibility of virus-containing aerosols inside the cabin, and (iii) present mitigation strategies, including the design of a novel HVAC system to reduce the risk of viral transmission inside urban subways. We assumed the complete efficiency of HEPA filters inside the recirculation ducts in eliminating airborne viruses and focus on the airflow pattern inside the cabin and exhaust duct configuration. In the current study, we performed the simulations at a constant supply velocity within the range of typical values used in practical settings with various configurations of the breathing source.

2. Numerical Method and Simulation Setup

2.1. Formulation of the Subway HVAC Air Flows in the Cabin

The ventilation process of the urban subway HVAC system was modeled assuming continuous and incompressible air flow. To track the viral dynamics, the concentration equation was added to the basic solver to model the interactions between the fluid flow of the HVAC system and created aerosol concentration. In other words, the movement of small respiratory particles (<5 μm) originate from the continuous deformation of the cloud of respiratory particles with a shear flow interface induced by the HVAC system air. By solving the concentration transport equation, we achieved appropriate results for small respiratory particles (neglecting coughing and sneezing), considering gravity, drag, and temperature-dependent thermophoresis. Recently, Zhang et al. [36] and Nazari et al. [28] applied a concentration equation to simulate particle movement during human breathing. The relevant equations of motion are the continuity equation, momentum conservation law, and temperature equation, in their incompressible forms [28,36,39] as expressed in the following equations:

$$div(\overline{u}_i) = 0, \tag{1}$$

$$\rho\frac{\partial(\overline{u}_i)}{\partial t} + \rho\frac{\partial(\overline{u}_i\overline{u}_j)}{\partial x_j} = -\frac{\partial\overline{p}}{\partial x_i} + g\cdot x\nabla(\frac{\rho}{\rho_0}) + \frac{\partial}{\partial x_j}[\mu(\frac{\partial\overline{u}_i}{\partial x_j} + \frac{\partial\overline{u}_j}{\partial x_i})] + \frac{\partial}{\partial x_j}(-\rho\overline{u}_i'\overline{u}_j'), \tag{2}$$

$$\rho\frac{\partial(\overline{T})}{\partial t} + \rho c_p(\frac{\partial}{\partial x_j}\overline{u}_i\overline{T}) = \nabla(k\nabla\overline{T}) + \frac{1}{2}\tau : (\nabla\overline{u}_i + \nabla\overline{u}_j^T) + \frac{\partial}{\partial x_j}(-\rho\overline{u}'_i\overline{T}), \tag{3}$$

$$\rho\frac{\partial(\overline{C})}{\partial t} + \rho\frac{\partial(\overline{u}_i\overline{C})}{\partial x_j} = \nabla(D_{eff}\nabla\overline{C}) + \frac{\partial}{\partial x_j}(\rho\overline{u}'_i\overline{C}), \tag{4}$$

where ρ, ρ_0, p, u, C, T, k, and D_{eff} are the density, reference density, dynamic pressure, velocity, aerosol concentration, temperature, thermal conductivity, and diffusivity of the fluid, respectively. In Equation (3), the *Boussinesq* approximation is used for the momentum equation. In addition, in Equation (4), $D_{eff} = \frac{\nu_t}{Sc_t} + \frac{\nu}{Sc}$, where Sc_t and Sc are the turbulent and laminar Schmidt numbers, respectively, and $Sc_t = Sc = 1$ [36]. In this study, a filter-based approach is used, which combines parts from both large eddy simulation (LES) and Reynolds-averaged Navier–Stokes equations (RANS) and compares the length of the characteristic with the mesh size-spatial filter to re-establish the turbulent viscosity as follows:

$$\mu_t = FC_\mu\rho\frac{k^2}{\varepsilon}, \tag{5}$$

where k and ε are the turbulent kinetic energy and the turbulent kinetic dissipation rate, respectively, and are derived directly from the standard k–ε two-equation turbulence closure model. Variable F is the filter function, which is defined in terms of filter size Δ as

$$F = \min(1, C_{3\varepsilon}\frac{\Delta\varepsilon}{k^{\frac{3}{2}}}). \tag{6}$$

Substituting Equation (6) into Equation (5), we have

$$\mu_t = C_\mu\min(1, C_{3\varepsilon}\frac{\Delta\varepsilon}{k^{\frac{3}{2}}})\rho\frac{k^2}{\varepsilon}. \tag{7}$$

The filter used in Equation (6) modifies the standard k-ε model for a coarse filter size. In the far-field zone, the filter produces a hybrid RANS- LES behavior by allowing the development of length scales comparable to the grid resolution, and the turbulent viscosity is given as follows:

$$\mu_t = C_\mu C_{3\varepsilon}\Delta\rho\sqrt{k}, \tag{8}$$

where $C_{3\varepsilon}$ = 1 in the present case. To ensure that the numerically resolvable scale is compatible with the filtering process, the lower bound of the filter is set as Δ_{grid}, where $\Delta_{grid} = (\Delta x \Delta y \Delta z)^{\frac{1}{3}}$. The standard k–ε turbulence model is used in the following equations:

$$\rho\frac{\partial(k)}{\partial t}+\rho\frac{\partial(k\overline{u}_i)}{\partial x_i}=\frac{\partial}{\partial x_j}[\frac{\mu_t}{\sigma_k}\frac{\partial k}{\partial x_j}]+2\mu_t E_{ij}E_{ij}-\rho\varepsilon, \tag{9}$$

$$\rho\frac{\partial(\varepsilon)}{\partial t}+\rho\frac{\partial(\varepsilon\overline{u}_i)}{\partial x_i}=\frac{\partial}{\partial x_j}[\frac{\mu_t}{\sigma_\varepsilon}\frac{\partial \varepsilon}{\partial x_j}]+C_{1\varepsilon}\frac{\varepsilon}{k}2\mu_t E_{ij}E_{ij}-C_{2\varepsilon}\rho\frac{\varepsilon^2}{k}, \tag{10}$$

where $C_\mu = 0.09$, $\sigma_k = 1.00$,$\sigma_\varepsilon = 1.30$, $C_{1\varepsilon} = 1.44$ and $C_{2\varepsilon} = 1.92$. The RSM turbulence model was used, as shown in equation (11). The following formula represents the partial differential equation (transport equation) for the stress tensor, which is derived from the Navier–Stokes equation as

$$\begin{aligned}
&\underset{C_{ij}}{\rho\frac{\partial(\overline{u}'_i\overline{u}'_j)}{\partial t}+\rho\frac{\partial(u_k\overline{u}'_i\overline{u}'_j)}{\partial x_k}}=\underset{D_{T,ij}}{-\frac{\partial}{\partial x_k}[\rho\overline{u}'_k\overline{u}'_i\overline{u}'_j+p(\delta_{kj}\overline{u}'_i+\delta_{ik}\overline{u}'_j)]}\\
&\underset{D_{L,ij}}{+\frac{\partial}{\partial x_k}[\mu\frac{\partial(\overline{u}'_i\overline{u}'_j)}{\partial x_k}]}\underset{P_{ij}}{-\rho(\overline{u}_k\overline{u}'_k\frac{\partial u_j}{\partial x_k}+\overline{u}'_j\overline{u}'_k\frac{\partial u_i}{\partial x_k})}\\
&\underset{G_{ij}}{-\rho\beta(g_i\overline{u}'_j\overline{\theta}+g_j\overline{u}'_i\overline{\theta})}\underset{\phi_{ij}}{+p(\frac{\partial\overline{u}'_i}{\partial x_j}+\frac{\partial\overline{u}'_j}{\partial x_i})}\underset{\varepsilon_{ij}}{-2\mu\frac{\partial\overline{u}'_i}{\partial x_k}\frac{\partial\overline{u}'_j}{\partial x_k}}\\
&\underset{F_{ij}}{-2\rho\Omega_k(\overline{u}'_j\overline{u}'_m e_{ikm}+\overline{u}'_i\overline{u}'_m e_{jkm})}
\end{aligned} \tag{11}$$

where C_{ij} is the convection term; $D_{T,ij}$ is the turbulent diffusion term; $D_{L,ij}$ is the molecular diffusion term; P_{ij} is the stress production term; G_{ij} is the buoyancy production term; φ_{ij} is the pressure strain term; ε_{ij} is the dissipation term; and F_{ij} is the production by the system rotation.

The particles potentially carrying SARS-CoV-2 have a size of 1–4 μm [57,58]. Aerosols within this size range have sufficient viral loads, and the effect of Brownian motion on aerosol movement is negligible [58]. Li et al. showed that Brownian force decreases rapidly with increasing particle diameter [59]. Moreover, they showed that the Brownian force becomes important only when the particle diameter is less than 0.3 μm. This feature can promote irregular movement of particles in space and enhance their diffusion ability. Some researchers have neglected Brownian motion to predict particle concentration distributions in rooms and obtained satisfactory results under flow settings similar to our current configuration [36,43]. To ignore the Brownian motion effect on aerosol diffusion, we assumed a Schmidt number equal to 1. This means that aerosols diffuse at the same rate as the momentum for 1–10 μm particles in the breathing model. This assumption is also used in the design of HVAC systems for clean rooms.

In the present study, we ignored the effects of spike-like structures (located at the viral surface) on viral transmission inside the urban subway because they are too small to affect the flow. Spike-like structures on the surface of SARS-CoV-2 may affect the dynamics of respiratory droplets if they are large [60]. Kanso et al. studied the effects of spike-like structures present on the surface of SARS-CoV-2 and their rotational diffusivity to attack a target cell. This study showed that the triangularity of the coronavirus spike bulb decreased its rotational diffusivity by 39%. They stated that SARS-CoV-2 particles rely on collisions with their surroundings to move into target cells. Kanso and co-workers studied a range of possible shapes of SARS-CoV-2 based on the virus size range and considered the rotational diffusivity and collision between them [60,61]. The applied concentration method for aerosol transmission inside an urban subway can encompass virus dispersion inside a confined space.

2.2. Numerical Technique and Boundary Conditions

In this study, we used the open-source field operation and manipulation (*OpenFOAM*) for computational fluid dynamics (CFD) software package version 5 to perform the numerical simulations. The *OpenFOAM* code is written in the C++ programming language and uses a finite-volume numerical technique to solve the conservation of mass, energy, momentum, and concentration equations, along with the equation of state, in their Reynolds-averaged forms.

Numerical simulations were performed for the configuration of the ducts of an urban subway car, as shown in Figure 1. We applied a three-dimensional Cartesian coordinate system to govern the equations. The y-direction was set parallel to the length of the subway car; the z-direction was set opposite the direction of gravity; and the x-direction was set perpendicular to the subway car length (parallel to the width). The origin was located at the left corner of the subway car. The length, width, and height of the subway car were 21.14 m, 2.95 m, and 2.87 m, respectively. The heights of the standing and sitting mannequins in the simulations were set as 2 m and 1.4 m, respectively.

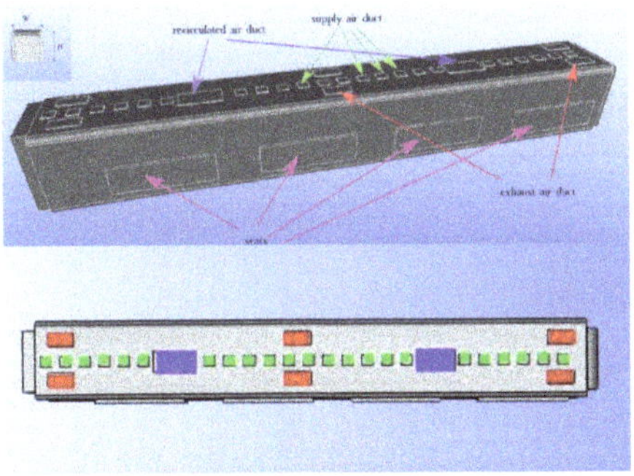

Figure 1. Schematic of a conventional urban subway car having supply air ducts, exhaust ducts, recirculated air ducts, and seats.

A Gaussian upwind scheme was used to handle the convective terms. A Gaussian linear-limited approach was employed to address the diffusion terms. The PIMPLE algorithm was applied to couple pressure and velocity. In addition, the maximum residuals for the convergence of pressure, velocity, temperature, and concentration were 10^{-5}, 10^{-8}, 10^{-7}, and 10^{-8}, respectively. To simulate aerosol movement within the urban subway, we ignored (i) the generation and (ii) interaction (collision) of particles due to the cloudy behavior of breath aerosols. The aerosol concentration is sensitive to the HVAC-induced airflow pattern, temperature, and ambient pressure. Among these environmental factors, relative humidity is a determinant parameter that strongly influences respiratory particle size during breathing. Various studies, such as those by Zhang et al. [36] and Liu et al. [29], demonstrated the traveling distance of respiratory particles based on their weight and size. The smallest particles were neutrally buoyant at a relative humidity of 20% and moved passively with the carrier fluid. Therefore, we ignored the Brownian motion of the particles and used a buoyant solver in *OpenFOAM*. In this method, gravity effects are considered instead of using the Lagrangian particle transport equation with Brownian motion. Zhang et al. [36] applied the buoyancy method to demonstrate respiratory particle transmission. However, both approaches are suitable.

Figure 1 represents the three-dimensional (3D) duct configuration of an urban subway. Figure 2 shows the standing and sitting mannequins within the urban subway cabin having continuous breathing. In this study, we determined the standing and sitting sampling surfaces to calculate aerosol encounter probability (Figure 2). The sampling surfaces cross the centers of the passenger's breathing zone hemispheres. The hemispherical breathing zone, having a radius of 300 mm, was defined as the area surrounding a healthy person's nose and mouth, from which most of the air is drawn into the lungs [62]. These surfaces were used to calculate the probability of an aerosol encounter, which was used as a metric to compare the performances of different duct configurations. To apply the breath source, we used *CreateBaffles* utility in *OpenFOAM*. The upper part of Figure 2 shows the uniformly structured elements for the course, fine, and fine meshes. Fixed-value porous and fixed-flux pressure conditions were imposed on the supply air ducts. The atmospheric value and inlet-outlet conditions were employed to model the exhaust and recirculated air ducts, respectively. Based on the DIN EN-14750-1:2006 standard [63], the vehicle classification for urban and suburban rolling stock was divided into two categories: A and B. In type A, the standing members were higher than 4 passengers/m^2; for category B, this value was lower or equal to 4 passengers/m^2. The DIN EN-14750-1:2006 standard notes that the minimum fresh air rate for categories A and B are 14 and 10 hr^{-1}, respectively. Furthermore, the Ashrae standard [64] determined that the air quality inside urban subway should be kept within the ranges of 10 to 13 hr^{-1} to provide adequate protection for public health. In this paper, category B satisfied our passengers' configurations. In this study, the air change rate (ACR) of the urban cabin was 10 hr^{-1}. The temperature values of the supply air, human breath, ambiance, and other cabin interior surfaces (such as mannequins, seats, and walls) were set at 20 °C, 30 °C, 25 °C, and 27 °C, respectively. The supply temperature was selected based on the standard ambient temperature for summer season. The temperature of a healthy human being is approximately 37 °C, while the exhaled breath temperature varies between 30 °C to 34 °C [65]; we set the human breath temperature to the lower temperature (30 °C) of this range to consider a lower evaporation of aerosols. The ambiance was set to the standard temperature and pressure (STP) conditions, and the thermally active surface temperature was set at 27 °C [66]. We defined thermally active surfaces in various HVAC systems as surfaces that are in direct contact with people or their respiratory flows. Figure 3 shows different views of the mesh configuration. Figure 4 shows mannequin meshing, which uses the snappyHexMesh utility in *OpenFOAM*. Figure 5 shows the schematic of the six cases. Red, yellow, and green mannequins represent infected individuals standing or sitting at the center (under combined exhaust and fresh air ducts), under the fresh air ducts, and under the recirculated ducts of the subway, respectively. The cell sizes of the mouth domain and ventilation ducts were 2×10^{-3} m and 5×10^{-2} m, respectively. The main input characteristics of the evaluated case studies are demonstrated in Table 1.

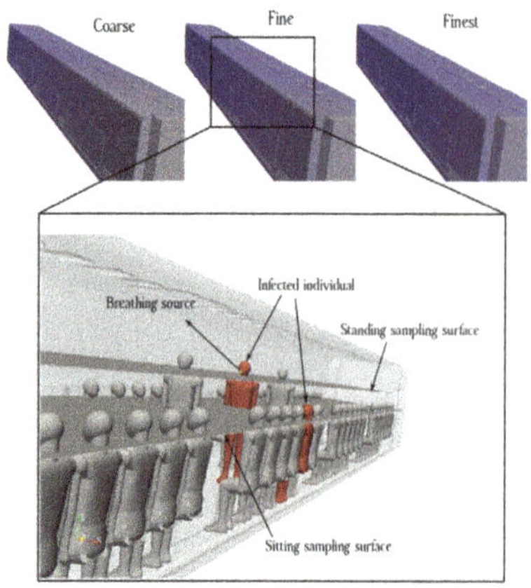

Figure 2. Schematic of a breathing source within the urban subway cabin, showing two mannequins (one standing and one seated) representing infected passengers.

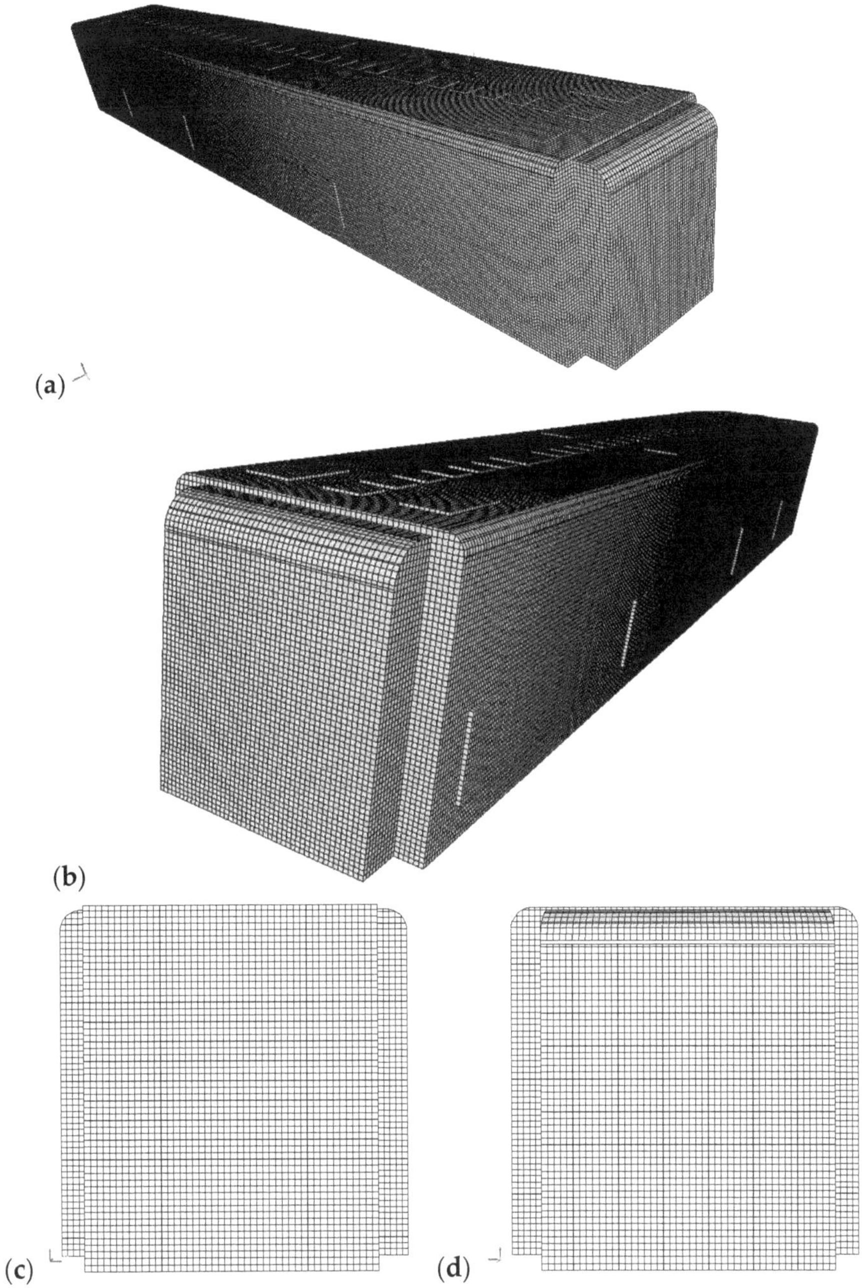

Figure 3. Schematic of mesh configuration for the present work. (**a**,**b**) are side views of urban subway fine mesh, (**c**) is the back view of fine mesh, and (**d**) is the front view of fine mesh.

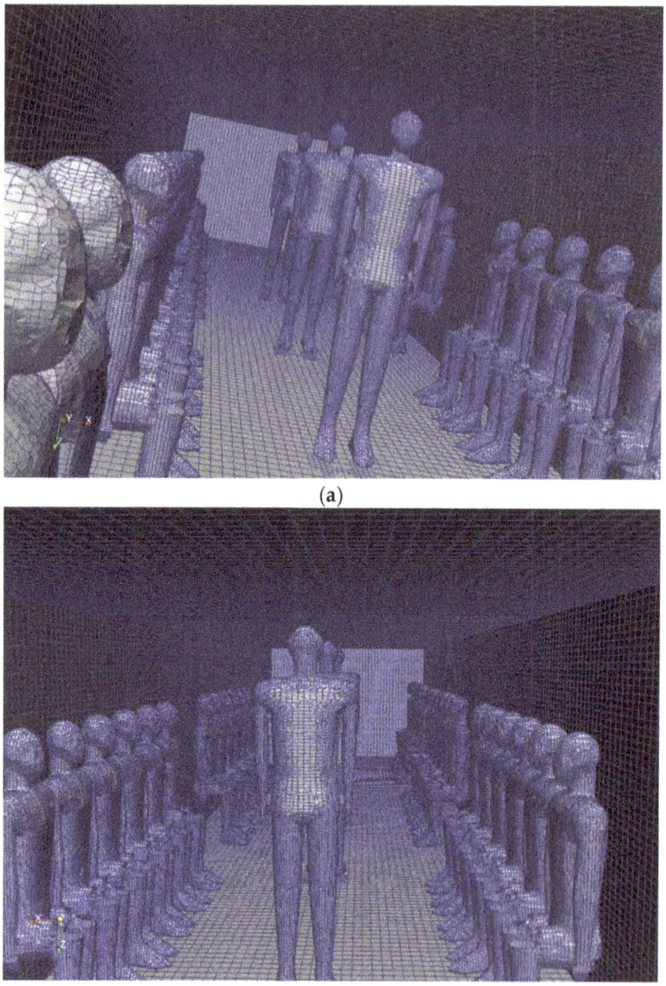

(**a**)

(**b**)

Figure 4. Schematic of mesh configuration for the present work, showing two mannequins. (**a**) the front view of the meshed mannequins, (**b**) the back view of meshed mannequins.

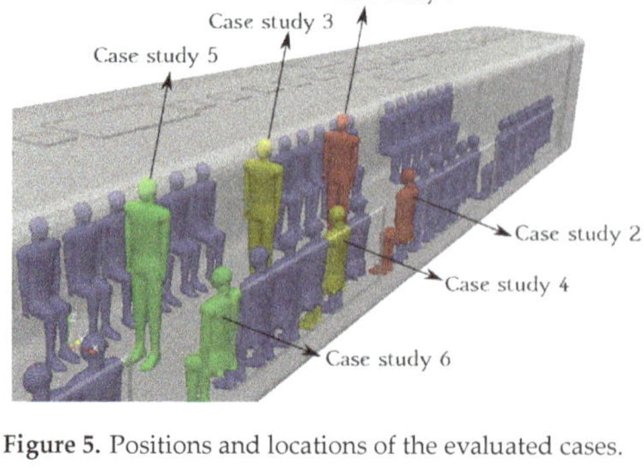

Figure 5. Positions and locations of the evaluated cases.

Table 1. The main characteristics of the evaluated case studies.

No. Case Study	Different Inputs	Common Inputs
The conventional HVAC system (Case studies 1, 2, 3, 4, 5 and 6)	The length and width of exhaust ducts are 2.5 m and 1.6 m, respectively. The number of exhaust ducts is 6,	Air change rate is 10 hr^{-1}. The temperature values of the supply air, human breath, ambiance, and other cabin interior surfaces (such as mannequins, seats, and walls) are set at 20 °C, 30 °C, 25 °C, and 27 °C, respectively. $\nu = 1.5 \times 10^{-5}$ m^2/s, T0 = 20 °C, $\beta = 3 \times 10^{-3}$ K^{-1}, Pr = 0.71, Prt = 0.9, g = 9.81 m/s^2, Sc = 1, Sct = 1, clothing insulation of mannequins is 0.60 clo.
The proposed HVAC system (Case studies 1, 2, 3, 4, 5 and 6)	The length and width of exhaust ducts are 4 m and 0.6 m, respectively. The number of ducts is 12.	

In this numerical work, we considered the sedentary activity for the external human body with no movement inside the urban subway. The value of clothing insulation was set at 0.60 clo. The *Clo* is a non-systematic unit for clothing insulation, determined as 0.155 $°Cm^2/W$ [67,68]. In other words, we supposed that mannequins' clothing level is trousers with a long-sleeve shirt. In this simulation, we used the fixed temperature boundary condition over the mannequins selected from Angelova's work [69].

2.3. Numerical Technique and Boundary Conditions

Table 2 compares our data with the experimental visualizations presented by Tao et al. for the 3D numerical simulation of the discharge velocity of the supply air [56]. The supply air of the urban subway passes through a porous media. The equation for the pressure drop through the porous media at the supply ducts is:

$$\Delta p = \frac{1}{2}C_2\rho\Delta n v^2 + \frac{\mu}{\alpha}\Delta n v, \tag{12}$$

where μ is the aerodynamic viscosity, which is set at 1.8×10^{-5}; v is the air velocity; and Δn represents the thickness of the supply panel. To obtain accurate results, a porous boundary condition was applied to the supply air ducts. The values of C_2 and $\frac{1}{\alpha}$ were set at 150,000 m^{-1} and 1.5×10^6 m^{-2}, respectively. The calculated points were positioned at the two ends and in the middle of the subway cabin and averaged in the z-direction. We calculated the average velocity and temperature in the passenger compartment of the subway at three horizontal measurement points and compared the findings with those from the experimental work of Tao et al. (v_a and T_a) [56]. Additionally, the maximum differences in the horizontal (v_x and T_x) and vertical (v_z and T_z) velocities and temperatures were calculated and compared. The maximum relative deviations of the velocity and temperature for the two cases in Table 2 were approximately 1.2% and 1.79%, respectively. The maximum temperature and velocity differences were used as evaluation metrics for the air distribution performance. These values were calculated using specific points along the x- and z-directions of the subway cabin. The European norm DIN EN-14750-1:2006 standard was applied to calculate the temperature and velocity differences and define specific points [63]. In addition, a new index, namely, the aerosol encounter probability inside an urban subway was defined to quantitatively evaluate the degree of the suppression effect of the HVAC system on airborne transmission. A grid independence test was performed to compute the required number of numerical cells to obtain convergent results. To obtain grid-independent results, simulations were performed on three different mesh topologies. A numerical grid with 900,000 cells for one core of the parallel solution was selected which had a convergent solution with an optimized computational cost (Figure 6). The maximum skewness of the grid was 0.3, which was suitable for obtaining accurate results. The number of grid points over the mannequin in the fine mesh was 8000. The Y^+ values were well below 1.0 over most regions of the mannequin surface and were limited to a maximum of 3.0, and the maximum value of x^+ and z^+ on any part of the mannequin was 2.0. The mesh quality influences the detailed flow in the vicinity of the complex geometry. However, as shown in our mesh independent test, selection of a suitable wall function near the wall limits the influence of mesh conditions on aerosol behavior. In addition, we selected a time step of 10^{-6} s to obtain an appropriate and convergent solution. The time-step size was set based on a maximum Courant number of 10. The results show that both the standard k–ε and RNG k–ε models can accurately simulate the fluid flow and temperature inside an urban subway HVAC system. The RSM turbulent approach had similar results to k-ε for the velocity parameter, but for temperature fields, the RSM results were significantly closer to Tao et al.'s study [56]. The results of the k-ε model were closer to the experimental data. A comparison of the four turbulence models is presented in Table 2. In addition, the standard wall function was applied to predict the near-wall turbulence. A no-slip boundary condition was applied to all solid surfaces (except the vents). The maximum velocity rate at the supply duct inlet was 4.2 m·s^{-1}, and recirculated air and exhaust ducts circulated 75% of aerosols exiting through the HVAC system. A *ZeroGradient* boundary condition was applied at the exhaust ducts.

Table 2. Comparison of the velocity and temperature with the results of the study by Tao et al. [56].

Values	Experimental Data	Present Work (RSM)	Present Work (k–ε)	Present Work (k-ε RNG)	Present Work (k–ω)	Present Work (k–ω SST)
v_a (average velocity)	0.170	0.168	0.168	0.168	0.160	0.161
v_x (horizontal velocity)	0.250	0.248	0.248	0.247	0.240	0.242
v_z (vertical velocity)	0.590	0.578	0.578	0.578	0.562	0.560
T_a (average temperature)	26.90	26.85	27.00	27.00	24.00	24.00
T_x (horizontal temperature)	06.68	06.65	06.80	06.80	05.71	05.81
T_z vertical temperature)	06.94	06.94	07.02	07.02	06.07	06.20

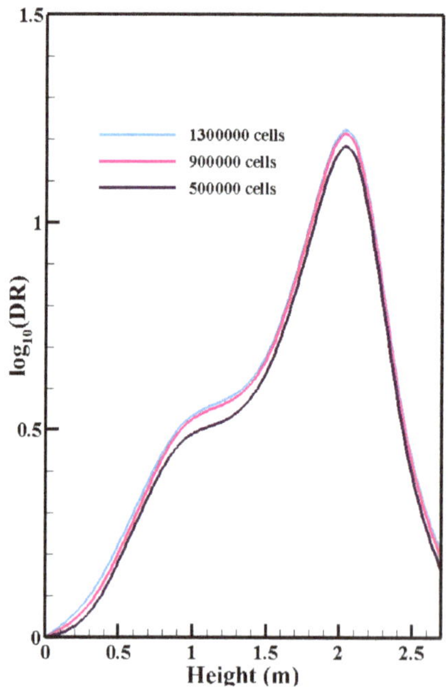

Figure 6. Log$_{10}$(DR) at 20 cm in front of mouth vs position along the height of urban subway obtained using coarse mesh 500,000, fine mesh 900,000, and finest mesh 1,300,000 (the structures of these meshes are shown in Figure 3). *DR* is the ratio of the continuous breathing source concentration to the initial concentration in the subway cabin.

2.4. Modelling of Aerosol

The temporal average was used to investigate the urban subway flow structure and is defined as [29]

$$\varphi = \frac{1}{t_2 - t_1}\int_{t_1}^{t_2} \varphi dt, \quad (13)$$

where t_1 is the time in which all aerosols reach a dynamic balance in the simulation, and t_2 is selected based on the traveling time of the urban subway between stations. The diffusion of aerosols from the infected person's mouth to the subway domain occurs close to equilibrium conditions only if the environmental concentration is minimal. Nonequilibrium diffusion occurs in environmental flows with high concentrations. A healthy individual inside an urban subway may create such environmental flows. We used a suitable initial condition to simulate aerosol transport to determine if all aerosols reached a dynamic balance. The equation $DR = \frac{C}{C_0}$, which is used in the probability of an aerosol encounter, is the ratio of continuous breathing source concentration to initial concentration in the subway cabin. The initial concentration cannot be equal to zero. The minimal concentration (equilibrium flow) value should be used to start the simulation. If we solved it with a higher concentration, the problem would not be related to the present work; it would be related to the science of aerosol interaction (nonequilibrium flow). Some researchers have used appropriate initial conditions in their simulations to ensure an equilibrium flow field and obtained satisfactory results [70,71]. Based on the typical temperature and ventilation velocity in an urban subway, the Wells–Riley equation,$P = 1 - e^{\frac{-qt}{DR}}$, may be used for aerosol encounter probability, where q is the quantum generation rate for an infected person (quanta·s^{-1}) [49]. A higher DR value in a certain zone indicates a higher risk of transmission [28]. The number of ejected particles is defined [36] as

$$N = \sum_{0}^{t} C\dot{V}, \quad (14)$$

where t and $\dot{V}$ are the exposure time and human breathing rate, assumed in this study to be equal to 180 s and 0.00033 $m^3 \cdot s^{-1}$, respectively [36]. In this study, we assumed continuous flow at the breathing sources inside the cabin. Besides, we assumed that the air at the individuals' mouths is constantly outward at a breathing frequency of 0.2 breaths/s (respiratory period of ~5 s and 12 breathing per minute). The simulation time is sufficiently long; such periodic effects can be neglected because our focus is on airborne transmission far from the emission source. The continuous breathing model is suitable for analyzing virus movement inside an urban subway cabin over exposure time. Therefore, the pulsatile nature of breathing was neglected. Instead, we used a steady flow velocity that was equivalent to the average velocity of the entire respiratory period. This length of time is mandatory for an exhalation-based viral load and essential for simulating aerosol dispersion in a specific confined space. Some researchers used this assumption (continuous flow at the breathing sources) to simulate aerosol extraction by breathing and achieved satisfactory results under flow settings similar to our current configuration [72,73]. The modeled breathing source had a hydraulic diameter of 0.04 m.

Experimental studies have shown that coughing and sneezing have a higher risk of causing infection owing to the intensity of the created particles [74]. However, sneezing or coughing was not simulated in this study. To date, no study has determined the number of viral particles required to infect an individual. Therefore, the number of particle encounters required to get infected was assumed to be 50, based on the studies by Zhang et al. [36] and Kolinski and Schneider [75]. To simulate the breath source, a source inside an urban subway cabin was applied for each case. Fixed flux and constant value conditions were used for velocity and concentration at the source, respectively. The variation in aerosol concentration with temperature was considered in this study. Given the complex velocity distributions and temperature gradients inside a cabin, averaging the temperature and velocity values is essential. The present analysis considered thermal gradients by solving the energy equation to refine the results. Solving the energy equation can influence the second term of the Navier–Stokes equation (Equation (2)/$g \cdot x\nabla(\frac{\rho}{\rho_0})$).

The density directly affects the aerosol behavior under gravitational force. Gravitational force is an important parameter that determines the traveling direction of the aerosol cloud inside the urban subway cabin, similar to buoyancy. Bhardwaj and Agrawal [76] introduced the diffusion coefficient as a function of environmental temperature as follows:

$$D_{eff} = 2.5 \times 10^{-4} \exp\left(-\frac{684.15}{T}\right). \tag{15}$$

The energy equation was solved to determine the thermal gradient and its effects on the aerosol motion to track aerosol clouds in the context of diffusion through the urban subway. The thermal gradients affect the density, hence, the gravitational force. The interplay between gravitational force, Stokes drag force, and Brownian motion affects the behavior of an aerosol cloud inside an urban subway. The induced ventilation force consists of the mean flow velocity, turbulent fluctuation, mean aerosol velocity, and fluctuating aerosol velocity effects [29]. The sum of these four effects is known as the Stokes drag force [18,29,77]. Moreover, because a buoyant solver was applied, the effect of Brownian motion was ignored in this study. Therefore, the aerosols resulting from continuous breathing were assumed to be transported by induced ventilation and gravitational forces.

3. Results and Discussion

The spread of particles (such as SARS-CoV-2-containing aerosols) increases when the infected individual breathes near the supply air ducts. In this section, we describe the concentration contours for six distinct breathing sources at different exposure times. A significant feature of subway transport is the regular stopping of subway trains for passenger boarding and disembarking, further increasing the risk of viral transmission. The proposed HVAC system for urban subways helps to reduce suspended aerosols in the cabin and mitigates the risk of viral transmission. The distance between the stations of the urban subway considerably affected the exposure time. A 3D schematic of the proposed HVAC system is depicted on the right side of Figure 7. According to this figure, the only difference between a conventional subway HVAC system (left side of Figure 7) and the proposed system is the application of a longitudinal exhaust duct along the cabin floor, instead of the ceiling, to reduce aerosol concentration.

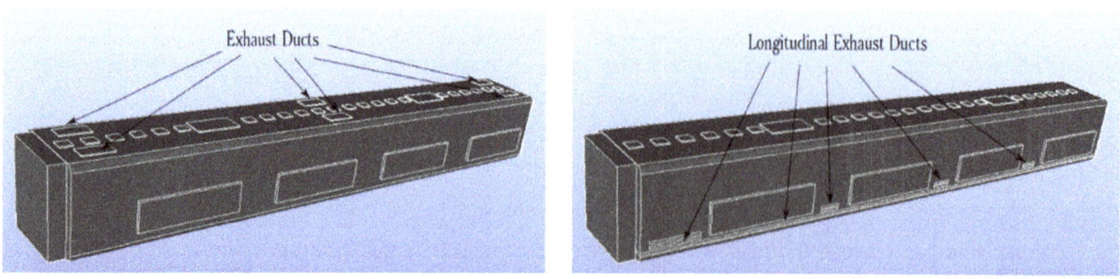

Figure 7. Comparison between a conventional (**left side**) and the proposed (**right side**) HVAC systems for an urban subway cabin.

Figure 7 shows a schematic of the computational domain, including the supply air ducts, exhaust ducts, and recirculated air ducts. In Figure 8, the airflow patterns of the conventional and proposed HVAC systems are compared. To compare the two HVAC systems, the areas of the exhaust ducts were made equal. As shown in Figure 8, the airflow of the proposed HVAC system was mostly downward, which did not permit the aerosols to spread horizontally.

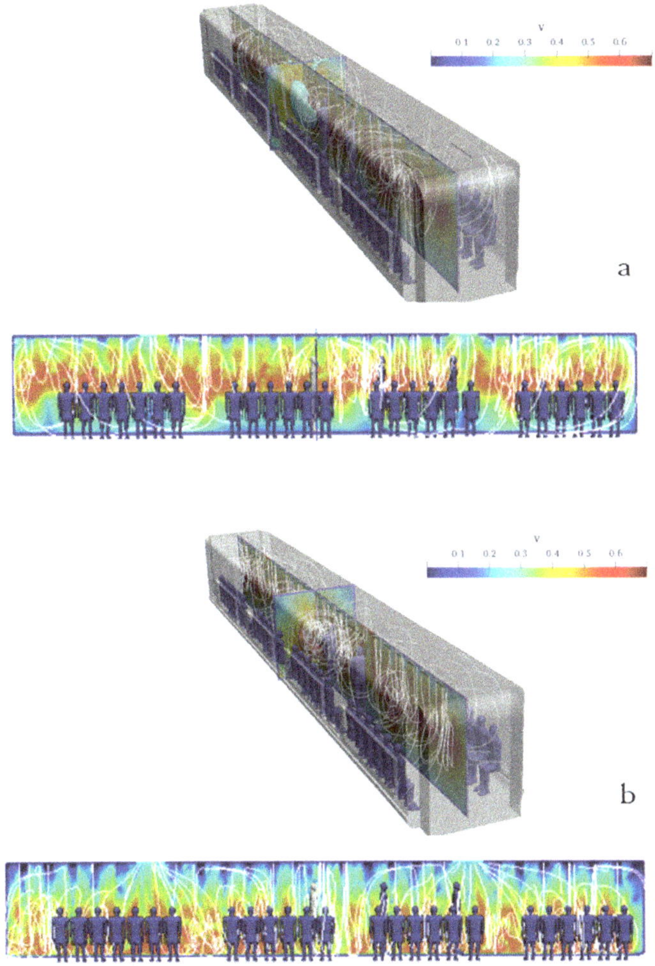

Figure 8. Comparison between the streamlines of a conventional (**a**) and the proposed (**b**) urban subway HVAC systems. The velocity contours shown are in the mid-planes.

3.1. Infected Individual Breathing near the Supply and Exhaust Ducts (Cases 1 and 2)

The flow pattern is complex at the center of an urban subway cabin because of the upward flow toward the exhaust ducts and downward flow of fresh air induced by the porous boundary. When an infected individual breathes near the exhaust ducts, the virus can readily exit the cabin. However, the downward flow of fresh air disperses aerosols to a certain extent. The flow pattern around the exhaust duct was not smooth, causing a more complex flow around the aerosol cloud. The breathing flow was much weaker than the ventilation flow; therefore, the ventilation flow can remove exhaled aerosols from the infected passenger. Figures 9 and 10 show the counters of $\log_{10}$(DR) when the infected individual was standing and sitting, respectively (Cases 1 and 2). The left column (a) shows the proposed HVAC system, and the right column (b) shows the conventional HVAC system of the urban subway. As shown in Figure 9, when the infected individual was standing and breathing near the exhaust and fresh-air ducts at the center of the cabin, the viral load was maintained than that of the proposed HVAC system over time. In the proposed HVAC system, the aerosol concentration near the floor of the subway cabin was, at times, higher than that of the conventional system.

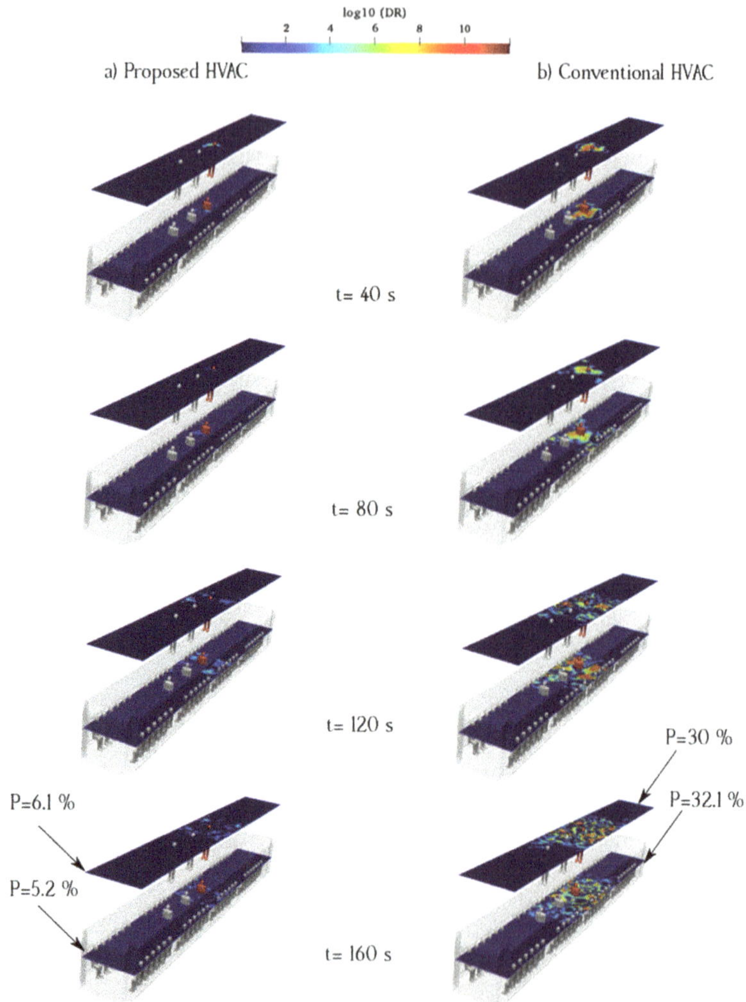

Figure 9. Time evolution of the counters of $\log_{10}$(DR) when the infected individual was standing near the supply and recirculated air ducts of the proposed (**a**) and conventional (**b**) HVAC systems (Case 1).

To better understand the behavior of aerosols inside the urban subway, the average value of $\log_{10}$(DR) along the sampling surfaces is shown in Figures 9 and 10. Logarithmic scales were used to delineate concentration figures. As shown in Figure 9, the average aerosol concentration along the standing sampling surface of the proposed HVAC system was lower than that of the conventional HVAC system. In addition, the average aerosol concentration along the sitting sampling surface for the proposed HVAC system was much lower than that of the conventional HVAC system. The main difference between the proposed HVAC system and conventional HVAC system was the flow pattern inside the urban subway. The longitudinal ventilation flows of the conventional HVAC system dispersed the aerosols more than those of the proposed HVAC system. The probability of aerosol encounters over the sampling surfaces for Case 1 is shown in Figure 9. The highest value of the aerosol encounter probability corresponding to the sitting sampling surface of the conventional HVAC system was 32.1%. This rate was 83.8% higher than that for the same surface in the proposed HVAC system.

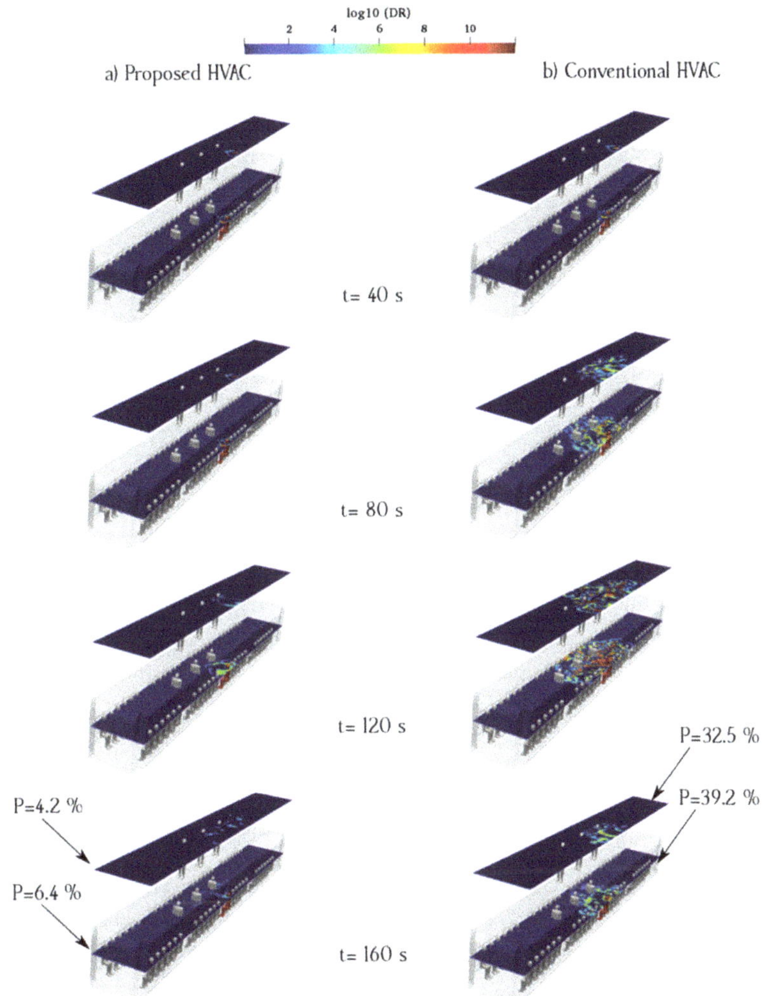

Figure 10. Time evolution of the counters of $\log_{10}$(DR) when the infected individual was sitting near the supply air and recirculated ducts of the proposed (**a**) and conventional (**b**) HVAC systems (Case 2).

Figure 10 shows the time evolution of the counters of $\log_{10}$(DR) for case 2, in which the infected individual was sitting near the exhaust and fresh-air ducts. The left column shows that in the case of the proposed system, a higher proportion of the aerosol cloud moved toward the cabin floor. The remaining part was dispersed in the surrounding spaces following the ventilation flows. However, in the case of the conventional HVAC system, aerosol clouds dispersed upward and affected a more extensive zone. The right side of the cabin, where the infected passenger was seated, contained a higher concentration of aerosols in both the HVAC systems. The aerosol concentrations in the standing/sitting surfaces of the conventional HVAC system were higher than those on the same surfaces in the proposed model. The probability of aerosol encounters at each sampling surface in Case 2 is shown in Figure 4. In the conventional model, the maximum aerosol encounter probability was 39.2% for the sitting sampling surface. This rate was 6.125 times higher than that of the proposed model. The lowest rates were on the surface in both models, and the rate for the proposed HVAC system was significantly lower, i.e., 4.2%. Impeding aerosols from rising inside a subway cabin is crucial to reduce the probability of aerosol encounters. By locating exhaust ducts near the floor of the subway, the overall aerosol concentration can be reduced through the suppression effect, which is defined as the rapid removal of indoor air pollutants by ventilation in a short pathway. This concept was considered in the development of the proposed model. However, some amount of aerosol continues to rise within the proposed HVAC system owing to the velocity distribution irregularities and position of the recirculated air ducts along the ceiling.

Different ventilation arrangements alter the flow circulation and therefore influence aerosol dispersion. The airflow circulation inside a subway cabin stems from the porous boundary of the HVAC system and secondary induced flow. Therefore, given a fixed position for the passengers, the smooth airflow part under the fresh-air ducts is an appropriate location for standing and sitting, whereas near the exhaust ducts is an inappropriate location. However, the movement of passengers inside the cabin may perturb the airflow pattern and disperse particles, thus increasing the risk of transmission.

The entering airflow pattern of an urban subway HVAC system depends on the porosity parameters. The interaction between the entering and exiting airflows with recirculated air ducts at the center of the urban subway increases the irregularity of the flow pattern at the aforementioned locations. The proposed HVAC system can suppress breath particles at the center of the urban subway HVAC system. In the proposed model, the concentration of particles and area of their spread were lower than those in a conventional HVAC system.

The conventional subway HVAC system can readily spread viruses in the top part of the subway cabin owing to its continuous flow and inconvenient exhaust duct configuration. For example, when an infected individual breathes near the supply air ducts, respiratory particles spread through the airflow, and a healthy standing passenger near the exhaust ducts can be infected.

3.2. Infected Individual Breathing near the Fresh-Air Ducts (Cases 3 and 4)

Aerosol dispersion inside an urban subway depends on the airflow pattern of the HVAC system. When an infected individual breathes near the supply ducts, viral particles take the longest path before being removed by the exhaust ducts. In this case, the viral particles remain suspended for an extended period, increasing the risk of infection (Figures 11 and 12). This risk is further increased if the infected person remains in the subway cabin for a longer time during the boarding and disembarking of passengers. In this case, we assumed that the infected individual and healthy individuals remain in a fixed location during the subway movement. By determining the safe zones and sitting on seats in those zones in the urban subway, the risk of transmission can be minimized. The crucial protective effect of face masks against SARS-CoV-2 has been highlighted by scientific authorities [18,20,57]. People inside the urban subway should wear face masks and sit on seats with appropriate social distancing. Airflow within the urban subway carries aerosols from breath source(s), which are then expelled from recirculated air ducts or atmospheric exhaust ducts. Furthermore, the application of HEPA filters and ultraviolet light emitters inside the recirculated air ducts decreased the virus concentration. In reality, aerosol movement resulting from ventilation flows may behave like a cloud flow, especially at small shear stresses of fresh air. The conventional HVAC system gives rise to higher $\log_{10}$ (DR) values than the proposed HVAC system. The distribution patterns of breath aerosols demonstrate that when an infected individual breathes under the fresh-air ducts of the cabin, the conventional HVAC system airflow carries these particles to the exhaust ducts. In this case, the traveling aerosol pathways are near the standing sampling surface and traverse a long distance toward the exhaust ducts, thus increasing the aerosol encounter probability.

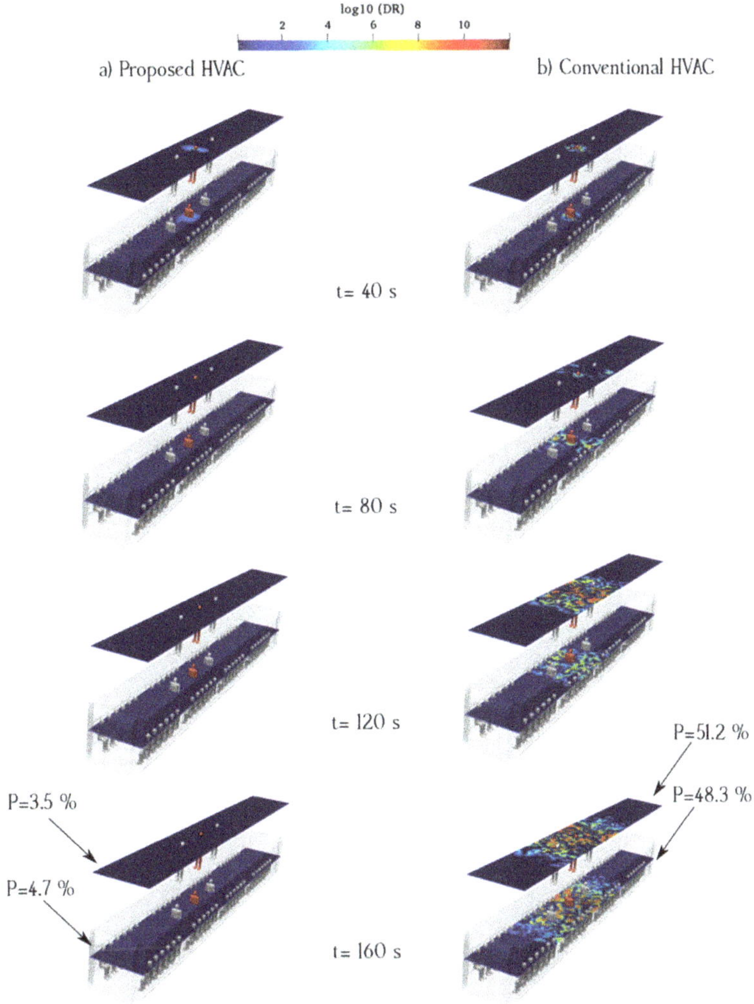

Figure 11. Time evolution of the counters of $\log_{10}$(DR) when the infected individual is standing near the supply air ducts of the proposed (**a**) and conventional (**b**) HVAC systems (Case 3).

Figure 12 shows the effect of the HVAC system type on aerosol dispersion for an infected passenger seated under a fresh-air ducts. With the conventional HVAC system configuration, the results show that aerosols spend a long time in suspended condition before reaching the exhaust duct when the infected person sits under the fresh-air duct. The aerosol concentration on the same side of the cabin as that of the infected passenger was higher than that of the other side. In contrast, for the proposed HVAC system, the aerosol concentration is considerably lower, given that exhaust ducts near the floor of the cabin can quickly remove aerosols and suppress suspended aerosols to reduce the spread of particles inside the cabin. Figure 12a shows the aerosol concentration counters when an infected individual was seated and breathing near the supply ducts in the proposed model. The aerosol path to the longitudinal exhaust ducts was narrow and continuous. Figure 11a shows the aerosol concentration for an infected individual standing and breathing continuously near the supply ducts in the proposed model, assuming a location similar to that in the sitting case. In this case, the flow pattern of aerosols is narrow and linear, which indicates that the aerosol spread mitigation efficiency is higher than that of the conventional HVAC system.

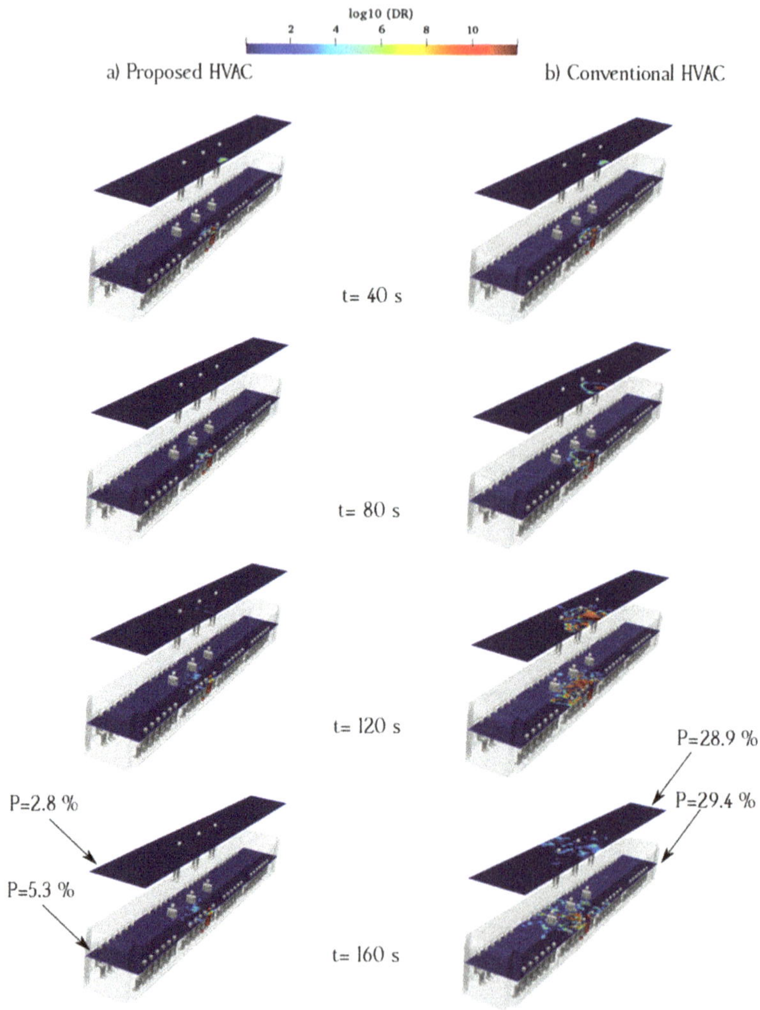

Figure 12. Time evolution of the counters of $\log_{10}$(DR) when the infected individual is sitting near the supply air ducts of the proposed (**a**) and conventional (**b**) HVAC systems (Case 4).

3.3. Infected Individual Breathes near the Recirculated Ducts (Cases 5 and 6)

Figures 13 and 14 show the aerosol concentrations in cases 5 and 6, in which the infected individual is standing and sitting near the recirculated air ducts, respectively. By comparing the $\log_{10}$(DR) counters of cases 5 and 6, we can conclude that a uniform air distribution reduces the suspension of aerosols under the recirculated and return ducts inside the subway cabin. The horizontal spreading of aerosols was minimal in the proposed model. The proposed HVAC system results in a lower aerosol concentration than that of the conventional model if the infected individual is in a standing position. The highest probability of aerosol encounters in the proposed HVAC system was 8.2% for the sitting sampling surface, which was 68.09% lower than that of the conventional HVAC system (Figure 13). Based on Figure 13, the highest aerosol encounter probabilities for the seated case near the recirculated ducts in the proposed and conventional HVAC systems were 8.2% and 25.7%, respectively. Our study shows that conventional HVAC systems can propagate airborne virus-bearing aerosols near recirculated ducts on the top side of the cabin, which can increase the transmission risk. In other words, the infected person may stand under the exhaust ducts, and the flow pattern expels particles in the conventional HVAC system. The $\log_{10}$(DR) variations over the sampling surfaces for Cases 5 and 6 are shown in Figures 13 and 14, respectively.

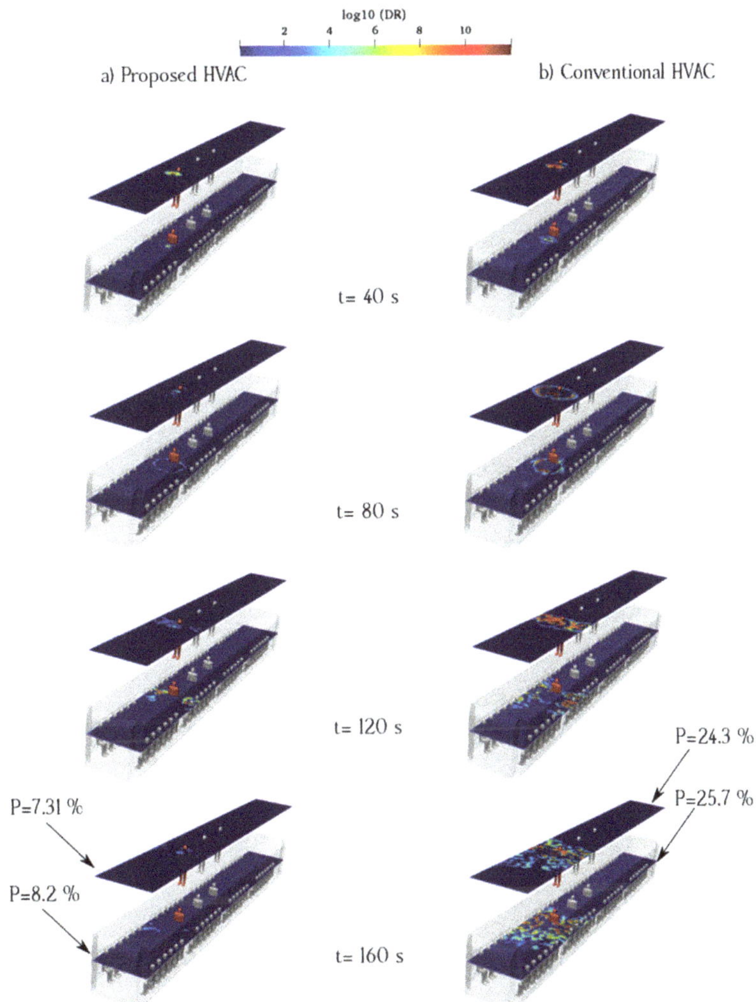

Figure 13. Time evolution of the counters of $\log_{10}$(DR) when the infected individual is standing near the recirculated ducts of the proposed (**a**) and conventional (**b**) HVAC systems (Case 5).

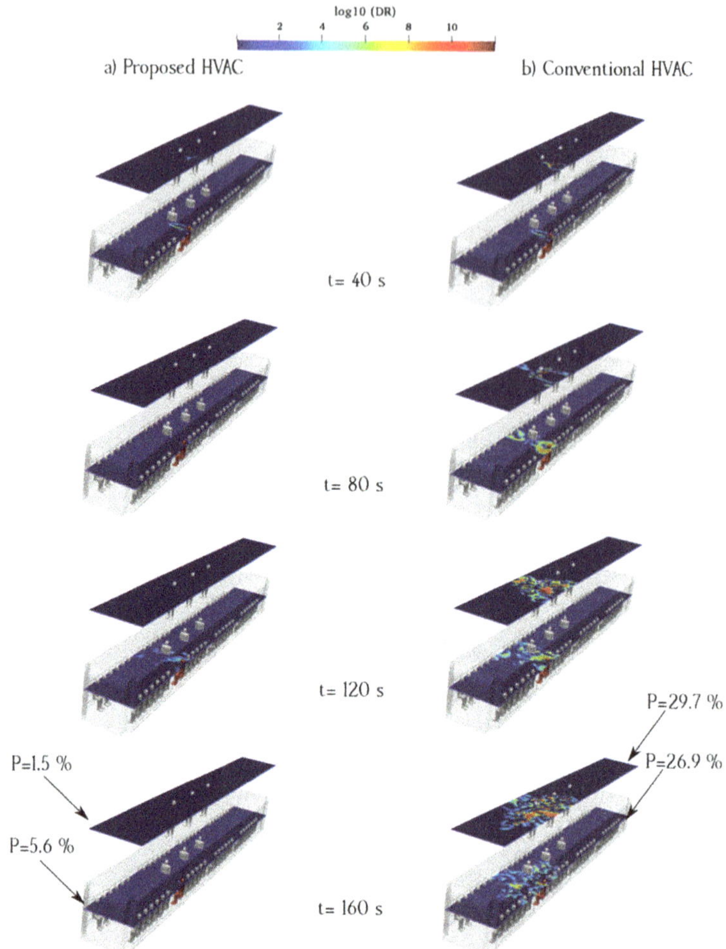

Figure 14. Time evolution of the counters of $\log_{10}$(DR) when the infected individual is sitting near the recirculated ducts of the proposed (**a**) and conventional (**b**) HVAC systems (Case 6).

3.4. General Comparison of Cases 1 to 6

Figure 15 summarizes the aerosol encounter probabilities (%) from the aforementioned simulated cases. The y-axis indicates the probability of an aerosol encounter, and the x-axis shows the sampling surfaces of both conventional and proposed HVAC systems. The vertical distribution of aerosols is shown for each case and can be used to understand the transmission risk of each sampling surface. The aerosol encounter probability is an index that can determine the exhaled aerosol behavior inside an urban subway. The values shown in Figure 9 are temporally averaged. The aerosol dynamics were plotted only for points along the sampling surfaces; otherwise, the values were set to 0. The spatial values of the surfaces were calculated for each time point. The average time for aerosol encounter probability was then calculated. The aerosol encounter probability at the standing surface of the urban subway for each HVAC model was higher than those for the other surfaces. This implies that when an infected person stands inside an urban subway cabin, the risk of aerosol transmission increases. The highest probability of aerosol encounter for the two sampling surfaces corresponds to the conventional HVAC system in Case 3. In this case, the infected individual stands under the cabin's fresh-air duct, where the incoming air flows with small shear stress and spreads the aerosol throughout the cabin toward the exhaust ducts. Irregularities in the velocity distribution caused by porous boundary conditions intensify the risk of transmission. However, for the proposed HVAC system, exhaust ducts were installed near the cabin floor, which could transfer aerosols via a short and safe pathway. Overall, the proposed HVAC system was more efficient in the rapid removal of aerosols through a safe pathway. The maximum viral transmission probability in the proposed HVAC system occurred in Case 5, with a rate of 8.2%, which was 68% less than that under similar conditions in the conventional model. In other words, by applying the introduced HVAC system to urban subways, the mean values of the aerosol encounter probability decreased by approximately 84% compared to those of the conventional system.

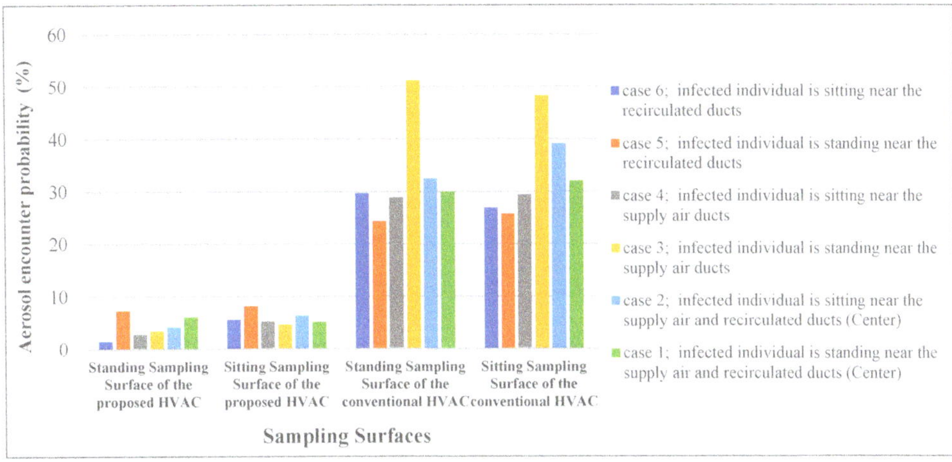

Figure 15. Comparison of aerosol encounter probability for cases 1 to 6.

An important result of this study was the determination of the transmission pathway of aerosols inside urban subways and the corresponding hazard zones (i.e., the space with higher probability of infection) under conventional subway HVAC settings. Assuming that an ample space of the urban subway is under fresh air injection, aerosols can spread both vertically and horizontally inside the conventional cabin. Based on our results, aerosols can move along a conventional urban subway and then rise to exit from the exhaust and recirculated ducts. During that time, healthy passengers who stand or sit near the exhaust and recirculated ducts are more vulnerable to viral particles compared to other locations. The unsafe zones of conventional urban subways with a higher probability of aerosol encounters can be calculated using equation (16) [28]

$$P_{tot} = \frac{\sum P_{unsafe,i} \times S_{unsafe,i}}{S_{tot}} \quad (16)$$

where P_{tot}, S_{tot}, $P_{unsafe,i}$ and $S_{unsafe,i}$ are the total probability of aerosol encounter, total sampling surface, unsafe probability of aerosol encounter of each case, and unsafe area of each case, respectively. Using Equation (16) and passenger breathing zones, unsafe areas (defined as zones with high aerosol concentration [28], high aerosol retention time [29], and high vertical aerosol spreading) for each source of breathing were determined. Notably, the region near the recirculated duct yielded a significantly higher probability of aerosol encounters. A healthy person standing in a hot spot under the recirculated duct is exposed to several times higher aerosol concentration than a person in a safe area (under fresh air ducts). The proposed HVAC system for an urban subway can eliminate unsafe hot spots existing in the conventional HVAC system by effectively minimizing the aerosol travel time and path lengths when those are being extracted by the ventilation system.

3.5. Effect of Supply Temperature of HVAC for Case 3

Based on the above analysis, when an infected individual is standing near the supply air ducts, the probability of aerosol encountering is high. In this study, we assumed four temperatures (18 °C, 20 °C, 26 °C, and 28 °C) to better understand aerosol transmission inside urban subways and calculate aerosol encounter probabilities. The present study considered the effects of temperature on aerosol behavior. The supply air temperature differed between winter and summer, with a higher thermal gradient in winter. The higher temperature of the induced air supply in winter leads to higher interaction between the thermal gradient and aerosol clouds. Under these conditions, the surficial evaporation of aerosol clouds inside urban subways decreases the probability of aerosol encounters. The aerosol concentrations over the sampling surfaces at various temperatures are shown in Figure 16. The calculated probabilities of aerosol encounter of standing sampling surface for 18 °C, 20 °C, 26 °C, and 28 °C temperatures of supply ducts were 55.32%, 51.2%, 39.2%, and 30.1%, respectively; the values for sitting sampling surface were 49.89%, 48.3%, 40.2%, and 33.2%, respectively. The interplay between the continuous temperature gradient of the induced air supply and buoyancy affects the aerosol cloud diffusion process inside the urban subway. The larger the difference between the ambient air and respiratory flow (containing aerosol cloud), the higher the buoyancy of the respiratory flow and lower the lateral dispersion of aerosols. The increase in aerosol-cloud buoyancy decreases virus transmission owing to the interaction between up-direction buoyant flow and down-direction shear flow of fresh air. The interaction between aerosol clouds and ambient air is an important mechanism for decreasing viral transmission. These results are consistent with those of He et al. [39], Nazari et al. [18], and Rezaei et al. [35]. At lower temperatures of the induced air flow, the effects of the mentioned aerosol buoyancy decrease, which can help disperse the viruses more with the induced air supply. We discuss the importance of this study from the perspective of reducing the aerosol encounter probability inside urban subways using HVAC system duct configuration, supply air temperature, and ACR. Breathing aerosol lifetime is an important parameter that reflects the duration over which a healthy individual can get infected if one encounters an induced aerosol.

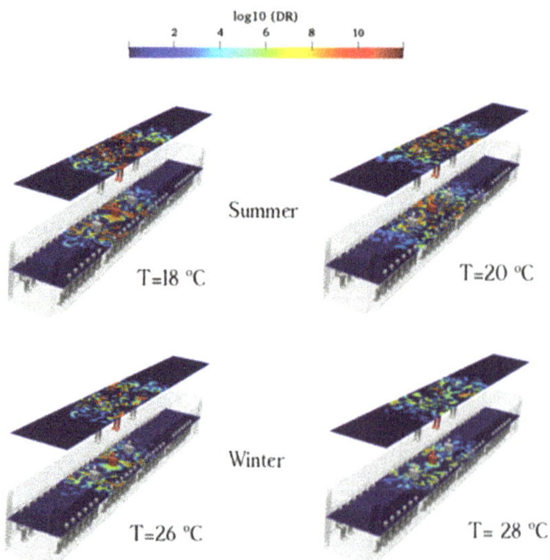

Figure 16. Evolution of the counters of $\log_{10}$(DR) when the infected individual is standing near the supply air ducts of the conventional HVAC systems (Case 3) at the time of 180 s for various temperatures (summer and winter season).

3.6. Effect of Supply Air change Rate (ACR) of HVAC for Case 3

This section investigates the effect of fresh air ACR inside urban subways on aerosol spread for case 3. Four ACRs were selected, and the aerosol encounter probabilities were calculated. To quantify the ACR effect, simulations were conducted using ACR values of 10, 11, 12, and 13 hr^{-1}. Based on Figure 11, an ACR of 10 hr^{-1} with an induced temperature of 20 °C represents the worst-case scenario. In this case, aerosol propagation on both sampling surfaces was higher than that on the same surfaces in the conventional HVAC system. An essential method to reduce aerosol encounter is to increase the fresh air velocity, which can be achieved by increasing the ACR. The HVAC system consumes higher energy with an increase in the ACR value. The calculated probabilities of the standing sampling surface aerosol encounters for 10, 11, 12, and 13 hr^{-1} were 51.2%, 42.3%, 35.2%, and 30.1%, respectively; the corresponding values for sitting sampling surface were 48.3%, 40.9%, 30.5%, and 25.4%, respectively. The aerosol concentrations over the sampling surfaces for the various ACRs are shown in Figure 17.

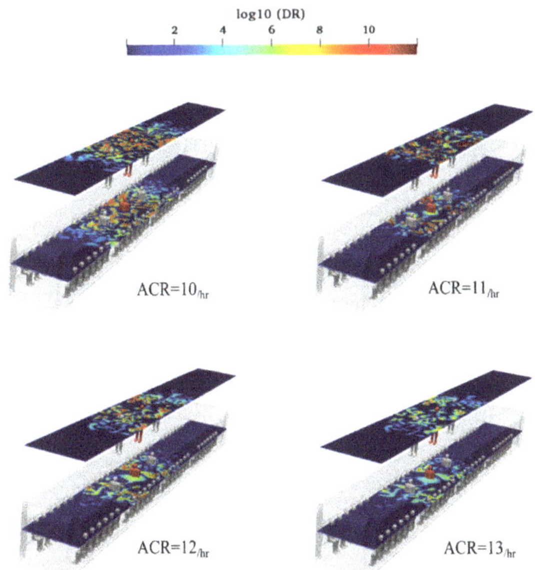

Figure 17. Evolution of the counters of $\log_{10}$(DR) when the infected individual was standing near the supply air ducts of the conventional HVAC systems (Case 3) at the time of 180 s for various well-ventilated air change rates.

In some cases, the supply air ducts cannot inject appropriate amount of fresh air into the subway cabin. To consider this scenario, poorly ventilated ACRs were investigated. Figure 18 shows the effects of ACRs of 4, 5, 6, and 7 hr^{-1} with an induced temperature of 20 °C on the breathing aerosol spreading inside the cabin. In a poorly ventilated ACR, the weak shear flow cannot overcome the aerosol cloud to change its path and direct it toward the exhaust ducts. The zone with high concentrations of aerosol encounter levels were created in the poorly ventilated scenario, which corresponded to the locations where virus transmission was greater than that in other places. These zones were spread in the longitudinal direction of the cabin, allowing the breathing aerosols to move in any desired direction. The calculated probabilities of the standing sampling surface aerosol encounters for 4, 5, 6, and 7 hr^{-1} were 76.1%, 74.4%, 68 %, and 65.1%, respectively; the corresponding values for sitting sampling surface were 71%, 70.2%, 62.2%, and 60.3%, respectively.

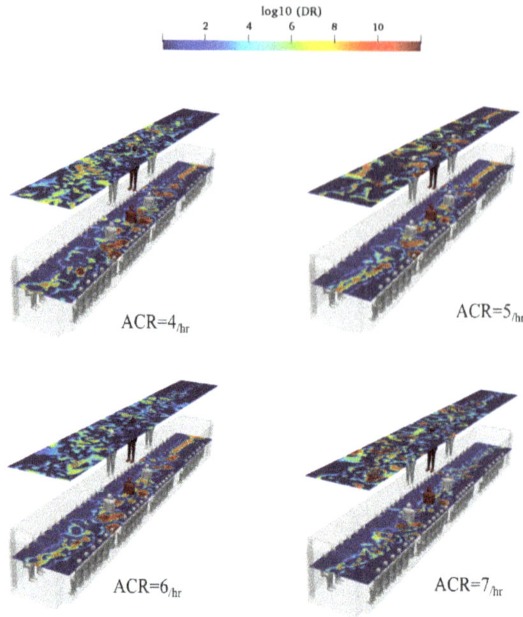

Figure 18. Evolution of the counters of $\log_{10}$(DR) when the infected individual was standing near the supply air ducts of the conventional HVAC systems (Case 3) at the time of 180 s for various poorly ventilated air change rates.

3.7. Effect of Imperfect Filtration of HVAC for Case 3

In all the sections mentioned above, filters inside the ducts were assumed to eliminate all airborne viruses. To simplify the simulations, we ignored the conditions that lack HEPA filters and assumed that the aerosol concentration in the injected fresh air was near zero. With the constant use of filters over time, some features of these filters change and lose their effectiveness, and aerosols can penetrate the cabin. A common index of HEPA filters, namely, removed particle present (RPP), was used to rate the efficiency of the filters. In the HVAC system of an urban subway, fresh outdoor air flows through the supply air duct, combines with recirculated airflow, and then flows through the supply air duct to the cabin. Part of the indoor air returns to the exhaust air duct and then enters the outdoor atmosphere. Because recirculated air and exhaust ducts circulate 75% of aerosols exiting through the HVAC system, HEPA filters are placed in all ducts to clean the air entering the cabin. To quantify the RPP effect, simulations were conducted with the RPP values of 50, 60, 70, and 80, and corresponding aerosol encounter probabilities were calculated. Notably, the worst case (case 3) with imperfect filtration in a conventional HVAC system was selected. The probability of aerosol encounters at each sampling surface in Case 3, with imperfect filtration, is shown in Figure 19. The highest aerosol encounter probability for an RPP of 50% was 89.9%. When the air supply blows at a particle number of 20, exhaled air interacts with fresh air. The mixing of these concentrated flows intensifies aerosol transmission in the vertical direction of the cabin. The effect of imperfect filtration was studied by changing the concentration in the fresh air ducts based on the following equation:

$$\mathrm{RPP} = \left(\frac{C_{\mathrm{beyond}} - C_{\mathrm{front}}}{C_{\mathrm{beyond}}} \right) \times 100, \tag{17}$$

where C_{beyond} and C_{front} are the concentrations of particulate matter outside and in front of the filter, respectively.

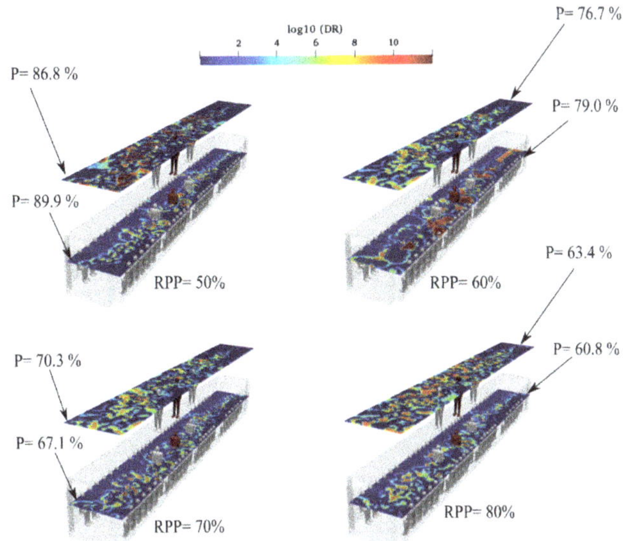

Figure 19. Evolution of the counters of $\log_{10}$(DR) when the infected individual was standing near the supply air ducts of the conventional HVAC systems (Case 3) with imperfect filtration at the time of 180 s for various removed particle present (RPP).

4. Conclusions

In this study, we computationally investigated the effects of a proposed HVAC system on the spread of viruses, such as SARS-CoV-2 inside an urban subway system. We employed a transport equation for the concentration, which was added to the basic buoyant solver to resolve aerosol transmission inside the urban subway by considering the thermal, turbulent, and induced ventilation flow effects. The mechanisms of virus spread through the movement of virus-containing aerosols were investigated using the *OpenFOAM* C++ libraries. Our simulation approach was validated using experimental data by applying a porous boundary condition to fresh-air ducts. Subsequently, we used a numerical approach to investigate the airborne transmission pattern and the associated infection risk for different ventilation designs and source locations. The exhaust ducts of a conventional HVAC system are located along the cabin ceiling and near the fresh-air ducts. These configurations create longitudinal ventilation flows inside the cabin and remove aerosols along long paths. This study introduced aerosol encounter probability to represent the probability of infection during airborne transmission. An increase in the aerosol encounter probability owing to the flow pattern depends on the exhaust configuration. To suppress aerosol encounter probability, we changed the exhaust duct location from near the cabin ceiling to near the floor. Our results show a remarkable relationship between regions with a high aerosol encounter probability and airflow patterns in urban subway cabins, adding to our understanding of aerosol transmission. The proposed ventilation design showed a significant improvement in mitigating the infection risk associated with airborne transmission in a subway, in comparison with the commonly used ventilation designs, for all the source locations simulated in this study. To calculate the aerosol encounter probability inside the urban subway, we selected two surfaces that crossed the center of the passengers' breathing zone hemisphere. For the case of the infected individual breathing near the fresh-air ducts, the conventional HVAC system design led to the largest spread and highest aerosol encounter probability (51.2% on the sitting sampling surface), but in the proposed design, this value was decreased to 3.5%. In the case of infected individuals breathing near the fresh-air ducts, the proposed design showed the most noticeable improvement compared to the conventional design. Based on our findings, we propose the following recommendations for risk mitigation in the subway considering that the ventilation design of the subway cannot be changed over a short period.

4.1. Concluding Remarks

1. Owing to the airflow directions, the safe zone is near the supply air ducts, and the hazardous zones are near the exhaust and recirculated ducts. Therefore, passengers should sit close to the air supply ducts and not stand inside the cabin. In other words, they should sit away from the exhaust and recirculate ducts.

2. With increased temperature and ACR of the supply air, the probabilities of aerosol encounter for all cases reduce.

3. Imperfect filtration of air in the HVAC system increase the probability for risk of aerosol encounter inside the urban subway.

4.2. Practical Recommendations

1. Based on the probabilities of encountering aerosols in the conventional HVAC system, authorities may provide alarm signs near an urban subway's exhaust and recirculated ducts. To minimize face-to-face contact and reduce the risk of cross-infection, we propose providing separate doors for passengers entering or exiting the cabin. In some parts of the study, we assumed that the HEPA filter could capture all the viral particles. Regularly changing the HEPA filters is suggested to maintain their efficacy.

2. The authors encourage wearing a face mask to prevent the spread of respiratory droplets and aerosols within the urban subway and maintain social distancing while sitting.

The current study has several limitations. For the proposed HVAC system, we did not consider the dispersion of aerosols because of sneezing or coughing inside the cabin. Further research is required to better understand virus transmission via respiratory droplets in both HVAC systems. Researchers may focus on the head's direction during coughing or sneezing inside the urban subway and study the high-value size of respiratory particles. An improved HVAC system model is required for future subway designs to manage viral transmission, keeping in mind the comfort of the passengers. Further research is needed to illustrate the energy consumption, thermal comfort, and aerodynamic behavior of urban subways. An improved HVAC system is necessary to mitigate aerosol transmission inside urban subways. To this end, this basic study must be experimentally validated.

Author Contributions: Conceptualization, A.N. and J.H.; methodology, A.N. and J.H.; software, A.N.; validation, A.N., J.H., F.T.-H. (Farhad Taghizadeh-Hesary) and F.T.-H. (Farzad Taghizadeh-Hesary); investigation, A.N., J.H., F.T.-H. (Farhad Taghizadeh-Hesary) and F.T.-H. (Farzad Taghizadeh-Hesary); resources, A.N., J.H., F.T.-H. (Farhad Taghizadeh-Hesary) and F.T.-H. (Farzad Taghizadeh-Hesary); data curation, A.N., J.H., F.T.-H. (Farhad Taghizadeh-Hesary) and F.T.-H. (Farzad Taghizadeh-Hesary); writing—original draft preparation, A.N.; writing—review and editing, J.H., F.T.-H. (Farhad Taghizadeh-Hesary) and F.T.-H. (Farzad Taghizadeh-Hesary); supervision, J.H. and F.T.-H. (Farhad Taghizadeh-Hesary); funding acquisition, F.T.-H. (Farhad Taghizadeh-Hesary). All authors have read and agreed to the published version of the manuscript.

Funding: This research was financially supported by the Grant-in-Aid for the Excellent Young Researcher of the Ministry of Education, Culture, Sports, Science and Technology of Japan (MEXT) and Grant-in-Aid for Young Scientists (No. 22K13432) of the Japan Society for the Promotion of Science (JSPS).

Institutional Review Board Statement: Not applicable.

Informed Consent Statement: Not applicable.

Data Availability Statement: The data that support the findings of this study are available from the corresponding author upon reasonable request.

Conflicts of Interest: The authors declare no conflict of interest.

References

1. Johns Hopkins University and Medicine. COVID-19 Mortality Rate. 2022. Available online: https://coronavirus.jhu.edu/map.html (accessed on 6 June 2022).
2. Rakhsha, A.; Azghandi, S.; Taghizadeh-Hesary, F. COVID-19 pandemic and patients with cancer: The protocol of a clinical oncology center in Tehran, Iran. *Rep. Pract. Oncol. Radiother.* **2020**, *25*, 765–767. [CrossRef] [PubMed]
3. Siavashpour, Z.; Taghizadeh-Hesary, F.; Rakhsha, A. Recommendations on management of locally advanced rectal cancer during the COVID-19 pandemic: An Iranian consensus. *J. Gastrointest. Cancer* **2020**, *51*, 800–804. [CrossRef] [PubMed]
4. Rakhsha, A.; Azghandi, S.; Taghizadeh-Hesary, F. Decision on chemotherapy amidst COVID-19 pandemic: A review and a practical approach from Iran. *Infect. Chemother.* **2020**, *52*, 496. [CrossRef] [PubMed]
5. Fazilat-Panah, D.; Fallah Tafti, H.; Rajabzadeh, Y.; Fatemi, M.A.; Ahmadi, N.; Jahansouz, D.; Tabasi, M.; Javadinia, S.A.; Joudi, M.; Harati, H.; et al. Clinical Characteristics and Outcomes of COVID-19 in 1290 New Cancer Patients: Single-center, Prospective Cohort Study from Iran. *Cancer Investig.* **2022**, *40*, 1–3. [CrossRef] [PubMed]
6. Rabaan, A.A.; Al-Ahmed, S.H.; Al-Malkey, M.K.; Alsubki, R.A.; Ezzikouri, S.; Al-Hababi, F.H.; Sah, R.; Al Mutair, A.; Alhumaid, S.; Al-Tawfiq, J.A.; et al. Airborne transmission of SARS-CoV-2 is the dominant route of transmission: Droplets and aerosols. *Infez Med.* **2021**, *29*, 10–19. [PubMed]
7. Agrawal, A.; Bhardwaj, R. Probability of COVID-19 infection by cough of a normal person and a super-spreader. *Phys. Fluids* **2021**, *33*, 031704. [CrossRef]
8. Avni, O.; Dagan, Y. Dynamics of Evaporating Respiratory Droplets in the Vicinity of Vortex Dipoles. *arXiv* **2021**, arXiv:210807068. [CrossRef]
9. Dbouk, T.; Drikakis, D. On respiratory droplets and face masks. *Phys. Fluids* **2020**, *32*, 063303. [CrossRef]
10. Bergmans, B.; Cattaneo, A.; Duarte, R.M.; Gomes, J.F.; Saraga, D.; Rodenas Garcia, M.; Querol, X.; Liotta, L.F.; Safell, J.; Spinazzé, A.; et al. Particulate matter indoors: A strategy to sample and monitor size-selective fractions. *Appl. Spectrosc. Rev.* **2022**, *57*, 675–704. [CrossRef]
11. Kaewrat, J.; Janta, R.; Sichum, S.; Rattikansukha, C.; Tala, W.; Kanabkaew, T. Human Health Risks and Air Quality Changes Following Restrictions for the Control of the COVID-19 Pandemic in Thailand. *Toxics* **2022**, *10*, 520. [CrossRef]
12. Hossain, M.; Faisal, N.H. Modeling aerosol cloud aerodynamics during human coughing, talking, and breathing actions. *AIP Adv.* **2021**, *11*, 045111. [CrossRef]
13. Manoj, M.G.; Satheesh Kumar, M.K.; Valsaraj, K.T.; Vijayan, S.K.; Nishanth, T. Exacerbation of Fatality Rates Induced by Poor Air Quality Due to Open-Air Mass Funeral Pyre Cremation during the Second Wave of COVID-19. *Toxics* **2022**, *10*, 306. [CrossRef] [PubMed]
14. Islam, M.S.; Larpruenrudee, P.; Paul, A.R.; Paul, G.; Gemci, T.; Gu, Y.; Saha, S.C. SARS CoV-2 aerosol: How far it can travel to the lower airways? *Phys. Fluids* **2021**, *33*, 061903. [CrossRef] [PubMed]
15. Jiang, G.; Li, F.; Hu, T. Transport Characteristics and Transmission Risk of Virus-Containing Droplets from Coughing in Outdoor Windy Environment. *Toxics* **2022**, *10*, 294. [CrossRef] [PubMed]
16. Liu, K.; Allahyari, M.; Salinas, J.S.; Zgheib, N.; Balachandar, S. Peering inside a cough or sneeze to explain enhanced airborne transmission under dry weather. *Sci. Rep.* **2021**, *11*, 1–9. [CrossRef]
17. Mohammadi Nafchi, A.; Blouin, V.; Kaye, N.; Metcalf, A.; Van Valkinburgh, K.; Mousavi, E. Room HVAC influences on the removal of airborne particulate matter: Implications for school reopening during the COVID-19 pandemic. *Energies* **2021**, *14*, 7463. [CrossRef]
18. Nazari, A.; Jafari, M.; Rezaei, N.; Arash-Azad, S.; Talati, F.; Nejad-Rahim, R.; Taghizadeh-Hesary, F.; Taghizadeh-Hesary, F. Effects of High-Speed Wind, Humidity, and Temperature on the Generation of a SARS-CoV-2 Aerosol; a Novel Point of View. *Aerosol Air Qual. Res.* **2021**, *21*, 200574. [CrossRef]
19. Stadnytskyi, V.; Anfinrud, P.; Bax, A. Breathing, speaking, coughing or sneezing: What drives transmission of SARS-CoV-2? *J. Intern. Med.* **2021**, *290*, 1010–1027. [CrossRef]
20. Ju, M.J.; Oh, J.; Choi, Y.H. Changes in air pollution levels after COVID-19 outbreak in Korea. *Sci. Total Environ.* **2021**, *750*, 141521. [CrossRef]
21. Zhou, M.; Zou, J. A dynamical overview of droplets in the transmission of respiratory infectious diseases. *Phys. Fluids* **2021**, *33*, 031301. [CrossRef]
22. Akagi, F.; Haraga, I.; Inage, S.-i.; Akiyoshi, K. Effect of sneezing on the flow around a face shield. *Phys. Fluids* **2020**, *32*, 127105. [CrossRef] [PubMed]
23. Wang, H.; Li, Z.; Zhang, X.; Zhu, L.; Liu, Y.; Wang, S. The motion of respiratory droplets produced by coughing. *Phys. Fluids* **2020**, *32*, 125102. [CrossRef] [PubMed]
24. Zeng, G.; Chen, L.; Yuan, H.; Yamamoto, A.; Maruyama, S. Evaporation flow characteristics of airborne sputum droplets with solid fraction: Effects of humidity field evolutions. *Phys. Fluids* **2021**, *33*, 123308. [CrossRef]
25. Li, L.; He, R.; Kong, M.; Eilts, S.M.; Hong, J.; Hogan Jr, C.J.; Pope, Z.C. Effect of low-cost recirculating portable air filtration on aerosol particle deposition and concentration in a conference room: Experiment, theory, and simulation comparison. *J. Aerosol Sci.* **2022**, *166*, 106048. [CrossRef]
26. He, R.; Gao, L.; Trifonov, M.; Hong, J. Aerosol generation from different wind instruments. *J. Aerosol Sci.* **2021**, *151*, 105669. [CrossRef] [PubMed]

27. Li, H.; Leong, F.Y.; Xu, G.; Ge, Z.; Kang, C.W.; Lim, K.H. Dispersion of evaporating cough droplets in tropical outdoor environment. *Phys. Fluids* **2020**, *32*, 113301. [CrossRef] [PubMed]
28. Nazari, A.; Jafari, M.; Rezaei, N.; Taghizadeh-Hesary, F.; Taghizadeh-Hesary, F. Jet fans in the underground car parking areas and virus transmission. *Phys. Fluids* **2021**, *33*, 013603. [CrossRef] [PubMed]
29. Liu, H.; He, S.; Shen, L.; Hong, J. Simulation-based study of COVID-19 outbreak associated with air-conditioning in a restaurant. *Phys. Fluids* **2021**, *33*, 023301. [CrossRef]
30. Li, Y.-y.; Wang, J.-X.; Chen, X. Can a toilet promote virus transmission? From a fluid dynamics perspective. *Phys. Fluids* **2020**, *32*, 065107. [CrossRef]
31. Schreck, J.H.; Lashaki, M.J.; Hashemi, J.; Dhanak, M.; Verma, S. Aerosol generation in public restrooms. *Phys. Fluids* **2021**, *33*, 033320. [CrossRef]
32. Wang, J.-X.; Li, Y.-Y.; Liu, X.-D.; Cao, X. Virus transmission from urinals. *Phys. Fluids* **2020**, *32*, 081703. [CrossRef] [PubMed]
33. Dbouk, T.; Drikakis, D. On airborne virus transmission in elevators and confined spaces. *Phys. Fluids* **2021**, *33*, 011905. [CrossRef] [PubMed]
34. Vranay, F.; Pirsel, L.; Kacik, R.; Vranayova, Z. Adaptation of HVAC Systems to Reduce the Spread of COVID-19 in Buildings. *Sustainability* **2020**, *12*, 9992. [CrossRef]
35. Rezaei, N.; Jafari, M.; Nazari, A.; Salehi, S.; Talati, F.; Torab, R.; Nejad-Rahim, R. A novel methodology and new concept of SARS-CoV-2 elimination in heating and ventilating air conditioning systems using waste heat recovery. *AIP Adv.* **2020**, *10*, 085308. [CrossRef]
36. Zhang, Z.; Han, T.; Yoo, K.H.; Capecelatro, J.; Boehman, A.L.; Maki, K. Disease transmission through expiratory aerosols on an urban bus. *Phys. Fluids* **2021**, *33*, 015116. [CrossRef]
37. Abuhegazy, M.; Talaat, K.; Anderoglu, O.; Poroseva, S.V. Numerical investigation of aerosol transport in a classroom with relevance to COVID-19. *Phys. Fluids* **2020**, *32*, 103311. [CrossRef]
38. Burgmann, S.; Janoske, U. Transmission and reduction of aerosols in classrooms using air purifier systems. *Phys. Fluids* **2021**, *33*, 033321. [CrossRef]
39. He, R.; Liu, W.; Elson, J.; Vogt, R.; Maranville, C.; Hong, J. Airborne transmission of COVID-19 and mitigation using box fan air cleaners in a poorly ventilated classroom. *Phys. Fluids* **2021**, *33*, 057107. [CrossRef]
40. Narayanan, S.R.; Yang, S. Airborne transmission of virus-laden aerosols inside a music classroom: Effects of portable purifiers and aerosol injection rates. *Phys. Fluids* **2021**, *33*, 033307. [CrossRef]
41. Foster, A.; Kinzel, M. Estimating COVID-19 exposure in a classroom setting: A comparison between mathematical and numerical models. *Phys. Fluids* **2021**, *33*, 021904. [CrossRef]
42. Wu, L.; Liu, X.; Yao, F.; Chen, Y. Numerical study of virus transmission through droplets from sneezing in a cafeteria. *Phys. Fluids* **2021**, *33*, 023311. [CrossRef] [PubMed]
43. Talaat, K.; Abuhegazy, M.; Mahfoze, O.A.; Anderoglu, O.; Poroseva, S.V. Simulation of aerosol transmission on a Boeing 737 airplane with intervention measures for COVID-19 mitigation. *Phys. Fluids* **2021**, *33*, 033312. [CrossRef] [PubMed]
44. Jia, D.; Lee Baker, J.; Rameau, A.; Esmaily, M. Simulation of a vacuum helmet to contain pathogen-bearing droplets in dental and otolaryngologic outpatient interventions. *Phys. Fluids* **2021**, *33*, 013307. [CrossRef] [PubMed]
45. Baudet, A.; Baurès, E.; Blanchard, O.; Le Cann, P.; Gangneux, J.P.; Florentin, A. Indoor carbon dioxide, fine particulate matter and total volatile organic compounds in private healthcare and elderly care facilities. *Toxics* **2022**, *10*, 136. [CrossRef]
46. Li, Z.; Zhang, X.; Wu, T.; Zhu, L.; Qin, J.; Yang, X. Effects of slope and speed of escalator on the dispersion of cough-generated droplets from a passenger. *Phys. Fluids* **2021**, *33*, 041701. [CrossRef] [PubMed]
47. Abraham, A.; He, R.; Shao, S.; Kumar, S.S.; Wang, C.; Guo, B.; Trifonov, M.; Placucci, R.G.; Willis, M.; Hong, J. Risk assessment and mitigation of airborne disease transmission in orchestral wind instrument performance. *J. Aerosol Sci.* **2021**, *157*, 105797. [CrossRef]
48. Duarte, R.M.; Gomes, J.F.; Querol, X.; Cattaneo, A.; Bergmans, B.; Saraga, D.; Maggos, T.; Di Gilio, A.; Rovelli, S.; Villanueva, F. Advanced instrumental approaches for chemical characterization of indoor particulate matter. *Appl. Spectrosc. Rev.* **2022**, *57*, 705–745. [CrossRef]
49. Shao, X.; Li, X. COVID-19 transmission in the first presidential debate in 2020. *Phys. Fluids* **2020**, *32*, 115125. [CrossRef] [PubMed]
50. Arias, F.J.; De Las Heras, S. The mechanical effect of moisturization on airborne COVID-19 transmission and its potential use as control technique. *Environ. Res.* **2021**, *197*, 110940. [CrossRef]
51. Baboli, Z.; Neisi, N.; Babaei, A.A.; Ahmadi, M.; Sorooshian, A.; Birgani, Y.T.; Goudarzi, G. On the airborne transmission of SARS-CoV-2 and relationship with indoor conditions at a hospital. *Atmos. Environ.* **2021**, 118563. [CrossRef]
52. Rowe, B.R.; Canosa, A.; Drouffe, J.-M.; Mitchell, J. Simple quantitative assessment of the outdoor versus indoor airborne transmission of viruses and covid-19. *Environ. Res.* **2021**, *198*, 111189. [CrossRef] [PubMed]
53. Moreno, T.; Pintó, R.M.; Bosch, A.; Moreno, N.; Alastuey, A.; Minguillón, M.C.; Anfruns-Estrada, E.; Guix, S.; Fuentes, C.; Buonanno, G.; et al. Tracing surface and airborne SARS-CoV-2 RNA inside public buses and subway trains. *Environ. Int.* **2021**, *147*, 106326. [CrossRef] [PubMed]
54. Chang, Z.; Yi, K.; Liu, W. A new ventilation mode of air conditioning in subway vehicles and its air distribution performance. *Energy Built Environ.* **2021**, *2*, 94–104. [CrossRef]

55. Gröndahl, M.; Goldbaum, C.; White, J. What Happens to Viral Particles on the Subway. 2021. Available online: www.nytimes.com/interactive/2020/08/10/nyregion/nyc-subway-coronavirus.html (accessed on 10 August 2020).
56. Tao, Y.; Yang, M.; Qian, B.; Wu, F.; Wang, T. Numerical and experimental study on ventilation panel models in a subway passenger compartment. *Engineering* **2019**, *5*, 329–336. [CrossRef]
57. Shao, S.; Zhou, D.; He, R.; Li, J.; Zou, S.; Mallery, K.; Kumar, S.; Yang, S.; Hong, J. Risk assessment of airborne transmission of COVID-19 by asymptomatic individuals under different practical settings. *J. Aerosol Sci.* **2021**, *151*, 105661. [CrossRef]
58. Chia, P.Y.; Coleman, K.K.; Tan, Y.K.; Ong, S.W.X.; Gum, M.; Lau, S.K.; Lim, X.F.; Lim, A.S.; Sutjipto, S.; Lee, P.H.; et al. Detection of air and surface contamination by SARS-CoV-2 in hospital rooms of infected patients. *Nat. Commun.* **2020**, *11*, 2800. [CrossRef]
59. Li, C.; Wang, H.; Yu, C.W.; Xie, D. Diffusion characteristics of the industrial submicron particle under Brownian motion and turbulent diffusion. *Indoor Built Environ.* **2022**, *31*, 17–30. [CrossRef]
60. Kanso, M.; Chaurasia, V.; Fried, E.; Giacomin, A. Peplomer bulb shape and coronavirus rotational diffusivity. *Phys. Fluids* **2021**, *33*, 033115. [CrossRef]
61. Kanso, M.; Naime, M.; Chaurasia, V.; Tontiwattanakul, K.; Fried, E.; Giacomin, A.J. Coronavirus Pleomorphism. *Phys. Fluids* **2022**, *34*, 063101. [CrossRef]
62. Yan, Y.; Li, X.; Shang, Y.; Tu, J. Evaluation of airborne disease infection risks in an airliner cabin using the Lagrangian-based Wells-Riley approach. *Build. Environ.* **2017**, *121*, 79–92. [CrossRef]
63. Amri, H.; Hofstädter, R.N.; Kozek, M. Energy efficient design and simulation of a demand-controlled heating and ventilation unit in a metro vehicle. In Proceedings of the 2011 IEEE Forum on Integrated and Sustainable Transportation Systems, Vienna, Austria, 29 June–1 July 2011.
64. Ventilation for Acceptable Indoor Air Quality. *Standard 62-1999*; American Society of Heating, Refrigerating and Air-Conditioning Engineers, Inc.: Atlanta, GA, USA, 1999.
65. Popov, T.; Kralimarkova, T.; Dimitrov, V. Measurement of exhaled breath temperature in science and clinical practice. *Breathe* **2012**, *8*, 186–192. [CrossRef]
66. Das Neves Almeida, M.; de Paula Xavier, A.A.; Michaloski, A.O. A Review of Thermal Comfort Applied in Bus Cabin Environments. *Appl. Sci.* **2020**, *10*, 8648. [CrossRef]
67. Ergonomics of the Thermal Environment: Estimation of the Thermal Insulation and Evaporative Resistance of a Clothing Ensembles. *ISO 9920*; International Standard Organization: Geneva, Switzerland, 1995.
68. ASHRAE. *Handbook of Fundamentals: Physiological Principles, Comfort Health*; ASHRAE: New York, NY, USA, 1997.
69. Angelova, R.A. Numerical simulation of the thermoregulation of clothed human body: Skin and clothing temperatures. *J. Braz. Soc. Mech. Sci. Eng.* **2015**, *37*, 297–303. [CrossRef]
70. Stefanitsis, D.; Koukouvinis, P.; Nikolopoulos, N.; Gavaises, M. Numerical Investigation of the Aerodynamic Droplet Breakup at Mach Numbers Greater Than 1. *J. Energy Eng.* **2021**, *147*, 04020077. [CrossRef]
71. Magnini, M.; Pulvirenti, B.; Thome, J. Characterization of the velocity fields generated by flow initialization in the CFD simulation of multiphase flows. *Appl. Math. Model.* **2016**, *40*, 6811–6830. [CrossRef]
72. Yin, Y.; Gupta, J.K.; Zhang, X.; Liu, J.; Chen, Q. Distributions of respiratory contaminants from a patient with different postures and exhaling modes in a single-bed inpatient room. *Build. Environ.* **2011**, *46*, 75–81. [CrossRef]
73. Zhu, S.; Kato, S.; Murakami, S.; Hayashi, T. Study on inhalation region by means of CFD analysis and experiment. *Build. Environ.* **2005**, *40*, 1329–1336. [CrossRef]
74. Gorbunov, B. Aerosol particles generated by coughing and sneezing of a SARS-CoV-2 (COVID-19) host travel over 30 m distance. *Aerosol Air Qual. Res.* **2021**, *21*, 200468. [CrossRef]
75. Kolinski, J.M.; Schneider, T.M. Superspreading events suggest aerosol transmission of SARS-CoV-2 by accumulation in enclosed spaces. *Phys. Rev. E* **2021**, *103*, 033109. [CrossRef]
76. Bhardwaj, R.; Agrawal, A. Tailoring surface wettability to reduce chances of infection of COVID-19 by a respiratory droplet and to improve the effectiveness of personal protection equipment. *Phys. Fluids* **2020**, *32*, 081702. [CrossRef]
77. Liu, Y.; Shen, L.; Zamansky, R.; Coletti, F. Life and death of inertial particle clusters in turbulence. *J. Fluid Mech.* **2020**, *902*. [CrossRef]

International Journal of
Environmental Research and Public Health

MDPI

Article

Chronic Home Radon Exposure Is Associated with Higher Inflammatory Biomarker Concentrations in Children and Adolescents

Brittany K. Taylor [1,2,*], OgheneTejiri V. Smith [1] and Gregory E. Miller [3]

1 Institute for Human Neuroscience, Boys Town National Research Hospital, Boys Town, NE 68010, USA
2 Department of Pharmacology and Neuroscience, Creighton University, Omaha, NE 68178, USA
3 Institute for Policy Research and Department of Psychology, Northwestern University, Evanston, IL 60208, USA
* Correspondence: brittany.taylor@boystown.org; Tel.: +1-531-355-8926

Abstract: Children are particularly vulnerable to the deleterious impacts of toxic environmental exposures, though the effects of some rather ubiquitous toxins have yet to be characterized in youths. One such toxin, radon gas, is known to accumulate to hazardous levels in homes, and has been linked with the incidence of lung cancer in aging adults. However, the degree to which chronic home radon exposure may impact risk for health problems earlier in life is unknown. Herein, we explored the degree to which chronic home radon exposure relates to biomarkers of low-grade inflammation in 68 youths ages 6- to 14 years old residing in an area of the United States prone to high home radon concentrations. Parents completed a home radon test kit, and youths provided a saliva sample to assess concentrations of five biomarkers. Using a multiple regression approach, we found that greater radon exposure was specifically associated with higher levels of C-reactive protein ($\beta = 0.31$, $p = 0.007$) and interleukin-1β ($\beta = 0.33$, $p = 0.016$). The data suggested specificity in associations between chronic home radon exposure and different biomarkers of inflammatory activity and highlight a pathway which may confer risk for future mental and physical health maladies.

Keywords: CRP; IL-1β; neurotoxicity; environmental exposure; immune dysregulation; youth

Citation: Taylor, B.K.; Smith, O.T.V.; Miller, G.E. Chronic Home Radon Exposure Is Associated with Higher Inflammatory Biomarker Concentrations in Children and Adolescents. *Int. J. Environ. Res. Public Health* **2023**, *20*, 246. https://doi.org/10.3390/ijerph20010246

Academic Editors: Esref Demir and Sam Kacew

Received: 20 October 2022
Revised: 19 December 2022
Accepted: 19 December 2022
Published: 23 December 2022

1. Introduction

Exposures to environmental toxins can cause a host of acute- and long-term consequences on brain structure and function, which can have cascading effects on cognition, behavior, and mental wellness. This is especially true for children and adolescents [1,2]. It is well-established that youths face unique, sometimes more severe costs relative to adults following exposure to toxicants due to modes of toxicant delivery and consumption, breathing and eating habits, and biological vulnerability during periods of rapid development [1,3]. Indeed, a growing body of work has linked exposure to various toxicants, ranging from ingested heavy metals (e.g., lead) to inhaled pollutants (e.g., particulate matter), with sensory dysfunction [4], atypical patterns of structural brain development [5], incidence of specific neurodevelopmental disorders like attention-deficit/hyperactivity disorder (ADHD; [6,7]), cognitive delays and deficits [8,9], and risk for mood and anxiety disorders [10,11]. The field of research understanding the neurotoxic effects of environmental exposures on children is rapidly growing; that said, certain toxins that are known to be readily present in commonly inhabited environments have been seldom studied for their impacts on youths.

One understudied, but ubiquitous environmental toxin is radon gas. Radon is a naturally occurring, radioactive byproduct of uranium decay in soil, and is a well-established carcinogen [12–14]. In fact, radon inhalation is a leading cause of lung cancer globally, second only to cigarette smoking [15]. In well-ventilated environments like outdoors, as radon gas rises from the soil and enters the air, it quickly dissipates and spreads to non-injurious levels [16]. However, it is surprisingly common for radon to accumulate to hazardous concentrations over time in less well-ventilated environments like inside homes and other structures [17,18]. This issue is critically important when considering the well-established linear dose–response effect of radon exposure, whereby increased dosage and chronicity of exposure predictably relates to increasing pathophysiological consequences [13,19].

Across the United States (US), 1 in every 15 homes is expected to have indoor radon concentrations exceeding the action limit for mitigation defined by the US Environmental Protection Agency; this limit, established in the 1980s at or above 4 pCi/L, is the carcinogenic equivalent of smoking 10 cigarettes per day [12,16]. That said, certain areas of the country where uranium deposits in soil are greater are expected to have even further elevated risks of high radon concentrations. For instance, across eastern Nebraska and through the entire state of Iowa, at least 50% of all homes are expected to test above the action limit for indoor radon concentrations [20]. Despite the known pathophysiological risks, and the well-established pattern of high radon concentrations in many parts of the country, much of the general public is unaware of the potential consequences of radon [15,21,22]. Thus, many domiciles remain untested and unmitigated for radon, leaving dwellers chronically exposed to potentially high doses of this ubiquitous environmental toxin.

Studies in rodent models and in adult humans have repeatedly shown that alpha radiation enters the respiratory tract when radon gas is inhaled, inducing mutagenic cellular changes to mucosal tissues lining the tract and subsequently promoting upregulation of proinflammatory activities [23,24]. Studies exploring the patterns of immune dysregulation associated with radon inhalation are varied and minimal to date, though researchers have reported upregulation of an array of inflammatory proteins including several interleukins [25–27], and tumor necrosis factor-alpha (TNF-α; [28]). Chronic increases in circulating proinflammatory biomarkers have been associated with cascading effects on the brain, cognition, behavior, and mental and physical illness [29–33]. These effects can be particularly detrimental among children and adolescents, for whom the brain is highly plastic and vulnerable to physiological stress signals [34–36]. These associations are well-established in many contexts, but research has yet to explore the degree to which exposure to radon gas may be associated with inflammatory activity in youths and potentially increasing their risk for downstream deficits in mental and physical wellness.

The present study explored the extent to which chronic home radon exposure was associated with alterations in inflammatory activity among children and adolescents. Herein, we focus on a sample of typically developing youths residing in eastern Nebraska and western Iowa, where radon concentrations are expected to exceed the US EPA action limit for mitigation in at least 50% of dwellings [16,20]. Families completed a home radon test kit to measure individual-level home radon concentrations, and they provided information about the dwelling itself and how long the family has lived in the tested residence. Youths also provided saliva samples from which we measured concentrations of five different inflammatory biomarkers. Because of our unique interest in potential impacts on the brain, we focused on biomarkers of inflammation that have been previously implicated in structural and functional neural outcomes. We also controlled for a number of anthropometric, maturational, and socioeconomic factors that are known to influence inflammatory activity. We anticipated that, above and beyond potential confounding factors, increasing radon exposure, defined as the combination of home radon concentrations and duration of exposure, would be associated with higher levels of in inflammatory biomarkers in youths. We did not have hypotheses about specific markers that may be more strongly linked with radon given the limited literature.

2. Materials and Methods

2.1. Participants

Children and adolescents who were part of an ongoing observational study of neurocognitive development were invited to participate in the current protocol. A total of

68 youths ages 6 to 14 years-old (M = 10.27 years, SD = 2.59; 33 male) consented to the study, and their families completed a home radon test kit. Youths included in this investigation did not differ from those in the larger study on the basis of age ($t = -1.93$, $p = 0.06$), SES ($t = -0.96$, $p = 0.34$), sex ($\chi^2 = 1.01$, $p = 0.32$), race ($\chi^2 = 4.60$, $p = 0.20$), or ethnicity ($\chi^2 = 3.03$, $p = 0.22$). Exclusion criteria included history of head trauma, neurological disorder or other medical illness affecting brain function, current substance abuse, and standard neuroimaging exclusions (e.g., dental braces, other non-removable ferromagnetic materials on the body). All parents of child participants provided signed informed consent, and youth participants gave signed assent to participate in the study. All procedures were approved by the local Institutional Review Board.

2.2. Home Radon Testing

Families were provided with a commercial short-term home radon testing kit (https://www.radon.com/, Accessed on 19 October 2022). The test kit is a standard carbon-based envelope that hangs on an interior wall on the lowest livable level of the home for three to seven days. After the testing period, the envelope is sealed and dropped in the mail for processing at the commercial lab. Parents were given the test kit along with instructions from the commercial vendor for proper exposure. We instructed families to leave the kit exposed for approximately four days. Our lab and the family each received a copy of the home radon results. In the case that a result exceeded the EPA action limit for mitigation ($\geq$4 pCi/L), the principal investigator called the family to ensure they understood the results and provided additional information on radon safety and local resources.

Given prior work suggesting that radon concentrations significantly vary seasonally, we provided a subset of 21 families with two radon kits. One kit was completed during summer months (June through September), and one during winter months (December through March). We compared the radon concentrations yielded from the two measurement periods to determine whether there was significant variability using a Wilcoxon Z test given the non-normal distribution of radon concentrations.

2.3. Questionnaires and Biometrics

Youths assessed for height and weight during a regularly scheduled lab visit. Height in centimeters and weight in kilograms were recorded for each child and used to compute each individual's body mass index (BMI) in accordance with standard procedures.

To assess pubertal development, youths or parents were asked to complete the Pubertal Development Scale (PDS; [37]). Specifically, if the child was under the age of 11 years old, we asked the parent to complete the survey. Youths who were 11 years old and older completed the survey themselves. Surveys were completed in a private room on a computer. A trained research assistant was available to answer any questions. We computed a pubertal development stage from the PDS in accordance with recommendations by Shirtcliff, Dahl, and Pollak [38], the end result being a score parallel to Tanner staging.

Parents were asked to complete a brief questionnaire when they began their home radon testing for the study. The custom-designed questionnaire asked for details about the construction of the home (e.g., how many stories, type of foundation, location of children's bedrooms), how long the family lived in the home, whether the home had ever been tested/mitigated for radon, and whether anyone in the home smoked cigarettes. The questionnaire was completed remotely at the parent's convenience via the Collaborative Informatics and Neuroimaging Suite (https://coins.trendscenter.org/, Accessed on 19 October 2022).

In addition to the custom radon questionnaire, parents completed the Barratt Simplified Measure of Social Status [39], which assesses parental education and occupation to provide a numeric index of socioeconomic status. Scores can range from 17 to 66, with higher scores indicating higher SES. This survey was completed during a visit to the laboratory.

2.4. Computing a Radon Exposure Index

Because the effects of radon exposure are cumulative, we computed a radon exposure index per participant. The index was defined as the child's home radon concentration (in pCi/L) obtained from the home testing kit multiplied by the amount of time they lived in that home (in years). That value was natural log transformed (see Equation (1)), providing us with a normally distributed index of chronic radon exposure in their current home. Of note, we achieved similar results when computing the radon index with exposure time computed in months versus years.

$$\text{Radon Exposure Index} = \ln\left([\text{radon concentration} \times \text{exposure time}] + 1\right) \quad (1)$$

2.5. Saliva Sample Acquisition and Analysis

During a visit to the laboratory, youths were asked to provide a saliva sample. As such, children were told to refrain from eating, drinking, or chewing gum for at least an hour prior to sample collection. Participants were instructed to passively drool into an Oragene DISCOVER (OGR-500; www.dnagenotek.com, Accessed on 19 October 2022) until liquid (not bubble) saliva reached the indicated fill line on the tube. Thus, we collected at least 2 mL of whole unstimulated saliva from each child. Samples were stored at −20 °C until processing at the University of Nebraska Lincoln Salivary Biosciences Laboratory (https://cb3.unl.edu/sbl/, Accessed on 19 October 2022). Samples were processed for concentrations of CRP, interleukin (IL)-1β, IL-6, IL-8, and TNF-α using commercially available assay kits (Salimetrics; www.salimetrics.com). We specifically used Meso Scale Discovery electrochemiluminescence cytokine assay kits for the interleukins and TNF-α, which provided additional sensitivity for cytokines that are naturally low in healthy populations. All assays were completed in duplicate, and the average of the two measures was used for analyses. Sensitivity, analytic range, and inter- and intra-assay coefficients of variability for each assay are listed in Table 1. In the case that a result was below the functional sensitivity for the assay, we replaced the value with half the lower limit of quantification (CRP: $n = 7$, IL-6: $n = 1$, and TNF-α: $n = 1$). The final values were then natural log transformed (ln [value + 1]) and inspected for normality. Log-transformed values exceeding three standard deviations above the group mean were excluded as outliers (CRP: $n = 1$, IL-8: $n = 1$).

Table 1. Characteristics of assays for inflammatory markers in the current study.

	Assay Range	Analytical Sensitivity	Functional Sensitivity	Intra-Assay CV	Inter-Assay CV
CRP	25–1600 pg/mL	0.042 pg/mL	19.44 pg/mL	1.93%	3.58%
IL-1β	0.646–375 pg/mL	0.05 pg/mL	0.646 pg/mL	2.65%	4.98%
IL-6	0.633–488 pg/mL	0.06 pg/mL	0.633 pg/mL	3.85%	2.88%
IL-8	0.591–375 pg/mL	0.07 pg/mL	0.591 pg/mL	1.31%	3.12%
TNF-α	0.690–248 pg/mL	0.04 pg/mL	0.690 pg/mL	4.90%	3.19%

2.6. Statistical Analysis

The main goal of the current study was to quantify the degree to which home radon exposure is associated with inflammation in children and adolescents. To address this aim, we constructed a multiple regression type model in which the radon exposure index was modeled as a predictor of each of the inflammatory markers (CRP, IL-1β, IL-6, IL-8, and TNF-α). All inflammatory markers were allowed to freely correlate. Statistical significance of hypothesized associations (i.e., those between radon exposure and each inflammatory biomarker) was determined using an $\alpha < 0.05$ threshold after correction for multiple comparisons via false discovery rate (FDR). We included age, sex, BMI, and SES as control variables on each of the inflammatory measures of interest. Of note, we did attempt an alternative model in which all control variables were enforced on the radon exposure index as well, but there were no significant effects of any control variables on radon exposure. Thus, we report the model without control variables imposed on radon exposure in favor of the more parsimonious model. The conceptual model is shown in Figure 1. Several inflammatory markers were missing across participants for various reasons including outlier data, contaminated saliva samples (i.e., food residue present in sample), or inadequate saliva sample volume to complete all assays. Thus, the model was tested with and without missing data estimation using full-information maximum likelihood (FIML). Because the conclusions were the same regardless, we report the results from the analysis using FIML for robustness. As an exploratory follow-up to compare parameter estimates, we computed Z scores based on the unstandardized beta and standard error of the associations of interest ($Z = [b_1 - b_2]/\sqrt{[SE_1^2 + SE_2^2]}$). All parameters were freely estimated using Mplus version 8.1. An a priori power analysis for a multiple regression with 6 predictors, power = 0.80, $\alpha = 0.05$, and small effect size of $f^2 = 0.15$ suggested a minimum sample size of 55 participants, thus we were sufficiently powered to conduct our analyses and detect relatively small effects.

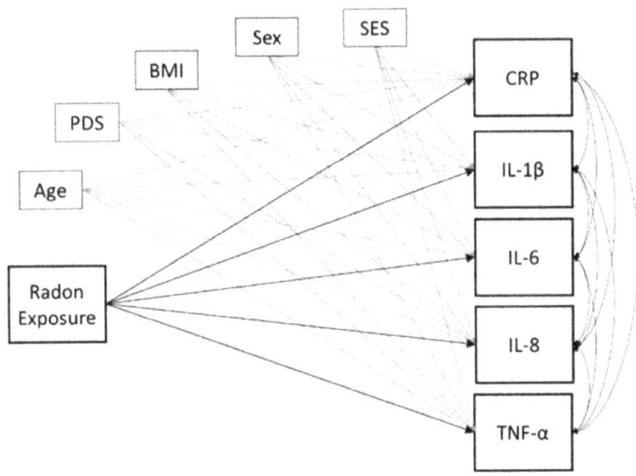

Figure 1. Conceptual model. Radon exposure is modeled as a predictor of each of the inflammatory markers of interest. Age, sex, SES, pubertal development, and BMI serve as control variables on all inflammatory markers. Single-headed arrows indicate directional, predictive relationships. Double-headed curved arrows show correlations. For clarity, control associations are shown in gray, and predictive relationships of interest are shown in black. Sex was dummy coded as 0 = "male", 1 = "female". "BMI" = body mass index; "CRP" = c-reactive protein; "IL" = interleukin; "PDS" = pubertal development stage; "SES" = socioeconomic status; "TNF-α" = tumor necrosis factor alpha.

3. Results

3.1. Sample Characteristics

Of the 68 youths who consented to the study, 10 did not successfully complete the home radon testing. Two of the tests were exposed for too long (i.e., longer than seven days), and the other eight took more than 11 days after completing the test exposure to arrive at the lab for processing. As such, the results from these test kits were unreliable. Additionally, one child who completed the radon testing did not provide a saliva sample. Thus, the final sample was comprised of 57 youths between the ages of 6 and 14 years-old ($M_{age} = 10.57$ years, $SD = 2.55$; 28 male). Youths' BMIs were generally healthy, though there was notable variability as would be expected in the study sample age range ($M = 19.04$, $SD = 5.07$). SES as indexed by the BSMSS suggested that the sample was largely comprised of middle- to upper-class families, with scores ranging from 31.67 to 64.50 ($M = 47.82$, $SD = 7.26$). Of note, one family did report that one person in the home smoked cigarettes.

3.2. Radon and Inflammation Descriptives

Home radon test kits were exposed in participants' homes for an average of 113.95 h ($SD = 26.25$). As expected, over half of homes tested in the current study (67%) had radon concentrations above the EPA action limit for mitigation (Figure 1). Radon levels ranged from <0.3 to 33.3 pCi/L (M = 6.60 pCi/L, SD = 7.43). As mentioned previously, a subset of 21 families completed two radon test kits: one during summer months, and one during winter months. Two of the families did take action to mitigate radon between test periods. However, even including these families in the analysis, we did not detect any significant difference in home radon concentrations acquired during summer (mean rank = 11.05, M = 8.53 pCi/L, SD = 8.10, range = 2.0–32.1 pCi/L) versus winter (mean rank = 9.95, M = 6.14 pCi/L, SD = 5.59, range = 0.3–18.0 pCi/L), Wilcoxon Z = −0.21, p = .84.

Youths in the study lived in their homes for 3.62 years on average (SD = 3.27). We used these data to compute our radon exposure index (ln ([radon concentration × exposure time] + 1)). The final index capturing chronic home radon exposure was normally distributed (M = 2.37, SD = 1.29; Figure 2). Finally, descriptive statistics for each of the inflammatory markers are reported in Table 2. Note that we report data prior to, as well as after natural log transformation.

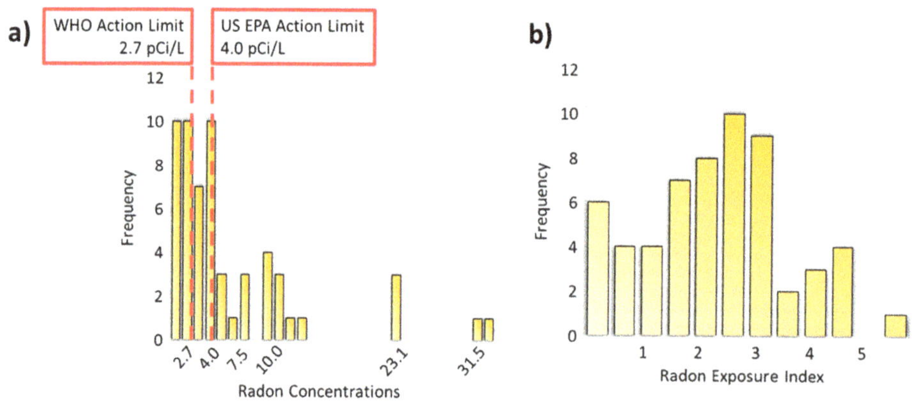

Figure 2. Histogram of acquired home radon concentrations (**a**) and of the computed radon exposure index (**b**). The left histogram is marked with the WHO and the US EPA action limits for indoor radon concentrations (at or above 2.7 and 4.0 pCi/L, respectively). The radon exposure index, calculated at the natural log of home radon concentrations multiplied by the amount of time the person has lived in their home (+1), is relatively normally distributed.

Table 2. Descriptive statistics for each of the inflammatory markers (pg/mL) before and after natural log transformation.

	Original (pg/mL)			**Natural Log Transformed**		
	M	*SD*	*Range*	*M*	*SD*	*Range*
CRP	319.74	1651.54	9.72–12,466.16	3.96	1.36	2.37–9.43
IL-1β	115.00	134.32	24.03–791.56	4.43	0.75	3.22–6.68
IL-6	7.21	10.23	0.32–58.79	1.67	0.86	0.27–4.09
IL-8	881.89	675.63	280.02–4292.42	6.62	0.54	5.64–8.36
TNF-α	5.10	5.18	0.35–25.76	1.58	0.65	0.30–3.29

3.3. Associations between Radon and Inflammation

We tested the degree to which the radon exposure index, a measure of chronic home radon exposure, was associated with each of the inflammatory markers while controlling for the potentially confounding effects of age, sex, and SES. A complete correlation matrix of the variables included in the final model is provided in Table 3. Above and beyond the effects of key demographic factors, radon exposure was significantly associated with both CRP (β = 0.31, b = 0.28, p = 0.007, p_{FDR} = 0.035) and IL-1β (β = 0.33, b = 0.19,

$p = 0.016$, $p_{FDR} = 0.040$; Figure 3 and Table 4). The data suggested that youths with greater radon exposure tended to have higher concentrations of these two inflammatory markers (Figure 3). There was a small, but non-statistically significant association between radon exposure and TNF-α ($\beta = 0.22$, $b = 0.11$, $p = 0.104$, $p_{FDR} = 0.173$), and associations with both IL-8 and IL-6 were near zero. Finally, follow-up comparison of the parameter estimates showed that the relationship between radon exposure and CRP was significantly stronger than the association between radon exposure and IL-8 ($Z = 2.10$, $p = 0.035$), and was marginally stronger than the association between radon exposure and IL-6 ($Z = 1.81$, $p = 0.070$).

Table 3. Correlations among all variables included in the model.

	1	2	3	4	5	6	7	8	9	10	11
1. Radon Exposure	–										
2. CRP	0.34	–									
3. IL-1β	0.28	0.32	–								
4. IL-6	0.10	0.20	0.40	–							
5. IL-8	0.12	0.33	0.53	0.45	–						
6. TNF-α	0.28	0.21	0.79	0.58	0.68	–					
7. Age	0.20	0.21	0.13	0.10	−0.08	−0.04	–				
8. PDS	0.13	0.36	0.17	−0.02	−0.05	−0.12	0.72	–			
9. Sex	−0.10	0.10	−0.03	−0.07	−0.27	−0.29	0.08	0.40	–		
10. SES	−0.18	−0.01	−0.28	−0.29	−0.08	−0.34	−0.14	−0.05	−0.01	–	
11. BMI	0.12	0.51	0.17	−0.04	−0.01	−0.06	0.46	0.64	0.25	−0.11	–

Note: All inflammatory markers were natural log transformed; Sex was dummy coded as 0 = "male", 1 = "female". "BMI" = body mass index; "CRP" = c-reactive protein; "IL" = interleukin; "PDS" = pubertal development stage; "SES" = socioeconomic status; "TNF-α" = tumor necrosis factor alpha.

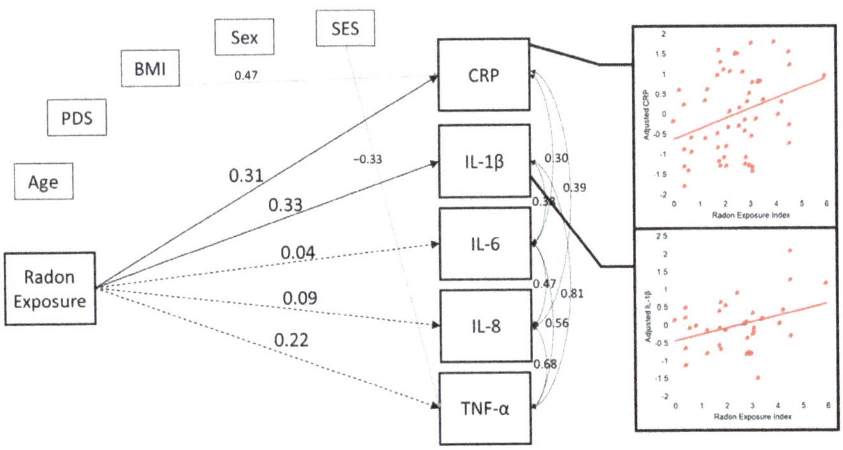

Figure 3. Model results. For simplicity, only associations of interest between radon exposure and inflammatory markers (regardless of statistical significance), and other effects significant at the $p_{FDR} < 0.05$ level are shown. All inflammatory markers were natural log transformed. All parameters reported in the figure are standardized coefficients. Solid lines indicate statistically significant associations at the $p_{FDR} < 0.05$ level, and dashed lines show non-statistically significant associations ($p_{FDR} > 0.05$). Scatterplots show the statistically significant associations between the radon exposure index and natural log transformed CRP (top), and IL-1β (bottom) after adjusting for the effects of age, sex, pubertal development, BMI, and SES. Sex was dummy coded as 0 = "male", 1 = "female". "BMI" = body mass index; "CRP" = c-reactive protein; "IL" = interleukin; "PDS" = pubertal development stage; "SES" = socioeconomic status; "TNF-α" = tumor necrosis factor alpha. Scatterplots of significant associations between radon exposure and inflammatory markers.

Table 4. Complete model results exploring associations between radon exposure and inflammatory markers, controlling for age, sex, SES, BMI, and PDS.

Path	β	b	SE	b/SE	*p*
		CRP On			
Radon Exposure	**0.307**	**0.277**	**0.103**	**2.682**	**0.007**
Age	−0.176	−0.080	0.076	−1.053	0.292
Sex	−0.040	−0.093	0.290	−0.322	0.747
SES	0.076	0.012	0.018	0.686	0.492
PDS	0.169	0.160	0.193	0.827	0.408
BMI	**0.466**	**0.106**	**0.032**	**3.264**	**0.001**
		IL-β on			
Radon Exposure	**0.328**	**0.191**	**0.079**	**2.405**	**0.016**
Age	−0.158	−0.046	0.079	−0.583	0.560
Sex	−0.102	−0.152	0.297	−0.512	0.609
SES	−0.225	−0.023	0.016	−1.455	0.146
PDS	0.246	0.150	0.192	0.783	0.434
BMI	0.047	0.007	0.025	0.274	0.784
		IL-6 on			
Radon Exposure	0.044	0.028	0.090	0.322	0.747
Age	0.133	0.045	0.066	0.675	0.499
Sex	−0.035	−0.061	0.262	−0.231	0.817
SES	**−0.279**	**−0.033**	**0.016**	**−2.121**	**0.034**
PDS	−0.061	−0.043	0.168	−0.254	0.799
BMI	−0.086	−0.014	0.028	−0.515	0.606
		IL-8 on			
Radon Exposure	0.087	0.030	0.057	0.532	0.595
Age	−0.287	−0.050	0.048	−1.044	0.297
Sex	−3.49	−0.313	0.186	−1.686	0.092
SES	−0.092	−0.006	0.010	−0.565	0.572
PDS	0.272	0.100	0.119	0.837	0.403
BMI	0.011	0.001	0.019	0.054	0.957
		TNF-α on			
Radon Exposure	0.217	0.105	0.064	1.626	0.104
Age	−0.146	−0.036	0.064	−0.559	0.576
Sex	−0.284	−0.352	0.230	−1.535	0.125
SES	**−0.326**	**−0.028**	**0.012**	**−2.241**	**0.025**
PDS	0.073	0.037	0.154	0.240	0.811
BMI	−0.029	−0.003	0.020	−0.173	0.863
		CRP with			
IL-1β	0.186	0.114	0.103	1.107	0.268
IL-6	**0.299**	**0.225**	**0.107**	**2.095**	**0.036**
IL-8	**0.390**	**0.152**	**0.068**	**2.232**	**0.026**
TNF-α	0.256	0.127	0.082	1.550	0.121
		IL-1β with			
IL-6	**0.376**	**0.201**	**0.095**	**2.111**	**0.035**
IL-8	**0.539**	**0.149**	**0.054**	**2.747**	**0.006**
TNF-α	**0.807**	**0.284**	**0.077**	**3.668**	**<0.001**
		IL-6 with			
IL-8	**0.471**	**0.161**	**0.063**	**2.539**	**0.011**
TNF-α	**0.561**	**0.243**	**0.079**	**3.091**	**0.002**
		IL-8 with			
TNF-α	**0.675**	**0.152**	**0.048**	**3.190**	**0.001**

Note: Sex was dummy coded as 0 = "male", 1 = "female"; "BMI" = body mass index; "CRP" = c-reactive protein; "IL" = interleukin; "PDS" = pubertal development stage; "SES" = socioeconomic status; "TNF-α" = tumor necrosis factor alpha all inflammatory markers were natural log transformed; statistically significant effects at the $p < 0.05$ level are bolded for clarity.

4. Discussion

The present investigation explored the association between chronic radon exposure and inflammatory activity in a sample of typically developing youths residing in a region of the US that is expected to have high indoor radon concentrations. In line with EPA estimates, over half of households tested in the present study did have radon concentrations exceeding the action limit at or above 4 pCi/L. Further, our key finding was that greater radon exposure, defined as the combination of home radon concentration and chronicity of exposure, was associated with higher concentrations of both CRP and IL-1β. We also detected a similar, trending association between radon exposure and TNF-α. We discuss the implications below.

Overall, the data at least partially supported our primary hypothesis that chronic radon exposure would be associated with higher inflammatory activity in children and adolescents. This was well-aligned with prior works showing that inhaled environmental toxicants (e.g., ultrafine particulate matter) tend to stimulate inflammatory activities [40,41]. Interestingly, we demonstrated specificity in the inflammatory markers with only CRP and IL-1β significantly increasing as a function of increasing home radon exposure. There was also a trending association with TNF-α, which may be significant with a larger study sample. Importantly, these associations were detected above and beyond the effects of socioeconomic, anthropometric, and maturational effects that are known to covary with levels of circulating inflammatory activity [42–44]. These findings generate many new hypotheses as they suggest a potential role of home radon exposure on oral inflammation in youths, which could have implications for neural development, psychological well-being, and overall physical health.

Both CRP and IL-1β have been implicated in the pathophysiology of psychological conditions, particularly major depressive disorders [45–47]. For instance, a longitudinal study in adults showed that individuals with greater levels of circulating CRP at the start of the study had more severe depressive symptomology, and were more likely to have experienced a major depressive relapse nine years later [47]. In another study using rodent models, Fourrier and colleagues [48] demonstrated that overexpression of IL-1β and TNF-α was associated with disrupted learning and memory, both of which play a major role in psychiatric health. The authors suggested that the effects may be explained by alterations to neural tissues, with demonstrated reductions in synaptic plasticity in the hippocampus [48]. Numerous studies have explored the links between neuroinflammatory activities and structural brain health. Indeed, research has shown that higher levels of both CRP and IL-1β are associated with atherosclerosis in the brain, reduced white matter fractional anisotropy (i.e., microstructural integrity), and impaired synaptic plasticity and neurogenesis [48–52]. Further, studies have also shown altered functional neural activity coupled with inflammatory activity in circulation [53–55], which is also associated with decrements in cognition and psychological wellness. Of note, all of these studies have focused on systemic inflammatory activity, and it remains unclear whether the patterns we observed here—in oral tissue—would have similar consequences for the brain.

Recent reviews do suggest that chronically elevated inflammatory activity early in life can and often does have long-term consequences [35,56]. This is critical when considering a new and burgeoning line of research linking chronic radon exposure to neurodegenerative diseases later in life, including multiple sclerosis, Alzheimer's Disease and related dementias, and Parkinson's Disease [57–60]. It is possible that chronic radon exposure starting early in life may, at least in part, contribute to later life neuropathologies due to continuously elevated inflammation. Further work is needed to understand if and how neuroinflammatory profiles may mediate long-term effects of radon exposure on neurological health.

With respect to physical health, inflammation induced by radon exposure, among other potential pollutants, is known to cause cascading effects on immune health overall and can confer risk for physical illnesses [61,62]. For instance, given the damage caused to respiratory epithelium, inhaled radon gas has been linked to the incidence of numerous respiratory diseases in childhood well before any type of carcinogenesis is detected; among the most common health maladies are asthma and chronic obstructive pulmonary disease (COPD) [62,63]. Such illnesses are associated with increasing lifetime burdens for the individual and for society, including school- and work-related losses, medical costs, limitation of physical activities, and overall lower quality of life [64]. Importantly, these chronic conditions are known to be exacerbated in the context of combined exposures to radon and other pollutants (e.g., phthalates, secondhand smoke; [65]). Given that radon is easily mitigated from homes, this seems an easy target to potentially quell the impacts of these lifelong conditions and improve quality of life for a large sect of the population.

In contrast to prior work [25–27], we did not detect associations between radon exposure and either IL-6 or IL-8. Typically, IL-6 has been associated with radon exposure under extreme circumstances. For example, Leng et al. [25] explored IL-6 as a promotor of lung squamous cell carcinoma in former uranium miners who would have been chronically exposed to extreme levels of radon and other potentially toxic compounds in their working environment. Similarly, Narayanan and colleagues [27] studies IL-8 in vitro in alpha-irradiated lung fibroblasts. Such conditions are not necessarily representative of effects in living humans, nor do they reflect the more normative, domestic radon exposure effects that we studied herein. Even so, future works should still consider IL-6 and IL-8 in the context of radon exposure given their known responses to other forms of ionizing radiation [66]. It is possible that the model we tested in our study did not adequately capture the roles of these interleukins. Van Der Meeren and colleagues [67] suggested that IL-6 and IL-8 might be better described as inflammatory mediators, serving as an important process linking environmental exposures and other downstream proinflammatory activities. More recent works have also demonstrated unique and interactive effects of different isoforms of CRP with both IL-6 and IL-8 [68]. Future works would require larger samples to adequately model the potentially complex mediating and moderating effects of different inflammatory activities following chronic home radon exposure.

Another contrasting body of work using controlled dosing in rodent models suggests that radon inhalation may actually inhibit proinflammatory activities by increasing the body's production of reactive oxygen species (ROS; e.g., [69,70]). Indeed, some work suggests potential therapeutic effects of brief, low-level radon exposures in specialized radon spas, particularly for relieving symptoms of inflammatory conditions like rheumatoid arthritis [71,72]. From this and dissenting literature, it is clear that the effects of radon on the body are complex in nature and may vary based on dosage, frequency, and mode of exposure (i.e., inhaled, ingested, etc). Further work is needed to better characterize the full effects of radon exposure on the human immune system.

Generally speaking, the literature provides sound evidence for the biological plausibility that cytokines may circulate in higher quantities in the presence of persistent environmental stressors (e.g., [73,74], including exposure to inhaled toxicants like radon (e.g., [25,27,28]). That said, increases in circulating cytokines is likely not the complete story. Biochemically, many cytokines, including those measured in the current investigation, have relatively short half-lives ranging from 18 min to roughly 30 h depending on the model organism and measurement modality [75,76]. Although the continued exposure to radon gas would likely perpetuate the production of these inflammatory biomarkers, their effects downstream on other biochemical processes like oxidative stress and damage to DNA [60,70,77,78] likely also play important roles in any noted effects of radon on the functioning system/ Future investigations should incorporate additional measurements of the biological impacts of radon exposure in vivo in children.

Before closing, it is important to recognize the limitations of the study. First off, radon was only assessed through short-term home radon test kits (three- to seven-day measurement period). Although the kits utilized herein are generally reliable, they are susceptible to inaccuracies due to intermediary factors including adverse weather at the time of testing, open doors or windows during the exposure period, and damp conditions in the location of the test. Further, historical ecidence suggests that radon concentrations vary across seasons, which would not be captured with such short-term measurements. Although we did not detect seasonal variability in the subset of participants who completed tests in both summer and winter months, this was a small sample and requires additional verification. Deploying multiple test kits simultaneously during multiple seasons would help solidify the reliability of measurements obtained, and help clarify the degree of seasonal variability in the local area; this is a study design limitation that could be easily addressed in future works. Alternatively, studies do suggest that longer-term kits exposed anywhere from 30 days to one year can more accurately measure radon concentrations within a dwelling [21], and these tests do tend to be quite robust against the aforementioned conditions. Second, the test kits used herein only assessed current home radon concentrations. Although we did approximate cumulative exposure based on the current home radon concentration and the amount of time the child had lived in the home, there may be better ways to assess lifetime exposure more accurately, especially if the child has lived in multiple homes for which we had no radon data. It is possible with specialized equipment to measure historic home radon concentrations if the family is able to provide a sample of glass that has been in the dwelling for the desired timeframe (i.e., the child's lifetime; [79–81]). Another limitation is that we only measured radon concentrations in homes. Although the home is the greatest source of indoor radon exposure for most people [82], children are exposed to radon in other venues including schools. Future studies may consider adding measurements of other commonly occupied facilities as well.

Aside from the radon measurements, there were several other limitations to this study. The investigation was cross-sectional, but a longitudinal design would allow for clearer linkages between the chronicity of radon exposure and inflammatory activity. Additionally, inflammatory markers were only assessed using saliva samples, which can be contaminated by oral health, and whose relevance for brain development and psychiatric illness is less clear. Future investigations can address these issues by measuring inflammatory biomarkers in circulating blood, or in biopsied tissue procured medical procedures (e.g., bronchoscopy). Finally, we did not have information on potential physical comorbidities that could have impacted inflammation and/or the effects of radon (e.g., asthma), and we did not control for other pollutants like secondhand smoke exposure (note that only one family reported any smokers in the home). Future investigations would benefit from additional controls for such comorbidities and concomitant exposure.

5. Conclusions

To conclude, the findings of the present study contribute to a growing body of work exploring the inflammatory consequences of an array of environmental toxic exposures. The work we presented herein focused on the inflammatory effects of a specific toxin, home radon, which has been sorely understudied for its potential effects on developing youths. We found significantly elevated concentrations of multiple inflammatory biomarkers as a function of increasing home radon exposure. These data provide insight into the biological effects of chronic radon exposure and support the need for further characterization of the effects of this ubiquitous environmental toxin on the developing body.

Author Contributions: Conceptualization, B.K.T.; methodology, B.K.T.; software, B.K.T.; validation, B.K.T.; formal analysis, B.K.T.; investigation, B.K.T.; resources, B.K.T.; data curation, B.K.T.; writing—original draft preparation, B.K.T. and O.T.V.S.; writing—review and editing, B.K.T., O.T.V.S. and G.E.M.; visualization, B.K.T.; supervision, B.K.T. and G.E.M.; project administration, B.K.T.; funding acquisition, B.K.T. All authors have read and agreed to the published version of the manuscript.

Funding: This work was supported by the National Institutes of Health (P20-GM144641 to BKT). The funders had no role in study design, data collection and analysis, decision to publish, or preparation of the manuscript.

Institutional Review Board Statement: The study was approved by the Institutional Review Board (IRB) at the University of Nebraska Medical Center (IRB# 521-19-EP), and at Pearl IRB and Boys Town National Research Hospital (IRB# 21-Boys-101). All procedures and protocols were in accordance with the Declaration of Helsinki.

Informed Consent Statement: Written informed consent was obtained from all parents of participants, and written assent was obtained from all child participants involved in the study.

Data Availability Statement: The data used herein will be made publicly available on request via the Collaborative Informatics and Neuroimaging Suite (COINS; www.coins.trendscenter.org, Accessed on 19 October 2022) after conclusion of the study data collection.

Acknowledgments: We would like to thank Tony Wilson, who served as an invaluable mentor during this project. We would also like to thank the members of the DICoN Laboratory who worked to acquire the data: Hallie Johnson, Madelyn Willett, Danielle Rice, Ryan Geslinger, and Michaela Frenzel.

Conflicts of Interest: The authors declare no conflict of interest.

References

1. Bearer, C.F. Environmental Health Hazards: How Children Are Different from Adults. *Future Child.* **1995**, *5*, 11–26. [CrossRef] [PubMed]
2. Samet, J.M. Risk Assessment and Child Health. *Pediatrics* **2004**, *113*, 952–956. [CrossRef] [PubMed]
3. Perera, F.; Viswanathan, S.; Whyatt, R.; Tang, D.; Miller, R.L.; Rauh, V. Children's Environmental Health Research—Highlights from the Columbia Center for Children's Environmental Health. *Ann. N. Y. Acad. Sci.* **2006**, *1076*, 15–28. [CrossRef] [PubMed]
4. Silver, M.K.; Li, X.; Liu, Y.; Li, M.; Mai, X.; Kaciroti, N.; Kileny, P.; Tardif, T.; Meeker, J.D.; Lozoff, B. Low-Level Prenatal Lead Exposure and Infant Sensory Function. *Environ. Health* **2016**, *15*, 65. [CrossRef]
5. Peterson, B.S.; Rauh, V.A.; Bansal, R.; Hao, X.; Toth, Z.; Nati, G.; Walsh, K.; Miller, R.L.; Arias, F.; Semanek, D.; et al. Effects of Prenatal Exposure to Air Pollutants (Polycyclic Aromatic Hydrocarbons) on the Development of Brain White Matter, Cognition, and Behavior in Later Childhood. *JAMA Psychiatry* **2015**, *72*, 531. [CrossRef] [PubMed]
6. Myhre, O.; Låg, M.; Villanger, G.D.; Oftedal, B.; Øvrevik, J.; Holme, J.A.; Aase, H.; Paulsen, R.E.; Bal-Price, A.; Dirven, H. Early Life Exposure to Air Pollution Particulate Matter (PM) as Risk Factor for Attention Deficit/Hyperactivity Disorder (ADHD): Need for Novel Strategies for Mechanisms and Causalities. *Toxicol. Appl. Pharmacol.* **2018**, *354*, 196–214. [CrossRef]
7. Sanders, A.P.; Claus Henn, B.; Wright, R.O. Perinatal and Childhood Exposure to Cadmium, Manganese, and Metal Mixtures and Effects on Cognition and Behavior: A Review of Recent Literature. *Curr. Environ. Health Rep.* **2015**, *2*, 284–294. [CrossRef]
8. Chiu, Y.-H.M.; Hsu, H.-H.L.; Coull, B.A.; Bellinger, D.C.; Kloog, I.; Schwartz, J.; Wright, R.O.; Wright, R.J. Prenatal Particulate Air Pollution and Neurodevelopment in Urban Children: Examining Sensitive Windows and Sex-Specific Associations. *Environ. Int.* **2016**, *87*, 56–65. [CrossRef]
9. Liu, J.; Lewis, G. Environmental Toxicity and Poor Cognitive Outcomes in Children and Adults. *J. Environ. Health* **2014**, *76*, 130–138.
10. Bornschein, S.; Hausteiner, C.; Konrad, F.; Förstl, H.; Zilker, T. Psychiatric Morbidity and Toxic Burden in Patients with Environmental Illness: A Controlled Study. *Psychosom. Med.* **2006**, *68*, 104–109. [CrossRef]
11. Lundberg, A. Psychiatric Aspects of Air Pollution. *Otolaryngol. Head Neck Surg.* **1996**, *114*, 227–231. [CrossRef] [PubMed]
12. Kang, J.-K.; Seo, S.; Jin, Y.W. Health Effects of Radon Exposure. *Yonsei Med. J.* **2019**, *60*, 597–603. [CrossRef] [PubMed]
13. Ruano-Ravina, A.; Wakeford, R. The Increasing Exposure of the Global Population to Ionizing Radiation. *Epidemiology* **2020**, *31*, 155–159. [CrossRef]
14. Sethi, T.K.; El-Ghamry, M.N.; Kloecker, G.H. Radon and Lung Cancer. *Clin. Adv. Hematol. Oncol.* **2012**, *10*, 157–164. [PubMed]
15. Vogeltanz-Holm, N.; Schwartz, G.G. Radon and Lung Cancer: What Does the Public Really Know? *J. Environ. Radioact.* **2018**, *192*, 26–31. [CrossRef]
16. United Stated Environmental Protection Agency A Citizens Guide to Radon 2016. Available online: https://www.epa.gov/sites/production/files/2016-12/documents/2016_a_citizens_guide_to_radon.pdf (accessed on 21 October 2022).
17. Laquarta, J. Indoor Air Quality. In *Indoor Environmental Quality*; IntechOpen: London, UK, 2019; p. 13.
18. Stanley, F.K.T.; Irvine, J.L.; Jacques, W.R.; Salgia, S.R.; Innes, D.G.; Winquist, B.D.; Torr, D.; Brenner, D.R.; Goodarzi, A.A. Radon Exposure Is Rising Steadily within the Modern North American Residential Environment, and Is Increasingly Uniform across Seasons. *Sci. Rep.* **2019**, *9*, 18472. [CrossRef] [PubMed]
19. Darby, S.; Hill, D.; Auvinen, A.; Barros-Dios, J.M.; Baysson, H.; Bochicchio, F.; Deo, H.; Falk, R.; Forastiere, F.; Hakama, M.; et al. Radon in Homes and Risk of Lung Cancer: Collaborative Analysis of Individual Data from 13 European Case-Control Studies. *BMJ* **2005**, *330*, 223. [CrossRef]
20. Nebraska Department of Health and Human Services. Radon. Available online: http://dhhs.ne.gov/Pages/Radon.aspx (accessed on 7 December 2020).
21. Novilla, M.L.B.; Johnston, J.D.; Beard, J.D.; Pettit, L.L.; Davis, S.F.; Johnson, C.E. Radon Awareness and Policy Perspectives on Testing and Mitigation. *Atmosphere* **2021**, *12*, 1016. [CrossRef]
22. Wang, Y.; Ju, C.; Stark, A.D.; Teresi, N. Radon Awareness, Testing, and Remediation Survey among New York State Residents. *Health Phys.* **2000**, *78*, 641–647. [CrossRef]
23. Al-Zoughool, M.; Krewski, D. Health Effects of Radon: A Review of the Literature. *Int. J. Radiat. Biol.* **2009**, *85*, 57–69. [CrossRef]
24. Chauhan, V.; Howland, M.; Mendenhall, A.; O'Hara, S.; Stocki, T.J.; McNamee, J.P.; Wilkins, R.C. Effects of Alpha Particle Radiation on Gene Expression in Human Pulmonary Epithelial Cells. *Int. J. Hyg. Environ. Health* **2012**, *215*, 522–535. [CrossRef] [PubMed]

25. Leng, S.; Thomas, C.L.; Snider, A.M.; Picchi, M.A.; Chen, W.; Willis, D.G.; Carr, T.G.; Krzeminski, J.; Desai, D.; Shantu, A.; et al. Radon Exposure, IL-6 Promoter Variants, and Lung Squamous Cell Carcinoma in Former Uranium Miners. *Environ. Health Perspect.* **2016**, *124*, 445–451. [CrossRef] [PubMed]
26. Li, B.-Y.; Tong, J. Adverse Effects Attributed to Long-Term Radon Inhalation in Rats. *J. Toxicol. Environ. Health Part A* **2007**, *70*, 925–930. [CrossRef] [PubMed]
27. Narayanan, P.K.; LaRue, K.E.A.; Goodwin, E.H.; Lehnert, B.E. Alpha Particles Induce the Production of Interleukin-8 by Human Cells. *Radiat. Res.* **1999**, *152*, 57–63. [CrossRef] [PubMed]
28. Robertson, A.; Allen, J.; Laney, R.; Curnow, A. The Cellular and Molecular Carcinogenic Effects of Radon Exposure: A Review. *Int. J. Mol. Sci.* **2013**, *14*, 14024–14063. [CrossRef]
29. Bollen, J.; Trick, L.; Llewellyn, D.; Dickens, C. The Effects of Acute Inflammation on Cognitive Functioning and Emotional Processing in Humans: A Systematic Review of Experimental Studies. *J. Psychosom. Res.* **2017**, *94*, 47–55. [CrossRef]
30. Danese, A.; Lewis, S.J. Psychoneuroimmunology of Early-Life Stress: The Hidden Wounds of Childhood Trauma? *Neuropsychopharmacology* **2017**, *42*, 99–114. [CrossRef]
31. DiSabato, D.J.; Quan, N.; Godbout, J.P. Neuroinflammation: The Devil Is in the Details. *J. Neurochem.* **2016**, *139* (Suppl. S2), 136–153. [CrossRef]
32. Furman, D.; Campisi, J.; Verdin, E.; Carrera-Bastos, P.; Targ, S.; Franceschi, C.; Ferrucci, L.; Gilroy, D.W.; Fasano, A.; Miller, G.W.; et al. Chronic Inflammation in the Etiology of Disease across the Life Span. *Nat. Med.* **2019**, *25*, 1822–1832. [CrossRef]
33. Miller, A.H.; Raison, C.L. The Role of Inflammation in Depression: From Evolutionary Imperative to Modern Treatment Target. *Nat. Rev. Immunol.* **2016**, *16*, 22–34. [CrossRef]
34. Ehrlich, K.B.; Miller, G.E.; Rogosch, F.A.; Cicchetti, D. Maltreatment Exposure across Childhood and Low-Grade Inflammation: Considerations of Exposure Type, Timing, and Sex Differences. *Dev. Psychobiol.* **2021**, *63*, 529–537. [CrossRef] [PubMed]
35. Loftis, J.M.; Byrne, M.L.; Measelle, J. Special Issue: Contexts and Consequences of Childhood Inflammation. *Brain Behav. Immun.* **2020**, *86*, 1–3. [CrossRef] [PubMed]
36. Miller, G.; Chen, E.; Cole, S.W. Health Psychology: Developing Biologically Plausible Models Linking the Social World and Physical Health. *Annu. Rev. Psychol.* **2009**, *60*, 501–524. [CrossRef] [PubMed]
37. Petersen, A.C.; Crockett, L.; Richards, M.; Boxer, A. A Self-Report Measure of Pubertal Status: Reliability, Validity, and Initial Norms. *J. Youth Adolesc.* **1988**, *17*, 117–133. [CrossRef] [PubMed]
38. Shirtcliff, E.A.; Dahl, R.E.; Pollak, S.D. Pubertal Development: Correspondence Between Hormonal and Physical Development: Hormonal Correlates of Pubertal Stage. *Child Dev.* **2009**, *80*, 327–337. [CrossRef] [PubMed]
39. Barratt, W. *The Barratt Simplified Measure of Social Status (BSMSS): Measuring SES*; Indiana State University: Terre Haute, IN, USA, 2006.
40. Guxens, M.; Sunyer, J. A Review of Epidemiological Studies on Neuropsychological Effects of Air Pollution. *Swiss Med. Wkly.* **2012**, *141*, w13322. [CrossRef]
41. Jankowska-Kieltyka, M.; Roman, A.; Nalepa, I. The Air We Breathe: Air Pollution as a Prevalent Proinflammatory Stimulus Contributing to Neurodegeneration. *Front. Cell. Neurosci.* **2021**, *15*, 647643. [CrossRef]
42. Decker, M.-L.; Grobusch, M.P.; Ritz, N. Influence of Age and Other Factors on Cytokine Expression Profiles in Healthy Children—A Systematic Review. *Front. Pediatr.* **2017**, *5*, 255. [CrossRef]
43. O'Connor, M.-F.; Bower, J.E.; Cho, H.J.; Creswell, J.D.; Dimitrov, S.; Hamby, M.E.; Hoyt, M.A.; Martin, J.L.; Robles, T.F.; Sloan, E.K.; et al. To Assess, to Control, to Exclude: Effects of Biobehavioral Factors on Circulating Inflammatory Markers. *Brain Behav. Immun.* **2009**, *23*, 887–897. [CrossRef]
44. Spyridaki, E.C.; Avgoustinaki, P.D.; Margioris, A.N. Obesity, Inflammation and Cognition. *Curr. Opin. Behav. Sci.* **2016**, *9*, 169–175. [CrossRef]
45. Farooq, R.K.; Asghar, K.; Kanwal, S.; Zulqernain, A. Role of Inflammatory Cytokines in Depression: Focus on Interleukin-1β (Review). *Biomed. Rep.* **2017**, *6*, 15–20. [CrossRef] [PubMed]
46. Tabatabaeizadeh, S.-A.; Abdizadeh, M.F.; Meshkat, Z.; Khodashenas, E.; Darroudi, S.; Fazeli, M.; Ferns, G.A.; Avan, A.; Ghayour-Mobarhan, M. There Is an Association between Serum High-Sensitivity C-Reactive Protein (Hs-CRP) Concentrations and Depression Score in Adolescent Girls. *Psychoneuroendocrinology* **2018**, *88*, 102–104. [CrossRef] [PubMed]
47. Zainal, N.H.; Newman, M.G. Increased Inflammation Predicts Nine-Year Change in Major Depressive Disorder Diagnostic Status. *J. Abnorm. Psychol.* **2021**, *130*, 829–840. [CrossRef] [PubMed]
48. Fourrier, C.; Singhal, G.; Baune, B.T. Neuroinflammation and Cognition across Psychiatric Conditions. *CNS Spectr.* **2019**, *24*, 4–15. [CrossRef]
49. Avci, A.Y.; Lakadamyali, H.; Arikan, S.; Benli, U.S.; Kilinc, M. High Sensitivity C-Reactive Protein and Cerebral White Matter Hyperintensities on Magnetic Resonance Imaging in Migraine Patients. *J. Headache Pain* **2015**, *16*, 9. [CrossRef]
50. Byrne, M.L.; Whittle, S.; Allen, N.B. The Role of Brain Structure and Function in the Association Between Inflammation and Depressive Symptoms: A Systematic Review. *Psychosom. Med.* **2016**, *78*, 389–400. [CrossRef]
51. Kempf, S.J.; Buratovic, S.; von Toerne, C.; Moertl, S.; Stenerlöw, B.; Hauck, S.M.; Atkinson, M.J.; Eriksson, P.; Tapio, S. Ionising Radiation Immediately Impairs Synaptic Plasticity-Associated Cytoskeletal Signalling Pathways in HT22 Cells and in Mouse Brain: An In Vitro/In Vivo Comparison Study. *PLoS ONE* **2014**, *9*, e110464. [CrossRef]

52. Kempf, S.J.; Casciati, A.; Buratovic, S.; Janik, D.; von Toerne, C.; Ueffing, M.; Neff, F.; Moertl, S.; Stenerlöw, B.; Saran, A.; et al. The Cognitive Defects of Neonatally Irradiated Mice Are Accompanied by Changed Synaptic Plasticity, Adult Neurogenesis and Neuroinflammation. *Mol. Neurodegener.* **2014**, *9*, 57. [CrossRef]
53. Felger, J.C.; Li, Z.; Haroon, E.; Woolwine, B.J.; Jung, M.Y.; Hu, X.; Miller, A.H. Inflammation Is Associated with Decreased Functional Connectivity within Corticostriatal Reward Circuitry in Depression. *Mol. Psychiatry* **2016**, *21*, 1358–1365. [CrossRef]
54. Inagaki, T.K.; Muscatell, K.A.; Irwin, M.R.; Cole, S.W.; Eisenberger, N.I. Inflammation Selectively Enhances Amygdala Activity to Socially Threatening Images. *Neuroimage* **2012**, *59*, 3222–3226. [CrossRef]
55. Nusslock, R.; Brody, G.H.; Armstrong, C.C.; Carroll, A.L.; Sweet, L.H.; Yu, T.; Barton, A.W.; Hallowell, E.S.; Chen, E.; Higgins, J.P.; et al. Higher Peripheral Inflammatory Signaling Associated with Lower Resting-State Functional Brain Connectivity in Emotion Regulation and Central Executive Networks. *Biol. Psychiatry* **2019**, *86*, 153–162. [CrossRef]
56. Jiang, N.M.; Tofail, F.; Ma, J.Z.; Haque, R.; Kirkpatrick, B.; Nelson, C.A.; Petri, W.A. Early Life Inflammation and Neurodevelopmental Outcome in Bangladeshi Infants Growing Up in Adversity. *Am. J. Trop. Med. Hyg.* **2017**, *97*, 974–979. [CrossRef] [PubMed]
57. Kempf, S.J.; Azimzadeh, O.; Atkinson, M.J.; Tapio, S. Long-Term Effects of Ionising Radiation on the Brain: Cause for Concern? *Radiat. Environ. Biophys.* **2013**, *52*, 5–16. [CrossRef] [PubMed]
58. Musella, A.; Gentile, A.; Rizzo, F.R.; De Vito, F.; Fresegna, D.; Bullitta, S.; Vanni, V.; Guadalupi, L.; Stampanoni Bassi, M.; Buttari, F.; et al. Interplay between Age and Neuroinflammation in Multiple Sclerosis: Effects on Motor and Cognitive Functions. *Front. Aging Neurosci.* **2018**, *10*, 238. [CrossRef] [PubMed]
59. Wang, Q.-Q.; Yin, G.; Huang, J.-R.; Xi, S.-J.; Qian, F.; Lee, R.-X.; Peng, X.-C.; Tang, F.-R. Ionizing Radiation-Induced Brain Cell Aging and the Potential Underlying Molecular Mechanisms. *Cells* **2021**, *10*, 3570. [CrossRef]
60. Zhang, Y.; Lu, L.; Chen, C.; Field, R.W.; D'Alton, M.; Kahe, K. Does Protracted Radon Exposure Play a Role in the Development of Dementia? *Environ. Res.* **2022**, *210*, 112980. [CrossRef]
61. Dozor, A.J. The Role of Oxidative Stress in the Pathogenesis and Treatment of Asthma. *Ann. N. Y. Acad. Sci.* **2010**, *1203*, 133–137. [CrossRef]
62. Mukharesh, L.; Greco, K.F.; Banzon, T.; Koutrakis, P.; Li, L.; Hauptman, M.; Phipatanakul, W.; Gaffin, J.M. Environmental Radon and Childhood Asthma. *Pediatr. Pulmonol.* **2022**, *57*, 3165–3168. [CrossRef]
63. Pando-Sandoval, A.; Ruano-Ravina, A.; Candal-Pedreira, C.; Rodríguez-García, C.; Represas-Represas, C.; Golpe, R.; Fernández-Villar, A.; Pérez-Ríos, M. Risk Factors for Chronic Obstructive Pulmonary Disease in Never-Smokers: A Systematic Review. *Clin. Respir. J.* **2022**, *16*, 261–275. [CrossRef]
64. Ferrante, G.; La Grutta, S. The Burden of Pediatric Asthma. *Front. Pediatr.* **2018**, *6*, 186. [CrossRef]
65. Kwan, W.S.; Nikezic, D.; Roy, V.A.L.; Yu, K.N. Multiple Stressor Effects of Radon and Phthalates in Children: Background Information and Future Research. *Int. J. Environ. Res. Public Health* **2020**, *17*, 2898. [CrossRef] [PubMed]
66. Di Maggio, F.M.; Minafra, L.; Forte, G.I.; Cammarata, F.P.; Lio, D.; Messa, C.; Gilardi, M.C.; Bravatà, V. Portrait of Inflammatory Response to Ionizing Radiation Treatment. *J. Inflamm.* **2015**, *12*, 14. [CrossRef]
67. Van DerMeeren, A.; Bertho, J.-M.; Vandamme, M.; Gaugler, M.-H. Ionizing Radiation Enhances IL-6 and IL-8 Production by Human Endothelial Cells. *Mediat. Inflamm.* **1997**, *6*, 185–193. [CrossRef]
68. Sproston, N.R.; Ashworth, J.J. Role of C-Reactive Protein at Sites of Inflammation and Infection. *Front. Immunol.* **2018**, *9*, 754. [CrossRef]
69. Kataoka, T. Study of Antioxidative Effects and Anti-Inflammatory Effects in Mice Due to Low-Dose X-Irradiation or Radon Inhalation. *J. Radiat. Res.* **2013**, *54*, 587–596. [CrossRef] [PubMed]
70. Kataoka, T.; Etani, R.; Kanzaki, N.; Kobashi, Y.; Yunoki, Y.; Ishida, T.; Sakoda, A.; Ishimori, Y.; Yamaoka, K. Radon Inhalation Induces Manganese-Superoxide Dismutase in Mouse Brain via Nuclear Factor-KB Activation. *J. Radiat. Res.* **2017**, *58*, 887–893. [CrossRef]
71. Zdrojewicz, Z.; Strzelczyk, J. (Jodi) Radon Treatment Controversy. *Dose-Response* **2006**, *4*, 106–118. [CrossRef]
72. Franke, A.; Reiner, L.; Pratzel, H.G.; Franke, T.; Resch, K.L. Long-term Efficacy of Radon Spa Therapy in Rheumatoid Arthritis—A Randomized, Sham-controlled Study and Follow-up. *Rheumatology* **2000**, *39*, 894–902. [CrossRef]
73. Gao, J.; Wei, Q.; Pan, R.; Yi, W.; Xu, Z.; Duan, J.; Tang, C.; He, Y.; Liu, X.; Song, S.; et al. Elevated Environmental PM2.5 Increases Risk of Schizophrenia Relapse: Mediation of Inflammatory Cytokines. *Sci. Total Environ.* **2021**, *753*, 142008. [CrossRef]
74. Prata, J.C.; da Costa, J.P.; Lopes, I.; Duarte, A.C.; Rocha-Santos, T. Environmental Exposure to Microplastics: An Overview on Possible Human Health Effects. *Sci. Total Environ.* **2020**, *702*, 134455. [CrossRef]
75. Kuribayashi, T. Elimination Half-Lives of Interleukin-6 and Cytokine-Induced Neutrophil Chemoattractant-1 Synthesized in Response to Inflammatory Stimulation in Rats. *Lab. Anim. Res.* **2018**, *34*, 80–83. [CrossRef] [PubMed]
76. Liu, C.; Chu, D.; Kalantar-Zadeh, K.; George, J.; Young, H.A.; Liu, G. Cytokines: From Clinical Significance to Quantification. *Adv. Sci.* **2021**, *8*, 2004433. [CrossRef]
77. Kobashi, Y. Comparison of Antioxidative Effects between Radon and Thoron Inhalation in Mouse Organs. *Radiation and Environmental Biophysics* **2020**, *59*, 473–482. [CrossRef] [PubMed]
78. Nie, J.-H.; Chen, Z.-H.; Liu, X.; Wu, Y.-W.; Li, J.-X.; Cao, Y.; Hei, T.K.; Tong, J. Oxidative Damage in Various Tissues of Rats Exposed to Radon. *J. Toxicol. Environ. Health Part A* **2012**, *75*, 694–699. [CrossRef]

79. Lagarde, F.; Falk, R.; Almrén, K.; Nyberg, F.; Svensson, H.; Pershagen, G. Glass-Based Radon-Exposure Assessment and Lung Cancer Risk. *J. Expo. Sci. Environ. Epidemiol.* **2002**, *12*, 344–354. [CrossRef] [PubMed]
80. Mahaffey, J.A.; Parkhurst, M.A.; James, A.C.; Cross, F.T.; Alavanja, M.C.R.; Boice, J.D.; Ezrine, S.; Henderson, P.; Brownson, R.C. Estimating Past Exposure to Indoor Radon from Household Glass. *Health Phys.* **1993**, *64*, 381–391. [CrossRef] [PubMed]
81. Samuelsson, C. Retrospective Determination of Radon in Houses. *Nature* **1988**, *334*, 338–340. [CrossRef]
82. Chen, J. Canadian Lung Cancer Relative Risk from Radon Exposure for Short Periods in Childhood Compared to a Lifetime. *Int. J. Environ. Res. Public Health* **2013**, *10*, 1916–1926. [CrossRef]

toxics

MDPI

Article

Organotin Antifouling Compounds and Sex-Steroid Nuclear Receptor Perturbation: Some Structural Insights

Mohd A. Beg [1,*,†], Md A. Beg [2,†], Ummer R. Zargar [3], Ishfaq A. Sheikh [1,4], Osama S. Bajouh [5], Adel M. Abuzenadah [4,6] and Mohd Rehan [4,6]

1 Reproductive Biology Laboratory, King Fahd Medical Research Center, King Abdulaziz University, Jeddah 21589, Saudi Arabia
2 Centre for Interdisciplinary Research in Basic Sciences, Jamia Millia Islamia University, New Delhi 110025, India
3 Department of Zoology, Government Degree College, Anantnag 192101, India
4 Department of Medical Laboratory Technology, Faculty of Applied Medical Sciences, King Abdulaziz University, Jeddah 21589, Saudi Arabia
5 Department of Obstetrics and Gynecology, Faculty of Medicine, King Abdulaziz University, Jeddah 21859, Saudi Arabia
6 King Fahd Medical Research Center, King Abdulaziz University, Jeddah 21589, Saudi Arabia
* Correspondence: mabeg51@gmail.com or mbeg@kau.edu.sa; Tel.: +966-552822451
† These authors contributed equally to this work.

Abstract: Organotin compounds (OTCs) are a commercially important group of organometallic compounds of tin used globally as polyvinyl chloride stabilizers and marine antifouling biocides. Worldwide use of OTCs has resulted in their ubiquitous presence in ecosystems across all the continents. OTCs have metabolic and endocrine disrupting effects in marine and terrestrial organisms. Thus, harmful OTCs (tributyltin) have been banned by the International Convention on the Control of Harmful Antifouling Systems since 2008. However, continued manufacturing by non-member countries poses a substantial risk for animal and human health. In this study, structural binding of common commercial OTCs, tributyltin (TBT), dibutyltin (DBT), monobutyltin (MBT), triphenyltin (TPT), diphenyltin (DPT), monophenyltin (MPT), and azocyclotin (ACT) against sex-steroid nuclear receptors, androgen receptor (AR), and estrogen receptors (ERα, ERβ) was performed using molecular docking and MD simulation. TBT, DBT, DPT, and MPT bound deep within the binding sites of AR, ERα, and Erβ, showing good dock score, binding energy and dissociation constants that were comparable to bound native ligands, testosterone and estradiol. The stability of docking complex was shown by MD simulation of organotin/receptor complex with RMSD, RMSF, Rg, and SASA plots showing stable interaction, low deviation, and compactness of the complex. A high commonality (50–100%) of interacting residues of ERα and ERβ for the docked ligands and bound native ligand (estradiol) indicated that the organotin compounds bound in the same binding site of the receptor as the native ligand. The results suggested that organotins may interfere with the natural steroid/receptor binding and perturb steroid signaling.

Keywords: organotins; butyltin; phenyltin; azocyclotin; androgen receptor; estrogen receptor; molecular docking; MD simulation; endocrine disruption

Citation: Beg, M.A.; Beg, M.A.; Zargar, U.R.; Sheikh, I.A.; Bajouh, O.S.; Abuzenadah, A.M.; Rehan, M. Organotin Antifouling Compounds and Sex-Steroid Nuclear Receptor Perturbation: Some Structural Insights. *Toxics* **2023**, *11*, 25. https://doi.org/10.3390/toxics11010025

Academic Editors: Esref Demir, Sam Kacew and Yang Song

Received: 5 December 2022
Accepted: 21 December 2022
Published: 27 December 2022

1. Introduction

Organotin compounds (OTCs), also called as stannanes, are a large class of organometallic compounds having at least one tin atom covalently bound to a carbon atom with a general formula 'RSnX' where 'R' represents an organic group, i.e., alkyl, phenyl, etc., and 'X' an anion such as chloride, fluoride, oxide, etc. [1–5]. Tin metal and its alloys have been used by humans for more than 6000 years, however, the first OTC was synthesized only about 175 years back in the year 1853, and, for nearly another 100 years, OTCs were not found to have any industrial utility for want of commercial applications [5,6]. As of now, there are more than 800 OTCs known, and, barring a few, all of them are anthropogenic [7]. OTCs constitute one the largest group of organometallic compounds in commercial use globally, and the United States and China are the largest consumers of OTCs, followed by Western Europe, the Middle East, and others [8]. The downside of the increased industrial applications and use has been that considerable amounts of OTCs have entered various ecosystems, and OTCs are now ubiquitous in the environment. The worldwide higher use and demand have resulted in an increase in global production of OTCs during the last several decades. Overall, the maximum utilization of OTCs is as polyvinyl chloride (PVC) stabilizers (about 70–80% of global consumption) and remaining as biocides [5,6,9].

OTCs are classified according to the number of organic functional groups on the tin atom or the number of 'carbon-tin' bonds, e.g., mono-, di-, tri-, and tetra-organotin compounds [5]. Generally, mono- and di-substituted OTCs, e.g., dibutyltin (DBT), diphenyltin (DPT), etc., are used as heat stabilizers for PVC plastics, lubricating oils, hydrogen peroxide, and polyolefins and as catalysts in the production of polyurethane foam, polymers, esters, plastisol prints, rubber, adhesives, etc., with applications in apparel and footwear industries [2,5]. Tri-substituted OTCs, e.g., tributyltin (TBT), triphenyltin (TPT), etc., have been used extensively as biocides, such as antifouling paints for boats, timber preservatives, preservatives in textiles, leathers and synthetic leathers, fungicides in crops including potatoes, sugar beets, and pecans, and as pesticides [2,5,10]. OTCs are also utilized in silicone-based finishes having elastomeric and water repellency properties. The seven most common OTCs commercially available for a variety of industrial applications are TBT, DBT, monobutyltin (MBT), TPT, DPT, monophenyltin (MPT), and azocyclotin (ACT).

Environmental contamination with OTCs has been reported to be mainly through agricultural runoffs and marine aquatic environment due to leaching from antifouling paints, treated timber, PVC pipes, and other contaminated material [4,5,10]. Biodegradation half-lives of organotins are generally shorter in tri-substituted organotins compared to di-substituted organotins, whereas monosubstituted organotins have the longest half-lives. The half-lives are longer in both seawater and soil/sediment than in fresh water and can range from six months to 15 years [5]. The 'C-Sn' bond of organotins is comparatively stable, as the anionic group is reported to dissociate easily on hydrolysis in water. Human exposure occurs through drinking water and eating food, especially aquatic food such as fish and other marine animals. The global terrestrial and aquatic contamination and the adverse effects on non-target organisms including abnormalities in shell calcification in oysters and the masculinization of female gastropods—imposex has resulted in outlawing the use of OTs, especially tributyltin in marine paints in 2008 under the International Convention on the Control of Harmful Antifouling Systems on Ships and World Health Organization [5,10,11]. However, developing nations that are not members of the International Maritime Organization are still permitted to use OTCs for industrial and agricultural purposes. Organotins have been found in household dust in many countries, including the United States and Germany [12,13]. Due to an increase in the usage of antifouling paints containing organotin compounds, organotin contaminations have also been observed in Saudi Arabia in recent years in fishing harbors of various coastal sites in the Eastern Province (Jubail, Khobar, and Qatif) [14,15]; the estimations of the contamination have shown more phenyltins than other organotins. The detection of high concentrations of hazardous triorganotins in the commercially important fish species caught from Arabian Gulf is of particular concern for human health in Saudi Arabia [16].

Limited epidemiological studies have been reported on the exposure of humans to organotin toxicity. In the United States, MBT, DBT, and TBT were detected in 70% of the human blood samples that were tested [17]. The potential toxicological significance of OTCs to humans on the basis of animal experiments indicates diverse health problems including immunosuppression, endocrine problems, neurotoxicity, metabolic and enzymatic problems, ocular, cardiovascular, gastrointestinal, reproductive, developmental, and several other problems [18]. OTCs cause morphological and functional changes in a number of tissues in animals that are involved in the regulation of endocrine function and metabolism in mammals, including the hypothalamus, pituitary, pancreas, gonads, adipose tissue, adrenal, and thyroid glands [19,20]. In regard to reproductive problems in humans, organotin exposure has been associated with congenital abnormalities in neonates such as cryptorchidism and hormone disbalance for LH and testosterone in newborn boys [21]. Organotins have been also reported to cause metabolic dysfunction in animal models similar to polycystic ovarian syndrome (PCOS) in women [22]. In animals, organotin exposure leads to phenotypic abnormalities, aberrant changes in hormones, transcriptome and proteome modifications, behavioral changes, and characteristic sexual organ masculinization called imposex, in which female gastropods develop penile structure and vas deferens [23,24]. In rats, organotins are associated with reduced sperm viability and motility, abnormalities of spermatids and spermatozoa, testicular necrosis, changes in the blood–testicular barrier, and hormone disbalance in males [24–26], as well as delayed follicular development, irregular estrous cycles, alterations in steroidogenic enzymes, abnormal adipogenesis, decline in fertility, failure of implantation, lower chances of conception, and other reproductive abnormalities in females [27–29].

The available literature indicates that very few epidemiological or clinical studies have been reported on the effects of organotins on human health despite the overwhelming evidence of the adverse effects on laboratory and marine species. Computational methods have been increasingly used for the prediction of binding pose and molecular interactions of ligands with their target protein molecules for the characterization of interactions, designing novel inhibitors, and/or as an aid for designing experimental and clinical studies [30–32]. The present study was performed to characterize the structural binding interactions of seven commonly available OTCs, viz., TBT, DBT, MBT, TPT, DPT, MPT, and ACT, against sex-steroid nuclear receptors, i.e., androgen receptor (AR) and estrogen receptors (ERα, ERβ). AR and ER are soluble proteins and function as intracellular transcription factors. Testosterone and dihydrotestosterone are the main steroid ligands of AR. In addition to the development of male sexual organs, maturation of male sexual organs and spermatogenesis, AR signaling plays a crucial role in a number of physiological and developmental processes related to male physiology. ER regulates the transcription of numerous genes, and its primary role is in the development of female phenotype, development of female reproductive tract, regulation of folliculogenesis, uterine development, and in other processes related to reproductive cycle. The suggested hypothesis is that OTCs may interfere with the natural interaction between ligands and proteins by binding to sex steroid nuclear receptors in the body, which might result in dysfunction of sex steroids target organs.

2. Materials and Methods

2.1. Data Retrieval

The three-dimensional crystal structures of the human sex-steroid hormone receptors were retrieved from the Protein Data Bank (PDB; https://www.rcsb.org, accessed on 4 December 2022) as AR (PDB ID: 2AM9), ERα (PDB ID: 5DXB), and ERβ (PDB ID: 5TOA). The two-dimensional structures of OTCs (TBT, DBT, MBT, TPT, DPT, MPT, and ACT) as illustrated (Figure 1) were retrieved from the PubChem database (https://pubchem.ncbi.nlm.nih.gov, accessed on 4 December 2022). PyMOL graphic interface was used for illustration and analysis of hormone receptors with their bound native ligands and binding pocket studies [33].

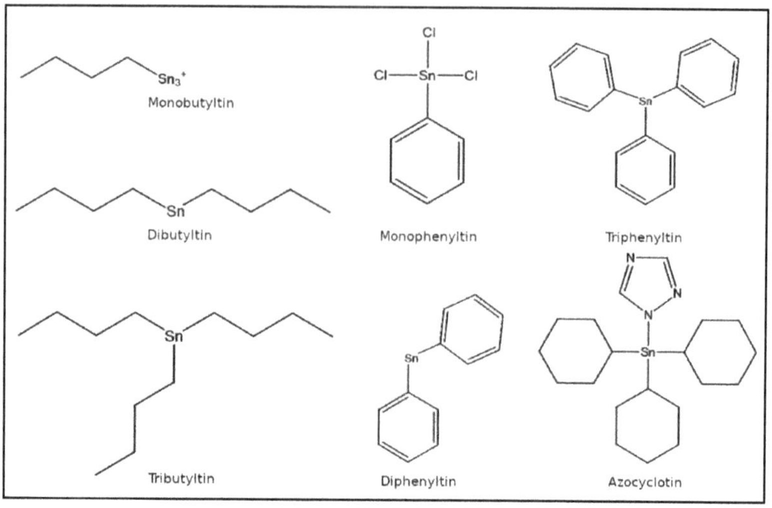

Figure 1. Two-dimensional structures of organotin compounds.

2.2. Molecular Docking

Molecular docking of the organotin compounds in the ligand binding sites of AR, ERα, and ERβ was carried out using Dock v.6.5 [34]. The initial structure preparation of the ligands and the receptors was performed using Chimera v.1.6.2 [35]. The native bound ligands were used as clue for the binding site and the residues around 10 Å of the native ligands were used as the binding site region for generating docking grid. Finally, the molecular docking of the compounds was performed using the rigid ligand docking option with default parameters and the docking poses and predicted results were generated, as detailed previously [36].

2.3. Protein–Ligand Complex Analysis

Discovery Studio (https://www.3ds.com/products-services/biovia/, accessed on 4 December 2022) was exploited for binding pose analysis and the illustrations of two-dimensional ligand-receptor interactions were prepared. The binding analysis of a two-dimensional interpretation of interacting residues and interfaces with high-quality protein and ligand contacts was generated in BIOVIA (https://www.3ds.com/products-services/biovia/, accessed on 4 December 2022) [37]. The result was an instructive depiction of the intermolecular interactions and their strengths, which included hydrogen bonds, hydrophobic contacts, and atom accessibilities.

2.4. Binding Energy and Dissociation Constant

The dock score is the score obtained directly from the docking software Dock v. 6.9 used in this study and demonstrates how fit the ligand is in the binding site. In addition to the dock score, the binding energy and dissociation constant terms were also evaluated for the ligand–protein complexes using another software, X-Score v. 1.2.11 [38].

2.5. Binding-Pose Comparison Analyses

The binding poses of the organotin compounds were compared to that of the native ligands in order to check if the docked organotin ligands were bound to the same ligand binding sites of AR, ERα, and ERβ where the respective native ligand were bound. The 3D structure of AR, ERα, and ERβ proteins had bound ligands (native) whose binding pocket ensured that the docking site chosen was unerring. Further, within the binding site, the interacting residues common for the organotin compounds and the native ligand were compared.

2.6. Molecular Dynamics (MD) Simulation Analysis

MD simulation was carried out with the Schrodinger Desmond tool [39] to obtain an insight into the binding stability of the docked complex of AR and TBT. Prior to MD simulation, docked complex structures were minimized using the protein preparation wizard of Schrodinger where the hydrogen bond network was optimized at pH 7.4, and final restrained minimization was performed using the OPLS3e force field [39]. Further minimized structures were incorporated within an orthorhombic box solvated with a TIP3P water model using a system builder module, and then 0.15 M NaCl counter ions were added to neutralize the system. All prepared systems were relaxed before the MD simulation by a series of energy minimization and short MD simulations. Finally, the MD simulations were subjected to a 100 ns time period for individual systems, and the coordinates were saved at an interval of 50 ps at 300 K temperature with 1.0325 bar pressure. The simulation event analysis module in Desmond was utilized further to analyze the simulation results. The dynamic profile of the AR/TBT complex was assessed by root mean square deviation (RMSD) from the 100 ns trajectory. Root mean square fluctuation, i.e., protein residue fluctuation, of the AR/TBT complex was monitored. In addition, a radius of gyration (Rg) calculation over MD simulation was performed, which is a major indication of structural compactness. Finally, an estimation of the total change in solvent accessibility surface area (SASA) of AR/TBT complex simulation was performed. The method has been detailed previously [40,41].

3. Results

3.1. Molecular Docking Analyses

The crystal structures of sex steroid hormone receptors, AR, ERα, and ERβ, containing the bound native ligands in the respective binding sites are presented in Figure 2. The analysis of binding strength scores of the organotins, i.e., TBT, DBT, MBT, TPT, DPT, MPT, and ACT in the present study with sex steroid receptors, AR, ERα, and ERβ and the interaction analysis of the ligand–protein complexes showed that four of seven organotin ligands bound well within the binding sites of the receptors (Table 1). The exception was MBT, which showed a negative but low value of dock score for all the three receptors, indicating a weak binding. Further analysis for other parameters for this organotin was not performed. In addition, TPT showed a high positive value of dock score for all the three receptors, indicating that it did bind to the receptors (a good binding is indicated by a higher negative dock score) and, lastly, ACT somehow could not be docked even after varying the default parameters. Therefore, MBT, TPT, and ACT were not considered for molecular interactions and binding pose analysis with the three receptors. Only four organotins, i.e., TBT, DBT, DPT, and MPT, were included in further analysis.

3.2. Molecular Docking of Organotin Compounds with AR

Intermolecular interactions of four organotin ligands, i.e., TBT, DBT, DPT, and MPT in complex with AR were analyzed and identified. All docked ligands were found to have similar binding poses to the native ligand, thus providing credence to the docking accuracy. All the organotin ligands bound well within the binding site of AR. The values for the dock score, binding energy, and dissociation constant were comparable among the ligands and were lower but close to native ligand, testosterone, indicating tight binding and similar binding strength to AR (Table 1). The two-dimensional interaction maps of amino acid residues of AR interacting with four organotin ligands and native bound ligand, testosterone, are illustrated in Figure 3.

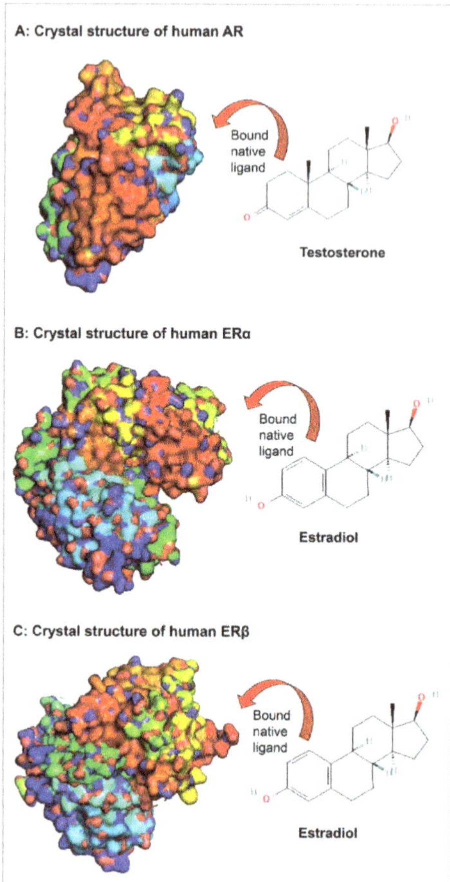

Figure 2. Surface representation of crystal structure of sex steroid hormone receptors containing the bound native ligands in the respective binding sites. Panel (**A**): crystal structure of human androgen receptor (AR) ligand binding domain in complex with testosterone; Panel (**B**): crystal structure of human estrogen receptor α (ERα) ligand binding domain in complex with estradiol; and Panel (**C**): crystal structure of human estrogen receptor β (ERβ) ligand binding domain in complex with estradiol.

The amino-acid residue interactions in the docking poses of organotin ligands, TBT, DBT, DPT, and MPT with AR are presented (Figure 3A–D). The docking pose analysis revealed that TBT interacted with nine amino acid residues of AR, i.e., Leu-701, Leu-707, Met-745, Val-746, Met-749, Phe-764, Met-780, Met-787, and Phe-876, and DBT interacted with six amino-acid residues of AR, i.e., Val-746, Met-749, Phe-764, Met-780, Leu-873, and Phe-876. In addition, DPT interacted with four amino acid residues of AR, i.e., Leu-704, Leu-707, Met-742, and Met-745, whereas MPT interacted with one amino acid residue (Leu-704) of AR. These residues were significantly important for the AR/organotin ligand bonding interaction in the active sites. The true conformation of bound native ligand testosterone in the binding site of AR is shown (Figure 2A), and the interacting amino acid residues are illustrated (Figure 3E). For testosterone, 10 amino acid residues were interacting with AR, i.e., Leu-704, Asn-705, Gln-711, Trp-741, Met-742, Met-745, Arg-752, Met-780, Leu-873, and Thr-877. Four hydrogen bonds were formed by Asn-705, Gln-711, Arg-752, and Thr-877, in addition to various other molecular interactions. When the interacting amino acid residues of AR for each of the four organotins were compared with those for native ligand (testosterone), two of nine interacting residues (Met-745, Met-780) for TBT, two of six interacting residues for DBT, three of four interacting residues for DPT, and the single interacting residue for MPT, were common between indicated organotin ligands and testosterone (Table 2). This suggested that the organotin ligands are bound in the same ligand binding pocket as the native ligand. Thus, on a preliminary basis, organotins have the potential to interfere with the binding of testosterone to AR and act as interfering compounds for androgen signaling.

Table 1. The binding strength scores, i.e., dock score, binding energy (BE), and dissociation constant (pK_d) for organotin compounds with androgen receptor, estrogen receptor α, and estrogen receptor β. The BE and pK_d values were calculated using X-Score. Lower (more negative) BE or dock score and higher pK_d denote better docking. Monobutyltin showed a weak binding (low dock score), and further analysis was not performed for its any other parameters. Triphenyltin did not bind to any of the receptors (high positive dock score), and azocyclotin could not be docked even after varying the default parameters.

Organotin Compounds	Androgen Receptor			Estrogen Receptor α			Estrogen Receptor β		
	Dock Score	BE (Kcal/mol)	pK_d	Dock Score	BE (Kcal/mol)	pK_d	Dock Score	BE (Kcal/mol)	pK_d
Tributyltin	−30.18	−7.82	5.73	−31.26	−7.67	5.62	−30.80	−7.69	5.64
Dibutyltin	−27.79	−6.81	4.99	−26.79	−6.72	4.93	−27.68	−6.78	4.97
Diphenyltin	−28.40	−8.17	5.99	−27.64	−8.10	5.94	−22.76	−8.13	5.96
Monophenyltin	−23.93	−7.60	5.57	−25.67	−7.63	5.59	−24.75	−7.66	5.61
Native ligand	-	−10.36	7.60	-	−9.78	7.17	-	−9.95	7.29
Monobutyltin	−16.96	No analysis was performed due to low dock score		−16.94	No analysis was performed due to low dock score		−16.67	No analysis was performed due to low dock score	
Triphenyltin	772.83	No binding to receptor (dock score high positive)		122.80	No binding to receptor (dock score high positive)		176.40	No binding to receptor (dock score high positive)	
Azocyclotin	Could not be docked even after varying the default parameters.								

3.3. Molecular Docking of Organotin Compounds with ERα

Intermolecular interactions of four organotin ligands, i.e., TBT, DBT, DPT, and MPT in complex with ERα were analyzed and identified. All docked ligands were found to have similar binding poses to the native ligand, thus providing credence to the docking accuracy. All the organotin ligands bound well within the binding site of ERα. The values for the dock score, binding energy, and dissociation constant were comparable among the ligands and were lower but close to native ligand, estradiol, indicating tight binding and similar binding strength to ERα (Table 1). The two-dimensional interaction maps of amino acid residues of ERα interacting with four organotin ligands and native bound ligand, estradiol, are illustrated in Figure 4.

The amino-acid residue interactions in the docking poses of organotin ligands, TBT, DBT, DPT, and MPT with ERα are presented (Figure 4A–D). The docking pose analysis revealed that TBT interacted with eight amino acid residues of ERα, i.e., Leu-349, Ala-350, Trp-383, Phe-404, Met-421, Ile-424, His-524, and Leu-525, and DBT interacted with six amino-acid residues of ERα, i.e., Met-343, Leu-349, Ala-350, Phe-404, Met-421, and His-524. In addition, DPT interacted with seven amino acid residues of ERα, i.e., Met-343, Leu-346, Ala-350, Leu-387, Leu-391, Phe-404, and Leu-525, whereas MPT interacted with five amino acid residues of ERα, i.e., Ala-350, Leu-384, Leu-387, Leu-391, and Phe-404. These residues were significantly important for the ERα/organotin ligand bonding interaction in the active sites. The true conformation of the bound native ligand estradiol in the binding site of ERα is shown (Figure 2B), and the interacting amino acid residues are illustrated (Figure 4E). For estradiol, 12 amino acid residues were interacting with ERα, i.e., Leu-346, Ala-350, Glu-353, Leu-384, Leu-38&, Met-388, Leu-391, Arg-394, Phe-404, Ile-424, His-524, and Leu-525. Three hydrogen bonds were formed by Arg-394, Gln-353, and His-524, in addition to various other molecular interactions. When the interacting amino acid residues of ERα for each of the four organotins were compared with those for bound native ligand (estradiol), five of eight (62%) interacting residues for TBT, three of six (50%) interacting residues for DBT, six of seven (86%) interacting residues for DPT, and the five of five (100%) interacting residue for MPT, were common between indicated organotin ligands and estradiol (Table 3). This suggested that the indicated four organotin ligands potentially bind in the same binding pocket of ERα as the native ligand estradiol. Thus, these compounds may interfere with the binding of estradiol to its receptor and cause dysfunction of estradiol signaling.

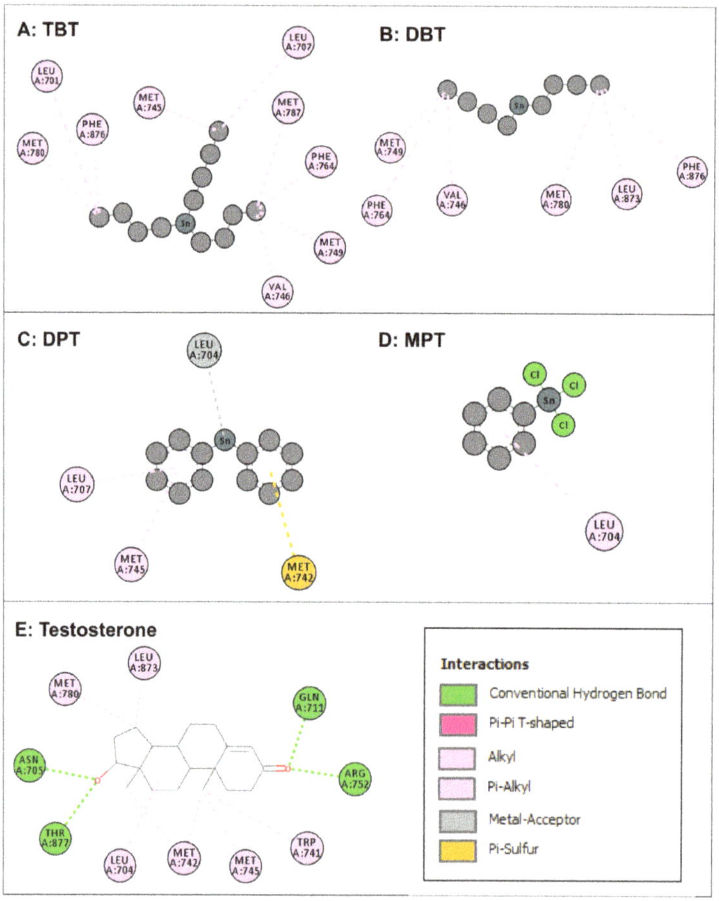

Figure 3. Two-dimensional interaction maps of amino acid residues of androgen receptor (AR) interacting with four organotin ligands during molecular docking in the ligand-binding site of AR. Amino acid residue interactions of the bound native ligand of AR (testosterone) are also shown. Tributyltin (TBT, Panel **A**); dibutyltin (DBT, Panel **B**); diphenyltin (DPT; Panel **C**); monophenyltin (MPT, Panel **D**); and testosterone (Panel **E**). Amino acid residues with alkyl or pi-alkyl bonds are indicated in light pink color, an amino acid with pi-sulfur bond is indicated in yellow color, and an amino acid with a metal acceptor bond is indicated in grey color. Interactions box: the depicted colors for interaction legends apply to all five interaction maps in the figure.

Table 2. Comparative list of amino acid residues of androgen receptor interacting with bound native ligand (testosterone), tributyltin, dibutyltin, diphenyltin, and monophenyltin. Residues within a row among native ligand and organotin ligands are overlapping interacting residues for the compounds.

Native (Testosterone)	Tributyltin	Dibutyltin	Diphenyltin	Monophenyltin
-	Leu-701	-	-	-
Leu-704	-	-	Leu-704	Leu-704
Asn-705	-	-	-	-
-	Leu-707	-	Leu-707	-
Gln-711	-	-	-	-
Trp-741	-	-	-	-
Met-742	-	-	Met-742	-
Met-745	Met-745	-	Met-745	-
-	Val-746	Val-746	-	-
-	Val-749	Met-749	-	-
Arg-752	-	-	-	-
-	Phe-764	Phe-764	-	-
Met-780	Met-780	Met-780	-	-
	Met-787			
Leu-873	-	Leu-873	-	-
-	Phe-876	Phe-876	-	-
Thr-877	-	-	-	-

3.4. Molecular Docking of Organotin Compounds with ERβ

Intermolecular interactions of four organotin ligands, i.e., TBT, DBT, DPT, and MPT in complex with ERβ were analyzed and identified. All docked ligands were found to have similar binding poses to the native ligand, thus providing credence to the docking accuracy. All the organotin ligands bound well within the binding site of ERβ. The values for the dock score, binding energy, and dissociation constant were comparable among the ligands and were lower but close to native ligand, estradiol, indicating tight binding and similar binding strength to ERβ (Table 1). The two-dimensional interaction maps of interacting amino acid residues of ERβ with four organotin ligands and native bound ligand, estradiol, are illustrated in Figure 5.

The amino-acid residue interactions in the docking poses of organotin ligands, TBT, DBT, DPT, and MPT with ERβ are presented (Figure 5A–D). The docking pose analysis revealed that TBT interacted with eight amino acid residues of ERβ, i.e., Met-295, Leu-298, Leu-301, Leu-339, Leu-343, Phe-356, Ile-373, His-475, and Leu-476, and DBT interacted with four amino-acid residues of ERβ, i.e., Leu-301, Leu-339, Phe-356, and His-475. In addition, DPT interacted with eight amino acid residues of ERβ, i.e., Met-295, Leu-298, Ala-302, Met-336, Leu-339, Leu-343, Phe-356, and Leu-476, whereas MPT interacted with four amino acid residues of ERβ, i.e., Ala-302, Met-336, Leu-339, and Phe-356. These residues were significantly important for the ERβ/organotin ligand bonding interactions in the active sites. The true conformation of the bound native ligand estradiol in the binding site of ERβ is shown (Figure 2C), and the interacting amino acid residues are illustrated (Figure 5E). For estradiol, 11 amino acids were interacting with ERβ, i.e., Leu-298, Ala-302, Glu-305, Met-336, Leu-339, Met-340, Phe-356, Ile-376, Gly-472, His-475, and Leu-476. Three hydrogen bonds were formed by Gln-305, Gly-472, and His-475, in addition to various other molecular interactions. When the interacting amino acid residues of ERβ for each of the four organotins were compared with those for bound native ligand (estradiol), five of nine (56%) interacting residues for TBT, three of four (75%) interacting residues for DBT, six of eight

(75%) interacting residues for DPT, and the four of four (100%) interacting residue for MPT, were common between indicated organotin ligands and estradiol (Table 4). This suggested that the indicated four organotin ligands potentially bind in the same binding pocket of ERβ as the bound native ligand estradiol. Thus, these compounds may interfere with the binding of estradiol to its receptor (ERβ) and cause dysfunction of estradiol signaling.

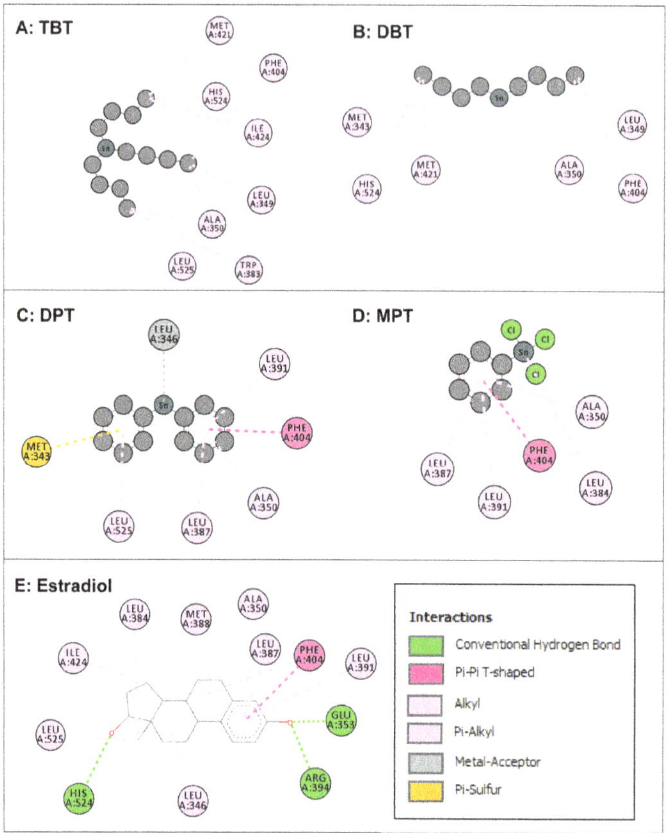

Figure 4. Two-dimensional interaction maps of amino acid residues of estrogen receptor (ERα) interacting with four organotin ligands during molecular docking in the ligand-binding site of ERα. Amino acid residue interactions of the bound native ligand of ERα (estradiol) are also shown. Tributyltin (TBT, Panel **A**); dibutyltin (DBT, Panel **B**); diphenyltin (DPT; Panel **C**); monophenyltin (MPT, Panel **D**); and estradiol (Panel **E**). Amino acid residues with alkyl or pi-alkyl bonds are indicated in light pink color, an amino acid with pi-sulfur bond is indicated in yellow color, amino acids with pi-pi T-shaped bond are indicated in dark pink color, and an amino acid with a metal acceptor bond is indicated in grey color. Interactions box: the depicted colors for interaction legends apply to all five interaction maps in the figure.

Table 3. Comparative list of amino acid residues of estrogen receptor α interacting with bound native ligand (estradiol), tributyltin, dibutyltin, diphenyltin, and monophenyltin. Residues within a row among native ligand and organotin ligands are overlapping interacting residues for the compounds.

Native (Estradiol)	Tributyltin	Dibutyltin	Diphenyltin	Monophenyltin
-	-	Met-343	Met-343	-
Leu-346	-	-	Leu-346	-
-	Leu-349	Leu-349	-	-
Ala-350	Ala-350	Ala-350	Ala-350	Ala-350
Glu-353	-	-	-	-
Leu-384	-	-	-	Leu-384
-	Trp-383	-	-	-
Leu-387	-	-	Leu-387	Leu-387
Met-388	-	-	-	-
Leu-391	-	-	Leu-391	Leu-391
Arg-394	-	-	-	-
Phe-404	Phe-404	Phe-404	Phe-404	Phe-404
-	Met-421	Met-421	-	-
Ile-424	Ile-424	-	-	-
His-524	His-524	His-524	-	-
Leu-525	Leu-525	-	Leu-525	-

3.5. Molecular Dynamics (MD) Simulation

In order to obtain an insight into protein–ligand complex stability, we performed an all-atomic molecular dynamics simulation of a 100 ns time period for AR and TBT complex. The dynamic profile of the complex was assessed by RMSD from the 100 ns trajectory. The results of the RMSD plot indicated that the AR/TBT ligand complex showed a stable interaction throughout MD simulation, and the system achieved equilibrium during the early period of MD simulation i.e., after 10 ns (Figure 6A). Both AR and TBT showed less deviation throughout MD simulation, which was 2.6 ± 0.22 and 2.6 ± 0.62, respectively. RMSF, i.e., protein residue fluctuation was also monitored during MD simulation (Figure 6B). Similarly, a Rg calculation over MD simulation, which is a major indication of structural compactness, was performed (Figure 6C). The Rg results showed that the AR/TBT complex maintains its compactness throughout and does not fluctuate much, and the average Rg was 19 ± 0.06 Å. Further, we estimated the total change in SASA of AR/TBT complex simulation (Figure 6D). The results showed that the SASA profile did not change during the whole simulation period indicating no huge structural deviation during MD simulation. Thus, the results indicate good stability of the AR/TBT protein–ligand complex.

Table 4. Comparative list of amino acid residues of estrogen receptor β interacting with bound native ligand (estradiol), tributyltin, dibutyltin, diphenyltin, and monophenyltin. Residues within a row among native ligand and organotin ligands are overlapping interacting residues for the compounds.

Native (Estradiol)	Tributyltin	Dibutyltin	Diphenyltin	Monophenyltin
-	Met-295	-	Met-295	-
Leu-298	Leu-298	-	Leu-298	-
-	Leu-301	Leu-301	-	-
Ala-302	-	-	Ala-302	Ala-302

Table 4. *Cont.*

Native (Estradiol)	Tributyltin	Dibutyltin	Diphenyltin	Monophenyltin
Glu-305	-	-	-	-
Met-336	-	-	Met-336	Met-336
Leu-339	Leu-339	Leu-339	Leu-339	Leu-339
Met-340	-	-	-	-
-	Leu-343	-	Leu-343	-
Phe-356	Phe-356	Phe-356	Phe-356	Phe-356
-	Ile-373	-	-	-
Ile-376		-	-	-
Gly-472	-	-	-	-
His-475	His-475	His-475	-	-
Leu-476	Leu-476	-	Leu-476	-

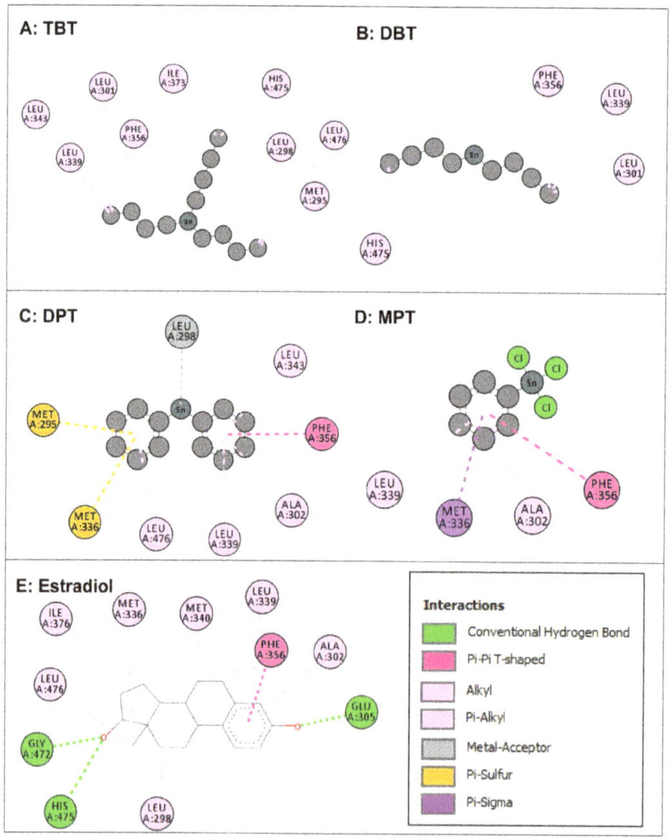

Figure 5. Two-dimensional interaction maps of amino acid residues of estrogen receptor β (ERβ) interacting with four organotin ligands during molecular docking in the ligand-binding site of ERβ. Amino acid residue interactions of the bound native ligand of ERβ (estradiol) are also shown. Tributyltin (TBT, Panel **A**); dibutyltin (DBT, Panel **B**); diphenyltin (DPT; Panel **C**); monophenyltin

(MPT, Panel **D**); and estradiol (Panel **E**). Amino acid residues with alkyl or pi-alkyl bonds are indicated in light pink color, an amino acid with pi-sulfur bond is indicated in yellow color, amino acids with pi-pi T-shaped bonds are indicated in dark pink color, an amino acid with a metal acceptor bond is indicated in grey color, and an amino acid with pi-sigma bond is indicated in purple color. Interactions box: the depicted colors for interaction legends apply to all five interaction maps in the figure.

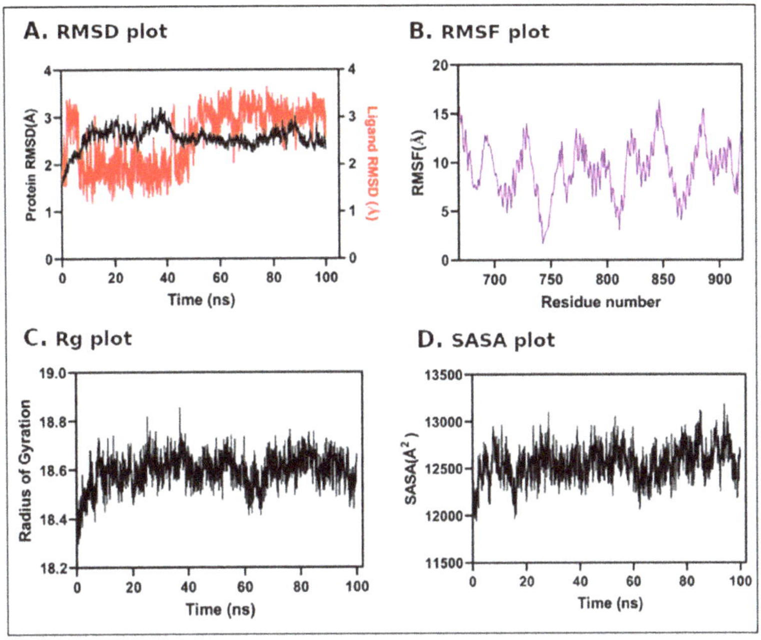

Figure 6. Molecular dynamics simulation analysis of androgen receptor in complex with tributyltin. Panel (**A**): root mean square deviation plot; Panel (**B**): root mean square fluctuation analysis; Panel (**C**): radius of gyration; and Panel (**D**): solvent accessible surface area.

4. Discussion

The objective of this study was to characterize the structural binding interactions of seven commonly available OTCs, viz., TBT, DBT, MBT, TPT, DPT, MPT, and ACT, against sex-steroid nuclear receptors, i.e., AR, ERα, and ERβ. As already indicated, docking analysis showed that four of seven organotin ligands, i.e., TBT, DBT, DPT, and MPT, bound well within the binding sites of the three receptors. The remaining three organotin ligands, MBT, TPT, and ACT, showed a weak binding or did not bind with the receptors and, hence, further analysis of these ligands for molecular interactions and binding pose analysis was not considered. The dock score, binding energy, and dissociation constants for the four organotins with AR, ERα, and ERβ indicated good docking and binding within the ligand binding sites of the receptors. Similar binding poses of the organotin ligands and the respective bound native ligands along with close similarity in binding energies indicated good binding and thus provided credence to the docking accuracy. In addition, high overlapping/commonality of interacting residues of each receptor for the docked ligands and respective native ligands provided further support for good docking. This was especially true for the interacting residues of ERα, and ERβ for the organotin ligands and native ligands as ERα, and ERβ showed better structural binding characteristics compared to AR. The good docking and stability of docking complex was further supported by the results of MD simulation of a representative organotin/receptor complex (TBT/AR), with a RMSD plot showing a stable interaction of TBT with AR with low deviation of ligand within good acceptable range. In addition, RMSF and Rg plots showed maintenance of compactness of the complex along with no change in the SASA profile during the whole simulation period. Thus, the MD simulation also provided support and indicated good stability of the receptor–ligand complex. Taken together, all the above results suggested that the organotin ligands are bound tightly in the same ligand binding pockets of AR, ERα, and ERβ as their respective bound native ligands, testosterone and estradiol, especially for the ERα, and ERβ. Thus, the suggested hypothesis that OTCs may interfere with the natural interaction between sex steroids and their receptors by binding to sex steroid nuclear receptors was supported. Hence, the indicated organotins have the potential to interfere with the binding of testosterone and estradiol17β to their respective receptors and result in dysfunction of steroid receptor signaling.

Previous reports on the in silico studies of organotins with sex steroid nuclear receptors are not available to the best of our knowledge. However, organotins, TBT and TPT, were shown to have competitive binding antagonistic activity against human ER in vitro [42]. The inhibitory effect on ER similar to other ER antagonists, such as 4-hydroxytamoxifen, was shown on ER-dependent reporter gene transcriptional activation by TBT and TPT at very low concentrations by interactions between human ERβ LBD and the co-activator SRC1 in a yeast two-hybrid detection system. In addition, TBT and TPT stimulated LA16 cells (that stably expressed androgen-responsive luciferase reporter gene and proliferates in response to androgen) and enhanced both AR-dependent transcription of luciferase gene and cell growth similar to dihydrotestosterone [43]. Simultaneous treatment of LA16 cells with dihydrotestosterone and TBT or TPT caused synergistic effects on AR activation, but an androgen antagonist, flutamide, did not inhibit the TBT- or TPT-induced AR activation, suggesting a novel mechanism other than the ligand-binding site of AR. TBT and TPT were also shown to interact with other nuclear receptors, e.g., as nanomolar agonists of retinoid X receptor (RXR) and peroxisome proliferator-activated receptor γ (PPARγ). In this regard, TBT and TPT exposure to a RXR-transfected human choriocarcinoma cell line (JEG-3 cells) stimulated luciferase expression, indicating activation of RXR [44]. In vivo and in vitro studies on organotins in marine invertebrates, fish, mammals, and laboratory animals have shown the profound endocrine disruption effects of organotins on reproductive and other functions. Development of imposex in mollusks even with exposure to very low TBT concentrations is the most sensitive and well known reproductive phenotypic effect of endocrine disruption [6]. The endocrine disruption by organotins in various species is believed to occur through three main mechanisms, i.e., (1) increased testosterone—organotins cause an androgen abundance by inhibiting the aromatase in females [45], (2) the APGWamide neuropeptide activation—organotins induce abnormally high levels of peptide APGWamide, which cause the development of male-like tissues subsequently producing androgens to promote the male-like sexual growth, and (3) the RXR agonism/activation—organotins cause abnormal activation of the RXR signaling pathway through ligand binding or by increasing the retinoid [23]. RXRs in association with other nuclear receptors, such as PPAR, etc., regulate cellular development and differentiation, metabolism, and cell death. In addition, recently inactivation of UDP-glucuronosyltransferases (UGTs) was also proposed to be another mechanism by which organotins can cause abnormal reproductive health effects; UGTs regulate metabolic inactivation of many endogenous hormones [46]. Inhibition of UGTs, including UGT2B15 (regulates catalysis of dihydrotestosterone glucuronidation) and UGT1A1 (regulates catalysis of estradiol-3-O-glucuronidation) by organotins such as TBT and TPT, may interfere with glucuronidation of endogenous sex hormones and thus inhibit their termination, resulting in abnormal endocrine actions [46]. The endocrine disruption mechanism of organotins in humans is not clear but has been postulated to most likely involve increased testosterone [4]. In addition, the potential for TBT and other organotins to act as obesogens in humans and to interfere with endocrine regulation of adipogenesis such as stimulation of preadipocyte differentiation into adipocytes in a PPARγ-dependent manner is also very concerning.

Very limited epidemiological studies have been reported on the association of organotin exposure and human reproductive problems. Several reports have mentioned general toxic effects associated with acute human organotin exposure, such as seizures, visual disturbances, paraparesis, forgetfulness, fatigue, weakness, loss of motivation, depression, and attacks of rage; some symptoms persist for at least three years [2,47,48]. Information associated with long term exposure is not reported. In other studies, severe toxicity of TPT showed neurotoxic symptoms, such as cerebellar syndrome, hearing impairment, and loss of consciousness with paroxysmal activity on electroencephalography [49]. In general population, significant levels of the organotins TBT and TPT have been detected in the blood of human male and female volunteers [50]. In another study, triorganotins, such as TBT and TPT, were detected in 37% and 99% of the placental tissues of women in Denmark and Finland, respectively, and a positive correlation between organotin (DBTCl) levels in the human placenta and incidence of cryptorchidism was reported in newborns in Denmark [21]. In addition, in the same study, the blood LH levels in four-month-old boys were negatively correlated, whereas the inhibin B levels were positively correlated with TBTCl in the placenta of women from Finland.

Several studies have reported adverse effects of organotins in male reproductive function in laboratory species [4,27]. In this regard, rat testes were reported to accumulate tin after three days of TBT exposure, resulting in reproductive abnormalities [25]. TBT induced weight gain and improved the food efficiency of male rats [51]. In another study [52], TPT resulted in decreased sperm count and motility in a dose-dependent manner, an impaired sperm histone–protamine replacement process, and significantly increased incidence of sperm deformities, as well as impaired proliferation of spermatogonia in adult male rats. In a recent study, TBT exposure decreased the number of Leydig cells and inhibited androgen production in rats [53]. In male mice, TBT exposure early after birth was associated with reduction in the weight of testis, epididymis, prostate, and seminal vesicles [54,55]. In zebra fish, TBT exposure during early development induced male bias and reduced or completely inhibited sperm motility, caused the absence of flagella, and resulted in only abnormal spermatozoa in semen in the exposed male offspring [56]. Similarly, exposure of ACT to adult male and female zebra fish altered the gene expression related to reproductive function, such as for LH in pituitary and for aromatase in gonads, along with reductions in estrogen in both sexes [57]. In addition, ACT exposure was associated with impaired spermatogenesis in males.

In addition to in vivo studies, in vitro studies have shown that TPT exerted inhibitory effects on the activity of human blood steroidogenic enzymes, i.e., 5-alpha-reductase type 2, cytochrome P450 aromatase, 17-beta-HSD type 3, 3-beta-HSD type 2, and 17-beta-HSD type 1 through the interaction with critical cysteine residues [50]. TBT, DBT and TPT salts were shown to inhibit human 5-alpha-reductase type 1 and 5-alpha-reductase type 2 enzymes, which are required for activation of androgens [58]. In vitro acute or short-term exposure of bovine sperm to TBT was associated with decrease in total motility, progressive motility, curvilinear velocity, and beat-cross frequency, along with a lower mitochondrial membrane potential [26]. In addition, TBT-exposed sperm resulted in a reduced cleavage rate and a lower rate of 8-16 cell morula development compared to embryos from unexposed sperms. DBT decreased the production of androgens from rat immature Leydig cells both under basal and LH stimulated conditions [59]. In another study [53], TBT exposure of immature rat Leydig cells in vitro reduced androgen production, cell viability, and cell cycle progression, while increasing reactive oxygen species (ROS) and apoptosis. Similarly, exposure of rat Leydig cells and Sertoli cells to TPT induced a significant decrease in the expression of steroidogenic and apoptotic indicators; similar exposure did not affect spermatogonia cells [52].

In female animals, organotin exposure was associated with toxicity of female reproductive system [3,27]. The most well known effect of many organotins, such as TBT and TPT, is the development of imposex in female gastropods, which refers to the development of male sexual organs in females, such as the penis and vas deferens [60–62]. In many laboratory mammalian animal models, TBT exposure was associated with reproductive, metabolic, and cardiovascular abnormalities, including hyperandrogenism, cystic ovarian follicles, irregular estrous cycle, elevated levels of LH, obesity, abnormal lipid profiles, glucose metabolism, and insulin resistance similar to those found in polycystic ovarian syndrome (PCOS) in women and animal models of PCOS [22,63–65]. In this regard, TBT exposure in female rats induced body weight gain and adiposity [66]. Exposure to TBT disrupts the proper functioning of the HPG axis of the female rats, probably in part through causing abnormal KISS and GnRH action, which regulate the reproductive axis of the hypothalamus and pituitary [64]. In this regard, TBT exposure in female rats caused irregular estrous cycles, downregulated hypothalamic GnRH mRNA expression, decreased exogenous KISS response, decreased basal and surge levels of LH, reduced exogenous GnRH responsiveness, decreased pituitary expression of both ERα and Erβ, increased testosterone, ovarian and uterine fibrosis, and decreased fertility. In mice and rats, TBT disrupts ovarian reserve, development of germ cells, folliculogenesis, steroidogenesis, ovulation, and CL formation [67]. In addition, in utero TBT exposure was associated with abnormal number and morphology of gonocytes with lipid droplets accumulating in the endoplasmic reticulum in the female rat offspring. In this regard, the placenta has been shown to accumulates the organotins, as exposure to TBT resulted in rat placental TBT levels that were five times higher than those found in maternal blood and ten times higher than those found in milk [68]. In other studies, TBT caused irregular estrous cycles, disturbed ovarian development, including increased presence of atretic and cystic follicles, fewer CLs, antral follicles and increased levels of atretic follicles, hyperandrogenism, high levels of serum LH, and decreased levels of serum sex hormone-binding globulin, in addition to an increase in the RXR/PPAR signaling pathway and other proteins that are involved in androgen biosynthesis [65,69]. Exposure to ACT in adult female zebra fish was associated with alterations in the reproductive related gene expression, reduced estrogen, and increased testosterone in females [57]. Increased accumulation of ACT in F1 eggs and embryonic abnormalities were also found after parental exposure.

In vitro studies have also demonstrated effects of organotin on female hypothalamic-gonadal-axis and ovarian cells similar to in vivo effects. TBT exposure of ovarian theca cells from five species (human, sheep, cow, pig, and mouse) affected cholesterol trafficking, luteinization, and steroidogenesis in all five species [70]. The effect was, in part, through modulation of RXR, as shown by RXR antagonist and RXRα knockdown. In human granulosa-like tumor cell line KGN, TBT exposure reduced the mRNA expression of aromatase and its activity by 30% compared to control cells [71]. In addition, exposure to higher TBT concentrations resulted in KGN cell death within 24 h, whereas lower TBT concentration promoted apoptosis. Exposure of TBT in bovine cumulus–oocyte complex cultures reduced estrogen and testosterone levels along with expression of aromatase and 3-beta-HSD mRNA [72]. In addition, TBT exposure inhibited LH stimulated estrogen synthesis in follicular granulosa cells. TBT and DBT acted as partial competitive inhibitors of aromatase enzyme in human placenta and showed inhibition of human 3-beta HSD type I activity [73].

Taken together, the results of our in silico structural interactions of organotin ligands with sex steroid receptors support the previously reported adverse effects, as discussed above, for in vivo and in vitro studies in human and laboratory animals. The morpho-physiological, hormonal, and molecular impairments discussed support our suggested hypothesis of perturbation of natural interaction between native ligands and sex steroid nuclear receptors leading to impaired reproductive function.

5. Conclusions

In the present study, molecular docking simulations of seven OTCs, viz., TBT, DBT, MBT, TPT, DPT, MPT, and ACT, against sex-steroid nuclear receptors, i.e., AR, ERα, and Erβ, were performed. The docking results showed that TBT, DBT, DPT, and MPT bound well within the binding sites of the three receptors. The remaining three organotin ligands, MBT,

TPT, and ACT, did not bind with the receptors. All the indicated four organotin compounds interacted with each of the sex steroid receptors and bound deep into their ligand-binding sites, showing good dock score, binding energy, and dissociation constants that were comparable to the bound native ligands, testosterone, and estradiol. The good docking and stability of docking complex was also shown by MD simulation of organotin/receptor complex, with RMSD, RMSF, Rg, and SASA plots showing stable interaction, low deviation, and compactness of the organotin/receptor complex during the whole simulation period. In addition, high commonality of interacting residues of each receptor for the docked ligands and respective bound native ligands, especially for ERα and Erβ, which indicated that the organotin compounds bound tightly in the same ligand binding site of the receptor as the bound native ligand. To conclude, the study suggested that the indicated organotin compounds may interfere with the natural interaction between sex steroids and their receptors and cause dysfunction of steroid receptor signaling.

Author Contributions: Conceptualization, M.A.B. (Mohd A. Beg); methodology, M.R. and M.A.B. (Md A. Beg); software, M.R. and M.A.B. (Md A. Beg); validation, M.R. and M.A.B. (Md A. Beg); formal analysis, M.R. and M.A.B. (Md A. Beg); data analysis, M.R. and M.A.B. (Md A. Beg); writing—original draft preparation, M.A.B. (Md A. Beg), U.R.Z. and M.A.B. (Mohd A. Beg); writing—review and editing, I.A.S., O.S.B., A.M.A. and M.R.; supervision, M.A.B. (Mohd A. Beg); project administration, M.A.B. (Mohd A. Beg); funding acquisition, M.A.B. (Mohd A. Beg). All authors have read and agreed to the published version of the manuscript.

Funding: Mohd Amin Beg received funding for this study from the Deputyship for Research & Innovation, Ministry of Education in Saudi Arabia through project number IFPIP: 1620-141-1442 at King Abdulaziz University, DSR, Jeddah, Saudi Arabia.

Institutional Review Board Statement: Not applicable.

Informed Consent Statement: This study did not involve humans.

Data Availability Statement: The majority of the data for the results of this study are provided in the main manuscript. In addition, any specific data are available from the corresponding author.

Acknowledgments: The authors extend their appreciation to the Deputyship for Research & Innovation, Ministry of Education in Saudi Arabia for funding this research work through project number IFPIP: 1620-141-1442, and King Abdulaziz University, DSR, Jeddah, Saudi Arabia. The authors also thank Sibasis Sahoo from the Membrane Protein Biology, International Centre for Genetic Engineering and Biotechnology, New Delhi India for helping with molecular dynamics simulation analysis.

Conflicts of Interest: The authors declare no conflict of interest.

References

1. Da Silva, I.F.; Lima, L.F.; Graceli, J.B.; Rodrigues, L.C.D.M. Organotins in Neuronal damage, brain function, and behavior: A short review. *Front. Endocrinol.* **2018**, *8*, 366. [CrossRef] [PubMed]
2. WHO. Organotins in Drinking-Water. Background Document for Development of Who Guidelines for Drinking-Water Quality. Geneva: World Health Organization (WHO/HEP/ECH/WSH/2020.7). Licence: CC BY-NC-SA 3.0 IGO. 2020. Available online: https://apps.who.int/iris/bitstream/handle/10665/338068/WHO-HEP-ECH-WSH-2020.7-eng.pdf (accessed on 4 December 2022).
3. Barbosa, K.L.; Dettogni, R.S.; da Costa, C.S.; Gastal, E.L.; Raetzman, L.T.; Flaws, J.A.; Graceli, J.B. Tributyltin and the female hypothalamic-pituitary-gonadal disruption. *Toxicol. Sci.* **2022**, *186*, 179–189. [CrossRef] [PubMed]
4. Beyer, J.; Song, Y.; Tollefsen, K.E.; Berge, J.A.; Tveiten, L.; Helland, A.; Øxnevad, S.; Schøyen, M. The ecotoxicology of marine tributyltin (TBT) hotspots: A review. *Mar. Envrion. Res.* **2022**, *179*, 105689. [CrossRef] [PubMed]
5. Metcalfe, C.D.; Bayen, S.; Desrosiers, M.; Muñoz, G.; Sauvé, S.; Yargeau, V. An introduction to the sources, fate, occurrence and effects of endocrine disrupting chemicals released into the environment. *Envrion. Res.* **2022**, *207*, 112658. [CrossRef] [PubMed]
6. Hoch, M. Organotin compounds in the environment—An overview. *Appl. Geochem.* **2001**, *16*, 719–743. [CrossRef]
7. Graceli, J.B. Editorial: Organotins as a complete physiologic and endocrine disruptor: Role of disease development. *Editor. Front. Endocrinol.* **2019**, *10*, 799. [CrossRef]
8. S&P Global. Organometallics: Chemical Economics Handbook. S&P Global Commodity Insights. 2022. Available online: https://www.spglobal.com/commodityinsights/en/ci/products/organometallics-chemical-economics-handbook.html (accessed on 4 December 2022).

9. Okoro, H.K.; Fatoki, O.S.; Adekola, F.A.; Ximba, B.J.; Snyman, R.G.; Opeolu, B. Human exposure, biomarkers, and fate of organotins in the environment. *Rev. Envrion. Contam. Toxicol.* **2011**, *213*, 27–54.
10. He, S.; Li, P.; Li, Z.H. Review on endocrine disrupting toxicity of triphenyltin from the perspective of species evolution: Aquatic, amphibious and mammalian. *Chemosphere* **2021**, *269*, 128711. [CrossRef]
11. Sadighara, P.; Jahanbakhsh, M.; Nazari, Z.; Mostashari, P. The organotin contaminants in food: Sources and methods for detection: A systematic review and meta-analysis. *Food Chem. X* **2021**, *12*, 100154. [CrossRef]
12. Fromme, H.; Mattulat, A.; Lahrz, T.; Rüden, H. Occurrence of organotin compounds in house dust in Berlin (Germany). *Chemosphere* **2005**, *58*, 1377–1383. [CrossRef]
13. Kannan, K.; Takahashi, S.; Fujiwara, N.; Mizukawa, H.; Tanabe, S. Organotin compounds, including butyltins and octyltins, in house dust from Albany, New York, USA. *Arch. Envrion. Contam. Toxicol.* **2010**, *58*, 901–907. [CrossRef] [PubMed]
14. Al-shatri, M.A.; Nuhu, A.A.; Basheer, C.; Al-Arfaj, A.; Al-Tawabini, B. Assessment of Tributyltin and triphenyltin compounds and their main degradation products in Saudi coastal waters. *Arab. J. Sci. Eng.* **2015**, *40*, 2959–2967. [CrossRef]
15. Hassan, A.T.; Qurban, M.; Manikandan, K.; Tawabini, B.; Basheer, C.; Periyadan, K. Assessment of the organotin pollution in the coastal sediments of the Western Arabian Gulf, Saudi Arabia. *Mar. Pollut. Bull.* **2019**, *139*, 174–180. [CrossRef] [PubMed]
16. Ashraf, M.W.; Salam, A.; Mian, A. Levels of organotin compounds in selected fish species from the arabian gulf. *Bull. Envrion. Contam. Toxicol.* **2017**, *98*, 811–816. [CrossRef] [PubMed]
17. Kannan, K.; Senthilkumar, K.; Giesy, J.P. Occurrence of butyltin compounds in human blood. *Envrion. Sci. Technol.* **1999**, *33*, 1776–1779. [CrossRef]
18. Nakanishi, T. Endocrine disruption induced by organotin compounds; organotins function as a powerful agonist for nuclear receptors rather than an aromatase inhibitor. *J. Toxicol. Sci.* **2008**, *33*, 269–276. [CrossRef]
19. Graceli, J.B.; Sena, G.C.; Lopes, P.F.; Zamprogno, G.C.; da Costa, M.B.; Godoi, A.F.; Dos Santos, D.M.; de Marchi, M.R.; Dos Santos Fernandez, M.A. Organotins: A review of their reproductive toxicity, biochemistry, and environmental fate. *Reprod. Toxicol.* **2013**, *36*, 40–52. [CrossRef]
20. Santos-Silva, A.P.; Andrade, M.N.; Pereira-Rodrigues, P.; Paiva-Melo, F.D.; Soares, P.; Graceli, J.B.; Dias, G.R.M.; Ferreira, A.C.F.; de Carvalho, D.P.; Miranda-Alves, L. Frontiers in endocrine disruption: Impacts of organotin on the hypothalamus-pituitary-thyroid axis. *Mol. Cell. Endocrinol.* **2018**, *460*, 246–257. [CrossRef]
21. Rantakokko, P.; Main, K.M.; Wohlfart-Veje, C.; Kiviranta, H.; Airaksinen, R.; Vartiainen, T.; Vartiainen, T.; Skakkebæk, N.E.; Toppari, J.; Virtanen, H.E. Association of placenta organotin concentrations with congenital cryptorchidism and reproductive hormone levels in 280 newborn boys from Denmark and Finland. *Hum. Reprod.* **2013**, *28*, 1647–1660. [CrossRef]
22. Merlo, E.; Silva, I.V.; Cardoso, R.C.; Graceli, J.B. The obesogen tributyltin induces features of polycystic ovary syndrome (PCOS): A review. *J. Toxicol. Environ. Healthy B. Crit. Rev.* **2018**, *21*, 181–206. [CrossRef]
23. Sternberg, R.M.; Gooding, M.P.; Hotchkiss, A.K.; LeBlanc, G.A. Environmental endocrine control of reproductive maturation in gastropods: Implications for the mechanism of tributyltin-induced imposex in prosobranchs. *Ecotoxicology* **2010**, *19*, 4–23. [CrossRef] [PubMed]
24. Lan, X.R.; Li, Y.W.; Chen, Q.L.; Shen, Y.J.; Liu, Z.H. Tributyltin impaired spermatogenesis and reproductive behavior in male zebrafish. *Aquat. Toxicol.* **2020**, *224*, 105503. [CrossRef] [PubMed]
25. Mitra, S.; Srivastava, A.; Khandelwal, S. Long term impact of the endocrine dis ruptor tributyltin on male fertility following a single acute exposure. *Environ. Toxicol.* **2017**, *32*, 2295–2304. [CrossRef] [PubMed]
26. Daigneault, B.W.; de Agostini Losano, J.D. Tributyltin chloride exposure to post-ejaculatory sperm reduces motility, mitochondrial function and subsequent embryo development. *Reprod. Fertil. Dev.* **2022**, *34*, 833–843. [CrossRef] [PubMed]
27. de Araújo, J.F.P.; Podratz, P.L.; Merlo, E.; Sarmento, I.V.; da Costa, C.S.; Niño, O.M.S.; Faria, R.A.; Freitas Lima, L.C.; Graceli, J.B. Organotin exposure and vertebrate reproduction: A review. *Front. Endocrinol.* **2018**, *9*, 64. [CrossRef]
28. da Costa, C.S.; Miranda-Alves, L.; La Merrill, M.A.; Silva, I.V.; Graceli, J.B. The tributyltin leads to obesogenic mammary gland abnormalities in adult female rats. *Toxicol. Lett.* **2019**, *307*, 59–71. [CrossRef]
29. Podratz, P.L.; Merlo, E.; de Araújo, J.; Ayub, J.; Pereira, A.; Freitas-Lima, L.C.; da Costa, M.B.; Miranda-Alves, L.; Cassa, S.; Carneiro, M.; et al. Disruption of fertility, placenta, pregnancy outcome, and multigenerational inheritance of hepatic steatosis by organotin exposure from contaminated seafood in rats. *Sci. Total Environ.* **2020**, *723*, 138000. [CrossRef]
30. Rehan, M.; Ahmad, E.; Sheikh, I.A.; Abuzenadah, A.M.; Damanhouri, G.A.; Bajouh, O.S.; AlBasri, S.F.; Assiri, M.M.; Beg, M.A. Androgen and progesterone receptors are targets for bisphenol A (BPA), 4-methyl-2,4-bis-(P-hydroxyphenyl)pent-1-ene—A potent metabolite of BPA, and 4-tert-octylphenol: A computational insight. *PLoS ONE* **2015**, *10*, e0138438. [CrossRef]
31. Beg, M.A.; Sheikh, I.A. Endocrine disruption: Molecular interactions of environmental bisphenol contaminants with thyroid hormone receptor and thyroxine-binding globulin. *Toxicol. Ind. Healthy* **2020**, *36*, 322–335. [CrossRef]
32. Rehan, M.; Zargar, U.R.; Sheikh, I.A.; Alharthy, S.A.; Almashjary, M.N.; Abuzenadah, A.M.; Beg, M.A. Potential disruption of systemic hormone transport by tobacco alkaloids using computational approaches. *Toxics* **2022**, *10*, 727. [CrossRef]
33. Seeliger, D.; de Groot, B.L. Ligand docking and binding site analysis with PyMOL and Autodock/Vina. *J. Comput. Aided Mol. Des.* **2010**, *24*, 417–422. [CrossRef] [PubMed]
34. Ewing, T.J.; Makino, S.; Skillman, A.G.; Kuntz, I.D. DOCK 4.0: Search strategies for automated molecular docking of flexible molecule databases. *J. Comput. Aided. Mol. Des.* **2001**, *15*, 411–428. [CrossRef] [PubMed]

35. Pettersen, E.F.; Goddard, T.D.; Huang, C.C.; Couch, G.S.; Greenblatt, D.M.; Meng, E.C.; Ferrin, T.E. UCSF Chimera–a visualization system for exploratory research and analysis. *J. Comput. Chem.* **2004**, *25*, 1605–1612. [CrossRef] [PubMed]
36. Rehan, M.; Ahmad, E.; Beg, M.A. Structural binding perspectives of a major tobacco alkaloid, nicotine, and its metabolite cotinine with sex-steroid nuclear receptors. *J. Appl. Toxicol.* **2020**, *40*, 1410–1420. [CrossRef] [PubMed]
37. Accelrys Software Inc. Discovery Studio Modeling Environment, Release 3.0, Accelrys Software Inc.: San Diego, CA, USA. 2018. Available online: www.accelrys.com (accessed on 4 December 2022).
38. Wang, R.; Lu, Y.; Wang, S. Comparative evaluation of 11 scoring functions for molecular docking. *J. Med. Chem.* **2003**, *46*, 2287–2303. [CrossRef]
39. Bowers, K.J.; Chow, E.; Xu, H.; Dror, R.O.; Eastwood, M.P.; Gregersen, B.A.; Klepeis, J.L.; Kolossvary, I.; Moraes, M.A.; Sacerdoti, F.D.; et al. Scalable algorithms for molecular dynamics simulations on commodity clusters. In Proceedings of the ACM/IEEE Conference on Supercomputing (SC06), Tampa, FL, USA, 11–17 November 2006.
40. Shivangi, E.M.K.; Meena, L.S. Essential biochemical, biophysical and computational inputs on efficient functioning of Mycobacterium tuberculosis H37Rv FtsY. *Int. J. Biol. Macromol.* **2021**, *171*, 59–73. [CrossRef]
41. Beg, M.A.; Sadaf; Shamsi, A.; Sahoo, S.; Yousuf, M.; Najm, M.Z.; Almutawif, Y.A.; Islam, A.; Aloliqi, A.A.; Athar, F. Mechanistic insight into the enzymatic inhibition of β-amyrin against mycobacterial Rv1636: In silico and in vitro approaches. *Biology* **2022**, *11*, 1214. [CrossRef]
42. Cho, E.M.; Lee, H.S.; Moon, J.S.; Kim, I.S.; Sim, S.; Ohta, A. Organotin compounds act as inhibitor of transcriptional activation with human estrogen receptor. *J. Microbiol. Biotechnol.* **2012**, *22*, 378–384. [CrossRef]
43. Yamabe, Y.; Hoshino, A.; Imura, N.; Suzuki, T.; Himeno, S. Enhancement of androgen-dependent transcription and cell proliferation by tributyltin and triphenyltin in human prostate cancer cells. *Toxicol. Appl. Pharmacol.* **2000**, *169*, 177–184. [CrossRef]
44. Nakanishi, T.; Nishikawa, J.; Hiromori, Y.; Yokoyama, H.; Koyanagi, M.; Takasuga, S.; Ishizaki, J.; Watanabe, M.; Isa, S.; Utoguchi, N.; et al. Trialkyltin compounds bind retinoid X receptor to alter human placental endocrine functions. *Mol. Endocrinol.* **2005**, *19*, 2502–2516. [CrossRef]
45. Fodor, I.; Urban, P.; Scott, A.P.; Pirger, Z. A critical evaluation of some of the recent so-called 'evidence' for the involvement of vertebrate-type sex steroids in the reproduction of mollusks. *Mol. Cell. Endocrinol.* **2020**, *516*, 110949. [CrossRef] [PubMed]
46. Lv, H.; Wang, J.J.; Wang, M.Y.; Shen, L.; Xiao, L.; Chen, T.J.; Sun, T.Z.; Li, W.J.; Zhu, L.L.; Zhang, X.K. Potent inhibition of tributyltin (TBT) and triphenyltin (TPT) against multiple UDP-glucuronosyltransferases (UGT): A new potential mechanism underlying endocrine disrupting actions. *Food Chem. Toxicol.* **2021**, *149*, 112039. [CrossRef] [PubMed]
47. Ross, W.D.; Emmett, E.A.; Steiner, J.; Tureen, R. Neurotoxic effects of occupational exposure to organotins. *Am. J. Psychiatry* **1981**, *138*, 1092–1095. [PubMed]
48. EFSA. Opinion of the scientific panel on contaminants in the food chain on a request from the commission to assess the health risks to consumers associated with exposure to organotins in foodstuffs. (European Food Safety Authority) (Question N° EFSA-Q-2003-110). *EFSA J.* **2004**, *102*, 1–119. Available online: https://www.efsa.europa.eu/en/efsajournal/pub/102 (accessed on 4 December 2022).
49. Lin, T.-J.; Hung, D.-Z.; Kao, C.-H.; Hu, W.-H.; Yang, D.-Y. Unique cerebral dysfunction following triphenyltin acetate poisoning. *Hum Exp Toxicol.* **1998**, *17*, 403–405. [CrossRef]
50. Lo, S.; Allera, A.; Albers, P.; Heimbrecht, J.; Jantzen, E.; Klingmuller, D.; Steckelbroeck, S. Dithioerythritol (DTE) prevents inhibitory effects of triphenyltin (TPT) on the key enzymes of the human sex steroid hormone metabolism. *J. Steroid Biochem. Mol. Biol.* **2003**, *84*, 569–576. [CrossRef]
51. Rodrigues-Pereira, P.; Andrade, M.N.; Santos-Silva, A.P.; Teixeira, M.P.; Soares, P.; Graceli, J.B.; de Carvalho, D.P.; Dias, G.; Ferreira, A.; Miranda-Alves, L. Subacute and low-dose tributyltin exposure disturbs the mammalian hypothalamus-pituitary-thyroid axis in a sex-dependent manner. *Comp. Biochem. Physiol. C Toxicol. Pharmacol.* **2022**, *254*, 109279. [CrossRef]
52. Lu, M.; Mu, Y.; Liu, Y. Triphenyltin disrupts the testicular microenvironment and reduces sperm quality in adult male rats. *Chemosphere* **2022**, *301*, 134726. [CrossRef]
53. Huang, X.; Ma, T.; Chen, X. Tributyltin inhibits development of pubertal Leydig cells in rats. *Reprod. Toxicol.* **2022**, *111*, 49–58. [CrossRef]
54. Chen, Y.; Zuo, Z.; Chen, S.; Yan, F.; Chen, Y.; Yang, Z.; Wang, C. Reduction of spermatogenesis in mice after tributyltin administration. *Toxicology* **2008**, *251*, 21–27. [CrossRef]
55. Mello, M.; Delgado, I.F.; Favareto, A.; Lopes, C.; Batista, M.M.; Kempinas, W.D.; Paumgartten, F. Sexual maturation and fertility of mice exposed to triphenyltin during prepubertal and pubertal periods. *Toxicol. Rep.* **2014**, *2*, 405–414. [CrossRef] [PubMed]
56. McAllister, B.G.; Kime, D.E. Early life exposure to environmental levels of the aromatase inhibitor tributyltin causes masculinisation and irreversible sperm damage in zebrafish (Danio rerio). *Aquat. Toxicol.* **2003**, *65*, 309–316. [CrossRef] [PubMed]
57. Ma, Y.N.; Cao, C.Y.; Wang, Q.W.; Gui, W.J.; Zhu, G.N. Effects of azocyclotin on gene transcription and steroid metabolome of hypothalamic–pituitary–gonad axis, and their consequences on reproduction in zebrafish (Danio rerio). *Aquat. Toxicol.* **2016**, *179*, 55–64. [CrossRef] [PubMed]
58. Doering, D.D.; Steckelbroeck, S.; Doering, T.; Klingmuller, D. Effects of butyltins on human 5alpha-reductase type 1 and type 2 activity. *Steroids* **2002**, *67*, 859–867. [CrossRef] [PubMed]
59. Li, G.; Chang, X.; Zhao, Y.; Li, D.; Kang, X. Dibutyltin (DBT) inhibits in vitro androgen biosynthesis of rat immature Leydig cells. *Toxicology* **2021**, *456*, 152779. [CrossRef]

60. Matthiessen, P.; Gibbs, P.E. Critical appraisal of the evidence for tributyltin-mediated endocrine disruption in mollusks. *Environ. Toxicol. Chem.* **1998**, *17*, 37–43. [CrossRef]
61. Lagadic, L.; Katsiadaki, I.; Biever, R.; Guiney, P.D.; Karouna-Renier, N.; Schwarz, T.; Meador, J.P. Tributyltin: Advancing the science on assessing endocrine disruption with an unconventional endocrine-disrupting compound. *Rev. Envrion. Contam. Toxicol.* **2018**, *245*, 65–127.
62. Vogt, É.L.; Model, J.F.A.; Vinagre, A.S. Effects of Organotins on crustaceans: Update and perspectives. *Front. Endocrinol.* **2018**, *9*, 65. [CrossRef]
63. de Araújo, J.; Podratz, P.L.; Sena, G.C.; Merlo, E.; Freitas-Lima, L.C.; Ayub, J.; Pereira, A.; Santos-Silva, A.P.; Miranda-Alves, L.; Silva, I.V.; et al. The obesogen tributyltin induces abnormal ovarian adipogenesis in adult female rats. *Toxicol. Lett.* **2018**, *295*, 99–114. [CrossRef]
64. Sena, G.C.; Freitas-Lima, L.C.; Merlo, E.; Podratz, P.L.; de Araújo, J.F.P.; Brandão, P.A.A.; Carneiro, M.T.W.D.; Zicker, M.C.; Ferreira, A.V.M.; Takiya, C.M.; et al. Environmental obesogen tributyltin chloride leads to abnormal hypothalamic- pituitary-gonadal axis function by disruption in kisspeptin/leptin signaling in female rats. *Toxicol. Appl. Pharmacol.* **2017**, *319*, 22–38. [CrossRef]
65. Yang, Z.; Shi, J.; Guo, Z.; Chen, M.; Wang, C.; He, C.; Zuo, Z. A pilot study on polycystic ovarian syndrome caused by neonatal exposure to tributyltin and bisphenol A in rats. *Chemosphere* **2019**, *231*, 151–160. [CrossRef] [PubMed]
66. Freitas-Lima, L.C.; Merlo, E.; Zicker, M.C.; Navia-Pelaez, J.M.; de Oliveira, M.; Capettini, L.D.S.A.; Nogueira, C.R.; Ferreira, A.V.M.; Santos, S.H.S.; Graceli, J.B. Tributyltin impacts in metabolic syndrome development through disruption of angiotensin II receptor signaling pathways in white adipose tissue from adult female rats. *Toxicol. Lett.* **2018**, *299*, 21–31. [CrossRef] [PubMed]
67. Podratz, P.L.; Filho, V.S.D.; Lopes, P.F.I.; Sena, G.C.; Matsumoto, S.T.; Samoto, V.Y.; Takiya, C.M.; Miguel, E.D.C.; Silva, I.V.; Graceli, J.B. Tributyltin Impairs the Reproductive Cycle in Female Rats. *J. Toxicol. Envrion. Healthy Part A* **2012**, *75*, 1035–1046. [CrossRef] [PubMed]
68. Cooke, G.M.; Forsyth, D.S.; Bondy, G.S.; Tachon, R.; Tague, B.; Coady, L. Organotin speciation and tissue distribution in rat dams, fetuses, and neonates following oral administration of tributyltin chloride. *J. Toxicol. Envrion. Healthy Part A* **2008**, *71*, 384–395. [CrossRef] [PubMed]
69. Lee, H.; Lim, S.; Yun, S.; Yoon, A.; Park, G.; Yang, H. Tributyltin increases the expression of apoptosis- and adipogenesis-related genes in rat ovaries. *Clin. Exp. Reprod. Med.* **2012**, *39*, 15–21. [CrossRef]
70. Pu, Y.; Pearl, S.; Gingrich, J.; Jing, J.; Martin, D.; Murga-Zamalloa, C.A.; Veiga-Lopez, A. Multispecies study: Lowdose tributyltin impairs ovarian theca cell cholesterol homeostasis through the RXR pathway in five mammalian species including humans. *Arch. Toxicol.* **2019**, *93*, 1665–1677. [CrossRef]
71. Saitoh, M.; Yanase, T.; Morinaga, H.; Tanabe, M.; Mu, Y.M.; Nishi, Y.; Nomura, M.; Okabe, T.; Goto, K.; Takayanagi, R.; et al. Tributyltin or triphenyltin inhibits aromatase activity in the human granulosa-like tumor cell line KGN. *Biochem. Biophys. Res. Commun.* **2001**, *289*, 198–204. [CrossRef]
72. Schoenfelder, M.; Schams, D.; Einspanier, R. Steroidogenesis during in vitro maturation of bovine cumulus oocyte complexes and possible effects of tri-butyltin on granulosa cells. *J. Steroid. Biochem. Mol. Biol.* **2003**, *84*, 291–300. [CrossRef]
73. Heidrich, D.D.; Steckelbroeck, S.; Klingmuller, D. Inhibition of human cytochrome P450 aromatase activity by butyltins. *Steroids* **2001**, *66*, 763–769. [CrossRef]

International Journal of
Environmental Research and Public Health

MDPI

Article

Health Risk Assessment for Human Exposure to Heavy Metals via Food Consumption in Inhabitants of Middle Basin of the Atrato River in the Colombian Pacific

Gabriel Caicedo-Rivas [1], Manuel Salas-Moreno [1,*] and José Marrugo-Negrete [2,*]

1 Biosistematic Research Group, Biology Department, Faculty of Natural Sciences, Universidad Tecnológica Del Chocó, Quibdó 270002, Chocó, Colombia

2 Faculty of Basic Sciences, Universidad de Córdoba, Carrera 6 No. 76-103, Montería 230002, Córdoba, Colombia

* Correspondence: d-manuel.salas@utch.edu.co (M.S.-M.); jmarrugo@correo.unicordoba.edu.co (J.M.-N.); Tel.: +57-3148056718 (M.S.-M.); +57-3233016454 (J.M.-N.)

Abstract: The Atrato river basin is one of the world's most biodiverse areas; however, it is highly impacted by mercury gold mining, which generates air, water, and soil pollution. (1) Background: The concentrations of persistent heavy metal pollutants, mercury (Hg), lead (Pb), cadmium (Cd), and arsenic (As) in the fish, fruits, and vegetables most consumed by the riverside inhabitants of the middle basin of the Atrato river represent a danger to public health; (2) Methods: A total of 154 samples of different fruits and vegetables and 440 samples of fish were analyzed by atomic absorption spectroscopy. A sample of 446 people were surveyed to evaluate food consumption and carcinogenic and non-carcinogenic risk; (4) Conclusions: High concentrations of As, Hg, Pb, and Cd were identified in fish, fruits-tubers, and vegetables-stems commonly consumed by inhabitants of the middle basin of the Atrato River, which exceeded the Codex limits and the limits established by the WHO/FAO, especially for carnivorous fish species. A high carcinogenic and non-carcinogenic risk was evidenced amongst inhabitants of the middle basin of the Atrato River due to the consumption of fish contaminated with high concentrations of As, MeHg, and THg. The risk due to the consumption of vegetables was very low.

Keywords: vegetables; fruits; fish; tubers; MeHg; As

Citation: Caicedo-Rivas, G.; Salas-Moreno, M.; Marrugo-Negrete, J. Health Risk Assessment for Human Exposure to Heavy Metals via Food Consumption in Inhabitants of Middle Basin of the Atrato River in the Colombian Pacific. *Int. J. Environ. Res. Public Health* **2023**, *20*, 435. https://doi.org/10.3390/ijerph20010435

Academic Editors: Esref Demir and Sam Kacew

Received: 8 October 2022
Revised: 4 November 2022
Accepted: 6 November 2022
Published: 27 December 2022

1. Introduction

The contamination of soil, air and water by heavy metals is a global concern due to its ability to affect different biological systems. Heavy metal contamination can result in bioaccumulation and biomagnification in the food chain affecting human health. Metal contamination in aquatic and terrestrial ecosystems is a serious environmental problem. Elements such as lead (Pb), arsenic (As), cadmium (Cd), mercury (Hg) and the organic form of the latter, methylmercury (MeHg), are toxic, even at low concentrations, and have a high capacity for bioaccumulation. The US Agency for Toxic Substances and Disease Registry (ATSDR) ranks these metals high on its 2017 Priority List of Hazardous Substances [1]. Contamination by these heavy metals can have various origins. However, this type of contamination is generally associated with anthropogenic activities, including gold mining, specifically the extraction of the precious metal by amalgamation with Hg. This activity generates significant levels of Hg contamination in soil and water; reports indicate that this type of activity produces emissions equivalent to about 880 tons of Hg per year [2–4]. A worrying aspect of these activities is that they generate heavy metals, such as Cd, Pb, and As, which are associated with gold minerals. These can be dispersed through erosion and chemical weathering of tailings from gold mining [5]. Some 200 tons of Hg are used in gold mining activities in Colombia, which generate emissions that range between 30 and 70 tons of Hg [6,7]. In the Chocó department, gold mining has been an important socioeconomic pillar in many communities for many years; it represents the base of the economy in many places within the department; however, these activities have generated environmental impacts on important water sources, such as the Atrato River [8]. Significant amounts of Hg are used in gold mining—approximately 150 tons—in this way a large amount of mining waste enters the Atrato river basin or its tributaries, contaminating the water and sediments [7,9]. In the sediments of the river basin, Hg is transformed into MeHg by the action of bacteria. This organic form of Hg is quite toxic—it can cross membranes and bioaccumulate in freshwater fish species in toxic concentrations [10,11]. The incorporation of heavy metals and MeHg into the food chain is one of the biggest concerns of these activities due to the inevitable transfer to human beings, with food intake as the main route of exposure. According to WHO reports, approximately 500,000 people perish annually across the world due to the consumption of contaminated food, 80,000 of which are associated with Colombia [12]. For this reason, the international standards that regulate the limits of tolerance or acceptance of the levels of heavy metals in food are important, with the objective of protecting human health in each region [13].

There have been several reports on different continents related to the risk assessments or quantification of heavy metals in plant and/or animal material arising from contamination as a direct result of the intervention of mining [14–17]. In Colombia, some investigations have reported high concentrations of these metals in fish commonly consumed in various communities; in addition, assessments of risk associated with the consumption of fish contaminated by heavy metals have been undertaken [18,19]. In the Colombian Pacific, Hg, Cd, Pb, and MeHg in fish from the Atrato river basin have been reported in high concentrations, particularly in fish with carnivorous habits [8,9,20,21]. In addition, Salazar-Camacho et al. [8] carried out a risk assessment of the consumption of fish in relation to levels of Hg, Cd, Pb, and MeHg in the riverside population of the Atrato River basin. It is important to note that most of the food consumed by families in the municipalities of the riverside areas of the Atrato River basin comes from crops and fishing in the territory itself. For this reason, the consumption of contaminated fish, fruits and vegetables is considered the main source of human exposure to heavy metals [22]. Therefore, the objectives of this study were: (1) to determine the concentrations of Hg, Cd, Pb and MeHg in fish, fruits and vegetables; (2) to determine the target hazard quotient (THQ) and total THQ to assess the carcinogenic and non-carcinogenic risk of metals from fish consumption; and (3) to evaluate the risk to human health due to the consumption of fish and vegetables (fruits-tubers and vegetables-stems) contaminated with heavy metals in residents of riverside municipalities of the middle basin of the Atrato River.

2. Materials and Methods

2.1. Study Area

This research was carried out in the Atrato river basin, a biodiverse lotic ecosystem located in the Colombian Pacific (Figure 1). It is made up of large bodies of water, forests, wetland swamps, grasslands, vast expanses of land used for agriculture, and many rural communities [23]. The basin has a depth of 31–38 m, an area of 35,700–36,400 km^2, a length of 750 km, and a width that varies between 150 and 500 m. It arises in the municipality of El Carmen de Atrato, specifically in the Cerro Plateado, and empties into the Gulf of Urabá, in the Caribbean Sea. The basin has a flow of approximately 4137 m^3/s, an annual precipitation between 5000- and 12,000- mm year^{-1}, and an average annual temperature of 26 °C [23,24]. It receives the flow of water from many rivers along its course, some of which have high levels of mining waste contaminated with Hg, Cd, Pb, and As. Among the most important of these are the Quito, Bebaramá, Bebará, Neguá, and Cabí [25,26].

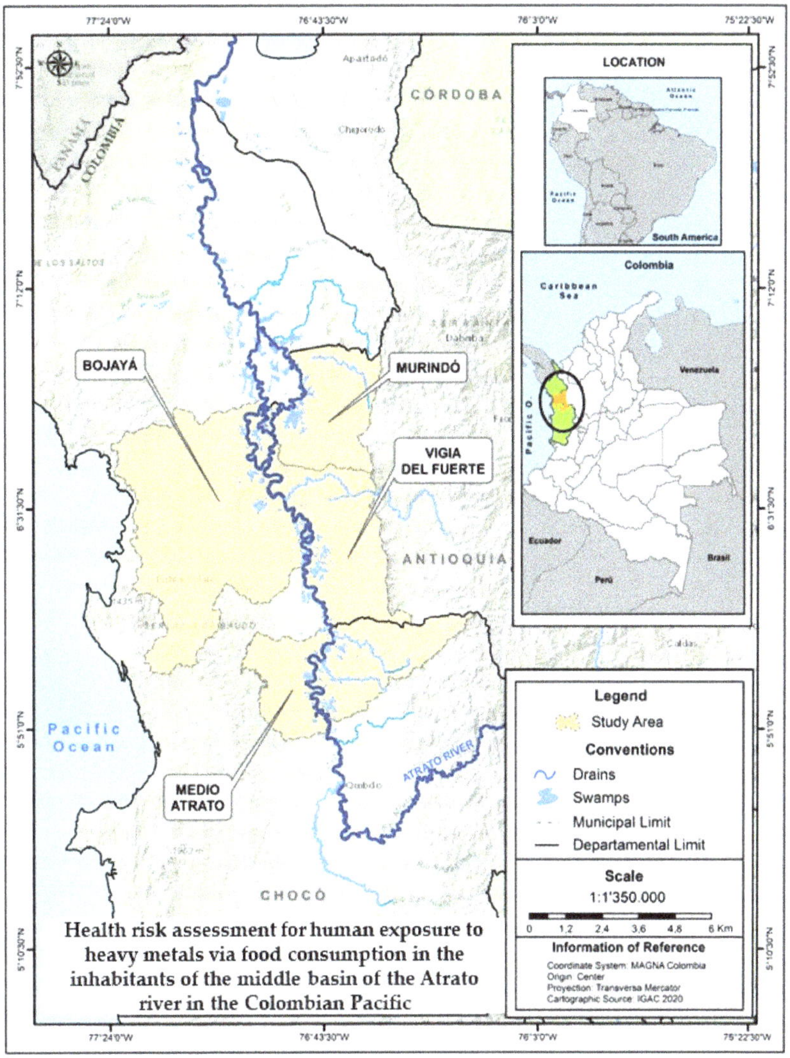

Figure 1. Map of the four municipalities studied belonging to the middle basin of the Atrato River (Chocó-Colombia): Medio Atrato, Bojaya, Vigía del Fuerte, and Murindó.

Our study was conducted in four areas of the tropical geographic basin of the Atrato River. The areas were the Medio Atrato, Bojaya, Murindó and Vigía del Fuerte, which were selected because they are important sites of gold mining [25,27]. The study was carried out on highly consumed fish, fruits and vegetables in the study areas. The fish samples were captured in the Atrato River within the study areas. Fruits and vegetables (fruits-tubers and vegetables-stems) are cultivated by the inhabitants of the communities for self-consumption and commercialization. These plants are cultivated in many cases in soils contaminated with heavy metals or, often, are treated with irrigation systems with water from the Atrato River.

2.2. Sampling

Fish: The capture of ichthyological material involved guides from each sampling point using artisanal fishing equipment, such as fishing nets, cast nets, and fishing rods. In total 440 individuals of 19 fish species were collected. The carnivorous species (330 individuals) were *Argeneiosus pardalis, Pimelodella chagresi, Rhamdia quelen, Trachelyopterus fisheri, Pimelodus punctatus, Andinoacara pulche, Leporinus muyscorum, Caquetaia kraussii, Caquetaia kraussii, Caquetaia kraussii, Caquetaia kraussii, Caquetaia kraussii*, and *Pimelodus punctatus, Leporinus muyscorum, Caquetaia kraussii, Ctenolucius beani, Pseudopimelodus schultzi, Sternopygus aequilabiatus, Hoplias malabaricus, Astyanax fasciatus, Caquetaia umbrifera, Cynopotamus atratoensis, Geophagus Pellegrini*, and *Pimelodus* sp. The non-carnivorous (110 individuals) were *Prochilodus magdalenae* and *Hypostomus hondae*. The fish samples were placed in polyethylene bags which had been previously labeled, placed in a polystyrene cooler to conserve the specimens, and transported to the Toxicology and Environmental Management laboratory of the University of Córdoba (Colombia). Subsequently, the total length of each fish was measured and, through an incision in the dorsal muscle, ten grams of tissue were extracted in one portion and again kept cold (4 °C) until the concentrations of the metals under study and the percentages of methylmercury (MeHg) were determined.

The fish species were identified using specialized taxonomic keys [28] with the help of the ichthyology team of the Technological University of Chocó (Colombia) and field guides.

Plants: The collection of vegetable material was carried out with the help of field guides at each sampling point. A total of 154 individual samples were collected and classified into two groups, fruits (12 species) and stems (one species). The fruit species were *Alibertia patinoi, Cocos nucifera, Citrus aurantifolia, Solanum sessilliflorum* Dunal, *Zea mays, Carica papaya, Musa balbisiana, Musa × paradisiac, Oryza sativa, Musa sapientum, Musa* sp., *Dioscorea trifida*; tubers (2): *Colocasia esculenta, Manihot esculenta*; vegetables (9): *Eryngium foetidum, Ocimum basilicum* L., *Ocimum tenuiflorum, Allium fistulosum* L., *Ocimum campechianum, Basella rubra* var, *Zingiber officinale, Origanum vulgare, Minthostachys mollis*. The stem species was *Saccharum officinalis*. The collections were georeferenced in situ, deposited in polyethylene bags and transported to the laboratory of Toxicology and Environmental Management of the University of Córdoba (Colombia). There, maceration was carried out followed by cold storage until the concentrations of the metals under study were determined.

The species were identified using specialized taxonomic keys [29] with the support of the biosystematics team of the Universidad Tecnológica del Chocó, the Chocó herbarium, the Universidad de Córdoba, and field description notebooks.

2.3. Analysis of THg, MeHg, As, Pb, and Cd in Fish Muscle, Fruits, and Vegetables

Quantities of 0.02 g of freeze-dried fish and plant material samples were analyzed for Hg concentration levels by atomic absorption spectrometry using a direct mercury analyzer (DMA-80 TRICELL, Milestone Inc, Italy) using the established EPA Method 7473. (EPA, 1998). For Cd and Pb analysis, Method 3051 A [30] and the procedure described by Karadede and ÜnlÜ [31] were used, respectively. The samples were digested with HNO_3/HCl (1:3 v/v) and Cd and Pb analyses were performed using a Thermo Elemental Solaar S4-graphite furnace method. Analysis was carried out by calcining a mixture of 1 g of each fish and plant material sample with Mg (NO_3) 2 at 550 °C in a muffle furnace, then 1 mL of concentrated HNO_3 was added and heated to dryness, subsequently dissolved with 4.5 N HCl, filtered through a 0.45 µm filter, then topped up to 25 mL with distilled water (Szkoda et al., 2006). A Thermo Scientific iCETM 3500 AAS atomic absorption spectrometer, coupled to a VP100 continuous flow steam generator (Waltham, MA, USA), was used for As analysis (HGAAS; standard Methods SM 31114, 2017). Certified µreference materials (CRM) IAEA 407 and DORM-4 and triplicate evaluation were used for quality control of the methods used. The recovery percentage was between 92 and 96% and the detection limits for the different metals were 0.014 µg g^{-1} for Hg, 0.006 µg g^{-1} for Cd, 0.010 µg g^{-1} for Pb, and 0.016 µg g^{-1} for As. MeHg analysis was only performed on the fish samples; for MeHg quantification, approximately 0.2–0.3 g of fresh fish were digested with hydrobromic acid and toluene. The resulting mixture was centrifuged and extracted several times with L-cysteine. Finally, a 100 μL aliquot of the aqueous phase was injected into DMA [32]. Quality control of the method was performed in triplicate using a CRM DORM-2 standard of dogfish muscle certificate (4.47 ± 0.32 μg g^{-1}). The percent recovery for MeHg was 99%. The limit of detection was 0.007 μg g^{-1}, while the limit of quantification was 0.023 μg g^{-1}.

2.4. Estimated Daily Intake (EDI)

A risk assessment of the estimated daily intake of fish, fruit, and vegetables in the municipalities of the middle Atrato basin was calculated using factors such as food consumption (μg kg^{-1} (bw) $week^{-1}$), the concentration of the metal, and the body weight (bw), necessary for the determination of the estimated daily intake (EDI). The average body weight in adults living in the middle basin of the Atrato river was 69.2 ± 3.3 kg (Table S1). On average, the inhabitants of the middle basin of the Atrato river consume 256 g/day of fish [8]. The average consumption of fruits and vegetables was established individually for each food; measurements were made by weighing the food portions from the information provided in the surveys (Table S1). The results obtained from the analysis of the concentrations of Hg, As, Cd, Pb, and MeHg in μg/kg of wet weight, and from the surveys, made it possible to calculate the necessary parameters to evaluate the risk of human exposure to these contaminants by the consumption of fish, fruits, and vegetables. For this calculation, the equation described by Chien et al. [33] was used.

$$\mathrm{EDI} = \frac{\mathrm{C} \times \mathrm{Cconc}}{\mathrm{BW}} \tag{1}$$

2.5. Determination of the Target Hazard Quotient (THQ)

The non-carcinogenic risk was calculated using the THQ formula (HQ/RfDo). When the THQ is less than one, it indicates that the hazard quotient (HQ) is below the reference dose (RfDo) and, therefore, that daily exposure at this level is unlikely to cause adverse effects over a person's lifetime. The THQ is a calculation using the assumptions of the US EPA Integrated Risk Analysis (USEPA, 2000). The THQ was determined using the following equation [34]:

$$\mathrm{THQ} = \frac{\mathrm{EFr} \times \mathrm{EDtot} \times \mathrm{FIR}}{\mathrm{RfDo} \times \mathrm{Bw} \times \mathrm{ATn}} \times \mathrm{C} - 10^{-3} \tag{2}$$

where EDtot is the exposure duration (30 years), EFr is the exposure frequency (350 days/year), FIR is the food ingestion rate (g/day), 10^{-3} is the unit conversion factor (kg/g); C is the element concentration in fish (μg/g ww), RfDo is the oral reference dose (mg/kg-day), Bw is the average adult body weight according to surveys in each municipality and ATn is the average exposure time for non-carcinogens (365 days/year × number of exposure years, assuming 30 years). The total THQ (TTHQ) was expressed as the sum of the THQ values for each studied element [34]:

$$\text{Total TQH (TTHQ)} = \text{TQH(Toxican 1)} + \text{TQH(Toxican 2)} + \text{TQH(Toxican 3)} + \ldots \tag{3}$$

2.6. Carcinogenic Risk Assessment (CR)

The contaminants associated with carcinogenic risk are As and Pb. The CR is defined as the lifetime chance of an individual developing any type of cancer due to exposure to carcinogenic hazards [35,36]. These carcinogenic health risks are calculated individually for each element throughout its lifetime according to the following equation [37,38]:

$$\mathrm{CDDingestion} = \frac{\mathrm{C} \times \mathrm{FIRing} \times \mathrm{ED} \times \mathrm{EF}}{\mathrm{Bw} \times \mathrm{AT} \times \mathrm{SF}} \tag{4}$$

$$\mathrm{CR} = \mathrm{CDDingestion} \times \mathrm{SF} \tag{5}$$

where CDDingestion is the chronic daily dose (mg/kg/day) established for potentially toxic heavy metals received by ingestion; C is the heavy metal content in fish and vegetables (μg/g); FIRingestion is the ingestion rate: 256 mg/day in fish [8], in vegetables the quantity varies (Table S1)—these amounts are for adult women and men who lived in the middle zone of the Atrato river basin; ED is the exposure duration, six years for children and 30 years for adults [35]; EF is the exposure frequency—in this study, 365 days/year; SF is the slope factor (kg/mg/Day)—SF is 1.5 for As and 0.0085 for Pb [38]; Bw is the mean body weight for each municipality; AT is the mean time for carcinogens (As and Pb) 70 × 365 days [38]; and CR is the carcinogenic risk—when the CR value is less than 1×10^{-6}, the risk is regarded as negligible, and if the CR value exceeds 1×10^{-4}, there is likely to be a risk to human health [37].

2.7. Assessment of Human Health Risk Related to MeHg

The risk assessment was carried out with a total of 769 voluntary respondents surveyed. Data was collected on educational level, average body weight, gender, frequency of fish consumption per week, number of times each participant ate fish per day in a week, and the type of fish. A total of 323 respondents were male and 446 were female. All respondents were aged ≥ 15 years. The respondent sample was divided into two groups: the first group, comprising children and women of childbearing age (WCHA) and the second group comprising the rest of the adult population (GP) (Table S1). The potential risk of human exposure to MeHg was assessed according to the estimated weekly intake (EWI-μg/bw/week) using the equation described by UNEP [39]:

$$\text{EWI} = \frac{\text{IR} \times \text{C}}{\text{Bw}} \tag{6}$$

where IR is the weekly intake (g/week) of fish, C is the median concentration of MeHg (μg/kg) in fish, and Bw is the bodyweight of the person (kg). The IRs were calculated taking into account the consumed portion of fish (g/day) and the frequency of consumption (days/week) in the four municipalities of the Atrato river basin.

The concentration of MeHg that the consumed fish species should contain to avoid exceeding the provisional tolerable weekly intake (PTWI) [34] was calculated using the following equation:

$$[\text{MeHg}]\text{permissible} = \frac{\text{C} \times \text{PTWI}}{\text{EDI}} \tag{7}$$

where PTWI is the reference value of 1.6 μg/kg bw/week for women of childbearing age and children, 3.2 μg/kg bw/week is the reference value for the adult population [34] and C is the median concentration of MeHg (μg/kg) in fish. The amount of fish consumption is crucial in risk assessment as it plays a key role in the generation of adverse effects on human health. For this reason, we estimated the maximum quantity of fish that a person could consume weekly (MFW) without adverse health effects, according to the following equation:

$$\text{MFW} = \frac{\text{PTWI} \times \text{IR}}{\text{EWI}} \tag{8}$$

Finally, to calculate the degree of Hg contamination in the most consumed fish species, we used the formula proposed by Zhang et al. [40]:

$$\text{Pi} = \frac{\text{Ci}}{\text{Si}} \tag{9}$$

where Ci and Si are the median concentration of the metal in the fish muscle and the value of the evaluation criteria, respectively, and Pi is the pollution index. Two reference limits were used: a threshold of 200 μg/kg ww [41] for vulnerable populations, such as children under 15 years of age and women of childbearing age, and a threshold of 500 μg/kg ww [42] for the adult population.

2.8. Data Analysis

Kolmogorov–Smirnov (n $\geq$ 50) and Shapiro–Wilk (n < 50) tests were used to assess whether data did or did not follow a normal distribution. The Kruskal–Wallis test was employed to evaluate the differences among Hg, As, Pb, and Cd concentrations between fish species. Spearman's test was performed to evaluate the correlation between the concentrations of the elements and the trophic level of the fish. A *p*-value of 0.05 was chosen to indicate statistical significance. THg, MeHg, As, Pb, and Cd concentrations were expressed as µg/kg ww of fish, fruits and vegetables. The statistical analyzes were carried out using the R Project statistical program version 3.6.1 (R Core Team, Vienna, Austria).

3. Results

3.1. Concentrations of Hg, Cd, Pb, As, and MeHg in Fish, Fruits, and Vegetables

The concentrations of Hg, Pb, Cd and As (µg kg^{-1}), the percentages of MeHg (%MeHg), the fish species consumed by the inhabitants and the trophic level of the fish species collected in the middle basin of the Atrato river are shown in Supplementary Materials Table S2. Based on the taxonomic identification, 19 species of fish were obtained, with a total of 440 individuals. Of these species, five were carnivorous (59 individuals), two were detritivorous (110 individuals), six were omnivorous species with a tendency to carnivory (120 individuals), one was omnivorous (14 individuals), one was an omnivorous species with a preference for fish and plant material (26 individuals), and one was a piscivorous species (111 individuals). The most common species were *Prochilodus magdalenae* (15.9%), *Hoplias malabaricus* (11.1%), *Rhamdia quelen* (10.0%), *Hypostomus hondae* (8.9%), and *Astyanax fasciatus* (8.6%). The concentrations of Hg were highest, followed by those of As, Pb, and Cd. (Table S2).

For the fish samples evaluated, the minimum values of Hg were found between two municipalities in Vigía del Fuerte, for *Leporinus muyscorum* with 44.5 $\pm$ 22.7 µg kg^{-1} and in Murindó for species *Hypostomus hondae* (41.5 $\pm$ 32.5 µg kg^{-1}) and *Andinoacara pulcher* (32.9 $\pm$ 4.3 µg kg^{-1}). Samples from the municipality of Vigía del Fuerte contained the highest concentration of As in *Ctenolucius beani* with 1008.0 $\pm$ 552.7 µg kg^{-1} (Table S2). Of the total fish samples for Hg and MeHg, 221 individuals (piscivores 111, omnivores with a tendency to carnivory 74, and carnivores 36) exceeded the limit for populations at risk, which was established at 200 µg kg $^{-1}$ [41]. Among these, 102 individuals (53 piscivores, 26 omnivores with a tendency to carnivory, and 25 carnivores) exceeded the maximum recommended limit for human consumption established in 500 µg kg^{-1} [42]. The species *C. beani* and *Ageneiosus pardalis* exceeded the maximum permissible limits for THg 500 µg kg $^{-1}$. Some species also exceeded the maximum permissible limits: *A. pardalis* (Murindó and Vigía del Fuerte), *Trachelyopterus fisheri* (Murindó and Vigía del Fuerte), *H. malabaricus* (Murindó), *Cynopotamus atratoensis* (Vigía del Fuerte), *C. beani* (Murindó and Bojayá), *Sternopygus aequilabiatus* (Murindó and Vigía del Fuerte), *Caquetaia kraussii* and *R. quelen* (Murindó). Five species exceeded the threshold for MeHg 500 µg kg^{-1}; these were: *A. pardalis* (Murindó and Vigía del Fuerte), *T. fisheri* (Murindó and Vigía del Fuerte), *C. beani* (Murindó and Bojayá), *C. atratoensis* (Vigía del Fuerte) and *S. aequilabiatus* (Vigía del Fuerte). *C. kraussii* and *H. malabaricus* were close to reaching this threshold, with concentrations of 473.05 and 485.71 µg kg^{-1}, respectively (Figure 2a). The species *P. magdalenae* was the most consumed in the middle basin of the Atrato river; its concentrations of THg and MeHg did not exceed the established thresholds of 200 µg kg^{-1} and 500 µg kg^{-1}.

None of the mean concentrations for the 19 species of fish reported in this study were higher than the maximum permissible levels of Cd, Pb, and As in the muscle of the fish, which were established at 300, 100, and 1000, µg kg^{-1}, respectively (Figure 2c–e) [43,44]. For As, two species *Geophagus Pellegrini* and *H. hondae* presented concentrations > 100 µg kg^{-1} < 300 µg kg^{-1} (Figure 2c).

The four metals studied were quantified in vegetables for two groups: fruits and vegetables; the species of vegetables with a total of 24 species were subdivided into 15 fruits and 9 vegetables. A total of 154 individual specimens are listed in Supplementary Materials,

Table S3. Hg presented the lowest concentration in the species *Citrus aurantifolia* for the municipalities of Murindó, Bojayá, and Vigía del Fuerte, with *O. vulgare* being the species with the highest mean concentration value (Figure 3a). The minimum values for As concentrations were in the species *Musa paradisiaca* and *Alibertia patinoi* (3.18–6.61 mg kg^{-1}). The maximum mean concentrations of As were found in Murindó (60.84 $\pm$ 107.8 mg kg^{-1}) and Bojayá (57.1 $\pm$ 99.6 mgkg^{-1}). The species with the highest concentrations for As were *Minthostachys mollis* and *Ocimum campechianum*. The concentrations for As observed in the food samples indicated that three species contained high concentrations and exceeded the permissible WHO limit of 100 mg kg^{-1} [41]; these were *Origanum vulgare, A. patinoi,* and *O. campechianum* (Figure 3b). Pb showed the lowest average concentration levels by municipality in Murindó (4.3 $\pm$ 0.0 mg kg^{-1}) and Bojayá (4.7 $\pm$ 1.5 mg kg^{-1}). For Cd, none of the observed concentrations were at a level above that recommended by the WHO (100 mg kg^{-1}) (Figure 3c). However, the mean concentrations of Hg, Pb, Cd, and As in fruits and vegetables in each municipality exceeded the maximum permissible levels of concentrations established by the Codex for Hg: 0.1 mg kg^{-1}, Pb: 0.1 mg kg^{-1}, Cd: 0.05 mg kg^{-1} and As: 0.1 mg kg^{-1} [45] and by the EU for Hg: 0.1 mg kg^{-1}, Pb: 0.2 mg kg^{-1}, Cd: 0.05 mg kg^{-1} and As: 0.2 mg kg^{-1} [46] (Table S3).

3.2. Species of Fish, Fruits, and Vegetables Most Consumed in the Middle Basin of the Atrato River

The data obtained for the average consumption of fish, fruits, and vegetables amongst inhabitants of the municipalities of Medio Atrato, Bojayá, Vigía del Fuerte, and Murindó, belonging to the middle basin of the Atrato river, indicated that, of the 19 species of fish captured, nine were associated with a high preference for consumption amongst riverside inhabitants of this middle area of the basin (Table S2). Among these, two species of non-carnivorous habits stood out: *P. magdalenae* and *H. hondae*. The results showed six species with high rates of population intake, including *P. magdalenae, L. muyscorum, P. schultzi, A. pardalis, H. Malabaricus* and *R. quelen* with 4.3, 4.3, 4.2, 3.5, 3.6 and 3.7 days/week reported mean consumption, respectively.

A total of 24 plant species were analyzed in the different municipalities studied, seven of which were associated with a high rate of preference for population consumption. Among the fruits, five stood out: *M. balbisiana, M. sapientum, C. aurantifolia, M. paradisiaca* and *O. sativa*; for vegetables, the data showed that only two species were associated with a high frequency of preference for population consumption: *E. foetidum* and *O. campechianum* (Figure 3).

3.3. Determination of Human Health Risk by Fish and Vegetables Consumption

The general characteristics of the population of the middle basin of the Atrato River are described in Supplementary Materials Table S1. A total of 57.9% (n = 446) of the respondents were women and 42.1% (n = 323) were men, with a mean age between 39 and 46 years (range: 15.2–88.8) and average weight between 68 and 72 kg.

The evaluation of the risk to health due to food consumption is important when considering the amounts consumed in the population under study. The average amount of fish consumed by inhabitants of the middle basin of the Atrato River was 256 g/day [8], with an average frequency of consumption of 3.4 days/week. The total weight of fish consumed was greater than 30 kg year^{-1}.

For fruits and vegetables, the amount consumed varied according to the food. The average frequency of consumption of fish, fruits, and vegetables for inhabitants of the middle basin of the Atrato river was 3.4 days/week.

The risk arising from the consumption of fish, fruits, and vegetables was calculated using the EDI, THQ, TTHQ, and CR metal contamination indices to estimate the accumulation and risk levels of metals for the most consumed species (Tables 1 and 2).

3.4. Determination of Health Risk from the Consumption of Fish, Fruits, and Vegetables

To assess the risks to human health based on frequent exposure of an individual through consumption of different species of vegetables and fish that accumulate different levels of contamination, the estimated daily intake (EDI) for each metal was determined. For the vegetables studied, the EDI values were 0.0135 µg/kg/day (range: 7.882×10^{-5}–0.068) for Hg, for As 0.0595 g/kg/day (range: 0.001–0.755), for Pb 0.0235 g/kg/day (range: 5.401–0.232) and for Cd 0.009 g/kg/day (range: 1.322×10^{-5}–0.860). The species *Musa paradisiaca* presented the lowest EDI values for Hg, Pb, Cd, and Pb for Medio Atrato. Only the species *Alibertia patinoi* (Medio Atrato) and *Ocimum campechianum* (Bojayá) exceeded the RfDo value limit for As (As 0.30 g/kg/day) (Table 1).

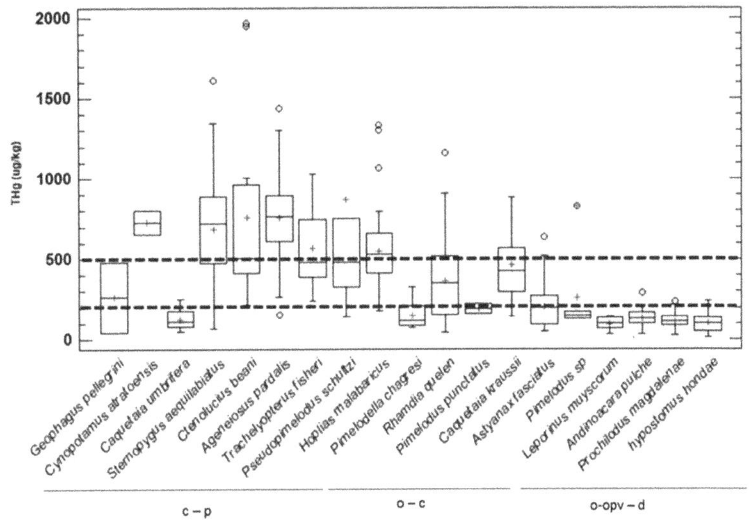

(**a**)

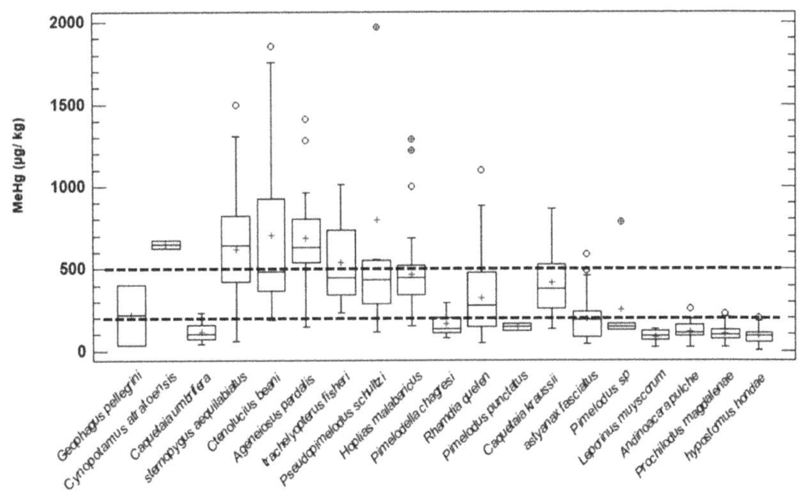

(**b**)

Figure 2. *Cont.*

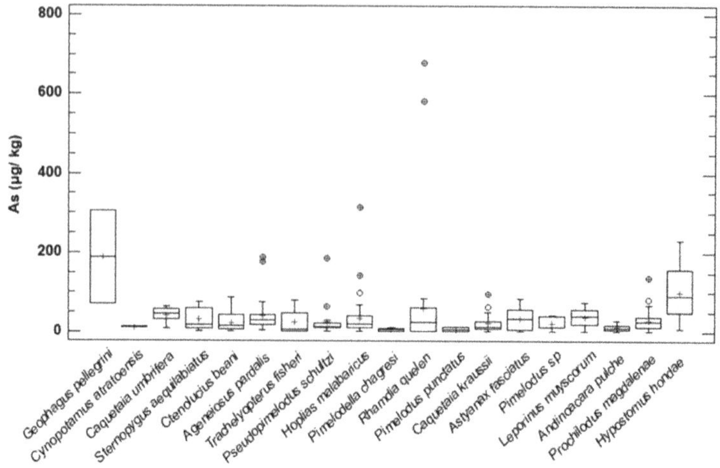

(c)

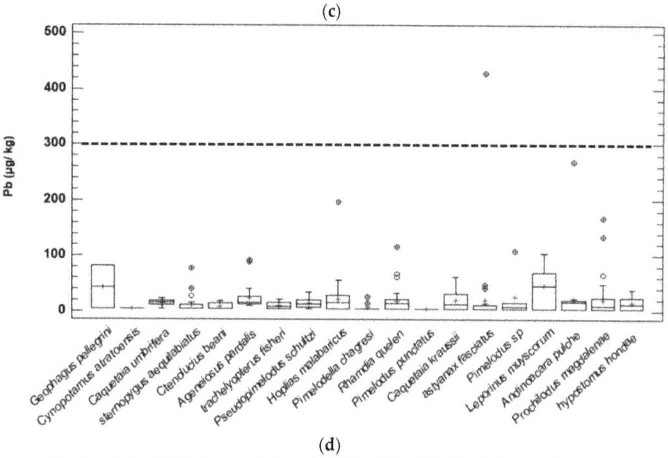

(d)

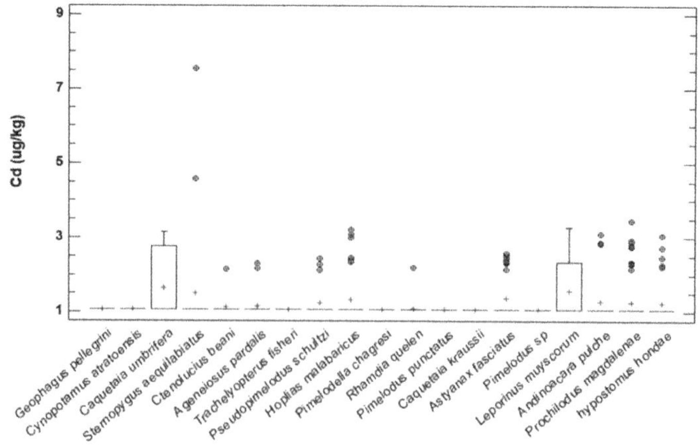

(e)

Figure 2. Concentrations of Hg (**a**), MeHg (**b**), As (**c**), Pb (**d**), and Cd (**e**) (µg kg^{-1} ww) in fish samples from the middle basin of the Atrato River. Consumption limits proposed by WHO [47,48] for these contaminants are as follows: for Hg and MeHg, threshold value of 500 ugkg^{-1} of body weight for the adult population; threshold value of 200 ugkg^{-1} of body weight for vulnerable populations (children, the elderly, and women of childbearing age). For As, Pb and Cd, 1000, 300 and 100 μg kg^{-1}, respectively [49,50]. Feeding habits: c-carnivore, p-piscivore, o-omnivore, d-detritivore, oc-omnivore/carnivore and opv-omnivore/piscivore. Scattered concentrations (°), mean concentrations (+).

For fish, the mean EDI values were 1.374 μg/kg/day (range: 0.122–3.556) for Hg, for As 0.119 g/kg/day (range: 0.016–0.672), for Pb 0.067 μg/kg/day (range: 0.015–0.067) and, for Cd, standard values of 0.003 g/kg/day were shown for all municipalities (Table 1). The species *Ctenolucius beani* presented the highest EDI values for Hg in the municipalities of Murindó (3.612 μg/kg/day), Bojayá (3.556 μg/kg/day), Medio Atrato (1.370 μg/kg/day) and Vigía del Fuerte. (1.655 μg/kg/day), followed by *Hoplias malabaricus* in Murindó (2.589 μg/kg/day), Bojayá (1.579 μg/kg/day), Medio Atrato (1.719 μg/kg/day), Vigía del Fuerte (1.694 μg/kg/day) and *Trachelyopterus fisheri* in Vigía del Fuerte (3.028 μg/kg/day), Murindó (2.673 μg/kg/day) and Medio Atrato (1.513 μg/kg/day). All the other species reported in this study presented values above what is recommended. However, species such as *Andinoacara pulcher* in Murindó and Vigía del Fuerte, *Leporinus muyscorum* in Murindó, *Hypostomus hondae* in Vigía del Fuerte, presented RfDo values below that established for Hg (0.16 μg/kg/day). In the case of As, the EDI values were above the established threshold (As 0.30 μg/kg/day) in species such as *H. hondae* (Murindó), and *Geophagus Pellegrini* (Vigía del Fuerte). In general, the EDI values for Pb remained below the limits; however, the species *L. muyscorum* in Murindó presented values above the RfDo limits (0.390 μg/kg/day) (Table 2).

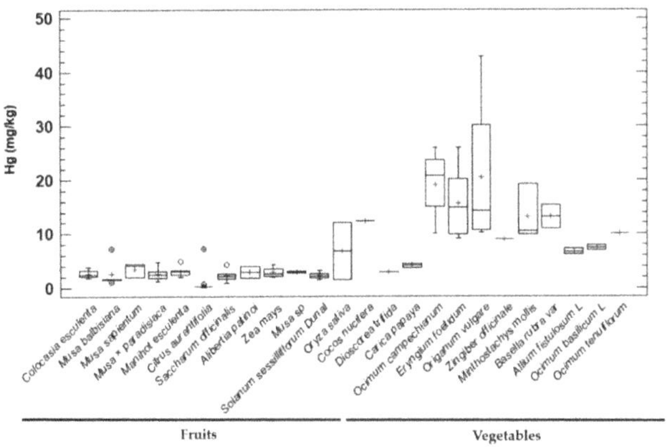

(a)

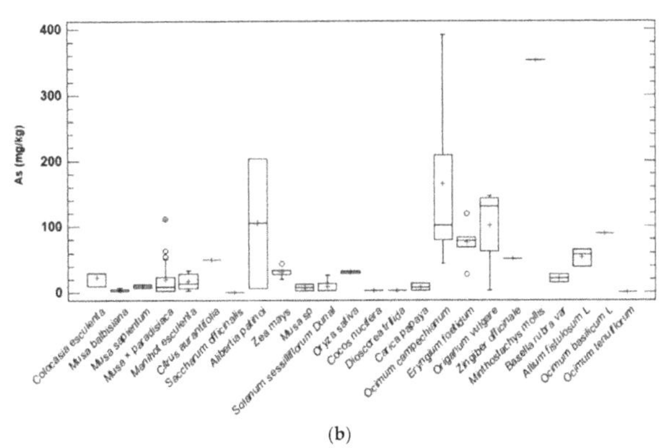

(b)

Figure 3. *Cont.*

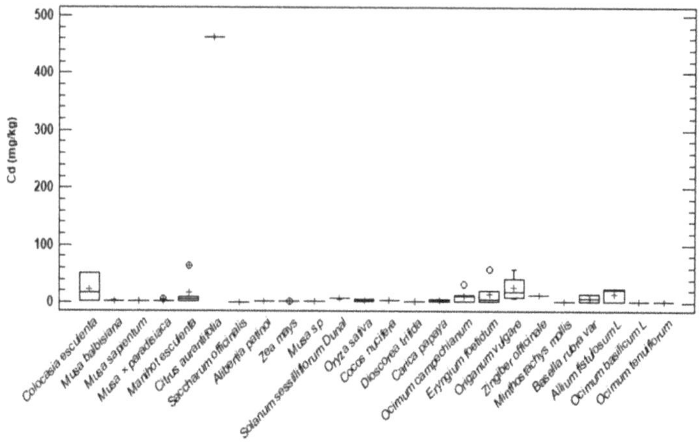

(**c**)

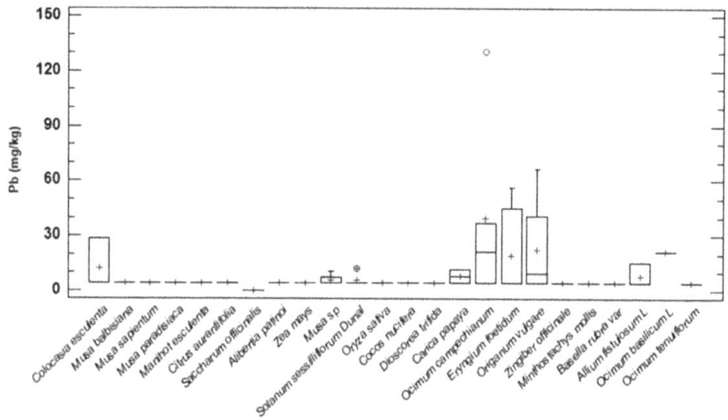

(**d**)

Figure 3. Concentrations of Hg (**a**), As (**b**), Pb (**c**), and Cd (**d**) in fruits and vegetables commonly consumed in municipalities of the middle basin of the Atrato River. Consumption limits proposed by the WHO: threshold value of 100 mg kg^{-1} of body weight for the adult population and for vulnerable populations (children, the elderly and women of childbearing age) for the heavy metals studied.

3.5. Assessment of Non-carcinogenic Health Risk

To establish the carcinogenic risk in the population exposed to the consumption of fish, fruit-bulbs, and vegetables-stems, the target hazard quotients (TQH and TTQH) were calculated. When the THQ value is less than one, it means that the exposure level is less than the RfDo, indicating that daily exposure at this level is unlikely to cause adverse effects during a person's lifetime [33,47]. For the present evaluation, the mean TQH values of fish were 3.700 for Hg (range: 0.643–21.668), for Cd 0.023 (range: 0.022–0.023), for As 0.228 (range: 0.053–1.398) and for Pb 0.012 (range: 0.004–0.140).

The THQ value for fish consumption for the metals Cd and Pb was less than one for all fish species in all the municipalities studied. Most of the fish species presented a TQH for Hg higher than one; the highest values were reported in *A. pardalis* (21.688) in Vigía del Fuerte, *C. beani* (21.649 and 21.313) in Murindó and Bojayá, and *H. malabaricus* (10.30) in Medio Atrato. However, species such as *Andinoacara pulcher* and *L. muyscuorum*, in the municipalities of Murindó and Vigía del Fuerte, and *H. hondae* in Murindó presented TQH values for Hg lower than one. For As, the TQH index was higher than one for

H. hondae and *T. fisheri* in Medio Atrato; for Cd and Pb all THQ values were less than one. The relative contributions to the total THQ score showed that Hg was the main contributor to risk. The average THQ for Hg far exceeded one (6.768). In general, the species with the highest values of EDI, THQ, and TTHQ for Hg were those of carnivorous, piscivorous and omnivorous habits with a tendency to carnivory including *H. malabaricus*, *A. pardalis*, *R. quelen*, *T. fisheri*, *S. aequilabiatus*, *C. kraussii*, *C. atratoensis*, *C. beani* and *P. schultzi* (Table 1).

For fruit and vegetables, the mean TQH values were 0.074 (range: 0.001–0.412) for Hg, for Cd 0.054 (range: 7.924×10^{-5}–1.322), for As 0.017 (range: 0.001–0.207) and for Pb 0.006 (range: 1.479×10^{-5}–0.063). The TQH values for vegetable consumption, based on the mean concentrations of Hg, Pb, and As for all vegetable species, were generally less than one in the four municipalities studied, except in species such as *Manihot esculenta* (1.322) in Murindó and *Colocasia esculenta* (2.405) in Bojayá. The relative contributions to total THQ showed that Cd was the main element contributing to the risk for vegetables (Table 1).

3.6. Assessment of Carcinogenic Health Risk

The calculation of the carcinogenic risk was applied only to the metals As and Pb due to their carcinogenic effect on humans. It was applied to the species of fish and vegetables commonly consumed by the inhabitants of the middle basin of the Atrato River. To do this, the USEPA has recommended using the carcinogenic risk index (CR). If the CR values are less than 1.0×10^{-6}, they are considered negligible, while a CR value greater than 1.0×10^{-4} indicates potential adverse effects in humans [37,38]. The data obtained for vegetables contaminated with As and Pb in the groups of fruit-bulbs and vegetables-stems collected in the municipalities of Medio Atrato, Bojayá, Vigía del Fuerte, and Murindó showed values lower than 1.0×10^{-6} (Table 1). Therefore, our results indicate that there was no carcinogenic risk to the health of the population from the consumption of these foods.

However, with respect to the carcinogenicity values of the metals Pb and As, the results showed, for all the species captured in the middle basin of the Atrato River, values lower than 1.0×10^{-6} for Pb. However, the results of the evaluation of the carcinogenic health risk of As through fish consumption showed that the CR values for the fish species *H. malabaricus*, *T. fisheri*, *P. schultzi*, *C. beani*, *Caquetaia umbrifera*, *Sternopygus aequilabiatus*, *A. fasciatus*, *L. muyscorum*, *Prochilodus magdalenae*, *H. hondae* in Medio Atrato; *A. pardalis*, *Geophagus pellegrini*, *Caquetaia kraussii*, *Rhamdia quelen*, *L. muyscorum*, *P. magdalenae*, *H. hondae* in Vigía del Fuerte; and *A. pardalis*, *H. malabaricus*, *C. beani*, *R. quelen*, *L. muyscorum*, *P. magdalenae*, *H. hondae* in Murindó were above the 1.0×10^{-6} limit, but below the 1.0×10^{-4} limit (Table 1). The results of this study indicate that there is a significant carcinogenic risk for the health of the inhabitants of the middle basin of the Atrtao River, mainly due to As, through the consumption of fish.

In general, the results of EDI, THQ, TTHQ, and CR show that the inhabitants of the middle basin of the Atrato River could present health problems during their lives, due to the consumption of fish, depending on the type of fish they consume.

Table 1. Estimated daily intake (EDI), estimated target hazard quotients (THQ), total estimated hazard quotient (TTHQ) and carcinogenic risk of the most consumed vegetables (fruits and vegetables) for individual metals from vegetables consumption.

Fish Species	Hg		Cd		As		Pb			As *	Pb *
	EDI+	THQ	EDI+	THQ	EDI+	THQ	EDI+	THQ	TTHQ	CR	CR
Medio Atrato											
Fruits-Tubers											
Colocasia esculenta	0.021	0.128	0.008	0.052	0.077	0.021	0.232	0.063	0.264	6.356×10^{-7}	1.894×10^{-6}
Oryza sativa	0.021	0.129	0.011	0.070	0.051	0.014	0.007	0.002	0.215	4.199×10^{-7}	6.324×10^{-8}
Alibertia patinoi	0.006	0.041	0.001	0.011	**0.356**	0.097	0.007	0.002	0.151	2.908×10^{-6}	6.198×10^{-8}
Musa × paradisiaca	0.013	0.080	0.006	0.036	0.018	0.004	0.024	0.006	0.126	1.486×10^{-7}	2.023×10^{-7}
Musa balbisiana	0.012	0.073	0.090	0.543	0.024	0.006	0.039	0.010	0.632	6.467×10^{-5}	2.026×10^{-7}
Cocos nucifera	0.035	0.021	0.008	0.049	0.007	0.002	0.012	0.003	0.075	6.333×10^{-8}	1.012×10^{-7}
Citrus aurantifolia	0.001	0.004	Nd	Nd	Nd	Nd	Nd	Nd	0.004	Nd	Nd
Musa sp	0.027	0.163	0.009	0.055	0.109	0.029	0.092	0.025	0.272	8.903×10^{-7}	7.577×10^{-7}
Musa sapientum	0.019	0.117	0.006	0.039	0.062	0.017	0.027	0.007	0.180	5.073×10^{-7}	2.206×10^{-7}
Solanum sessilliflorum	0.014	0.084	0.034	0.208	0.011	0.003	0.018	0.005	0.300	9.500×10^{-8}	1.518×10^{-7}
Zea mays	0.011	0.067	0.003	0.018	0.109	0.029	0.012	0.003	0.117	8.918×10^{-7}	1.012×10^{-7}
Dioscorea trifida	0.020	0.122	0.007	0.045	0.019	0.005	0.030	0.008	0.180	1.567×10^{-7}	2.504×10^{-7}
Carica papaya	0.004	0.302	0.040	0.243	0.034	0.009	0.034	0.009	0.563	2.795×10^{-7}	2.851×10^{-7}
Manihot esculenta	0.012	0.074	0.006	0.038	0.016	0.004	0.026	0.007	0.123	1.330×10^{-7}	2.215×10^{-7}
Vegetables-Stems											
Eryngium foetidum	0.015	0.090	0.007	0.042	0.014	0.054	0.028	0.007	0.193	4.447×10^{-7}	2.285×10^{-7}
Ocimum campechianum	0.047	0.028	0.022	0.132	0.221	0.060	0.126	0.034	0.254	1.806×10^{-6}	1.032×10^{-6}
Origanum vulgare	0.018	0.110	0.010	0.065	0.063	0.017	0.028	0.007	0.199	5.143×10^{-7}	2.338×10^{-7}
Saccharum officinalis	0.015	0.094	Nd	Nd	Nd	Nd	Nd	Nd	0.094	Nd	Nd
Bojayá											
Fruits-tubers											
Colocasia esculenta	0.014	0.088	0.401	**2.405**	0.231	0.063	0.034	0.009	**2.565**	1.887×10^{-6}	2.782×10^{-7}
Musa balbisiana	0.013	0.081	0.037	0.225	0.023	0.006	0.038	0.010	0.322	1.948×10^{-7}	3.112×10^{-7}
Manihot esculenta	0.013	0.083	0.012	0.075	0.075	0.020	0.025	0.006	0.184	6.166×10^{-7}	2.042×10^{-7}
Musa × paradisiaca	0.011	0.068	0.019	0.114	0.099	0.027	0.023	0.006	0.215	8.137×10^{-7}	1.945×10^{-7}
Citrus aurantifolia	0.002	0.013	0.860	0.476	0.093	0.025	0.008	0.002	0.516	7.591×10^{-7}	6.566×10^{-8}
Alibertia patinoi	0.003	0.018	0.001	0.010	0.011	0.003	0.007	0.002	0.033	9.258×10^{-8}	5.958×10^{-8}
Musa sp	0.022	0.136	0.008	0.053	0.022	0.006	0.036	0.009	0.204	1.856×10^{-7}	2.967×10^{-7}
Solanum sessilliflorum	0.010	0.065	0.033	0.203	0.011	0.003	0.017	0.004	0.275	9.138×10^{-8}	1.459×10^{-7}
Zingiber officinale	0.004	0.024	0.005	0.034	0.022	0.006	0.001	0.001	0.065	1.873×10^{-7}	1.605×10^{-8}
Vigía del fuerte											
Fruits-tubers											
Colocasia esculenta	0.019	0.118	Nd	Nd	Nd	Nd	Nd	Nd	0.118	Nd	Nd
Oryza sativa	0.002	0.015	0.001	0.011	0.057	0.015	0.007	0.002	0.043	4.653×10^{-7}	6.191×10^{-8}
Citrus aurantifolia	0.001	0.003	Nd	Nd	Nd	Nd	Nd	Nd	0.003	Nd	Nd
Musa balbisiana	0.015	0.090	0.009	0.057	0.024	0.006	0.038	0.010	0.163	1.984×10^{-7}	3.170×10^{-7}
Zea mays	0.005	0.032	0.002	0.017	0.075	0.020	0.012	0.003	0.072	6.120×10^{-7}	9.906×10^{-8}
Musa sp	0.014	0.087	0.009	0.054	0.107	0.029	0.037	0.026	0.196	8.757×10^{-7}	7.747×10^{-7}
Musa × paradisiaca	0.012	0.073	0.005	0.035	0.215	0.059	0.024	0.006	0.173	1.759×10^{-6}	1.981×10^{-7}
Manihot esculenta	0.019	0.116	0.033	0.199	0.170	0.046	0.025	0.006	0.367	1.390×10^{-6}	2.080×10^{-7}
Solanum sessilliflorum	0.007	0.047	0.020	0.121	0.070	0.019	0.029	0.008	0.195	5.717×10^{-7}	2.426×10^{-7}

Table 1. *Cont.*

Fish Species	Hg		Cd		As		Pb			As *	Pb *
	EDI⁺	THQ	EDI⁺	THQ	EDI⁺	THQ	EDI⁺	THQ	TTHQ	CR	CR
Vegetables-Stems											
Ocimum campechianum	0.029	0.176	0.024	0.148	0.199	0.054	0.008	0.002	0.380	1.626×10^{-6}	6.934×10^{-8}
Eryngium foetidum	0.006	0.040	0.004	0.028	0.052	0.014	0.003	0.001	0.083	4.309×10^{-7}	2.476×10^{-8}
Origanum vulgare	0.007	0.043	0.005	0.031	0.051	0.014	0.006	0.001	0.089	4.173×10^{-7}	1.486×10^{-8}
Ocimum basilicum L.	0.015	0.092	0.002	0.012	0.175	0.048	0.042	0.011	0.163	1.434×10^{-6}	3.443×10^{-7}
Minthostachys mollis	0.006	0.041	Nd	Nd	Nd	Nd	Nd	Nd	0.041	Nd	Nd
Basella rubra var	0.009	0.055	0.001	0.005	0.012	0.003	0.003	0.001	0.064	1.012×10^{-7}	2.972×10^{-8}
Saccharum officinalis	0.014	0.085	Nd	Nd	Nd	Nd	Nd	Nd	0.085	Nd	Nd
Murindó											
Fruits-tubers											
Colocasia esculenta	0.032	0.192	0.138	0.832	0.251	0.068	0.036	0.009	1.101	2.047×10^{-6}	2.962×10^{-7}
Musa balbisiana	0.068	0.412	0.009	0.059	0.064	0.017	0.040	0.012	0.500	5.274×10^{-7}	3.314×10^{-7}
Musa sapientum	0.029	0.174	0.006	0.040	0.062	0.017	0.027	0.007	0.238	5.100×10^{-7}	2.258×10^{-7}
Musa × paradisiaca	0.016	0.099	0.011	0.068	0.231	0.063	0.025	0.006	0.236	1.885×10^{-6}	2.071×10^{-7}
Manihot esculenta	0.024	0.147	0.220	1.322	0.122	0.033	0.026	0.007	1.509	9.967×10^{-7}	2.175×10^{-7}
Citrus aurantifolia	0.001	0.003	Nd	Nd	Nd	Nd	Nd	Nd	0.003	Nd	Nd
Alibertia patinoi	0.003	0.022	0.006	0.037	0.011	0.003	0.007	0.002	0.064	9.679×10^{-8}	6.344×10^{-8}
Zea mays	0.006	0.041	0.004	0.028	0.077	0.021	0.012	0.003	0.093	6.322×10^{-7}	1.035×10^{-7}
Vegetables-stems											
Eryngium foetidum	0.007	0.043	0.002	0.012	0.057	0.015	0.003	0.001	0.071	4.661×10^{-7}	2.589×10^{-8}
Ocimum basilicum L.	0.014	0.085	Nd	Nd	Nd	Nd	Nd	Nd	0.085	Nd	Nd
Minthostachys mollis	0.014	0.084	0.001	0.004	0.258	0.070	0.003	0.001	0.159	2.110×10^{-6}	2.589×10^{-8}
Ocimum tenuiflorum	0.020	0.123	0.002	0.013	Nd	Nd	0.008	0.002	0.138	Nd	7.251×10^{-8}
Allium fistulosum L.	8.324×10^{-5}	0.001	1.322×10^{-5}	7.924×10^{-5}	0.001	0.001	5.401×10^{-5}	1.479×10^{-5}	9.405×10^{-5}	3.937×10^{-9}	4.402×10^{-10}
Saccharum officinalis	0.008	0.053	Nd	Nd	Nd	Nd	Nd	Nd	0.053	Nd	Nd

The order of the municipalities is from upstream to downstream on the Atrato River. * Carcinogenic risk. EDI⁺ values in (μg/kg/day). Oral reference dose (RfD) values used were: Hg = 0.16 μg/kg/day; Cd = 1.00 μg/kg/day; Pb = 3.50 μg/kg/day; and As = 0.30 μg/kg/day. The values highlighted in bold are above the established limits.

Table 2. Estimated daily intake (EDI), estimated target hazard quotients (THQ), total estimated hazard quotient (TTHQ) and carcinogenic risk of the most consumed fish for individual metals from fish consumption.

Fish Species	Hg		Cd		As		Pb			As *	Pb *
	EDI+	THQ	EDI+	THQ	EDI+	THQ	EDI+	THQ	TTHQ	CR	CR
Medio Atrato											
Hoplias malabaricus	**1.719**	**10.30**	0.003	0.023	0.117	0.376	0.074	0.020	10.719	$\mathbf{1.695 \times 10^{-3}}$	6.057×10^{-7}
Trachelyopterus fisheri	**1.513**	**9.070**	0.003	0.023	0.223	**1.340**	0.047	0.013	10.446	$\mathbf{3.217 \times 10^{-3}}$	3.890×10^{-7}
Pseudopimelodus schultzi	**1.622**	**9.722**	0.003	0.023	0.062	0.200	0.045	0.012	9.957	9.003×10^{-5}	3.703×10^{-7}
Ctenolucius beani	**1.370**	**8.215**	0.003	0.023	0.156	0.501	0.015	0.004	8.743	$\mathbf{2.256 \times 10^{-3}}$	1.295×10^{-7}
Caquetaia umbrifera	**0.401**	**2.407**	0.003	0.023	0.169	0.541	0.058	0.016	2.987	$\mathbf{2.436 \times 10^{-3}}$	4.777×10^{-7}
Sternopygus aequilabiatus	**0.617**	**3.700**	0.003	0.023	0.237	0.757	0.024	0.006	4.486	$\mathbf{3.409 \times 10^{-3}}$	2.003×10^{-7}
Rhamdia quelen	**0.608**	**3.646**	0.003	0.023	0.067	0.214	0.060	0.016	3.899	9.674×10^{-5}	4.958×10^{-7}

Table 2. *Cont.*

Fish Species	Hg		Cd		As		Pb			As *	Pb *
	EDI+	THQ	EDI+	THQ	EDI+	THQ	EDI+	THQ	TTHQ	CR	CR
Astyanax fasciatus	**0.781**	**4.681**	0.003	0.023	0.172	0.550	0.015	0.004	5.258	**2.476×10^{-3}**	1.295×10^{-7}
Andinoacara pulcher	**0.522**	**3.129**	0.003	0.023	0.041	0.133	0.064	0.017	3.302	6.003×10^{-5}	5.256×10^{-7}
Leporinus muyscorum	**0.431**	**2.585**	0.003	0.023	0.155	0.497	0.051	0.014	3.119	**2.237×10^{-3}**	4.207×10^{-7}
Prochilodus magdalenae	**0.292**	**1.752**	0.003	0.023	0.140	0.450	0.034	0.009	2.234	**2.026×10^{-3}**	2.797×10^{-7}
Hypostomus hondae	**0.487**	**2.922**	0.003	0.023	**0.437**	**1.398**	0.082	0.022	4.365	**6.291×10^{-3}**	6.723×10^{-7}
Bojayá											
Hoplias malabaricus	**1.579**	**9.465**	0.003	0.022	0.063	0.201	0.090	0.024	9.532	9.070×10^{-5}	7.374×10^{-7}
Ctenolucius beani	**3.556**	**21.313**	0.003	0.022	0.025	0.080	0.015	0.004	21.419	3.610×10^{-5}	1.245×10^{-7}
Prochilodus magdalenae	**0.501**	**3.003**	0.003	0.022	0.056	0.179	0.057	0.015	3.219	8.090×10^{-5}	4.697×10^{-7}
Vigía del fuerte											
Ageneiosus pardalis	**3.615**	**21.668**	0.003	0.022	0.120	0.384	0.049	0.013	22.087	**1.728×10^{-3}**	4.033×10^{-7}
Hoplias malabaricus	**1.694**	**10.155**	0.003	0.022	0.042	0.136	0.015	0.004	10.317	6.131×10^{-5}	1.268×10^{-7}
Trachelyopterus fisheri	**3.028**	**18.149**	0.003	0.022	0.009	0.031	0.004	0.004	18.206	1.400×10^{-5}	1.268×10^{-7}
Pseudopimelodus schultzi	**2.105**	**12.619**	0.003	0.022	0.043	0.139	0.050	0.013	12.793	6.272×10^{-5}	4.076×10^{-7}
Ctenolucius beani	**1.655**	**9.923**	0.003	0.022	0.044	0.143	0.026	0.007	10.095	6.441×10^{-5}	2.132×10^{-7}
Sternopygus aequilabiatus	**2.776**	**16.641**	0.003	0.022	0.039	0.126	0.015	0.004	16.793	5.705×10^{-5}	1.268×10^{-7}
Cynopotamus atratoensis	**2.621**	**15.710**	0.003	0.022	0.044	0.143	0.015	0.004	15.879	6.467×10^{-5}	1.268×10^{-7}
Geophagus Pellegrini	**0.943**	**5.653**	0.003	0.022	**0.672**	**2.149**	0.154	0.042	7.866	**9.674×10^{-3}**	1.256×10^{-6}
Caquetaia kraussii	**1.446**	**8.667**	0.003	0.022	0.072	0.230	0.015	0.004	8.923	**1.036×10^{-3}**	1.268×10^{-7}
Rhamdia quelen	**1.639**	**9.824**	0.003	0.022	0.187	0.598	0.015	0.004	10.448	**2.694×10^{-3}**	1.268×10^{-7}
Pimelodus sp	**0.531**	**3.185**	0.003	0.022	0.052	0.167	0.032	0.008	3.382	7.536×10^{-5}	2.632×10^{-7}
Andinoacara pulcher	0.149	0.893	0.003	0.022	0.068	0.217	0.039	0.010	1.142	9.788×10^{-5}	3.251×10^{-7}
Leporinus muyscorum	**0.347**	**2.080**	0.003	0.022	0.209	0.668	0.144	0.039	2.809	**3.006×10^{-3}**	1.179×10^{-6}
Prochilodus magdalenae	**0.347**	**2.083**	0.003	0.022	0.097	0.310	0.015	0.004	2.419	**1.398×10^{-3}**	1.268×10^{-7}
Hypostomus hondae	0.107	0.643	0.003	0.022	0.079	0.253	0.015	0.004	0.922	**1.142×10^{-3}**	1.268×10^{-7}
Murindó											
Ageneiosus pardalis	**2.633**	**15.781**	0.003	0.023	0.100	0.321	0.064	0.017	16.142	**1.448×10^{-3}**	5.244×10^{-7}
Hoplias malabaricus	**2.589**	**15.521**	0.003	0.023	0.082	0.262	0.045	0.012	15.818	**1.182×10^{-3}**	3.712×10^{-7}
Trachelyopterus fisheri	**2.673**	**16.025**	0.003	0.023	0.016	0.053	0.038	0.010	16.111	2.409×10^{-5}	3.117×10^{-7}
Pseudopimelodus schultzi	**1.310**	**7.855**	0.003	0.023	0.022	0.071	0.016	0.004	7.953	3.197×10^{-5}	1.325×10^{-7}
Ctenolucius beani	**3.612**	**21.649**	0.003	0.023	0.071	0.228	0.016	0.004	21.904	**1.028×10^{-3}**	1.325×10^{-7}
Sternopygus aequilabiatus	**2.468**	**14.796**	0.003	0.023	0.010	0.032	0.057	0.015	14.866	1.464×10^{-5}	4.704×10^{-7}
Caquetaia kraussii	**1.904**	**11.412**	0.003	0.023	0.037	0.119	0.091	0.025	11.579	5.383×10^{-5}	7.443×10^{-7}
Rhamdia quelen	**1.970**	**11.808**	0.003	0.023	0.077	0.247	0.069	0.019	12.097	**1.113×10^{-3}**	5.633×10^{-7}
Astyanax fasciatus	**0.318**	**1.908**	0.003	0.023	0.060	0.193	0.158	0.043	2.167	8.729×10^{-5}	1.293×10^{-6}
Pimelodus punctatus	**0.726**	**4.356**	0.003	0.023	0.030	0.096	0.016	0.004	4.479	4.328×10^{-5}	1.325×10^{-7}
Pimelodella chagresi	**0.443**	**2.657**	0.003	0.023	0.023	0.075	0.016	0.004	2.759	3.382×10^{-5}	1.325×10^{-7}
Andinoacara pulcher	0.123	0.740	0.003	0.023	0.029	0.093	0.513	0.140	0.996	4.195×10^{-5}	4.183×10^{-6}
Leporinus muyscorum	0.122	0.736	0.003	0.023	0.106	0.341	0.390	0.106	1.206	**1.535×10^{-3}**	3.181×10^{-6}
Prochilodus magdalenae	**0.468**	**2.807**	0.003	0.023	0.091	0.293	0.080	0.021	3.144	**1.322×10^{-3}**	6.537×10^{-7}
Hypostomus hondae	**0.165**	0.994	0.003	0.023	**0.312**	0.997	0.044	0.012	2.006	**4.490×10^{-3}**	3.619×10^{-7}

The order of the municipalities is from upstream to downstream on the Atrato River. * Carcinogenic risk. EDI^{+} values in (μg/kg/day). Oral reference dose (RfD) values used are: Hg = 0.16 μg/kg/day; Cd = 1.00 μg/kg/day; Pb = 3.50 μg/kg/day; and As = 0.30 μg/kg/day. The values highlighted in bold are above the established limits.

3.7. Risk Assessment by MeHg in Most Consumed Fish

The mean concentrations of MeHg (μg kg^{-1}) and the percentage of MeHg (%MeHg) in the fish species consumed by the inhabitants of the middle basin of the Atrato River are presented in Supplementary Materials, Table S2. Of the total fish species studied, 221 individuals exceeded the limit for populations at risk, which was established at 200 μg kg^{-1} of MeHg [41]. Among these 102 individuals exceeded the maximum recommended limit for human consumption, established at 500 μg kg^{-1} of MeHg (Table S2). All the municipalities studied, except Medio Atrato, reported concentrations that exceeded the permissible limits of 500 μg kg^{-1}. The species with the highest concentrations were: in Murindó, *C. beani*, *A. pardalis*, *T. fisheri*, *H. malabaricus*, *S. aequilabiatus*, *R. quelen*, *C. kraussii*; in Bojaya, *C. beani*, *A. pardalis*, *T. fisheri*, *C. atratoensis*; and in Vigía del Fuerte, *P. schultzi*. Other species, such as *P. schultzi* (Murindó and Medio Atrato), *C. kraussii* (Vigia del Fuerte), *H. malabaricus* (Bojayá, Medio Atrato, and Vigía del Fuerte), *T. fisheri*, *A. fasciatus*, *G. Pellegrini* (Medio Atrato), and *C. beani* (Medio Atrato and Vigía del Fuerte) exceeded the limit for vulnerable populations (WCHA) of 200 μg kg^{-1}. In general, the carnivorous species with the highest concentrations of MeHg was *A. pardalis* with 956.82 μg kg^{-1} in the municipality of Vigía del Fuerte, which represents an important species in the food security of the population under study. However, for the species *P. magdalenae*, the highest consumption for which was reported in the middle basin of the Atrato river, the concentrations of MeHg did not exceed any of the thresholds established by the WHO (200 μg kg^{-1} and 500 μg kg^{-1}) (Table S2).

When the daily intake rate (RI) was estimated for the fish species commonly consumed in the studied sites, six species presented values higher than 500 g/week and lower than 700 g/week (which is of particular relevance when assessing vulnerable populations, e.g., children and women of childbearing age), including *C. atratoensis* (Vigía del Fuerte), *S. aequilabiatus*, *C. umbrifera* (Medio Atrato), *T. fisheri* (Medio Atrato and Murindó), *P. punctatus* (Murindó) and *C. beani* (Murindó and Bojayá). In the municipality of Murindó, the species *R. quelen*, *L. muyscorum* and *P. schultzi*, presented the highest levels of IR, with consumption levels of 1254.4 g/week, 1433.6 g/week and 1587.2 g/week, respectively (Table S4). Regarding the frequency of consumption (FIR), for the species *P. magdalenae* weekly consumption of four or more times was reported in all the municipalities studied; however, the municipality of Murindó presented the highest FIR values for *P. schultzi* (6.2 days/week), *L. muyscorum* (5.6 days/week), and *R. quelen* (4.9 days/week). Values close to this limit were observed in *R. quelen* (2.7 days/week) and *T. fisheri* (2.7 days/week), both in the municipality of Vigía del Fuerte (Table S4). In relation to the estimated weekly intake (EWI), the results showed that 10 fish species exceeded the potential weekly intake threshold (PTWI) for the GP group (3.2 μg kg bw/week). For the WCHA group (1.6 μg kg bw/week), it was also shown that the species with piscivorous, carnivorous and omnivorous habits with a carnivorous tendency, had the highest EWI values. These species were *A. pardalis* (Vigía del Fuerte and Murindó), *C. beani* (Medio Atrato, Bojayá, and Murindó) and *R. quelen* (Murindó) with values of 4.7, 2.8 and 2.7 times the PTWI for the GP group and 9.1, 5.4 and 5.3 times the PTWI for the WCHA group, respectively. The results showed that fish consumption limits higher than those recommended (MFW) were obtained in all the municipalities studied. In Medio Atrato, the species presented values between 0.3 to 1.6 and 0.5 to 3.3 times the PTWI for the GP and WCHA group, respectively; in Bojayá, between 0.6 to 2.1 and 1.2 to 4.4 times, respectively; in Vigía del Fuerte between 0.1 to 9.1 and 0.2 to 9.1 times, respectively; and in the municipality of Murindó between 0.1 to 3.0 and 0.2 to 6.0 times, respectively. The lowest values recorded for *Hypostomus hondae* (Vigía del Fuerte) were about 10.3- and 5.5-fold lower than the PTWI for the GP and WCHA, respectively. High consumption species, such as *A. pardalis*, *H. malabaricus*, *C. beani* and *P. schultzi* represent a serious risk to the health of the inhabitants in the studied areas due to the high concentrations of MeHg in their tissues, such that it is recommended for riverside populations to reduce or eliminate the consumption of these fish. Similarly, species of high consumption preference, such as *P. magdalenae*, presented values 2.7 times below the recommended consumption in Medio Atrato, 2.6 times below in Vigía del Fuerte, 1.7 times below in Bojayá, and 1.5 below times in Murindó, and for *L. muyscorum*, 5.4 times below in the municipality of Medio Atrato. These results suggest that these species can be consumed frequently by the inhabitants of the studied areas as their content does not exceed recommended MFW consumption limits. In addition, these species could be important for replacement of species that present high concentrations in the diet.

3.8. Diagnosis of the Population

The contamination index (Pi) [40] was used to show the degree of contamination with Hg for each species of fish in each municipality, taking as a reference the permissible limits established by the WHO (500 $\mu g\ kg^{-1}$ and 200 $\mu g\ kg^{-1}$) [47,48]. Table S5 shows that, when the Pi values were calculated according to the WHO limit [42], the fish species *A. pardalis* in the municipalities of Murindó and Vigía del Fuerte, *T. fisheri* in Vigía del Fuerte and Murindó, *C. beani* in Bojayá, and *C. atratoensis* in Vigía del Fuerte presented slight contamination ($1 < Pi \leq 2$) (Table S5). The contamination index was also calculated based on the WHO threshold [47] (2008) indicating that *P. schultzi* in Murindó, *H. malabaricus* in Bojayá and Vigía del Fuerte, *C. beani* and *T. fisheri* in Medio Atrato, and *C. kraussii* and *G. Pellegrini* in Vigía del Fuerte showed a slight degree of contamination ($1 < Pi \leq 2$) (Table S5). The species *Sternopygus aequilabiatus* (Murindó), *A. pardalis* (Murindó), *R. quelen* (Murindó and Vigía del Fuerte), *C. kraussii* (Medio Atrato), *Hoplias malabaricus* (Murindó and Medio Atrato), *P. schultzi* (Medio Atrato and Vigia del Fuerte) and *C. beani* and *R. quelen* (Vigia del Fuerte) showed a moderate degree of contamination ($2 < Pi < 3$). Similarly, *T. fisheri* in Murindó and Vigia del Fuerte, *C. beani* in Murindó and Bojayá, and *S. aequilabiatus*, *C. atratoensis* and *A. pardalis* in Vigia del Fuerte, presented a high degree of contamination (Table S5).

4. Discussion

The middle basin of the Atrato river of the Colombian Pacific is within an area where the Murindó, Bebará, Bebaramá, and Neguá rivers flow—these tributaries are associated with high levels of gold-mining activity. This area has suffered significant impacts on the rivers and surrounding soils due to indiscriminate gold mining, resulting in contamination with significant concentrations of heavy metals of the fish in the area which are important for food security. This contamination has also affected the crops that play an important role in the food security of the inhabitants of this area. However, there are no records of metal concentrations in this type of food, nor of any risk assessments. Based on ruling T-622 and the Minamata agreement, it is important to assess the risk to human health from the consumption of food contaminated with these heavy metals by the inhabitants of the middle basin of the Atrato River.

In this study, the concentrations of metals in fruits-tubers and vegetable-stems followed the order As > Pb > Hg > Cd. In Medio Atrato, fruits-tubers, in general, presented low levels of metals, except for As in the species *Alibertia patinoi* (203.16) and Zea mays (38.16). Vegetable-stems showed higher levels of As, especially *Ocimum campechianum* (110.41), *Eryngium foetidum* (76.12), and *Origanum vulgare* (146.72); in Bojaya, Vigia del Fuerte, and Murindó, the metal concentrations were found to be the same. It is important to highlight that the concentrations of As in vegetables in all the municipalities studied were between 3.18–391.75 mg/kg, which is well above the maximum standards established internationally for vegetables (0.5 mg/kg) [45,46]. These results are similar to those observed in studies carried out on *Daucus carota*, *Cynara scolymus*, and *Petroselinum crispum* in the city of Sibaté (Colombia) [48].

The concentrations of As in these foods could be derived from the original soil material or the application of fertilizers and pesticides still used in agricultural activities [49–51]. In addition, studies have reported that there are differences in the concentrations in vegetables. Leafy vegetables are a group of plants recognized for having a high capacity for heavy metal accumulation [51,52], with plant species having different capacities for the absorption and accumulation of metal(oids), associated with factors such as the different characteristics of the soil or the growth period of each plant [53,54]. It is evident that these plants have good translocation characteristics for As, either derived from mining waste sources from soil removal in gold extraction activities or from natural sources in the Earth's crust [51]. The concentrations of Hg in vegetables-stems were at higher levels compared to fruit-tubers; this accumulation behavior by this type of plant species has been reported in research carried out in China [52]. However, the concentrations of Hg reported in the municipality of Lloró for vegetables were found to be lower than those in fruits and tubers [51]. These concentrations could be related to low translocation factors and common mining processes in the studied areas, with atmospheric deposition of Hg not having a significant influence on crops due to high precipitation in these areas, causing Hg to be deposited on the ground [51,55]. Pb concentrations in fruits and vegetables from the middle Atrato basin did not exceed WHO permissible consumption limits [41,42]. These results are in agreement with studies carried out on fruits and vegetables from the African countries of South Africa and Mozambique [14].

Pb levels were higher in the vegetables-stems, consistent with the findings of investigations carried out on vegetables in Baiyin, China [56]. The municipalities of Vigía del Fuerte and Medio Atrato reported higher concentration levels in vegetables-stems above those allowed by the Codex [45,46], especially for *E. foetidum* (cilantro), *O. campechianum* (basil), *O. vulgare* (oregano), *B. rubra var* (spinach), *M. mollis* (pennyroyal) and *O. basilicum L.* (white basil) with concentrations between 4.33–66.71 mg/kg. These Pb concentrations in vegetables, in general, showed values above the maximum standards established internationally by the Codex for vegetables (0.1 mg/kg) in all the studied municipalities [45,46]. The concentration levels observed in our study were very high compared to those carried out in Arequipa (Peru) in quinoa, corn and rice products, with high Pb concentrations observed of 0.55, 0.75 and 5.08 mg kg^{-1}, respectively [57]. They are consistent with Pb concentrations reported in cabbage (23.1 $\pm$ 1.5), lettuce (17.2 $\pm$ 2.7), and tomato (15.0 $\pm$ 1.1) from the city of Arba Minch (Ethiopia) [58] and concentrations of between 0.84 and 12.5 for vegetables of dietary importance in the city of Sibaté (Colombia) [48]. In the case of fruit-tubers, such as *C. esculenta* (28.34 $\pm$ 0.9), Pb levels in Medio Atrato were high, contrary to the results reported for concentrations in this tuber grown in the Canary Islands (Spain) [59].The above is possibly related to the growth of these species in contaminated soils as a result of soil removal for mining activities and the use of irrigation water contaminated with metals, something which is very common in the studied areas of the Atrato environment [60]. Another reason could be the high density of stomata in the leaves of these plants, which allows for the accumulation of atmospheric Pb [61]. Pb can be adsorbed and fixed in the clay material of soils. It is characterized by being a highly mobile metal, with mobility increasing with pH, particularly in acid conditions, such as those found in the soils of the Middle Atrato, which could favor its adsorption and accumulation [62]. In addition, the results show that these plants have the potential and capacity to bioaccumulate and translocate concentrations of Pb, as well as the other metals evaluated in this study. The levels of Cd in the fruits-bulbs of the municipalities of Murindó and Bojayá were high compared to the other two municipalities, particularly for *C. esculenta*, which is frequently consumed by the inhabitants, which presented concentrations of 1.06–50.92 mg/kg, exceeding the Codex limit [45,46]. However, the levels of Pb reported in *C. esculenta* cultivated in the Canary Islands (Spain) showed levels below the Codex limits [59]. These plants are characterized by being tubers; the bulbs grow underground and are in direct contact with the contaminated soil present in the study area and, as a result, can accumulate significant concentrations of Cd. There are reports of the large capacity for accumulation of heavy metals from this type of plant because its parts grow under the earth's surface comprising large tubers for consumption [63,64]. In general, Cd, like other metals, can be absorbed by the pores of the stomata of the leaves. However, unlike other toxic metals, such as Pb, it has high mobility in the soil, is easily absorbed by the roots, and transported to the shoots, and is uniformly distributed in the plants [64]. Its high degree of bioaccumulation is due more to soil contamination than to atmospheric deposition [65]. In our research, the concentrations of Cd did not exceed the permissible limits of consumption specified by the WHO [41,42], in contrast to the findings of Genthe et al. [14] for fruits and vegetables from African countries. In addition, the bioaccumulation ranges observed were above those reported by Real et al. [66], in which the established range was 0.003–1.616 for *O. sativa*. On the other hand, the data from our research showed concentrations higher than those reported for *Solanum lycopersicum* [58], *Daucus carota*, *Cynara escolymus* and *Petroselinum crispum* [48], *M. paradisiaca* and *C. aurantifolia* [67], *Apium graveolens*, *Lepidium sativum* and *Porrum de Alliuml* [68], and *Z. mays* and *O. sativa* [57]. Subsistence agriculture for the inhabitants of the Atrato river basin has developed on the ground and the river is the main source of water irrigation for crops. However, the high impact of gold mining in the Atrato river basin and its tributaries has generated high levels of contamination, which has contributed to the contamination of crops with Hg, Cd, Pb, and As. Therefore, these anthropogenic activities constitute an imminent risk to human health in the riverside populations of the basin. The estimated daily intake rates (EDI) of fruits-tubers and vegetables-stems for all the inhabitants of the middle basin of the Atrato river are shown in Table 1, as well as data for the average body weight by population group, the intake by food category, and the reference doses (RfDs) for Hg, Cd, As, and Pb [47]. In general, the EDI values for none of the vegetable groups exceeded the oral RfD for metals in the studied areas, except for the *Alibertia patinoi* fruit in Medio Atrato for As. Therefore, exposure to Hg, Cd, As, and Pb through the consumption of fruits-tubers and vegetables-stems does not represent a threat to the health of the inhabitants of the municipalities of the middle basin of the Atrato River since the calculated EDI values do not exceed the reference dose tolerable intake values (RfD: 0.16 μg Hg/kg/day, 1.0 μg Cd/kg/day, 0.3 μg As/kg/day, 3.5 μg Pb/kg/day) [47]. Our results are in contrast to results obtained for fruits and tubers in the municipality of Lloró (Chocó-Colombia) by Marrugo-Madrid et al. [51], where the exposure of As through the consumption of fruits could represent a threat to the health of all population groups studied since the calculated EDI values exceed the tolerable consumption reference dose by 10 times.

The non-carcinogenic health risk was also evaluated based on THQ. With a THQ < 1, the exposed population should not experience any adverse risk, but if THQ > 1, the population could experience health risks not related to cancer. Table 1 shows that most of the THQ values did not exceed one, except for *Colocasia esculenta* (HQ = 2.405) in the municipality of Bojayá for As. These results indicate that, in general, people would not experience significant health risks from the ingestion of individual meta(loids) through vegetable consumption. Carcinogenic risk (CR) evaluation was also carried out for As and Pb using the same method. For this investigation values greater than 1.0×10^{-4} were taken as indicators of risk for vegetables [37]. In our study, none of the plant species evaluated presented a carcinogenic risk for As and Pb. These results were in contrast to those of similar investigations of vegetables showing carcinogenic risk associated with Pb and As concentrations in Bangladesh and Ireland [66,69], and observations from Peru, where some species, such as *O. sativa*, exposed to As showed evidence of possible risk to the population [57]. Similarly, in the municipality of Lloró (Colombia), the fruits and tubers showed a possible carcinogenic risk by exposure to As. However, in all the municipalities, the studied vegetables presented levels above the Codex. Therefore, considering the accumulative properties of heavy metals, these may represent a health hazard for the riverside populations of the middle basin of the Atrato River.

The results showed that the concentrations in fish followed the order Hg > As > Pb > Cd, with high concentration levels of Hg, MeHg, and As observed. *C. beani* presented the highest concentrations of Hg (1008.0 ± 552.7 g kg^{-1}) in the municipality of Bojayá. However, the municipality of Vigía del Fuerte had a greater number of species with high concentrations, including *C. kraussii*, *R. quelen*, *C. beani*, *H. malabaricus*, *P. schultzi*, *C. atratoensis*, *S. aequilabiatus*, *T. fisheri*, and *A. pardalis*. These areas are characterized by extensive mining activities on the Murrí River and other tributaries [21]. In the Atrato river basin, other studies reported similar data for *A. pardalis, H. malabaricus*, and *C. beani*; the observed concentrations were related to extractive mining of the Atrato river and its tributaries [9,21]. The high concentrations of THg found were consistent with observations by Vargas-Licona and Marrugo-Negrete [70] who warned about the toxicological risk due to mining in some ecosystems in Colombia. The concentrations of MeHg in fish from the middle basin of the Atrato, especially those with a carnivorous habit, exceeded the permissible safe consumption limits by 500 μg kg^{-1} [42]. These results were similar to the findings of Salazar-Camacho et al. [21] in this area of the basin. The concentrations of THg and MeHg in fish reported in this investigation were similar to those reported in other investigations in the Atrato river basin [8,9,21]. These results are also consistent with other investigations, where species of carnivorous habit c-p and oc, such as *H. malabaricus, C. kraussii*, and *A. pardalis*, presented greater bioaccumulation of THg and MeHg [71].

The mean concentrations of Pb and Cd in fish were 12.03 $\pm$ 24.4 and 1.1 $\pm$ 2.2 g kg^{-1}, respectively. These concentrations were higher than those for investigations of fish from the Buriganga River (Bangladesh), specifically, the species *Heteropneustes fossilis, Channa striata, Labeo rohita*, and *Catla catla* [66]; however, other studies have reported lower concentrations in fish from the Ciénaga Grande de Santa Marta (Colombia) [18].

Concentrations of the metals Pb and Cd for the fish species in the middle basin exceeded the limits established by the Codex [45,46]; however, they did not exceed the permissible intake limits defined by the WHO [41,42]. In general, the concentrations of As in fish for the basin were above what is allowed [45,46]. The municipalities of Medio Atrato and Vigia del Fuerte recorded the highest levels of average concentrations in the basin (134.2 $\pm$ 61.0 and 187.2 $\pm$ 164.8 ug kg^{-1}). The species *P. magdalenae* and *Leporinus muyscorum*, which are of gastronomic importance, did not exceed the permissible intake limits defined by the WHO [41,42], which is a reflection of the preferred feeding habit of the species in the ecosystem. The EDI of fish for all the inhabitants of the middle basin of the Atrato river are shown in Table 2, as well as the intake by food category, and the RfDs for Hg, Cd, As, and Pb [38]. In general, the EDI values for Hg and As for the vegetable groups exceeded the oral RfD for metals in the studied areas, especially for Hg, where the values were between 0.107–3.615; only four fish species had concentrations below the RfD. The EDI values for As exceeded the RfD only for *H. hondae* (Medio Atrato and Mutindó) and *G. Pellegrini* (Vigia del Fuerte). Therefore, the exposure to Hg and As through the consumption of fish could represent a threat to the health of all the inhabitants of the municipalities of the middle basin of the Atrato River since the calculated EDI values exceeded the reference dose tolerable intake. Our results were similar to the findings of studies of fish in the municipality of Lloró (Chocó-Colombia), where the exposure to As and Hg through the consumption of fish could represent a threat to the health of all the studied population groups, especially for Hg [51].

The TQH data for Hg in our study were very high 0.736–21.68, which represents a risk for the coastal populations through consumption of fish contaminated with Hg. These data were similar to those reported for the municipalities of Vigía del Fuerte, Murindó, Bojayá, Medio Atrato [8], and Lloró [51]. Other studies have shown that species such as *P. schultzi, A. pardalis, S. aequilabiatus, R. quelen, H. malabaricus, Cathorops melanopus, Centrpomo undecimalis, C. umbrifera, C. kraussii, Prochilodus* sp., *Prochilodus punctatus, Prochilodus magdalenae, Spatuloricaria atratoensis, Leporinus muyscorum, Hemiancistrus wilsoni*, and *Cyphochara Magdalena* presented values greater than one for the THQ index [9]. Our results showed (Table 2) that the CR values through As exposure from fish consumption were well above 1×10^{-4} for all the municipalities (except Bojayá), especially in Medio Atrato (CR: 9.674×10^{-5}–6.291×10^{-3}), indicating that the local population should reduce its intake of fish. Species such as *A. pardalis, P. magdalenae, H. malabaricus, L. muyscorum*, and *R. quelen*, which are of gastronomic importance, showed levels of contamination of carcinogenic importance for As, except for the municipality of Bojayá. Salazar-Camacho et al. [8], reported similar data for carcinogenic risk through As exposure. The Colombian National Institute of Cancerology (INC) has estimated that there are about 74.8 new cases of cancer per 100,000 inhabitants, although in the Pacific Region the increase in national carcinogenic risk is not considered to be decisive; however, heavy metal contamination and excessive intake of contaminated food may affect local and national carcinogenic risk in the future. Genes are influenced by the environment and their modification can lead to many types of cancer—there is a direct relationship between environmental contaminants and the increase in many types of cancer.

MeHg is the most toxic form of Hg and exposure to this pollutant is associated with the consumption of fish. In this study, it represented between 71.08–99.21% of the THg. Therefore, it is important to assess the potential risk of exposure to MeHg in the population based on the estimated weekly intake (EWI), corresponding to the maximum quantity of fish that can be consumed weekly (MFW) per person without harmful effects on health, and the permissible safety level concentration of MeHg in fish for human consumption. Worryingly, our study showed that the WCHA group had a fish intake with a frequency from 0.2 to 9.1 times more than recommended, compared to the 1.3–2.1 times reported by Salazar-Camacho et al. [8] who also presented MFW values higher than those recommended. This report shows that women of childbearing age or pregnant women (WCHA group), especially in the municipalities of Medio Atrato, Vigía del Fuerte, and Munrindo, were at risk of having high concentrations of MeHg in placental tissue, blood, and cord blood, which could affect the health of the mother, fetus, and newborn.

Table S4 shows that the highest EWI corresponded to *A. pardalis* in Vigía del Fuerte (10.65 µg/kg/week). According to the recommended limits for the consumption (MFW) of fish, in the municipalities of Medio Atrato, Bojayá, Vigía del Fuerte, and Murindó, consumption of species such as *C. beani*, *P. schultzi*, *H. malabaricus*, *T. fisheri*, *A. pardalis*, *S. aequilabiatus*, *R. quelen*, *C. atratoensis*, and *C. kraussii* exceeded the weekly recommended intake by residents. Our results agree with previous reports for some of these fish species in the Atrato river basin [8,51]. The inhabitants (including children, pregnant women, and women of childbearing age, the elderly, and adults) of all the studied areas presented EWI values higher than the reference (PTWI) and the estimated values (MFW), respectively. As such, there is a potential risk to the health of these inhabitants. The results of this study are important because they show the impact of fish consumption on the most vulnerable population (WCHA) of the middle basin of the Atrato River. Local and national authorities need to implement strategies to prevent children and the WCHA group from consuming fish containing high levels of Hg, such as *H. malabaricus*, *A. pardalis*, *P. schultzi*, *C. kraussi* and *R. quelen*. They should, alternatively, recommend that these groups eat fish with low concentrations of MeHg, such as *Andinoacara pulcher*, *Leporinus muyscorum*, and *Hypostomus hondae*. Therefore, it is recommended to establish continuous monitoring of the content of metal(oids) in the riverside populations of the basin and their food, and to implement bioremediation strategies to decontaminate the soil and water in these areas, to guarantee the consumption of safe foods with respect to heavy metal content and improve the health security of the inhabitants who depend on the aquatic resources and crops of the area as the basis of food security and economic sustenance.

5. Conclusions

High concentrations of As, Hg, Pb, and Cd were identified in fish, fruits-tubers and vegetables-stems commonly consumed by inhabitants of the middle basin of the Atrato River, which exceeded Codex limits. Similarly, in fish, the concentrations of MeHg and THg exceeded the limits established by the WHO/FAO for vulnerable populations and the rest of the adult population, especially for carnivorous fish species.

A high carcinogenic and non-carcinogenic risk was evidenced for the inhabitants of the middle basin of the Atrato River due to the consumption of fish contaminated with high concentrations of As, MeHg, and THg. However, the risk associated with consumption of vegetables was very low with only a couple of species showing some degree of risk for As.

Health risks from consuming MeHg-contaminated fish are a matter of concern since many fish species were consumed at levels exceeding the recommended weekly intake

(MFW) and the PTWI for all population groups, in all areas studied. Thus, it is recommended that the consumption of carnivorous species is reduced or replaced, and that consumption of non-carnivorous species, such as *P. magdalenae*, occurs instead.

The combined exposure to the four metals through the consumption of fish, fruits, and vegetables would probably result in exceeding the RfD for the population of the middle Atrato basin. It is important that further studies of multiple exposure to toxins found in the foods most consumed by the inhabitants are undertaken. In addition, it is necessary that periodic monitoring of heavy metals is carried out in riverside populations and their food and that bioremediation strategies for soils and water sources are implemented to reduce the concentration of these pollutants and to improve food production and quality.

Supplementary Materials: The following supporting information can be downloaded at: https://www.mdpi.com/article/10.3390/ijerph20010435/s1, Table S1: General characteristics of the population from the middle basin of the Atrato river (Chocó-Colombia) (n = 446). Table S2: Concentrations of Hg, MeHg, As, Pb, and Cd (μg kg−1 ww) in fish of the Atrato region. Table S3: Concentrations of heavy metals (mg/kg) and consumption (days/week) of fruits and vegetables in the middle basin of the Atrato river. Table S4: Estimation of the potential risk to the population from fish consumption in the Atrato river basin. Table S5: Classification of pollution index of metals (Zhang et al., 2019) [40].

Author Contributions: G.C.-R. performed the experiments, data/evidence collection, writing—review, and editing of the initial draft, data curation, and did the critical review; M.S.-M. provided reagents, materials, atomic absorption spectrometer analysis, data curation, and did the critical review, commentary, and revision for the pre-publication stage, and provided oversight and leadership for the research activity planning and execution; J.M.-N. undertook oversight and leadership responsibility for the research activity planning and execution, developed and designed the methodology, and formulated the evolution of overarching research goals and aims, provided reagents, materials, instrumentation and computing resources for atomic absorption spectrometer analysis, and did the critical review, commentary, and revision for the pre-publication stage. All authors have read and agreed to the published version of the manuscript.

Funding: This work was supported by the Colombian Ministry of Science, Technology, and Innovation (Minciencias) and the University of Córdoba for the financing of the contract 849-2018. Additionally, funding was received from the University of Córdoba project FCB-01-19.

Institutional Review Board Statement: The study did not require ethical approval.

Informed Consent Statement: Informed consent was obtained from all subjects involved in the study, a format endorsed by the Ministry of Science, Technology and Innovation of Colombia was designed for the parents of the children and another for adults.

Data Availability Statement: In this study, the data are for the exclusive use of the Ministry of Science, Technology and Innovation of Colombia and no data was reported in the study.

Acknowledgments: The authors are grateful to the members of the Laboratory of Toxicology and Environmental Management of the University of Córdoba (Colombia) and the Biosystematic Research Group of the Technological University of Chocó, (Colombia). We also thank Minciencias, the Technological University of Chocó 'DLC'.

Conflicts of Interest: The authors declare no conflict of interest.

References

1. ATSDR (Agency for Toxic Substances and Disease Registry). *Priority List of Hazardous Substances;* ATSDR: Atlanta, GA, USA, 2017.
2. UNEP (United Nations Environment Programme). *Mercury–Time to Act;* UNEP: Athens, Greece, 2013.
3. Kocman, D.; Wilson, S.J.; Amos, H.M.; Telmer, K.H.; Steenhuisen, F.; Sunderland, E.M.; Mason, R.P.; Outridge, P.; Horvat, M. Toward an assessment of the global inventory of present-day mercury releases to freshwater environments. *Int. J. Environ. Res. Public Health* **2017**, *14*, 138. [CrossRef] [PubMed]
4. Obrist, D.; Kirk, J.L.; Zhang, L.; Sunderland, E.M.; Jiskra, M.; Selin, N.E. A review of global environmental mercury processes in response to human and natural perturbations: Changes of emissions, climate, and land use. *Ambio* **2018**, *47*, 116–140. [CrossRef] [PubMed]
5. da Silva, E.; Zhang, C.; Pinto, L.; Patinha, C.; Reis, P. Hazard assessment on arsenic and lead in soils of Castromil gol mining area. *Portugal. Appl. Geochem.* **2004**, *19*, 887–898. [CrossRef]

6. Telmer, K.; Veiga, M. World emissions of mercury from artisanal and small scale gold mining. In *Mercury Fate and Transport in the Global Atmosphere*; Pirrone, N., Mason, R., Eds.; Springer: New York, NY, USA, 2009; Chapter 6; pp. 131–172.
7. UPME (Unidad de Planeación Minero Energética); Ministerio de Minas y Energía—Universidad de Córdoba. *Estudio de la Cadena de Mercurio en Colombia con Énfasis en la Actividad Minera del oro*; UPME (Unidad de Planeación Minero Energética): Bogotá DC, Colombia; Ministerio de Minas y Energía—Universidad de Córdoba: Córdoba, Spain, 2014.
8. Salazar-Camacho, C.; Salas-Moreno, M.; Marrugo-Madrid, S.; Paternina-Uribe, R.; Marrugo-Negrete, J.; Díez, S. A human health risk assessment of methylmercury, arsenic and metals in a tropical river basin impacted by gold mining in the Colombian Pacific región. *Environ. Res.* **2022**, *212*, 113120. [CrossRef] [PubMed]
9. Palacios, Y.; Caballero, K.; Olivero-Verbel, J. Mercury pollution by gold mining in a global biodiversity hotspot, the Choco biogeographic region, Colombia. *Chemosphere* **2018**, *193*, 421–430. [CrossRef]
10. Buck, D.; Evers, D.; Adams, E.; DiGangi, J.; Beeler, B.; Samánek, J.; Petrlik, J.; Turnquist, M.; Speranskaya, O.; Regan, K.; et al. A global-scale assessment of fish mercury concentrations and the identification of biological hotspots. *Sci. Total Environ.* **2019**, *687*, 956–966. [CrossRef] [PubMed]
11. Gutiérrez, H.; Marrugo, J.; Díez, S.; Morales, G.; Montoya, L.; Jonathan, M.P. Distribution of chemical forms of mercury in sediments from abandoned ponds created during former gold mining operations in Colombia. *Chemosphere* **2020**, *258*, 127319. [CrossRef]
12. Organización Mundial de la Salud (OMS). *Evaluación de Seguridad de Ciertos Aditivos Alimentarios y Contaminantes; 53ª reunión del Comité Mixto FAO/OMS de Expertos en Aditivos Alimentarios (JECFA)*; Serie de aditivos alimentarios de la OMS No. 44; OMS: Ginebra, Switzerland, 2000.
13. Barraza-Martín, H. *Análisis Cuantitativo de Metales Pesados en Pescados para Exportación a la Unión Europea*; Trabajo de grado de Maestría; Facultad de Ciencias veterinarias, UNCPBA: Tandil, Argentina, 2018.
14. Genthe, B.; Kapwata, T.; Le Roux, W.; Chamier, J.; Wright, C.Y. The reach of human health risks associated with metals/metalloids in water and vegetables along a contaminated river catchment: South Africa and Mozambique. *Chemosphere* **2018**, *199*, 19. [CrossRef]
15. Wei, J.; Gao, J.; Cen, K. Levels of eight heavy metals and health risk assessment considering food consumption by China's residents based on the 5th China total diet study. *Sci. Total Environ.* **2019**, *689*, 1141–1148. [CrossRef]
16. Solgi, E.; Beigzadeh-Shahraki, F. Accumulation and Human Health Risk of Heavy Metals in Cultured Rainbow Trout (*Oncorhynchus mykiss*) Form Different Fish Farms of Eight Cities of Chaharmahal and Bakhtiari Province, Iran. *Thalassas: Int. J. Mar. Sci.* **2019**, *35*, 305–317. [CrossRef]
17. Erasmus, J.H.; Zimmermann, S.; Smit, N.J.; Malherbe, W.; Nachev, M.; Sures, B.; Wepener, V. Human health risks associated with consumption of fish contaminated with trace elements from intensive mining activities in a peri-urban region. *Sci. Total Environ.* **2022**, *825*, 154011. [CrossRef] [PubMed]
18. Pinzón-Bedoya, C.H.; Pinzón-Bedoya, M.L.; Pinedo-Hernández, J.; Urango-Cardenas, I.; Marrugo-Negrete, J. Assessment of Potential Health Risks Associated with the Intake of Heavy Metals in Fish Harvested from the Largest Estuary in Colombia. *Int. J. Environ. Res. Public Health* **2020**, *17*, 2921. [CrossRef] [PubMed]
19. Alcala-Orozco, M.; Caballero-Gallardo, K.; Olivero-Verbel, J. Biomonitoring of Mercury, Cadmium and Selenium in Fish and the Population of Puerto Nariño, at the Southern Corner of the Colombian Amazon. *Arch. Environ. Contam. Toxicol.* **2020**, *79*, 354–370. [CrossRef] [PubMed]
20. Salazar, C.; Salas-Moreno, M.; Marrugo, S.; Marrugo, J.; Díez, S. Dietary human exposure to mercury in two artisanal small-scale gold mining communities of northwestern Colombia. *Environ. Int.* **2017**, *107*, 47–54. [CrossRef]
21. Salazar, C.; Salas-Moreno, M.; Paternina, R.; Marrugo, J.; Díez, S. Mercury species in fish from a tropical river highly impacted by gold mining at the Colombian Pacific region. *Chemosphere* **2021**, *264*, 128478. [CrossRef]
22. Rice, K.; Walker, E.; Wu, M.; Gillette, C.; Blough, E. Environmental mercury and its toxic effects. *J. Prev. Med. Public Health* **2014**, *47*, 74–83. [CrossRef]
23. Ángel, S.P.; Anaya-Acevedo, J.A.; Botero, B.A. Evaluation of 3B42V7 and IMERG daily-precipitation products for a very high-precipitation region in northwestern South America. *Atmos. Rest.* **2019**, *217*, 3748.
24. Velásquez, M.; Poveda, G. Estimación del balance hídrico de la región Pacífica Colombiana. *Dyna* **2019**, *86*, 297–306. [CrossRef]
25. CODECHOCÓ (Corporación Autónoma Regional para el Desarrollo Sostenible del Chocó). *Situación Actual de la Minería en el Departamento del Chocó Informe de Gestión*; Subdirección de Calidad y Control Ambiental; CODECHOCÓ: Quibdo, DC, USA, 2012.
26. UNODOC. *(Oficina de las Naciones Unidas Contra La Droga y el Delito). Explotación de oro de Aluvión. Evidencias a Partir de Percepción Remota*; UNODOC: Washington, DC, USA, 2016.
27. Río Atrato Como Sujeto de Derechos. *Sentencia T-622 de 2016*. 2016. Available online: https://redjusticiaambientalcolombia.files.wordpress.com/2017/05/sentencia-t-622-de-2016-rio-atrato.pdf (accessed on 7 November 2022).
28. FishBase. Available online: https://www.fishbase.se/search.php (accessed on 7 November 2022).
29. Instituto de Ciencias Naturales, Facultad de Ciencias, Universidad Nacional de Colombia. 2004 y Continuamente Actualizado. Colecciones en Línea. Publicado en Internet. Available online: http://www.biovirtual.unal.edu.co (accessed on 29 March 2011).
30. EPA. *Microwave Assisted Acid Digestion of Sediments, Sludges, Soils, and Oils*; Revision 1; EPA: Washington, DC, USA, 2007; pp. 1–30.
31. Karadede, H.; Unlu, E. Heavy metal concentrations in water, sediment, fish and some benthic organisms from Tigris River, Turkey. *Environ. Monit. Assess.* **2007**, *131*, 323–337. [CrossRef]

32. Cordeiro, F.; Gonçalves, S.; Caldéron, J.; Robouch, P.; Emteborg, H.; Coneely, P.; Tumba-Tshilumba, M.F.; Kortsen, B.; de la Calle, B. *IMEP-115: Determination of Methylmercury in Seafood—A Collaborative Trial Report*; EUR 25830; The Publications Office: Luxembourg, 2013.
33. Chien, L.; Hung, T.; Choang, K.; Yeh, C.; Meng, P.; Shieh, M.; Han, B. Daily intake of TBT, Cu, Zn, Cd and as for fishermen in Taiwan. *Sci. Total Environ.* **2002**, *285*, 177–185. [CrossRef]
34. FAO (Food and Agriculture Organization)/WHO. *The State of Food Security and Nutrition in the World 2017*; Building Resilience for Peace and Food Security; FAO: Rome, Italy; WHO: Geneva, Switzerland, 2017.
35. Adimalla, N. Spatial distribution, exposure, and potential health risk assessment from nitrate in drinking water from semi-arid region of South India. *Hum. Ecol. Risk Assess. Int. J.* **2019**, *26*, 310–334. [CrossRef]
36. Zhaoyong, Z.; Mamat, A.; Simayi, Z. Pollution assessment and health risks evaluation of (metalloid) heavy metals in urban street dust of 58 cities in China. *Environ. Sci. Pollut. Res.* **2019**, *26*, 126–140. [CrossRef] [PubMed]
37. USEPA. *Risk Assessment Guidance for Superfund. Human Health Evaluation Manual (Part A)*; Office of Emergency and Remedial Response: Washington, DC, USA, 1989; Volume 1.
38. USEPA (United States Environmental Protection Agency). *Treatment Technologies for Mercury in Soil, Waste, and Water*; Office of Solid Waste and Emergency Response: Washington, DC, USA, 2002.
39. UNEP (United Nations Environment Programme). *Guidance for Identifying Population at Risk from Mercury Exposure*; UNEP (DTIE)/Hg/INC.2/INF/3; UNEP: Geneva, Switzerland, 2010.
40. Zhang, T.; Xu, W.; Lin, X.; Yan, H.; Ma, M.; He, Z. Assessment of heavy metals pollution of soybean grains in North Anhui of China. *Sci. Total Environ.* **2019**, *646*, 914–922. [CrossRef] [PubMed]
41. WHO (World Health Organization). *Guidance for Identifying Populations at Risk from Mercury Exposure*; UNEP DTIE Chemicals Branch and WHO Department of Food Safety, Zoonoses and Foodborne Diseases; UNEP; WHO: Geneva, Switzerland, 2008.
42. WHO. *Methylmercury in Environmental Health Criteria 101*; World Health Organization: Geneva, Switzerland, 1990; p. 19.
43. FAO/OMS (FAO/Organización Mundial de la Salud). *Codex Alimentarius—Normas Generales de Contaminantes y Toxinas en los Alimentos. Anexo 1 Niveles máximos y de Referencia para Contaminantes y Toxinas en Los Alimentos*; Referencia CX/FAC 02/16; Alimentación Conjunta FAO/OMS Programa de Normas, Comité del Codex: Rotterdam, The Netherlands, 2002.
44. FAO. *Recopilación de Límites Legales para Sustancias Peligrosas en el Pescado y los Productos Pesqueros (Organización de las Naciones Unidas para la Agricultura y la Alimentación)*; FAO: Rome, Italy, 2009.
45. *Codex Stand 193-1995*; Norma General del codex para los Contaminantes y las Tóxinas Presentes en los Alimentos y Piensos. FAO: Rome, Italy, 1995.
46. Unión Europea (UE). *Contenidos Máximos en Métales Pesados en Productos Alimenticios*; Unión Europea (UE): Maastricht, The Netherlands, 2016.
47. USEPA. *Supplementary Guidance for Conducting Health Risk Assessment of Chemical Mixtures*; USEPA: Washinton, DC, USA, 2000; p. 20460.
48. Lizarazo, M.F.; Herrera, C.D.; Celis, C.A.; Pombo, L.M.; Teherán, A.A.; Piñeros, L.G.; Rodríguez, O.E. Contamination of staple crops by heavy metals in Sibaté, Colombia. *Heliyon* **2020**, *6*, e04212. [CrossRef] [PubMed]
49. Islam, M.S.; Ahmed, M.K.; Al-mamun, M.H.; Masunaga, S. Trace metals in soil and vegetables and associated health risk assessment. *Environ. Monit. Assess.* **2014**, *186*, 8727–8739. [CrossRef]
50. UPME-MME-UC (Unidad de Planeación Minero Energética, Ministerio de Minas y Energía, Universidad de Córdoba). *Identificación y Caracterización de Unidades Básicas de Beneficio Aurífero en Municipios de Colombia*; Informe Técnico (Reporte Final); UPME-MME-UC: Bogotá DC, Colombia, 2016.
51. Marrugo-Madrid, S.; Pinedo-Hernández, J.; Paternina-Uribe, R.; Marrugo-Negrete, J.; Díez, S. Health risk assessment for human exposure to mercury species and arsenic via consumption of local food in a gold mining area in Colombia. *Environ. Res.* **2022**, *215*, 113950. [CrossRef]
52. Zhong, T.; Xue, D.; Zhao, L.; Zhang, X. Concentration of heavy metals in vegetables and potential health risk assessment in China. *Environ. Geochem. Health* **2017**, *40*, 313–322. [CrossRef]
53. Saha, N.; Zaman, M.R. Evaluation of possible health risks of heavy metals by consumption of foodstuffs available in the central market of Rajshahi City, Bangladesh. *Environ. Monit. Assess.* **2013**, *185*, 3867–3878. [CrossRef]
54. Margenat, A.; You, R.; Cañameras, N.; Carazo, N.; Díez, S.; Bayona, J.M.; Matamoros, V. Occurrence and human health risk assessment of antibiotics and trace elements in *Lactuca sativa* amended with different organic fertilizers. *Environ. Res.* **2020**, *190*, 109946. [CrossRef]
55. Siudek, P.; Kurzyca, I.; Siepak, J. Atmospheric deposition of mercury in central Poland: Sources and seasonal trends. *Atmos. Res.* **2016**, *170*, 14–22. [CrossRef]
56. Herforth, A.; Arimond, M.; Alvarez-S-anchez, C.; Coates, J.; Christianson, K.; Muehlhoff, E.A. A Global Review of Food-Based Dietary Guidelines. *Adv. Nutr.* **2019**, *10*, 590–605. [CrossRef] [PubMed]
57. Román-Ochoa, Y.; Delgado, G.T.C.; Tejada, T.R.; Yucra, H.R.; Durand, A.E.; Hamaker, B.R. Heavy metal contamination and health risk assessment in grains and grain-based processed food in Arequipa region of Peru. *Chemosphere* **2021**, *274*, 129792. [CrossRef] [PubMed]

58. Guadie, A.; Yesigat, A.; Gatew, S.; Worku, A.; Liu, W.; Ajibade, F.O.; Wang, A. Evaluating the health risks of heavy metals from vegetables grown on soil irrigated with untreated and treated wastewater in Arba Minch, Ethiopia. *Sci. Total Environ.* **2020**, *761*, 143302. [CrossRef] [PubMed]
59. Luis-González, G.; Rubio, C.; Gutiérrez, Á.; González-Weller, D.; Revert, C.; Hardisson, A. Essential and toxic metals in taros (*Colocasia esculenta*) cultivated in the Canary Islands (Spain): Evaluation of content and estimate of daily intake. *Environ. Monit. Assess.* **2015**, *187*, 4138. [CrossRef]
60. Zwolak, A.; Sarzyaska, M.; Szpyrka, E.; Stawarczyk, K. Sources of Soil Pollution by Heavy Metals and Their Accumulation in Vegetables: A Review. *Wat. Air Soil Poll.* **2019**, *230*, 164. [CrossRef]
61. Gan, Y.; Wang, L.; Yang, G.; Dai, J.; Wang, R.; Wang, W. Multiple factors impact the contents of heavy metals in vegetables in high natural background area of China. *Chemophere* **2017**, *184*, 1388–1395. [CrossRef]
62. Prieto, G.; González, L.; Vargas, O.; Matamoros, A. Geoquímica de Suelos de la Cuenca del Río Bogotá-a. *Bol. Geol.* **2018**, *41*, 41–56. [CrossRef]
63. Gomez, G.; Fisberg, R.M.; Nogueira Previdelli, A.; Hermes Sales, C.; Kovalskys, I.; Fisberg, M.; Herrera-Cuenca, M.; Cortés Sanabria, L.Y.; Yépez García, M.C.; Pareja Torres, R.G.; et al. Diet Quality and Diet Diversity in Eight Latin American Countries: Results from the Latin American Study of Nutrition and Health (ELANS). *Nutrients* **2019**, *11*, 1605. [CrossRef]
64. Bi, C.J.; Zhou, Y.; Chen, Z.L.; Jia, J.P.; Bao, X. Heavy metals and lead isotopes in soils, road dust and leafy vegetables and health risks via vegetable consumption in the industrial areas of Shanghai, China. *Sci. Total Environ.* **2018**, *619*, 1349–1357. [CrossRef]
65. Zhou, Y.; Bi, C.J.; Zhou, X.X.; Zhang, H.H.; Chen, Z.L.; Bao, X.Y. Distribution characteristics and health risk of heavy metals in vegetables near the industrial area in Shanghai. *Environ. Sci.* **2017**, *38*, 5292–5298.
66. Real, M.I.H.; Azam, H.M.; Majed, N. Consumption of heavy metal contaminated foods and associated risks in Bangladesh. *Environ. Monit. Assess.* **2017**, *189*, 651. [CrossRef]
67. Marini, M.; Angouria-Tsorochidou, E.; Caro, D.; Thomsen, M. Daily intake of heavy metals and minerals in food—A case study of four Danish dietary profiles. *J. Clean. Prod.* **2020**, *280*, 124279. [CrossRef]
68. Tariq, F.S. Heavy metals concentration in vegetables irrigated with municipal wastewater and their human daily intake in Erbil city. *Environ. Nanotechnol. Monit. Manag.* **2021**, *16*, 100475. [CrossRef]
69. Nag, R.; Cummins, E. Human health risk assessment of lead (Pb) through the environmental-food pathway. *Sci. Total Environ.* **2022**, *810*, 151168. [CrossRef]
70. Vargas-Licona, S.P.; Marrugo-Negrete, J.L. Mercurio, metilmercurio y otros metales pesados en peces de Colombia: Riesgo por ingesta. *Acta Biol. Colomb.* **2019**, *24*, 232–242. [CrossRef]
71. Marrugo-Negrete, J.; Ruiz-Guzmán, J.; Ruiz-Fernández, A. Biomagnification of Mercury in Fish from Two Gold Mining-Impacted Tropical Marshes in Northern Colombia. *Arch. Environ. Contam. Toxicol.* **2018**, *74*, 121–130. [CrossRef] [PubMed]

toxics

MDPI

Article

Exposure to Polypropylene Microplastics via Oral Ingestion Induces Colonic Apoptosis and Intestinal Barrier Damage through Oxidative Stress and Inflammation in Mice

Rui Jia [1,2], Jie Han [2], Xiaohua Liu [2], Kang Li [2], Wenqing Lai [2], Liping Bian [2], Jun Yan [2,*] and Zhuge Xi [1,2,*]

1 College of Marine Ecology and Environment, Shanghai Ocean University, Shanghai 201306, China
2 Tianjin Institute of Environmental and Operational Medicine, Tianjin 300050, China
* Correspondence: yanjying@sina.cn (J.Y.); zhugexi2003@sina.com (Z.X.)

Abstract: Extensive environmental pollution by microplastics has increased the risk of human exposure to plastics. However, the biosafety of polypropylene microplastics (PP-MPs), especially of PP particles < 10 μm, in mammals has not been studied. Thus, here, we explored the mechanism of action and effect of exposure to small and large PP-MPs, via oral ingestion, on the mouse intestinal tract. Male C57BL/6 mice were administered PP suspensions (8 and 70 μm; 0.1, 1.0, and 10 mg/mL) for 28 days. PP-MP treatment resulted in inflammatory pathological damage, ultrastructural changes in intestinal epithelial cells, imbalance of the redox system, and inflammatory reactions in the colon. Additionally, we observed damage to the tight junctions of the colon and decreased intestinal mucus secretion and ion transporter expression. Further, the apoptotic rate of colonic cells significantly increased after PP-MP treatment. The expression of pro-inflammatory and pro-apoptosis proteins significantly increased in colon tissue, while the expression of anti-inflammatory and anti-apoptosis proteins significantly decreased. In summary, this study demonstrates that PP-MPs induce colonic apoptosis and intestinal barrier damage through oxidative stress and activation of the TLR4/NF-κB inflammatory signal pathway in mice, which provides new insights into the toxicity of MPs in mammals.

Keywords: polypropylene microplastics (PP-MPs); intestinal barrier; oxidative stress; inflammatory reaction; apoptosis

Citation: Jia, R.; Han, J.; Liu, X.; Li, K.; Lai, W.; Bian, L.; Yan, J.; Xi, Z. Exposure to Polypropylene Microplastics via Oral Ingestion Induces Colonic Apoptosis and Intestinal Barrier Damage through Oxidative Stress and Inflammation in Mice. *Toxics* **2023**, *11*, 127. https://doi.org/10.3390/toxics11020127

Academic Editors: Esref Demir and Sam Kacew

Received: 24 December 2022
Revised: 23 January 2023
Accepted: 26 January 2023
Published: 28 January 2023
Corrected: 25 August 2023

1. Introduction

The output of plastic and related products is continuously increasing because of its widespread use and low cost. It is estimated that the total output of plastic waste will reach 12 billion tons in 2050 [1]. Plastic use in the past few decades has caused environmental issues through the lack of sound recycling and treatment measures. Widespread plastic pollution is considered to be a global threat to human and animal health, especially during the period of the coronavirus disease 2019 (COVID-19) when increased use of masks and surgical gloves led to the generation of a large amount of medical waste [2–5].

Plastics are further degraded into fragments in the environment. Thompson et al. [6] first proposed microplastics (MPs), which were later defined as "plastic particles smaller than 5 mm" [7]. Meanwhile, plastic particles below 0.1 μm are called nanoplastics (NPs) [8]. MPs have been detected in water [9,10], soil [11,12], air [13–15], tap water [16–18], drinking water [19,20], and even human food [21–25]. The main types of MPs are polyethylene (PE), polypropylene (PP), and polystyrene (PS) [26]. MPs can move along the food chain to higher-level organisms, including humans. Oral intake is a major route for human exposure to MPs, and they have been detected in adult and infant feces [27–29] and human placenta [30,31], among which, PP constituted the highest proportion of MPs, and its relative mass abundance in adult feces was up to 61.0% [28]. In addition, infants who are fed formula from PP infant feeding bottles are exposed to PP-MPs ranging from 14,600–4,550,000 particles per capita per day, depending on the region [32]. Therefore, more attention should be paid to the influence of MPs (especially PP) on human health.

Most current reports on the toxicity of MPs focus on PS, while the biological safety assessment of PP is mainly concentrated on aquatic organisms [33–35] and plants [36,37]. Previous studies have reported that PP-MPs can reduce the thickness of the intestinal mucosa and intestinal muscle layer, cause oxidative stress and inflammation in intestinal tissue, and seriously interfere with lipid metabolism in zebrafish [38,39]. Ten micrograms per liter of 70 μm PP-MP significantly reduced the survival rate of zebrafish [40]. However, few studies have measured the effects of PP-MPs on mammalian and cell models, although PP-MPs contribute to pulmonary inflammation in vivo and affect the level of immune cytokines in vitro [41,42]. Meanwhile, PP-MPs with an average particle size of >50 μm do not exhibit acute toxicity in rats [43–45]. However, there is no report on the biosafety of PP particles < 10 μm in the environment, thus warranting further study.

The intestinal tract may be the primary target organ after oral ingestion of PP-MPs. The intestinal barrier plays a key role in evaluating PP-MP intestinal toxicity and its toxic effects on distal tissues and organs. Orally ingested MPs can accumulate in organisms through the intestinal barrier and cause adverse effects such as flora imbalance and metabolic changes, which may lead to damage in multiple systems and organs [46–48]. Thus, MP exposure can cause intestinal barrier dysfunction or even damage, which may be a key factor for MP biotoxicity. However, the effect of orally ingested PP-MPs on the intestinal tract (especially on the intestinal barrier) remains unknown.

The mechanical and chemical barriers are two important aspects of the intestinal barrier. The mechanical barrier consists of intact intestinal epithelial cells and tight connections between cells. Goblet cells in the intestinal epithelium secrete mucus to form the intestinal mucus barrier, which is an important component of the chemical barrier [49,50]. The intestinal mucus layer covers almost the entire intestinal cavity surface; it lubricates, resists bacterial invasion, and protects the intestinal tract from mechanical damage and pathogenic bacteria [51,52]. Therefore, both the intestinal epithelial cell layer and mucus layer form the first line of defense against external factors and play an important role in maintaining the balance of the intestinal environment and blocking intestinal pathogens and toxins [53,54]. However, pathogenic factors, such as stress and inflammation, can destroy intestinal barrier function and cause intestinal mucosal barrier injury [55,56].

This study aimed to clarify the mechanism and effect of orally ingested PP-MPs on the intestinal tract. To this end, two kinds of PP-MP particles with different scales were selected, the relevant doses of environmental exposure were adopted, and a subacute oral ingestion model in mice was established to study the effects of PP-MPs on the intestinal mucosal barrier and the changes in related signaling pathways. Ultimately, we aimed to evaluate the intestinal toxicity caused by PP-MPs. This is the first study on the intestinal toxicity of ~10 μm PP-MPs in mice, which provides a toxicological reference for the biosafety assessment of environmental PP-MPs and their risk of exposure to humans.

2. Materials and Methods

2.1. Characterization of PP Particles and Suspension Preparation

PP was purchased from Shanghai Macklin Biochemical Co., Ltd. (Shanghai, China), and PP-MP particles were prepared according to a previous method [41,57]. PP material was frozen in liquid nitrogen for 10 min and the powder was ground with a temperature-controlled three-dimensional vibration ball mill (TJSKW; Techin, Tianjin, China) at 4 °C for 25 min. The powder was collected and separated with an electromagnetic vibration sieve (TJ-TAS; Techin). The PP particles were separated by size using sieves with ~10 μm and 20–100 μm pore sizes. All of the PP particles were irregular blocks according to scanning electron microscopy (SEM) images (Tescan VEGA3; Tescan, Brno, Czech Republic) (Figure 1A,B). The particle size ranges of PP are 1–10 μm (50% are 8 μm) and 40–90 μm (60% are 70 μm) (NIH, Bethesda, MD, USA) [58] (n = 100) (Figure 1C,D). Therefore, to express the results of this study, we used 8 μm to represent the size of smaller PP particles and 70 μm to represent the size of larger PP particles.

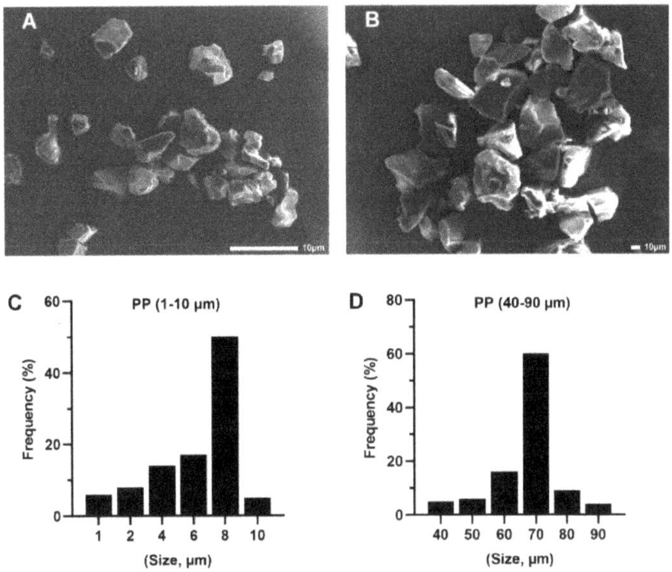

Figure 1. Characterization and scanning electron microscopy (SEM) images of polypropylene (PP) particles. (**A**) SEM images of 8 μm PP particles (scale bar = 10 μm) determined with a particle size analyzer (**C**). (**B**) SEM images of 70 μm PP particles (scale bar = 10 μm) determined with a particle size analyzer (**D**).

A 10 mg/mL suspension of PP particles was prepared in pure water, followed by the addition of Tween 80 to 0.01% v/v. The mixture was ultrasonicated at 40K Hz for 1 hr using ultrasonic cleaner (DS-2510DTH; Shanghai Aucy Scientific Instrument Co., Ltd., Shanghai, China), then diluted to the desired concentration. The suspension was ultrasonicated for 30 min and fully mixed in the eddy current oscillator before administration to animals.

2.2. Animals and Experimental Design

Seventy-two male C57BL/6 mice of 22–26 g body weight were purchased from Beijing Vital River Laboratory Animal Technology Co., Ltd., China (animal production license number: SCXK (Beijing) 2016–0006). All animal experiments were approved by the Experimental Animal Welfare Ethics Committee of the Tianjin Institute of Environmental and Operational Medicine, IACUC approval code AMMS-04-2021-014. The mice were reared in a specific pathogen-free (SPF) animal room with an ambient temperature of 23 ± 2 °C, a photoperiod of 12 h light/12 h darkness, and a relative humidity of 40–60%. The diet contained bran, soybean meal, corn, flour, sorghum flour, fish meal, calcium hydrogen phosphate, and salt. All ingredients complied with the corresponding national food hygiene standards. Food and sterilized water were provided ad libitum. The exposure dose of PP-MPs (8 and 70 μm) was 1, 10, and 100 mg/kg/d; that is, the exposure concentration was 0.1, 1, and 10 mg/mL, respectively. This exposure dose is based on the previous literature: the intake of MPs for adult humans (calculated for 70 kg) is 0.1–5 g per week, or 0.2–10.2 mg/kg body weight (bw)/d [59]. Therefore, the PP dose selected for the test reflects the real range of human MP intake.

Mice were acclimatized for one week, then randomly divided into eight groups: blank control (pure water, BC), solvent control (pure water containing 0.01% v/v Tween-80, SC), 8 μm PP at 0.1 mg/mL (Ls), 8 μm PP at 1.0 mg/mL (Ms), 8 μm PP at 10 mg/mL (Hs),

70 μm PP at 0.1 mg/mL (Lb), 70 μm PP at 1.0 mg/mL (Mb), and 70 μm PP at 10 mg/mL (Hb). Each mouse group was housed in two cages, four or five in each cage. The mice were given 0.1 mL/10 g bw of PP-MPs suspension by oral gavage for 28 days. At the end of the experiment, mice were fasted for 12 h and anesthetized. Colon tissues were carefully isolated, colon segments of about 4 cm were taken, and then the intestinal contents were washed with aseptic PBS buffer. Some of the colon samples were fixed with different solutions for subsequent staining sections, and the remaining samples were frozen in liquid nitrogen for Western blotting and enzymatic analysis.

2.3. Histopathological Examination and Electron Microscopy Analysis

The colonic tissues of mice in each group were fixed in 4% w/v paraformaldehyde solution at 4 °C for 24 h, then dehydrated in gradient ethanol from 75% to 100% for 40 min, respectively, immersed in xylene to make it transparent, and embedded in paraffin wax. The embedded samples were cut into 3 μm thick sections with a microtome (RM2245; Leica, Nussloch, Germany) and stained with hematoxylin–eosin (H&E), and the histopathological changes were observed under an Olympus DP26 microscope (Tokyo, Japan).

The colonic tissues were soaked in 2.5% w/v glutaraldehyde at 4 °C for 24 h and fixed in 1% w/v osmic acid at 20 °C for 2 h. The samples were dehydrated with gradient ethanol from 30% to 100% for 20 min, respectively, and embedded in epoxy resin. Ultrathin sections (50 nm) were prepared with a microtome (UC7; Leica). The sections were double-stained with 2% w/v uranium acetate and lead citrate for 15 min at 20 °C and dried overnight. Ultrastructural changes were observed with a transmission electron microscope (Fei Tecnai G20 TWIN, FEI, Hillsboro, OR, USA).

2.4. Alcian Blue/Periodic Acid–Schiff (AB-PAS) Staining

Paraffin sections of the colonic tissues were prepared according to the method above. The sections were dewaxed in xylene, hydrated in a graded ethanol series (100%, 85%, and 75%), and stained with alcian blue/periodic acid–Schiff (AB-PAS). The dye solution was prepared according to the requirements of the kit (Nanjing JianCheng Bioengineering Institute, Nanjing, China). Then dye solution was added to the glass slide sample, incubated at room temperature (about 20 °C) for 8–15 min, washed with running water, dried naturally, and examined under a microscope. The mucus coverage ratio was calculated as the pixels in the mucus area to the total pixel area of the gut section. The pixels were determined using Image Pro Plus 6.0 software (Media Cybernetics, Rockville, MD, USA).

2.5. Immunohistochemical Analysis

Colon paraffin sections were taken for immunohistochemical analysis. The sections were dewaxed in xylene and hydrated in a graded ethanol series (100%, 85%, and 75%). Antigen retrieval was performed using sodium citrate antigen retrieval solution (pH 6.0), following the manufacturer's instructions (Beijing Solarbio Science & Technology Co., Ltd., Beijing, China). The sections were incubated with 3% v/v hydrogen peroxide in the dark for 25 min to quench endogenous peroxidase activity. After blocking with normal goat serum (Solarbio) at room temperature (about 20 °C) for 30 min, sections were separately incubated with the following antibodies at 4 °C overnight: anti-solute carrier family 26 member 6 (SLC26A6, 1:400; Bioss, Beijing, China), anti-Na-K-2Cl cotransporter 1 (NKCC1, 1:200; CST, Boston, MA, USA), and anti-cystic fibrosis transmembrane conductance regulator (CFTR, 1:100; CST). Then, the sections were incubated with goat anti-rabbit IgG HRP secondary antibody (1:500; Sino Biological, Beijing, China) for 50 min at room temperature, stained with diaminobenzidine (DAB; Solarbio) for color development, and counterstained with hematoxylin (cell nucleus) for 8 min. Finally, the sections were dehydrated in ascending ethanol (75%, 85%, and 100%), hyalinized in xylene, and sealed with neutral gum. The positive areas of the sections were observed and photographed using an Olympus BX51 fluorescence microscope, and the fluorescence intensity was analyzed using with Image Pro Plus 6.0 software.

2.6. Detection of Oxidative Stress Markers

The colon tissues of each group of mice were prepared according to the kit requirements to detect the levels of colonic oxidative stress markers. The concentrations of reduced glutathione (GSH) and oxidized glutathione (GSSG) (GSH and GSSG Assay Kit, S0053, Beyotime Biotechnology Co., Ltd., Shanghai, China), malondialdehyde (MDA) (Lipid Peroxidation MDA Assay Kit, S0131S, Beyotime), and catalase (CAT) (Catalase assay kit (Visible light), A007-1-1, Jiancheng) were detected using colorimetric kits. Glutathione peroxidase (GSH-Px) expression was detected using a Total Glutathione Peroxidase Assay Kit with NADPH (S0058, Beyotime). Superoxide dismutase (SOD) content was determined using a Total Superoxide Dismutase Assay Kit with WST-8 (S0101S, Beyotime). Protein concentration was determined with a BCA Protein Assay Kit (Beyotime).

2.7. Enzyme-Linked Immunosorbent Assay (ELISA)

ELISA kits were used to determine the expression of inflammatory factors and intestinal-barrier-related proteins in colon tissues. Inflammatory factors included mouse tumor necrosis factor alpha (TNF-α; Mouse TNFα ELISA Kit, JL10484, Shanghai Jianglai Biotechnology Co., Ltd., Shanghai, China), interleukin 1 beta (IL-1β; Mouse IL-1β ELISA Kit, JL18442, Jianglai), interleukin 6 (IL-6; Mouse IL-6 ELISA Kit, JL20268, Jianglai), and interleukin 10 (IL-10; Mouse IL-10 ELISA Kit, JL20242, Jianglai). Intestinal-barrier-related proteins included mouse zonula occludens 1 (ZO-1; Mouse ZO-1 ELISA Kit, JL20409, Jianglai), occludin (Mouse Occludin ELISA Kit, JL20408, Jianglai), mucin-1 (MUC1; Mouse MUC1 ELISA Kit, JL26951, Jianglai), and claudin-1 (CLDN1; ELISA Kit for CLDN1, SEC388Mu, Wuhan Cloud Clone Technology Co., Ltd., Wuhan, China). The extracted colon homogenate was mixed with the reagent in the kits, followed by the addition of a chromogenic agent for color development, and finally treated with a stop solution according to the manufacturer's protocols. The optical density (OD) of each sample was measured at 450 nm with a microplate reader (Molecular Devices M5E, Silicon Valley, Calif., USA). The levels of inflammatory factors and intestinal-barrier-related proteins were calculated using a standard curve. Protein concentration was determined with a BCA Protein Assay Kit (Beyotime).

2.8. Apoptosis Detection in the Colon

The paraffin sections of colonic tissue were generated as previously stated. The sections were dewaxed in xylene and hydrated with a graded ethanol series (100%, 85%, and 75%). A membrane-breaking working solution (PBS containing 0.5% Triton X-100) was added to cover the tissues. After being stained with TUNEL fluorescence staining solution (In Situ Cell Death Detection Kit-POD; Roche, Basel, Switzerland) at 37 °C for 2 h in the dark, the sections were washed with PBS (pH 7.4) three times, counterstained with DAPI (4′,6-diamidino-2-phenylindole) (cell nucleus) and sealed with anti-fluorescence quenching sealing tablets. Images of the sections were observed with an Olympus BX51 fluorescence microscope, and the number of apoptotic cells was counted. The TUNEL positive cell rate = number of apoptotic cells/total cells × 100% in each visual field.

2.9. Western Blot Analysis

Total protein extraction of colonic tissue was performed by treatment with RIPA buffer, followed by centrifugation at 12000 r/min for 5 min at 4 °C, and the concentration was measured with a BCA Protein Assay Kit (Beyotime). Proteins (50 μg) were electrophoresed in an 8–10% gradient sodium dodecyl sulfate polyacrylamide gel and transferred to polyvinylidene fluoride (PVDF) membranes (Millipore, Billerica, MA, USA). After blocking with 5% *w*/*v* nonfat milk dispersed in Tris-buffered saline containing 0.05% Tween-20 (TBST) for 1 h, the membranes were incubated with primary antibodies against IκBα (CST), p105 (CST), p50 (Abcam, Cambridge, UK), p65 (CST), p-p65 (CST), Toll-like receptor 4 (TLR4; Abcam), Bax (CST), Bcl-2 (CST), caspase-9 (CST), caspase-3 (CST), cleaved caspase-9 (CST) (all 1:1000), and cleaved caspase-3 (1:500; Bioss), as well as the internal control glyceraldehyde 3-phosphate dehydrogenase (GAPDH) antibody (1:2000; Tianjin UtiBody Biotechnology

Co., Ltd., Tianjin, China) overnight at 4 °C. The membranes were washed with TBST (10 min wash × 3), followed by incubation with goat anti-rabbit IgG HRP secondary antibody (1:3000; Bioss) and goat anti-mouse IgG HRP (1:3000; Bioss) secondary antibody at room temperature (about 20 °C) for 1 hr. Membranes were washed three times with TBST, stained using ECL chemiluminescence reagent (Solarbio), and imaged in a fully automatic chemiluminescence image analysis system (Tanon 5200, Shanghai, China). The grayscale value of protein imprinting was analyzed using Gel-Pro analyzer software.

2.10. Statistical Analysis

Data are expressed as the mean ± standard deviation (SD). Statistical analyses were performed using SPSS 25.0 software. Differences between groups were determined according to one-way analysis of variance (ANOVA) followed by the least post-squares post hoc test for equal variances or Dunnett's T3 post hoc test for unequal variances. Any p-value of <0.05 was considered to be statistically significant.

3. Results

3.1. Polypropylene Microplastic Exposure Changes Colonic Histopathology and Ultrastructure

The whole-colon tissue structure of mice in the blank control group and solvent control group was basically normal, and there were no abnormalities in the mucosal layer and muscular layer. However, mild edema of the submucosa was found in the colon of mice in all PP-MP treatment groups. In addition, a large number of lymphocytes gathered in the colon of mice in the 1.0 mg/mL treatment group using 8 μm PP-MPs. A small amount of inflammatory cells infiltrated the colon of mice in the 10 mg/mL treatment group. Glandular hyperplasia of the mucosal layer appeared in the colon of mice in the 10 mg/mL treatment group using 70 μm PP-MPs (Figure 2A).

In addition, the ultrastructure of the colonic epithelial cells of mice was basically normal in the blank control group and the solvent control group; also, the nucleus was clearly visible, and the mitochondrial structure was normal. The ultrastructure of colonic epithelial cells of mice in all PP-MP treatment groups was also basically normal, and the nucleus was clearly visible. Meanwhile, PP granules were found in colonic epithelial cells of mice in all PP-MPs treatment groups. The ultrastructural changes of colonic epithelial cells of mice treated with 8 μm PP-MPs (1 and 10 mg/mL) and 70 μm PP-MPs (10 mg/mL) included mild abnormalities in the mitochondrial structure: loose ridge arrangement, reduced matrix, partial swelling, and vacuolation (Figure 2B).

3.2. Polypropylene Microplastic Exposure Induces Colonic Oxidative Stress and Inflammation

There was no significant difference in the levels of MDA, GSSG, SOD, GSH, GSH-Px, and CAT in the colon of mice in the blank control group compared with that of the solvent control group (Figure 3A–F). The levels of SOD, GSH, GSH-Px, and CAT in all PP-MP treatment groups significantly decreased ($p < 0.05$) compared with that of the solvent control group, and the MDA level significantly increased ($p < 0.05$); each of these relationships showed a certain concentration dependence. Exposure to 8 μm and 70 μm PP-MPs at 1 mg/mL and 10 mg/mL significantly increased the level of GSSG compared with that in the solvent control group ($p < 0.05$). In addition, the expression of CAT in the 8 μm PP-MP treatment group was significantly lower than that in the 70 μm PP-MP treatment group ($p < 0.05$). The MDA level in the 8 μm PP-MP treatment group was significantly higher than that in the 70 μm PP-MP (1 and 10 mg/mL) treatment groups ($p < 0.05$). The levels of SOD and GSH-Px in the 8 μm PP-MP treatment group were significantly lower than those in the 70 μm PP-MP treatment group ($p < 0.05$) at the exposure concentration of 10 mg/mL. These results indicate that PP-MP exposure caused an imbalance in the redox system in the intestinal tissues of mice, resulting in oxidative damage.

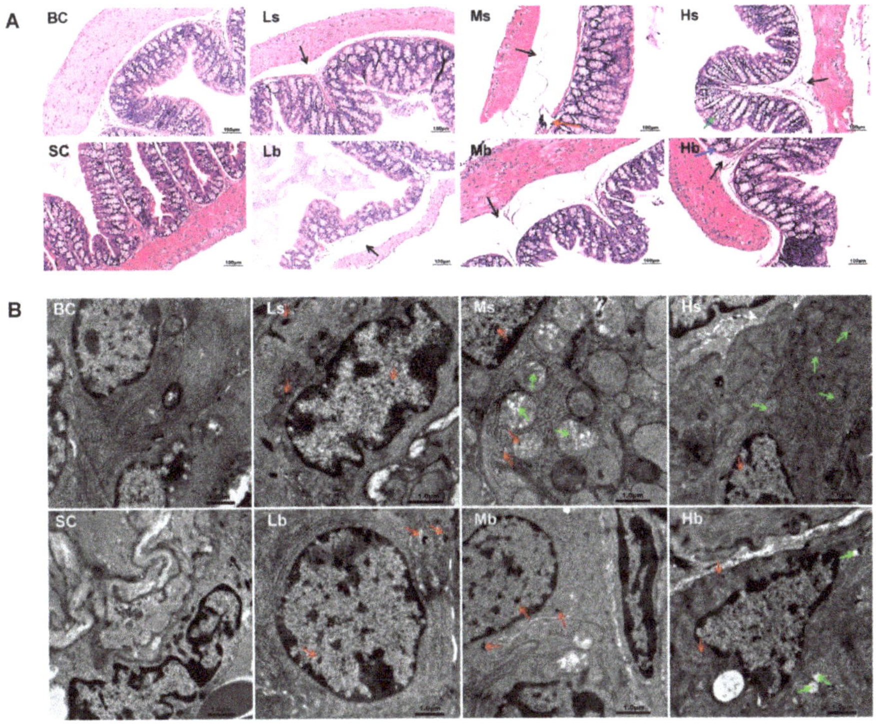

Figure 2. Effects of exposure to PP-MPs on colonic histopathology and ultrastructure in mice. (**A**) Images of H&E-stained colon sections of mice treated with PP-MPs. Black arrow: submucosal edema; red arrow: massive aggregation of lymphocytes; blue arrow: glandular hyperplasia of mucosal layer; green arrow: inflammatory cell infiltration. (**B**) Ultrastructural changes of colon after exposure to PP-MPs. Red arrows: PP particles; green arrow: slight abnormality of mitochondrial structure, with loose ridge arrangement, less matrix, partial swelling, and vacuolation. $n = 3$. BC: blank control (pure water); SC: solvent control (pure water containing 0.01% v/v Tween-80); PP-MPs: polypropylene microplastics. Ls: 8 μm PP at 0.1 mg/mL; Ms: 8 μm PP at 1.0 mg/mL; Hs: 8 μm PP at 10 mg/mL; Lb: 70 μm PP at 0.1 mg/mL; Mb: 70 μm PP at 1.0 mg/mL; Hb: 70 μm PP at 10 mg/mL. scale bar = 1 μm.

There was no significant difference in the levels of TNF-α, IL-1β, IL-6, and IL-10 in the colon of mice in the blank control group compared with that of the solvent control group (Figure 3G–J). The levels of TNF-α, IL-1β, and IL-6 significantly increased in all PP-MP treatment groups, and IL-10 expression significantly decreased ($p < 0.05$) compared with that of the solvent control group. The effect was concentration-dependent. In addition, the levels of TNF-α, IL-1β, and IL-6 were significantly higher in the 8 μm PP-MP treatment group than those in the 70 μm PP-MP (1 and 10 mg/mL) treatment groups ($p < 0.05$). The level of IL-10 in the 8 μm PP-MP treatment group was significantly lower than that in the 70 μm PP-MP (10 mg/mL) treatment group ($p < 0.05$). These results suggest that inflammatory reactions occur in the colon of mice exposed to PP-MPs.

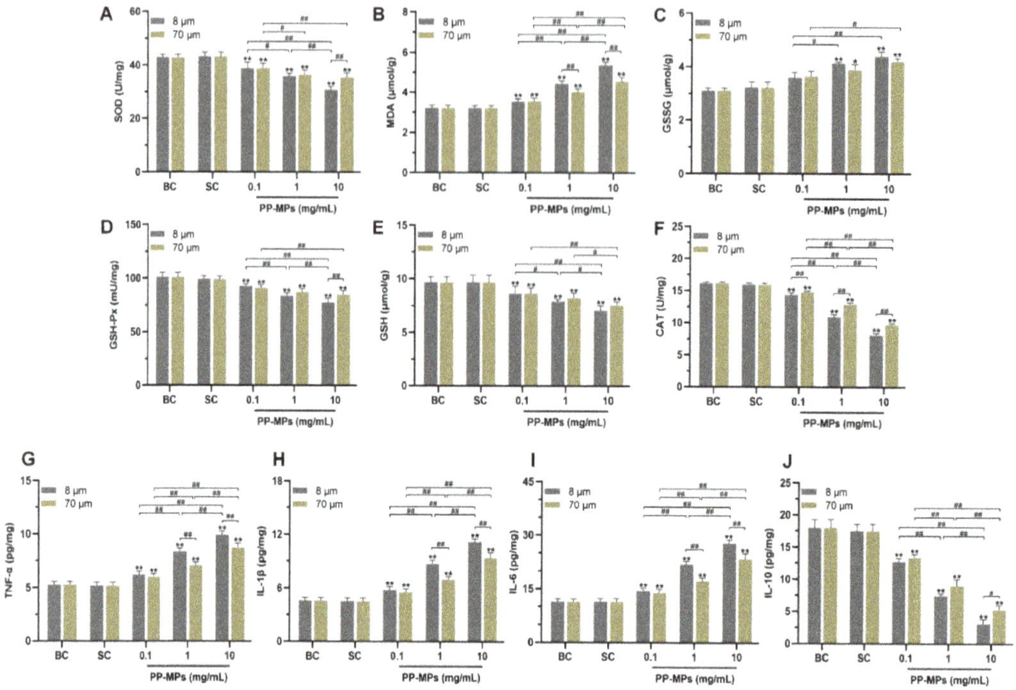

Figure 3. Oxidative stress and inflammation occurred in the colon of mice after PP-MP exposure. (**A–F**) Colonic levels of SOD (**A**), MDA (**B**), GSSG (**C**), GSH-Px (**D**), GSH (**E**), and CAT (**F**). (**G–J**) Colonic levels of TNF-α (**G**), IL-1β (**H**), IL-6 (**I**), and IL-10 (**J**). The presented values are the means ± SD ($n = 5$). $^{\#}$ $p < 0.05$, $^{\#\#}$ $p < 0.01$ between PP-MP treated groups, and * $p < 0.05$, ** $p < 0.01$ vs. SC as determined by one-way analysis of variance (ANOVA). BC: blank control (pure water); SC: solvent control (pure water containing 0.01% v/v Tween-80); PP-MPs: polypropylene microplastics.

3.3. Polypropylene Microplastic Exposure Activates the TLR4/NF-κB Signaling Pathway

The levels of TLR4, p105, p50, p65, *p*-p65, and IκBα were not significantly different in the colon of the blank control mice from those in the solvent control group (Figure 4). The levels of TLR4, p50, and *p*-p65 significantly increased in all PP-MP treatment groups, and IκBα expression significantly decreased ($p < 0.05$) compared with that of the solvent control group. The differences were all concentration-dependent. In addition, the level of IκBα was significantly lower in the 8 μm PP-MP treatment group than that in the 70 μm PP-MP (1 and 10 mg/mL) treatment groups ($p < 0.05$). Also, the levels of p50 and *p*-p65 in the 8 μm PP-MP treatment group were significantly higher than that in the 70 μm PP-MP treatment group ($p < 0.05$). The level of TLR4 in the 8 μm PP-MP treatment group was significantly higher than that in the 70 μm PP-MP (10 mg/mL) treatment group at ($p < 0.05$).

3.4. Polypropylene Microplastic Exposure Destroys the Intestinal Mucosal Barrier in Mice

AB-PAS staining showed that there was no significant difference in the colonic mucus coverage rate between the blank control group and the solvent control group (Figure 5A). The colonic mucus coverage rate of mice treated with 8 μm and 70 μm PP-MPs (1 and 10 mg/mL) significantly decreased compared with that of the solvent control group ($p < 0.05$).

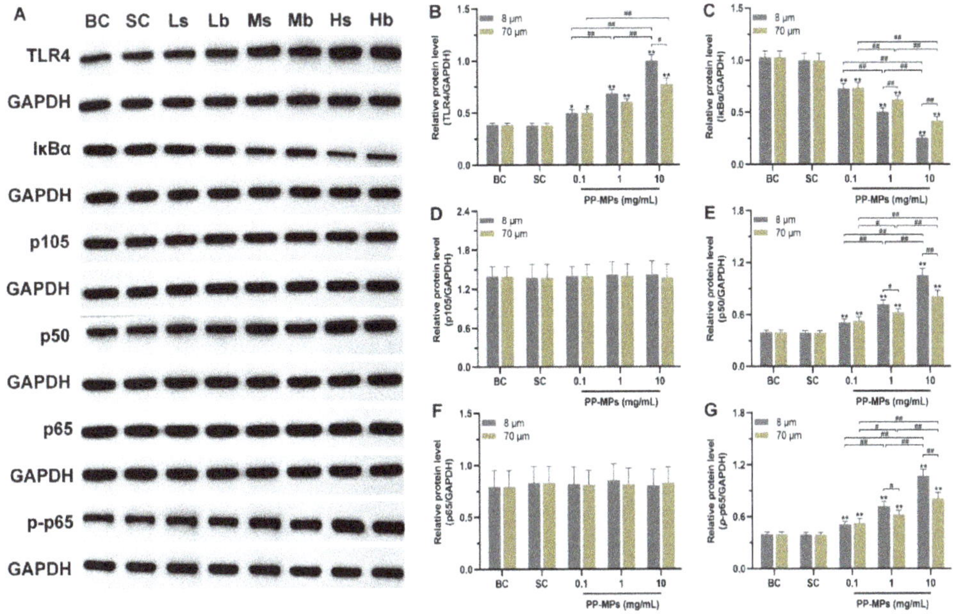

Figure 4. Effects of PP-MP exposure on protein expression of the TLR4/NF-κB signaling pathway. (**A**) Western blot of TLR4, p105, p50, p65, *p*-p65, and IκBα protein expression. GAPDH was used as an internal reference. (**B–G**) Quantitative expression of these proteins. The values are the means ± SD ($n = 4$). # $p < 0.05$, ## $p < 0.01$ between PP-MP treated groups, and * $p < 0.05$, ** $p < 0.01$ vs. SC as determined by one-way analysis of variance (ANOVA). BC: blank control (pure water); SC: solvent control (pure water containing 0.01% *v*/*v* Tween-80); PP-MPs: polypropylene microplastics.

The levels of ZO-1, claudin-1, occludin, and mucin (MUC1) were not significantly different in the colon of mice in the blank control group from those in the solvent control group (Figure 5B–E). The levels of occludin and MUC1 in all PP-MP treatment groups significantly decreased ($p < 0.05$) in a concentration-dependent manner compared with that of the solvent control group. The levels of ZO-1 and claudin-1 significantly decreased in the 8 μm PP-MP (1 and 10 mg/mL) treatment groups and 70 μm PP-MP (10 mg/mL) treatment group compared with that of the solvent control group ($p < 0.05$). In addition, the expression of ZO-1, claudin-1, and MUC1 were significantly lower in the 8 μm PP-MP treatment group than those in the 70 μm PP-MP treatment group ($p < 0.05$) at 1 mg/mL and 10 mg/mL exposure concentrations. Occludin expression was significantly lower in the 8 μm PP-MP treatment group than that in the 70 μm PP-MP treatment group ($p < 0.05$) at the exposure concentration of 10 mg/mL.

Immunohistochemical analysis of the colon showed that there was no significant difference in the levels of CFTR, SLC26A6, and NKCC1 between the blank control group and the solvent control group (Figure 5F). The level of NKCC1 significantly decreased in all PP-MP treatment groups ($p < 0.05$) compared with that of the solvent control group. The level of SLC26A6 significantly decreased in the 8 μm PP-MP (0.1, 1, and 10 mg/mL) treatment groups and 70 μm PP-MP (1 and 10 mg/mL) treatment groups compared with that of the solvent control group ($p < 0.05$). The level of CFTR was significantly lower in the 8 μm and 70 μm PP-MP treatment groups than that in the solvent control group ($p < 0.05$) at the exposure concentration of 10 mg/mL.

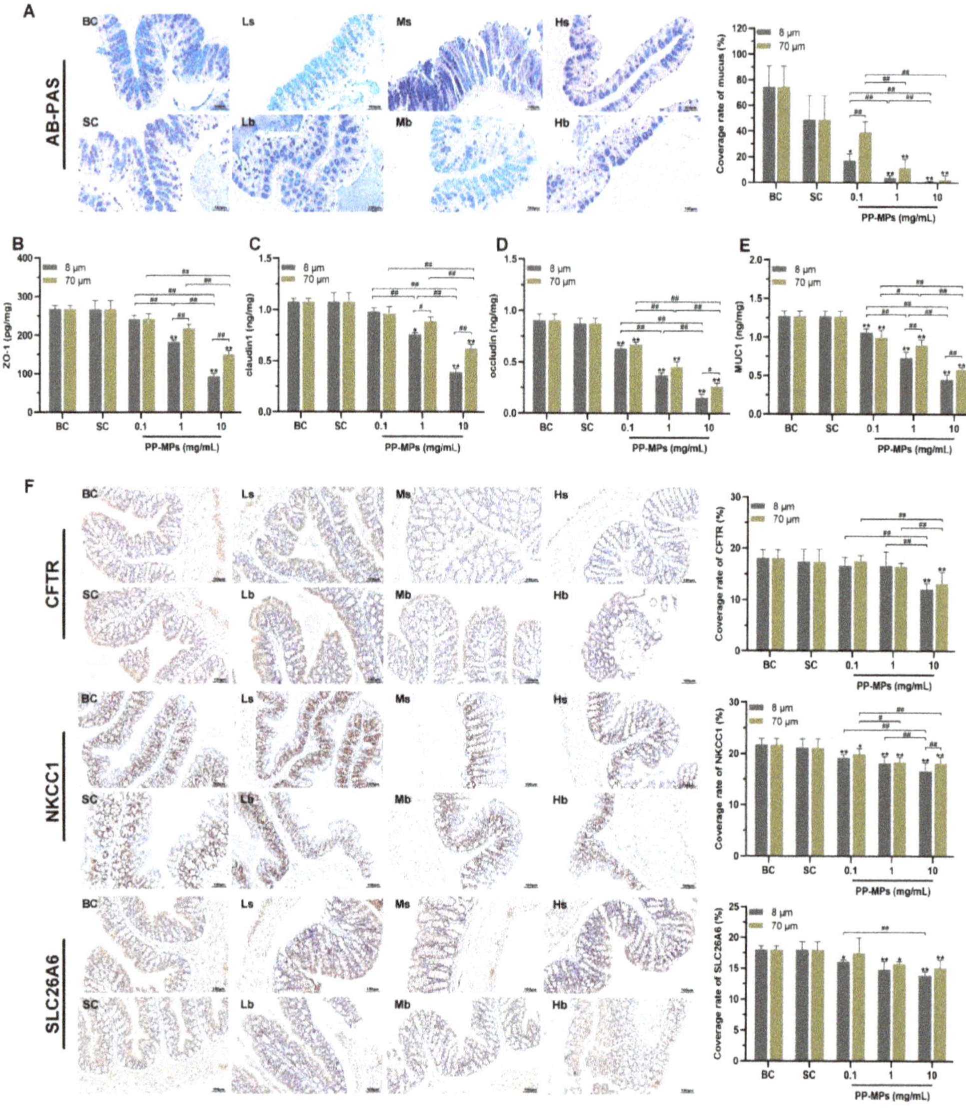

Figure 5. Polypropylene microplastics affect the intestinal barrier in mice. (**A**) AB-PAS staining and the ratio of the mucus coverage area to the entire colon area. Neutral mucin is purplish red, and acidic mucin is blue; the tissues and cells containing both neutral and acidic mucins show different degrees of purple. (**B–E**) The expression of ZO-1 (**B**), claudin-1 (**C**), occludin (**D**), and MUC1 (**E**) in the colon. (**F**) Immunohistochemistry of the colonic ion channel transport proteins CFTR, NKCC1, and SLC26A6. The nucleus stained with hematoxylin is blue and the positive expression is brown. The values are the means ± SDs (normalized amounts of AB-PAS and immunohistochemistry, $n = 3$; intestinal-barrier-related protein level, $n = 5$). $^{\#}$ $p < 0.05$, $^{\#\#}$ $p < 0.01$ between PP-MP treated groups, and * $p < 0.05$, ** $p < 0.01$ vs. SC was determined via one-way analysis of variance (ANOVA). BC: blank control (pure water); SC: solvent control (pure water containing 0.01% v/v Tween-80); PP-MPs: polypropylene microplastics.

3.5. Polypropylene Microplastic Exposure Promotes Apoptosis of Colonic Cells

The TUNEL experiment showed that there was no significant difference in the apoptotic rate of colonic cells between the blank control group and the solvent control group (Figure 6A,B). The apoptotic rate of colonic cells significantly increased ($p < 0.05$) in all PP-MP treatment groups in a concentration-dependent manner compared with that of the solvent control group. In addition, the apoptotic rate of colonic cells was significantly higher in the 8 μm PP-MP treatment group than that in the 70 μm PP-MP (1 and 10 mg/mL) treatment groups ($p < 0.05$).

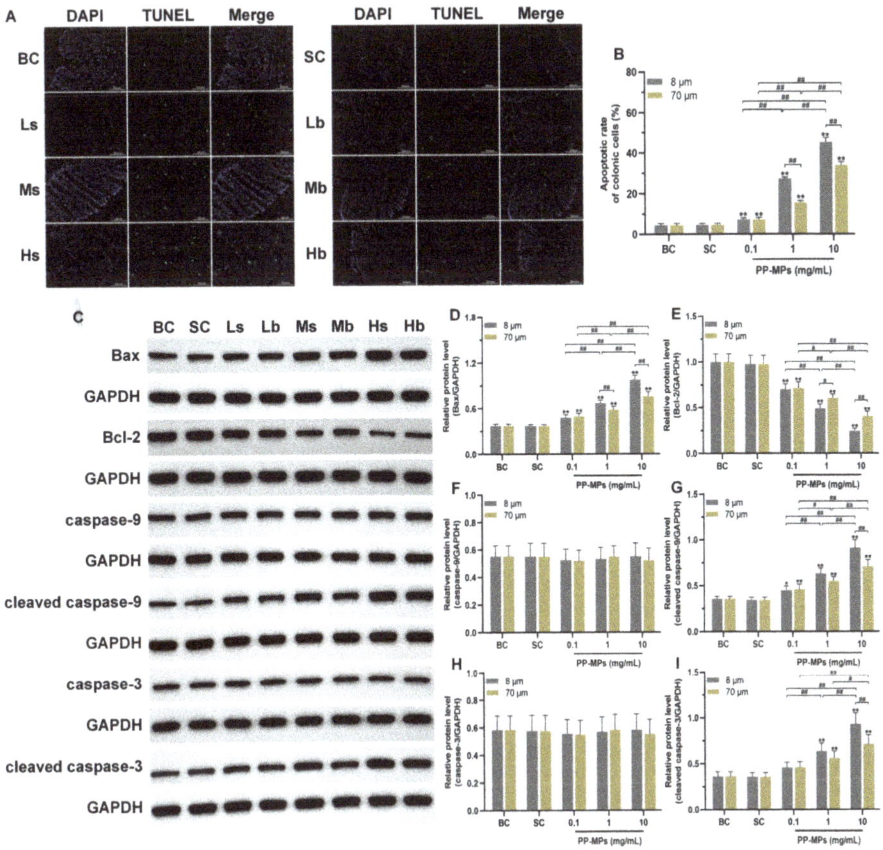

Figure 6. Effects of exposure to PP-MPs on apoptosis and the apoptosis pathway in colon tissue. (**A**) Images of colon sections stained with TUNEL to assess cell apoptosis after PP-MP exposure (200×, scale bar = 50 μm). Apoptotic cells are green, and the nucleus was stained with DAPI in blue. (**B**) Quantification of the apoptotic rate of colonic cells. (**C**) Western blot of Bax, Bcl-2, caspase-9, caspase-3, cleaved caspase-9, and cleaved caspase-3. (**D–I**) Quantitative expression of these proteins. The presented values are the means ± SDs (TUNEL staining: $n = 3$; Western blot analysis: $n = 4$). $^{\#} p < 0.05$, $^{\#\#} p < 0.01$ between the PP-MP treated groups, and $^{*} p < 0.05$, $^{**} p < 0.01$ vs. SC was determined via one-way analysis of variance (ANOVA). BC: blank control (pure water); SC: solvent control (pure water containing 0.01% v/v Tween-80); PP-MPs: polypropylene microplastics.

Furthermore, there was no significant difference in the levels of Bax, Bcl-2, caspase-9, cleaved caspase-9, caspase-3, and cleaved caspase-3 (all proteins related to apoptosis) in the colon of mice in the blank control group compared with the solvent control group

(Figure 6C–I). Bcl-2 levels significantly decreased in all PP-MP treatment groups, while the levels of Bax and cleaved caspase-9 significantly increased ($p < 0.05$) compared with the solvent control group; each change was concentration-dependent. The levels of cleaved caspase-3 were significantly higher ($p < 0.05$) in the 8 μm and 70 μm PP-MP treatment groups (1 and 10 mg/mL) than those in the solvent control group. At the same time, the Bax level was significantly higher after 8 μm PP-MP treatment than after 70 μm PP-MP treatment, while the Bcl-2 level was significantly lower after 8 μm PP-MP treatment compared with that of the 70 μm PP-MP treatment group ($p < 0.05$) at exposure concentrations of 1 mg/mL and 10 mg/mL. The levels of cleaved caspase-9 and cleaved caspase-3 were significantly higher in the 8 μm PP-MP treatment group than those in the 70 μm PP-MP treatment group ($p < 0.05$) at the exposure concentration of 10 mg/mL.

4. Discussion

There is little research on the biosafety of PP-MPs in mammals. This study evaluated the intestinal toxicity and possible mechanism of subacute oral ingestion of PP-MPs (8 μm and 70 μm) in mice. These sizes were selected because MPs with a particle size of 1–50 μm have been found to be dominant in tap water samples collected from consumer homes [18]. Moreover, MPs (<10 μm) can be absorbed by intestinal epithelial cells and accumulate in the intestine, after which they pass through the intestinal barrier and enter the systemic circulation to reach other organs [60,61]. MP fragments with sizes ranging between 5–10 μm were found in the placenta after vaginal delivery; the most common type of detected MP was PP [30], while those with sizes above 50 μm were found in the placenta and meconium post-cesarean section delivery [31].

According to our pathological and electron microscopy results, we speculate that PP-MPs may enter the intestinal epithelial cells through the damaged intestinal mucosal barrier, after which they occur in the colon. In addition, we found that the 10 mg/mL treatment group exhibited the most severe colon injury at 8 μm PP-MP or 70 μm PP-MP. Inflammatory and mucosal injury in colon tissue and mitochondrial injury in the epithelial cell ultrastructure are typically related to oxidative stress [62,63]. Therefore, we measured indicators of oxidative stress and inflammation in the colon. Notably, damage caused by 8 μm PP-MP to the colon tissue of mice was more severe than that by 70 μm PP-MP at the exposure concentration of 1 mg/mL, and we propose that the particle size of PP affects its toxicity.

Our study also found that the levels of the antioxidant GSH and the antioxidant enzymes SOD, CAT, and GSH-Px significantly decreased in the colon of mice after PP-MP exposure, while MDA levels significantly increased. Among them, GSH plays an important role in scavenging reactive oxygen species, and GSH-Px directly or indirectly plays the role of the antioxidant [64]. MDA is the product of lipid peroxidation which can be used as an index to measure oxidative damage [65]. These results show that PP-MP exposure reduces the level of antioxidant defense in the colon tissue of mice, resulting in increased oxidative stress reactions. Simultaneously, the levels of the pro-inflammatory factors TNF-α, IL-1β, and IL-6 significantly increased in colon tissue after oral ingestion of PP-MPs, while the level of anti-inflammatory factor IL-10 significantly decreased, indicating the occurrence of inflammatory injury in the colon of mice. Our results are consistent with those of previous studies showing that MPs induce oxidative stress and inflammation [66–68]. In addition, the colonic oxidative stress and inflammatory damage was more severe in mice treated with 8 μm PP-MPs than in those treated with 70 μm PP-MPs at an exposure concentration of 10 mg/mL. Therefore, we speculate that oxidative stress and inflammatory injury play a key role in intestinal injury caused by oral ingestion of PP-MPs.

Pathological conditions such as inflammation or disorder of the redox system can increase tyrosine phosphorylation of occludin protein and change the interaction between occludin and ZO-1 protein, thus breaking the tight junction [69]. Tight junctions are mainly composed of occludin and claudin proteins, which are anchored to the actin cytoskeleton by the scaffold protein Zos [70]. ZO-1 is closely related to other tight junction proteins and functions as an adaptor that connects transmembrane proteins to the perijunctional actomyosin ring [71,72]. Occludin is a tight junction transmembrane protein that plays a key role in maintaining the tight junction barrier [73]. Claudin is the main determinant of tight junction paracellular permeability, and changes in claudin protein may lead to intestinal instability, inflammation, and the progression of necrotizing enterocolitis [74,75]. Destruction of the tight junction increases intestinal permeability, and harmful substances such as bacteria and endotoxins enter the systemic milieu, adversely affecting the health of the organism [76,77]. Rao et al. [78] believe that oxidative stress induces tight-junction destruction and increases intestinal permeability by mediating tyrosine phosphorylation and the redistribution of occluding–ZO-1 and E-cadherin–beta-catenin complexes from the intracellular junctions. In addition, TNF and interleukin (IL) can significantly disrupt the expression and distribution of tight junction proteins, thus damaging intestinal barrier function [79,80]. TNF-α increases intestinal permeability by affecting occludin internalization and interferes with ZO-1 subcellular localization and protein expression by stimulating NF-κB signal transduction, resulting in damage to intestinal barrier function [81,82]. Incubation of the intestinal epithelial cell monolayer (Caco-2 and T84) with IFN-γ and TNF-α promotes the recombination of tight junction proteins including ZO-1, occludin, and claudin-1, and reduces epithelial barrier function [83]. IL-1, IL-6, and IL-10 have a great influence on epithelial and endothelial paracellular permeability [82]. IL-1 intervention in vitro increases the permeability of intestinal epithelial tight junctions and downregulates occludin expression by activating NF-κB. In vitro IL-6 intervention increases the permeability across endothelial cells and leads to ZO-1 mislocalization. Meanwhile, IL-10 blocks epithelial permeability induced by IFN-γ to maintain the epithelial barrier and chloride secretion function; IL-10-deficient mice exhibit increased intestinal permeability [84–87]. We found that occludin expression significantly decreased in all PP-MP treatment groups, and the expression of ZO-1 and claudin-1 significantly decreased in the 8 μm and 70 μm PP-MP treatment groups at 10 mg/mL exposure concentration. This indicates that tight junctions were broken. In addition, the colonic tight junction injury in mice treated with 8 μm PP-MPs was more severe than in mice treated with 70 μm PP-MPs at 10 mg/mL exposure concentration. Therefore, we conclude that PP-MPs destroy the tight junctions of mice intestinal epithelium by mediating colonic oxidative stress and inflammation and induce intestinal mucosal barrier injury.

The mucus layer on the surface of the intestinal cavity (and secreted mucins) protects the intestinal mucosa and resists bacterial invasion [88]. MUC1 is a cell-surface mucin of intestinal epithelial cells that protects epithelial cells and participates in signal transduction [89,90]. Our results show that colonic mucus secretion and MUC1 expression significantly decreased in the 8 μm and 70 μm PP-MP treatment groups at exposure concentrations of 1 mg/mL and 10 mg/mL. Similarly, oral ingestion of PS nano/microplastics reduces mucus secretion and the level of *Muc1* gene transcription in the colon of mice [55,91], and exposure to PS-MPs through drinking water also resulted in such effects in the colon of mice [47,92]. Oxidative damage can induce mucin degradation, and higher ROS levels can reduce mucus barrier thickness [93]. TNF-α causes the loss of mucin-producing goblet cells by inducing the death of intestinal epithelial cells, while also affecting mucus layer composition by regulating the expression of other components in the mucus [94]. IL-10 promotes the production of intestinal mucus by inhibiting protein misfolding and endoplasmic reticulum stress in goblet cells, thus maintaining the mucus barrier [95]. Our results indicate that oral ingestion of PP-MPs causes oxidative damage and changes the levels of inflammatory cytokines in the intestinal tract of mice which further induces destruction of the intestinal mucus barrier.

Tight junctions maintain the intestinal epithelial barrier and regulate the osmotic gradients required for ion transport [96]. In particular, epithelial anion secretion can ensure the correct ion environment needed for the formation of a normal mucus layer [97–99]. We speculate that the ion transport process may be affected after intestinal barrier injury in mice. Our results show that the expression of chloride channel proteins in the colon significantly changed after PP-MP exposure, and NKCC1 expression significantly decreased in all PP-MP treatment groups. The levels of CFTR and SLC26A6 significantly decreased in the 8 μm and 70 μm PP-MP treatment groups at 10 mg/mL exposure concentration. CFTR is a chloride channel located in the apical membrane of cells; it plays a key role in regulating intestinal epithelial secretion, and its dysfunction leads to abnormal bacterial colonization [96,100,101]. NKCC1 provides chlorine for apical secretion through CFTR by mediating chloride uptake at the basolateral pole of intestinal epithelial cells [96]. SLC26 is a conservative anion transporter family, and SLC26A6 has the most extensive exchange function in the family; it mediates the transport of Cl^-/HCO_3^- and other anions, and plays an important role in ion homeostasis and acid-base balance [102]. Therefore, our results show that PP-MP exposure affects ion transport in the intestinal tract of mice. This is suggestive of damage to the intestinal barrier and mucus layer, which may cause pathogens and other harmful substances to enter blood circulation through the intestinal wall [103–105]. PP particles likely enter intestinal epithelial cells through the damaged intestinal barrier and deposit in the cells. This is consistent with the results observed using electron microscopy in this study.

TLR4 is the key receptor of intestinal innate immunity. It is expressed on a variety of cell surfaces of the intestinal mucosa and plays a key role in inducing inflammatory responses and producing inflammatory mediators [106]. NF-κB participates in the expression and regulation of various genes, plays a role in inflammation and immune response, and is a common transcription factor in the process of inflammation mediated by TLR4. NF-κB is composed of five members: p50 (p105), p52 (p100), p65, RelB, and c-Rel. The active form of p50 is formed by p105 proteolysis, and the signal transduction pathway triggered by inflammatory stimulation usually leads to the release and nuclear translocation of the NF-κB p50/p65 dimer [107–109]. In this study, there was no significant change in p105 and p65 levels in the colon of all PP-MP treated mice. However, the levels of TLR4, p50, and *p*-p65 significantly increased, and the level of IκBα significantly decreased. This indicates that PP-MPs induce a colonic inflammatory response through the TLR4/NF-κB signaling pathway. In addition, the level of IκBα was significantly lower in the 8 μm PP-MP treatment group than in the 70 μm PP-MP treatment group at an exposure concentration of 10 mg/mL, while the levels of TLR4, p50, and *p*-p65 were significantly higher than those in the 70 μm PP-MP treatment group. This indicates that 8 μm PP-MPs had a stronger effect on inducing intestinal inflammation in mice. TLR4 recruited MyD88 and further recruited the downstream kinase IRAK4 and the ubiquitin ligase TRAF6. This led to the activation of TAK1 and IκB phosphorylation, thus releasing the NF-κB p50/p65 dimer. The released dimer is translocated into the nucleus through modification processes such as phosphorylation. This induces NF-κB-mediated transcription of inflammatory cytokines, especially the increase in pro-inflammatory cytokines TNF-α, IL-1β, and IL-6 [109–113], which is consistent with the changes in inflammatory factors in this study.

Disorder of the redox system and inflammation can promote apoptosis [114,115]. In this study, colonic epithelial cells of mice showed mitochondrial damage after PP-MP exposure, and TUNEL results suggested apoptosis of colonic epithelial cells. We further detected proteins related to apoptosis in the mitochondrial pathway. PP-MP exposure significantly upregulated the levels of the apoptosis markers Bax, cleaved caspase-9, and cleaved caspase-3, and significantly downregulated the level of anti-apoptosis factor Bcl-2. Intracellular stress induces intrinsic apoptosis which depends on the process of mitochondrial outer membrane permeabilization. The process is mediated by the Bcl-2 family (including the pro-apoptotic proteins Bax and Bak and anti-apoptotic proteins Bcl-2, Bcl-xL, and MCL-1), which can release the contents of the mitochondrial intermembrane space (such as cytochrome C) into the cytoplasm. Cytochrome C interacts with apoptosis protease-activating factor 1 (Apaf-1), thus activating caspase-9 and triggering the cascade activation of caspase-3 and caspase-7 [116–120]. Caspase-3 is a terminal caspase protein that plays a vital role in cell apoptosis [121]. NF-κB can also activate apoptosis by upregulating the pro-apoptotic factor, Bax [122,123]. Further, we found that the levels of Bax, cleaved caspase-9, and cleaved caspase-3 were significantly higher in the 8 μm PP-MP treatment group than in the 70 μm PP-MP treatment group at an exposure concentration of 10 mg/mL. Moreover, the Bcl-2 level was significantly lower in the 8 μm PP-MP treatment group than in the 70 μm PP-MP treatment group. This indicates that 8 μm PP-MPs had a stronger effect on inducing colonic cell apoptosis in the mitochondrial pathway. These results indicate that PP-MPs induce endogenous apoptosis in the colon via the Bax/Bcl-2/Caspase-3 signaling pathway. Abnormal apoptosis in colonic cells (especially epithelial cells) causes damage to the intestinal mucosal barrier by reducing barrier components, and may reduce mucus production by reducing the number of goblet cells, which further destroys the intestinal barrier.

The particle size of MPs may affect their biotoxicity. Our results indicated that the toxicity of 8 μm PP-MP to the colon of mice was stronger than that of 70 μm PP-MP, especially at an exposure concentration of 10 mg/mL. Similarly, PP particles with diameters < 20 μm (dispersed in DMSO) are more toxic to macrophages than PP particles with diameters of 25–200 μm [41]. Compared with larger particles, smaller particles have a higher bioreactivity due to their higher specific surface area [60], which may be the main reason for the stronger toxic effects of smaller PP particles. This means that 8 μm PP-MP is more likely to come into contact with the intestine and exerts stronger irritation effects on the colon, which may lead to more severe damage to the intestinal mucosal barrier. Moreover, smaller MP particles may elicit stronger cytotoxicity compared to the larger particles due to the different absorption modes of intestinal epithelial cells [124]. A comparison of the toxic mechanism of MPs with different particle size scales requires further study. PS particles (<10 μm) can activate p38, MAPK, Wnt/β-catenin, and other signaling pathways to exhibit toxicity [125–128]. However, there are few reports on the mechanism of toxicity of larger MP particles. In addition, with respect to the shape of MPs, the PP particles used in this study were irregular. Irregular and regularly shaped (spherical) MPs, such as PP and PS, can exert a variety of adverse effects by disrupting the balance of the redox system of organisms [34,39,125,126]. Therefore, the contribution of the shape to the biotoxicity of PP particles requires further research.

5. Conclusions

This study preliminarily demonstrated that oral ingestion of PP-MPs induces an imbalance in the redox system and activates the TLR4/NF-κB inflammatory signal pathway in the mouse intestine. This induces oxidative stress, inflammation, and apoptosis of intestinal epithelial cells through the mitochondrial pathway, which leads to intestinal barrier dysfunction and intestinal mucosal barrier damage, resulting in intestinal toxicity. Smaller PP-MPs exhibit higher intestinal toxicity compared with larger PP-MPs at the same exposure dose. This study provides data and reference for biological risk assessment of exposure to environmental PP-MPs via oral ingestion in mice. The findings can further aid in the formulation of pollution prevention and control policies for plastics.

Author Contributions: Conceptualization and supervision, X.L.; conception and design, Z.X.; investigation, data curation, and formal analysis, J.H.; writing—original draft preparation, R.J.; writing—review and editing, J.Y.; visualization, W.L., L.B., and K.L. All authors have read and agreed to the published version of the manuscript.

Funding: This research was funded by the National Natural Science Foundation of China [grant number 21607172].

Institutional Review Board Statement: The animal study protocol was approved by the Experimental Animal Welfare Ethics Committee of the Tianjin Institute of Environmental and Operational Medicine, IACUC approval code AMMS-04-2021-014. Approval date: 11 April 2021.

Informed Consent Statement: Not applicable.

Data Availability Statement: The data presented in this study are available on request from the corresponding author. The data are not publicly available due to privacy.

Conflicts of Interest: The authors declare no conflict of interest.

References

1. Geyer, R.; Jambeck, J.R.; Law, K.L. Production, use, and fate of all plastics ever made. *Sci. Adv.* **2017**, *3*, e1700782. [CrossRef] [PubMed]
2. Zarus, G.M.; Muianga, C.; Hunter, C.M.; Pappas, R.S. A review of data for quantifying human exposures to micro and nanoplastics and potential health risks. *Sci. Total Environ.* **2021**, *756*, 144010. [CrossRef] [PubMed]
3. Aragaw, T.A. Surgical face masks as a potential source for microplastic pollution in the COVID-19 scenario. *Mar. Pollut. Bull.* **2020**, *159*, 111517. [CrossRef] [PubMed]
4. Fadare, O.O.; Okoffo, E.D. COVID-19 face masks: A potential source of microplastic fibers in the environment. *Sci. Total Environ.* **2020**, *737*, 140279. [CrossRef]
5. Saadat, S.; Rawtani, D.; Hussain, C.M. Environmental perspective of COVID-19. *Sci. Total Environ.* **2020**, *728*, 138870. [CrossRef]
6. Thompson, R.C.; Olsen, Y.; Mitchell, R.P.; Davis, A.; Rowland, S.J.; John, A.W.G.; McGonigle, D.; Russell, A.E. Lost at sea: Where is all the plastic? *Science* **2004**, *304*, 838. [CrossRef]
7. Courtney, A.; Baker, J.; Bamford, H. (Eds.) *Proceedings of the International ReseArch. Workshop on the Occurrence, Effects, and Fate of Microplastic Marine Debris, September 9–11, 2008, University of Washington Tacoma, Tacoma, WA, USA*; NOAA Technical Memorandum NOS-OR&R-30; NOAA: Silver Spring, MD, USA, 2009.
8. EFSA Panel on Contaminants in the Food Chain (CONTAM). Presence of microplastics and nanoplastics in food, with particular focus on seafood. *EFSA J.* **2016**, *14*, e04501. [CrossRef]
9. He, D.; Chen, X.; Zhao, W.; Zhu, Z.; Qi, X.; Zhou, L.; Chen, W.; Wan, C.; Li, D.; Zou, X.; et al. Microplastics contamination in the surface water of the Yangtze River from upstream to estuary based on different sampling methods. *Environ. Res.* **2021**, *196*, 110908. [CrossRef]
10. Lindeque, P.K.; Cole, M.; Coppock, R.L.; Lewis, C.N.; Miller, R.Z.; Watts, A.J.R.; Wilson-McNeal, A.; Wright, S.L.; Galloway, T.S. Are we underestimating microplastic abundance in the marine environment? A comparison of microplastic capture with nets of different mesh-size. *Environ. Pollut.* **2020**, *265*, 114721. [CrossRef]
11. Cheng, Y.; Song, W.; Tian, H.; Zhang, K.; Li, B.; Du, Z.; Zhang, W.; Wang, J.; Wang, J.; Zhu, L. The Effects of High-Density Polyethylene and Polypropylene Microplastics on the Soil and Earthworm Metaphire Guillelmi Gut Microbiota. *Chemosphere* **2021**, *267*, 129219. [CrossRef]
12. Zhu, B.K.; Fang, Y.M.; Zhu, D.; Christie, P.; Ke, X.; Zhu, Y.G.; Christie, X.K.; Zhu, Y.G. Exposure to Nanoplastics Disturbs the Gut Microbiome in the Soil Oligochaete Enchytraeus Crypticus. *Environ. Pollut.* **2018**, *239*, 408–415. [CrossRef] [PubMed]
13. Kelly, F.J.; Fussell, J.C. Toxicity of Airborne Particles-Established Evidence, Knowledge Gaps and Emerging Areas of Importance. *Philos. Trans. A Math. Phys. Eng. Sci.* **2020**, *378*, 20190322. [CrossRef] [PubMed]
14. Zhang, Q.; Zhao, Y.; Du, F.; Cai, H.; Wang, G.; Shi, H. Microplastic fallout in different indoor environments. *Environ. Sci. Technol.* **2020**, *54*, 6530–6539. [CrossRef] [PubMed]
15. Sridharan, S.; Kumar, M.; Singh, L.; Bolan, N.S.; Saha, M. Microplastics as an emerging source of particulate air pollution: A critical review. *J. Hazard. Mater.* **2021**, *418*, 126245. [CrossRef] [PubMed]
16. Kosuth, M.; Mason, S.A.; Wattenberg, E.V. Anthropogenic contamination of tap water, beer, and sea salt. *PLoS ONE* **2018**, *13*, e0194970. [CrossRef] [PubMed]
17. Wang, Z.; Lin, T.; Chen, W. Occurrence and removal of microplastics in an advanced drinking water treatment plant (ADWTP). *Sci. Total Environ.* **2020**, *700*, 134520. [CrossRef] [PubMed]
18. Tong, H.; Jiang, Q.; Hu, X.; Zhong, X. Occurrence and identification of microplastics in tap water from China. *Chemosphere* **2020**, *252*, 126493. [CrossRef] [PubMed]
19. Koelmans, A.A.; Mohamed Nor, N.H.; Hermsen, E.; Kooi, M.; Mintenig, S.M.; De France, J. Microplastics in freshwaters and drinking water: Critical review and assessment of data quality. *Water Res.* **2019**, *155*, 410–422. [CrossRef] [PubMed]
20. Oßmann, B.E. Microplastics in Drinking Water? Present State of Knowledge and Open Questions. *Curr. Opin. Food Sci.* **2021**, *41*, 44–51. [CrossRef]
21. Liebezeit, G.; Liebezeit, E. Non-pollen particulates in honey and sugar. *Food Addit. Contam. Part A Chem. Anal. Control Exp. Risk Assess* **2013**, *30*, 2136–2140. [CrossRef]
22. Liebezeit, G.; Liebezeit, E. Origin of synthetic particles in honeys. *Pol. J. Food Nutr. Sci.* **2015**, *65*, 143–147. [CrossRef]
23. Zhu, J.; Zhang, Q.; Li, Y.; Tan, S.; Kang, Z.; Yu, X.; Lan, W.; Cai, L.; Wang, J.; Shi, H. Microplastic pollution in the Maowei Sea, a typical mariculture bay of China. *Sci. Total Environ.* **2019**, *658*, 62–68. [CrossRef] [PubMed]
24. Yang, D.; Shi, H.; Li, L.; Li, J.; Jabeen, K.; Kolandhasamy, P. Microplastic pollution in table salts from China. *Environ. Sci. Technol.* **2015**, *49*, 13622–13627. [CrossRef] [PubMed]
25. Lee, H.; Kunz, A.; Shim, W.J.; Walther, B.A. Microplastic contamination of table salts from Taiwan, including a global review. *Sci. Rep.* **2019**, *9*, 10145. [CrossRef] [PubMed]
26. Zolotova, N.; Kosyreva, A.; Dzhalilova, D.; Fokichev, N.; Makarova, O. Harmful effects of the microplastic pollution on animal health: A literature review. *PeerJ* **2022**, *10*, e13503. [CrossRef] [PubMed]
27. Schwabl, P.; Köppel, S.; Königshofer, P.; Bucsics, T.; Trauner, M.; Reiberger, T.; Liebmann, B. Detection of various microplastics in human stool: A prospective case series. *Ann. Intern. Med.* **2019**, *171*, 453–457. [CrossRef]

28. Zhang, N.; Li, Y.B.; He, H.R.; Zhang, J.F.; Ma, G.S. You are what you eat: Microplastics in the feces of young men living in Beijing. *Sci. Total Environ.* **2021**, *767*, 144345. [CrossRef]
29. Zhang, J.; Wang, L.; Trasande, L.; Kannan, K. Occurrence of polyethylene terephthalate and polycarbonate microplastics in infant and adult feces. *Environ. Sci. Technol. Lett.* **2021**, *8*, 989–994. [CrossRef]
30. Ragusa, A.; Svelato, A.; Santacroce, C.; Catalano, P.; Notarstefano, V.; Carnevali, O.; Papa, F.; Rongioletti, M.C.A.; Baiocco, F.; Draghi, S.; et al. Plasticenta: First evidence of microplastics in human placenta. *Environ. Int.* **2021**, *146*, 106274. [CrossRef]
31. Braun, T.; Ehrlich, L.; Henrich, W.; Koeppel, S.; Lomako, I.; Schwabl, P.; Liebmann, B. Detection of microplastic in human placenta and meconium in a clinical setting. *Pharmaceutics* **2021**, *13*, 921. [CrossRef]
32. Li, D.; Shi, Y.; Yang, L.; Xiao, L.; Kehoe, D.K.; Gun'ko, Y.K.; Boland, J.J.; Wang, J.J. Microplastic release from the degradation of polypropylene feeding bottles during infant formula preparation. *Nat. Food* **2020**, *1*, 746–754. [CrossRef]
33. Jemec Kokalj, A.; Dolar, A.; Drobne, D.; Marinšek, M.; Dolenec, M.; Škrlep, L.; Strmljan, G.; Mušič, B.; Škapin, A.S. Environmental Hazard. of polypropylene microplastics from disposable medical masks: Acute toxicity towards *Daphnia magna* and current knowledge on other polypropylene microplastics. *Microplast. Nanoplast.* **2022**, *2*, 1. [CrossRef]
34. Jeyavani, J.; Sibiya, A.; Gopi, N.; Mahboob, S.; Riaz, M.N.; Vaseeharan, B. Dietary consumption of polypropylene microplastics alter the biochemical parameters and histological response in freshwater benthic mollusc *Pomacea paludosa*. *Environ. Res.* **2022**, *212*, 113370. [CrossRef] [PubMed]
35. Piccardo, M.; Bertoli, M.; Pastorino, P.; Barceló, D.; Provenza, F.; Lesa, D.; Anselmi, S.; Elia, A.C.; Prearo, M.; Pizzul, E.; et al. Lethal and sublethal responses of *Hydropsyche pellucidula* (Insecta, Trichoptera) to commercial polypropylene microplastics after different preconditioning treatments. *Toxics* **2021**, *9*, 256. [CrossRef] [PubMed]
36. Esterhuizen, M.; Kim, Y.J. Effects of polypropylene, polyvinyl chloride, polyethylene terephthalate, polyurethane, high-density polyethylene, and polystyrene microplastic on *Nelumbo nucifera* (Lotus) in water and sediment. *Environ. Sci. Pollut. Res. Int.* **2022**, *29*, 17580–17590. [CrossRef]
37. Shi, R.; Liu, W.; Lian, Y.; Wang, Q.; Zeb, A.; Tang, J. Phytotoxicity of polystyrene, polyethylene and polypropylene microplastics on tomato (*Lycopersicon esculentum* L.). *J. Environ. Manag.* **2022**, *317*, 115441. [CrossRef]
38. Zhao, Y.; Qiao, R.; Zhang, S.; Wang, G. Metabolomic profiling reveals the intestinal toxicity of different length of microplastic fibers on zebrafish (*Danio rerio*). *J. Hazard. Mater.* **2021**, *403*, 123663. [CrossRef]
39. Sheng, C.; Zhang, S.; Zhang, Y. The influence of different polymer types of microplastics on adsorption, accumulation, and toxicity of triclosan in zebrafish. *J. Hazard. Mater.* **2021**, *402*, 123733. [CrossRef]
40. Lei, L.; Wu, S.; Lu, S.; Liu, M.; Song, Y.; Fu, Z.; Shi, H.; Raley-Susman, K.M.; He, D. Microplastic particles cause intestinal damage and other adverse effects in zebrafish *Danio rerio* and nematode *Caenorhabditis elegans*. *Sci. Total Environ.* **2018**, *619–620*, 1–8. [CrossRef]
41. Hwang, J.; Choi, D.; Han, S.; Choi, J.; Hong, J. An assessment of the toxicity of polypropylene microplastics in human derived cells. *Sci. Total Environ.* **2019**, *684*, 657–669. [CrossRef]
42. Woo, J.H.; Seo, H.J.; Lee, J.Y.; Lee, I.; Jeon, K.; Kim, B.; Lee, K. Polypropylene microplastic exposure leads to lung inflammation through p38-mediated NF-κB pathway due to mitochondrial damage. *Part. Fibre Toxicol.* **2022**, *20*, 2. [CrossRef]
43. Stock, V.; Laurisch, C.; Franke, J.; Dönmez, M.H.; Voss, L.; Böhmert, L.; Braeuning, A.; Sieg, H. Uptake and cellular effects of PE, PP, PET and PVC microplastic particles. *Toxicol. Vitr.* **2021**, *70*, 105021. [CrossRef]
44. Lehner, R.; Wohlleben, W.; Septiadi, D.; Landsiedel, R.; Petri-Fink, A.; Rothen-Rutishauser, B. A novel 3D intestine barrier model to study the immune response upon exposure to microplastics. *Arch. Toxicol.* **2020**, *94*, 2463–2479. [CrossRef] [PubMed]
45. Kim, J.; Maruthupandy, M.; An, K.S.; Lee, K.H.; Jeon, S.; Kim, J.S.; Cho, W.S. Acute and subacute repeated oral toxicity study of fragmented microplastics in Sprague-Dawley rats. *Ecotoxicol. Environ. Saf.* **2021**, *228*, 112964. [CrossRef] [PubMed]
46. Deng, Y.; Yan, Z.; Shen, R.; Wang, M.; Huang, Y.; Ren, H.; Zhang, Y.; Lemos, B. Microplastics release phthalate esters and cause aggravated adverse effects in the mouse gut. *Environ. Int.* **2020**, *143*, 105916. [CrossRef] [PubMed]
47. Jin, Y.; Lu, L.; Tu, W.; Luo, T.; Fu, Z. Impacts of polystyrene microplastic on the gut barrier, microbiota and metabolism of mice. *Sci. Total Environ.* **2019**, *649*, 308–317. [CrossRef] [PubMed]
48. Djouina, M.; Vignal, C.; Dehaut, A.; Caboche, S.; Hirt, N.; Waxin, C.; Himber, C.; Beury, D.; Hot, D.; Dubuquoy, L.; et al. Oral exposure to polyethylene microplastics alters gut morphology, immune response, and microbiota composition in mice. *Environ. Res.* **2022**, *212*, 113230. [CrossRef] [PubMed]
49. Cui, Y.; Wang, Q.; Chang, R.; Zhou, X.; Xu, C. Intestinal barrier function-non-alcoholic fatty liver disease interactions and possible role of gut microbiota. *J. Agric. Food Chem.* **2019**, *67*, 2754–2762. [CrossRef]
50. Huang, L.; He, F.; Wu, B. Mechanism of effects of nickel or nickel compounds on intestinal mucosal barrier. *Chemosphere* **2022**, *305*, 135429. [CrossRef] [PubMed]
51. Johansson, M.E.; Ambort, D.; Pelaseyed, T.; Schütte, A.; Gustafsson, J.K.; Ermund, A.; Subramani, D.B.; Holmén-Larsson, J.M.; Thomsson, K.A.; Bergström, J.H.; et al. Composition and functional role of the mucus layers in the intestine. *Cell Mol. Life Sci.* **2011**, *68*, 3635–3641. [CrossRef]
52. Schroeder, B.O. Fight them or feed them: How the intestinal mucus layer manages the gut microbiota. *Gastroenterol. Rep.* **2019**, *7*, 3–12. [CrossRef] [PubMed]
53. Chelakkot, C.; Ghim, J.; Ryu, S.H. Mechanisms regulating intestinal barrier integrity and its pathological implications. *Exp. Mol. Med.* **2018**, *50*, 1–9. [CrossRef] [PubMed]

54. Catalioto, R.M.; Maggi, C.A.; Giuliani, S. Intestinal epithelial barrier dysfunction in disease and possible therapeutical interventions. *Curr. Med. Chem.* **2011**, *18*, 398–426. [CrossRef] [PubMed]
55. Liang, B.; Zhong, Y.; Huang, Y.; Lin, X.; Liu, J.; Lin, L.; Hu, M.; Jiang, J.; Dai, M.; Wang, B.; et al. Underestimated health risks: Polystyrene micro- and nanoplastics jointly induce intestinal barrier dysfunction by ROS-mediated epithelial cell apoptosis. *Part. Fibre Toxicol.* **2021**, *18*, 20. [CrossRef]
56. Wang, K.; Qiu, L.; Zhu, J.; Sun, Q.; Qu, W.; Yu, Y.; Zhao, Z.; Yu, Y.; Shao, G. Environmental contaminant BPA causes intestinal damage by disrupting cellular repair and injury homeostasis in vivo and in vitro. *Biomed. Pharmacother.* **2021**, *137*, 111270. [CrossRef]
57. Stock, V.; Fahrenson, C.; Thuenemann, A.; Dönmez, M.H.; Voss, L.; Böhmert, L.; Braeuning, A.; Lampen, A.; Sieg, H. Impact of artificial digestion on the sizes and shapes of microplastic particles. *Food Chem. Toxicol.* **2020**, *135*, 111010. [CrossRef]
58. Schneider, C.A.; Rasband, W.S.; Eliceiri, K.W. NIH Image to ImageJ: 25 years of image analysis. *Nat. Methods* **2012**, *9*, 671–675. [CrossRef] [PubMed]
59. Senathirajah, K.; Attwood, S.; Bhagwat, G.; Carbery, M.; Wilson, S.; Palanisami, T. Estimation of the mass of microplastics ingested—A pivotal first step towards human health risk assessment. *J. Hazard. Mater.* **2021**, *404*, 124004. [CrossRef]
60. Stock, V.; Böhmert, L.; Lisicki, E.; Block, R.; Cara-Carmona, J.; Pack, L.K.; Selb, R.; Lichtenstein, D.; Voss, L.; Henderson, C.J.; et al. Uptake and effects of orally ingested polystyrene microplastic particles in vitro and in vivo. *Arch. Toxicol.* **2019**, *93*, 1817–1833. [CrossRef]
61. Mu, Y.; Sun, J.; Li, Z.; Zhang, W.; Liu, Z.; Li, C.; Peng, C.; Cui, G.; Shao, H.; Du, Z. Activation of pyroptosis and ferroptosis is involved in the hepatotoxicity induced by polystyrene microplastics in mice. *Chemosphere* **2022**, *291*, 132944. [CrossRef]
62. Bao, W.; Liu, M.; Meng, J.; Liu, S.; Wang, S.; Jia, R.; Wang, Y.; Ma, G.; Wei, W.; Tian, Z. MOFs-based nanoagent enables dual mitochondrial damage in synergistic antitumor therapy via oxidative stress and calcium overload. *Nat. Commun.* **2021**, *12*, 6399. [CrossRef] [PubMed]
63. Wu, G.; Wei, P.; Chen, X.; Zhang, Z.; Jin, Z.; Liu, J.; Liu, L. Less is more: Biological effects of NiSe(2)/rGO nanocomposites with low dose provide new insight for risk assessment. *J. Hazard. Mater.* **2021**, *415*, 125605. [CrossRef] [PubMed]
64. Yun, S.H.; Moon, Y.S.; Sohn, S.H.; Jang, I.S. Effects of cyclic heat stress or vitamin C supplementation during cyclic heat stress on HSP70, inflammatory cytokines, and the antioxidant defense system in Sprague Dawley rats. *Exp. Anim.* **2012**, *61*, 543–553. [CrossRef] [PubMed]
65. Impellizzeri, D.; Cordaro, M.; Campolo, M.; Gugliandolo, E.; Esposito, E.; Benedetto, F.; Cuzzocrea, S.; Navarra, M. Anti-inflammatory and antioxidant effects of flavonoid-rich fraction of bergamot juice (BJe) in a mouse model of intestinal ischemia/reperfusion injury. *Front. Pharmacol.* **2016**, *7*, 203. [CrossRef] [PubMed]
66. Hou, J.; Lei, Z.; Cui, L.; Hou, Y.; Yang, L.; An, R.; Wang, Q.; Li, S.; Zhang, H.; Zhang, L. Polystyrene microplastics lead to pyroptosis and apoptosis of ovarian granulosa cells via NLRP3/Caspase-1 signaling pathway in rats. *Ecotoxicol. Environ. Saf.* **2021**, *212*, 112012. [CrossRef]
67. Yin, K.; Lu, H.; Zhang, Y.; Hou, L.; Meng, X.; Li, J.; Zhao, H.; Xing, M. Secondary brain injury after polystyrene microplastic-induced intracerebral hemorrhage is associated with inflammation and pyroptosis. *Chem. Biol. Interact.* **2022**, *367*, 110180. [CrossRef]
68. Hu, M.; Palić, D. Micro- and nano-plastics activation of oxidative and inflammatory adverse outcome pathways. *Redox Biol.* **2020**, *37*, 101620. [CrossRef]
69. Kale, G.; Naren, A.P.; Sheth, P.; Rao, R.K. Tyrosine phosphorylation of occludin attenuates its interactions with ZO-1, ZO-2, and ZO-3. *Biochem. Biophys. Res. Commun.* **2003**, *302*, 324–329. [CrossRef]
70. Salim, S.Y.; Söderholm, J.D. Importance of disrupted intestinal barrier in inflammatory bowel diseases. *Inflamm. Bowel Dis.* **2011**, *17*, 362–381. [CrossRef]
71. Ulluwishewa, D.; Anderson, R.C.; McNabb, W.C.; Moughan, P.J.; Wells, J.M.; Roy, N.C. Regulation of tight junction permeability by intestinal bacteria and dietary components. *J. Nutr.* **2011**, *141*, 769–776. [CrossRef]
72. Suzuki, T.; Tanabe, S.; Hara, H. Kaempferol enhances intestinal barrier function through the cytoskeletal association and expression of tight junction proteins in Caco-2 cells. *J. Nutr.* **2011**, *141*, 87–94. [CrossRef] [PubMed]
73. Furuse, M.; Hirase, T.; Itoh, M.; Nagafuchi, A.; Yonemura, S.; Tsukita, S.; Tsukita, S. Occludin: A novel integral membrane protein localizing at tight junctions. *J. Cell Biol.* **1993**, *123*, 1777–1788. [CrossRef] [PubMed]
74. Lynn, K.S.; Peterson, R.J.; Koval, M. Ruffles and spikes: Control of tight junction morphology and permeability by claudins. *Biochim. Biophys. Acta Biomembr.* **2020**, *1862*, 183339. [CrossRef] [PubMed]
75. Griffiths, V.; Al Assaf, N.; Khan, R. Review of claudin proteins as potential biomarkers for necrotizing enterocolitis. *Ir. J. Med. Sci.* **2021**, *190*, 1465–1472. [CrossRef] [PubMed]
76. Chen, Y.; Zhang, H.S.; Fong, G.H.; Xi, Q.L.; Wu, G.H.; Bai, C.G.; Ling, Z.Q.; Fan, L.; Xu, Y.M.; Qin, Y.Q.; et al. PHD3 stabilizes the tight junction protein occludin and protects intestinal epithelial barrier function. *J. Biol. Chem.* **2015**, *290*, 20580–20589. [CrossRef] [PubMed]
77. Camilleri, M. Leaky gut: Mechanisms, measurement and clinical implications in humans. *Gut* **2019**, *68*, 1516–1526. [CrossRef]
78. Rao, R.K.; Basuroy, S.; Rao, V.U.; Karnaky, K.J., Jr.; Gupta, A. Tyrosine phosphorylation and dissociation of occludin-ZO-1 and E-cadherin-beta-catenin complexes from the cytoskeleton by oxidative stress. *Biochem. J.* **2002**, *368*, 471–481. [CrossRef] [PubMed]

79. Kim, K.Y.; Oh, T.W.; Do, H.J.; Yang, J.H.; Yang, I.J.; Jeon, Y.H.; Go, Y.H.; Ahn, S.C.; Ma, J.Y.; Park, K.I. *Acer palmatum thumb.* Ethanol extract alleviates interleukin-6-induced barrier dysfunction and dextran sodium sulfate-induced colitis by improving intestinal barrier function and reducing inflammation. *J. Immunol. Res.* **2018**, *2018*, 5718396. [CrossRef]
80. Zhou, H.Y.; Zhu, H.; Yao, X.M.; Qian, J.P.; Yang, J.; Pan, X.D.; Chen, X.D. Metformin regulates tight junction of intestinal epithelial cells via MLCK-MLC signaling pathway. *Eur. Rev. Med. Pharmacol. Sci.* **2017**, *21*, 5239–5246. [CrossRef] [PubMed]
81. Marchiando, A.M.; Shen, L.; Graham, W.V.; Weber, C.R.; Schwarz, B.T.; Austin, J.R., 2nd; Raleigh, D.R.; Guan, Y.; Watson, A.J.; Montrose, M.H.; et al. Caveolin-1-dependent occludin endocytosis is required for TNF-induced tight junction regulation in vivo. *J. Cell Biol.* **2010**, *189*, 111–126. [CrossRef] [PubMed]
82. Capaldo, C.T.; Nusrat, A. Cytokine regulation of tight junctions. *Biochim. Biophys. Acta* **2009**, *1788*, 864–871. [CrossRef] [PubMed]
83. Zolotarevsky, Y.; Hecht, G.; Koutsouris, A.; Gonzalez, D.E.; Quan, C.; Tom, J.; Mrsny, R.J.; Turner, J.R. A membrane-permeant peptide that inhibits MLC kinase restores barrier function in in vitro models of intestinal disease. *Gastroenterology* **2002**, *123*, 163–172. [CrossRef] [PubMed]
84. Al-Sadi, R.M.; Ma, T.Y. IL-1beta causes an increase in intestinal epithelial tight junction permeability. *J. Immunol.* **2007**, *178*, 4641–4649. [CrossRef] [PubMed]
85. Desai, T.R.; Leeper, N.J.; Hynes, K.L.; Gewertz, B.L. Interleukin-6 causes endothelial barrier dysfunction via the protein kinase C pathway. *J. Surg. Res.* **2002**, *104*, 118–123. [CrossRef] [PubMed]
86. Madsen, K.L.; Lewis, S.A.; Tavernini, M.M.; Hibbard, J.; Fedorak, R.N. Interleukin 10 prevents cytokine-induced disruption of T84 monolayer barrier integrity and limits chloride secretion. *Gastroenterology* **1997**, *113*, 151–159. [CrossRef]
87. Madsen, K.L.; Malfair, D.; Gray, D.; Doyle, J.S.; Jewell, L.D.; Fedorak, R.N. Interleukin-10 gene-deficient mice develop a primary intestinal permeability defect in response to enteric microflora. *Inflamm. Bowel Dis.* **1999**, *5*, 262–270. [CrossRef]
88. Linden, S.K.; Sutton, P.; Karlsson, N.G.; Korolik, V.; McGuckin, M.A. Mucins in the mucosal barrier to infection. *Mucosal Immunol.* **2008**, *1*, 183–197. [CrossRef]
89. Sheng, Y.H.; Davies, J.M.; Wang, R.; Wong, K.Y.; Giri, R.; Yang, Y.; Begun, J.; Florin, T.H.; Hasnain, S.Z.; McGuckin, M.A. MUC1-mediated macrophage activation promotes colitis-associated colorectal cancer via activating the interleukin-6/ signal transducer and activator of transcription 3 axis. *Cell Mol. Gastroenterol. Hepatol.* **2022**, *14*, 789–811. [CrossRef]
90. Gendler, S.J. MUC1, the renaissance molecule. *J. Mammary Gland Biol. Neoplasia* **2001**, *6*, 339–353. [CrossRef]
91. Lu, L.; Wan, Z.; Luo, T. Polystyrene microplastics induce gut microbiota dysbiosis and hepatic lipid metabolism disorder in mice. *Sci. Total Environ.* **2018**, *631–632*, 449–458. [CrossRef]
92. Luo, T.; Wang, C.; Pan, Z.; Jin, C.; Fu, Z.; Jin, Y. Maternal polystyrene microplastic exposure during gestation and lactation altered metabolic homeostasis in the dams and their F1 and F2 offspring. *Environ. Sci. Technol.* **2019**, *53*, 10978–10992. [CrossRef] [PubMed]
93. Brownlee, I.A.; Knight, J.; Dettmar, P.W.; Pearson, J.P. Action of reactive oxygen species on colonic mucus secretions. *Free Radic. Biol. Med.* **2007**, *43*, 800–808. [CrossRef] [PubMed]
94. Leppkes, M.; Roulis, M.; Neurath, M.F.; Kollias, G.; Becker, C. Pleiotropic functions of TNF-α in the regulation of the intestinal epithelial response to inflammation. *Int. Immunol.* **2014**, *26*, 509–515. [CrossRef]
95. Hasnain, S.Z.; Tauro, S.; Das, I.; Tong, H.; Chen, A.C.; Jeffery, P.L.; McDonald, V.; Florin, T.H.; McGuckin, M.A. IL-10 promotes production of intestinal mucus by suppressing protein misfolding and endoplasmic reticulum stress in goblet cells. *Gastroenterology* **2013**, *144*, 357–368.e9. [CrossRef]
96. Murek, M.; Kopic, S.; Geibel, J. Evidence for intestinal chloride secretion. *Exp. Physiol.* **2010**, *95*, 471–478. [CrossRef]
97. Muchekehu, R.W.; Quinton, P.M. A new role for bicarbonate secretion in cervico-uterine mucus release. *J. Physiol.* **2010**, *588*, 2329–2342. [CrossRef] [PubMed]
98. Garcia, M.A.; Yang, N.; Quinton, P.M. Normal mouse intestinal mucus release requires cystic fibrosis transmembrane regulator-dependent bicarbonate secretion. *J. Clin. Investig.* **2009**, *119*, 2613–2622. [CrossRef] [PubMed]
99. Gustafsson, J.K.; Lindén, S.K.; Alwan, A.H.; Scholte, B.J.; Hansson, G.C.; Sjövall, H. Carbachol-induced colonic mucus formation requires transport via NKCC1, K^+ channels and CFTR. *Pflug. Arch.* **2015**, *467*, 1403–1415. [CrossRef]
100. Li, C.; Krishnamurthy, P.C.; Penmatsa, H.; Marrs, K.L.; Wang, X.Q.; Zaccolo, M.; Jalink, K.; Li, M.; Nelson, D.J.; Schuetz, J.D.; et al. Spatiotemporal coupling of cAMP transporter to CFTR chloride channel function in the gut epithelia. *Cell* **2007**, *131*, 940–951. [CrossRef]
101. Lynch, S.V.; Goldfarb, K.C.; Wild, Y.K.; Kong, W.; De Lisle, R.C.; Brodie, E.L. Cystic fibrosis transmembrane conductance regulator knockout mice exhibit aberrant gastrointestinal microbiota. *Gut Microbes* **2013**, *4*, 41–47. [CrossRef]
102. Wang, J.; Wang, W.; Wang, H.; Tuo, B. Physiological and pathological functions of SLC26A6. *Front. Med.* **2020**, *7*, 618256. [CrossRef] [PubMed]
103. Zheng, J.; Ahmad, A.A.; Yang, Y.; Liang, Z.; Shen, W.; Feng, M.; Shen, J.; Lan, X.; Ding, X. *Lactobacillus rhamnosus* CY12 enhances intestinal barrier function by regulating tight junction protein expression, oxidative stress, and inflammation response in lipopolysaccharide-induced caco-2 cells. *Int. J. Mol. Sci.* **2022**, *23*, 11162. [CrossRef] [PubMed]
104. Wang, G.; Sun, S.; Wu, X.; Yang, S.; Wu, Y.; Zhao, J.; Zhang, H.; Chen, W. Intestinal environmental disorders associate with the tissue damages induced by perfluorooctane sulfonate exposure. *Ecotoxicol. Environ. Saf.* **2020**, *197*, 110590. [CrossRef] [PubMed]

105. Huang, Z.; Weng, Y.; Shen, Q.; Zhao, Y.; Jin, Y. Microplastic: A potential threat to human and animal health by interfering with the intestinal barrier function and changing the intestinal microenvironment. *Sci. Total Environ.* **2021**, *785*, 147365. [CrossRef] [PubMed]
106. Yao, H.; Hu, C.; Yin, L.; Tao, X.; Xu, L.; Qi, Y.; Han, X.; Xu, Y.; Zhao, Y.; Wang, C.; et al. Dioscin reduces lipopolysaccharide-induced inflammatory liver injury via regulating TLR4/MyD88 signal pathway. *Int. Immunopharmacol.* **2016**, *36*, 132–141. [CrossRef] [PubMed]
107. Gao, K.; Dai, D.L.; Martinka, M.; Li, G. Prognostic significance of nuclear factor-kappaB p105/p50 in human melanoma and its role in cell migration. *Cancer Res.* **2006**, *66*, 8382–8388. [CrossRef]
108. Ryseck, R.P.; Bull, P.; Takamiya, M.; Bours, V.; Siebenlist, U.; Dobrzanski, P.; Bravo, R. RelB, a new Rel family transcription activator that can interact with p50-NF-kappa B. *Mol. Cell Biol.* **1992**, *12*, 674–684. [CrossRef] [PubMed]
109. Li, S.; Yang, F.; Ma, C.; Cao, W.; Yang, J.; Zhao, Z.; Tian, H.; Zhu, Z.; Zheng, H. Porcine epidemic diarrhea virus nsp14 inhibits NF-κB pathway activation by targeting the IKK complex and p65. *Anim. Dis.* **2021**, *1*, 24. [CrossRef]
110. Matelski, L.; Morgan, R.K.; Grodzki, A.C.; Van de Water, J.; Lein, P.J. Effects of cytokines on nuclear factor-kappa B, cell viability, and synaptic connectivity in a human neuronal cell line. *Mol. Psychiatry* **2021**, *26*, 875–887. [CrossRef] [PubMed]
111. Yang, T.N.; Li, X.N.; Wang, Y.X.; Ma, X.Y.; Li, J.L. Disrupted microbiota-barrier-immune interaction in phthalates-mediated barrier defect in the duodenum. *Chemosphere* **2022**, *308*, 136275. [CrossRef]
112. Hu, X.; Liu, S.; Zhu, J.; Ni, H. Dachengqi decoction alleviates acute lung injury and inhibits inflammatory cytokines production through TLR4/NF-κB signaling pathway in vivo and in vitro. *J. Cell Biochem.* **2019**, *120*, 8956–8964. [CrossRef]
113. Chang, Y.; Yuan, L.; Liu, J.; Muhammad, I.; Cao, C.; Shi, C.; Zhang, Y.; Li, R.; Li, C.; Liu, F. Dihydromyricetin attenuates *Escherichia coli* lipopolysaccharide-induced ileum injury in chickens by inhibiting NLRP3 inflammasome and TLR4/NF-κB signalling pathway. *Vet. Res.* **2020**, *51*, 72. [CrossRef] [PubMed]
114. Bai, Q.; Wang, Z.; Piao, Y.; Zhou, X.; Piao, Q.; Jiang, J.; Liu, H.; Piao, H.; Li, L.; Song, Y.; et al. Sesamin alleviates asthma airway inflammation by regulating mitophagy and mitochondrial apoptosis. *J. Agric. Food Chem.* **2022**, *70*, 4921–4933. [CrossRef] [PubMed]
115. Li, W.J.; Zhang, L.; Wu, H.X.; Li, M.; Wang, T.; Zhang, W.B.; Du, Z.Y.; Zhang, M.L. Intestinal microbiota mediates gossypol-induced intestinal inflammation, oxidative stress, and apoptosis in fish. *J. Agric. Food Chem.* **2022**, *70*, 6688–6697. [CrossRef] [PubMed]
116. Zhang, H.; Huang, J.; Yang, J.; Cai, J.; Liu, Q.; Zhang, X.; Bao, J.; Zhang, Z. Cadmium induces apoptosis and autophagy in swine small intestine by downregulating the PI3K/Akt pathway. *Environ. Sci. Pollut. Res. Int.* **2022**, *29*, 41207–41218. [CrossRef] [PubMed]
117. Spierings, D.; McStay, G.; Saleh, M.; Bender, C.; Chipuk, J.; Maurer, U.; Green, D.R. Connected to death: The (unexpurgated) mitochondrial pathway of apoptosis. *Science* **2005**, *310*, 66–67. [CrossRef]
118. Oberst, A.; Bender, C.; Green, D.R. Living with death: The evolution of the mitochondrial pathway of apoptosis in animals. *Cell Death Differ.* **2008**, *15*, 1139–1146. [CrossRef]
119. Tang, D.; Kang, R.; Berghe, T.V.; Vandenabeele, P.; Kroemer, G. The molecular machinery of regulated cell death. *Cell Res.* **2019**, *29*, 347–364. [CrossRef]
120. Hou, L.; Wang, D.; Yin, K.; Zhang, Y.; Lu, H.; Guo, T.; Li, J.; Zhao, H.; Xing, M. Polystyrene microplastics induce apoptosis in chicken testis via crosstalk between NF-κB and Nrf2 pathways. *Comp. Biochem. Physiol. C Toxicol. Pharmacol.* **2022**, *262*, 109444. [CrossRef]
121. Zhou, W.J.; Wang, S.; Hu, Z.; Zhou, Z.Y.; Song, C.J. Angelica sinensis polysaccharides promotes apoptosis in human breast cancer cells via CREB-regulated caspase-3 activation. *Biochem. Biophys. Res. Commun.* **2015**, *467*, 562–569. [CrossRef]
122. Shou, Y.; Li, N.; Li, L.; Borowitz, J.L.; Isom, G.E. NF-kappaB-mediated up-regulation of Bcl-X(S) and Bax contributes to cytochrome c release in cyanide-induced apoptosis. *J. Neurochem.* **2002**, *81*, 842–852. [CrossRef] [PubMed]
123. Gill, J.S.; Windebank, A.J. Ceramide initiates NFkappaB-mediated caspase activation in neuronal apoptosis. *Neurobiol. Dis.* **2000**, *7*, 448–461. [CrossRef] [PubMed]
124. Visalli, G.; Facciolà, A.; Pruiti Ciarello, M.; De Marco, G.; Maisano, M.; Di Pietro, A. Acute and sub-chronic effects of microplastics (3 and 10 μm) on the Human Intestinal Cells Ht-29. *Int. J. Environ. Res. Public Health* **2021**, *18*, 5833. [CrossRef] [PubMed]
125. Li, Z.; Zhu, S.; Liu, Q.; Wei, J.; Jin, Y.; Wang, X.; Zhang, L. Polystyrene microplastics cause cardiac fibrosis by activating Wnt/β-catenin signaling pathway and promoting cardiomyocyte apoptosis in rats. *Environ. Pollut.* **2020**, *265*, 115025. [CrossRef]
126. Liu, Z.; Huang, Y.; Jiao, Y.; Chen, Q.; Wu, D.; Yu, P.; Li, Y.; Cai, M.; Zhao, Y. Polystyrene nanoplastic induces ROS production and affects the MAPK-HIF-1/NFkB-mediated antioxidant system in *Daphnia pulex*. *Aquat. Toxicol.* **2020**, *220*, 105420. [CrossRef]
127. Jeong, C.B.; Won, E.J.; Kang, H.M.; Lee, M.C.; Hwang, D.S.; Hwang, U.K.; Zhou, B.; Souissi, S.; Lee, S.J.; Lee, J.S. Microplastic size-dependent toxicity, oxidative stress induction, and p-JNK and p-p38 activation in the Monogonont Rotifer (*Brachionus Koreanus*). *Environ. Sci. Technol.* **2016**, *50*, 8849–8857. [CrossRef]
128. Qiang, L.; Cheng, J. Exposure to polystyrene microplastics impairs gonads of zebrafish (*Danio Rerio*). *Chemosphere* **2021**, *263*, 128161. [CrossRef]

toxics

MDPI

Article

Y-27632 Impairs Angiogenesis on Extra-Embryonic Vasculature in Post-Gastrulation Chick Embryos

Johannes W. Duess [1,2,3,*], Jan-Hendrik Gosemann [1,2], Anna Kaskova Gheorghescu [3], Prem Puri [2,3] and Jennifer Thompson [2,3]

1 Department of Pediatric Surgery, University of Leipzig, 04103 Leipzig, Germany
2 National Children's Research Centre, Our Lady's Children's Hospital, Crumlin, 12 Dublin, Ireland
3 School of Medicine and Medical Science, University College Dublin, Belfield, 4 Dublin, Ireland
* Correspondence: johannes.duess@medizin.uni-leipzig.de; Tel.: +49-341-97-26400; Fax: +49-34-97-26409

Abstract: Y-27632 inhibits Rho-associated coiled-coil-containing protein kinase (ROCK) signaling, which is involved in various embryonic developmental processes, including angiogenesis, by controlling actin cytoskeleton assembly and cell contractility. Administration of Y-27632 impairs cytoskeletal arrangements in post-gastrulation chick embryos, leading to ventral body wall defects (VBWDs). Impaired angiogenesis has been hypothesized to contribute to VBWDs. ROCK is essential in transmitting signals downstream of vascular endothelial growth factor (VEGF). VEGF-mediated angiogenesis induces gene expressions and alterations of the actin cytoskeleton upon binding to VEGF receptors (VEGFRs). The aim of this study was to investigate effects of Y-27632 on angiogenesis in post-gastrulation chick embryos during early embryogenesis. After 60 h incubation, embryos in shell-less culture were treated with Y-27632 or vehicle for controls. Y-27632-treated embryos showed reduced extra-embryonic blood vessel formation with impaired circulation of the yolk sac, confirmed by fractal analysis. Western blot confirmed impaired ROCK downstream signaling by decreased expression of phosphorylated myosin light chain. Interestingly, RT-PCR demonstrated increased gene expression of VEGF and VEGFR-2 1 h post-treatment. Protein levels of VEGF were higher in Y-27632-treated embryos at 8 h following treatment, whereas no difference was seen in membranes. We hypothesize that administration of Y-27632 impairs vessel formation during angiogenesis, which may contribute to failure of VWB closure, causing VBWDs.

Keywords: Y-27632; ROCK; post-gastrulation; chick embryo; angiogenesis; ventral body wall defect

Citation: Duess, J.W.; Gosemann, J.-H.; Kaskova Gheorghescu, A.; Puri, P.; Thompson, J. Y-27632 Impairs Angiogenesis on Extra-Embryonic Vasculature in Post-Gastrulation Chick Embryos. *Toxics* **2023**, *11*, 134. https://doi.org/10.3390/toxics11020134

Academic Editors: Esref Demir and Sam Kacew

Received: 1 December 2022
Revised: 14 January 2023
Accepted: 16 January 2023
Published: 30 January 2023

1. Introduction

Ventral body wall defects (VBWDs) are human birth anomalies reportedly present in about one in two thousand live births, with omphalocele and gastroschisis being the most frequently observed defects [1,2]. Although similar in that both conditions involve herniation of abdominal contents through the ventral body wall, they differ in that omphalocele occurs through the base of the umbilical cord and has a peritoneal covering, whereas gastroschisis occurs usually to the right of the umbilical cord and herniated organs have no peritoneal covering. Both conditions are relatively infrequent, and their exact causes are still unknown [3,4].

Omphalocele may occur as an isolated lesion but may also occur as part of various syndromes associated with chromosomal abnormalities, such as Beckwith–Wiedmann, Down's, and Patau syndromes. The VBWD in omphalocele is thought to result from failure of the midgut loop to rotate and return to the abdominal cavity from the physiological hernia of early pregnancy. Incomplete development and migration of the lateral plate mesoderm to form the body wall and incomplete differentiation and migration of somitic myotomes to form precursors of the body wall musculature may also contribute. Gene mutations and environmental factors have also been implicated in omphalocele formation [3,5–7].

The etiology of gastroschisis is reported to be the result of a disruption to blood flow to the developing body wall, leading to deficient development of the affected tissues and subsequently to an intestinal herniation through the defect [3,8]. It has been hypothesized that the abnormal involution of the right umbilical vein in particular leads to impaired growth and viability of the surrounding mesenchyme, which then results in a defect of skin formation in the periumbilical area and subsequent rupture of the intestines through the defect [1,7,9,10]. In addition to the latter, Hoyme stated that disruption of the omphalomesenteric artery leads to infarction and necrosis at the base of the umbilical cord with subsequent gut herniation [1,11]. The importance of the appropriate progress of angiogenic activities during embryogenesis is highlighted in studies on vasoactive substances. Mothers who took pseudoephedrines and/or analgesics (such as aspirin) during the first trimester of pregnancy were more likely to have a baby that had gastroschisis [12]. Smoking cigarettes during pregnancy is also linked with VBWDs during embryogenesis [13,14]. In addition, alcohol consumption and recreational drug use, particularly cocaine, has been correlated with a higher incidence of gastroschisis [15].

Since a direct correlation between an environmental factor and body wall malformation is difficult to verify, animal models have been used to investigate hypotheses on the relationship between VBWDs and angiogenesis. In chick embryos, the administration of the heavy metal cadmium (Cd) following completion of gastrulation at Hamburger–Hamilton (H-H) stages 16–17 altered the extra-embryonic vascular branching pattern, leading to an isolated morphological defect, similar to human VBWDs [10,16–18]. Rho-associated coiled-coil-containing protein kinase (ROCK) gene expression levels were found to be significantly decreased in chick embryos following Cd exposure [19]. In other studies, chick embryos exposed at H-H stages (16–17) to the pyridine derivative Y-27632 formed abnormal somites and kinking in the lumbosacral region, findings similar to those observed following Cd exposure. Y-27632 inhibits ROCK signaling, which is involved in numerous developmental processes during embryogenesis, such as cell adhesion, motility, proliferation, differentiation, and apoptosis, primarily by controlling actin cytoskeleton assembly and cell contractility. Western blot analysis implicated interference with ROCK downstream signaling in affected embryos. In addition, administration of Y-27632 at H-H stages 16–17 has also been shown to lead to increased cell death in somites, the neural tube, and ectoderm, as well as to alteration in cytoskeletal arrangements during early chick embryogenesis, all of which may contribute to failure of ventral body wall closure at later stages of development in chick embryos [20–22].

Angiogenesis, also known as neovascularization, is the process of new capillary formation from pre-existing vessels through endothelial cell proliferation and stem cell recruitment in combination with morphogenesis [23–25]. It must be distinguished from vasculogenesis, a process whereby vessels are formed de novo from endothelial cell precursors termed angioblasts and which occurs in the extra- and intra-embryonic tissues of embryos [26–29]. Angioblasts and newly formed endothelial cells proliferate, migrate, and remodel into tubular structures and fuse into larger vessels to create an interconnecting network of vessels referred to as the primitive capillary plexus [26,30]. With subsequent growth, this primitive vascular plexus remodels during the process of angiogenesis [31].

Vascular endothelial growth factor (VEGF) is one of the most important factors regulating angiogenesis during embryogenesis, including vascular endothelial cell migration, proliferation, and permeability, as well as skeletal growth and reproductive functions [32]. The biological effects of VEGF are mediated through three structurally related VEGF receptor tyrosine kinases (VEGFR), namely VEGFR-1, VEGFR-2, and VEGFR-3 [32–34]. VEGFR-2, also designated as KDR or Flk1, is the predominant receptor in angiogenic signaling and the main VEGF receptor on endothelial cells. It is essential for the regulation of endothelial cell migration, proliferation, differentiation, and survival as well as vessel permeability and dilatation [33,35]. VEGF induces hyperpermeability by enhancing nitric oxide (NO) production, which promotes endothelial NO synthase (eNOS) and inducible NOS (iNOS) expression in endothelial cells [27]. In addition, VEGFR-2 expression levels are particularly increased during embryonic vasculogenesis and angiogenesis [33].

Although the importance of VEGF signaling during angiogenesis is well established, the precise morphogenetic processes involved in the organization of proliferating endothelial cells into new blood vessels are still unknown. It has been hypothesized that major mechanisms are controlled through cytoskeletal elements [23]. Therefore, the role of members of the Rho family of small GTPases as key regulators of angiogenesis has been investigated, as they modulate the extra-cellular matrix and regulate migration, proliferation, morphogenesis, and survival through the cytoskeleton. RhoA signaling and its downstream target Rho-associated coiled-coil-containing protein kinase (ROCK) have been shown to play an important role in VEGF-dependent in vivo angiogenesis and in initial steps of in vitro endothelial cord assembly [23,36–41]. ROCK proteins phosphorylate various substrates, including myosin light chain (MLC) phosphatase and LIM-kinases (LIMK-1, LIMK-2) and activate them to phosphorylate and inactivate cofilin-1 (also known as non-muscle cofilin) or cofilin-2 (also known as muscle cofilin), which leads to inhibition of actin depolymerization and collectively to a reorganization of the actin cytoskeleton [42,43]. It was supposed that Rho proteins are downstream targets of VEGF-mediated signaling, leading to the activation of Rho GTPases after VEGF and VEGFR-2 stimulation, therefore controlling endothelial cell migration [30,36,44,45]. However, several studies have identified a critical and selective role for RhoA in the regulation of endothelial cell assembly into new blood vessels [23].

The research goal of the current study is largely mechanistic in terms of the effect of Y-27632 on angiogenesis, focusing on the extra-embryonic vascular branching pattern in chick embryos following exposure to teratogenic doses of this drug after 60 h incubation [21]. Moreover, the impact of Y-27632 on VEGF signaling during angiogenesis and different genes involved in vessel formation are investigated. The incubation time of 60 h is a point in chick development at which gastrulation is completing and body wall folding is initiating. Embryos were explanted using the Dugan method [46], which permits staging prior to treatment and intermittent review of the extra-embryonic vasculature and developmental progress prior to harvesting embryos.

2. Materials and Methods

Fertilized unincubated eggs of the Ross strain were obtained from a commercial hatchery (Enfield Broiler Breeders Ltd., Maynooth, Co. Kildare, Ireland). They were incubated at 38 °C in a forced draught incubator (Shell Lab) at 65–75% humidity until they reached the required stage for treatment. All animal studies were carried out in accordance with the established protocols of the institute. After 60 h of incubation, chick embryos that had attained H-H stages 16–17 (distinguished by the presence of the cervical flexure, absent or rudimentary limb buds, blood islands below the tail region, and otherwise well-developed extra-embryonic vasculature) were explanted into shell-less culture using a method adapted from Dugan et al. [46].

This method for ex ovo culture of chick embryos involves the explantation of the embryo and egg contents into a concave well made with cling film, which is attached to the exterior of a polystyrene cup using an elastic band, ballasted with 30–50 mL of water. The eggs were removed from the incubator and left for 5–10 min at room temperature, allowing them to cool down and thus reducing the possibility of yolk rupture upon explantation. Subsequently, the shells were cracked, and the contents were emptied into the wells with the blastodisc facing upwards. Embryos were staged according to H-H parameters prior to treatment. Using an adjustable volume pipette, 50 µL of 500 µM of ROCK inhibitor Y-27632 (Y-27632 dihydrochloride, Sigma-Aldrich, cat.# Y0503, Ireland), diluted in chick saline (0.72 g NaCl/100 mL dH_2O), was applied to the center of the blastodisc, and entered the embryo and vitelline membranes by a process of diffusion. Embryos in the control group received 50 µL of vehicle. The dose of 500 µM of Y-27632 has recently been shown to result in the optimal balance between mortality and morphological abnormality rate [21].

Embryos were assessed 8 h, 24 h, 48 h, and 4 d after treatment at 60 h for development of extra-embryonic vasculature. Measurements, including length of sinus terminalis (ST), size of area vasculosa (AV), and absolute area vasculosa (AOV), as well as vascular index (VI), were performed at 60 + 24 h to analyze differences in formation of extra-embryonic membrane (EEM) vasculature between controls and Y-27632-treated embryos. Fractal dimensions were calculated to study the branch morphological pattern of right- and left-sided extra-embryonic vessels (D-right, D-left) at 4 d following treatment. Western blot was performed at 8 h post-treatment to determine whether there was interference with ROCK signaling. Relative gene expression levels of pro- and anti-angiogenic factors were analyzed 1 h, 4 h, and 8 h (for membranes) or 24 h (for embryos) after treatment at 60 h. Enzyme-Linked Immunosorbent Assays (ELISA) for VEGF and VEGFR-2 were performed at 8 h and 24 h following treatment. These time points are summarized in Table 1.

Table 1. Overview of procedures as described in Material and Methods.

Procedure	Time Point
Treatment	at 60 h incubation
Gross Morphology of extra-embryonic vasculature	60 + 8 h, +24 h, +48 h, +4 d
Measurements of EEM	60 + 24 h
Fractal analysis	60 + 4 d
Western Blot	60 + 8 h
Real-Time qRT-PCR	60 + 1 h, +4 h, +8 h/+24 h
ELISA	60 + 8 h, +24 h

Embryos and EEM were harvested and assessed for stage development and anomalies using the dissection microscope (Leica S6E). Dead embryos were recorded and discarded. Surviving chicks were photographed with a single-lens reflex camera to study the formation of the extra-embryonic vasculature or with the stereomicroscope (Olympus SZX12) attached to a camera port (Olympus DP70) to analyze the morphological development of chick embryos.

2.1. Measurements of EEM Formation

Eight chick embryos were randomly collected from the control and experimental group 24 h following treatment at 60 h to measure parameters of the extra-embryonic vasculature (Figure 1). The length of ST (in cm) and the size of the area vasculosa (AV, in cm^2) were determined using Fiji® image processing software. The size of AV was calculated by subtracting the chick embryo expanse from the area surrounded by the ST. The area occupied by vessels of the AV (AOV in cm^2) was measured using Adobe Photoshop (Adobe Systems). VI (in %) is expressed as the percentage of AOV from AV. Measurements were performed double-blinded by two independent investigators.

2.2. Fractal Analysis

Eight chick embryos were randomly collected from the control and experimental groups 4 d following treatment at 60 h to calculate fractal dimensions for the study of the branch morphological pattern of right- and left-sided extra-embryonic vessels (D_f-right, D_f-left). Fractal dimensions (D_f) were determined using the "box counting" method in Fiji® image processing software. An area covering extra-embryonic vessels on one side was randomly selected at 1, 3, 5, 7, 9, and 11 o'clock and converted to an 8-bit image. To analyze the pattern branching, the images were converted to dichotomic binarized (black and white) pictures and skeletonized to obtain a final picture where each vessel had the structural shape of the object, being in this case a direct representation of total vessel length [47]. D_f was then calculated from the skeletonized line tracing using the box counting method [47–49]. The logarithm of the number of boxes containing a segment is plotted against the logarithm of the number of boxes covering the image, and the D_f is calculated from the slope by the computer program [50]. Fractal analysis is thought to be more sensitive to changes in vascular patterns when skeletons rather than segmented images are considered [47,50]. Measurements were performed double-blinded by two independent investigators.

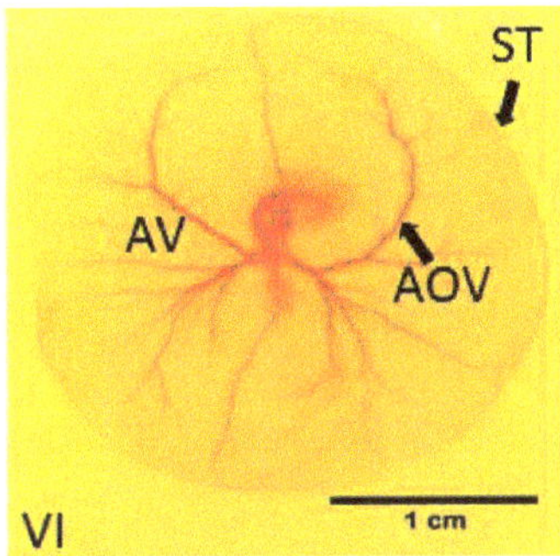

Figure 1. Parameters for measurement of extra-embryonic vasculature formation: sinus terminalis (ST, in cm), size of area vasculosa (AV, in cm^2), absolute area vasculosa (AOV, in cm^2), and vascular index (VI, in %). Scale bar 1 cm.

2.3. Western Blot

Western blot (WB) was performed in triplicate to confirm inhibition of ROCK downstream signaling in embryos and EEM at the teratogenic dose of 500 μM of Y-27632.

Following treatment with the aforementioned dose of Y-27632 or vehicle, chick embryos were harvested and sonicated in 200 μL RIPA (Radio Immuno Precipitation Assay) buffer. Protein concentrations were then measured using Bradford assays and diluted with gel loading buffer Laemmli (Sigma-Aldrich, Arklow, Co. Wicklow, Ireland) prior to gel loading. Proteins were separated electrophoretically in 10% SDS-PAGE and transferred to a 0.45 μm nitrocellulose membrane by Western blotting for 90 min at 30 V. Following blotting, the membranes were blocked in 3% BSA-0.05% Tween for 30 min to overnight. Primary antibody (Anti-Myosin light chain (phospho S20), 1:500, Abcam, Cambridge, UK) incubation was performed from 90 min to overnight. After extensive washing (four times in PBS-0.05% Tween 20 or 4 h for overnight incubation) the membranes were incubated with the secondary antibody (Anti-rabbit IgG, HRP-linked Antibody, 1:5000, Cell Signaling Technology-Isis Ltd., Stillorgan, Co. Dublin, Ireland) for 90 min followed by extensive washing. Protein detection was performed with PIERCE chemiluminescence kit (ECL Western blotting substrate Thermo Scientific Pierce, Fisher Scientific, Dublin, Ireland). Beta-actin (Anti-beta Actin antibody, Abcam, Cambridge, UK) was used to control for equal loading and transfer of the samples. Bands of WB were quantified using densitometry.

2.4. Real-Time Quantitative Reverse Transcription Polymerase Chain Reaction (RT-PCR)

RT-PCR was performed to evaluate the relative mRNA levels of VEGF, VEGFR-2, Ang-2, and iNOS in embryos and EEM harvested 1 h, 4 h, and 8 h (for membranes) or 24 h (for embryos) after treatment in shell-less culture. Total RNA of twelve embryos (six controls, six Y-27632-treated embryos) at each time point was isolated using TRIzol® reagent (Bio-Sciences, cat.# 15596-026, Dublin, Ireland) according to the recommended protocol. For reverse transcription (RT), first strand cDNA was synthesized from RNA by using a Transcriptor High Fidelity cDNA Synthesis Kit (Roche, cat.# 05091284001, UK). Following RT at 44° for 60 min, polymerase chain reaction (PCR) was performed using a LightCycler® 480 SYBR Green I Master (Roche, cat.# 04887352001, West Sussex, UK). After an initial denaturation step of 5 min at 95°, PCR cycles of amplification for target primers were carried out. Relative levels of gene expression were measured using a LightCycler

480 (Roche, UK). The relative mRNA expression levels of target genes were normalized against the level of GAPDH gene expression in each sample. Experiments were carried out in triplicate for each data point. The specific primer pairs used in this study are listed in Table 2.

Table 2. Oligonucleotide primers used for RT-PCR analysis.

Gene	Sequence (5′-3′)
GAPDH	
Forward	cctctctggcaaagtccaag
Reverse	ggtcacgctcctggaagata
VEGF	
Forward	caattgagaccctggtggac
Reverse	catcagaggcacacaggatg
VEGFR-2	
Forward	gacagtggcatggtgttcag
Reverse	gtgcagtttccttcctggag
Ang-2	
Forward	ttgaggaggttggacagttc
Reverse	gcttcatttccttcccagtc
iNOS	
Forward	agtggtatgctctgcctgct
Reverse	ccagtcccattcttcttcc

2.5. Enzyme-Linked Immunosorbent Assay (ELISA)

In embryos and EEM harvested at 8 h and 24 h following treatment, ELISA was performed in 16 embryos and membranes (eight from each group for one time point) using ELISA kits for VEGF-A and VEGFR-2 (Chicken vascular endothelial growth factor A ELISA Kit, MyBioSource, Inc., USA; Chicken VEGFR-2/KDR ELISA Kit, MyBioSource, Inc., San Diego, CA, USA) according to the manufacturer's protocol. Experiments were carried out in technical duplicate for each data point. The results were measured at 450 nm using the Synerg Mx microplate reader (BioTek, Winooski, VT, USA) immediately after adding the stop solution.

2.6. Statistical Analysis

All numerical data are shown as number of cases (%) or mean (± standard deviation, SD). Differences in extra-embryonic measurements, fractal dimension, densitometry, and relative mRNA and protein expressions were compared using 2-tailed unpaired t test when data were normally distributed or Mann–Whitney U test when data deviated from normal distribution. A p-value < 0.05 was considered statistically significant.

3. Results

3.1. Effects of Y-27632 on Angiogenic Development

At 60 h incubation, embryos in both groups showed blood islands in the tail region and well-developed extra-embryonic vasculature (Figure 2A–D).

At 8 h following treatment, all controls (n = 54) and 78 (97.5%) out of 80 Y-27632-treated embryos were alive. In the experimental group, 67 (85.8%) embryos demonstrated improper formation of extra-embryonic vasculature. Omphalomesenteric vessels appeared to be disrupted or strikingly underdeveloped, and immaturely formed blood islands were also evident in the tail region at 60 + 8 h in comparison to controls (Figure 3A,B). The thin vitelline vessels and pre-existing vascular architecture appeared severely impaired in the experimental group post-treatment (Figure 3C,D). Out of the 67 embryos with impaired blood vessel formation, 60 (89.6%) revealed morphological abnormalities, such as improperly developed somites in the tail region. A significant weight difference between controls and Y-27632-treated embryos with abnormalities in vessel formation and morphology was also found (0.0151 g ± 0.003 g vs. 0.0067 g ± 0.002 g, $p < 0.001$) at 60 + 8 h.

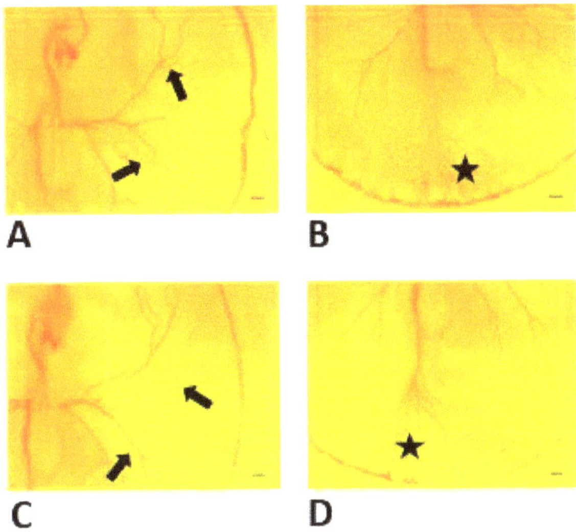

Figure 2. Development of extra-embryonic vasculature of chick embryos prior to treatment at 60 h incubation. Well-developed extra-embryonic vasculature (**A**,**C**; arrows) and blood islands in the tail region (**B**,**D**; asterisks) were evident prior to treatment. Control embryos were treated with vehicle (**A**,**B**), and embryos in the experimental group (**C**,**D**) received 500 µM of Y-27632. Scale bars 200 µm.

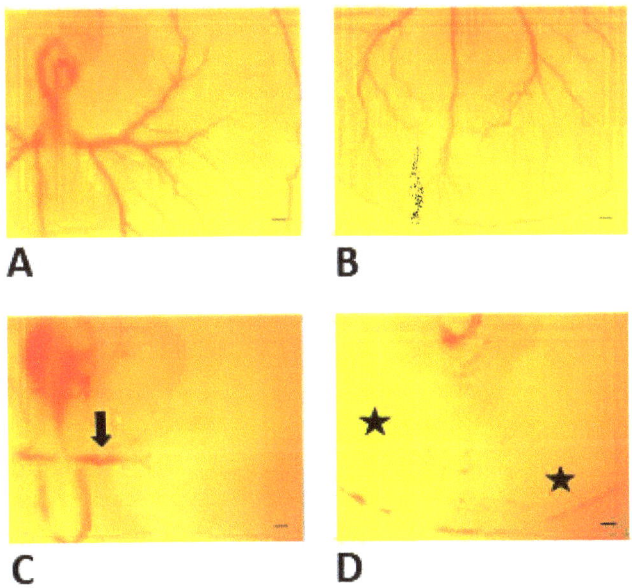

Figure 3. Development of extra-embryonic vasculature of control embryo (**A**,**B**) and Y-27632 (500 µM)-treated embryo (**C**,**D**) at 8 h following treatment at 60 h. In the experimental group, the majority of embryos revealed improper formation of omphalomesenteric vessels (**C**, arrow) and vitelline vessels in the tail region (**D**, asterisks) when compared with controls (**A**,**B**). Scale bars 200 µm.

For the analysis of extra-embryonic angiogenesis at later stages of development, 38 embryos were treated with vehicle for the control group and 69 embryos with 500 µM of Y-27632 for the experimental group at 60 h incubation. Following 24 h, 48 h, and 4 d in the incubator, embryos were assessed for the formation of extra-embryonic vasculature at each time point. Prior to treatment, embryos in both groups demonstrated well-developed extra-embryonic vasculature and blood islands in the tail region (Figure 4A,B). After 24 h, 36 (94.7%) controls and 43 (62.3%) Y-27632-treated embryos were alive at examination. In controls, 33 (91.7%) embryos showed normal extra-embryonic vasculature. In contrast, 23 (53.5%) chick embryos in the experimental group showed reduced extra-embryonic blood vessel formation with impaired circulation of the yolk sac (Figure 4D,F,H) when compared with controls (Figure 4C,E,G) during the time period of investigation. Out of these 23 embryos, morphological abnormalities in the somites and the neural tube were detected in 21 (91.3%) embryos. A significant weight difference between controls and Y-27632-treated embryos with abnormalities in vessel formation and morphology was also found (0.0263 g ± 0.005 g vs. 0.0185 g ± 0.0052 g, $p < 0.05$) at 60 + 24 h.

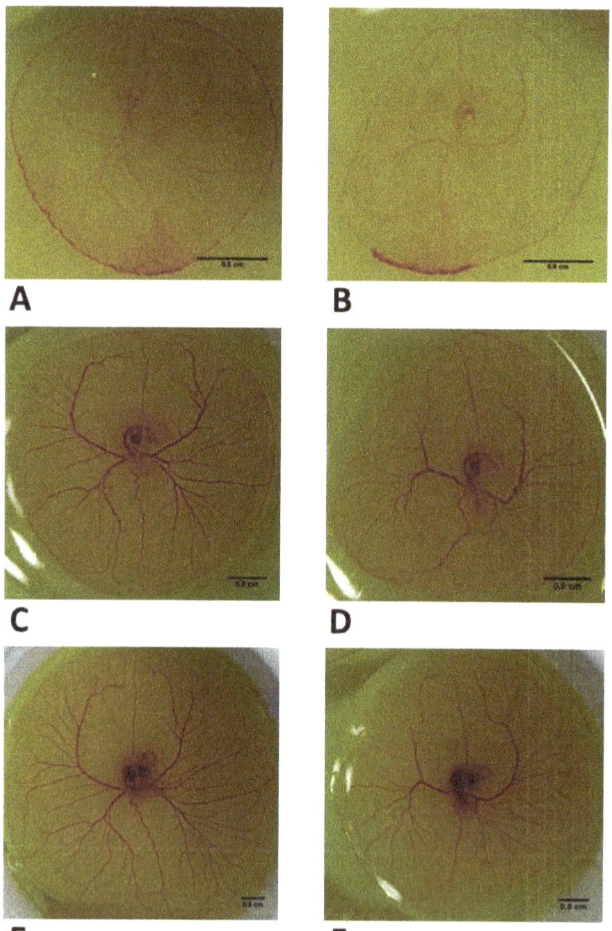

Figure 4. *Cont.*

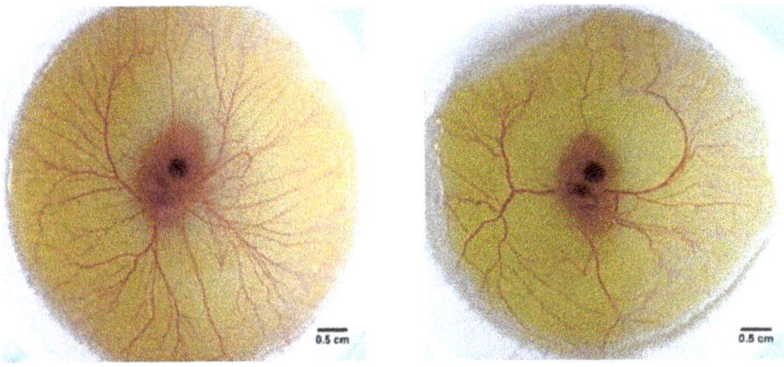

G H

Figure 4. Development extra-embryonic vasculature of control embryo (**A**,**C**,**E**,**G**) and Y-27632 (500 μM)-treated embryo (**B**,**D**,**F**,**H**) at 60 h ((**A**,**B**)–prior to treatment), 60 + 24 h (**C**,**D**), 60 + 48 h (**E**,**F**), and 60 h + 4 d (**G**,**H**). Following administration of agents, embryos in the experimental group revealed strikingly impaired neovascularization with reduced formation of blood vessels and reduced circulation of the yolk sac when compared with controls at 24 h (**C**,**D**), 48 h (**E**,**F**), and 4 d (**G**,**H**) post-treatment. Scale bars 0.5 cm.

3.2. Measurements of EEM Formation

The mean length of ST was 9.7 cm (±1.1) in controls vs. 8 cm (±0.6) in Y-27632-treated embryos ($p < 0.05$, Figure 5A). The mean size of AV was 7.1 cm^2 (± 1.1) vs. 5 cm^2 (±1.0) ($p < 0.05$, Figure 5B). There was also a significant difference in the mean AOV between the two groups (6 cm^2 (±0.4) vs. 3.6 cm^2 (±0.6), $p < 0.05$) (Figure 5C). The mean VI was 85.3% (±10.0) in controls compared to 72.5% (± 12.2) in Y-27632-treated embryos ($p < 0.05$, Figure 5D).

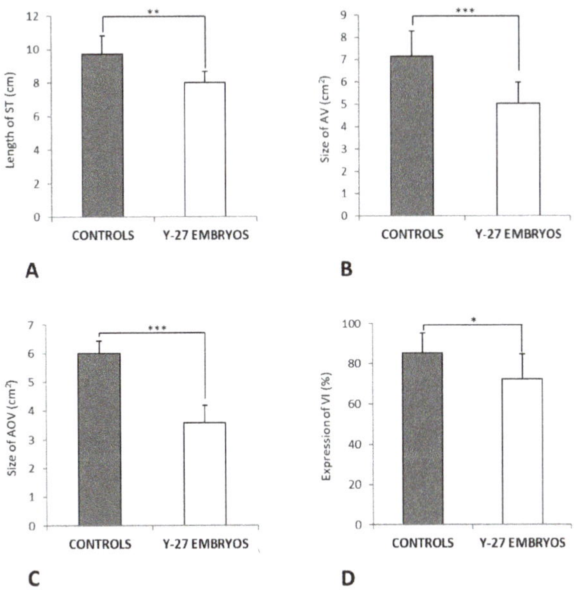

Figure 5. Measurements of EEM formation at 60 + 24 h. There was a significant difference in the mean length of ST (**A**, ** $p = 0.001$), mean size of AV (**B**, *** $p < 0.001$), mean size of AOV (**C**, *** $p < 0.001$), and mean VI (**D**, * $p < 0.05$) between controls and Y-27632 (500 μM)-treated embryos. The vertical bars represent the SD. *y-axis*: Length of ST in cm (**A**), Size of AV and AOV in cm^2 (**B**,**C**), and Expression of VI in % (**D**).

3.3. Fractal Analysis

At examination, controls embryos showed a more complex branching pattern from 60 h incubation (Figure 6A) to 4 d following treatment (Figure 6B–D) in comparison to a paucity of vessels in Y-27632-treated embryos (Figure 6E–H) during the time period of investigation. A significant difference in the mean D_f-left was found between controls and Y-27632-treated embryos (1.232 ± 0.035 vs. 1.064 ± 0.043, $p < 0.05$) (Figure 6I). The mean D_f-right was also significantly lower in control embryos compared with embryos in the experimental group (1.227 ± 0.044 vs. 1.077 ± 0.029, $p < 0.05$) (Figure 6J).

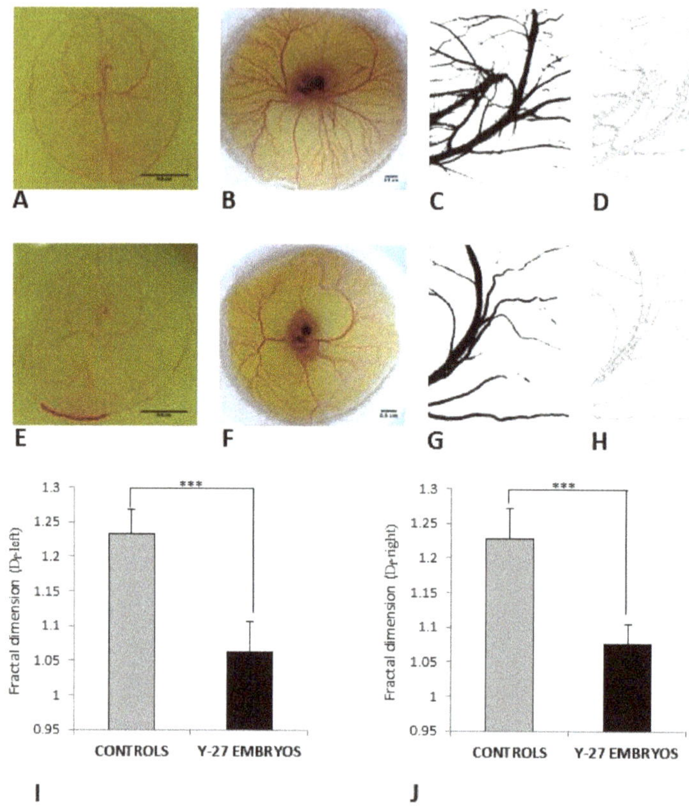

Figure 6. Fractal analysis of control embryo (**A–D**) and Y-27632 (500 μM)-treated embryo (**E–H**) with D_f-left (**I**) and D_f-right (**J**) at 60 h + 4 d. Prior to treatment, embryos in both groups showed blood islands in the tail region and well-developed extra-embryonic vasculature (**A**,**E**). Control embryos developed a more complex branching pattern (**B–D**) in comparison to embryos in the experimental group (**F–H**). D_f-left was significantly lower in Y-27632 (500 μM)-treated embryos in comparison to controls (**I**, *** $p < 0.001$). There was also a significant difference in D_f-right between controls and embryos in the experimental group (**J**, *** $p < 0.001$). The color image version (**B**,**F**). The segmented version (**C**,**G**). The skeletonized version (**D**,**H**). The vertical bars represent the SD. *y-axis*: Fractal dimension (D_f-left, (**I**); D_f-right, (**J**)). Scale bars: 0.5 cm.

3.4. Western Blot

Western blot analysis confirmed Y-27632 interference with ROCK downstream signaling in embryos treated with the teratogenic dose of 500 μM of Y-27632, with significant reduction in the expression of phosphorylated MLC (p-MLC) at 8 h following treatment ($p < 0.05$, Figure 7A). In addition, ROCK signaling was also impaired in EEM of Y-27632-

treated embryos, leading to a significant reduction in the expression of p-MLC at the same time point ($p < 0.05$, Figure 7B).

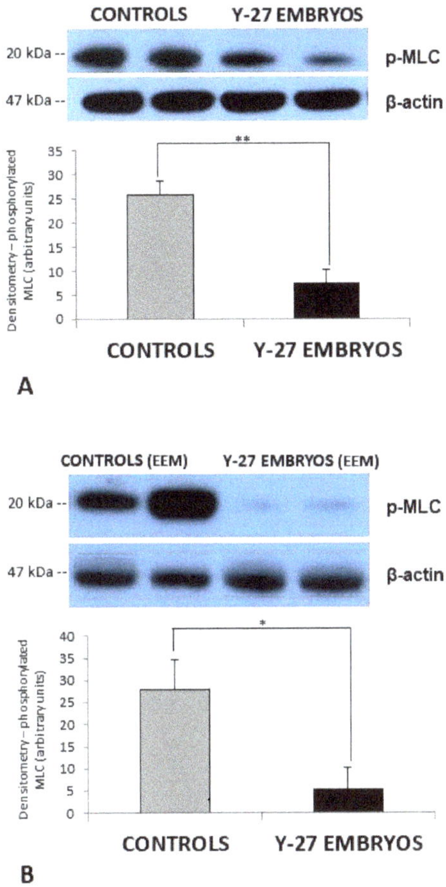

Figure 7. Embryos treated with vehicle (controls, **A**) or Y-27632 (500 μM, **A**) and EEM of controls (**B**) or Y-27632-treated embryos (**B**) were subjected to Western blot at 60 + 8 h to detect phosphorylated MLC. The results show that the amount of phosphorylated MLC was significantly reduced in Y-27632-treated embryos in comparison to controls (**A**, ** $p < 0.001$ and **B**, * $p < 0.05$). The vertical bars represent the SD. y-axis: Densitometry (grey levels, arbitrary units).

3.5. RT-PCR of VEGF and VEGFR-2

Effects of Y-27632 on VEGF and VEGFR-2 signaling were analyzed at 1 h, 4 h, and 8 h (for membranes) or 24 h (for embryos) following treatment at 60 h incubation.

In Y-27632-treated embryos, relative mRNA expression levels of *VEGF* were significantly higher 1 h post-treatment in comparison to controls (0.123 ± 0.03 vs. 0.083 ± 0.011, $p < 0.05$) (Figure 8A). However, gene expression levels of *VEGF* were not different at 4 h and 24 h following treatment between controls and embryos treated with the inhibitor ($p > 0.05$, Figure 8B,C). The relative mRNA expression levels of *VEGF* were also significantly higher at 60 + 1 h in EEM of Y-27632-treated embryos compared with controls (1.159 ± 0.339 vs. 0.673 ± 0.135, $p < 0.05$) (Figure 8D). Gene expression levels in EEM at 4 h and 8 h following treatment were not different between the control and experimental group ($p > 0.05$, Figure 8E,F).

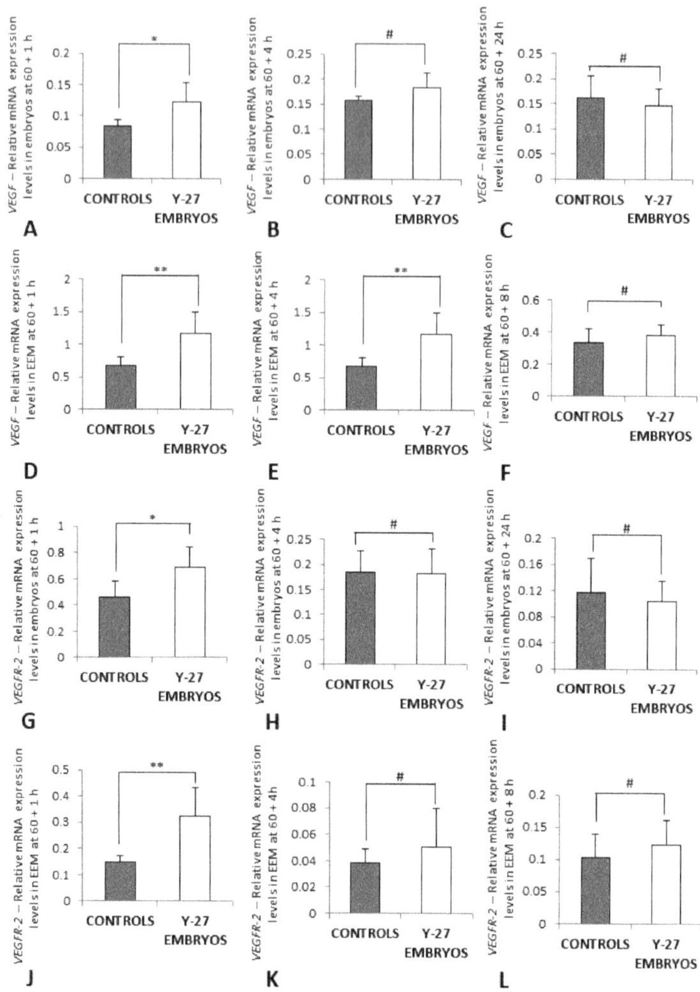

Figure 8. The relative mRNA expression levels of *VEGF* (**A–F**) and *VEGFR-2* (**G–L**) in controls and embryos treated with Y-27632 (500 μM) at 60 h incubation. Embryos in (**A–C**,**G–I**), EEM in (**D**,**F**,**J–L**). The expression levels of *VEGF* and *VEGF*R-2 at 60 + 1 h were significantly higher in the treatment group when compared to controls (**A**,**G**, * $p < 0.05$ and **D**,**J**, ** $p = 0.009$). The expression levels of *VEGF* and *VEGFR-2* at other time points (**B**,**C**,**E**,**F**,**H**,**I**,**K**,**L**) were not different between controls and Y-27632-treated embryos (# $p > 0.05$). The vertical bars represent the SD. *y*-axis: Relative mRNA Expression Levels.

Relative mRNA expression levels of *VEGFR-2* were also significantly higher in Y-27632-treated embryos 1 h post-treatment in comparison to controls (0.692 ± 0.155 vs. 0.458 ± 0.127, $p < 0.05$) (Figure 8G). At 4 h and 24 h following treatment, gene expression levels of *VEGFR-2* were not different between controls and embryos treated with the inhibitor ($p > 0.05$, Figure 8H,I). In EEM of Y-27632-treated embryos, the relative mRNA expression levels of *VEGFR-2* were significantly higher at 60 + 1 h as well compared with controls (0.324 ± 0.109 vs. 0.147 ± 0.026, $p < 0.05$) (Figure 8J). In EEM at 4 h and 8 h following treatment, gene expression levels were not different between the control and experimental group ($p > 0.05$, Figure 8K,L).

3.6. Protein Expression of VEGF and VEGFR-2 (ELISA)

Protein expression of VEGF and VEGFR-2 in embryos and membranes was analyzed 8 h and 24 h following treatment at 60 h incubation with vehicle for the control group or 500 μM of Y-27632 for the experimental group.

The protein levels of VEGF were significantly higher in embryos in the experimental group when compared with controls at 60 + 8 h (973.0 ± 120.1 vs. 331.0 ± 52.5, $p < 0.05$) (Figure 9A). However, there was no difference between embryos and EEM of controls and Y-27632-treated embryos at other time points ($p > 0.05$, Figure 9B–H).

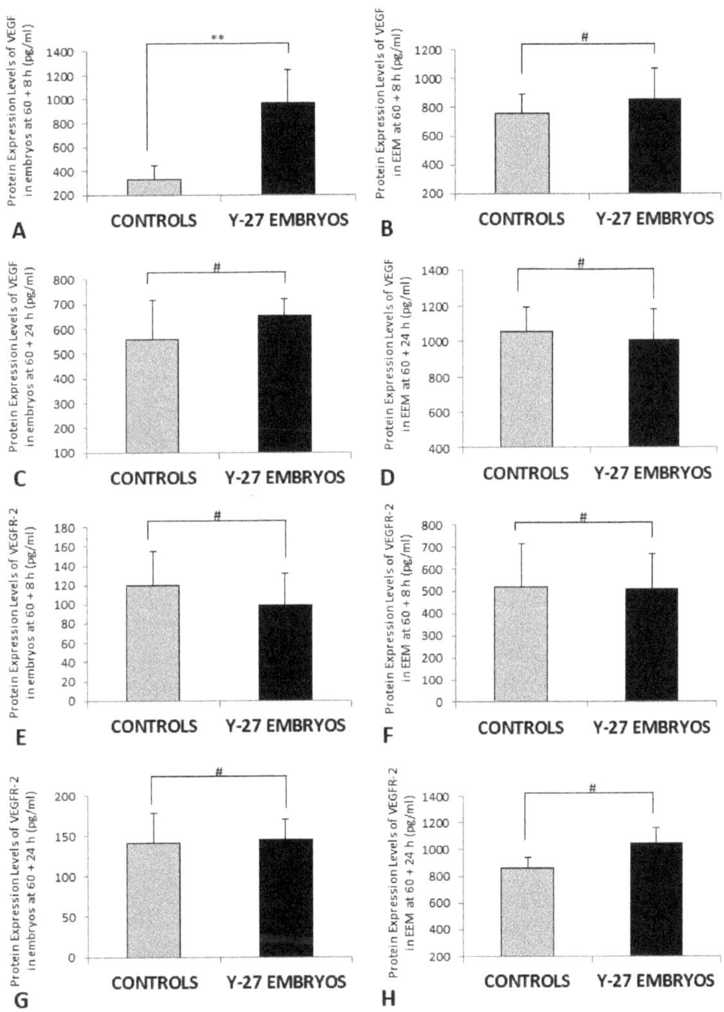

Figure 9. Protein expression of VEGF and VEGFR-2 in embryos (**A**,**C**,**E**,**G**) and EEM (**B**,**D**,**F**,**H**) of the control group and Y-27632 (500 μM) treatment group at 60 + 8 h and 60 + 24 h. Protein expression of VEGF was significantly higher in Y-27632-treated embryos in comparison to controls at 60 + 8 h (**A**, ** $p = 0.001$). Protein levels of VEGF and VEGFR-2 were not different between embryos and membranes of the control and treatment group at other time points (**B–H**, # $p > 0.05$). The vertical bars represent the SD. *y-axis*: Protein Expression Levels (pg/mL).

3.7. RT-PCR of Ang-2 and iNOS

Gene expression of Ang-2 and iNOS was analyzed at 60 + 1 h (for Ang-2 and iNOS) and 60 + 8 h (for iNOS) to investigate their contribution in angiogenesis in chick embryos treated with vehicle or 500 μM of ROCK inhibitor Y-27632 at 60 h incubation.

The relative mRNA expression levels of Ang-2 at 60 + 1 h were not different between controls and embryos in the experimental group ($p > 0.05$, Figure 10A) but were significantly upregulated in EEM of Y-27632-treated embryos at 1 h following treatment in comparison with controls (0.160 ± 0.021 vs. 0.128 ± 0.017, $p < 0.05$) (Figure 10B).

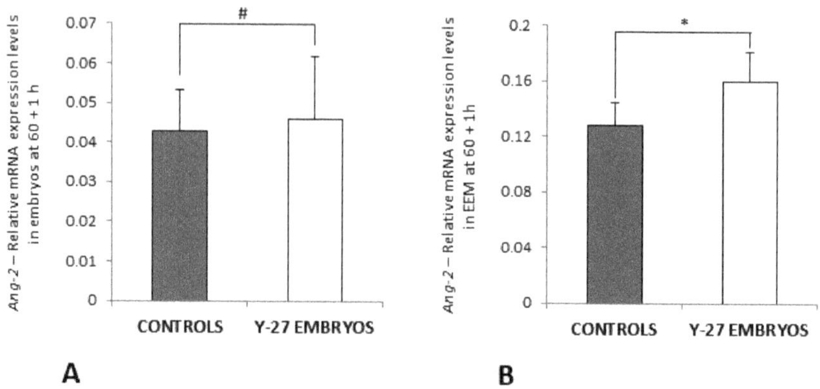

Figure 10. The relative mRNA expression levels of *Ang-2* in embryos (**A**) and EEM (**B**) of the control group and Y-27632 (500 μM) treatment group at 60 + 1 h. The expression levels of *Ang-2* at 60 + 1 h were not different between controls and Y-27632-treated embryos (**A**, # $p > 0.05$) but were significantly higher in membranes of the treatment group when compared to the control group (**B**, * $p < 0.05$). The vertical bars represent the SD. *y-axis*: Relative mRNA Expression Levels.

Gene expression levels of iNOS were significantly higher in Y-27632-treated embryos compared with controls at 60 + 1 h (0.202 ± 0.056 vs. 0.132 ± 0.075, $p < 0.05$) (Figure 11A), whereas relative mRNA expression of iNOS was significantly lower in embryos in the experimental group in comparison with controls at 60 + 8 h (0.316 ± 0.029 vs. 0.434 ± 0.068, $p < 0.05$) (Figure 11B). The relative mRNA expression levels of iNOS were not different in membranes of the control group and treatment group at 1 h and 8 h following treatment at 60 h incubation ($p > 0.05$, Figure 11C,D).

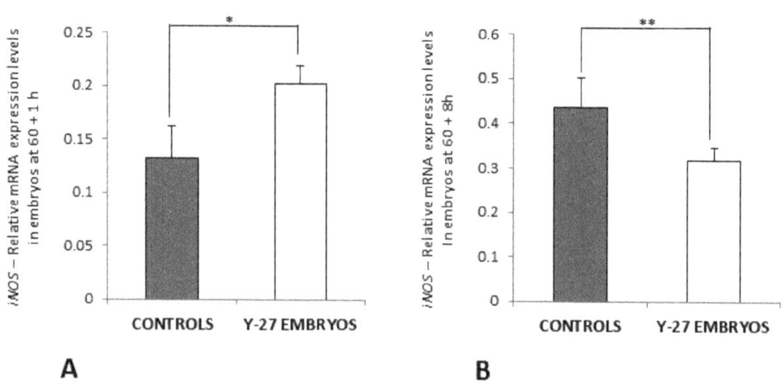

Figure 11. *Cont.*

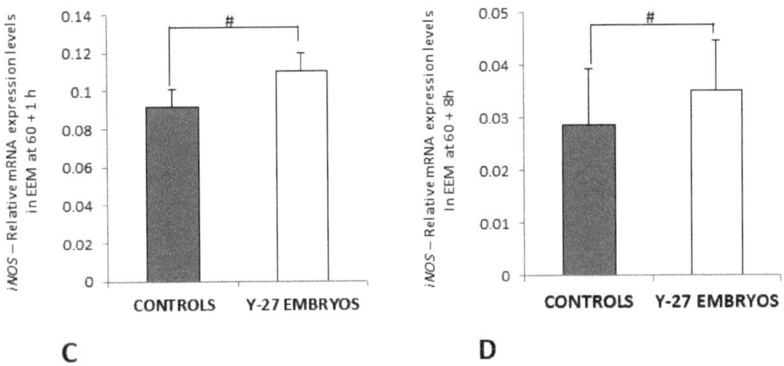

Figure 11. The relative mRNA expression levels of *iNOS* in embryos (**A**,**B**) and EEM (**C**,**D**) of the control group and Y-27632 (500 μM) treatment group at 60 + 1 h and 60 + 8 h. The expression levels of *iNOS* at 60 + 1 h were significantly higher in the treatment group when compared to controls (**C**, * $p < 0.05$). The expression levels of *iNOS* at 60 + 8 h were significantly lower in the treatment group when compared to controls (**D**, ** $p = 0.004$). The expression levels of *iNOS* were not different between membranes of the control and treatment group at 60 + 1 h and 60 + 8 h (**C**,**D**, # $p > 0.05$). The vertical bars represent the SD. *y-axis*: Relative mRNA Expression Levels.

4. Discussion

The major finding of this study is that the exposure of post-gastrulation chick embryos to ROCK inhibitor Y-27632 at 60 h incubation significantly impaired neovascularization, with reduced formation of extra-embryonic blood vessels and reduced circulation of the yolk sac. At very early stages of embryogenesis (60 + 8 h), Y-27632-treated embryos already revealed improper formation of the omphalomesenteric and vitelline vessels in the tail region of the yolk sac compared to controls. Although the formation of extra-embryonic vasculature could be observed in the experimental group throughout development, the process of neovascularization was significantly diminished throughout the period of investigation as confirmed by measurements of ST, the AV, the AOV, and the VI, in addition to fractal analysis at a more advanced developmental stage. Furthermore, the inhibition of ROCK downstream signaling was demonstrated in membranes and embryos when the analysis of the effects on the extra-embryonic angiogenesis and morphology was conducted at 8 h following treatment, as shown by the decreased expression of p-MLC in both tissues. Finally, Y-27632 caused an increased gene expression of VEGF and VEGFR-2 in embryos and membranes at 1 h post-treatment, whereas a significant difference was not seen at later time points. Protein levels of VEGF were higher in Y-27632-treated embryos at 8 h following treatment, whereas no difference was seen in membranes at that time point. Protein expression of VEGF at 60 + 24 h and of VEGFR-2 at 60 + 8 h and 60 + 24 h was not different between embryos and membranes of the control and treatment groups.

Impaired angiogenesis has been connected with multiple pathological conditions in humans, including failure of ventral body wall closure. It has been hypothesized that gastroschisis, one of the most frequently observed VBWDs, results from the premature atrophy or abnormal persistence of the right umbilical vein causing mesenchymal damage to the abdominal wall and intestinal herniation. Alternatively, it has been proposed that the disruption of the right omphalomesenteric artery can lead to the development of gastroschisis [51,52].

VEGF is the crucial regulatory growth factor orchestrating vascular development during embryogenesis (vasculogenesis) as well as blood vessel formation (angiogenesis) in later stages of development [53,54]. During angiogenesis, VEGF is involved in the main processes of neovascularization, mainly by regulation of gene expression and alterations of the actin cytoskeleton [55,56]. The latter, in association with cell–matrix adhesion, are the key functions in endothelial cell migration in angiogenesis [56]. The biological functions of VEGF are exerted upon binding to three receptor tyrosine kinases, known as VEGFR [35,54]. VEGFR-2, the main receptor on endothelial cells, is predominantly described as the leading receptor in angiogenic signaling [35]. Mice deficient in VEGFR-2 have been reported to die in utero at 8.5 days post-coitum from early defects in the development of hematopoietic and endothelial cells. The authors demonstrated the absence of yolk sac blood islands and organized blood vessels as well as reduction in hematopoietic progenitors [57,58]. This phenotype is similar to that of VEGFA$^{-/-}$ mice revealing impaired angiogenesis and blood island formation leading to developmental anomalies [59,60]. Therefore, it has been agreed that VEGFR-2 is the major transducer of the mitogenic, angiogenic, and permeability-enhancing effects of VEGF on endothelial cell differentiation, proliferation, migration, and formation of the vascular tube [32,34,35].

The role of the Rho family of small GTPases as key regulators of angiogenesis has been well established, and it also plays an essential role in transmitting signals downstream of VEGF [41,61,62]. Specifically, Rho proteins appear to be direct downstream targets of VEGF-mediated endothelial cell signaling, leading to activation of RhoA, Rac1, and Cdc42 within several minutes after VEGF stimulation [36,44,45]. Following this rapid increase, Rho proteins return to basal levels within 30 min, which emphasizes their importance during very early stages of VEGF-controlled blood vessel formation [36]. In chick embryos, Uchida et al. have shown that spontaneous angiogenesis in vivo is dependent on RhoA using a chick chorioallantoic membrane (CAM) assay [63]. In particular, the RhoA/ROCK pathway is crucial for the fundamental processes in angiogenesis, inducing capillary permeability, migration, capillary cord formation, and the proliferation of endothelial cells, mainly by its effects on the actin cytoskeleton and cell contractility [23,55]. Due to the fact that this pathway is involved in the cytoskeletal function of the vasculature, it has therefore been suggested that the RhoA/ROCK signaling is also related to vascular diseases. However, studies investigating the function of ROCK during angiogenesis and vascular remodeling using inhibitors such as Y-27632 or H-1152 have shown contradictory and inconclusive results so far [38,64].

VEGF-induced endothelial cell migration has been shown to involve dynamic, coordinate changes in cell adhesion, signal transduction, and cytoskeletal dynamics and organization [30]. Van NieuwAmerongen et al. have studied the effects of VEGF on the endothelial cytoskeleton in human foreskin microvascular and human umbilical vein endothelial cells. Stimulation of cells with VEGF was described to cause an increase in cytoplasmic F-actin staining and stress fiber (SF) formation, accompanied by the loss of peripheral F-actin and an increase in active RhoA and its recruitment to the cell membrane. Treatment with Y-27632 disrupted the endothelial F-actin cytoskeleton and VEGF-induced SF formation. The authors demonstrated that the inhibitor abolished VEGF-enhanced endothelial cell migration and reduced the mean tube length of the capillary-like tubular structure formed in response to VEGF and TNF-α in a dose-dependent manner [56]. Similar results were also demonstrated by Hoang et al. investigating RhoA function during neovascularization in a model of mouse skin angiogenesis using VEGF transfectants. Bryan et al. analyzed the role of RhoA/ROCK signaling in various essential processes during angiogenesis, such as endothelial cell migration, survival, and cell permeability. Using retinal explant cultures from bovine eyes, the authors demonstrated that VEGF strongly promotes vessel outgrowth, which was significantly disrupted by the addition of Y-27632. ROCK inhibition was shown to reduce VEGF-induced permeability, but it did not affect VEGF-mediated proliferation, which is similar to findings presented by Hoang et al. The compound also impaired VEGF-mediated tube morphogenesis, leading to an improper vascular network consisting of flattened multicellular structures [53].

All in all, these aforementioned results obtained from various cell types suggest that ROCK signaling is heavily involved in numerous and fundamental mechanisms during angiogenesis, resulting in immature or reduced blood vessel formation following treatment with agents inhibiting the ROCK pathway, which is in agreement with our findings. Treatment of chick embryos at 60 h incubation with ROCK inhibitor Y-27632 significantly decreased the formation and amount of blood vessels in the extra-embryonic membranes throughout the embryonic development. At 60 + 8 h, the most striking effects on the vasculature could be seen in the experimental group, revealing impaired development of omphalomesenteric and vitelline vessels in the tail region. During the investigation period, vessel formation appeared to proceed but was strongly delayed and decreased compared to the control group, as confirmed by calculating the length of ST and areas of vasculature as wells as fractal analysis [26].

VEGF-induced migration of endothelial cells involves actin reorganization. Beside phosphorylation of MLC, VEGF has been stated to regulate the phosphorylation and activity of focal adhesion kinase (FAK) as well, which is important for focal adhesion assembly and disassembly as well as actin organization [35,65]. Hence, VEGF in association with paxillin and talin or vinculin leads to the recruitment of FAK to focal adhesions. VEGFR-2 also transduces its mitogenic functions by influencing FAK, thus regulating cytoskeleton organization and migration [35]. Kroll et al. have recently provided important in vivo evidence for the interaction of VEGF and VEGFR-2 during angiogenesis and have even postulated a potentiating effect of Rho kinase inhibition using H-1152 on VEGF-driven retinal neovascularization and sprouting angiogenesis in human umbilical vein endothelial cells. Pharmacological inhibition as well as expression silencing of ROCK I and ROCK II in endothelial cells activated VEGF-stimulated activation of VEGFR-2, which is necessary to induce VEGF-induced migration, proliferation, NO release, and permeability. The H-1152-driven increase in blood vessel growth has been stated to be completely dependent on VEGFR-2 activation, suggesting a negative feedback loop in which VEGF-induced ROCK I and ROCK II activation counteracts VEGFR-2 activation [64]. Van NieuwAmerongen at et al. have seized on the Kroll conclusion and postulated a dual role for ROCK in the regulation of angiogenesis (Figure 12) [38].

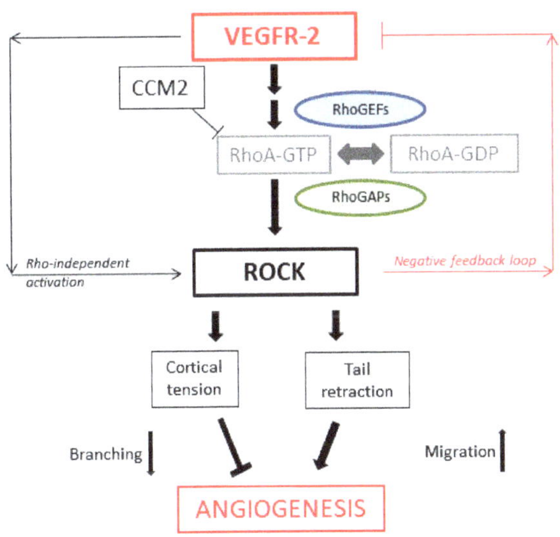

Figure 12. Dual role of ROCK signaling in regulation of angiogenesis; adapted from [38].

Although we recorded a significant decrease in blood vessel formation following ROCK inhibition in our experiments, which is contrary to the previously mentioned results, we still wanted to investigate the potential effects of Y-27632 on VEGF and VEGFR-2 activity based on Kroll's assumption [64]. At 60 + 1 h, gene expression levels of VEGF and VEGFR-2 were significantly upregulated in membranes and embryos in the experimental group, which is in line with the negative feedback loop presented by Kroll et al. This may reflect an early inhibitory effect of Y-27632 on ROCK signaling in angiogenesis and subsequently explain the activating effect on VEGFR-2 and VEGF mRNA levels. Y-27632 seems to have a very rapid inhibiting effect on vessel formation during angiogenesis when compared to the effect on morphologic development in the embryos themselves. The ROCK inhibitor may affect neovascularization soon after administration, hence disrupting the underlying processes and mechanism necessary for further vessel formation, which may then lead to the reduced extra-embryonic vascularization at later stages of embryonic development [66,67]. It has been shown that during development, the first hematopoietic cells arise in the yolk sac from mesodermal structures and differentiate into endothelial and blood cells, forming blood islands [26,68]. Therefore, the effects of ROCK inhibitor Y-27632 on the extra-embryonic membranes may also contribute to abnormalities in the morphology of the embryos, as described by the authors in recent studies [20–22]. The impact and chronology of this improper supply of nutrients provided by the yolk sac may be highlighted by the early effect of Y-27632 on the extra-embryonic vasculature recorded at 60 + 1 h and the delayed morphological anomalies seen at 60 + 8 h. The mRNA levels of VEGF and VEGFR-2 were not different at 4 h and 8 h in Y-27632-treated membranes and embryos. This may indicate an early compensatory effect in the experimental group. To further examine possible post-translational effects of Y-27632 on VEGF and VEGFR-2 in later stages of development, we studied protein levels in membranes and embryos at 8 and 24 h following the administration of vehicle or the inhibitor. Protein expression of VEGF was only higher in Y-27632-treated embryos at 60 + 8 h, which may reflect a compensatory effect in protein levels following ROCK inhibition. However, no significant differences in protein expression of VEGF and VEGFR-2 were recorded in the remaining ELISA procedures. Hyvelin et al. have found similar results in their study investigating the role of the RhoA/ROCK pathway in the development of chronic hypoxia-induced pulmonary hypertension and the associated structural changes in the pulmonary vasculature. Inhibition of ROCK signaling with Y-27632 abrogated the development of hypoxic pulmonary hypertension by suppressing sustained pulmonary vasoconstriction and hypoxia-induced vascular remodeling revealing improperly formed intra-acinar vessels. Moreover, the compound did not alter the expression of the pro-angiogenic genes VEGF-A and Ang-1 after one week of hypoxia. However, the authors did suggest that the expression of these genes might have been affected by Y-27632 at earlier time points, thus inhibiting angiogenesis [69].

We also tested other genes which might have influenced the impaired vessel formation in treated embryos. Ang-2 is a complex regulator of angiogenesis that plays a central role in both vessel sprouting and vessel regression. It is almost exclusively produced by endothelial cells and functions mainly as a vessel-destabilizing molecule by regulating the vessel maturation and remodeling of Ang-1/Tie2 in an antagonistic manner [70–72]. It has been demonstrated in previous studies that Ang-2 prevented the formation of capillary-like structures and impaired angiogenesis in a CAM chick model [71]. However, in contrast, Ang-2 has also been shown to phosphorylate the integrin adaptor protein FAK inducing sprouting angiogenesis in a Tie2-independent manner [70]. Maisonpierre et al. generated transgenic mice overexpressing Ang-2 in their blood vessels and found that the animals were smaller and exhibited an abnormal vascular phenotype. The authors suggested that this factor may play an early role at the sites of vessel invasion [72]. We found an upregulation of Ang-2 at 60 + 1 h in membranes of Y-27632-treated embryos and found that treated embryos were smaller and exhibited abnormalities. The upregulation may contribute to the disruption of vascular formation by blocking the Ang-1/Tie2 signaling axis or may reflect a compensatory effect, as it has been stated that in the presence of abundant VEGF, as seen at 60 + 1 h in our study, Ang-2 can promote vessel sprouting by inhibiting a stabilizing Ang-1 signal [73].

Next, NO is known to play an integral role in the development and maintenance of the microvascular network, including angiogenesis and vascular permeability [38,74,75]. VEGF enhances NO production and promotes eNOS and iNOS expression in endothelial cells in vitro [74]. Inhibition of ROCK, in turn, has been described to induce increased expression and activity of eNOS [38,64]. Fukumura et al. have analyzed the relative contribution of eNOS and iNOS in VEGF-induced angiogenesis and vascular permeability and found that eNOS plays the predominant role in these processes. Nevertheless, the authors have postulated that iNOS selective inhibitor and iNOS gene deletion slowed angiogenesis in VEGF-containing gels, although it was less effective than eNOS gene deletion. Activation of VEGFR-2 has also been shown to lead to an upregulation of iNOS protein in human umbilical vein endothelial cells [76]. The iNOS gene expression was not altered in membranes treated with Y-27632 compared with controls in our analyses, whereas an upregulation of iNOS mRNA levels was noted at 60 + 1 h in treated embryos, which seems to be associated with the noted upregulation of VEGF and VEGFR-2 at that time point. In contrast, gene expression levels of iNOS decreased in the experimental group during development and were significantly different compared with controls at 60 + 8 h, possibly contributing to the delayed angiogenesis as a consequence.

As already mentioned, endothelial cell migration involves cytoskeletal dynamics and organization, mechanisms to which normal ROCK signaling is fundamental. ROCK regulates actomyosin contractility, an essential process in endothelial cell migration, by direct phosphorylation of MLC and phosphorylation and inactivation of myosin phosphatase, thereby contributing to the formation of multicellular, precapillary cord-like structures arranged by endothelial cells [26,53]. The effects of Y-27632 on the cytoskeletal arrangements in chick embryos have recently been shown by the authors to lead to the failure of ventral body wall closure at later stages of development [20]. In the present study, Western blot analysis showed inhibition of ROCK downstream signaling in both membranes and embryonic tissue in the experimental group at 8 h following treatment with Y-27632, with significantly decreased expression of phosphorylated MLC when compared with controls.

Taken together, the major impact of ROCK inhibitor Y-27632 in angiogenesis seems to be on the cell migration of endothelial cells. The impaired regulation of the cytoskeletal rearrangement may affect and delay the process of organization of endothelial cells into capillary structures, subsequently leading to the attenuated vascularization observed in the experimental group. The vascular defect in the yolk sac and regression of pre-existing vessels noted at very early stages during development impede the maintenance of proper circulation, possibly causing developmental delays and morphological abnormalities in Y-27632-treated embryos, all of which were noted in follow-up investigations at more advanced stages of development [21].

5. Conclusions

In conclusion, our experiments have shown that exposure to ROCK inhibitor Y-27632 in the critical period of ventral body wall formation in chick embryos strikingly impairs extra-embryonic angiogenesis, leading to delayed and reduced vessel formation in further development. We speculate that Y-27632 also disrupts the rearrangement process of the cytoskeleton required for VEGF/VEGFR-2-induced endothelial cell migration to establish the regular vasculature in membranes and embryos. Hence, defects in vascularization may contribute to morphological abnormalities during early development and subsequently to a deficient formation of the embryonic tissue, which may result in the rupture of the body wall and intestinal herniation through the defect. Further work to determine the contribution of ROCK signaling or its inhibition in angiogenic processes will be subject of future studies.

Author Contributions: Conceptualization: J.W.D. and J.T.; methodology: J.W.D., J.-H.G., A.K.G. and J.T.; validation: P.P., J.-H.G. and J.T.; formal analysis: J.W.D.; investigation: J.W.D.; resources: J.W.D. and A.K.G.; data curation: P.P. and J.T.; writing—original draft preparation: J.W.D.; writing—review and editing: P.P. and J.T.; visualization: J.W.D. and A.K.G.; supervision: P.P. and J.T.; project administration: P.P. and J.T.; funding acquisition: P.P. and J.T. All authors have read and agreed to the published version of the manuscript.

Funding: This work was funded by the Children's and Medical Research Foundation, Our Lady's Children's Hospital, Crumlin [E/12/1].

Institutional Review Board Statement: Not applicable for this study.

Informed Consent Statement: Not applicable.

Data Availability Statement: Not applicable.

Conflicts of Interest: The authors declare no conflict of interest.

References

1. Feldkamp, M.L.; Carey, J.C.; Sadler, T.W. Development of gastroschisis: Review of hypotheses, a novel hypothesis, and implications for research. *Am. J. Med. Genet. A* **2007**, *143A*, 639–652. [CrossRef] [PubMed]
2. Williams, T. Animal models of ventral body wall closure defects: A personal perspective on gastroschisis. *Am. J. Med. Genet. C Semin. Med. Genet.* **2008**, *148C*, 186–191. [CrossRef] [PubMed]
3. Pechriggl, E.; Blumer, M.; Tubbs, R.S.; Olewnik, L.; Konschake, M.; Fortelny, R.; Stofferin, H.; Honis, H.R.; Quinones, S.; Maranillo, E.; et al. Embryology of the Abdominal Wall and Associated Malformations—A Review. *Front. Surg.* **2022**, *9*, 891896. [CrossRef]
4. Revels, J.W.; Wang, S.S.; Nasrullah, A.; Revzin, M.; Iyer, R.S.; Deutsch, G.; Katz, D.S.; Moshiri, M. An Algorithmic Approach to Complex Fetal Abdominal Wall Defects. *AJR Am. J. Roentgenol.* **2020**, *214*, 218–231. [CrossRef]
5. Sadler, T.W. Embryology of neural tube development. *Am. J. Med. Genet. C Semin. Med. Genet.* **2005**, *135C*, 2–8. [CrossRef] [PubMed]
6. Sadler, T.W. The embryologic origin of ventral body wall defects. *Semin. Pediatr. Surg.* **2010**, *19*, 209–214. [CrossRef] [PubMed]
7. Sadler, T.W.; Feldkamp, M.L. The embryology of body wall closure: Relevance to gastroschisis and other ventral body wall defects. *Am. J. Med. Genet. C Semin. Med. Genet.* **2008**, *148C*, 180–185. [CrossRef] [PubMed]
8. Brewer, S.; Williams, T. Finally, a sense of closure? Animal models of human ventral body wall defects. *Bioessays* **2004**, *26*, 1307–1321. [CrossRef]
9. deVries, P.A. The pathogenesis of gastroschisis and omphalocele. *J. Pediatr. Surg.* **1980**, *15*, 245–251. [CrossRef]
10. Ledbetter, D.J. Congenital abdominal wall defects and reconstruction in pediatric surgery: Gastroschisis and omphalocele. *Surg. Clin. N. Am.* **2012**, *92*, 713–727. [CrossRef]
11. Hoyme, H.E.; Jones, M.C.; Jones, K.L. Gastroschisis: Abdominal wall disruption secondary to early gestational interruption of the omphalomesenteric artery. *Semin. Perinatol.* **1983**, *7*, 294–298. [PubMed]
12. Torfs, C.P.; Katz, E.A.; Bateson, T.F.; Lam, P.K.; Curry, C.J. Maternal medications and environmental exposures as risk factors for gastroschisis. *Teratology* **1996**, *54*, 84–92. [CrossRef]
13. Haddow, J.E.; Palomaki, G.E.; Holman, M.S. Young maternal age and smoking during pregnancy as risk factors for gastroschisis. *Teratology* **1993**, *47*, 225–228. [CrossRef]
14. Haustein, K.O. Cigarette smoking, nicotine and pregnancy. *Int. J. Clin. Pharmacol. Ther.* **1999**, *37*, 417–427. [PubMed]
15. Torfs, C.P.; Velie, E.M.; Oechsli, F.W.; Bateson, T.F.; Curry, C.J. A population-based study of gastroschisis: Demographic, pregnancy, and lifestyle risk factors. *Teratology* **1994**, *50*, 44–53. [CrossRef] [PubMed]
16. Hamburger, V.; Hamilton, H.L. A series of normal stages in the development of the chick embryo. 1951. *Dev. Dyn.* **1992**, *195*, 231–272. [CrossRef]
17. Gheorghescu, A.; Thompson, J. Delayed vasculogenesis and impaired angiogenesis due to altered Ang-2 and VE-cadherin levels in the chick embryo model following exposure to cadmium. *Pediatr. Surg. Int.* **2016**, *32*, 175–186. [CrossRef] [PubMed]
18. Gheorghescu, A.K.; Tywoniuk, B.; Duess, J.; Buchete, N.V.; Thompson, J. Exposure of chick embryos to cadmium changes the extra-embryonic vascular branching pattern and alters expression of VEGF-A and VEGF-R2. *Toxicol. Appl. Pharmacol.* **2015**, *289*, 79–88. [CrossRef]
19. Doi, T.; Puri, P.; Bannigan, J.; Thompson, J. Downregulation of ROCK-I and ROCK-II gene expression in the cadmium-induced ventral body wall defect chick model. *Pediatr. Surg. Int.* **2008**, *24*, 1297–1301. [CrossRef]
20. Duess, J.W.; Puri, P.; Thompson, J. Impaired cytoskeletal arrangements and failure of ventral body wall closure in chick embryos treated with rock inhibitor (Y-27632). *Pediatr. Surg. Int.* **2016**, *32*, 45–58. [CrossRef]
21. Duess, J.W.; Gosemann, J.H.; Puri, P.; Thompson, J. Teratogenesis in the chick embryo following post-gastrulation exposure to Y-27632 -effect of Y-27632 on embryonic development. *Toxicol. Appl. Pharmacol.* **2020**, *409*, 115277. [CrossRef] [PubMed]
22. Duess, J.W.; Fujiwara, N.; Corcionivoschi, N.; Puri, P.; Thompson, J. ROCK inhibitor (Y-27632) disrupts somitogenesis in chick embryos. *Pediatr. Surg. Int.* **2013**, *29*, 13–18. [CrossRef] [PubMed]
23. Hoang, M.V.; Whelan, M.C.; Senger, D.R. Rho activity critically and selectively regulates endothelial cell organization during angiogenesis. *Proc. Natl. Acad. Sci. USA* **2004**, *101*, 1874–1879. [CrossRef] [PubMed]
24. Moses, M.A. The regulation of neovascularization of matrix metalloproteinases and their inhibitors. *Stem Cells* **1997**, *15*, 180–189. [CrossRef]
25. Eelen, G.; Treps, L.; Li, X.; Carmeliet, P. Basic and Therapeutic Aspects of Angiogenesis Updated. *Circ. Res.* **2020**, *127*, 310–329. [CrossRef] [PubMed]

26. Kamijo, H.; Matsumura, Y.; Thumkeo, D.; Koike, S.; Masu, M.; Shimizu, Y.; Ishizaki, T.; Narumiya, S. Impaired vascular remodeling in the yolk sac of embryos deficient in ROCK-I and ROCK-II. *Genes Cells* **2011**, *16*, 1012–1021. [CrossRef]
27. Papetti, M.; Herman, I.M. Mechanisms of normal and tumor-derived angiogenesis. *Am. J. Physiol. Cell Physiol.* **2002**, *282*, C947–C970. [CrossRef]
28. Risau, W. Mechanisms of angiogenesis. *Nature* **1997**, *386*, 671–674. [CrossRef]
29. Schmidt, A.; Brixius, K.; Bloch, W. Endothelial precursor cell migration during vasculogenesis. *Circ. Res.* **2007**, *101*, 125–136. [CrossRef]
30. Lamalice, L.; Le Boeuf, F.; Huot, J. Endothelial cell migration during angiogenesis. *Circ. Res.* **2007**, *100*, 782–794. [CrossRef]
31. Eichmann, A.; Simons, M. VEGF signaling inside vascular endothelial cells and beyond. *Curr. Opin. Cell Biol.* **2012**, *24*, 188–193. [CrossRef]
32. Ferrara, N.; Gerber, H.P.; LeCouter, J. The biology of VEGF and its receptors. *Nat. Med.* **2003**, *9*, 669–676. [CrossRef] [PubMed]
33. Koch, S.; Claesson-Welsh, L. Signal transduction by vascular endothelial growth factor receptors. *Cold Spring Harb. Perspect. Med.* **2012**, *2*, a006502. [CrossRef] [PubMed]
34. Koch, S.; Tugues, S.; Li, X.; Gualandi, L.; Claesson-Welsh, L. Signal transduction by vascular endothelial growth factor receptors. *Biochem. J.* **2011**, *437*, 169–183. [CrossRef] [PubMed]
35. Cebe-Suarez, S.; Zehnder-Fjallman, A.; Ballmer-Hofer, K. The role of VEGF receptors in angiogenesis; complex partnerships. *Cell. Mol. Life Sci.* **2006**, *63*, 601–615. [CrossRef]
36. Bryan, B.A.; D'Amore, P.A. What tangled webs they weave: Rho-GTPase control of angiogenesis. *Cell. Mol. Life Sci.* **2007**, *64*, 2053–2065. [CrossRef]
37. Park, H.J.; Kong, D.; Iruela-Arispe, L.; Begley, U.; Tang, D.; Galper, J.B. 3-hydroxy-3-methylglutaryl coenzyme A reductase inhibitors interfere with angiogenesis by inhibiting the geranylgeranylation of RhoA. *Circ. Res.* **2002**, *91*, 143–150. [CrossRef] [PubMed]
38. van Nieuw Amerongen, G.P.; van Hinsbergh, V.W. Role of ROCK I/II in vascular branching. *Am. J. Physiol. Heart Circ. Physiol.* **2009**, *296*, H903–H905. [CrossRef]
39. Varon, C.; Basoni, C.; Reuzeau, E.; Moreau, V.; Kramer, I.J.; Genot, E. TGFbeta1-induced aortic endothelial morphogenesis requires signaling by small GTPases Rac1 and RhoA. *Exp. Cell Res.* **2006**, *312*, 3604–3619. [CrossRef]
40. Liu, J.; Wada, Y.; Katsura, M.; Tozawa, H.; Erwin, N.; Kapron, C.M.; Bao, G.; Liu, J. Rho-Associated Coiled-Coil Kinase (ROCK) in Molecular Regulation of Angiogenesis. *Theranostics* **2018**, *8*, 6053–6069. [CrossRef]
41. Urade, R.; Chiu, Y.H.; Chiu, C.C.; Wu, C.Y. Small GTPases and Their Regulators: A Leading Road toward Blood Vessel Development in Zebrafish. *Int. J. Mol. Sci.* **2022**, *23*, 4991. [CrossRef] [PubMed]
42. Riento, K.; Ridley, A.J. Rocks: Multifunctional kinases in cell behaviour. *Nat. Rev. Mol. Cell Biol.* **2003**, *4*, 446–456. [CrossRef] [PubMed]
43. Scott, R.W.; Olson, M.F. LIM kinases: Function, regulation and association with human disease. *J. Mol. Med. (Berl.)* **2007**, *85*, 555–568. [CrossRef] [PubMed]
44. Montalvo, J.; Spencer, C.; Hackathorn, A.; Masterjohn, K.; Perkins, A.; Doty, C.; Arumugam, A.; Ongusaha, P.P.; Lakshmanaswamy, R.; Liao, J.K.; et al. ROCK1 & 2 perform overlapping and unique roles in angiogenesis and angiosarcoma tumor progression. *Curr. Mol. Med.* **2013**, *13*, 205–219.
45. Sun, H.; Breslin, J.W.; Zhu, J.; Yuan, S.Y.; Wu, M.H. Rho and ROCK signaling in VEGF-induced microvascular endothelial hyperpermeability. *Microcirculation* **2006**, *13*, 237–247. [CrossRef]
46. Dugan, J.D., Jr.; Lawton, M.T.; Glaser, B.; Brem, H. A new technique for explantation and in vitro cultivation of chicken embryos. *Anat. Rec.* **1991**, *229*, 125–128. [CrossRef]
47. Parsons-Wingerter, P.; Elliott, K.E.; Clark, J.I.; Farr, A.G. Fibroblast growth factor-2 selectively stimulates angiogenesis of small vessels in arterial tree. *Arterioscler. Thromb. Vasc. Biol.* **2000**, *20*, 1250–1256. [CrossRef]
48. Masters, B.R. Fractal analysis of the vascular tree in the human retina. *Annu. Rev. Biomed. Eng.* **2004**, *6*, 427–452. [CrossRef]
49. Stosic, T.; Stosic, B.D. Multifractal analysis of human retinal vessels. *IEEE Trans. Med. Imaging* **2006**, *25*, 1101–1107. [CrossRef]
50. Liew, G.; Mitchell, P.; Rochtchina, E.; Wong, T.Y.; Hsu, W.; Lee, M.L.; Wainwright, A.; Wang, J.J. Fractal analysis of retinal microvasculature and coronary heart disease mortality. *Eur. Heart J.* **2011**, *32*, 422–429. [CrossRef]
51. Lammer, E.J.; Iovannisci, D.M.; Tom, L.; Schultz, K.; Shaw, G.M. Gastroschisis: A gene-environment model involving the VEGF-NOS3 pathway. *Am. J. Med. Genet. C Semin. Med. Genet.* **2008**, *148C*, 213–218. [CrossRef]
52. Khan, F.A.; Raymond, S.L.; Hashmi, A.; Islam, S. Anatomy and embryology of abdominal wall defects. *Semin. Pediatr. Surg.* **2022**, *31*, 151230. [CrossRef] [PubMed]
53. Bryan, B.A.; Dennstedt, E.; Mitchell, D.C.; Walshe, T.E.; Noma, K.; Loureiro, R.; Saint-Geniez, M.; Campaigniac, J.-P.; Liao, J.K. D'Amore, P.A. RhoA/ROCK signaling is essential for multiple aspects of VEGF-mediated angiogenesis. *FASEB J.* **2010**, *24*, 3186–3195. [CrossRef] [PubMed]
54. Olsson, A.K.; Dimberg, A.; Kreuger, J.; Claesson-Welsh, L. VEGF receptor signalling—In control of vascular function. *Nat. Rev. Mol. Cell Biol.* **2006**, *7*, 359–371. [CrossRef] [PubMed]
55. Chen, W.; Mao, K.; Liu, Z.; Dinh-Xuan, A.T. The role of the RhoA/Rho kinase pathway in angiogenesis and its potential value in prostate cancer (Review). *Oncol. Lett.* **2014**, *8*, 1907–1911. [CrossRef]

56. van Nieuw Amerongen, G.P.; Koolwijk, P.; Versteilen, A.; van Hinsbergh, V.W. Involvement of RhoA/Rho kinase signaling in VEGF-induced endothelial cell migration and angiogenesis in vitro. *Arterioscler. Thromb. Vasc. Biol.* **2003**, *23*, 211–217. [CrossRef]
57. Shalaby, F.; Ho, J.; Stanford, W.L.; Fischer, K.D.; Schuh, A.C.; Schwartz, L.; Bernstein, A.; Rossant, J. A requirement for Flk1 in primitive and definitive hematopoiesis and vasculogenesis. *Cell* **1997**, *89*, 981–990. [CrossRef]
58. Shalaby, F.; Rossant, J.; Yamaguchi, T.P.; Gertsenstein, M.; Wu, X.F.; Breitman, M.L.; Schuh, A.C. Failure of blood-island formation and vasculogenesis in Flk-1-deficient mice. *Nature* **1995**, *376*, 62–66. [CrossRef]
59. Carmeliet, P.; Ferreira, V.; Breier, G.; Pollefeyt, S.; Kieckens, L.; Gertsenstein, M.; Fahrig, M.; Vandenhoeck, A.; Harpal, K.; Eberhardt, C.; et al. Abnormal blood vessel development and lethality in embryos lacking a single VEGF allele. *Nature* **1996**, *380*, 435–439. [CrossRef]
60. Ferrara, N.; Carver-Moore, K.; Chen, H.; Dowd, M.; Lu, L.; O'Shea, K.S.; Powell-Braxton, L.; Hillan, K.J.; Moore, M.W. Heterozygous embryonic lethality induced by targeted inactivation of the VEGF gene. *Nature* **1996**, *380*, 439–442. [CrossRef]
61. Eckenstaler, R.; Hauke, M.; Benndorf, R.A. A current overview of RhoA, RhoB, and RhoC functions in vascular biology and pathology. *Biochem. Pharmacol.* **2022**, *206*, 115321. [CrossRef]
62. Combedazou, A.; Gayral, S.; Colombie, N.; Fougerat, A.; Laffargue, M.; Ramel, D. Small GTPases orchestrate cell-cell communication during collective cell movement. *Small GTPases* **2020**, *11*, 103–112. [CrossRef]
63. Uchida, S.; Watanabe, G.; Shimada, Y.; Maeda, M.; Kawabe, A.; Mori, A.; Arii, S.; Uehata, M.; Kishimoto, T.; Oikawa, T.; et al. The suppression of small GTPase rho signal transduction pathway inhibits angiogenesis in vitro and in vivo. *Biochem. Biophys. Res. Commun.* **2000**, *269*, 633–640. [CrossRef]
64. Kroll, J.; Epting, D.; Kern, K.; Dietz, C.T.; Feng, Y.; Hammes, H.P.; Wieland, T.; Augustin, H.G. Inhibition of Rho-dependent kinases ROCK I/II activates VEGF-driven retinal neovascularization and sprouting angiogenesis. *Am. J. Physiol. Heart Circ. Physiol.* **2009**, *296*, H893–H899. [CrossRef]
65. Yin, L.; Morishige, K.; Takahashi, T.; Hashimoto, K.; Ogata, S.; Tsutsumi, S.; Takata, K.; Ohta, T.; Kawagoe, J.; Takahashi, K.; et al. Fasudil inhibits vascular endothelial growth factor-induced angiogenesis in vitro and in vivo. *Mol. Cancer Ther.* **2007**, *6*, 1517–1525. [CrossRef]
66. Yadgary, L.; Kedar, O.; Adepeju, O.; Uni, Z. Changes in yolk sac membrane absorptive area and fat digestion during chick embryonic development. *Poult. Sci.* **2013**, *92*, 1634–1640. [CrossRef] [PubMed]
67. Yadgary, L.; Yair, R.; Uni, Z. The chick embryo yolk sac membrane expresses nutrient transporter and digestive enzyme genes. *Poult. Sci.* **2011**, *90*, 410–416. [CrossRef] [PubMed]
68. Minko, K.; Bollerot, K.; Drevon, C.; Hallais, M.F.; Jaffredo, T. From mesoderm to blood islands: Patterns of key molecules during yolk sac erythropoiesis. *Gene Expr. Patterns* **2003**, *3*, 261–272. [CrossRef]
69. Hyvelin, J.M.; Howell, K.; Nichol, A.; Costello, C.M.; Preston, R.J.; McLoughlin, P. Inhibition of Rho-kinase attenuates hypoxia-induced angiogenesis in the pulmonary circulation. *Circ. Res.* **2005**, *97*, 185–191. [CrossRef]
70. Felcht, M.; Luck, R.; Schering, A.; Seidel, P.; Srivastava, K.; Hu, J.; Bartol, A.; Kienast, Y.; Vettel, C.; Loos, E.K.; et al. Angiopoietin-2 differentially regulates angiogenesis through TIE2 and integrin signaling. *J. Clin. Investig.* **2012**, *122*, 1991–2005. [CrossRef] [PubMed]
71. Lee, O.H.; Fueyo, J.; Xu, J.; Yung, W.K.; Lemoine, M.G.; Lang, F.F.; Bekele, B.N.; Zhou, X.; Alonso, M.A.; Aldape, K.D.; et al. Sustained angiopoietin-2 expression disrupts vessel formation and inhibits glioma growth. *Neoplasia* **2006**, *8*, 419–428. [CrossRef] [PubMed]
72. Maisonpierre, P.C.; Suri, C.; Jones, P.F.; Bartunkova, S.; Wiegand, S.J.; Radziejewski, C.; Compton, D.; McClain, J.; Aldrich, T.H.; Papadopoulos, N.; et al. Angiopoietin-2, a natural antagonist for Tie2 that disrupts in vivo angiogenesis. *Science* **1997**, *277*, 55–60. [CrossRef] [PubMed]
73. Davis, S.; Yancopoulos, G.D. The Angiopoietins: Yin and Yang in Angiogenesis. In *Vascular Growth Factors and Angiogenesis*; Claesson-Welsh, L., Ed.; Springer: Berlin/Heidelberg, Germany, 1999; pp. 173–185.
74. Fukumura, D.; Gohongi, T.; Kadambi, A.; Izumi, Y.; Ang, J.; Yun, C.O.; Buerk, D.G.; Huang, P.L.; Jain, R.K. Predominant role of endothelial nitric oxide synthase in vascular endothelial growth factor-induced angiogenesis and vascular permeability. *Proc. Natl. Acad. Sci. USA* **2001**, *98*, 2604–2609. [CrossRef] [PubMed]
75. Peng, J.F.; Ling, Y.; Gou, W.Y.; Zhang, H.; Wu, C.X. Identification of chicken eNOS gene and differential expression in highland versus lowland chicken breeds. *Poult. Sci.* **2012**, *91*, 2275–2281. [CrossRef]
76. Kroll, J.; Waltenberger, J. VEGF-A induces expression of eNOS and iNOS in endothelial cells via VEGF receptor-2 (KDR). *Biochem. Biophys. Res. Commun.* **1998**, *252*, 743–746. [CrossRef]

toxics

Article

Downregulation of LncRNA GCLC-1 Promotes Microcystin-LR-Induced Malignant Transformation of Human Liver Cells by Regulating GCLC Expression

Xinglei Huang [1,†], Zhaohui Su [1,†], Jiangheng Li [1], Junquan He [1], Na Zhao [1], Liyun Nie [1], Bin Guan [1], Qiuyue Huang [1], Huiliu Zhao [2], Guo-Dong Lu [3] and Qingqing Nong [1,4,5,*]

1 Department of Environmental Health, School of Public Health, Guangxi Medical University, Nanning 530021, China
2 Department of Clinical Laboratory, The Affiliated Tumor Hospital of Guangxi Medical University, Nanning 530021, China
3 Department of Toxicology, School of Public Health, Guangxi Medical University, Nanning 530021, China
4 Guangxi Colleges and Universities Key Laboratory of Prevention and Control of Highly Prevalent Diseases, Guangxi Medical University, Nanning 530021, China
5 Guangxi Key Laboratory of Environment and Health Research, Guangxi Medical University, Nanning 530021, China
* Correspondence: nnqq@gxmu.edu.cn; Tel.: +86-771-5358146
† These authors contributed equally to this work.

Abstract: Microcystin-LR (MCLR) is an aquatic toxin, which could lead to the development of hepatocellular carcinoma (HCC). Long non-coding RNAs (lncRNAs) are considered important regulatory elements in the occurrence and development of cancer. However, the roles and mechanisms of lncRNAs during the process of HCC, induced by MCLR, remain elusive. Here, we identified a novel lncRNA, namely lnc-GCLC-1 (lncGCLC), which is in close proximity to the chromosome location of glutamate–cysteine ligase catalytic subunit (GCLC). We then investigated the role of lncGCLC in MCLR-induced malignant transformation of WRL68, a human hepatic cell line. During MCLR-induced cell transformation, the expression of lncGCLC and GCLC decreased continuously, accompanied with a consistently high expression of miR-122-5p. Knockdown of lncGCLC promoted cell proliferation, migration and invasion, but reduced cell apoptosis. A xenograft nude mouse model demonstrated that knockdown of lncGCLC promoted tumor growth. Furthermore, knockdown of lncGCLC significantly upregulated miR-122-5p expression, suppressed GCLC expression and GSH levels, and enhanced oxidative DNA damages. More importantly, the expression of lncGCLC in human HCC tissues was significantly downregulated in the high-microcystin exposure group, and positively associated with GCLC level in HCC tissues. Together, these findings suggest that lncGCLC plays an anti-oncogenic role in MCLR-induced malignant transformation by regulating GCLC expression.

Keywords: microcystin-LR; lncRNA GCLC-1; GCLC; carcinogenesis

Citation: Huang, X.; Su, Z.; Li, J.; He, J.; Zhao, N.; Nie, L.; Guan, B.; Huang, Q.; Zhao, H.; Lu, G.-D.; et al. Downregulation of LncRNA GCLC-1 Promotes Microcystin-LR-Induced Malignant Transformation of Human Liver Cells by Regulating GCLC Expression. *Toxics* **2023**, *11*, 162. https://doi.org/10.3390/toxics11020162

Academic Editors: Esref Demir and Sam Kacew

Received: 21 December 2022
Revised: 3 February 2023
Accepted: 5 February 2023
Published: 9 February 2023

1. Introduction

Microcystins (MCs), a group of cyclic heptapeptide compounds, are secondary metabolites produced by freshwater cyanobacteria [1]. So far, over 279 structural variants of MCs have been identified, among which microcystin-LR (MCLR) is regarded as the most toxic [2]. Since current tap water treatment technologies are not effective in removing MCs, MCs contamination has become an imperative threat to people's drinking water safety. Epidemiological studies have suggested that chronic exposure to MCLR via consumption of contaminated water was associated with an increased incidence of hepatocellular carcinoma (HCC) [3–7]. MCLR is currently classified as a group 2B carcinogen by the International

Agency for Research on Cancer [8]. However, the underlying mechanism of MCLR-induced HCC remains elusive.

One of the predominant forms of free radical-induced oxidative lesions, 8-hydroxy-2′-deoxyguanosine (8-OHdG), has been widely used as a biomarker for oxidative stress and carcinogenesis [9]. Long-term and persistent exposure to MCLR increased the 8-OHdG levels of DNA in liver cells, damaged the integrity of mitochondrial DNA (mtDNA) and nuclear DNA, and altered the mtDNA content [10,11]. A large number of proteins has been identified to be involved in the DNA damage process and response. Glutamate–cysteine ligase (GCL) is one of the key enzymes of oxidative stress and functions as a rate-limiting enzyme of glutathione (GSH) synthesis. This enzyme is a heterodimer, consisting of a catalytic subunit (GCLC) and a modifier subunit (GCLM) [12]. Downregulation of GCLC was observed in multiple types of human cancer cell lines and tumors [13–15]. Previously, we have found that the expression of GCLC and activity of GCL decreased continuously, accompanied with consistently low levels of GSH, but high levels of oxidative DNA damages during MCLR-induced cell transformation, suggesting that downregulation of GCLC participates in MCLR-induced oxidative DNA damages and malignant transformation in human liver cells [16].

Long non-coding RNAs (lncRNAs) are a group of non-coding transcripts, longer than 200 nucleotides. Abnormalities in lncRNAs have been confirmed to exhibit tumor suppressor or carcinogenic effects, and play an important role in the development of tumors. For instance, the expression of lncRNA-DQ786227 was significantly increased during the transformation of BEAS-2B cells induced by benzo(a)pyrene, while silencing of lncRNA-DQ786227 expression inhibited cell proliferation and colony formation, and promoted cell apoptosis [17]. LncRNA linc00152 was upregulated in transformed 16HBE cells, induced by cigarette smoke extract (CSE), and expedited cell transformation by regulating cyclin D1 [18]. Cadmium (Cd) exposure-caused lncRNA MEG3 downregulation leads to enhanced cell cycle progression and apoptosis resistance, promoting Cd-induced cell transformation and cancer stem cell (CSC)-like property [19]. Furthermore, numerous lncRNAs, including lncRNA-LET [20], lncRNA-PRAL [21] and lnc-DILC [22] were found to be downregulated in HCC tissues and dramatically inhibited HCC growth by regulating histone acetylation, p53 ubiquitination, or autocrine interleukin-6/STAT3 signaling. However, the roles of lncRNAs, in the occurrence of cancer induced by MCLR, are poorly reported.

It is well established that the function of lncRNAs is frequently associated with their chromosomal location. Accumulating evidence suggests that numerous lncRNA loci act locally (in cis) to regulate the expression of neighbouring genes through functions of the lncRNA promoter, transcription, or transcript itself [23]. Recently, we identified a novel lncRNA named lnc-GCLC-1 (defined as lncGCLC here: https://lncipedia.org/db/transcript/lnc-GCLC-1:7, accessed on 20 December 2022), of which the chromosome location (chr6: 53,561,289–53,617,007) is near that of GCLC (chr6: 53,497,341–53,616,970). To date, the biological function and expression of lncGCLC remains unexplored. In this study, we investigated the potential roles of lncGCLC in MCLR-induced hepatocarcinogenesis, based on the previously established MCLR-induced malignant transformation model in human liver cell lines WRL68 and the MC-exposed human samples. Further, we analyzed the relationship among lncGCLC, redox and hepatocarcinogenesis induced by MCLR.

2. Materials and Methods

2.1. Cell Lines and MCLR-Induced Malignant Transformation Model

The human normal liver cell lines WRL68 and human hepatoma cell lines (HepG2 and SMMC7721) were purchased from the American Type Culture Collection (Manassas, VA, USA). WRL68, HepG2 and SMMC7721 cells were cultured in Dulbecco's Modified Eagle's Medium (DMEM, Gibco, USA), supplemented with 10% fetal bovine serum (FBS, Gibco, USA) and 1% penicillin/streptomycin (Gibco, USA) in a cell incubator, with 5% carbon dioxide at 37 °C. The MCLR-induced malignant transformation model of WRL68 cells was established as described in the previous study [24,25]. In brief, the cells were continuously exposed to phosphate-buffered saline (PBS; negative control) or 10 μg/L of MCLR (Alexis, Switzerland) for 72 h per passage. The process was continued for about 11 weeks (25 passages). The soft agar assay and in vivo tumorigenicity assays in nude mice were performed to detect the malignancy of MCLR-treated WRL68 cells [16].

2.2. Construction of WRL68 Cells with Stable Knockdown of LncGCLC

Small hairpin RNA (shRNA) was synthesized against lncGCLC by iGeneBio (Guangzhou, China). The shRNA, or scrambled control shRNA of lncGCLC, was cloned into vector psi-LVRH1GPF, named sh-lncGCLC and sh-NC, respectively. Then, the constructed plasmids were transfected into 293T cells, to collect viral particles using Lenti-Pac™ HIV packaging mix (GeneCopoeia, Inc., Guangzhou, China). The viral particles were then used to infect WRL68 cells, and screened by adding puromycin (10 μg/mL). The stably transfected cell line was confirmed by a fluorescence microscope and quantitative real-time polymerase chain reaction (qRT-PCR). The interference sequence of sh-lncGCLC and sh-NC are as follows: sh-lncGCLC, 5′-GCTTCTCCTCACTCCCAATTA-3′; sh-NC, 5′-GCTTCGCGCCGTAGTCTTA-3′. Then, the vector, or sh-lncGCLC-transfected cells, was separately treated with PBS or 10 μg/L of MCLR for 25 passages.

2.3. Tumor Tissue of HCC Patients with MC Exposure

HCC patients with an MC exposure were selected from our previous case-control study that involved 541 participants [26]. Briefly, thirty pairs of liver tumor tissue and adjacent normal tissue were collected during surgery on patients diagnosed with HCC from the Affiliated Tumor Hospital of Guangxi Medical University. Tissue samples were used to detect the expressions of lncGCLC and GCLC using qRT-PCR. Serum MC levels were determined using enzyme-linked immunosorbent assay (ELISA) kits (Beacon Analytical Systems Inc., Saco, ME, USA). The 30 participants were matched by age, sex, and hepatitis B virus infection status, and were divided into two groups according to the median values of serum MCs (0.14 μg/L) from all 541 HCC patients, termed the low MC exposure group and the high MC exposure group, respectively. Informed consent was obtained from patients prior to specimen collection, and all samples were snap-frozen in liquid nitrogen, immediately after excision, and stored at −80 °C until use.

2.4. RNA Extraction and qRT-PCR Analysis

Total RNA was isolated from cells and tissues using a TRIzol reagent (Invitrogen, Carlsbad, CA, USA), according to the manufacturer's instructions. RNA quality and concentration were measured using NanoDrop ND-1000 Spectrophotometer (Agilent, Santa Clara, CA, USA). Total RNA was reverse-transcribed using a PrimeScript1 RT Reagent Kit (TaKaRa, Kusatsu, Japan), and the SYBR Premix Ex Taq™ Kit (TaKaRa, Kusatsu, Japan) was then used to determine gene expression with gene-specific primers, according to the manufacturer's instructions. PCR reactions were performed on the StepOnePlus™ Real-Time PCR System (Applied Biosystems, Foster City, CA, USA), and the cycling conditions were as follows: 95 °C for 30 s, followed by 40 cycles of 95 °C for 5 s and 60 °C for 30 s. The PCR products were identified using a melting-curve analysis. Relative fold changes were calculated using the $2^{-\triangle\triangle Ct}$ method. The sequences of primers are shown in Supplementary Table S1.

2.5. Cell Proliferation Assay

A cell proliferation assay was performed using a Cell Counting Kit-8 (CCK-8; Dojindo, Japan), abiding by the manufacturer's protocols. Cells were seeded into 96-well plates at 5×10^3 cells per well. After incubation for 24, 48, 72, or 96 h, the culture medium was discarded and 110 μL of fresh medium containing 10 μL of CCK-8 reagent was added to each well. After incubation for 2 h, A450 was measured using a microplate reader (Thermo Fisher Scientific, Rochester, NY, USA).

2.6. Flow Cytometry Assay

For cell cycle analysis, cells were harvested and fixed overnight in 70% (*v*/*v*) ethanol at 4 °C. Subsequently, cells were stained with 1 mg/mL propidium iodide at 37 °C for 30 min after mixing with 10 mg/mL RNase and assessed immediately by flow cytometry (FACScan; BD Biosciences, Shanghai, China). To analyse cell apoptosis, cells were harvested and washed twice with ice-cold PBS, followed by resuspension in the binding buffer. Subsequently, 5 μL of Annexin V-APC and 10 μL of 7-AAD were added to 100 μL of the cell suspension at 4 °C for 15 min in the dark. Cell apoptosis was determined by flow cytometry (FACScan; BD Biosciences, Shanghai, China). Data on the cell cycle and apoptosis were analysed by FlowJo 7.6 (Treestar, Woodburn, OR, USA).

2.7. Transwell Migration and Invasion Assays

Migration or invasion of cells in each group was evaluated using 24-well Transwell chambers (BD Biosciences, Becton, NJ, USA) without or with Matrigel-coating, according to the manufacturer's instructions. Cells were harvested and adjusted to 1×10^5/mL of cell suspension using serum-free medium, containing 0.08% bovine serum albumin. Briefly, 200 μL cell suspension was added to the upper chamber of the Transwell (8-μm pore size). Subsequently, 2 × DMEM medium was poured into the bottom well. After incubating at 37 °C for 60 h, cells were fixed with 4% paraformaldehyde, and stained with 0.1% crystal violet. After removing non-migrated or non-invaded cells from the top of the membrane with cotton swabs, migrated or invaded cells from the bottom of the membrane were counted using an inverted microscope (Olympus, Tokyo, Japan), in three randomly selected fields (magnification; 100×).

2.8. Soft Agar Assay

The cells in the logarithmic growth phase were harvested and adjusted to a density of 5×10^4/mL. The agar (1.2% *w*/*v*, Sigma-Aldrich, St. Louis, MO, USA), in a small beaker, was completely dissolved using a microwave oven, cooled to 60 °C, and rapidly mixed with the same volume of 2 × DMEM medium (containing 20% FBS). The mixture was then injected into each well of 6-well plates. After solidification at room temperature, 5×10^3 cells per well of each group were suspended in 2 mL of 0.3% (*w*/*v*) agar medium, supplemented with 10% FBS and then seeded in the plate (three plates per group). These plates were incubated at 37 °C under 5% CO_2 for 15 days and stained with 0.005% crystal violet for 30 min. The colony formation efficiency was measured when clones contained more than 50 cells under an inverted microscope (Olympus, Tokyo, Japan). The rate of soft agar colony formation was calculated as follows: (number of colonies)/(number of cells seeded) × 100%.

2.9. Tumorigenicity Assays in Nude Mice

Thirty-six BALB/c athymic nude mice (4–5 weeks, male: female = 1:1) were purchased from the Guangxi Medical University Laboratory Animals Center. After a week of adaptive feeding in a pathogen-free room, mice were randomly divided into six groups (n = 3 for both male and female nude mice per group). The 25th passage of sh-lncGCLC-transfected cells, vector-transfected cells, and control cells (with or without MCLR treatment) were harvested after adjusting the cell suspension to 1×10^7/mL with PBS, then, 100 μL was injected subcutaneously into nude mice to evaluate the tumorigenicity of transformed cells. The tumor volume was measured every three days and calculated using the equation: $V = (LW^2)/2$ (L, longitudinal diameter; W, width diameter) [25,27]. After 22 days of injection, the mice were sacrificed, and tumors were harvested, weighed, and fixed with 4% paraformaldehyde for pathological examination.

2.10. Relative Expression of LncGCLC in the Nucleus and Cytoplasm

To establish the subcellular localization of lncGCLC, nuclear and cytoplasmic fractions were isolated from WRL68 cells using the Nuclear/Cytosol Fractionation Kit (BioVision,

Milpitas, MA, USA), and nuclear and cytoplasmic RNA was extracted. Nuclear and cytoplasmic RNA (800 ng) was then converted to cDNA and analyzed for the expression of lncGCLC by qRT-PCR.

2.11. Western Blots Analysis

Briefly, cells were harvested and lysed using the RIPA buffer (Beyotime, Shanghai, China). Total proteins were collected from the supernatant liquid of the cell homogenate after centrifuging, and the concentrations were measured by a BCA protein assay kit (Beyotime, Shanghai, China). The proteins were subjected to sodium dodecyl sulfate-polyacrylamide gel electrophoresis (SDS-PAGE) and were electrophoretically transferred to nitrocellulose filter membranes. Then, the membranes were blocked in 5% fat-free milk for 2 h and incubated with the appropriate primary antibodies overnight at 4 °C with gentle shaking. The primary antibodies used were GCLC (1:400, Abcam, Cambridge, UK) and β-actin (1:5000, Abcam, Cambridge, UK). Subsequently, the membranes were incubated with secondary antibodies (1:10^4; Beijing Bioss Biotechnology Co., Ltd., Beijing, China) for 2 h at room temperature. Finally, the protein bands were determined by Odyssey two-color infrared fluorescence imaging system (Li-cor, Ltd., Lincoln, NE, USA).

2.12. Determination of GCL, GSH, and 8-OHdG

The activity of GCL and the content of GSH were determined by the kits obtained from Nanjing Jiancheng Biology Engineering Institute (Nanjing, China). The content of 8-OHdG was measured by an ELISA kit, supplied by BlueGene Biotech Co., Ltd. (Shanghai, China). All procedures were performed according to the corresponding manufacturer's protocols.

2.13. Statistical Analysis

All statistical analyses were performed using SPSS 19.0 software (IBM Corp., Armonk, NY, USA). Quantitative data are presented as the mean ± standard deviation (SD). Differences between any two groups were analyzed by a paired *t*-test or an independent *t*-test. Differences among multiple groups were assessed using one-way analysis of variance analysis (ANOVA). If the ANOVA was statistically significant, the student-Newman-Keuls test would be used to analyze the differences between the two groups. A value of $p < 0.05$ was statistically significant.

3. Results

3.1. LncGCLC Is Downregulated in MCLR-Transformed WRL68 Cells and HCC Tissues with MC Exposure

We investigated whether lncGCLC is differentially expressed in MCLR-transformed cells, liver cancer cell lines, and patient samples. Continuously decreased levels of lncGCLC were observed in WRL68 cells after exposure to MCLR for 10 passages, compared to the passage-matched control cells (Figure 1A). Using qRT-PCR, we examined the expression of lncGCLC in two cell lines derived from human liver cancers (HepG2 and SMMC7721). LncGCLC expression of the two liver cancer cell lines was much lower than that in normal human liver cell lines WRL68 (Figure 1B). Next, the detection of lncGCLC in liver tissue of HCC patients with MC exposure was performed to further determine the relationship between MC exposure and lncGCLC expression. The results are shown in Figure 1C,D; the lncGCLC expression of tumor tissue was significantly lower than that of the matched normal tissues. Compared to the low MC exposure group, the lncGCLC expression of tumor tissue was significantly downregulated in high MC exposure group. As shown in Supplementary Table S2, there was no significant difference between the groups divided by the clinicopathologic factors, including tumor size, differentiation, lymph node metastasis, and Barcelona Clinic Liver Cancer (BCLC) stages.

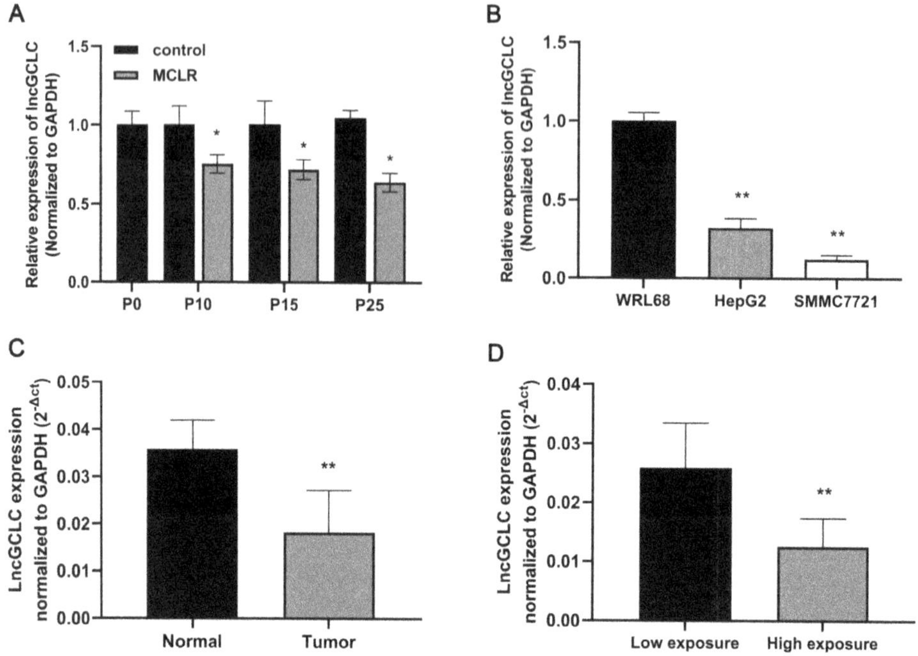

Figure 1. The lncGCLC expression in MCLR-exposed cells and population samples. (**A**) Changes of lncGCLC expression in WRL68 cells after exposure to 0 or 10 μg/L of MCLR for 0, 10, 15, and 25 passages. * $p < 0.05$ compared with passage-matched control cells. (**B**) Comparison of lncGCLC expression in WRL68, HepG2, and SMMC7721 cells. ** $p < 0.01$ compared with WRL68 cells. (**C**) Comparison of lncGCLC expression in tumor tissue and adjacent normal tissue from HCC patients with MC exposure (n = 30), ** $p < 0.01$. (**D**) LncGCLC expression in HCC patients with high MC exposure (serum MCs ≥ 0.14 μg/L, n = 17) was lower than that in those with low MC exposure (serum MCs < 0.14 μg/L, n = 13), ** $p < 0.01$. Data are presented as the mean ± SD of three independent experiments.

3.2. Knockdown of lncGCLC Promoted MCLR-Induced Cell Proliferation

To further clarify the role of lncGCLC in MCLR-induced malignant transformation, lncGCLC was continuously knocked down in WRL68 cells (passage 0) using shRNA. Because lncGCLC is a 3653-bp non-coding RNA, the success rate of construction of lentiviral vector overexpressing lncGCLC is relatively low, thus knockdown of lncGCLC was used to confirm its function. As shown in Figure 2A, a low expression level of lncGCLC was detected in the WRL68 cells stably transfected with lncGCLC shRNA (sh-lncGCLC), using qRT-PCR. In contrast, there was no significant difference in the expression level of lncGCLC between the control cells and empty vector-transfected cells. Next, the vector- or sh-lncGCLC-transfected cells were separately exposed to 10 μg/L of MCLR for 25 passages. MCLR downregulated the expression of lncGCLC in the empty vector-transfected cells, and the alteration was more pronounced by lncGCLC knockdown. CCK-8 assays indicated that the proliferation rate of the MCLR-treated cells and lncGCLC knockdown cells at 25 passages were significantly faster than that of the control (Figure 2B). Moreover, knockdown of lncGCLC significantly promoted cell proliferation induced by MCLR. These results further proved the effect of lncGCLC knockdown on cell proliferation.

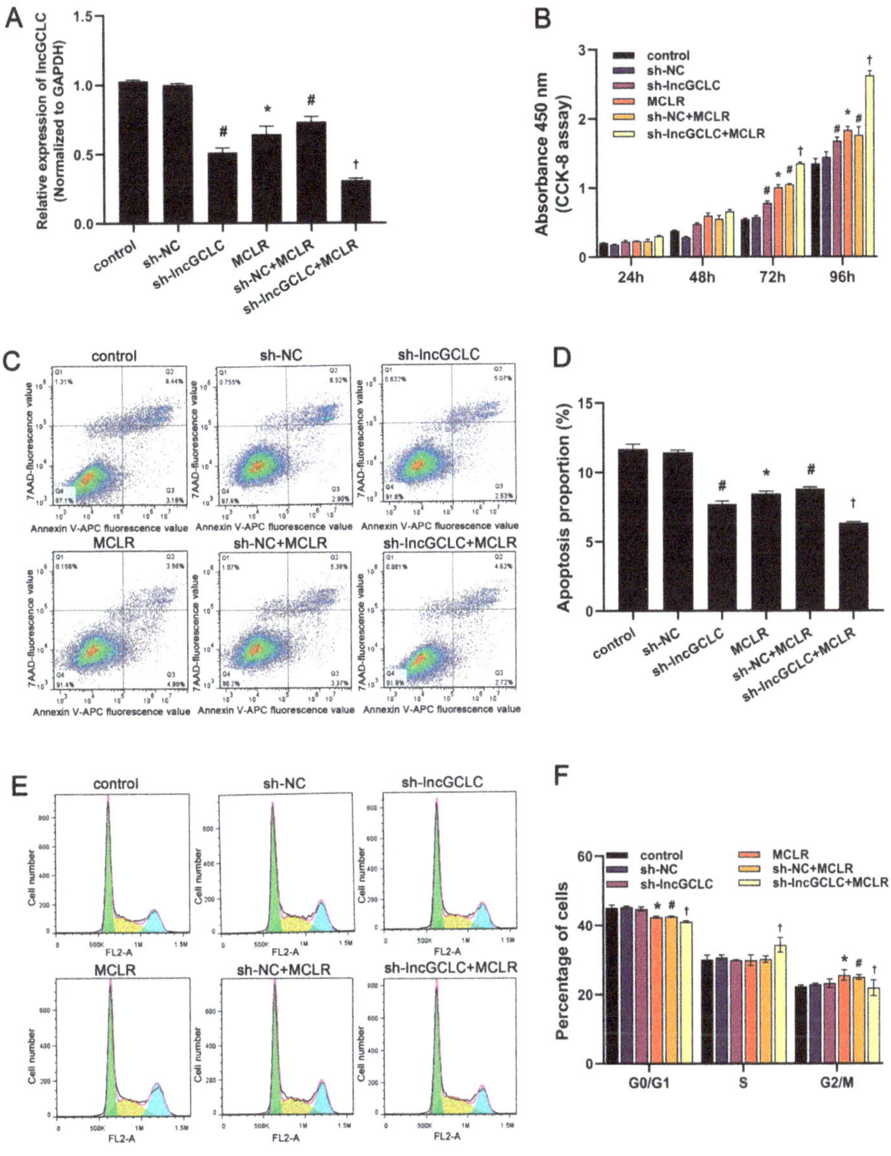

Figure 2. Knockdown of lncGCLC promoted the proliferation of MCLR-treated WRL68 cells. Vector- or sh-lncGCLC-transfected cells were treated with or without 10 μg/L of MCLR for 25 passages. (**A**) The expression of lncGCLC was detected using qRT-PCR. (**B**) Cell proliferation activity was detected by CCK-8. (**C**) Cell apoptosis was analyzed using flow cytometry. (**D**) Apoptosis rate in each group. (**E**) The percentage distribution of cells in the G1/G0, S, and G2/M phases of the cell cycle was determined by flow cytometry. (**F**) Cell cycle distribution quantification. Data are presented as the means ± SD of three independent experiments in each group. * $p < 0.05$ compared with the control group; # $p < 0.05$ compared with the sh-NC group; † $p < 0.05$ compared with the sh-NC + MCLR group.

Because lncGCLC affects cell proliferation, we hypothesized that it may do so by affecting the cell cycle and/or apoptosis. The apoptotic rate in sh-lncGCLC-transfected cells, following treatment with MCLR, was significantly lower than in vector-transfected cells treated with MCLR, whereas the frequency of apoptosis in vector-transfected cells was similar to that of the control cells (Figure 2C,D). Knockdown of lncGCLC significantly inhibited cell apoptosis. As shown in Figure 2E,F, the percentage of WRL68 cells in the G0/G1 phase was significantly reduced, while that in the G2/M phases was significantly increased after treatment with MCLR for 25 passages. Examination of cell cycle changes in sh-lncGCLC-transfected cells, following treatment with MCLR, disclosed an evident increase in the percentage of cells in the S phase and a concomitant decrease in the percentage of cells in the G0/G1 and G2/M phases. These results suggested that lncGCLC affected cell proliferation by functioning as a cell cycle checkpoint.

3.3. Knockdown of lncGCLC Promoted MCLR-Induced Cell Migration and Invasion

To determine whether lncGCLC has any effect on the metastatic capacity of MCLR-induced malignant transformed cells, we examined the effects of lncGCLC knockdown on the motility of MCLR-untreated or -treated cells with a Transwell assay. The Transwell experiment found that the number of cells passing through the transwell membranes significantly increased after knockdown of lncGCLC (Figure 3A–D). These results further indicated that the expression of lncGCLC could promote the migration and invasion of malignantly transformed WRL68 cells induced by MCLR.

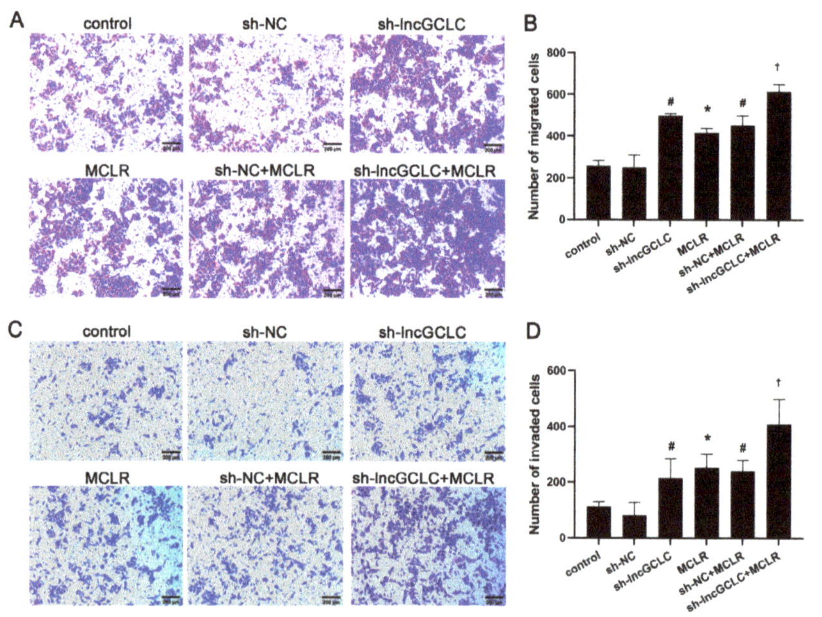

Figure 3. Knockdown of lncGCLC promoted the migration and invasion of MCLR-treated WRL68 cells. Vector- or sh-lncGCLC-transfected cells were treated with or without 10 μg/L of MCLR for 25 passages. (**A**) Representative images of a cell migration assay (100×), scale bar = 200 μm. (**B**) Quantification of cell migration. (**C**) Representative images of a cell invasion assay (Original magnification ×100, scale bar = 200 μm). (**D**) Quantification of cell invasion. Data are presented as the means ± SD of three independent experiments in each group. * $p < 0.05$ compared with the control group; # $p < 0.05$ compared with the sh-NC group; † $p < 0.05$ compared with the sh-NC + MCLR group.

3.4. Knockdown of lncGCLC Promoted the Growth of Malignantly Transformed WRL68 Cells, Induced by MCLR in Nude Mice

The results of the colony forming assay experiment showed that anchorage -independent growth capacity of MCLR-transformed cells was significantly increased after knockdown of lncGCLC (Figure 4A,B). We performed a tumor xenograft experiment in nude mice to verify whether lncGCLC can promote the growth of tumor in vivo. In the group injected with sh-lncGCLC-transfected cells, MCLR-treated cells, vector-transfected cells with MCLR treatment or sh-lncGCLC-transfected cells with MCLR treatment, tumor incidence was 100%. In contrast, in the group injected with control cells or vector-transfected cells, none of the mice developed tumors, with 0% of tumor incidence. Moreover, the tumors of the mice injected with sh-lncGCLC-transfected cells, following treatment with MCLR, were significantly larger than those with vector-transfected cells, following treatment with MCLR (Figure 4C–F). This promoted the effect of lncGCLC knockdown on tumor growth consistent with our in vitro results. Tumors in nude mice induced by MCLR-transformed cells were composed of poorly differentiated hepatocytes (Figure 4G). Though the tumors tended to grow faster in male compared to female mice in the group injected with MCLR-transformed cells, the differences did not reach statistically significant level (Supplementary Figure S1).

3.5. lncGCLC Regulates the Expression of GCLC in MCLR-Induced Malignant Transformation

Following the observation of the role of lncGCLC in MCLR-induced malignant transformation of WRL68 cells, we attempted to analyze the underlying mechanism. Recent studies have found that some lncRNAs act in cis, therefore they can regulate the expression of one or more nearby genes on the same chromosome [23,28]. We investigated whether lncGCLC acts in cis. The expression of five nearby genes, GCLC, ELOVL5, TMEM14A, ICK, and FBXO9, extending across approximately 1.5 Mb downstream of lncGCLC, was detected by qRT-PCR after the knockdown of lncGCLC by shRNA. As shown in Figure 5A, only GCLC mRNA expression in sh-lncGCLC-transfected cells was significantly lower than that in vector control cells, suggesting that lncGCLC may regulate GCLC expression.

We also determined the cytoplasmic/nuclear distributions of lncGCLC. As shown in Figure 5B, the lncGCLC was located both in the cytoplasm and the nucleus, and the expression level in the cytoplasm was higher than that in the nucleus, indicating that lncGCLC may function as a regulator at the post-transcriptional level. Bioinformatics analysis employed to predict miRNAs that potentially interact with lncGCLC (http://starbase.sysu.edu.cn and http://bioinfo.life.hust.edu.cn/LNCediting, accessed on 20 December 2022) revealed direct interactions with miR-122-5p.

To further validate interactions among lncGCLC, miR-122-5p and GCLC, we analyzed lncGCLC and GCLC expression patterns in HCC tissue with MC exposure. Our data suggested that lncGCLC and GCLC were lowly expressed in hepatocellular carcinoma but also positively correlated (Figure 5C). The GCLC expression significantly decreased in tumor tissue compared with normal liver tissue from HCC patients with MC exposure (Figure 5D). Furthermore, the GCLC expression in the high MC exposure group was significantly lower than those in the low MC exposure group (Figure 5E). With prolonged exposure time, expression of miR-122-5p increased and that of GCLC mRNA decreased in MCLR-treated cells (Figure 5F). Western blot data showed that GCLC protein expression was gradually decreased with cell passage during MCLR-induced malignant transformation (Figure 5G,H). As shown in Figure 5I–L, the knockdown of lncGCLC significantly upregulated the expression of miR-122-5p and suppressed MCLR-induced inhibition of GCLC at both mRNA and protein levels.

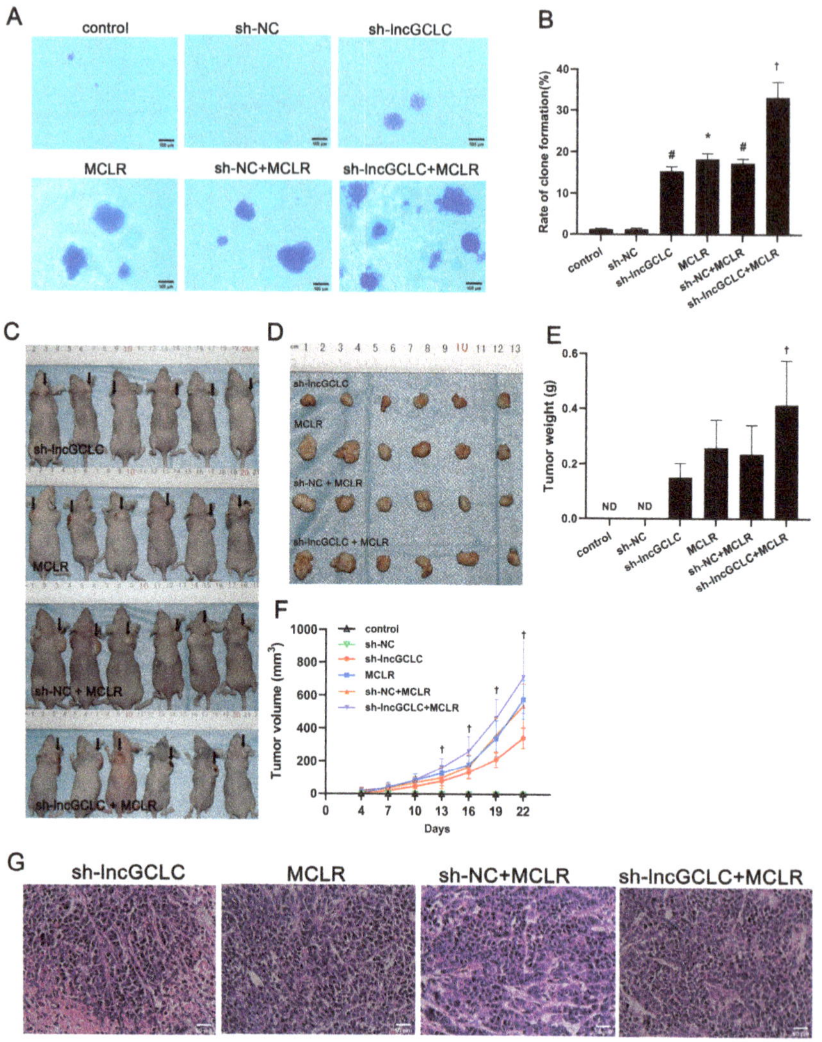

Figure 4. Knockdown of lncGCLC promoted the growth of MCLR-induced malignantly transformed WRL68 cells in nude mice. Vector- or sh-lncGCLC-transfected cells were treated with or without 10 μg/L of MCLR for 25 passages. (**A**) Representative images of a soft agar assay (Original magnification ×200, scale bar = 100 μm). (**B**) Quantification of colony formation in soft agar. (**C**) Tumorigenicity test of BALB/c nude mice was performed. Solid tumors were removed after the sacrifice at 22 days. (**D**) Representative image showing subcutaneous tumor size. (**E**) The weights of solid tumors. The values given are mean ± SD (n = 3 for both male and female nude mice per group). (**F**) Tumor volume was monitored every 3 days after injection of WRL68 cells. (**G**) Pathological changes of the tumor tissue in nude mice in each group (HE stains, original magnification ×400, scale bar = 50 μm). ND—not detected. Data are presented as the means ± SD of three independent experiments in each group. * $p < 0.05$ compared with the control group; # $p < 0.05$ compared with the sh-NC group; † $p < 0.05$ compared with the sh-NC + MCLR group.

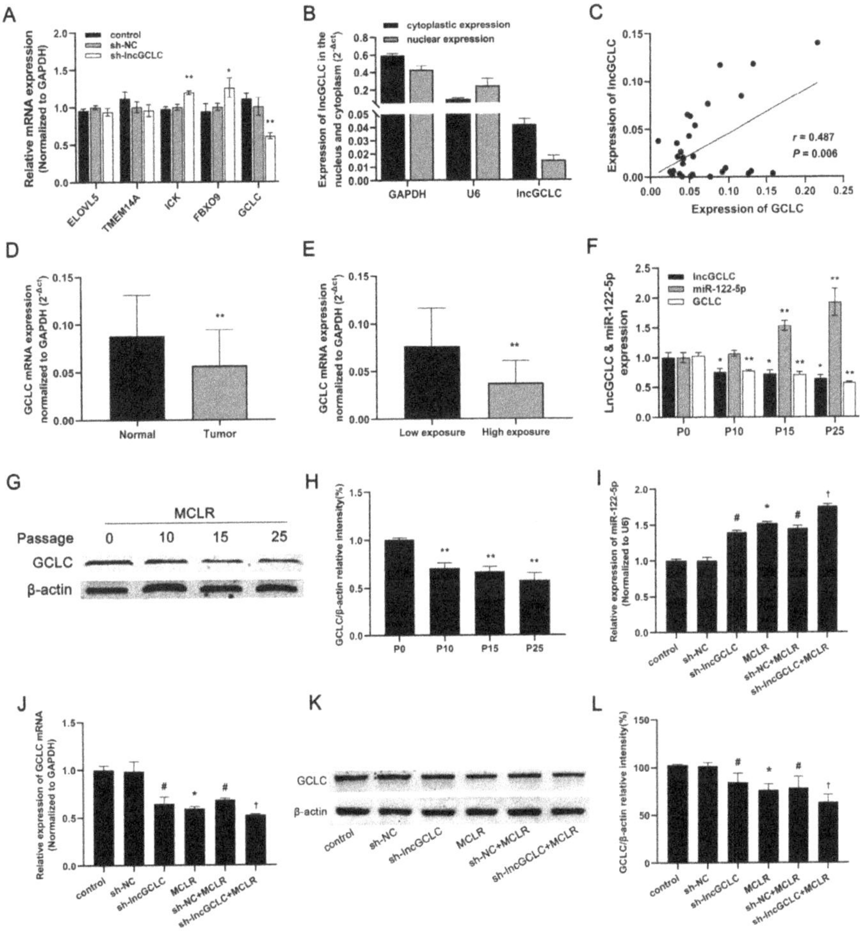

Figure 5. Verification of the interrelationships among lncGCLC, miR-122-5p, and GCLC. (**A**) The mRNA expression level of neighboring genes which are located nearly 1.5 Mb downstream of lncGCLC in lncGCLC knockdown cells. * $p < 0.05$ and ** $p < 0.01$ compared with vector control cells. (**B**) The expression level of lncGCLC in the nuclear and cytoplasmic fractions of WRL68 cells. (**C**) The correlation between the lncGCLC expression and the GCLC expression in HCC tissues (n = 30). Correlation coefficient (r) and P value were calculated by Pearson correlation analysis. (**D**) Comparison of GCLC expression in tumor tissue and adjacent normal tissue from HCC patients with MC exposure (n = 30), ** $p < 0.01$. (**E**) GCLC expression in HCC patients with high MC exposure (serum MCs $\geq$ 0.14 μg/L, n =17) was lower than in those with low MC exposure (serum MCs < 0.14 μg/L, n = 13), ** $p < 0.01$. (**F**) After exposure to 10 μg/L of MCLR for 0, 10, 15, and 25 passages, expression of lncGCLC, miR-122-5p and GCLC mRNA was detected in WRL68 cells. * $p < 0.05$ and ** $p < 0.01$ compared with passage-matched control cells. (**G**,**H**) Expression of GCLC protein in P0, P10, P15 and P25 MCLR-induced malignantly transformed WRL68 cells. ** $p < 0.01$ compared with passage-matched control cells. (**I**) Changes of miR-122-5p in both vector- and sh-lncGCLC-transfected WRL68 cells treated with 0 or 10 μg/L of MCLR for 25 passages. (**J–L**) Changes of GCLC mRNA (**J**) and protein (**K**,**L**) expression in both vector- and sh-lncGCLC-transfected WRL68 cells treated with or without 10 μg/L of MCLR for 25 passages. Data are presented as the means ± SD of three independent experiments in each group. * $p < 0.05$ compared with the control group; $^{\#}$ $p < 0.05$ compared with the sh-NC group; † $p < 0.05$ compared with the sh-NC + MCLR group.

3.6. Knockdown of lncGCLC Reduces GSH Levels and Enhances Oxidative DNA Damages

To further test whether knockdown of lncGCLC is involved in redox regulation, we measured the GCL activity, GSH, and 8-OHdG contents in vector- or sh-lncGCLC-transfected cells exposed to MCLR for 25 passages. As shown in Figure 6A,B, continuous treatment of MCLR for 25 passages led to an evident decrease in GCL activities and GSH levels in the empty vector-transfected cells. On the contrary, the 8-OHdG levels were increased after exposure to MCLR for 25 passages (Figure 6C). Knockdown of lncGCLC significantly reduced GCL activities and GSH levels, and increased 8-OHdG levels.

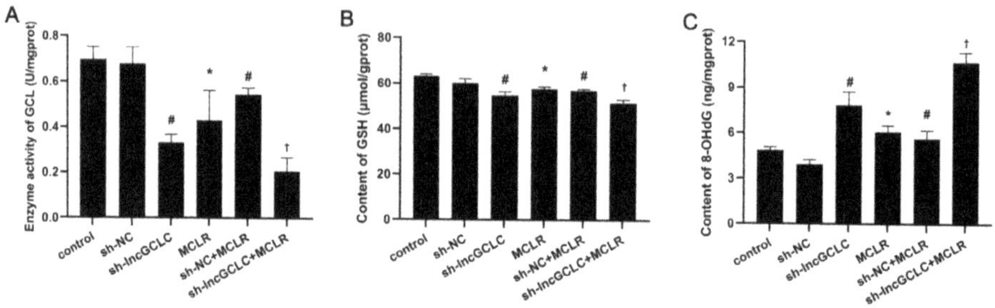

Figure 6. Knockdown of lncGCLC reduced GSH levels and induced oxidative DNA damage in MCLR-treated WRL68 cells. Alterations in GCL activity (**A**), GSH (**B**), and 8-OHdG content (**C**) in vector- or sh-lncGCLC-transfected cells exposed to 0 or 10 μg/L of MCLR for 25 passages. Data are presented as the means ± SD of three independent experiments in each group. * $p < 0.05$ compared with the control group; # $p < 0.05$ compared with the sh-NC group; † $p < 0.05$ compared with the sh-NC + MCLR group.

4. Discussion

Although many lncRNAs have been reported to link with liver cancer development, little is known about their roles in MCLR-induced cell transformation. In this study, we conducted bioinformatics analysis and identified lncGCLC, a 3653 bp transcript without protein-coding potency, located on chromosome 6p12.1 (chr6: 53561289-53617007), upstream of the GCLC gene. Zhou et al. [29] previously reported that nickel exposure downregulated the expression of lncRNA MEG3 in a time- and dose-dependent manner, and low lncRNA MEG3 expression was shown to promote the malignant transformation of human bronchial epithelial cells induced by nickel. Reportedly, lncRNA-Dreh, lnc-DILC, and lncRNA-LET are poorly expressed in HCC cell lines and tissues, and act as tumor suppressors in the development of liver cancer [20,22,30]. We have shown that the expression level of lncGCLC in the liver cancer cell lines HepG2 and SMMC7721 was significantly lower than that in WRL68 cells. LncGCLC was significantly downregulated in HCC tissues compared with the adjacent non-HCC tissues. These results indicated that the downregulation of lncGCLC is a frequent event in liver cancer and lncGCLC may have an anti-oncogene-like function. In this study, we found that the expression of lncGCLC was continuously reduced during the process of MCLR-induced malignant transformation. The knockdown of lncGCLC caused a cell cycle arrest at the S Phase, inhibited cell apoptosis, and promoted cell proliferation, induced by MCLR. This finding implies that lncGCLC is involved in tumorigenesis through the inhibition of cell apoptosis and accelerated cell cycle progress, which is consistent with the knockdown of lncRNA MEG3, inhibiting apoptotic activity [31]. In addition, we found that knockdown of lncGCLC could effectively increase the capacities of invasion and migration of MCLR-transformed cells. More importantly, knockdown of lncGCLC could increase the degree of malignancy in vitro and effectively promote tumor growth in vivo. These results suggest that lncGCLC might play a pivotal role in the regulation of MCLR-induced malignant transformation.

LncRNAs could act as sponges to compete miRNAs with mRNAs, participating in various biological processes [32]. For instance, lncRNA BCAR4 may promote the proliferation, migration, and invasion of liver cancer cells by directly binding to miR-1261 and targeting the anaphase-promoting complex subunit 11 (ANAPC11) gene [33]. GCLC, the most important subunit of GCL, contains all substrate-binding sites, and catalytic activity of GCL and plays an important role in GSH biosynthesis [34]. Mougiakakos et al. [14] reported that the malignant phenotype of melanoma cells, including survival, invasiveness, and switch from E-cadherin to N-cadherin expression, was found at significantly higher levels in cells with a lower GCLC expression. Previously, we have found that GCLC expression was progressively reduced during MCLR-induced cell transformation. Overexpression of GCLC decreased the capacities of proliferation, migration and invasion of MCLR-transformed cells [16]. In the present study, we found that lncGCLC exerted an inhibitory action on carcinogenesis, similar to that exhibited by GCLC. The expressions of lncGCLC and GCLC in human HCC tissue were significantly downregulated in the high MC exposure group. LncGCLC expression level showed a positive relation with GCLC levels in HCC tissue. During MCLR-induced cell transformation, the expression of lncGCLC decreased continuously, accompanied with consistent low expression of GCLC, but high expression of miR-122-5p. Knockdown of lncGCLC significantly upregulated miR-122-5p expression, and suppressed GCLC expression. MiR-122-5p is one of the most abundant miRNAs in the liver, constituting 70% of all hepatic miRNAs [35]. It has been reported that the expression of miR-122-5p was significantly higher in the alpha-fetoprotein (AFP)-producing gastric cancer tissues and plasma samples [36]. MiR-122-5p promotes aggression and epithelial–mesenchymal transition in triple-negative breast cancer, by suppressing charged multivesicular body protein 3 (CHMP3) through mitogen-activated protein kinase (MAPK) signaling [37]. Therefore, we speculate that lncGCLC may regulate GCLC expression by acting as an endogenous competitive RNA for miR-122-5p. Further studies are required to clarify these mechanisms.

Recently, it has been reported that long non-coding RNAs are involved in the regulation of redox reactions. Zhang et al. [38] revealed that downregulation of lncRNA MAGI2-AS3 decreased the H_2O_2 content and delayed cell senescence, by stabilizing the HSPA8 protein level. LncRNA HCP5 was found to promote the stemness and chemoresistance of gastric cancer cells by driving fatty acid oxidation [39]. LncRNA LCPAT1 was reported to be involved in DNA damage induced by cigarette smoke extract through the RCC2 gene [40]. It has been shown that the GCLC primarily regulates de novo synthesis of glutathione and is central to the antioxidant capacity of the cell [41]. In the present study, the levels of GCL and GSH were significantly decreased in WRL68 cells after exposure to the low concentration of MCLR for 25 passages. On the contrary, levels of 8-OHdG were increased after exposure to MCLR for 25 passages. Knockdown of lncGCLC significantly reduced GCL activities and GSH levels, and increased 8-OHdG levels, suggesting that lncGCLC may downregulate GCLC expression, reduce GSH levels, and subsequently induce oxidative DNA damage, which may contribute to oncogenesis caused by low-level MCLR.

Among over 279 identified variants of MCs, MCLR is the most abundant and toxic. Several studies have also found that it is rather difficult to accurately determine the variants of MCs in human serum using the method of ELISA [2,42]. Currently, the commercial microcystin ELISA kit does not distinguish between MCLR and other MC variants (microcystin-RR, -YR, etc), but detects their presence at varying degrees. In this study, we used an ELISA method to detect serum MC levels for a rough estimation of MCLR exposure, which may not be able to accurately assess the internal exposure level of MCLR. Thus, there is a pronounced need to further investigate a more sensitive serum MCLR detection method and more appropriate biomarkers in exposed populations.

5. Conclusions

In conclusion, we identified a novel lncRNA, lncGCLC, which was downregulated in MCLR-transformed cells and tumor tissue of HCC patients. MCLR exposure-caused lncGCLC downregulation enhanced miR-122-5p expression, suppressed GCLC expression and GSH levels, and promoted oxidative DNA damages. Furthermore, the downregulation of lncGCLC enhanced cell proliferation, apoptosis resistance, migration and invasion, all of which promote malignant transformation of liver cells (Figure 7). More importantly, the expression of lncGCLC in human HCC tissues was significantly downregulated in the high MC exposure group, and positively associated with GCLC level in HCC tissues. Taken together, the current results suggest that lncGCLC plays an anti-oncogenic role in MCLR-induced malignant transformation by regulating GCLC expression. Thus, increasing the expression of lncGCLC by pharmacological agents may be a potential strategy to prevent the liver carcinogenesis of MCLR.

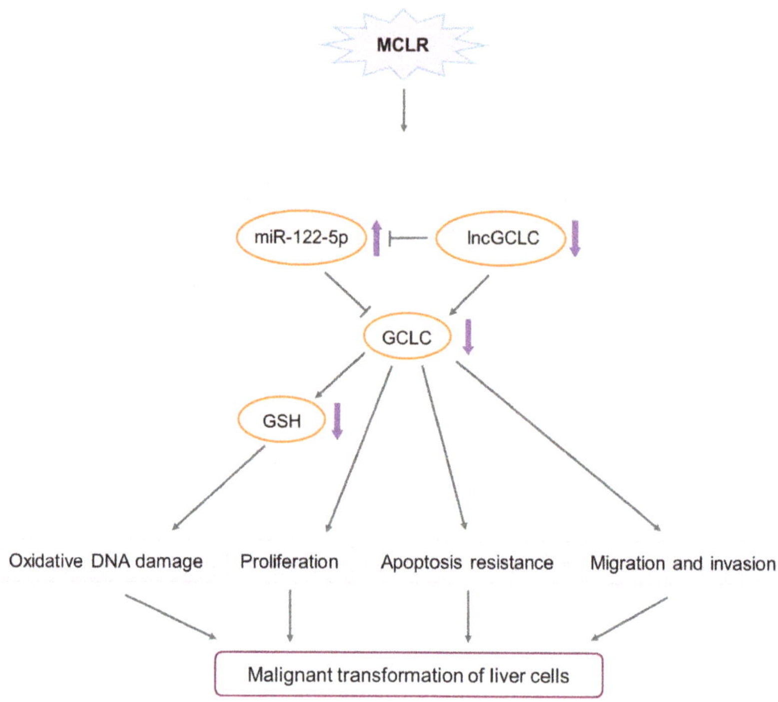

Figure 7. Schematic diagram for the role of lncGCLC pathway in MCLR-induced malignant transformation.

Supplementary Materials: The following supporting information can be downloaded at: https://www.mdpi.com/article/10.3390/toxics11020162/s1, Table S1: Primer sequences used in the experiment; Table S2: Clinicopathologic factors of hepatocellular carcinoma patients with MC exposure; Figure S1: Comparison of tumor weight (A) or volume (B) between male and female nude mice in each group after inoculation for 22 days. ND—not detected.

Author Contributions: Conceptualization, J.L. and Q.N.; Data curation, X.H., Z.S., J.H., N.Z. and L.N.; Funding acquisition, Q.N.; Investigation, J.H., N.Z., L.N. and H.Z.; Methodology, B.G. and Q.H.; Project administration, Q.N.; Supervision, Q.N.; Writing—original draft, X.H., Z.S., J.L. and Q.N.; Writing—review and editing, G.-D.L. and Q.N. All authors have read and agreed to the published version of the manuscript.

Funding: The research was funded by the National Natural Science Foundation of China (grant numbers: 81960582 and 81660529).

Institutional Review Board Statement: The study was conducted according to the guidelines of the Declaration of Helsinki, and was approved by the Medical Ethics Committee of Guangxi Medical University (GXMU-20160303-9).

Informed Consent Statement: Informed consent was obtained from all subjects involved in the study.

Data Availability Statement: The data are currently not publicly available, as the data also form part of an ongoing study.

Acknowledgments: We thank all study participants, research staff, and students who participated in this study.

Conflicts of Interest: The authors declare that they have no competing financial interest.

References

1. Chen, L.; Chen, J.; Zhang, X.; Xie, P. A review of reproductive toxicity of microcystins. *J. Hazard. Mater.* **2016**, *301*, 381–399. [CrossRef] [PubMed]
2. Chen, L.; Giesy, J.P.; Adamovsky, O.; Svirčev, Z.; Meriluoto, J.; Codd, G.A.; Mijovic, B.; Shi, T.; Tuo, X.; Li, S.; et al. Challenges of using blooms of *Microcystis* spp. in animal feeds: A comprehensive review of nutritional, toxicological and microbial health evaluation. *Sci. Total Environ.* **2021**, *764*, 142319. [CrossRef] [PubMed]
3. McGlynn, K.A.; Petrick, J.L.; El-Serag, H.B. Epidemiology of Hepatocellular Carcinoma. *Hepatology* **2021**, *73* (Suppl. 1), 4–13. [CrossRef] [PubMed]
4. Ueno, Y.; Nagata, S.; Tsutsumi, T.; Hasegawa, A.; Watanabe, M.F.; Park, H.D.; Chen, G.C.; Chen, G.; Yu, S.Z. Detection of microcystins, a blue-green algal hepatotoxin, in drinking water sampled in Haimen and Fusui, endemic areas of primary liver cancer in China, by highly sensitive immunoassay. *Carcinogenesis* **1996**, *17*, 1317–1321. [CrossRef]
5. Zheng, C.; Zeng, H.; Lin, H.; Wang, J.; Feng, X.; Qiu, Z.; Chen, J.; Luo, J.; Luo, Y.; Huang, Y.; et al. Serum microcystin levels positively linked with risk of hepatocellular carcinoma: A case-control study in southwest China. *Hepatology* **2017**, *66*, 1519–1528. [CrossRef]
6. Svircev, Z.; Drobac, D.; Tokodi, N.; Luzanin, Z.; Munjas, A.M.; Nikolin, B.; Vuleta, D.; Meriluoto, J. Epidemiology of cancers in Serbia and possible connection with cyanobacterial blooms. *J. Environ. Sci. Health C Environ. Carcinog. Ecotoxicol. Rev.* **2014**, *32*, 319–337. [CrossRef]
7. Zhang, F.; Lee, J.; Liang, S.; Shum, C.K. Cyanobacteria blooms and non-alcoholic liver disease: Evidence from a county level ecological study in the United States. *Environ. Health* **2015**, *14*, 41. [CrossRef]
8. Grosse, Y.; Baan, R.; Straif, K.; Secretan, B.; El Ghissassi, F.; Cogliano, V. Carcinogenicity of nitrate, nitrite, and cyanobacterial peptide toxins. *Lancet Oncol.* **2006**, *7*, 628–629. [CrossRef]
9. Valavanidis, A.; Vlachogianni, T.; Fiotakis, C. 8-hydroxy-2′ -deoxyguanosine (8-OHdG): A critical biomarker of oxidative stress and carcinogenesis. *J. Environ. Sci. Health C Environ. Carcinog. Ecotoxicol. Rev.* **2009**, *27*, 120–139. [CrossRef]
10. Li, X.; Zhao, Q.; Zhou, W.; Xu, L.; Wang, Y. Effects of chronic exposure to microcystin-LR on hepatocyte mitochondrial DNA replication in mice. *Environ. Sci. Technol.* **2015**, *49*, 4665–4672. [CrossRef]
11. Liu, K.; Zhao, X.; Guo, M.; Zhu, J.; Li, D.; Ding, J.; Han, X.; Wu, J. Microcystin-leucine arginine (MC-LR) induces mouse ovarian inflammation by promoting granulosa cells to produce inflammatory cytokine via activation of cGAS-STING signaling. *Toxicol. Lett.* **2022**, *358*, 6–16. [CrossRef]
12. Chen, X.; Xue, H.; Fang, W.; Chen, K.; Chen, S.; Yang, W.; Shen, T.; Chen, X.; Zhang, P.; Ling, W. Adropin protects against liver injury in nonalcoholic steatohepatitis via the Nrf2 mediated antioxidant capacity. *Redox Biol.* **2019**, *21*, 101068. [CrossRef]
13. Kim, S.H.; Yoon, D.; Lee, Y.-H.; Lee, J.; Kim, N.D.; Kim, S.; Jung, Y.-S. Transformation of liver cells by 3-methylcholanthrene potentiates oxidative stress via the downregulation of glutathione synthesis. *Int. J. Mol. Med.* **2017**, *40*, 2011–2017. [CrossRef]
14. Mougiakakos, D.; Okita, R.; Ando, T.; Dürr, C.; Gadiot, J.; Ichikawa, J.; Zeiser, R.; Blank, C.; Johansson, C.C.; Kiessling, R. High expression of GCLC is associated with malignant melanoma of low oxidative phenotype and predicts a better prognosis. *J. Mol. Med.* **2012**, *90*, 935–944. [CrossRef]
15. Sun, Y.; Kojima, C.; Chignell, C.; Mason, R.; Waalkes, M.P. Arsenic transformation predisposes human skin keratinocytes to UV-induced DNA damage yet enhances their survival apparently by diminishing oxidant response. *Toxicol. Appl. Pharmacol.* **2011**, *255*, 242–250. [CrossRef]
16. Jia, X.; Guan, B.; Liao, J.; Hu, X.; Fan, Y.; Li, J.; Zhao, H.; Huang, Q.; Ma, Z.; Zhu, X.; et al. Down-regulation of GCLC is involved in microcystin-LR-induced malignant transformation of human liver cells. *Toxicology* **2019**, *421*, 49–58. [CrossRef]
17. Gao, L.; Mai, A.; Li, X.; Lai, Y.; Zheng, J.; Yang, Q.; Wu, J.; Nan, A.; Ye, S.; Jiang, Y. LncRNA-DQ786227-mediated cell malignant transformation induced by benzo(a)pyrene. *Toxicol. Lett.* **2013**, *223*, 205–210. [CrossRef]
18. Liu, Z.; Liu, A.; Nan, A.; Cheng, Y.; Yang, T.; Dai, X.; Chen, L.; Li, X.; Jia, Y.; Zhang, N.; et al. The linc00152 Controls Cell Cycle Progression by Regulating CCND1 in 16HBE Cells Malignantly Transformed by Cigarette Smoke Extract. *Toxicol. Sci.* **2019**, *167*, 496–508. [CrossRef]
19. Lin, H.-P.; Rea, M.; Wang, Z.; Yang, C. Down-regulation of lncRNA MEG3 promotes chronic low dose cadmium exposure-induced cell transformation and cancer stem cell-like property. *Toxicol. Appl. Pharmacol.* **2021**, *430*, 115724. [CrossRef]
20. Yang, F.; Huo, X.-S.; Yuan, S.-X.; Zhang, L.; Zhou, W.-P.; Wang, F.; Sun, S.-H. Repression of the Long Noncoding RNA-LET by Histone Deacetylase 3 Contributes to Hypoxia-Mediated Metastasis. *Mol. Cell* **2013**, *49*, 1083–1096. [CrossRef]
21. Zhou, C.-C.; Yang, F.; Yuan, S.-X.; Ma, J.-Z.; Liu, F.; Yuan, J.-H.; Bi, F.-R.; Lin, K.-Y.; Yin, J.-H.; Cao, G.-W.; et al. Systemic genome screening identifies the outcome associated focal loss of long noncoding RNA PRAL in hepatocellular carcinoma. *Hepatology* **2016**, *63*, 850–863. [CrossRef] [PubMed]

22. Wang, X.; Sun, W.; Shen, W.; Xia, M.; Chen, C.; Xiang, D.; Ning, B.; Cui, X.; Li, H.; Li, X.; et al. Long non-coding RNA DILC regulates liver cancer stem cells via IL-6/STAT3 axis. *J. Hepatol.* **2016**, *64*, 1283–1294. [CrossRef] [PubMed]
23. Joung, J.; Engreitz, J.M.; Konermann, S.; Abudayyeh, O.O.; Verdine, V.K.; Aguet, F.; Gootenberg, J.S.; Sanjana, N.E.; Wright, J.B.; Fulco, C.P.; et al. Genome-scale activation screen identifies a lncRNA locus regulating a gene neighbourhood. *Nature* **2017**, *548*, 343–346. [CrossRef] [PubMed]
24. Xu, L.; Qin, W.; Zhang, H.; Wang, Y.; Dou, H.; Yu, D.; Ding, Y.; Yang, L.; Wang, Y. Alterations in microRNA expression linked to microcystin-LR-induced tumorigenicity in human WRL-68 Cells. *Mutat. Res.* **2012**, *743*, 75–82. [CrossRef] [PubMed]
25. Chen, H.-Q.; Zhao, J.; Li, Y.; He, L.-X.; Huang, Y.; Shu, W.-Q.; Cao, J.; Liu, W.-B.; Liu, J.-Y. Gene expression network regulated by DNA methylation and microRNA during microcystin-leucine arginine induced malignant transformation in human hepatocyte L02 cells. *Toxicol. Lett.* **2018**, *289*, 42–53. [CrossRef]
26. Wang, Y.; Huang, Q.; Huang, X.; Zhao, H.; Guan, B.; Ban, K.; Zhu, X.; Ma, Z.; Tang, Y.; Su, Z.; et al. Genetic Variant of PP2A Subunit Gene Confers an Increased Risk of Primary Liver Cancer in Chinese. *Pharm. Pers. Med.* **2021**, *14*, 1565–1574. [CrossRef]
27. Du, X.; Liu, H.; Tian, Z.; Zhang, S.; Shi, L.; Wang, Y.; Guo, X.; Zhang, B.; Yuan, S.; Zeng, X.; et al. PI3K/AKT/mTOR pathway mediated-cell cycle dysregulation contribute to malignant proliferation of mouse spermatogonia induced by microcystin-leucine arginine. *Environ. Toxicol.* **2022**, *38*, 343–358. [CrossRef]
28. Rinn, J.L.; Chang, H.Y. Genome Regulation by Long Noncoding RNAs. *Annu. Rev. Biochem.* **2012**, *81*, 145–166. [CrossRef]
29. Zhou, C.; Huang, C.; Wang, J.; Huang, H.; Li, J.; Xie, Q.; Liu, Y.; Zhu, J.; Li, Y.; Zhang, D.; et al. LncRNA MEG3 downregulation mediated by DNMT3b contributes to nickel malignant transformation of human bronchial epithelial cells via modulating PHLPP1 transcription and HIF-1alpha translation. *Oncogene* **2017**, *36*, 3878–3889. [CrossRef]
30. Huang, J.-F.; Guo, Y.-J.; Zhao, C.-X.; Yuan, S.-X.; Wang, Y.; Tang, G.-N.; Zhou, W.-P.; Sun, S.-H. Hepatitis B virus X protein (HBx)-related long noncoding RNA (lncRNA) down-regulated expression by HBx (Dreh) inhibits hepatocellular carcinoma metastasis by targeting the intermediate filament protein vimentin. *Hepatology* **2013**, *57*, 1882–1892. [CrossRef]
31. Zhang, X.; Wu, N.; Wang, J.; Li, Z. LncRNA MEG3 inhibits cell proliferation and induces apoptosis in laryngeal cancer via miR-23a/APAF-1 axis. *J. Cell. Mol. Med.* **2019**, *23*, 6708–6719. [CrossRef]
32. Zhou, S.; Wang, L.; Yang, Q.; Liu, H.; Meng, Q.; Jiang, L.; Wang, S.; Jiang, W. Systematical analysis of lncRNA–mRNA competing endogenous RNA network in breast cancer subtypes. *Breast Cancer Res. Treat.* **2018**, *169*, 267–275. [CrossRef]
33. Zhang, Y.; Zhou, H. LncRNA BCAR4 promotes liver cancer progression by upregulating ANAPC11 expression through sponging miR-1261. *Int. J. Mol. Med.* **2020**, *46*, 159–166. [CrossRef]
34. Lu, S.C. Regulation of glutathione synthesis. *Mol. Asp. Med.* **2009**, *30*, 42–59. [CrossRef]
35. Jopling, C. Liver-specific microRNA-122: Biogenesis and function. *RNA Biol.* **2012**, *9*, 137–142. [CrossRef]
36. Maruyama, S.; Furuya, S.; Shiraishi, K.; Shimizu, H.; Akaike, H.; Hosomura, N.; Kawaguchi, Y.; Amemiya, H.; Kawaida, H.; Sudo, M.; et al. miR-122-5p as a novel biomarker for alpha-fetoprotein-producing gastric cancer. *World J. Gastrointest. Oncol.* **2018**, *10*, 344–350. [CrossRef]
37. Wang, Z.; Wang, X. miR-122-5p promotes aggression and epithelial-mesenchymal transition in triple-negative breast cancer by suppressing charged multivesicular body protein 3 through mitogen-activated protein kinase signaling. *J. Cell. Physiol.* **2020**, *235*, 2825–2835. [CrossRef]
38. Zhang, Y.; Qiao, X.; Liu, L.; Han, W.; Liu, Q.; Wang, Y.; Xie, T.; Tang, Y.; Wang, T.; Meng, J.; et al. Long noncoding RNA MAGI2-AS3 regulates the H_2O_2 level and cell senescence via HSPA8. *Redox Biol.* **2022**, *54*, 102383. [CrossRef]
39. Wu, H.; Liu, B.; Chen, Z.; Li, G.; Zhang, Z. MSC-induced lncRNA HCP5 drove fatty acid oxidation through miR-3619-5p/AMPK/PGC1α/CEBPB axis to promote stemness and chemo-resistance of gastric cancer. *Cell Death Dis.* **2020**, *11*, 233. [CrossRef]
40. Gao, S.; Lin, H.; Yu, W.; Zhang, F.; Wang, R.; Yu, H.; Qian, B. LncRNA LCPAT1 is involved in DNA damage induced by CSE. *Biochem. Biophys. Res. Commun.* **2019**, *508*, 512–515. [CrossRef]
41. Dickinson, D.A.; Levonen, A.-L.; Moellering, D.; Arnold, E.K.; Zhang, H.; Darley-Usmar, V.; Forman, H.J. Human glutamate cysteine ligase gene regulation through the electrophile response element. *Free. Radic. Biol. Med.* **2004**, *37*, 1152–1159. [CrossRef] [PubMed]
42. Chen, J.; Xie, P.; Li, L.; Xu, J. First identification of the hepatotoxic microcystins in the serum of a chronically exposed human population together with indication of hepatocellular damage. *Toxicol. Sci.* **2009**, *108*, 81–89. [CrossRef] [PubMed]

toxics

MDPI

Systematic Review

Association of Dietary Nitrate, Nitrite, and N-Nitroso Compounds Intake and Gastrointestinal Cancers: A Systematic Review and Meta-Analysis

Monireh Sadat Seyyedsalehi [1,2], Elham Mohebbi [2,3], Fatemeh Tourang [2], Bahareh Sasanfar [2], Paolo Boffetta [1,4] and Kazem Zendehdel [2,5,*]

1 Department of Medical and Surgical Sciences, University of Bologna, 40138 Bologna, Italy
2 Cancer Research Center, Cancer Institute, Tehran University of Medical Sciences, Tehran 1419733141, Iran
3 Department of Oncology, Lombardi Comprehensive Cancer Center, Georgetown University, Washington, DC 20007, USA
4 Stony Brook Cancer Center, Stony Brook University, Stony Brook, NY 11794, USA
5 Cancer Biology Research Center, Cancer Institute, Tehran University of Medical Sciences, Tehran 1419733141, Iran
* Correspondence: kzendeh@sina.tums.ac.ir

Abstract: N-nitroso compounds (NOCs) are a class of chemical carcinogens found in various environmental sources such as food, drinking water, cigarette smoke, the work environment, and the indoor air population. We conducted a systematic review and meta-analysis to investigate the links between nitrate, nitrite, and NOCs in food and water and the risk of gastrointestinal (GI) cancers, including esophageal cancer (EC), gastric cancer (GC), colorectal cancer (CRC), and pancreatic cancer (PC). A systematic search of the literature in Scopus, PubMed, Google Scholar, Web of Science, ScienceDirect, and Embase was performed for studies on the association between NOCs in drinking water and food sources and GI cancers. Forest plots of relative risk (RR) were constructed for all the cancer sites and the intake sources. The random-effects model was used to assess the heterogeneity between studies. Forty articles were included after removing duplicate and irrelevant articles. The meta-analysis indicated that the intake of high dose vs. low dose of these compounds was significantly associated with the overall GI cancer risk and nitrite (RR = 1.18, 95% CI = 1.07–1.29), and N-nitrosodimethylamine (NDMA) (RR = 1.32, 95% CI = 1.06–1.65). We found that dietary nitrite intake increased GC (RR = 1.33, 95% CI = 1.02–1.73), and EC (RR = 1.38, 95% CI = 1.01–1.89). Additionally, dietary NDMA intake increased the risk of CRC (RR = 1.36, 95% CI = 1.18–1.58). This meta-analysis provides some evidence that the intake of dietary and water nitrate, nitrite, and NOCs may be associated with GI cancers. In particular, dietary nitrite is linked to GC and EC risks and dietary NDMA intake is associated with CRC.

Keywords: neoplasms; organ; nitrate; nitrite; esophageal neoplasms; gastric neoplasms; intestinal neoplasms; pancreatic neoplasms; organic chemicals

Citation: Seyyedsalehi, M.S.; Mohebbi, E.; Tourang, F.; Sasanfar, B.; Boffetta, P.; Zendehdel, K. Association of Dietary Nitrate, Nitrite, and N-Nitroso Compounds Intake and Gastrointestinal Cancers: A Systematic Review and Meta-Analysis. *Toxics* **2023**, *11*, 190. https://doi.org/10.3390/toxics11020190

Academic Editors: Esref Demir and Sam Kacew

Received: 27 November 2022
Revised: 16 January 2023
Accepted: 27 January 2023
Published: 17 February 2023

1. Introduction

In 2020, cancer accounted for nearly 10 million deaths and nearly one in six deaths worldwide. Gastrointestinal (GI) cancers, with an estimation of approximately 5 million new cases and 3.5 million deaths worldwide, accounted for half of the cancer burden in 2020 [1,2]. The age-adjusted incidence rate (ASR) and mortality rate (ASMR) were equal to 19.5 and 9.7 per 100,000 for colorectal cancer (CRC) in both genders, respectively. The corresponding rates were 11.1 and 7.7 per 100,000 for gastric cancers (GC), 6.3 and 5.6 for esophageal cancer (EC), and 4.9 and 4.5 for pancreatic cancer (PC), respectively [1].

In addition to non-modifiable risk factors such as the demographic characteristics, family history, and genetic predisposition, some preventable risk factors are well established for GI cancer, including tobacco smoking, alcohol consumption, *H pylori* infection, high body mass index, low physical activity, and dietary factors [3–6].

The N-nitroso compounds (NOCs) are a broad class of chemical carcinogens that exist in various environmental sources such as food, drinking water, cigarette smoke, the work environment, and the indoor air population, although intrinsic sources are also available for crucial needs of our bodies [7–12]. The NOCs act as alkylating agents and may react with DNA to cause mutations leading to carcinogenicity [13–15]. N-nitrosodimethylamine (NDMA) is one of the NOCs found in human food, predominantly in processed/cured meats and smoked/salted fish [13]. Additionally, ingested nitrate is reduced to nitrite by the bacterial flora in the mouth and digestive tract. Subsequently, nitrite may react with amines, amides, and other nitrosation precursors in the gastrointestinal tract to form the NOCs under physiological conditions [11]. The main sources of nitrate are vegetables (beetroot, spinach, and cabbages, etc.) and the main sources of nitrite are animal foods (processed and smoked food) [8,16]. Drinking water is another source of nitrate in most countries, resulting from the overuse of chemicals or the improper disposal of human and animal waste, including fertilizers, feedlots, industrial and food processing waste [9].

Several experimental and epidemiological studies have shown that nitrate, nitrite and NDMA can affect human health [10], including an association cancer [17–22]. However, the results on the associations between GI cancers and these compounds are inconsistent. We performed a systematic review and meta-analysis to study the associations between the intake of these three compounds from food and water and the risk of GI cancers, including esophageal, gastric, colorectal, and pancreatic cancers.

2. Materials and Methods

2.1. Data Sources, Search Strategy, and Selection Criteria

The online database searches were performed in January 2022. Searches were undertaken for English-language peer-reviewed publications on the association of nitrate, nitrite, and NOCs and the risk of GI cancers between 1990 to the present. The databases included Scopus, PubMed, Google Scholar, Web of Science, ScienceDirect, and Embase. The search strategy was designed using MeSH terms like "Gastrointestinal Cancers", "Digestive System Cancers", "Organic Chemicals", "Nitroso Compounds", "Nitrate", "Nitrite","NDMA", and "Epidemiologic Studies". Based on our searches on the databases, a total of 12,750 articles were retrieved, including 4539 articles for GC, 2204 for EC, 6673 for CRC, and 2243 for PC. A total of 40 studies were included after the title, abstract, full-text evaluations, and quality assessment [23–62] were undertaken. Figure 1 shows the PRISMA flow diagram of the literature searches and the study selection process. The inclusion criteria were as follows; case-control and cohort studies reporting either a relative risk (RR) or an odds ratio (OR) for the associations between GI cancers and the consumption of nitrate, nitrite, and NOCs from drinking water or food sources.

2.2. Data Extraction and Quality Assessment

The study screening and quality assessment were conducted by three researchers (MSS, FT, and BS), and in the event of any discrepancies, a referee (EM) intervened. The data extraction file contained the demographic characteristics of the article such as the author's name, the year of publication, the title, the type of study, the country, as well as the design characteristics, including the sample size, the sampling method, the source of the population, the type of controls, the type of cancer, the source of nitrate, nitrite, and the NOCs (water/food/vegetable/fruit/all), and the method of data collection (e.g., food frequency questionnaire). Finally, the effect size measures, including the relative risks (RRs) for the cohort studies and the odds ratios (ORs) for the case-control studies and their 95% CI were abstracted.

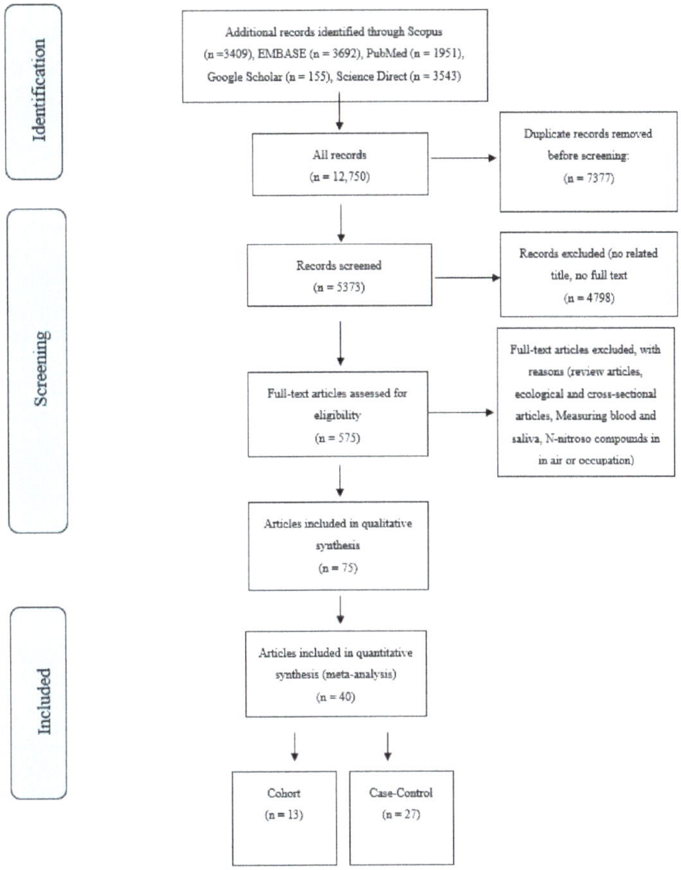

Figure 1. Selection process for studies included in the meta-analysis for the intake of nitrate, nitrite, and N-nitroso compounds from diet and water and the risk of gastrointestinal cancers.

2.3. Quality Assessment

Quality assessment of the included articles (case-control or cohort studies) was accomplished using the Newcastle-Ottawa Scale (NOS) [63]. For case-control studies, the checklist contained the definition and selection of cases and controls, comparability, ascertainment of exposure, and the method of ascertainment for the cases and the controls. The checklist of cohort studies included the situation of the exposed and non-exposed cohort, comparability, assessment of outcome, and the duration and adequacy of follow-up of the cohort study. Based on the NOS, the selection score could result in a maximum of four stars, the comparability scores a maximum of two stars, and the outcome/exposure score a maximum of three stars [64].

2.4. Statistical Analysis

The relationships between nitrite, nitrate, or NMDA intake and the risk of GI cancers were examined based on the effect size measurements and the corresponding 95% CIs of each study. Because of the rare disease assumption, the ORs are assumed to approximate the RRs [65]. Heterogeneity (Het.) among studies was evaluated by the Q test, based on the variation across studies rather than within studies, and the I^2 statistic (the percentage of variance in a meta-analysis that is attributable to study heterogeneity) [66]. The pooled effect of nitrate, nitrite, and NOCs, in the case of significant heterogeneity (PQ test > 0.10 and

$I^2 > 60\%$), was estimated using a random-effects model [67]. All analyses were completed using the STATA version 14.0 (Stata, College Station, TX, USA). Given that approximately 82% of the studies reported combined results for both genders, we focused our analysis on these results. Due to the high variation of nitrogen types, we conducted separate analyses on nitrate, nitrite, and NMDA intake among the studies. In order to estimate the impact of the dose-response relationship and to describe the magnitude of the response, we calculated the high/moderate doses of the nitrogen types and compared it to the low dose as a reference group in each study. A stratified analysis was conducted by gender (male and female), food sources (plant and animal), study design (case-control and cohort studies), and sub-sites of CRC (colon and rectum). Since most studies have been conducted in the United States (48%), stratification based on the country was ineffective. The *p*-value of heterogeneity was assessed using a sub meta-analysis. Moreover, publication bias was examined by creating a funnel plot and a regression asymmetry test [68,69].

3. Results

A total of 40 independent studies were included in the meta-analysis, including 27 case-control studies [23–28,31,33–39,41–43,45,51,52,54–57,60,61] and 13 cohort studies [29,30,32,40,46–50,53,58,59,62]. Details on these studies were provided in Supplementary Tables S1 and S2. These studies reported a total of 13 risk estimates for CRC [34,36,42,46,49,51,53–56,59,62], 22 for GC [23–25,27–35,38–41,44,45,47,49,50,57], seven for EC [26,35,44,47,49,50,60], and five for PC [37,48,52,58,61]. The studies included in the meta-analysis had a minimum score of 7 out of 9 stars in the NOS.

Meta-Analysis

The meta-analysis was repeated for each cancer site based on the nitrate, nitrite, NDMA and the source intake.

The findings revealed a statistically significant relationship between the highest vs. the lowest level of nitrite intake from the food sources and EC risk (RR = 1.38, 95% CI = 1.01–1.89). There were no statistically significant excess risks from the food sources of nitrate, NDMA, or the water sources of nitrate intake. We observed significant heterogeneity for studies of the food sources of NDMA (I^2 = 80.2%, P-heterogeneity = 0.025), the food intake of nitrate (I^2 = 89.4%, P-heterogeneity = 0.000), and the food intake of nitrite (I^2 = 73.8%, P-heterogeneity = 0.004) (Figure 2). There were no significant results according to gender and food source (plant or animal) among the nitrate and nitrite intake groups (Tables 1 and 2).

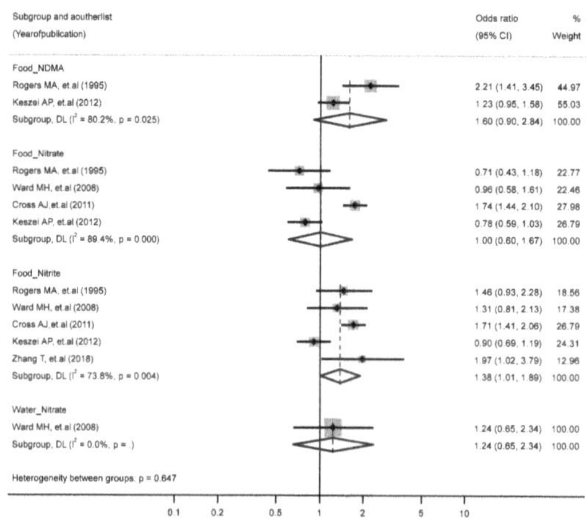

Figure 2. Forest plot (random-effects model) quantifying the relationships between NDMA, nitrite, and nitrate intake and esophagus cancer risk stratified by sources. List of references includes: [26,44,47,50,60].

Table 1. Association between nitrate intake and selected gastrointestinal cancers stratified by gender, food sources, study type, and topography.

		Colorectal Cancer				Gastric Cancer				Esophageal Cancer				Pancreatic Cancer			
		N	RR (95% CI)	I^2	P_b	N	RR (95% CI)	I^2	P_b	N	RR (95% CI)	I^2	P_b	N	RR (95% CI)	I^2	P_b
gender	Male	0	-	-	-	1	0.78 (0.63–0.97)	-	-	1	0.84 (0.61–1.18)	-	-	2	0.86 (0.45–1.66)	9.82	0.002
	Female	0	-	-	-	1	1.08 (0.79–1.47)	-	-	1	0.62 (0.37–1.05)	-	-	2	1.50 (1.27–1.77)	0.01	0.917
P-heterogeneity			-				0.513				0.066				0.400		
Food Sources	Animal	3	1.12 (0.93–1.35)	11.70	0.003	3	1.06 (0.76–1.48)	6.28	0.043	1	1.73 (1.43–2.10)	-	-	2	0.83 (0.41–1.69)	12.12	0.000
	Plant	2	0.99 (0.76–1.29)	5.34	0.021	3	0.99 (0.77–1.28)	0.00	0.999	1	0.96 (0.57–1.60)	-	-	1	1.23 (0.99–1.53)	-	-
P-heterogeneity			0.368				0.884				0.294				0.206		
Study Type	Cohort	4	1.19 (1.10–1.29)	3.96	0.266	4	1.08 (0.83–1.39)	11.36	0.010	2	1.17 (0.53–2.57)	21.83	0.000	2	0.92 (0.82–1.04)	0.54	0.461
	Case_ control	2	1.01 (0.81–1.26)	5.12	0.024	8	1.14 (0.65–1.99)	148.92	0.000	2	0.82 (0.57–1.18)	0.67	0.412	3	1.00 (0.62–1.61)	17.46	0.000
P-heterogeneity			0.093				0.469				0.419				0.182		
Topography	Colon	5	1.07 (0.94–1.21)	14.93	0.005												
	Rectum	5	1.13 (0.96–1.33)	10.87	0.028												
p-heterogeneity			0.084														

N = number of studies; P_b = p value for heterogeneity.

Table 2. Association between nitrite intake and selected gastrointestinal cancers stratified by gender, food sources, study type, and topography.

		Colorectal Cancer				Gastric Cancer				Esophagus Cancer				Pancreatic Cancer			
		N	RR (95% CI)	I^2	P_b	N	RR (95% CI)	I^2	P_b	N	RR (95% CI)	I^2	P_b	N	RR (95% CI)	I^2	P_b
Food Sources	Animal	4	1.03 (0.89–1.19)	20.80	0.000	4	1.00 (0.81–1.23)	7.73	0.052	2	1.25 (0.67–2.33)	14.17	0.000	3	1.36 (0.88–2.10)	28.11	0.000
	Plant	2	0.95 (0.86–1.05)	0.01	0.937	3	1.10 (0.86–1.40)	0.24	0.888	1	1.31 (0.80–2.13)	-	-	2	0.85 (0.75–0.96)	0.86	0.353
P-heterogeneity			0.546				0.618				0.199				0.916		
Study Type	Cohort	4	1.02 (0.88–1.19)	10.40	0.015	3	1.30 (0.94–1.81)	6.62	0.036	1	1.70 (1.41–2.06)	-	-	2	0.91 (0.81–1.03)	0.04	0.835
	Case_ Control	2	1.38 (1.20–1.59)	0.68	0.409	9	1.42 (0.99–2.02)	111.4	0.000	3	1.49 (1.11–2.00)	0.97	0.616	2	1.39 (0.66–2.93)	18.75	0.000
P-heterogeneity			0.254				0.014				0.000				0.711		
Topography	Colon	4	1.08 (0.97–1.20)	6.47	0.091												
	Rectum	4	1.04 (0.85–1.29)	9.13	0.028												
P-heterogeneity			0.153														

N = number of studies; P_b = p value for heterogeneity.

We found a significant association between the highest vs. the lowest level of nitrite consumption and GC (RR = 1.33, 95% CI = 1.02–1.73) (I^2 = 92.2%, P-heterogeneity = 0.000). Additionally, there were significant heterogeneities for studies of the food intake of NDMA (I2 = 94.4%, P-heterogeneity = 0.000), the food intake of nitrate (I^2 = 93.2%, P-heterogeneity = 0.000), and the water intake of nitrate (I^2 = 93.5%, P-heterogeneity = 0.000) (Figure 3). We found non-significant differences in the association of GC and nitrite and nitrate intakes stratified for other factors (Tables 1 and 2).

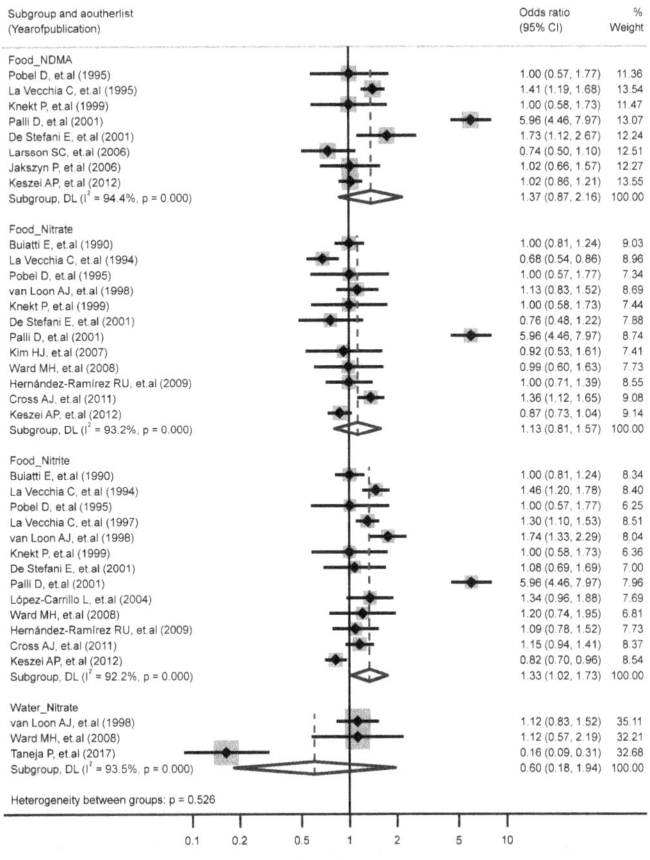

Figure 3. Forest plot (random-effects model) quantifying the relationships between NDMA, nitrite, and nitrate intake and gastric cancer risk stratified by sources. List of references includes: [23–25,27,30,32–34,38–41,44,45,47,50,57].

The association between the highest vs. the lowest levels of nitrate, nitrite, and NDMA and CRC risk only showed that the intake of NDMA from the food increased the risk of CRC significantly (RR = 1.36, 95% CI = 1.18–1.58). No statistically significant heterogeneity across studies was observed for CRC in relation to NDMA intake, but heterogeneity was statistically significant for studies of the food nitrate intake (I^2 = 73.4%, P-heterogeneity = 0.002), the food nitrite intake (I^2 = 77.2%, P-heterogeneity = 0.001), and the water intake of nitrate (I^2 = 98%, P-heterogeneity = 0.000) (Figure 4). Stratified analyses according to the study type, the gender, the topography and the food sources (plant, animal) revealed no significant differences due to these variables (Tables 1 and 2).

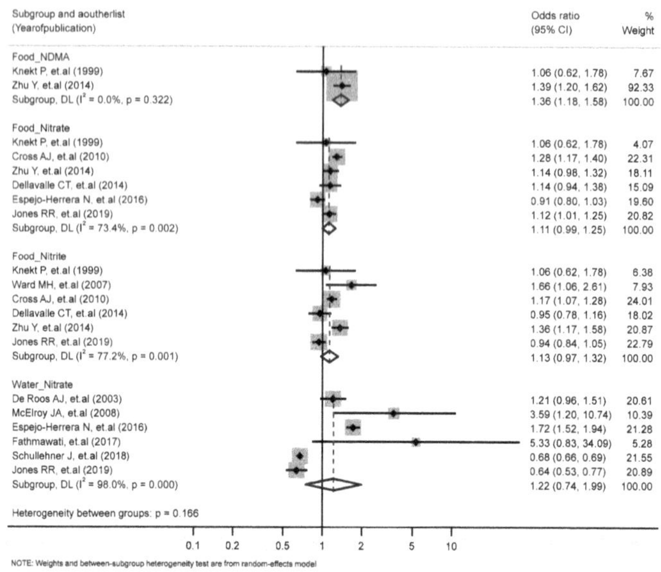

Figure 4. Forest plot (random-effects model) quantifying the relationships between NDMA, nitrite, and nitrate intake and colorectal cancer risk stratified by sources. List of references includes: [32,36,42,43,46,53–56,59,62].

We found no significant associations between PC in the highest vs. the lowest levels of nitrate, nitrite, and NDMA. However, we observed significant heterogeneity for the food intake of nitrate (I^2 = 80.7%, P-heterogeneity = 0.000), and nitrite (I^2 = 93.5%, P-heterogeneity = 0.000) (Figure 5). No significant differences were found by study types, genders, topographies, and food sources (plants, animals) (Tables 1 and 2).

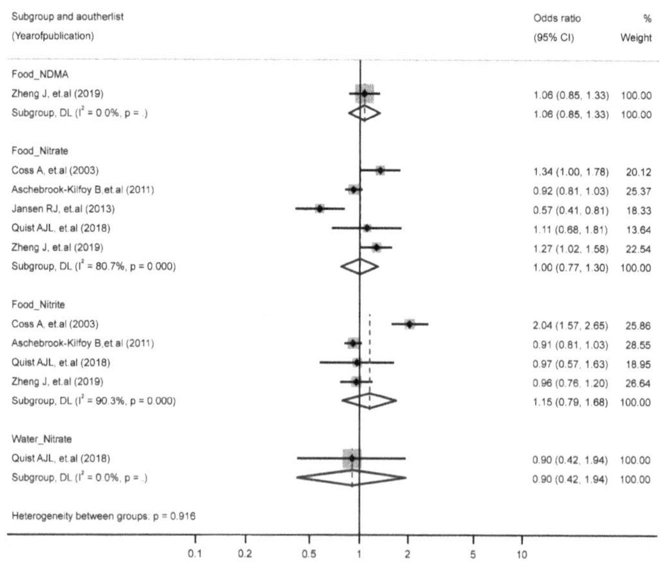

Figure 5. Forest plot (random-effects model) quantifying the relationships between NDMA, nitrite, and nitrate intake and pancreatic cancer risk stratified by sources. List of references includes: [37,48,52,58,61].

4. Discussion

We found limited high-quality research on nitrate, nitrite, NOCs, and cancer risk, most likely due to the difficulty of quantifying these compounds from different sources worldwide. Despite this limitation, the literature suggests that dietary and water sources may be risk factors for GI cancers, in particular, nitrate intake from water sources and diet for CRC, nitrite intake from diet for GC, and NDMA intake for CRC and EC [14,15,56,59,70].

Nitrate and nitrite as precursors of the NOCs (e.g., NDMA) are suspected of playing a key role in cancer carcinogenesis through the induction of DNA-damaging metabolites, like aldehydes and alkyldiazonium ions, which could cause cancerous lesions in cells [14–16,58,70]. Several in vivo and epidemiological studies reported that different factors might contribute to the carcinogenic effect of nitrate, nitrite, and NOCs, including the dose and the type of the compounds, the cancer sub-site, the morphology of the tumor, the type of diet source (animal, plant, or water), the gender, the exposure time, the cooking methods, the effect of seasonal rainfall on water composition and the different fertilizer compounds used in different areas [62]. For example, consuming fresh fruits and vegetables high in vitamins, essential minerals, and antioxidants such as vitamins C and E, as well as reducing meat, fatty foods, and processed foods were found to improve health. This may be important in modifying any harmful effects of dietary nitrates and nitrites on particularly susceptible tissues in the digestive system [58,71].

There was heterogeneity in the study designs and the methodologies, which may explain the differences in the results of the published studies. The high heterogeneity observed in our meta-analyses may be due to the variety of confounders used in the models as well as the differences between the study settings, the population characteristics, the compound ranges, and the estimation methods. Except for age and gender, other confounders were less consistently controlled for, particularly dietary factors and specific risk factors, such as *H pylori* infection in the GC studies [72].

4.1. Esophageal Cancer

The number of reviews for EC was limited. A meta-analysis study by Lie et al. [19] in 2016 found no significant association between dietary nitrite and nitrate and EC, and a study conducted by Essien et al. [18] in 2020 reported similar results based on water nitrate. We found that dietary nitrite and animal sources of nitrate were positively associated with a higher risk of EC, but NDMA was not associated with EC. These results may indicate that the source and the amount of the intake are important determinants of cancer risk. Animal sources of food (mainly processed meats) contain amines and amides that are necessary precursors for endogenous nitrosation [19]. Although other sources like vegetables may contain these compounds, they are also rich in antioxidants, which may decrease their harmful effect and explain their protective role in cancer risk [71].

4.2. Gastric Cancer

In 2016, Lie and others found an inverse association between dietary nitrate consumption and GC, and a borderline association between dietary nitrite consumption and GC [19]. In 2015, Peng Song et al. [22] found that nitrates from the diet can reduce GC risk, but nitrites and NDMA increased the risk. Fei-Xiong and colleagues reported that dietary nitrate decreased, and nitrite intake increased GC risk [21]. A meta-analysis carried out in 2020 by Essien et al. found no significant association between water-source nitrate intake and GC risk. [18]. In the present study, we found that dietary nitrite intake increased the risk of GC, but the associations were not significant for all kinds of dietary and water sources of nitrate and NDMA intake. Unexpectedly, we reported a significant inverse association between dietary nitrate and GC in males. In addition to the source of compounds, other GC risk factors such as *H.pylori* infection may increase the NOCs product level and increase the risk of GC [39].

4.3. Colorectal Cancer

Previous studies on association between the NOCs and CRC are inconsistent. In 2016, one systematic review reported that dietary nitrite/nitrate intake was not significantly linked with CRC [19], however, another review in 2020 reported that dietary nitrate consumption was linked to a higher risk of CRC [18]. However, the intake of dietary nitrite and nitrate from the drinking water was not associated with CRC risk. Another review in 2020 revealed that nitrate from water sources was significantly associated with CRC [20]. Our study showed that dietary NDMA intake increased the risk of CRC, but nitrite and nitrate sources were not significantly associated with CRC. The possible reasons for the association between NDMA and CRC may relate to the main source of intake, including meats, particularly bacon, hot dogs, and sausage, which are naturally high in amines. The association may be explained by the synergy between NDMA and amine groups which are found in protein structures [54].

4.4. Pancreatic Cancer

None of the previous reviews reported a significant relationship between dietary or water intake of nitrate, nitrite, and NDMA and the risk of PC [18,19]. Our results of the relation between water and dietary sources of nitrate, nitrite, and NDMA showed no significant association with PC.

The geographical distribution of the studies plays an important role in achieving reliable results. A majority of the studies included in our review were conducted in Europe (30%) and North America (48%), and we found no reliable studies from low- and middle-income countries (LMICs). As a result, our findings cannot be generalized to LMICs, where the amounts of nitrate, nitrite, and NOCs may vary based on the dietary patterns, the cooking methods, the water sources, the effect of seasonal rainfall on water compounds, and the fertilizer compounds [73]. Also, most studies on the water sources were of ecological design and therefore were not included in our analysis. A small number of studies in relation to several exposure/outcome combinations, and high heterogeneity in most of our meta-analyses were additional limitations. Furthermore, possible residual confounding like consuming diets and lifestyle factors and inadequate adjustment in several studies might have hampered the results of the published studies. However, the present systematic review and meta-analysis have several important strengths. We included the most updated articles that reported the associations for GC, EC, CRC, and PC exclusively. We also studied different sources of nitrate, nitrite, and NOCs (i.e., animal, plant, and water sources) intake.

5. Conclusions

In conclusion, our meta-analysis shows modest evidence on the association between dietary and water nitrate, nitrite, and NOCs intake and certain types of GI cancer risk. The dietary intake of nitrite is associated with GS and EC and the dietary NDMA intake is associated with CRC. In future studies, the influence of different compound types on cancer should be explored by considering the source intake, and gender differences, particularly in less studied geographical regions.

Supplementary Materials: The following supporting information can be downloaded at: https://www.mdpi.com/article/10.3390/toxics11020190/s1. Table S1. Characteristics of cohort studies in the systematic review; Table S2. Characteristics of case-control studies in the systematic review.

Author Contributions: K.Z and M.S.S.: Conceived and designed. B.S., F.T. and M.S.S.: Searches and screening process. M.S.S. and E.M.: Analyzed the data. M.S.S.: Wrote the article. P.B., K.Z. and E.M.: Contributed to the editing of the tables and text. P.B.: Critical appraisal of analyses and results. Interpretation of the findings and improving the paper. All authors have read and agreed to the published version of the manuscript.

Funding: This research received no external funding.

Institutional Review Board Statement: The study was conducted in accordance with the Declaration of Helsinki and approved by the Ethics Committee of the Tehran University of Medical Sciences (TUMS) (Code: IR.TUMS.IKHC.REC.1400.382). All participants provided written informed consent.

Informed Consent Statement: Not applicable.

Data Availability Statement: Not applicable.

Conflicts of Interest: The authors declare no conflict of interest.

References

1. Sung, H.; Ferlay, J.; Siegel, R.L.; Laversanne, M.; Soerjomataram, I.; Jemal, A.; Bray, F. Global Cancer Statistics 2020: GLOBOCAN Estimates of Incidence and Mortality Worldwide for 36 Cancers in 185 Countries. *CA Cancer J. Clin.* **2021**, *1*, 209–249. [CrossRef] [PubMed]
2. Tabatabaiefar, M.A.; Moridnia, A. Gastrointestinal Cancers. In *Cancer Genetics and Psychotherapy*; Mehdipour, P., Ed.; Springer: Cham, Switzerland, 2017. [CrossRef]
3. Neagoe, A.; Molnar, A.M.; Acalovschi, M.; Seicean, A.; Serban, A. Risk factors for colorectal cancer: An epidemiologic descriptive study of a series of 333 patients. *Rom. J. Gastroenterol.* **2004**, *13*, 187–193. [PubMed]
4. Capasso, M.; Franceschi, M.; Rodriguez-Castro, K.I.; Crafa, P.; Cambiè, G.; Miraglia, C.; Barchi, A.; Nouvenne, A.; Leandro, G.; Meschi, T.; et al. Epidemiology and risk factors of pancreatic cancer. *Acta Biomed.* **2018**, *89*, 141–146. [PubMed]
5. Domper Arnal, M.J.; Ferrández Arenas, Á.; Lanas Arbeloa, Á. Esophageal cancer: Risk factors, screening and endoscopic treatment in Western and Eastern countries. *World J. Gastroenterol.* **2015**, *21*, 7933–7943. [CrossRef] [PubMed]
6. Yusefi, A.R.; Bagheri Lankarani, K.; Bastani, P.; Radinmanesh, M.; Kavosi, Z. Risk Factors for Gastric Cancer: A Systematic Review. *Asian Pac. J. Cancer Prev.* **2018**, *19*, 591–603.
7. Tricker, A.R.; Preussmann, R. Carcinogenic N-nitrosamines in the diet: Occurrence, formation, mechanisms and carcinogenic potential. *Mutat. Res.* **1991**, *259*, 277–289. [CrossRef]
8. Griesenbeck, J.S.; Steck, M.D.; Huber, J.C., Jr.; Sharkey, J.R.; Rene, A.A.; Brender, J.D. Development of estimates of dietary nitrates, nitrites, and nitrosamines for use with the short willet food frequency questionnaire. *Nutr. J.* **2009**, *8*, 16. [CrossRef]
9. Mahvi, A.A.H.; Nouri, J.; Babaei, A.A.; Nabizadeh, R. Agricultural activities impact on groundwater nitrate pollution. *Int. J. Environ. Sci. Technol.* **2005**, *2*, 41–47. [CrossRef]
10. Wogan, G.; Tannenbaum, S. Environmental N-Nitroso Compounds: Implications for Public Health. *Toxicol. Appl. Pharmacol.* **1975**, *31*, 375–383. [CrossRef]
11. IARC Working Group on the Evaluation of Carcinogenic Risks to Humans. Ingested Nitrate and Nitrite, and Cyanobacterial Peptide Toxins. In *IARC Monographs on the Evaluation of Carcinogenic Risks to Humans*; International Agency for Research on Cancer: Lyon, France, 2010; Volume 94, p. 448.
12. Li, Y.; Hecht, S.S. Metabolism and DNA Adduct Formation of Tobacco-Specific N-Nitrosamines. *Int. J. Mol. Sci.* **2022**, *23*, 5109. [CrossRef]
13. Li, Y.; Hecht, S.S. Metabolic Activation and DNA Interactions of Carcinogenic N-Nitrosamines to Which Humans Are Commonly Exposed. *Int. J. Mol. Sci.* **2022**, *23*, 4559. [CrossRef]
14. Lijinsky, W. Carcinogenicity and mutagenicity of N-nitroso compounds. *Mol. Toxicol.* **1987**, *1*, 107–119.
15. Jain, D.; Chaudhary, P.; Varshney, N.; Janmeda, P. Carcinogenic effects of N-nitroso compounds in the environment. *Environ. Conserv. J.* **2020**, *21*, 25–41. [CrossRef]
16. Wylie, L.J.; Park, J.W.; Vanhatalo, A.; Kadach, S.; Black, M.I.; Stoyanov, Z.; Schechter, A.N.; Jones, A.M.; Piknova, B. Human skeletal muscle nitrate store: Influence of dietary nitrate supplementation and exercise. *J. Physiol.* **2019**, *597*, 5565–5576. [CrossRef] [PubMed]
17. SeyyedSalehi, M.S.; Mohebbi, E.; Sasanfar, B.; Toorang, F.; Zendehdel, K. Dietary N-nitroso compounds intake and bladder cancer risk: A Systematic Review and Meta-Analysis. *Nitric Oxide* **2021**, *115*, 1–7. [CrossRef] [PubMed]
18. Essien, E.E.; Abasse, K.S.; Côté, A.; Mohamed, K.S.; Ashraf Baig, M.M.F.; Habib, M.; Naveed, M.; Yu, X.; Xie, W.; Jinfang, S.; et al. Drinking-Water Nitrate and Cancer Risk: A Systematic Review and Meta-analysis. *Arch. Environ. Occup. Health* **2020**, *3*, 51–67. [CrossRef] [PubMed]
19. Xie, L.; Mo, M.; Jia, H.X.; Liang, F.; Yuan, J.; Zhu, J. Association between dietary nitrate and nitrite intake and specifc cancer risk: Evidence from observational studies. *Oncotarget* **2016**, *7*, 56915–56932. [CrossRef] [PubMed]
20. Hosseini, F.; Majdi, M.; Naghshi, S.; Sheikhhossein, F.; Djafarian, K.; Shab-Bidar, S. Nitrate-nitrite exposure through drinking water and diet and risk of colorectal cancer: A systematic review and meta-analysis of observational studies. *Clin. Nutr.* **2020**, *40*, 3073–3081. [CrossRef] [PubMed]
21. Zhang, F.X.; Miao, Y.; Ruan, J.G.; Meng, S.P.; Dong, J.D.; Yin, H.; Huang, Y.; Chen, F.R.; Wang, Z.C.; Lai, Y.F. Association Between Nitrite and Nitrate Intake and Risk of Gastric Cancer: A Systematic Review and Meta-Analysis. *Med. Sci. Monit.* **2019**, *25*, 1788–1799. [CrossRef] [PubMed]
22. Song, P.; Wu, L.; Guan, W. Dietary Nitrates, Nitrites, and Nitrosamines Intake and the Risk of Gastric Cancer: A Meta-Analysis. *Nutrients* **2015**, *7*, 9872–9895. [CrossRef]
23. Buiatti, E.; Palli, D.; Decarli, A.; Amadori, D.; Avellini, C.; Bianchi, S.; Bonaguri, C.; Cipriani, F.; Cocco, P.; Giacosa, A.; et al. A case-control study of gastric cancer and diet in italy: 11. association with nutrients. *Int. J. Cancer* **1990**, *45*, 896–901. [CrossRef] [PubMed]

24. La Vecchia, C.; Ferraroni, M.; D'Avanzo, B.; Decarli, A.; Franceschi, S. Selected micronutrient intake and the risk of gastric cancer. *Cancer Epidemiol. Biomark. Prev.* **1994**, *3*, 393–3988.
25. La Vecchia, C.; D'Avanzo, B.; Airoldi, L.; Braga, C.; Decarli, A. Nitrosamine intake and gastric cancer risk. *Eur. J. Cancer Prev.* **1995**, *4*, 469–474. [CrossRef] [PubMed]
26. Rogers, M.A.; Vaughan, T.L.; Davis, S.; Thomas, D.B. Consumption of nitrate, nitrite, and nitrosodimethylamine and the risk of upper aerodigestive tract cancer. *Cancer Epidemiol. Biomark. Prev.* **1995**, *4*, 29–36.
27. Pobel, D.D.; Riboli, E.; Cornée, J.; Hémon, B.; Guyader, M. Nitrosamine, nitrate and nitrite in relation to gastric cancer: A case-control study in Marseille, France. *Eur. J. Epidemiol.* **1995**, *11*, 67–73. [CrossRef] [PubMed]
28. La Vecchia, C.; Negri, E.; Franceschi, S.; Decarli, A. Case-control study on influence of methionine, nitrite, and salt on gastric carcinogenesis in northern Italy. *Nutr. Cancer* **1997**, *27*, 65–68. [CrossRef] [PubMed]
29. van Loon, A.J.; Botterweck, A.A.; Goldbohm, R.A.; Brants, H.A.; van den Brandt, P.A. Nitrate intake and gastric cancer risk: Results from the Netherlands cohort study. *Cancer Lett.* **1997**, *114*, 259–261. [CrossRef]
30. van Loon, A.J.; Botterweck, A.A.; Goldbohm, R.A.; Brants, H.A.; van Klaveren, J.D.; van den Brandt, P.A. Intake of nitrate and nitrite and the risk of gastric cancer: A prospective cohort study. *Br. J. Cancer* **1998**, *78*, 129–135. [CrossRef]
31. De Stefani, E.; Boffetta, P.; Mendilaharsu, M.; Carzoglio, J.; Deneo-Pellegrini, H. Dietary nitrosamines, heterocyclic amines, and risk of gastric cancer: A case-control study in Uruguay. *Nutr. Cancer* **1998**, *30*, 158–162. [CrossRef]
32. Knekt, P.; Järvinen, R.; Dich, J.; Hakulinen, T. Risk of colorectal and other gastrointestinal cancers after exposure to nitrate, nitrite and N-nitroso compounds: A follow-up study. *Int. J. Cancer* **1999**, *80*, 852–856. [CrossRef]
33. Palli, D.; Russo, A.; Decarli, A. Dietary patterns, nutrient intake and gastric cancer in a high-risk area of Italy. *Cancer Causes Control.* **2001**, *12*, 163–172. [CrossRef] [PubMed]
34. De Stefani, E.; Ronco, A.; Brennan, P.; Boffetta, P. Meat consumption and risk of stomach cancer in Uruguay: A case-control study. *Nutr. Cancer* **2001**, *40*, 103–107. [CrossRef] [PubMed]
35. Mayne, S.T.; Risch, H.A.; Dubrow, R.; Chow, W.H.; Gammon, M.D.; Vaughan, T.h.L.; Farrow, D.C.; Schoenberg, J.B.; Stanford, J.L.; Ahsan, H.; et al. Nutrient intake and risk of subtypes of esophageal and gastric cancer. *Cancer Epidemiol. Biomark. Prev.* **2001**, *10*, 1055–1062.
36. De Roos, A.J.; Ward, M.H.; Lynch, C.F.; Cantor, K.P.; Charles, F.L.; Kenneth, P.C. Nitrate in public water supplies and the risk of colon and rectum cancers. *Epidemiology* **2003**, *14*, 640–649. [CrossRef]
37. Coss, A.; Cantor, K.P.; Reif, J.S.; Lynch, C.F.; Ward, M.H. Pancreatic cancer and drinking water and dietary sources of nitrate and nitrite. *Am. J. Epidemiol.* **2003**, *159*, 693–701. [CrossRef]
38. López-Carrillo, L.; Torres-López, J.; Galván-Portillo, M.; Muñoz, L.; López-Cervantes, M. Helicobacter pylori-CagA seropositivity and nitrite and ascorbic acid food intake as predictors for gastric cancer. *Eur. J. Cancer Prev.* **2004**, *40*, 1752–1759. [CrossRef]
39. Jakszyn, P.; Bingham, S.; Pera, G.; Agudo, A.; Luben, R.; Welch, A.; Boeing, H.; Del Giudice, G.; Palli, D.; Saieva, C.; et al. Endogenous versus exogenous exposure to N-nitroso compounds and gastric cancer risk in the European Prospective Investigation into Cancer and Nutrition (EPIC-EURGAST) study. *Carcinogenesis* **2006**, *27*, 1497–1501. [CrossRef]
40. Larsson, S.C.; Bergkvist, L.; Wolk, A. Processed meat consumption, dietary nitrosamines and stomach cancer risk in a cohort of Swedish women. *Int. J. Cancer* **2006**, *119*, 915–919. [CrossRef]
41. Kim, H.J.; Lee, S.S.; Choi, B.Y.; Kim, M.K. Nitrate intake relative to antioxidant vitamin intake affects gastric cancer risk: A case-control study in Korea. *Nutr. Cancer* **2007**, *59*, 185–191. [CrossRef]
42. Ward, M.H.; Cross, A.J.; Divan, H.; Kulldorff, M.; Nowell-Kadlubar, S.; Kadlubar, F.F.; Sinha, R. Processed meat intake, CYP2A6 activity and risk of colorectal adenoma. *Carcinogenesis* **2007**, *28*, 1210–1216. [CrossRef]
43. McElroy, J.A.; Trentham-Dietz, A.; Gangnon, R.E.; Hampton, J.M.; Bersch, A.J.; Kanarek, M.S.; Newcomb, P.A. Nitrogen-nitrate exposure from drinking water and colorectal cancer risk for rural women in Wisconsin, USA. *J. Water Health* **2008**, *6*, 399–409. [CrossRef] [PubMed]
44. Ward, M.H.; Heineman, E.F.; Markin, R.S.; Weisenburger, D.D. Adenocarcinoma of the stomach and esophagus and drinking water and dietary sources of nitrate and nitrite. *Int. J. Occup. Environ. Health* **2008**, *14*, 193–197. [CrossRef] [PubMed]
45. Hernández-Ramírez, R.U.; Galván-Portillo, M.V.; Ward, M.H.; Agudo, A.; González, C.A.; Oñate-Ocaña, L.F.; Herrera-Goepfert, R.; Palma-Coca, O.; López-Carrillo, L. Dietary intake of polyphenols, nitrate and nitrite and gastric cancer risk in Mexico City. *Int. J. Cancer* **2009**, *125*, 1424–1430. [CrossRef] [PubMed]
46. Cross, A.J.; Ferrucci, L.M.; Risch, A.; Graubard, B.I.; Ward, M.H.; Park, Y.; Hollenbeck, A.R.; Schatzkin, A.; Sinha, R. A large prospective study of meat consumption and colorectal cancer risk: An investigation of potential mechanisms underlying this association. *Cancer Res.* **2010**, *70*, 2406–2414. [CrossRef] [PubMed]
47. Cross, A.J.; Freedman, N.D.; Ren, J.; Ward, M.H.; Hollenbeck, A.R.; Schatzkin, A.; Sinha, R.; Abnet, C.C. Meat consumption and risk of esophageal and gastric cancer in a large prospective study. *Am. J. Gastroenterol.* **2011**, *106*, 432–442. [CrossRef]
48. Aschebrook-Kilfoy, B.; Cross, A.J.; Stolzenberg-Solomon, R.Z.; Schatzkin, A.; Hollenbeck, A.R.; Sinha, R.; Ward, M.H. Pancreatic cancer and exposure to dietary nitrate and nitrite in the NIH-AARP Diet and Health Study. *Am. J. Epidemiol.* **2011**, *174*, 305–315. [CrossRef]
49. Loh, Y.H.; Jakszyn, P.; Luben, R.N.; Mulligan, A.A.; Mitrou, P.N.; Khaw, K.T. N-Nitroso compounds and cancer incidence: The European Prospective Investigation into Cancer and Nutrition (EPIC)-Norfolk Study. *Am. J. Clin. Nutr.* **2011**, *93*, 1053–1061. [CrossRef]
50. Keszei, A.P.; Goldbohm, R.A.; Schouten, L.J.; Jakszyn, P.; van den Brandt, P.A. Dietary N-nitroso compounds, endogenous nitrosation, and the risk of esophageal and gastric cancer subtypes in the Netherlands Cohort Study. *Am. J. Clin. Nutr.* **2012**, *97*, 135–146. [CrossRef]

51. Miller, P.E.; Lazarus, P.; Lesko, S.M.; Cross, A.J.; Sinha, R.; Laio, J.; Zhu, J.; Harper, G.; Muscat, J.E.; Hartman, T.J. Meat-related compounds and colorectal cancer risk by anatomical subsite. *Nutr. Cancer* **2013**, *65*, 202–226. [CrossRef]
52. Jansen, R.J.; Robinson, D.P.; Frank, R.D.; Stolzenberg-Solomon, R.Z.; Bamlet, W.R.; Oberg, A.L.; Rabe, K.G.; Olson, J.E.; Petersen, G.M.; Sinha, R.; et al. Meat-Related Mutagens and Pancreatic Cancer: Null Results from a Clinic-Based Case–Control Study. *Cancer Epidemiol. Biomark. Prev.* **2013**, *22*, 1336–1339. [CrossRef]
53. Dellavalle, C.T.; Xiao, Q.; Yang, G.; Shu, X.O.; Aschebrook-Kilfoy, B.; Zheng, W.; Lan, L.H.; Ji, B.T.; Rothman, N.; Chow, W.H.; et al. Dietary nitrate and nitrite intake and risk of colorectal cancer in the Shanghai Women's Health Study. *Int. J. Cancer* **2014**, *134*, 2917–2926. [CrossRef]
54. Zhu, Y.; Wang, P.P.; Zhao, J.; Green, R.; Sun, Z.; Roebothan, B.; Squires, J.; Buehler, S.; Dicks, E.; Zhao, J.; et al. Dietary N-nitroso compounds and risk of colorectal cancer: A case-control study in Newfoundland and Labrador and Ontario, Canada. *Br. J. Nutr.* **2014**, *111*, 1109–1117. [CrossRef]
55. Espejo-Herrera, N.; Gràcia-Lavedan, E.; Boldo, E.; Aragonés, N.; Pérez-Gómez, B.; Pollán, M.; Molina, A.J.; Fernández, T.; Martín, V.; La Vecchia, C.; et al. Colorectal cancer risk and nitrate exposure through drinking water and diet. *Int. J. Cancer* **2016**, *139*, 334–346. [CrossRef] [PubMed]
56. Fathmawati, F.J.; Gravitiani, E.; Sarto Husodo, A.H. Nitrate in drinking water and risk of colorectal cancer in Yogyakarta, Indonesia. *J. Toxicol. Environ. Health A* **2017**, *80*, 120–128. [CrossRef] [PubMed]
57. Taneja, P.; Labhasetwar, P.; Nagarnaik, P.; Ensink, J.H.J. The risk of cancer as a result of elevated levels of nitrate in drinking water and vegetables in Central India. *J. Water Health* **2017**, *15*, 602–614. [CrossRef] [PubMed]
58. Quist, A.J.L.; Inoue-Choi, M.; Weyer, P.J.; Anderson, K.E.; Cantor, K.P.; Krasner, S.; Freeman, L.E.B.; Ward, M.H.; Jones, R.R. Ingested nitrate and nitrite, disinfection by-products, and pancreatic cancer risk in postmenopausal women. *Int. J. Cancer* **2018**, *142*, 251–261. [CrossRef] [PubMed]
59. Schullehner, J.; Hansen, B.; Thygesen, M.; Pedersen, C.B.; Sigsgaard, T. Nitrate in drinking water and colorectal cancer risk: A nationwide population-based cohort study. *Int. J. Cancer* **2018**, *143*, 73–79. [CrossRef]
60. Zhang, T.; Pan, D.; Su, M.; Fu, L.M.; Miao, C.Y.; Yan, Q.Y.; Wang, J.; Yang, L.G.; Wang, S.K.; Sun, G.J. Determination of dietary nitrite in patients with esophageal pre-cancerous lesion and normal people: A duplicate diet study. *Food Addit. Contam.* **2018**, *35*, 2298–2308. [CrossRef]
61. Zheng, J.; Stuff, J.; Tang, H.; Hassan, M.M.; Daniel, C.R.; Li, D. Dietary N-nitroso compounds and risk of pancreatic cancer: Results from a large case-control study. *Carcinogenesis* **2019**, *40*, 254–262. [CrossRef]
62. Jones, R.R.; DellaValle, C.T.; Weyer, P.J.; Robien, K.; Cantor, K.P.; Krasner, S.; Beane Freeman, L.E.; Ward, M.H. Ingested nitrate, disinfection by-products, and risk of colon and rectal cancers in the Iowa Women's Health Study cohort. *Environ. Int.* **2019**, *126*, 242–251. [CrossRef]
63. Wells, G.A.; Shea, N.D.; O'Connell, J.P.; Welch, V.M.; Losos, P.T. *The Newcastle-Ottawa Scale (NOS) for Assessing the Quality of Nonrandomised Studies in Meta-Analyses*; The Ottawa Health Research Institute: Ottawa, ON, USA, 2006.
64. Stang, A. Critical evaluation of the Newcastle-Ottawa scale for the assessment of the quality of nonrandomized studies in meta-analyses. *Eur. J. Epidemiol.* **2010**, *25*, 603–605. [CrossRef]
65. Greenland, S. Quantitative methods in the review of epidemiologic literature. *Epidemiol. Rev.* **1987**, *9*, 1–30. [CrossRef] [PubMed]
66. Higgins, J.P.; Thompson, S.G. Quantifying heterogeneity in a meta-analysis. *Stat. Med.* **2002**, *21*, 1539–1558. [CrossRef] [PubMed]
67. Ades, A.E.; Lu, G.; Higgins, J.P. The interpretation of random-effects meta-analysis in decision models. *Med. Decis. Mak.* **2005**, *25*, 646–654. [CrossRef] [PubMed]
68. Egger, M.; Davey Smith, G.; Schneider, M.; Minder, C. Bias in meta›analysis detected by a simple, graphical test. *BMJ* **1997**, *315*, 629–634. [CrossRef]
69. Duval, S.; Tweedie, R. Trim and Fill: A Simple Funnel-Plot-Based Method of Testing and Adjusting for Publication Bias in Meta-Analysis. *Biometrics* **2000**, *56*, 455–463. [CrossRef]
70. Bartsch, H.; O'Neill, L.K. Ninth International Meeting on N-NitrosoCompounds: Exposures, Mechanisms, and Relevance to Human Cancer. *Cancer Res.* **1988**, *48*, 4711–4714.
71. Mirvish, S.S. Effects of vitamins C and E on N-nitroso compound formation, carcinogenesis, and cancer. *Cancer* **1986**, *58*, 1842–1850. [CrossRef]
72. Yang, L.; Kartsonaki, C.; Yao, P.; de Martel, C.; Plummer, M.; Chapman, D.; Guo, Y.; Clark, S.; Walters, R.G.; Chen, Y.; et al. China Kadoorie Biobank Collaborative Group. The relative and attributable risks of cardia and non-cardia gastric cancer associated with Helicobacter pylori infection in China: A case-cohort study. *Lancet Public Health* **2021**, *6*, e888–e896. [CrossRef]
73. Jones, R.R.; Weyer, P.J.; DellaValle, C.T.; Inoue-Choi, M.; Anderson, K.E.; Cantor, K.P.; Krasner, S.; Robien, K.; Freeman, L.E.; Silverman, D.T.; et al. Nitrate from Drinking Water and Diet and Bladder Cancer Among Postmenopausal Women in Iowa. Environ. *Health Perspect.* **2016**, *124*, 11. [CrossRef]

toxics

MDPI

Article

Subchronic Low-Dose Methylmercury Exposure Accelerated Cerebral Telomere Shortening in Relevant with Declined Urinary aMT6s Level in Rats

Xi Wu [1], Ping Li [2], Junyan Tao [1], Xiong Chen [1,*] and Aihua Zhang [1,*]

[1] The Key Laboratory of Environmental Pollution Monitoring and Disease Control, Ministry of Education, School of Public Health, Guizhou Medical University, Guiyang 550025, China
[2] State Key Laboratory of Environmental Geochemistry, Institute of Geochemistry, Chinese Academy of Sciences, Guiyang 550081, China
* Correspondence: scicelnat_cx@163.com (X.C.); aihuagzykd@163.com (A.Z.)

Abstract: Methylmercury (MeHg) is a global pollutant with established toxic effects on the central nervous system (CNS). However, early events and early-warning biomarkers of CNS damage following exposure to low-dose MeHg are still lacking. This study aimed to investigate whether subchronic low-dose MeHg exposure had adverse effects on the cerebral telomere length, as well as serum melatonin and its urinary metabolite 6-sulfatoxymelatonin (aMT6s) in rats. Sixteen male Sprague Dawley rats were divided into two groups. Group I was the control group. In group II, rats were exposed to MeHg by gavage at a dose of 0.1 mg/kg/day for 3 months. This study revealed that MeHg exposure resulted in impairment of learning and memory ability, a slightly reduced number of neurons and an irregular arrangement of neurons in the hippocampus. It also significantly accelerated telomere shortening in the cerebral cortex, hippocampus and hypothalamus. Moreover, MeHg exposure decreased the levels of melatonin in serum and aMT6s in urine, partly by suppressing the synthesis of 5-hydroxytryptamine (5-HT) in the brain but promoted the expression of melatonin-catalyzing AANAT and ASMT. Importantly, cerebral telomere length was positively correlated with MT and aMT6s after MeHg exposure. These results suggested that the shortened telomere length in the brain may be an early event in MeHg-induced CNS toxicity, and the level of aMT6s in urine may serve as an early-warning biomarker for MeHg-induced CNS damage.

Keywords: methylmercury; telomere shortening; melatonin; 6-sulfatoxymelatonin; 5-hydroxytryptamine

Citation: Wu, X.; Li, P.; Tao, J.; Chen, X.; Zhang, A. Subchronic Low-Dose Methylmercury Exposure Accelerated Cerebral Telomere Shortening in Relevant with Declined Urinary aMT6s Level in Rats. *Toxics* **2023**, *11*, 191. https://doi.org/10.3390/toxics11020191

Academic Editors: Esref Demir and Sam Kacew

Received: 25 January 2023
Revised: 16 February 2023
Accepted: 17 February 2023
Published: 18 February 2023

1. Introduction

Mercury is known to be a widespread environmental contaminant and one of the most toxic heavy metals detectable in the environment [1]. Even at low concentrations, mercury exposure can cause a variety of health problems [2,3]. It has been widely used in metallurgy, scientific measuring instruments (such as thermometers and barometers), dental amalgam fillers and other manufacturing activities, and can enter the human body in various ways, such as the atmosphere, soil, water and food [4,5], seriously threatening human health. Methylmercury (MeHg) is one of the most toxic forms of mercury, with particularly adverse effects on the central nervous system (CNS). Subchronic MeHg exposure has been reported to affect brain development and cause motor and cognitive impairments in children, as well as neurological damage and neurodegenerative disease in adults [6–8]. Nervous system damage caused by MeHg is related to the fact that the brain is a highly oxygen-consuming organ and, therefore, more prone to oxidative stress [9]. Moreover, MeHg can also reduce the antioxidant activity of cells by interacting directly with antioxidants or selenium [10].

Telomeres are sequences of genes found at the ends of chromosomes and are responsible for maintaining genome integrity [11]. A shorter telomere length predicts an increasing risk of disease [12]. An increasing body of evidence suggests that the shortened telomere length may reflect the adverse health effects of environmental pollutants [13,14]. Studies have shown that the generation of reactive oxygen species and consequent oxidative stress is an important toxic mechanism of MeHg [4]. Excessive oxidative stress can induce telomere damage, especially to the repetitive series structure, which is rich in base G and causes telomeres to break and shorten under non-replication conditions [15]. However, whether MeHg has a negative effect on cerebral telomere length is unclear.

Melatonin is a neuroendocrine hormone that is primarily synthesized in the pineal gland and mainly synthesized by 5-hydroxytryptamine (5-HT) under the continuous catalysis of arylalkylamine-N-acetyltransferase (AANAT) and acetylserotonin O-methyltransferase (ASMT) [16,17]. It is a powerful endogenous antioxidant that is effective in preventing oxidative-stress-induced cellular oxidative damage [18]. Importantly, melatonin, when synthesized under physiological conditions with a circadian rhythm, has been shown to have a protective effect regarding mercury-induced CNS damage [19–21]. Therefore, it is significant to investigate the potential impact of MeHg on endogenous melatonin secretion for the prevention and treatment of MeHg-induced CNS damage.

Melatonin synthesis and secretion are reflected by detecting melatonin levels in peripheral blood; this collection is difficult and invasive. In vivo, melatonin is catalyzed into 6-hydroxymelatonin by liver microsomal hydroxylase, and nearly 60–80% of 6-hydroxymelatonin binds to sulfate to form 6-sulfatoxymelatonin (aMT6s) [22]. At present, there is a growing body of evidence suggesting that urinary aMT6s levels could reflect the circulating melatonin level [23,24]. Thus, melatonin secretion can also be evaluated by detecting the urinary aMT6s level.

Therefore, this study aimed to investigate the adverse effect of subchronic low-dose MeHg exposure on CNS and its underlying mechanism in rats, including cerebral telomere length and melatonin secretion level, and to analyze the association between urinary aMT6s level and cerebral telomere length. This study provides evidence of urinary aMT6s level as an effective early-warning biomarker to assess MeHg-induced CNS damage, as well as providing insight regarding the prevention of CNS damage in populations exposed to MeHg.

2. Materials and Methods

Warning: The use of methyl mercury is extremely hazardous and requires special precautions during handling to reduce the risk of harm.

2.1. Animals and Experimental Design

Male Sprague Dawley rats weighing 180–200 g were purchased from and maintained in the Experimental Animal Center of Guizhou Medical University (Guiyang, China) The rats were provided with commercial feed obtained from Henan Huanyu Hekang Biotechnology Co., Ltd (Anyang, China). The feed contained corn, fish meal, flour, vegetable oil, vitamins and trace elements, amino acids, etc. During the feeding process, the rats drank and ate ad libitum under a 12-h light/12-h dark cycle; the temperature of the rats' housing environment was 22 ± 2 °C. All animals were acclimatized to the facility for 7 days before the experiment began. Sixteen rats were divided into two groups, with eight rats in each group: a control group and MeHg exposure group. The MeHg (CH_3HgCl, Sigma-Aldrich, Helsinki, Finland) was administered by gavage at a dose of 0.1 mg/kg/day (equivalent to 0.116 mg/kg/day of CH_3HgCl). The dose of MeHg (0.1 mg/kg/day) was selected based on a previous study of the daily MeHg ingestion by Brazilian Riparian communities exposed to MeHg through contaminated fish intake [25]. Although the dose is not exactly comparable to that in fish-eating communities, it is much more representative of environmental MeHg exposure than traditional studies of the effects of MeHg exposure in animals. Exposure lasted 3 months, which is representative of subchronic exposure in rats.

During the experiments, the weights of rats were recorded weekly. Rat urine was collected at 19:00–22:00/22:00–1:00/1:00–4:00/4:00–7:00 time periods for aMT6s analysis one week before the rats were sacrificed. After feeding for 3 months, rats were anesthetized with 3% pentobarbital sodium, and blood samples were taken from the heart. Half of each rat's blood was collected and placed in the anticoagulant tube to detect Blood-Hg concentrations; the rest was placed in a non-anticoagulant tube, which was stored at room temperature for 1 h and centrifuged for 20 min at 1000× g at 4 °C; then, the supernatant was removed to test the serum melatonin level. Pineal glands were homogenized to detect the mRNA level of AANAT and ASMT. Rat brains were weighed, washed and dissected into two portions. The brain somatic index was calculated using the formula (100 × [brain weight (g)/body weight (g)]). Half of the brain was fixed in 4% paraformaldehyde solution for later histological analysis. The rest of the brain tissue was separated into three portions (cortex, hippocampus and hypothalamus), half of each portion was used to detect telomeres, and the other half was used to detect the level of 5-HT.

All experiments and procedures associated with this study were performed in accordance with guidelines for animal care and use and approved by the ethics committee of the Guizhou Medical University (Ethics No. 2200493).

2.2. Morris Water Maze

Behavioral testing was performed using the Morris water maze. The assay was performed based on the methods reported in a previous study [26]. The black circular pool (120 cm in diameter and 60 cm in height) was filled with tap water (22 ± 1 °C) until the escape platform (10 cm diameter) was 2 cm below the surface.

Positioning and navigation experiment: First, the pool was divided into four quadrants; the platform was located in the fourth quadrant, 2 cm below the water level. The time provided to the rat to find the hidden platform was 120 s, and if it failed, it was gently guided to the platform and was allowed to remain there for 20 s. The training lasted for 4 days. Space search experiment: On the 5th day, the platform was removed, one quadrant was chosen, and the rat surface wall went into the water. In the absence of the platform, the time taken and the first time reached the platform of the rats were recorded.

2.3. ELISA

After adding RIPA lysis buffer, the cortex, hippocampus and hypothalamus tissue were, respectively, placed into the grinder to be grinded. The supernatant was taken, and total protein concentration was determined with the BCA kit (Solarbio, Beijing, China). The sample was diluted with PBS to ensure that the total protein in each test hole of 5-HT did not exceed 0.3 mg. The urine collected at each time was centrifuged at 4000 rpm for 10 min, and the supernatant was then taken. MT levels (Elabscience, Wuhan, China) in serum, aMT6s levels (RunYu, Shanghai, China) in urine and 5-HT levels (FineTest, Wuhan, China) in brain were measured using commercially available enzyme-linked immunosorbent assay (ELISA) kits according to the manufacturer's instructions.

2.4. Hematoxylin–Eosin (HE) Staining

After fixation with 4% paraformaldehyde for 24 h, the brain tissue was embedded in paraffin and sliced to 4 μm. Paraffin sections were dewaxed twice in xylene, dyed with hematoxylin solution and washed with distilled water to remove floating color. The differentiation solution was rinsed twice with tap water after differentiation. Then, each section was dyed with eosin solution, dehydrated with anhydrous ethanol, made transparent with xylene, and sealed with a neutral gum. The images were obtained using a microscope (Nikon Eclipse E100, Tokyo, Japan).

2.5. Nissl Staining

The fixed brain tissue was washed with running water and immersed in paraffin for embedding. Slice thickness was 5 μm, and slicers were dewaxed to water. The sections were placed in Cresyl violet stain, and the dye tank was immersed in a 56 °C box for 1 h. Each section was placed in Nissl Differentiation for a few seconds (until the background was nearly colorless). Each section was dehydrated, made transparent and sealed. Image acquisition was finished using a microscope (Nikon Eclipse E100, Tokyo, Japan).

2.6. Determine Blood-Hg Concentrations

Whole-blood samples were taken and placed in a 25 mL borosilicate glass colorimetric tube. Then, 5 mL of HNO_3 was added, acid-treated glass spheres were placed in the colorimetric tube, and samples were digested in water bath at 95 °C for 3 h. After cooling, a small amount of ultra-pure water was added first, and then 0.5 mL of BrCl was added. About 24 h later, 2–3 drops of $NH_2OH{\cdot}HCl$ solution were added to create yellow recede, and ultra-pure water was added at a constant volume to 25 mL. The method was reduced by $SnCl_2$ and determined by cold atomic fluorescence spectrometry (AF-630A, Beijing, China).

2.7. Real-Time Quantitative PCR

Total RNA was extracted from the pineal tissue using TRIzol reagent (Invitrogen, Carlsbad, CA, USA), according to the manufacturer's instructions. Complementary DNA was synthesized using a RevertAid First Strand cDNA Synthesis Kit (Thermo Fisher Scientifific, Vilnius, Lithuania). qPCR for AANAT, ASMT and GAPDH was performed using a Real-Time PCR Detection System. Primer sequences used for qPCR were as follows: forward AANAT 5′-GTG GCT GCT GAC CCA AG-3′, reverse AANAT 5′-TGC TGT CTC CCT TCA TGC T-3′, forward ASMT 5′-GTG CCT GCG TGG AGT TG-3′, reverse ASMT 5′-CCA TGA CCC TGT GAC CCT-3′, forward GAPDH 5′-TCT CTG CTC CTC CCT GTT C-3′, and reverse GAPDH 5′-ACA CCG ACC TTC ACC ATC T-3′. PCR amplifications were performed in a total volume of 20 μL using SuperReal PreMix (SYBR Green) (Tiangen) with an iCycler thermal cycler (Bio-Rad, CFX96™ Optics Module). The gene expression level was calculated using the $2^{-\Delta\Delta Ct}$ method, and the relative AANAT and ASMT level was normalized to that of GAPDH.

2.8. Genomic DNA Isolation and Telomere Length Analysis

Genomic DNA was extracted from the PBMCs isolated from whole blood using TIANamp Genomic DNA Kit (Tiangen, Beijing, China). DNA concentration was measured using the microplate reader. Samples were diluted to a final concentration of 25 ng/1.5 μL to measure telomere length. qPCR was performed using SuperReal PreMix (SYBR Green) (Tiangen, Beijing, China). Primers used were as follows: forward TEL 5′-GGT TTT TGA GGG TGA GGG TGA GGG TGA GGG TGA GGG t-3′, reverse TEL 5′-TCC CGA CTA TCC CTA TCC CTA TCC CTA TCC CTA TCC CTA-3′, forward AT1 5′-ACG TGT TCT CAG CAT CGA CCG CTA CC-3′, and reverse AT1 5′-AGA ATG ATA AGG AAA GGG AAC AAG AAG CCC-3′. The relative telomere length was measured by comparing the ratio of telomere repeat copy number (T as Tel1) and single gene copy number (S as AT1), expressed as telomere length (T/S) ratio. Each individual value obtained by qPCR was processed through the formula $T/S = 2^{-\Delta Ct}$, where $\Delta CT = CT_{telomere} - CT_{AT1}$. This ratio was then compared with the ratio of the reference DNA. Each DNA sample collected was measured in duplicate.

2.9. Statistical Analysis

For parametric data, an unpaired Student's t-test was used when there were two groups. For nonparametric data, a Mann–Whitney test was performed when there were two groups. The Spearman correlation coefficient test was used to analyze the correlation between all parameters. The data were expressed as mean ± standard deviation (SD). Differences between groups were considered significant at $p < 0.05$. Statistical analyses were performed using GraphPad Prism software (version 6.0; GraphPad Software Inc., San Diego, CA, USA).

3. Results

3.1. Influence of MeHg on Blood Mercury Concentration and Brain Somatic Index in Rats

To determine the internal load of MeHg in rats, the blood mercury concentration was detected. Compared with the control group, MeHg exposure for 3 months induced a significant increase in the mercury concentration in rat blood ($p < 0.05$; Figure 1A). To assess the effect of MeHg on body growth and brain development, the body weight and brain weight were measured. The body weight of MeHg-exposed rats was significantly lower than that of the control group from the 10th week ($p < 0.05$; Figure 1B). MeHg exposure did not show an obvious effect on brain weight ($p > 0.05$; Figure 1C) but increased brain somatic index ($p < 0.05$; Figure 1D).

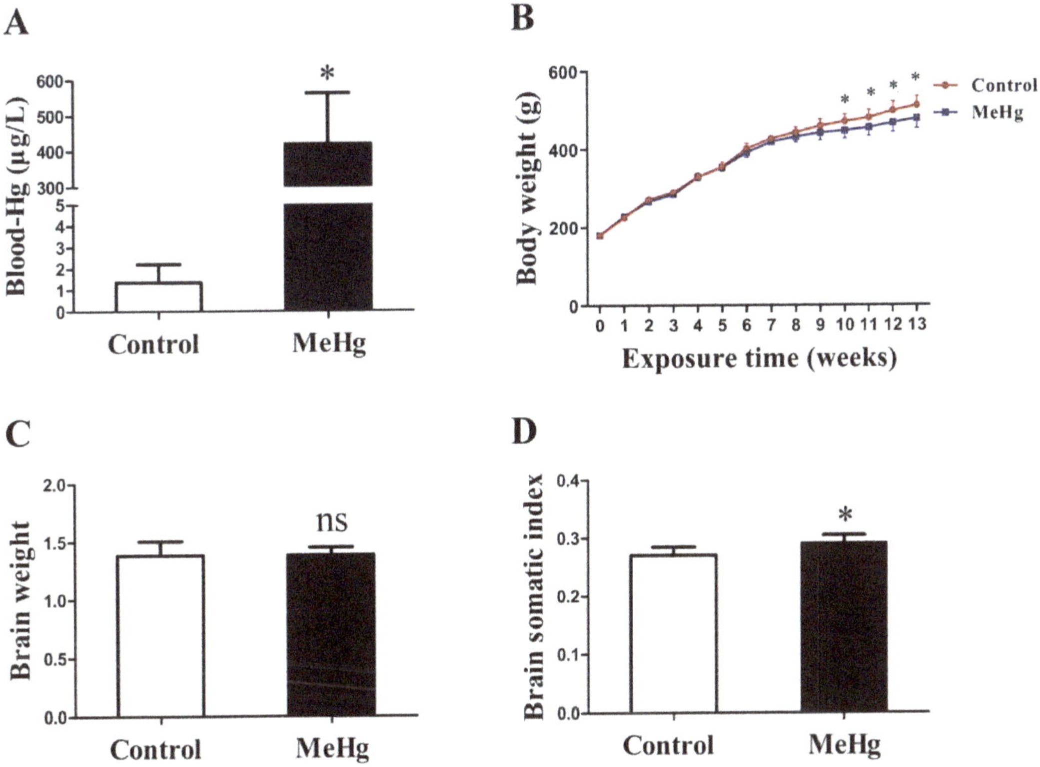

Figure 1. Impact of 0.1 mg/kg/day MeHg exposure for 3 months on mercury concentrations in blood (**A**), body weight (**B**), brain weight (**C**) and brain somatic index (**D**) in rats. Blood-Hg, blood mercury concentration; MeHg, methylmercury. The data are expressed as mean ± SD for eight rats per group. ns, no significance compared to the control group, $p > 0.05$. * $p < 0.05$, significant difference compared to the control group.

3.2. The Effect of MeHg on Spatial Learning and Memory in Rats

To explore the impacts of MeHg on spatial learning and memory, Morris water maze tests were performed. There was no statistically significant difference between the escape latency result of the two groups ($p > 0.05$; Figure 2A). The probe test was performed on the last day of MeHg exposure. The results showed that MeHg-treated rats spent less time in the target quadrant than the control group ($p < 0.05$; Figure 2B), but the difference in swimming speed between the two groups was not statistically significant ($p > 0.05$; Figure 2C). However, MeHg significantly decreased the frequency of entering the target quadrant ($p < 0.05$; Figure 2D). These results suggest that subchronic low-dose MeHg exposure could impair spatial memory performance in rats.

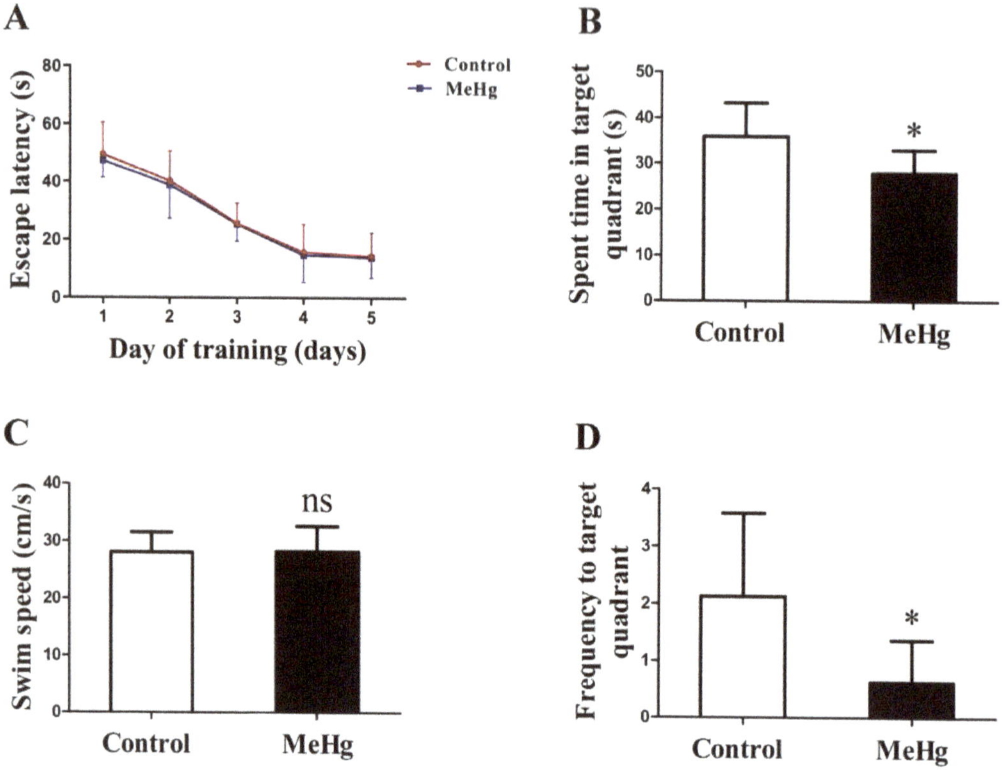

Figure 2. The influence of MeHg exposure on spatial learning and memory in the Morris water maze, including the escape latencies of the groups during the training trials (**A**), time spent in target quadrant (**B**), swimming speed (**C**), and the frequency of entering the target quadrant (**D**). The data are expressed as mean ± SD for eight rats per group. ns, no significance compared to the control group, $p > 0.05$. * $p < 0.05$, significant difference compared to the control group.

3.3. Impact of MeHg on Pathological Morphology and Telomere Length of Rat Brain Tissue

HE-staining and Nissl-staining results showed that, in the control group, the morphology of CA1 and CA3 region in hippocampus was regular and well-organized (Figure 3A). However, in the MeHg exposure group, the layer of pyramidal cells was slightly thinner in the CA1 and CA3 regions of the hippocampus (Figure 3A), and the number of neurons in the CA1 and CA3 regions of the hippocampus were slightly reduced and showed an irregular arrangement (Figure 3B). Histograms of the Nissl body counts in the hippocampus quantified from Nissl staining analysis (Figure 3C). The telomere length in cerebral cortex, hippocampus and hypothalamus were significantly shortened after exposure to MeHg compared with the control ($p < 0.05$; Figure 3D–F).

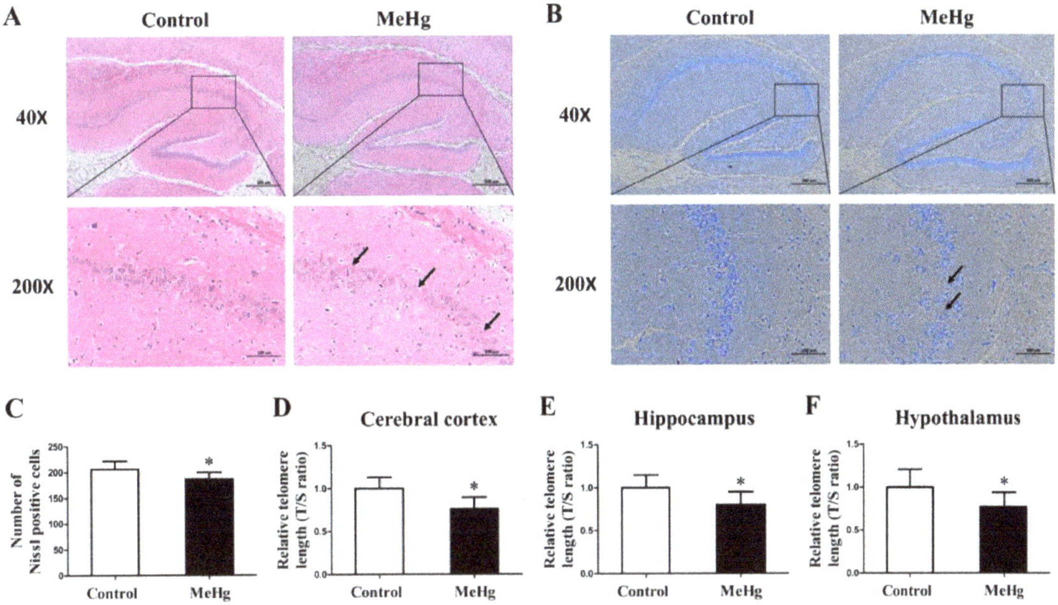

Figure 3. The influence of MeHg exposure on pathological morphology of rat brain tissue detected by HE staining (**A**) and Nissl staining (**B**,**C**) as well as the telomere length in cerebral cortex (**D**), hippocampus (**E**) and hypothalamus (**F**). The pyramidal cells in the regions indicated by black arrows was decreased and showed an irregular arrangement. The data are expressed as mean ± SD for eight rats per group. * $p < 0.05$, significant difference compared to the control group.

3.4. MeHg-Induced Alterations in the Levels of Serum Melatonin and Urinary aMT6s in Rats

The serum level of melatonin in the rats was significantly decreased after MeHg exposure for 3 months ($p < 0.05$; Figure 4A). To further illustrate the impact of MeHg on melatonin metabolism, the aMT6s level in the urine was measured. The results showed that the urinary aMT6s level in rats was also statistically significantly declined in the MeHg-exposure group compared to the control group ($p < 0.05$; Figure 4B).

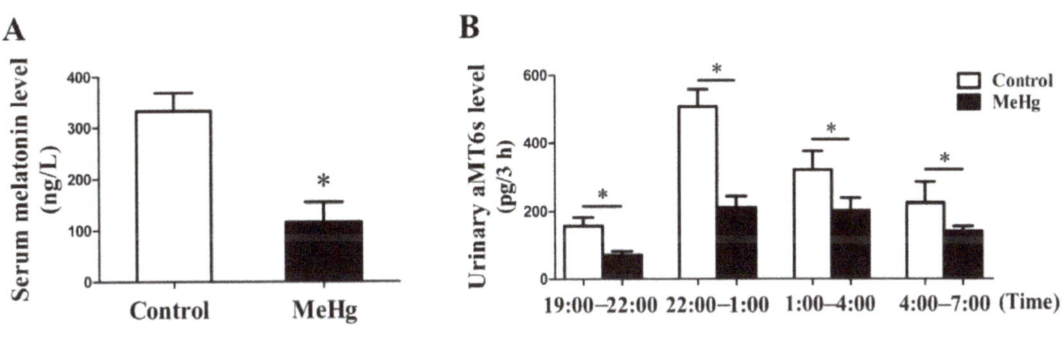

Figure 4. Impact of MeHg exposure on serum melatonin (**A**) and urinary aMT6s (**B**) in rats. MT, melatonin. The data are expressed as mean ± SD for eight rats per group. * $p < 0.05$, significant difference compared to the control group.

3.5. Influence of MeHg on the Expression of Melatonin Synthetase and 5-HT

To elucidate the cause of declining serum melatonin levels under low-dose MeHg exposure, we further detected the expression of melatonin synthetase in the hippocampus and 5-HT in selected brain regions. The results showed that the mRNA expression levels of melatonin-catalyzing enzymes AANAT and ASMT in the pineal tissue of rats were increased in the MeHg-exposure group compared to the control group ($p < 0.05$; Figure 5A,B). The 5-HT level in the cerebral cortex and hippocampus ($p < 0.05$; Figure 5C,D) was significantly decreased, but there was no obvious difference in the hypothalamus ($p > 0.05$; Figure 5E) after exposure to MeHg when compared with the control.

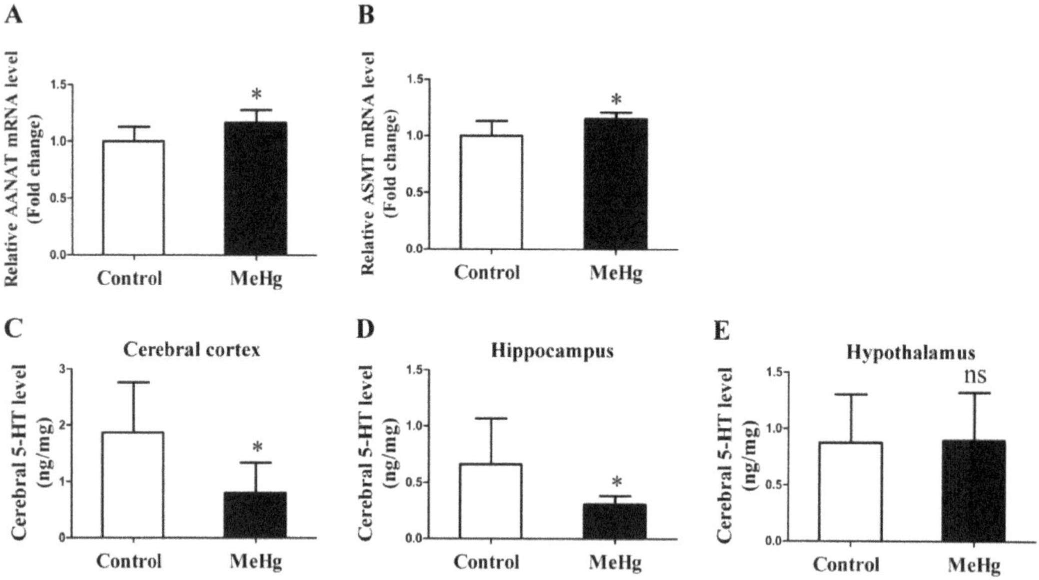

Figure 5. The influence of MeHg exposure on AANAT (**A**) and ASMT (**B**) mRNA expression level and in hippocampal tissue and 5-HT level in cerebral cortex (**C**), hippocampus (**D**) and hypothalamus (**E**). The data are expressed as mean ± SD for eight rats per group. ns, no significance compared to the control group, $p > 0.05$. * $p < 0.05$, significant difference compared to the control group.

3.6. Correlations of the Level of Serum Melatonin and Urinary aMT6s with Cerebral Telomere Length in Rats

To elucidate the possible causal relationship between cerebral telomere shortening and declined levels of serum melatonin and urinary aMT6s, a correlation analysis was conducted. The results of the correlation analysis showed a positive correlation between serum melatonin level and telomere length in the cerebral cortex, hippocampus and hypothalamus ($p < 0.05$; Figure 6A–C) in the two experiment groups ($n = 16$). There was also a positive correlation between urinary aMT6s level and telomere length in the cerebral cortex, hippocampus and hypothalamus ($p < 0.05$; Figure 6D–F) in the two experiment groups ($n = 16$).

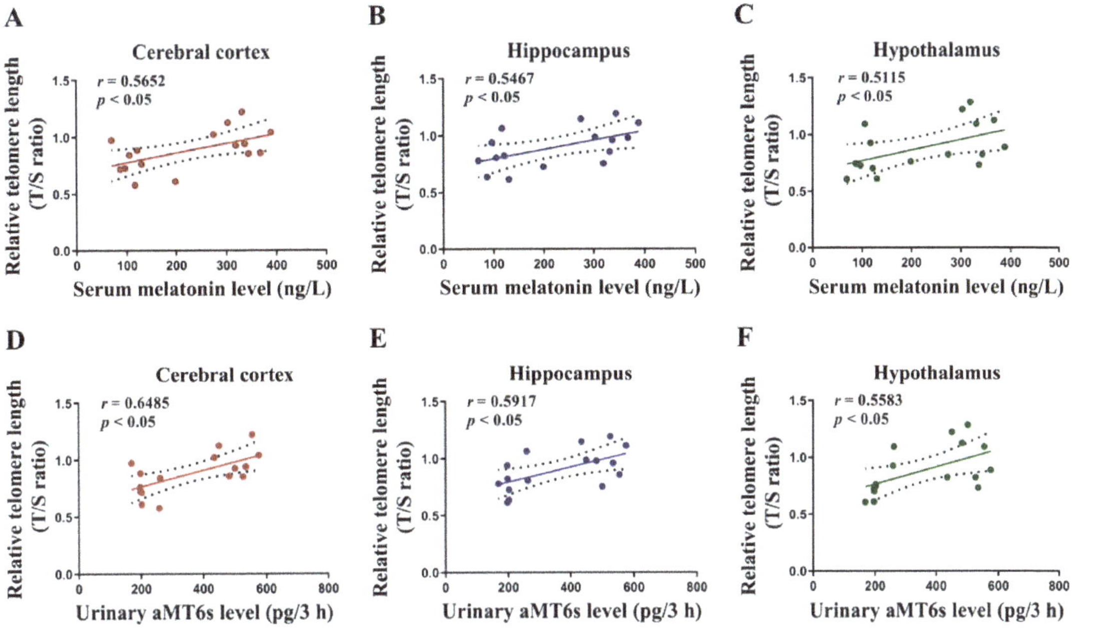

Figure 6. Correlations of serum melatonin and urinary aMT6s level with cerebral telomere length in rats. The serum levels of melatonin were positively correlated with telomere length in cerebral cortex ($r = 0.5652$, $p < 0.05$), hippocampus ($r = 0.5467$, $p < 0.05$) and hypothalamus ($r = 0.5115$, $p < 0.05$) (**A–C**), and the urinary levels of aMT6s were positively correlated with telomere length in cerebral cortex ($r = 0.6485$, $p < 0.05$), hippocampus ($r = 0.5917$, $p < 0.05$) and hypothalamus ($r = 0.5583$, $p < 0.05$) (**D–F**).

3.7. The Hypothesis of Adverse Impacts of MeHg on Cerebral Telomere Length and Causal Relationship with Reduced Melatonin Synthesis and Metabolites

As shown in Figure 7, these results indicate that subchronic low-dose MeHg exposure could cause neuronal damage and telomere shortening, resulting in decreased neuronal synthesis of 5-HT, which, in turn, reduces the level of melatonin synthesis and secretion, as well as the level of aMT6s in urine.

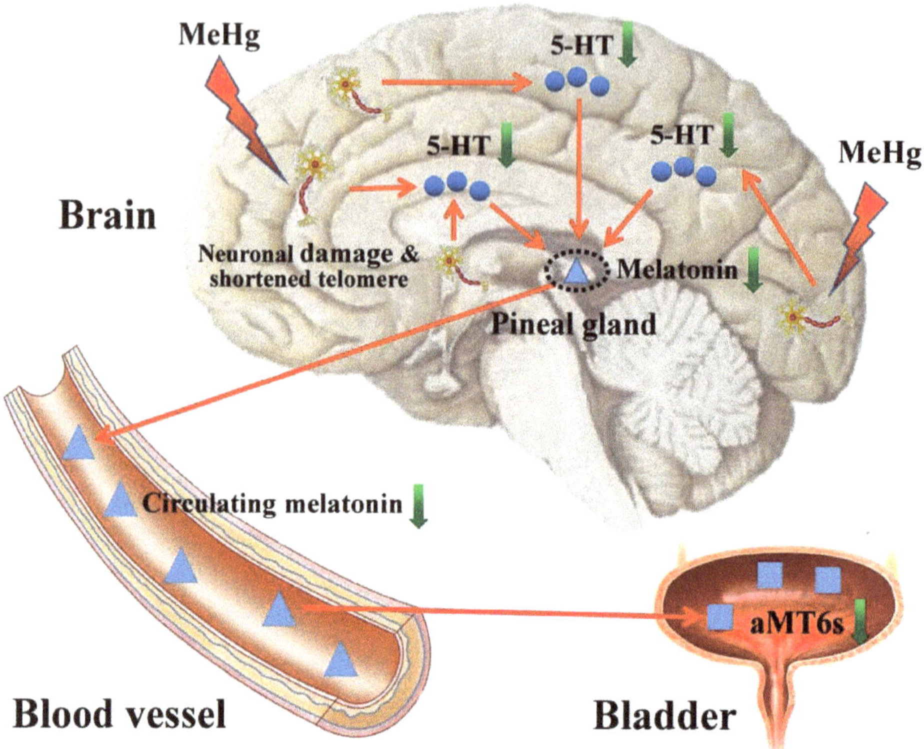

Figure 7. Proposed model of adverse impacts of MeHg on cerebral telomere length and causal relationship with reduced melatonin synthesis and metabolites. The blue dots indicate 5-HT, the blue triangles indicate melatonin and the blue squares indicate aMT6s.

4. Discussion

MeHg exposure has been identified as a neurotoxic substance [27] and a risk factor for neurodegenerative disease in humans [7]. Despite knowledge of the adverse effects of high-dose MeHg exposure on the CNS, evidence of the neurotoxic effects of low-dose and continual MeHg exposure is limited. In addition, there is a lack of effective early-warning biomarkers to assess low-dose MeHg-induced CNS damage.

The neurotoxic effects of MeHg may be significantly related to the exposure dose. Previous studies have indicated that MeHg exposure at high doses ($\geq$1 mg/kg/day) induces neurodevelopmental toxicity and behavioral disorders [28,29]. Valentini and Grotto showed that subchronic MeHg exposure at low doses (0.1 mg/kg/day) decreased the activities of butyrylcholinesterase and antioxidant enzymes in plasma [30,31]. In the present study, subchronic MeHg exposure at dose of 0.1 mg/kg/day decreased the frequency of entering the target quadrant and spending time in the target quadrant in Sprague Dawley rats undergoing a Morris water maze experiment. However, the escape latency showed no statistical significance between the two experimental groups, as shown in Figure 2. Furthermore, low-dose MeHg exposure reduced the number of neurons in hippocampal CA1 and CA3 regions slightly and affected neuronal arrangement in rats. These results suggest that subchronic low-dose MeHg exposure induces a kind of early and indiscoverable neurotoxic effect.

Telomere shortening is a hallmark of cell senescence [32]. However, accelerated telomere shortening and cell senescence in brain cells has been associated with several neurodegenerative diseases, including Parkinson's and Alzheimer's diseases [33–35]. Investigating the effect of MeHg on the telomere length of brain cells may provide a basis for the prevention of neurological diseases. Although an epidemiological study has shown that MeHg exposure was not associated with leukocyte telomere length in both mothers and their children [36], an animal study revealed that MeHg reduced the expression of telomerase reverse transcriptase (TERT) in mice brains [37]. Our results further suggest that low-dose MeHg exposure for 3 months could significantly reduce the telomere length in the cerebral cortex, hippocampus and hypothalamus in rats. Therefore, the shortened telomere length of brain cells may be an important early manifestation of CNS damage induced by MeHg at low levels.

Melatonin is an important endogenous hormone secreted by the pineal gland [16]. Numerous studies have shown that exogenous melatonin could antagonize mercury-induced neurotoxicity [19–21]. To date, however, the effect of MeHg on endogenous melatonin secretion is unknown. In general, the level of serum melatonin was a good indicator, reflecting the secretion of melatonin and, hence, the need to detect the level of serum melatonin. Our results showed that the serum melatonin level in rats exposed to MeHg was significantly lower than those in the control group, indicating that low-dose MeHg may inhibit endogenous melatonin synthesis.

There is a growing body of evidence suggesting that the level of aMT6s in urine could reflect the circulating melatonin [23,24]. In addition, detection biomarkers in urine are non-invasive and more readily available compared to blood; therefore, we further examined the effect of MeHg exposure on the level of melatonin urinary metabolite aMT6s. Previous research has shown that the level of aMT6s in urine reached a peak at 22:00–1:00 a.m. in rats [38,39]. We examined urinary aMT6s levels in four consecutive time periods in rats. In accordance with previous studies, the results showed that the aMT6s level at 22:00–1:00 a.m. was the highest and then declined. Moreover, the urinary aMT6s level in the MeHg-exposure group was lower than that in the control group at the four selected time periods. Therefore, aMT6s level in urine may reflect the circulating melatonin level and melatonin secretion to some extent following MeHg exposure.

To explain the reason for the decline in melatonin secretion induced by MeHg, we focused on two aspects. On the one hand, studies have demonstrated that AANAT and ASMT are two key catalyzing enzymes in the synthesis of melatonin [40] and can catalyze the synthesis of 5-HT into melatonin [41]. Thus, the effect of MeHg on the expression of AANAT and ASMT was detected in the pineal gland. Contrary to expectation, however, the expression of both enzymes increased rather than decreased under MeHg exposure. It is speculated that the increased expression of AANAT and ASMT may be related to the decreased level of melatonin through a type of negative feedback regulation. On the other hand, previous studies have shown that MeHg exposure caused a marked reduction in 5-HT in rat brains [42,43]. The expression of 5-HT in rat brain tissue was further detected. In accordance with previous studies, our results showed that 5-HT level in the cerebral cortex and hippocampus significantly decreased but that there was no obvious difference in the hypothalamus after exposure to MeHg. Since 5-HT was primarily synthesized in neurons and concentrated in synapses in the CNS [44], we hypothesized that the MeHg-induced decline in 5-HT could be attributed to neuronal damage.

Based on the above findings, it was speculated that there may be a correlation between neuron damage and the secretion and metabolism of melatonin under MeHg exposure. Therefore, the correlation between brain tissue telomere length and the level of serum melatonin and aMT6s in urine at peak time (namely, 22:00–1:00 a.m.) was further analyzed. The results showed that the telomere lengths of the three brain regions were all positively correlated with the level of serum melatonin and urinary aMT6s, suggesting that aMT6s in urine may act as an early sensitive indicator of the CNS damage caused by MeHg.

5. Conclusions

Taken together, this study reveals that subchronic low-dose MeHg exposure accelerated cerebral telomere shortening in rats. Moreover, low-dose MeHg exposure led to declines in melatonin synthesis and metabolism, possibly due to the reduced 5-HT synthesis as a result of neuronal damage. Importantly, the urinary aMT6s level was positively correlated with a shortened cerebral telomere length, indicating that urinary aMT6s may be an early-warning biomarker to assess low-dose MeHg-induced CNS damage. This study will provide insight into the prevention of CNS damage for populations exposed to MeHg. As the results were from animal experimental research, the association between the degree of CNS damage and urinary aMT6s level in populations exposed to environmental MeHg needs further investigation.

Author Contributions: Methodology, X.W.; data curation, X.W.; writing—original draft preparation, X.W.; writing—review and editing, J.T., X.C. and A.Z.; visualization, X.W. and J.T.; supervision, P.L., X.C. and A.Z.; project administration, X.C. and A.Z.; funding acquisition, X.C. and A.Z. All authors have read and agreed to the published version of the manuscript.

Funding: This research was funded by Joint Fund of the National Natural Science Foundation of China-Guizhou Province Karst Scientific Research Center Project (U1812403).

Institutional Review Board Statement: All experiments and procedures associated with this study were performed in accordance with guidelines for animal care and use approved by the ethics committee of the Guizhou Medical University (Ethics No. 2200493).

Informed Consent Statement: Not applicable.

Data Availability Statement: The data presented in this study are available upon request from the corresponding author.

Conflicts of Interest: The authors declare that they have no known competing financial interest or personal relationships that could have appeared to influence the work reported in this paper.

References

1. Wei, Y.; Lyu, Y.; Cao, Z.; Zhao, F.; Liu, Y.; Chen, C.; Li, C.; Gu, H.; Lu, F.; Zhou, J.; et al. Association of low cadmium and mercury exposure with chronic kidney disease among Chinese adults aged ≥80 years: A cross-sectional study. *Chin. Med. J.* 2022, *in press*. [CrossRef]
2. Zahir, F.; Rizwi, S.J.; Haq, S.K.; Khan, R.H. Low dose mercury toxicity and human health. *Environ. Toxicol. Pharmacol.* **2005**, *20*, 351–360. [CrossRef]
3. Kim, S.; Song, S.H.; Lee, C.W.; Kwon, J.T.; Park, E.Y.; Oh, J.K.; Kim, H.J.; Park, E.; Kim, B. Low-Level Environmental Mercury Exposure and Thyroid Cancer Risk Among Residents Living Near National Industrial Complexes in South Korea: A Population-Based Cohort Study. *Thyroid* **2022**, *32*, 1118–1128. [CrossRef]
4. Rice, K.M.; Walker, E.M., Jr.; Wu, M.; Gillette, C.; Blough, E.R. Environmental mercury and its toxic effects. *J. Prev. Med. Public Health* **2014**, *47*, 74–83. [CrossRef]
5. Renzoni, A.; Zino, F.; Franchi, E. Mercury levels along the food chain and risk for exposed populations. *Environ. Res.* **1998**, *77*, 68–72. [CrossRef]
6. Gustin, K.; Tofail, F.; Mehrin, F.; Levi, M.; Vahter, M.; Kippler, M. Methylmercury exposure and cognitive abilities and behavior at 10 years of age. *Environ. Int.* **2017**, *102*, 97–105. [CrossRef]
7. Weiss, B.; Clarkson, T.W.; Simon, W. Silent latency periods in methylmercury poisoning and in neurodegenerative disease. *Environ. Health Perspect.* **2002**, *110*, 851–854. [CrossRef]
8. Bittencourt, L.O.; Matta, P.P.M.; Nascimento, P.C.; Eiró-Quirino, L.; Aragão, W.A.B.; Dionizio, A.; Fernandes, L.M.P.; Silva, M.C.F.; Buzalaf, M.A.R.; Aschner, M.; et al. Deciphering the Global Proteomic Profile Involved in Methylmercury-Induced Cerebellar Neurodegeneration and Motor Dysfunction in Adult Rats. *Toxics* **2022**, *10*, 531. [CrossRef]
9. Li, X.; Pan, J.; Wei, Y.; Ni, L.; Xu, B.; Deng, Y.; Yang, T.; Liu, W. Mechanisms of oxidative stress in methylmercury-induced neurodevelopmental toxicity. *Neurotoxicology* **2021**, *85*, 33–46. [CrossRef]
10. Fujimura, M.; Usuki, F. Methylmercury-Mediated Oxidative Stress and Activation of the Cellular Protective System. *Antioxidants* **2020**, *9*, 1004. [CrossRef]
11. O'Sullivan, R.J.; Karlseder, J. Telomeres: Protecting chromosomes against genome instability. *Nat. Rev. Mol. Cell Biol.* **2010**, *11*, 171–181. [CrossRef]
12. Aviv, A.; Shay, J.W. Reflections on telomere dynamics and ageing-related diseases in humans. *Philos. Trans. R. Soc. Lond B Biol. Sci.* **2018**, *373*, 20160436. [CrossRef]
13. Møller, P.; Wils, R.S.; Jensen, D.M.; Andersen, M.H.G.; Roursgaard, M. Telomere dynamics and cellular senescence: An emerging field in environmental and occupational toxicology. *Crit. Rev. Toxicol.* **2018**, *48*, 761–788. [CrossRef]
14. Louzon, M.; Coeurdassier, M.; Gimbert, F.; Pauget, B.; de Vaufleury, A. Telomere dynamic in humans and animals: Review and perspectives in environmental toxicology. *Environ. Int.* **2019**, *131*, 105025. [CrossRef]

15. Henle, E.S.; Han, Z.; Tang, N.; Rai, P.; Luo, Y.; Linn, S. Sequence-specific DNA cleavage by Fe2+-mediated fenton reactions has possible biological implications. *J. Biol. Chem.* **1999**, *274*, 962–971. [CrossRef]
16. Gurunathan, S.; Kang, M.H.; Kim, J.H. Role and Therapeutic Potential of Melatonin in the Central Nervous System and Cancers. *Cancers* **2020**, *12*, 1567. [CrossRef]
17. Pomianowski, K.; Gozdowska, M.; Burzyński, A.; Kalamarz-Kubiak, H.; Sokołowska, E.; Kijewska, A.; Kulczykowska, E. A study of aanat and asmt expression in the three-spined stickleback eye and skin: Not only "on the way to melatonin". *Comp. Biochem. Physiol. A Mol. Integr. Physiol.* **2020**, *241*, 110635. [CrossRef]
18. He, R.; Cui, M.; Lin, H.; Zhao, L.; Wang, J.; Chen, S.; Shao, Z. Melatonin resists oxidative stress-induced apoptosis in nucleus pulposus cells. *Life Sci.* **2018**, *199*, 122–130. [CrossRef]
19. Kim, C.Y.; Nakai, K.; Kameo, S.; Kurokawa, N.; Liu, Z.M.; Satoh, H. Protective effect of melatonin on methylmercury-Induced mortality in mice. *Tohoku J. Exp. Med.* **2000**, *191*, 241–246. [CrossRef]
20. Said, E.S.; Ahmed, R.M.; Mohammed, R.A.; Morsi, E.M.; Elmahdi, M.H.; Elsayed, H.S.; Mahmoud, R.H.; Nadwa, E.H. Ameliorating effect of melatonin on mercuric chloride-induced neurotoxicity in rats. *Heliyon* **2021**, *7*, e07485. [CrossRef]
21. Rao, M.V.; Purohit, A.; Patel, T. Melatonin protection on mercury-exerted brain toxicity in the rat. *Drug Chem. Toxicol.* **2010**, *33*, 209–216. [CrossRef]
22. Minami, M.; Takahashi, H.; Inagaki, H.; Yamano, Y.; Onoue, S.; Matsumoto, S.; Sasaki, T.; Sakai, K. Novel tryptamine-related substances, 5-sulphatoxydiacetyltryptamine, 5-hydroxydiacetyltryptamine, and reduced melatonin in human urine and the determination of those compounds, 6-sulphatoxymelatonin, and melatonin with fluorometric HPLC. *J. Chromatogr. B Analyt. Technol. Biomed. Life Sci.* **2009**, *877*, 814–822. [CrossRef]
23. Lv, X.D.; Liu, S.; Cao, Z.; Gong, L.L.; Feng, X.P.; Gao, Q.F.; Wang, J.; Hu, L.; Cheng, X.C.; Yu, C.H.; et al. Correlation between serum melatonin and aMT6S level for age-related macular degeneration patients. *Eur. Rev. Med. Pharmacol. Sci.* **2016**, *20*, 4196–4201.
24. Yang, T.H.; Chen, Y.C.; Ou, T.H.; Chien, Y.W. Dietary supplement of tomato can accelerate urinary aMT6s level and improve sleep quality in obese postmenopausal women. *Clin. Nutr.* **2020**, *39*, 291–297. [CrossRef]
25. Passos, C.J.; Da Silva, D.S.; Lemire, M.; Fillion, M.; Guimarães, J.R.; Lucotte, M.; Mergler, D. Daily mercury intake in fish-eating populations in the Brazilian Amazon. *J. Expo. Sci. Environ. Epidemiol.* **2008**, *18*, 76–87. [CrossRef]
26. Wang, W.; Liu, L.; Jiang, P.; Chen, C.; Zhang, T. Levodopa improves learning and memory ability on global cerebral ischemia-reperfusion injured rats in the Morris water maze test. *Neurosci. Lett.* **2017**, *636*, 233–240. [CrossRef]
27. Novo, J.P.; Martins, B.; Raposo, R.S.; Pereira, F.C.; Oriá, R.B.; Malva, J.O.; Fontes-Ribeiro, C. Cellular and Molecular Mechanisms Mediating Methylmercury Neurotoxicity and Neuroinflammation. *Int. J. Mol. Sci.* **2021**, *22*, 3101. [CrossRef]
28. Maia, C.S.; Lucena, G.M.; Corrêa, P.B.; Serra, R.B.; Matos, R.W.; Menezes, F.C.; Santos, S.N.; Sousa, J.B.; Costa, E.T.; Ferreira, V.M. Interference of ethanol and methylmercury in the developing central nervous system. *Neurotoxicology* **2009**, *30*, 23–30. [CrossRef]
29. Reardon, A.J.F.; Karathra, J.; Ribbenstedt, A.; Benskin, J.P.; MacDonald, A.M.; Kinniburgh, D.W.; Hamilton, T.J.; Fouad, K.; Martin, J.W. Neurodevelopmental and Metabolomic Responses from Prenatal Coexposure to Perfluorooctanesulfonate (PFOS) and Methylmercury (MeHg) in Sprague-Dawley Rats. *Chem. Res. Toxicol.* **2019**, *32*, 1656–1669. [CrossRef]
30. Valentini, J.; Vicentini, J.; Grotto, D.; Tonello, R.; Garcia, S.C.; Barbosa Jr, F. Sub-Chronic Exposure to Methylmercury at Low Levels Decreases Butyrylcholinesterase Activity in Rats. *Basic Clin. Pharmacol. Toxicol.* **2010**, *106*, 95–99. [CrossRef]
31. Grotto, D.; de Castro, M.M.; Barcelos, G.R.; Garcia, S.C.; Barbosa Jr, F. Low level and sub-chronic exposure to methylmercury induces hypertension in rats: Nitric oxide depletion and oxidative damage as possible mechanisms. *Arch Toxicol.* **2009**, *83*, 653–662. [CrossRef]
32. Bernadotte, A.; Mikhelson, V.M.; Spivak, I.M. Markers of cellular senescence. Telomere shortening as a marker of cellular senescence. *Aging* **2016**, *8*, 3–11. [CrossRef]
33. Wang, J.; Liu, Y.; Xia, Q.; Xia, Q.; Wang, B.; Yang, C.; Liang, J.; Liu, X. Potential roles of telomeres and telomerase in neurodegenerative diseases. *Int. J. Biol. Macromol.* **2020**, *163*, 1060–1078. [CrossRef]
34. Liu, M.; Huo, Y.R.; Wang, J.; Wang, C.; Liu, S.; Liu, S.; Wang, J.; Ji, Y. Telomere Shortening in Alzheimer's Disease Patients. *Ann. Clin. Lab. Sci.* **2016**, *46*, 260–265.
35. Baker, D.J.; Petersen, R.C. Cellular senescence in brain aging and neurodegenerative diseases: Evidence and perspectives. *J. Clin. Investig.* **2018**, *128*, 1208–1216. [CrossRef]
36. Yeates, A.J.; Thurston, S.W.; Li, H.; Mulhern, M.S.; McSorley, E.M.; Watson, G.E.; Shamlaye, C.F.; Strain, J.J.; Myers, G.J.; Davidson, P.W.; et al. PUFA Status and Methylmercury Exposure Are Not Associated with Leukocyte Telomere Length in Mothers or Their Children in the Seychelles Child Development Study. *J. Nutr.* **2017**, *147*, 2018–2024. [CrossRef]
37. Crespo-López, M.E.; Soares, E.S.; Macchi, B.M.; Santos-Sacramento, L.; Takeda, P.Y.; Lopes-Araújo, A.; Paraense, R.S.O.; Souza-Monteiro, J.R.; Augusto-Oliveira, M.; Luz, D.A.; et al. Towards Therapeutic Alternatives for Mercury Neurotoxicity in the Amazon: Unraveling the Pre-Clinical Effects of the Superfruit Açaí (Euterpe oleracea, Mart.) as Juice for Human Consumption. *Nutrients* **2019**, *11*, 2585. [CrossRef]
38. Price, M.R.; Kruse, J.A.; Galvez, M.E.; Lorincz, A.M.; Avigdor, M.; Heideman, P.D. Failure to respond to endogenous or exogenous melatonin may cause nonphotoresponsiveness in Harlan Sprague Dawley rats. *J. Circadian Rhythms.* **2005**, *3*, 12. [CrossRef]
39. Yie, S.M.; Liu, G.Y.; Johansson, E.; Brown, C.; Brown, G.M. Age-associated changes and sex differences in urinary 6-sulphatoxymelatonin circadian rhythm in the rat. *Life Sci.* **1992**, *50*, 1235–1242. [CrossRef]

40. do Amaral, F.G.; Cipolla-Neto, J.; Afeche, S.C. Melatonin Synthesis Enzymes Activity: Radiometric Assays for AANAT, ASMT, and TPH. *Methods Mol. Biol.* **2022**, *2550*, 33–43.
41. Kamal, M.; Gbahou, F.; Guillaume, J.L.; Daulat, A.M.; Benleulmi-Chaachoua, A.; Luka, M.; Chen, P.; Kalbasi Anaraki, D.; Baroncini, M.; Mannoury la Cour, C.; et al. Convergence of melatonin and serotonin (5-HT) signaling at MT2/5-HT2C receptor heteromers. *J. Biol. Chem.* **2015**, *290*, 11537–11546. [CrossRef]
42. Ben Bacha, A.; Norah, A.O.; Al-Osaimi, M.; Harrath, A.H.; Mansour, L.; El-Ansary, A. The therapeutic and protective effects of bee pollen against prenatal methylmercury induced neurotoxicity in rat pups. *Metab. Brain Dis.* **2020**, *35*, 215–224. [CrossRef]
43. Tsuzuki, Y. Effect of methylmercury exposure on different neurotransmitter systems in rat brain. *Toxicol. Lett.* **1982**, *13*, 159–162.
44. De-Miguel, F.F.; Trueta, C. Synaptic and extrasynaptic secretion of serotonin. *Cell Mol. Neurobiol.* **2005**, *25*, 297–312. [CrossRef]

International Journal of
Molecular Sciences

MDPI

Article

Triazine Herbicides Risk Management Strategies on Environmental and Human Health Aspects Using In-Silico Methods

Tianfu Yao †, Peixuan Sun † and Wenjin Zhao *

College of New Energy and Environment, Jilin University, Changchun 130012, China
* Correspondence: zhaowj@jlu.edu.cn
† These authors contributed equally to this work.

Abstract: As an effective herbicide, 1, 3, 5-Triazine herbicides (S-THs) are used widely in the pesticide market. However, due to their chemical properties, S-THs severely threaten the environment and human health (e.g., human lung cytotoxicity). In this study, molecular docking, Analytic Hierarchy Process—Technique for Order Preference by Similarity to the Ideal Solution (AHP-TOPSIS), and a three-dimensional quantitative structure-active relationship (3D-QSAR) model were used to design S-TH substitutes with high herbicidal functionality, high microbial degradability, and low human lung cytotoxicity. We discovered a substitute, Derivative-5, with excellent overall performance. Furthermore, Taguchi orthogonal experiments, full factorial design of experiments, and the molecular dynamics method were used to identify three chemicals (namely, the coexistence of aspartic acid, alanine, and glycine) that could promote the degradation of S-THs in maize cropping fields. Finally, density functional theory (DFT), Estimation Programs Interface (EPI), pharmacokinetic, and toxicokinetic methods were used to further verify the high microbial degradability, favorable aquatic environment, and human health friendliness of Derivative 5. This study provided a new direction for further optimizations of novel pesticide chemicals.

Keywords: triazine herbicides; cleaner production; 3D-QSAR; molecular docking; molecular dynamics; microbial degradation pathways; field application program

Citation: Yao, T.; Sun, P.; Zhao, W. Triazine Herbicides Risk Management Strategies on Environmental and Human Health Aspects Using In-Silico Methods. *Int. J. Mol. Sci.* **2023**, *24*, 5691. https://doi.org/10.3390/ijms24065691

Academic Editors: Esref Demir and Sam Kacew

Received: 17 February 2023
Revised: 9 March 2023
Accepted: 12 March 2023
Published: 16 March 2023

1. Introduction

Triazine herbicides (THs) have long held an important position in the pesticide market. They are applied primarily to maize cropping fields [1], owing to their broad spectrum, high performance, and low cost [1], but they also have high toxicity, environmental persistence, and endocrine disrupting effect [1,2]. THs target the D1 protein (D1-PSII) of the photosynthetic system II (PSII) and act as herbicides by inhibiting plant photosynthesis [3]. However, only 10–30% of THs are currently absorbed by target plants or adsorbed by soil particles, with the majority polluting water bodies, such as surface water, via surface runoff and irrigation, eventually reaching the ocean [4]. In 2020, the US Environmental Protection Agency (EPA) designated atrazine (ATZ), promazine, and simazine as pesticides "likely to adversely affect (LAA)" in species and ecosystems [5]. Furthermore, the introduction of bans or restrictions on the use of 1, 3, 5-Triazine herbicides (S-THs) in the EU, Ulaanbaatar, Nigeria, and India highlights the continued residues and biohazards of S-THs in the environment [6]. The continuing residues and biohazards of S-THs in the environment have attracted widespread attention.

S-THs are highly persistent in soil and aqueous sediment environments, with half-lives of 4–12 weeks for prometryne (PRT) [7] and 4–57 weeks for ATZ in soil environments, respectively [8]. In addition, S-THs can pollute the aquatic environments through rainfall, irrigation, and surface runoff, causing long-term damage to aquatic organisms. For example, S-THs are acutely toxic to fish, and 1200 μg/L PRT can significantly reduce embryo hatching and survival in carp [9]. Hao et al. [10] discovered that ATZ reduced zebrafish hatching rates and increased the incidence of malformations and embryo mortality with increasing ATZ exposure, which also caused necrosis and congestion in carp gill epithelial cells [11], oxidative stress in catfish liver and gills [12], and ovarian lesions in female blackhead dull fish [13]. Furthermore, S-THs can be harmful to humans, with PRT being toxic to human pulmonary adenocarcinoma cell lines and human bronchial epithelial cell lines [14], and ATZ causing carcinogenesis, teratogenesis, and mutagenesis after long-term exposure [15,16].

Soil microorganisms such as *Acinetobacter* spp. Brisou and Prévot (Moraxellales: Moraxellaceae), *Arthrobacter* sp. Conn and Dimmick (Micrococcales: Micrococcaceae), *Agrobacterium* sp. Conn (Hyphomicrobiales: Rhizobiaceae), *Bacillus* sp. Cohn (Bacillales: Bacillaceae), *Deinococcus* sp. Brooks and Murray (Deinococcales: Deinococcaceae), *Microbacterium* sp. Orla-Jensen (Micrococcales: Mycobacteriaceae), *Nocardioides* sp. Prauser (Propionibacteriales: Nocardioidaceae), and *Rhodococcus rhodochrous* Tsukamura (Mycobacteriales: Nocardiaceae), among others, have been shown to degrade S-THs partially [17–24]. White-rot fungi (*Phanerochaete chrysosporium* Burds. (Polyporales: Phanerochaetaceae)) and lignocellulose-degrading fungi (*Pleurotus pulmonarius* Fr. (Agaricales: Pleurotaceae)) can dechlorinate ATZ water to produce hydroxylated ATZ and dealkylated ATZ metabolites with nitrogen [25,26]. Although various microorganisms can degrade S-THs, none of the triazine rings are broken, resulting in the persistence of S-THs in the soil. Vonberg et al. [27] showed that, although ATZ has been banned in Germany for 31 years, residue can still be detected in groundwater, surface water, and soil. Therefore, developing functional and environmentally friendly substitutes for S-THs is crucial for the ecological environment and human health.

The three main research objectives of this study are as follows: (1) create S-THs substitutes (with high herbicidal functionality, high microbial degradability, and high environmental friendliness but low human lung cytotoxicity). (2) to design and screen optimal field application schemes (with high microbial degradation promotion in maize cropping fields. (3) to further verify and assess the excellent comprehensive performance of S-THs substitutes and their degradation products (with high microbial degradability, and favorable aquatic environment, and human health friendliness).

2. Results

2.1. Construction and Evaluation of the Single-Effect and Comprehensive-Effect 3D-QSAR Models of Herbicidal Functionality Properties, Microbial Degradability, and Human Lung Cytotoxicity of S-THs

The structural information of 26 S-THs was used as an independent variable, and the docking score (LibDock Score, LDS) values of herbicidal functionality properties, microbial degradability, and human lung cytotoxicity (hereafter referred to as herbicidal functionality properties, degradability, and toxicity) of S-THs were adopted as the dependent variables to build the single-effect CoMSIA models for herbicidal properties, degradability, and toxicity of S-THs (Table S1). In addition, the final weight results of herbicidal functionality properties, degradability, and toxicity of the comprehensive value (CV) are shown in Figure S1, and the results of the CV calculations are shown in Table S1. The CV was adopted as a dependent variable to construct the comprehensive-effect CoMSIA model of herbicidal functionality properties, degradability, and toxicity of S-THs. Table S2 lists the relevant modeling materials.

In addition, the evaluation parameters of the single-effect CoMSIA and comprehensive-effect CoMSIA models of herbicidal functionality properties, degradability, and toxicity of S-THs are shown in Table 1. The comprehensive-effect CoMSIA model of herbicidal functionality properties, degradability, and toxicity of S-THs was used as an example; the model cross-validation coefficient q^2 was 0.751 (>0.5), the best principal component n was 10, the non-cross-validation coefficient R^2 was 0.998, and the standard deviation was 0.008, manifesting that the constructed model had an excellent internal prediction and fitting ability [28]. In addition, the model test set external validation interaction test coefficient

r^2_{pred} was 0.678 (>0.6), manifesting that the constructed model had favorable external prediction ability [29]. The model $(R^2 - q^2)/R^2$ (<30%) manifested that the constructed model was not over-fitted [30].

Table 1. CoMSIA model evaluation parameters for the herbicidal functionality properties, degradability, and toxicity of S-THs and their comprehensive effects.

CoMSIA Models	Enzymes	q^2	n	R^2	SEE	F	r^2_{pred}	$(R^2 - q^2)/R^2$ (%)
Comprehensive	ALL	0.789	8	0.993	0.007	132.292	0.615	20.54
Herbicide	1FC9	0.751	9	0.997	0.393	207.333	0.785	24.67
Degradation	4L9X	0.757	5	0.986	1.529	157.196	0.713	23.23
Toxicity	6K1J	0.706	10	1.000	0.285	1660.695	0.792	29.40

2.2. Design of S-TH Substitutes Based on the 3D Isopotential Diagrams of the CoMSIA Model

ATZ, which is primarily used in agriculture, was chosen as the target molecule to analyze the three-dimensional (3D) isopotential diagrams of single-effect and comprehensive-effect CoMSIA models of herbicidal functionality properties, degradability, and toxicity and to design the substitutes. Figure S2 depicts the molecular structure and proposed modification sites for ATZ.

The 3D isopotential diagrams of the hydrophobic (H), hydrogen-bonded acceptor (A), hydrogen-bonded donor (D), electrostatic (E), and steric (S) fields for the single-effect and comprehensive-effect CoMSIA models of herbicidal functionality properties, degradability, and toxicity of S-THs are shown in Figure 1.

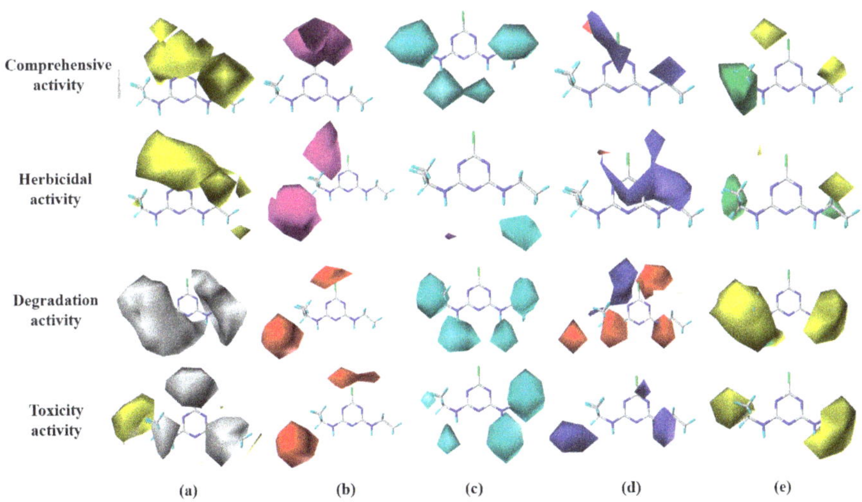

Figure 1. The 3D isopotential diagrams of the single-effect and comprehensive-effect CoMSIA models of herbicidal functionality properties, degradability, and toxicity of S-THs: (**a**) hydrophobic field; (**b**) hydrogen-bonded acceptor field; (**c**) hydrogen-bonded donor field; (**d**) electrostatic field, and (**e**) steric field.

In this study, we aimed to reduce the cytotoxicity of S-THs in the human lung by designing ATZ substitutes based on the reverse law of the substitution principle of 3D isopotential diagrams. According to the single-effect and comprehensive-effect CoMSIA models of herbicidal functionality properties, degradability, and toxicity of S-THs, and the contribution ratio of each force field in the 3D isopotential diagrams (Table 2), single, double, and multiple substitutions could be performed at the 1–5 point sites of ATZ (Table S3).

Table 2. The proportion of each force field in the single-effect and comprehensive-effect CoMSIA models of herbicidal functionality properties, degradability, and toxicity of S-THs based on the 3D isopotential diagrams.

Fields	Proportion of Fields (%) CoMSIA Models			
	For Comprehensive Activity	For Herbicidal Activity	For Degradation Activity	For Toxicity Activity
Hydrophobic (H)	37.4	50	29.3	32.5
Hydrogen-bond acceptor (A)	4.7	6.9	7.1	6.9
Hydrogen-bond donor (D)	26.2	6.8	27.6	22.8
Electrostatic (E)	16.8	18.8	17.8	19.4
Steric (S)	14.9	16.5	18.1	18.4

Therefore, hydrophobic groups (-F, -Cl, -Br, -SH, -C≡C, $-OCH_3$, and $-CF_3$) were introduced to site 1, a more electronegative group ($-CF_3$) was introduced to site 3, and a small volume of group ($-CH_3$) was introduced to site 4, to design and screen a total of S-THs with improved single- and comprehensive-effects of S-TH substitutes (Table S3).

2.3. *Prediction and Evaluation of the Single-Effect and Comprehensive-Effect 3D-QSAR Models of Herbicidal Properties, Microbial Degradability, and Human Lung Cytotoxicity of S-THs*

In this study, the herbicidal functionality properties, degradability, and toxicity of the 40 designed substitutes were predicted using four constructed CoMSIA models, and the predicted values were normalized using Formula (2). The functionality and degradability (positive indices) were normalized by "bigger, better type," while toxicity (negative indices) was normalized by "smaller, better type" (Table S4). The comprehensive effect of the 40 S-TH substitutes ranged from -26.79% to 70.44%, while the eight S-THs, D-3, D-4, D-5, D-18, D-29, D-30, D-31, and D-35, were consistent with the weighted values of the comprehensive-effect model (49.31%:25.17%:25.52%). The results verified the effectiveness of the comprehensive-effect CoMSIA model of herbicidal functionality properties, degradability, and toxicity of S-THs, the reasonableness of the molecular design of the substitutes, and verified that the hydrophobic, electrostatic, and steric fields of the comprehensive-effect CoMSIA model were the primary factors influencing the comprehensive effects of S-THs.

2.4. *Evaluation of the Microbial Degradability Universality and Toxicity of Antioxidant Systems in Fish of S-TH Substitutes*

2.4.1. Evaluation of the Microbial Degradability Universality of S-TH Substitutes

We selected three other target proteins for the microbial degradation of S-THs in addition to triazine hydrolase (TrzN), namely AtzC (PDB ID: 2QT3), LiP (PDB ID: 1B85), and MnP (PDB ID: 1MNP) [31,32] using the Protein Data Bank (PDB) database [33]. Molecular docking of the S-TH substitutes with the above three proteins was carried out. The LDS was used as an evaluation index to assess the microbial degradability of the S-TH substitutes. It was found that the microbial degradability of the eight S-TH substitutes, including D-3, D-4, D-5, D-18, D-29, D-30, D-31, and D-35, improved to varying degrees compared to ATZ or remained essentially unchanged (Table S5).

2.4.2. Evaluation of the Toxicity of Antioxidant Systems in Fish of S-TH Substitutes

We selected two antioxidant proteins from carp, SOD (UniProt ID: Q8JFG7) and CAT (UniProt ID: E2CWE8), using the UniProt database [34]. Eight S-TH substitutes previously screened were molecularly docked to the two antioxidant proteins. The LDS of the two proteins were added together using a 1:1 weighting, and the CV of the toxicity of the antioxidant system in fish of S-TH substitutes was calculated (Table S6). Compared to ATZ, the toxicity of five S-TH substitutes (D-4, D-5, D-19, D-21, and D-25) was lower in the antioxidant system (range 0.12–23.98%), with substitute D-5 showing a significant reduction.

2.5. Screen of Optimal Field Application Schemes to Promote the Microbial Degradation of S-TH Substitutes in Maize Cropping Fields

Compared to the blank control group (group 1, with a binding energy of −93.414 kJ/mol), the results of the Taguchi orthogonal experiment (Table S7) revealed that the binding energy values of groups 2, 10, 14, 15, 19, 21, 26, 27, 28, and 29 all showed varying degrees of reduction (1.51–58.32%), with group 2 (binding energy of −147.893 kJ/mol) showing the most significant reduction. Therefore, the external conditions from group 2 (aspartic acid (Q), alanine (R), and glycine (S)) were chosen as a field application scheme to perform a full factorial design experiment used with the 3-factors (Q, R, and S) and 2-levels (0 for no addition and 1 for addition), with a total of 8 different sets of external conditions schemes. The absolute values of the binding energy of the eight schemes were calculated and used as the response values for the factor analysis about the main-, second-, and third-order interaction effects among the three factors (Figure 2, Table S8).

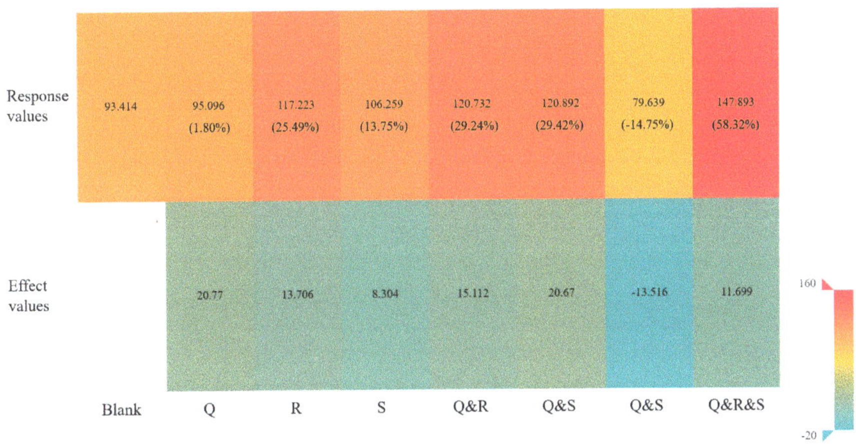

Figure 2. Response and effect values for the 8 external conditions schemes (3-factors/2-levels).

The results of the main-, second-, and third-order interaction effects of each scheme (Figure 2) showed that the main-effect values of Q, R, and S (groups 2, 3, and 4) were positive, indicating that the main-effect factors (Q, R, and S) in the optimal field application schemes could promote the microbial degradation of S-TH substitutes in maize cropping fields. In the second-order interaction effects (groups 5, 6, and 7), the second-effect values of Q and R, and Q and S were positive, while the second-effect value of R and S was negative, indicating that the coexistence of aspartic acid and alanine, and aspartic acid and glycine exhibited synergistic effects in promoting the microbial degradation of S-TH substitutes in maize cropping fields. In contrast, the coexistence of alanine and glycine exhibited antagonistic effects. In the third-order interaction effects (group 8), the third-effect value of Q, R, and S was positive, indicating that aspartic acid inhibited the antagonistic effects of alanine and glutamic acid on the microbial degradation of S-TH substitutes, indicating significant synergistic effects of aspartic acid, alanine, and glycine. Consequently, combined with the maximum response value of Q, R, and S in Group 8 (value 147.893), the coexistence of Q, R, and S could be screened as the optimal field application schemes to promote the microbial degradation of S-TH substitutes in maize cropping fields.

2.6. Simulation of Microbial Degradation Pathways of S-TH Substitutes

Figure 3 shows that the microbial degradation pathways of ATZ and substitute D-5 have the same process, Stage 2, with the only difference being the reactants and products of Stage 1. Therefore, the microbial degradation of ATZ and substitute D-5 was analyzed by comparing the differences of the two-step (Steps 1 and 2) reaction energy barrier (ΔE) in Stage 1 (Table 3).

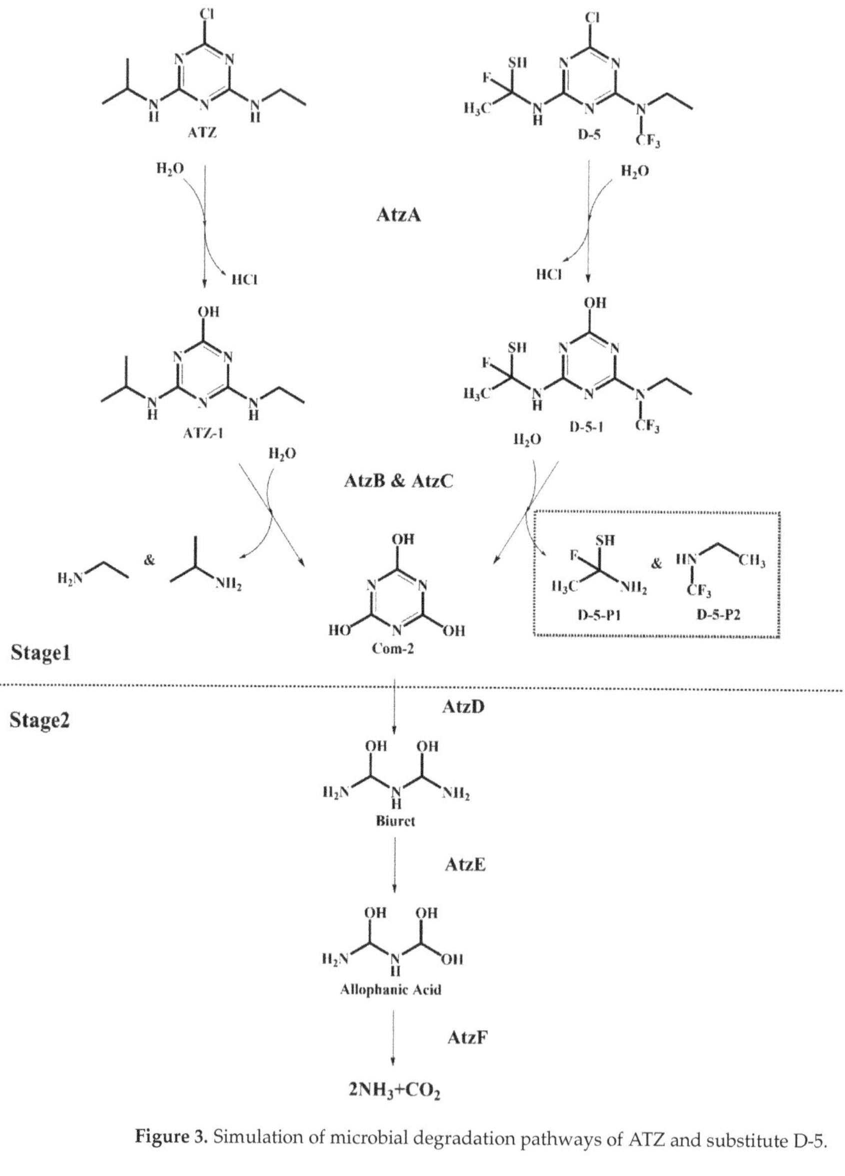

Figure 3. Simulation of microbial degradation pathways of ATZ and substitute D-5.

Table 3. Calculation of reaction energy barrier and change rates of the microbial degradation of ATZ and substitute D-5 in Stage 1.

Compounds	Steps	Reactants	Products	ΔE (kJ/mol)	ΔE (Total) (kJ/mol)	Change Rate (%)
ATZ	1	ATZ	ATZ-1	160.893	200.806	-
	2	ATZ-1	Com-1	39.913		
D-5	1	D-5	D-5-1	60.147	94.533	−62.62
	2	D-5-1	Com-1	34.386		−13.85

It has been shown that the ΔE (>0) represents the difficulty of reaction occurrence, with smaller values indicating that the reaction is more likely to occur [35]. Compared to the ΔE values of Stage 1 (Steps 1 and 2) of the ATZ, that of the substitute D-5 were reduced by 62.62% and 13.85%, respectively, indicating that the groups on the modified S-THs triazine ring (-C_2H_5SNF, and -$C_3H_5NF_3$) were more susceptible to be hydrolyzed to triuric acid by microorganisms. In addition, the above analysis results demonstrated that the microbial degradability of S-TH substitute D-5 was significantly enhanced, confirming the rationality of the comprehensive-effect CoMSIA model of herbicidal functionality properties, degradability, and toxicity of S-THs and the precision of the molecular design of substitutes constructed in this study.

2.7. Evaluation of Aquatic Biotoxicity and Human Health Risks of Microbial Degradation Products of S-TH Substitutes

As shown in Tables S9 and S10, regarding aquatic toxicity, the intermediate microbial degradation products D-5-P1 and D-5-P2 were significantly less toxic to green algae and fish and, to a lesser extent, Daphnia than ATZ. Regarding human health risks, compared to that of ATZ, the hepatotoxicity of D-5-P1 and D-5-P2 was significantly reduced, whereas the maximum tolerated dose was significantly increased. The carcinogenicity of D-5-P1 in male mice was reduced to nontoxic levels. The toxicity levels of the five toxicity models, including skin irritation, sensitization, and carcinogenicity, remained unchanged in male and female rats and female mice. The skin sensitization level of D-5-P2 was reduced to a low toxicity level, and the toxicity levels of the five toxicity models, including skin irritation and rodent carcinogenicity, remained unchanged.

3. Discussion

In the present work, the single-effect and comprehensive-effect CoMSIA models of herbicidal functionality properties, microbial degradability, and human lung cytotoxicity of S-THs were constructed, which showed excellent stability, predictability, and fitting, and the S-TH substitutes with excellent comprehensive performance were designed based on the 3D isopotential diagrams of the above models.

Overall, the hydrophobic field contributed the highest proportion to the molecular effect of S-THs in the four CoMSIA models, which was regarded as the main modifying force field to improve the herbicidal functionality properties of the S-TH substitutes. The single-effect CoMSIA model of herbicidal functionality properties contributed the highest proportion among the four models. The analysis of the 3D isopotential diagrams of the single-effect CoMSIA models of degradability and toxicity showed that the electrostatic fields have the largest and most comprehensive difference in the color block distribution, with the largest contribution of the E fields to the S-THs performance (17.8% and 19.4%, respectively). Therefore, the E field was regarded as the main modifying force field to improve the degradability of the S-TH substitutes and avoid increased toxicity. Furthermore, in the single-effect CoMSIA model of toxicity, the stereoscopic field accounted for the highest proportion, which was regarded as the main modifying force field to reduce the toxicity of substitutes for reverse design. Similar to our previous study, based on the constructed plant-microbial synergistic degradation CoMISA model of quinolones (QNs), the hydrophobic field and electrostatic field of this model were regarded as the main modifying force field. By introducing groups with hydrophobicity (-SH, -Cl, and -F), as well as groups with strong electronegativity ($-CF_3$, $-CH_3$, and $-CH_2F$), QNs substitutes with enhanced plant-microbial synergistic degradation effect were designed reasonably [36].

In addition, the optimal field application scheme of aspartic acid, alanine, and glycine indicated the effective role of amino acids in the promotion of S-THs microbial degradation. Shen et al. [37] found that the stress of organic pollutants, such as petroleum hydrocarbons, could lead to a positive plant response and, to some extent, promote the secretion of amino acids in soil inter-roots. Furthermore, amino acid content had been correlated with soil N effectiveness, which improved the activity and respiration of soil microorganisms, and further enhanced the degradation of organic pollutants by soil inter-rooted microorganisms [38,39]. In addition, Li et al. [40] discovered that root secretions could effectively stimulate microbial degradation of organic pollutants within plant roots, which was consistent with the results of the present study that the coexistence of the three root secretions, aspartic acid, alanine, and glycine, could promote the microbial degradation of S-THs substitutes in maize cropping fields.

The simulation of microbial degradation pathways of S-TH substitutes indicated that, compared to the ΔE values of Stage 1 (Steps 1 and 2) of the ATZ, the substitute D-5 were reduced by 62.62% and 13.85%, respectively, indicating that the groups on the modified S-THs triazine ring ($-C_2H_5SNF$, and $-C_3H_5NF_3$) were more susceptible to be hydrolyzed to triuric acid by microorganisms. In addition, the above analysis results demonstrated that the microbial degradability of substitute D-5 was significantly enhanced, confirming the rationality of the comprehensive-effect CoMSIA model of herbicidal functionality properties, degradability and toxicity of S-THs, and the precision of the molecular design of substitutes constructed in this study. Furthermore, Fu et al. [41], Li et al. [42], and Xue et al. [43] all used pharmacokinetic and toxicokinetic methods to predict and evaluate the human health risks of designed substitutes. The predicted results of indictors (hepatotoxicity, maximum tolerated dosage, skin sensitization, skin irritation, and rodent carcinogenicity) all indicated that the designed molecules had a low risk to human health. According to the above studies, the human health risk assessment based on pharmacokinetic and toxicokinetic methods had certain rationality and reliability. Therefore, the aquatic biotoxicity and human health risks of the microbial degradation products of the substitute D-5 designed in the present work were significantly reduced.

The in-silico methods used in this study have certain efficiency, rationality, and convenience, which could provide a new direction for the research and development of more similar functional chemicals. However, in the future, we still need to combine as much experimental data as possible. The environmental and human health hazards, as well as economic applicability and other aspects, should be taken into more comprehensive consideration in order to obtain more efficient and environmentally friendly new chemical substitutes.

4. Materials and Methods

4.1. Characterization of Herbicidal Functionality Properties, Microbial Degradability, and Human Lung Cytotoxicity of S-THs—Molecular Docking Method

This study selected 26 S-THs, including ATZ. First, the molecular structures were drawn and optimized using the Sketch Molecule, Minimize and Align Database modules of SYBYL-X2.0 software (Tripos, Inc.: St. Louis, MO, USA). Then, the molecular structures were optimized using Tripos force fields, Gasteiger-Huckel charges, and 10,000 iterations [44] to achieve the optimal conformation with the lowest molecular energy.

The PDB database [33] was used to identify and select the three target proteins mentioned above (Figure 4): the D1 protein of *Tetradesmus obliquus* Turpin (Sphaeropleales: Scenedesmaceae) photosynthetic system II (D1-PSII, PDB ID:1FC9) [45], the triazine hydrolyzable protein of *Paenarthrobacter aurescens* Phillips (Micrococcales: Micrococcaceae) (TrzN, PDB ID:4L9X) [46], and the human H2AX protein (H2AX, PDB ID:6K1J) [47] (the rationale for the selection of the above three proteins is shown in the appendix). The Discovery Studio (DS) 2020 software (BIOVIA Inc.: Shenzhen, Guangdong, China) used the three proteins mentioned above as receptor proteins. Furthermore, the structurally optimized S-THs molecules were used as ligand molecules in the LibDock module for rapid ligand-receptor docking to characterize the ligand-receptor binding ability using LDS. These include herbicidal functionality properties, degradability, and toxicity.

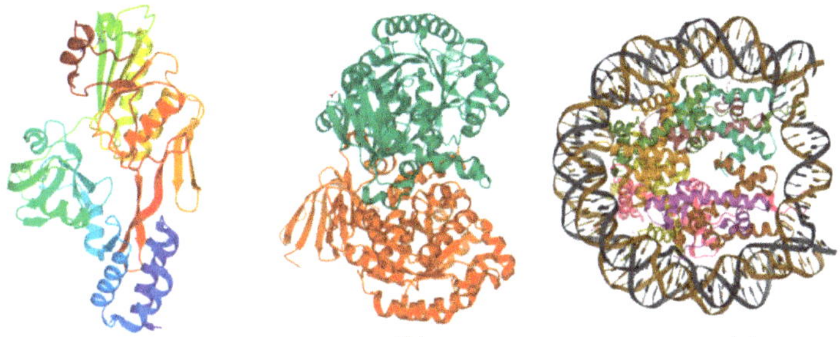

(a) (b) (c)

Figure 4. Schematic structure of the herbicidal, degradative, and toxic receptor proteins of S-THs: (**a**) 1FC9; (**b**) 4L9X; (**c**) 6K1J.

4.2. Characterization of the Comprehensive Effects of Herbicidal Functionality Properties, Microbial Degradability, and Human Lung Cytotoxicity of S-THs—AHP-TOPSIS Method

In this study, we used the Analytic Hierarchy Process—Technique for Order Preference by Similarity to the Ideal Solution (AHP-TOPSIS) [48] to normalize the herbicidal functionality properties, degradation, and toxicity indices of S-THs and calculated the comprehensive herbicidal functionality properties, degradation, toxicity, and CV of S-THs molecules according to the weighting ratios of the AHP-TOPSIS method. In addition, the CV was normalized using the AHP-TOPSIS method. The equations are as follows:

(1) Subjective weighting W1 (j) of herbicidal functionality properties, degradability, and toxicity of S-THs—SPSSAU software method

This study is primarily concerned with the herbicidal functionality properties of S-THs. The Analytic Hierarchy Process (AHP) module of SPSSAU software (QingSi Technology Ltd.: Beijing, China) was used to calculate the weight W1 (j) (the calculation and weighting methods are shown in the appendix).

(2) Objective weighting W_2 (j) of herbicidal functionality properties, degradability, and toxicity of S-THs—TOPSIS weighting method

The docking scores of herbicidal functionality properties, degradability, and toxicity of S-THs were normalized in this study based on the type of indicator (positive and negative indicators, respectively), where herbicidal functionality properties and degradability are positive indicators, and toxicity is a negative indicator, and are calculated as follows:

Normalization of positive indicators of herbicidal functionality properties or degradability:

$$Z_{ij}^{+} = \frac{X_{ij} - \min\{X_{ij}\}}{\max\{X_{ij}\} - \min\{X_{ij}\}} \tag{1}$$

Normalization of negative indicators of toxicity:

$$Z_{ij}^{-} = \frac{\max\{X_{ij}\} - X_{ij}}{\max\{X_{ij}\} - \min\{X_{ij}\}} \tag{2}$$

where i denotes the S-THs molecule (i = 1, 2, . . . , 26), j denotes the receptor protein (j = 1, 2, 3 for 1FC9, 4L9X, and 6K1J, respectively), $Z_{ij}{}^{+}$ denotes the positive indicator of the S-THs molecule normalized for herbicidal functionality properties or degradability, $Z_{ij}{}^{-}$ denotes the negative indicator of the S-THs molecule normalized for toxicity, and X_{ij} denotes the molecular docking scoring value of the ith molecule to the jth protein.

The docking scores of herbicidal functionality properties, degradability, and toxicity of S-THs were used as column vectors a_1, a_2, and a_3 to construct a 3 × 26 normalized data matrix (a_1, a_2, a_3), and the best and worst values of each column were selected to construct the best vector $A^+ = (a_1{}^+, a_2{}^+, a_3{}^+)$ and the worst vector $A^- = (a_1{}^-, a_2{}^-, a_3{}^-)$, respectively. The normalized data vector of each S-THs molecule was then compared to the best and worst vectors. Closer to the best vector indicates that the S-THs molecule's comprehensive effect is better. In contrast, closer to the worst vector indicates that the S-THs molecule's comprehensive effect is worse. Therefore, the elements of the optimal and worst vectors were calculated as follows:

$$a_j^+ = \begin{cases} \max(X_{ij}), X_{ij} \text{ is a positive indicator} \\ \min(X_{ij}), X_{ij} \text{ is a negative indicator} \end{cases} \quad (3)$$

$$a_j^- = \begin{cases} \max(X_{ij}), X_{ij} \text{ is a positive indicator} \\ \min(X_{ij}), X_{ij} \text{ is a negative indicator} \end{cases} \quad (4)$$

where $a_j{}^+$ is the jth column element of the best vector and $a_j{}^-$ is the jth column element of the worst vector.

Based on the optimal and inferior vectors A^+ and A^-, the distances of each S-THs numerator from the optimal and inferior vectors were calculated, and a positive relative error matrix $R^+ = (r)_{ij}{}^+{}_{3\times 26}$ and a negative relative error matrix $R^- = (r_{ij}{}^-)_{3\times 26}$ were constructed based on the ratio between them and the maximum distance.

$$r_{ij}^+ = \frac{\left|X_{ij} - a_j^+\right|}{\max(X_{ij}) - \min(X_{ij})} \quad (5)$$

$$r_{ij}^- = \frac{\left|X_{ij} - a_j^-\right|}{\max(X_{ij}) - \min(X_{ij})} \quad (6)$$

where r_{ij}^+ denotes the element in row i, column j of the positive relative error matrix, and r_{ij}^- denotes the element in row i, column j of the negative relative error matrix.

The cosine of the relative error angle between the herbicidal functionality properties, degradability, and toxicity indicators of the S-THs was calculated based on the relative error matrix θ_j:

$$\theta_j = \cos < r_{ij}^+, r_{ij}^- > = \frac{\sum_{i=1}^{m} r_{ij}^+ \cdot r_{ij}^-}{\sqrt{\sum_{i=1}^{m} r_{ij}^{+2}} \cdot \sqrt{\sum_{i=1}^{m} r_{ij}^{-2}}} \quad (7)$$

The objective TOPSIS weights $W_2(j)$ for herbicidal functionality properties, degradability, and toxicity of the comprehensive effect of S-THs were calculated by normalizing the cosine of the relative error clincher.

$$W_2(j) = \frac{\theta_j}{\sum_{j=1}^{m} \theta_j} \quad (8)$$

(3) The compound weighting of subjective and objective

The minF optimization problem was designed using the minimum entropy principle. The following are the calculated comprehensive weights w(j) of the herbicidal functionality properties, degradability, and toxicity of S-THs.

$$\min F = \sum_{j=1}^{m} w(j) \ln \frac{w(j)}{W_1(j)} + \sum_{j=1}^{m} w(j) \ln \frac{w(j)}{W_2(J)} \tag{9}$$

$$s.t. \sum\nolimits_{j=1}^{m} w(j) = 1, w(j) > 0 \tag{10}$$

The comprehensive weights w (j) for herbicidal properties, degradability, and toxicity of the comprehensive effects of S-THs were calculated using the Lagrange multiplier method to solve the above equation. The subjective weights W_1 (j) and objective weights W_2 (j) was substituted to calculate the comprehensive weights.

$$w(j) = \sqrt{W_1(j)W_2(j)} / \sum\nolimits_{j=1}^{3} \sqrt{W_1(j)W_2(j)}, j = 1,2,3 \tag{11}$$

4.3. Construction of a Model for the Comprehensive Effects of Herbicidal Functionality Properties, Microbial Degradability, and Human Lung Cytotoxicity of S-THs—3D-QSAR Model

The optimized molecules from the SYBYL-X2.0 software [49] Minimize module were classified into training and test sets in a 3:1 ratio randomly (template molecules were present at both the training and test levels; the molecular distribution of the model training and test sets is shown in Table S2), with the more widely used Propazine (PRZ) chosen as the template molecule. The Align Database module [50] was used for molecular stacking (the molecular structure and common backbone are shown in Figure S3).

The stacked training set molecules and docking scores were imported using the SYBYL-X2.0 software (where the comprehensive value was imported to construct the CoMSIA model [36,50] for the comprehensive effect of herbicidal properties, degradability and toxicity of S-THs). The Calculate Properties module output the calculated values for the hydrophobic (H), hydrogen bond acceptor (A), hydrogen bond donor (D), electrostatic (E), and steric fields (S) were output using the Calculate Properties module, and cross-verification and non-cross-verification [36,50]. The molecules from the superimposed test set were then imported into SYBYL-X2.0 software. Based on the analysis results of the CoMSIA model constructed using the training set molecules, the predicted value of the test set molecule output was obtained using the Predict function under the Add a Computed Column module [36,50]. Finally, using the Calculate Properties module [36,50], the predicted values were externally validated against the original scoring values (the comprehensive model for the comprehensive values). After passing all the above validations, the built model of the herbicidal, degradation, and toxicity effects of S-THs and their comprehensive effects proved stable, predictive, and well-fitting. The parameters and model evaluation criteria for the training sets and test sets of the CoMSIA model developed in this study are listed in Table S11 [36].

4.4. Design of S-TH Substitutes—SYBYL-X2.0 Software

In this study, using the SYBYL-X2.0 software, we propose the selection of the most widely used ATZ as a template molecule and determine the substitutable group sites and substitution groups based on the 3D isopotential diagrams of each force field (including hydrophobic field (H), hydrogen bond acceptor field (A), hydrogen bond donor field (D), electrostatic field (E), and steric field (S)) of the constructed CoMSIA model of the comprehensive effect of herbicidal properties, degradability, and toxicity of S-THs. The introduction of hydrophobic substituents (-CH_3, CF_3, -F/Cl/Br, -OCH_3, and -SH) near the yellow region of the hydrophobic field (H) and hydrophilic substituents (-OH, -CHO, -COOH, and -NH_2) near the white region of the hydrophobic field (A) improved the activities of the compounds. The addition of hydrogen bond acceptors (-NO_2, NF_2, and -$COCF_3$) in the purple region of the hydrogen bond acceptor field (A) and hydrogen bond donors (-NH_2, -OH, -$COCH_3$, and -$CONH_2$) in the red region of the field improved the activities of the compounds. The addition of hydrogen bond donors in the cyan region of the hydrogen bond donor field (D) and hydrogen bond acceptors in the purple region of the field improved the activities of the compounds. Introducing less electronegative substituents

(-H, -OH, -COOH, and –NH) in the blue region of the electrostatic field (E) improved the activities of the compounds. The introduction of less electronegative substituents (-H, -OH, and -CH_3) in the blue region of the electrostatic field (E) and more electronegative substituents (-CF_3, -F, -CH_2 F, and -CHF_2) in the red region of the field improved the activities of the compounds. The activities of the compounds were improved by increasing the size of the substituents in the green region of the steric field (S) and decreasing the size in the yellow region [51]. The Sketch Molecule, Minimize, and Align Database modules of the SYBYL-X2.0 software were used to map and optimize substitutes [36].

4.5. Evaluation of the Microbial Degradability Universality and Toxicity of Antioxidant Systems in Fish of S-THs Substitutes—Molecular Docking Method

4.5.1. Evaluation of the Microbial Degradability Universality of S-THs Substitutes in the Soil Environment

The PDB database was used to search and select three proteins as the target proteins of microbial degradation (Figure 5, the rationale for the selection of the above three proteins is shown in the appendix), which are AtzC of *Pseudomonas* sp. Migula (Pseudomonadales: Pseudomonadaceae) ADP (PDB ID: 2QT3), LiP of *Phanerodontia chrysosporium* Burds. (Polyporales: Phanerochaetaceae) (PDB ID: 1B85), and MnP (PDB ID: 1MNP). The LibDock module of DS software was used to perform molecular docking between the designed S-TH substitutes and the above proteins. LDS was used to characterize the microbial degradability of S-TH substitutes, which was used as an evaluation index of the microbial degradability universality of S-TH substitutes in the the soil environment.

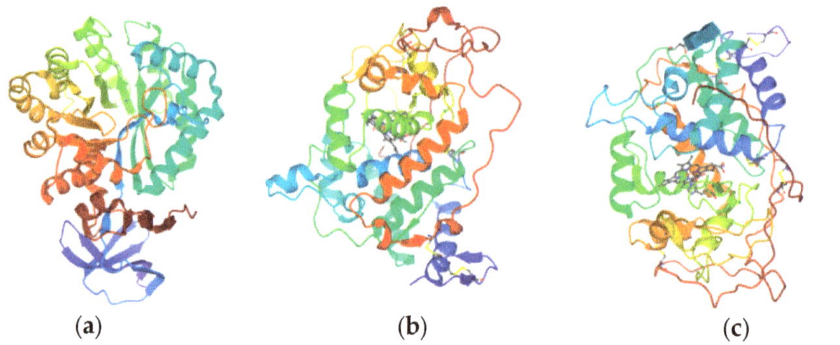

Figure 5. Schematic diagram of the structure of a universal degrading protein of S-TH substitutes in the soil environment: (**a**) 2QT3; (**b**) 1B85; (**c**) 1MNP.

4.5.2. Evaluation of the Toxicity of Antioxidant Systems in Fish of S-TH Substitutes in the Aquatic Environment

The UniProt database [34] was used to identify superoxide dismutase (SOD, UniProt ID: Q8JFG7), and hydrogen peroxide proteins (CAT, UniProt ID: E2CWE8) (Figure 6) as receptor proteins in carp (the rationale for the selection of the above three proteins is shown in the appendix). The above proteins were molecularly docked with S-TH substitutes with universal microbial degradability using the LibDock module of the DS software. Protein activity was more likely to be inhibited when the docking score was higher. The docking scores of the two proteins were added at a weight of 1:1, and the comprehensive value was used to characterize the toxic effect of S-TH substitutes on the antioxidant system of fish. The higher the comprehensive value, the stronger the toxicity of S-TH substitutes on the antioxidant system of fish.

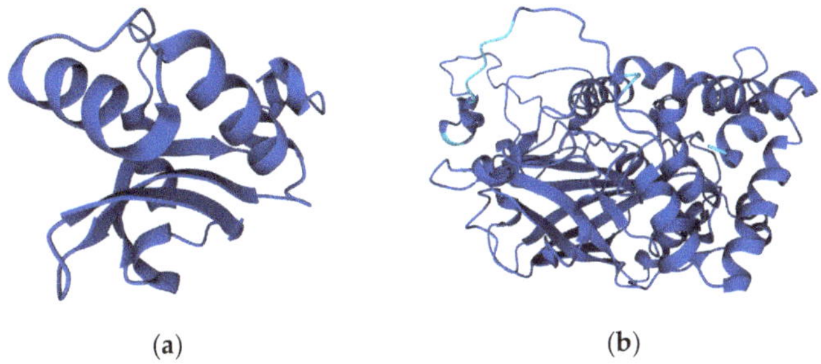

Figure 6. Schematic representation of antioxidant proteins in carp: (**a**) Q8JFG7; (**b**) E2CWE8.

4.6. Screen of Optimal Field Application Schemes to Promote the Microbial Degradation of S-TH Substitutes in Maize Cropping Fields—Taguchi Orthogonal Experiments, Full Factorial Design of Experiments, and Molecular Dynamics Methods

THs are applied primarily to maize [1]. In this study, the maize cropping field was selected as the primary research area to explore the influence of different external stimulus conditions (such as fertilizer and crop root secretion) of the maize cropping field on the microbial degradability of S-TH substitutes in the soil environment and to screen and determine the best field application scheme of S-TH substitutes. Furthermore, root exudates can promote microbial degradation near the rhizosphere [40]. In this study, 18 compounds, including the most widely used nitrogen fertilizer [52] and 17 kinds of rhizosphere secretory substances of maize, were selected (Table S12) [53,54] as external addictive conditions to investigate their influence on the degradability of S-TH substitutes.

In this study, 18 compounds (Table S12) were chosen as external conditions to design an 18-factors and 2-levels (0 for no addition, 1 for addition) Taguchi orthogonal experiment, with a total of 32 different sets of external conditions schemes. In addition, the binding energy of the substitute D-5, triazine hydrolysate protein 4L9X, and the complex system of external conditions (Table S7) were calculated to screen the optimal field application schemes to promote the microbial degradation of S-TH substitutes preliminary.

The Taguchi orthogonal experimental design is a method for independently evaluating the single-factor-level effects [55]. Eighteen compounds were chosen as external additives in this study. The Taguchi (T) module under the design of the experiment module in Minitab 20 software was used to construct a Taguchi orthogonal experimental design with 18 factors and two levels (0 represents no addition, 1 represents addition). A total of 32 groups of external condition-adding schemes were included. Among them, the S-TH substitutes and 4L9X proteins selected as fixed conditions were added to this scheme's molecular dynamics calculation system.

The Gromacs 4.6.5 software (GROMACS development team: Stockholm, Sweden) [56] was used to simulate the molecular dynamics of the 32 groups of schemes. After docking, the compound system of S-TH substitutes, external condition compounds, and 4L9X protein were placed in a periodic cubic aqueous solution with a side length of 15 nm. The GROMOS96 43a1 force field was utilized for molecular restraint. In addition, a positively charged Na^+ neutralizing system was added. The binding energy between S-TH substitutes, external condition compounds, and 4L9X proteins in each group was calculated using the Molecular Mechanics/Poisson-Boltzmann Surface Area (MMPBSA) method (the smaller the binding energy, the stronger the promoting effect of the scheme on the microbial degradation of S-TH substitutes). Finally, the external condition addition scheme with the minimum binding energy was used as the initial screening scheme for the subsequent total factorial experimental design. Furthermore, the best field application scheme is conducive to degrading S-TH substitutes by soil rhizosphere microorganisms in maize cropping fields.

The full factorial design of experiments is a method that allows rapid screening of multilevel, multifactorial, and correlated vital factors, thereby reducing experimental workload and increasing efficiency [57]. In this study, we used the Factor (F) module under the design of the experiment module of the Minitab 20 software (Minitab LLC.: Centre County, Pennsylvania, United States) to construct an n-factor 2-level full factorial design of the experiment based on the initial screening scheme of the Taguchi orthogonal experiment (assuming the number of external conditions in the scheme is n). Molecular dynamics was used to calculate the binding energies of S-THs, external conditions, and 4L9X protein. In addition, using Minitab 20 software, the absolute values of binding energies were entered as response values into the constructed design of the experimental table, and the factorial design was analyzed using the Factorial (F) module under the design of the experimental module. The interaction mechanisms (synergistic and antagonistic) between the n-factors were further analyzed to verify the reliability of the screened optimal field application options.

4.7. Simulation of Microbial Degradation Pathways of S-TH Substitutes—DFT and Microbial Degradation Pathway Simulation

In this study, the microbial degradation pathways of S-THs before and after molecular modification were simulated and inferred from that of ATZ indicated in the literature (Figure S4) [58]. Furthermore, the possible intermediate and final products produced during the degradation and the change of microbial degradation degree molecules before and after modification were analyzed based on DFT. Gaussian09 software (Gaussian Inc. Wallingford, Connecticut, United States) was used to optimize and calculate the reaction energy barrier ($\Delta E > 0$, with a smaller energy barrier indicating a more accessible reaction) of the microbial degradation of S-THs before and after molecular modification at the B3LYP/6-31G* unit level [35,59]. It has been shown that the ΔE (>0) represents the difficulty of reaction occurrence, with smaller values indicating that the reaction is more likely to occur [35]. Therefore, the ΔE results were used to evaluate the microbial degradation of S-THs before and after molecular modification.

4.8. Evaluation of Aquatic Biotoxicity and Human Health Risks of Microbial Degradation Products of S-TH Substitutes—EPI Software Method, Pharmacokinetic and Toxicokinetic Methods

Estimation Programs Interface (EPIWEB 4.1) software (SRC Inc.: Syracuse, New York, NY, USA) [60] was used in this study to predict and evaluate the aquatic toxicity of intermediate microbial degradation products by calculating the toxicity of the aquatic organisms (green algae (EC_{50}), Daphnia (LC_{50}), and fish (LC_{50})). In addition, the pharmacokinetic and toxicokinetic methods in the ADMET module of the DS software were used to calculate the human health risks (hepatotoxicity, maximum tolerated dosage, skin sensitization, skin irritation, and rodent carcinogenicity) of S-THs and their substitutes' microbial degradation products. These indicators can be used to predict and evaluate the human health risks posed by microbial degradation products.

5. Conclusions

This study developed an ecological and sustainable S-THs control scheme that can effectively reduce the environmental and human health impacts of S-THs application in maize cropping fields through molecular source prevention, field application process control, and end-of-soil degradation evaluation. The main findings were as follows: (1) the design of a substitute to S-THs with high herbicidal functionality and microbial degradability, low human health risk, and environmental friendliness using three-dimensional quantitative structure-activity relationship (3D-QSAR) modeling and molecular docking methods; (2) the simulation and screening of fertilizer and soil secretion that can promote microbial degradation of the substitute to S-THs in maize cropping fields using Taguchi orthogonal experiments and full factorial design of experiments; and (3) Based on the DFT, we simulated and calculated the degradation pathways and reaction energy barrier of S-TH molecules before and after modification, and confirmed that the designed S-TH substitutes have stronger microbial degradability. This study developed a source modification scheme for S-TH substitutes, an optimal application process control scheme for S-TH substitutes in maize cropping fields, and an environmental and human health evaluation scheme for their terminal potential degradation products, which provides theoretical guidance for minimizing the risk of S-THs application to the environment and human health in maize cropping fields.

Supplementary Materials: The following supporting information can be downloaded at: https://www.mdpi.com/article/10.3390/ijms24065691/s1, Figure S1: The outcomes of a comprehensive weighting of herbicidal, degradability, and toxicity for the comprehensive effect of S-TH molecules; Figure S2: Molecular structure of ATZ with proposed modification sites (Site 1: -CH_3; Site 2: -H; Site 3: -H; Site 4: -CH_3, -CH_3; Site 5: -Cl); Figure S3: The structure and the common backbone of PRZ; Figure S4: Microbial degradation pathway of ATZ; Table S1: Docking scores for herbicidal, degradative, and toxic effects of S-TH molecules and calculation of their comprehensive value; Table S2: Molecular combination of the CoMSIA model training sets and test sets; Table S3: Substitution sites and substituent groups for the design of molecular substitutes to S-THs; Table S4: Predicted values of herbicidal, degradative, and toxic effects of molecular substitutes to S-THs, their comprehensive effects, and rates of change; Table S5: The LDS values and their rates of change for the microbial degradation pervasiveness of molecular substitutes to S-THs; Table S6: Comprehensive values of the toxic effects of molecular substitutes to S-THs on the antioxidant system of fish; Table S7: Molecular dynamics simulations of the external condition addition scheme (18 factors/2 levels) based on the Taguchi orthogonal experiments; Table S8: Response and effect values for the external condition addition scenario (3 factors/2 levels); Table S9: Evaluation of the aquatic toxicity of D-5 microbial degradation products based on the EPI software method; Table S10: Human health risk assessment of D-5 microbial degradation products based on pharmacokinetic and toxicokinetic data; Table S11: CoMSIA model evaluation methods and model evaluation criteria; Table S12: Nitrogen fertilizer and maize root secretion and their corresponding factors.

Author Contributions: Conceptualization, T.Y. and P.S.; methodology, T.Y. and P.S.; software, T.Y. and P.S.; validation, W.Z.; formal analysis, T.Y.; investigation, P.S. and W.Z.; resources, P.S.; data curation, T.Y. and P.S.; writing—original draft preparation, T.Y.; writing—review and editing, P.S.; visualization, W.Z.; supervision, W.Z.; project administration, W.Z. All authors have read and agreed to the published version of the manuscript.

Funding: This research received no external funding.

Institutional Review Board Statement: Not applicable.

Informed Consent Statement: Not applicable.

Data Availability Statement: Data is contained within the article or supplementary material.

Conflicts of Interest: The authors declare no conflict of interest.

References

1. Manousi, N.; Kabir, A.; Zachariadis, G.A. Recent advances in the extraction of triazine herbicides from water samples. *J. Sep. Sci.* **2021**, *45*, 113–133. [CrossRef] [PubMed]
2. Yu, Q.Q.; Lu, F.F.; Ma, L.Y.; Yang, H.; Song, N.H. Residues of Reduced Herbicides Terbuthylazine, Ametryn, and Atrazine and Toxicology to Maize and the Environment through Salicylic Acid. *ACS Omega* **2021**, *6*, 27396–27404. [CrossRef] [PubMed]
3. Funar-Timofei, S.; Borota, A.; Crisan, L. Combined molecular docking and QSAR study of fused heterocyclic herbicide inhibitors of D1 protein in photosystem II of plants. *Mol. Divers.* **2017**, *21*, 437–454. [CrossRef] [PubMed]
4. Li, D.J.; Daler, D. Ocean pollution from land-based sources: East China sea, China. *Ambio* **2004**, *33*, 107–113. [CrossRef] [PubMed]
5. EPA Releases Draft Biological Evaluations for Atrazine, Simazine and Propazine. Available online: https://www.epa.gov/pesticides/epa-releases-draft-biological-evaluations-atrazine-simazine-and-propazine (accessed on 15 May 2022).
6. Gu, L.L. New progress of research and development of triazine herbicides. *World Pestic.* **2021**, *43*, 12–23. [CrossRef]
7. Galvín, R.M.; Rodríguez Mellado, J.M.; Higuera, M.J. Reductive deactivation of some s-triazine herbicides: Prometryne, desmetryne and terbutryne. *J. Serb. Chem. Soc.* **2002**, *67*, 381–392. [CrossRef]
8. Zhao, Y.H. Biodegradation and Regulation of the s-Triazine Herbicides. Ph.D. Thesis, Zhejiang University, Hangzhou, China, 2002.
9. Velisek, J.; Stara, A.; Koutnik, D.; Machova, J. Effects of prometryne on early life stages of common carp (*Cyprinus carpio* L.). *Pestic. Biochem. Physiol.* **2015**, *118*, 58–63. [CrossRef]
10. Hao, M.J.; Wang, J.H.; Dong, W.; Duan, L.M. Study on the toxicity of herbicide atrazine to zebrafish embryos. *Sci. Technol. Vis.* **2018**, *22*, 68–69. [CrossRef]

11. Khoshnood, Z.; Jamili, S.; Khodabandeh, S. Histopathological effects of atrazine on gills of caspian kutum rutilus frisii kutum fingerlings. *Dis. Aquat. Organ.* **2015**, *113*, 227–234. [CrossRef]
12. Mela, M.; Guiloski, I.C.; Doria, H.B.; Randi, M.A.F.; Ribeiro, C.A.D.; Pereira, L.; Maraschi, A.C.; Prodocimo, V.; Freire, C.A.; de Assis, H.C.S. Effects of the herbicide atrazine in neotropical catfish (*Rhamdia quelen*). *Ecotoxicol. Environ. Saf.* **2013**, *93*, 13–21. [CrossRef]
13. Tillitt, D.E.; Papoulias, D.M.; Whyte, J.J.; Richter, C.A. Atrazine reduces reproduction in fathead minnow (*Pimephales promelas*). *Aquat. Toxicol.* **2010**, *99*, 149–159. [CrossRef]
14. Liu, Q.Y. Research on the Cytotoxicity and Related Mechanisms of Herbicide Prometrvn In Vitro. Master's Thesis, Zhejiang University, Hangzhou, China, 2017.
15. Joo, H.; Choi, K.; Hodgson, E. Human metabolism of atrazine. *Pestic. Biochem. Physiol.* **2010**, *98*, 73–79. [CrossRef]
16. Zeljezic, D.; Garaj-Vrhovac, V.; Perkovic, P. Evaluation of DNA damage induced by atrazine and atrazine-based herbicide in human lymphocytes in vitro using a comet and DNA diffusion assay. *Toxicol. Vitro* **2006**, *20*, 923–935. [CrossRef] [PubMed]
17. Wang, Q.F.; Xie, S.G. Isolation and characterization of a high-efficiency soil atrazine-degrading Arthrobacter sp. strain. *Int. Biodeterior. Biodegrad.* **2012**, *71*, 61–66. [CrossRef]
18. Li, Q.Y.; Li, Y.; Zhu, X.K.; Cai, B.L. Isolation and characterization of atrazine-degrading Arthrobacter sp. AD26 and use of this strain in bioremediation of contaminated soil. *J. Environ. Sci.* **2008**, *20*, 1226–1230. [CrossRef]
19. Vargha, M.; Takats, Z.; Marialigeti, K. Degradation of atrazine in a laboratory scale model system with Danube River sediment. *Water Res.* **2005**, *39*, 1560–1568. [CrossRef]
20. Singh, P.; Suri, C.R.; Cameotra, S.S. Isolation of a member of Acinetobacter species involved in atrazine degradation. *Biochem. Biophys. Res. Commun.* **2004**, *317*, 697–702. [CrossRef]
21. Strong, L.C.; Rosendahl, C.; Johnson, G.; Sadowsky, M.J.; Wackett, L.P. Arthrobacter aurescens TC1 metabolizes diverse s-triazine ring compounds. *Appl. Environ. Microbiol.* **2002**, *68*, 5973–5980. [CrossRef]
22. Katz, I.; Dosoretz, C.G.; Mandelbaum, R.T.; Green, M. Atrazine degradation under denitrifying conditions in continuous culture of Pseudomonas ADP. *Water Res.* **2001**, *35*, 3272–3275. [CrossRef]
23. Jones, L.R.; Owen, S.A.; Horrell, P. Bacterial inoculation of granular activated carbon filters for the removal of atrazine from surface water. *Water Res.* **1998**, *32*, 2542–2549. [CrossRef]
24. Rousseaux, S.; Hartmann, A.; Lagacherie, B.; Piutti, S.; Andreux, F.; Soulas, G. Inoculation of an atrazine-degrading strain, Chelatobacter heintzii Cit1, in four different soils: Effects of different inoculum densities. *Chemosphere* **2003**, *51*, 569–576. [CrossRef] [PubMed]
25. Mougin, C.; Laugero, C.; Asther, M.; Dubroca, J.; Frasse, P.; Asther, M. Biotransformation of the herbicide atrazine by the white rot rungus *Phanerochaete chrysosporium*. *Appl. Environ. Microbiol.* **1994**, *60*, 705–708. [CrossRef] [PubMed]
26. Masaphy, S.; Levanon, D.; Vaya, J.; Henis, Y. Isolation and characterization of a novel atrazine metabolite produced by the fungus *Pleurotus pulmonarius*, 2-chloro-4-ethylamino-6-(1-hydroxyisopropyl) amino-1,3,5-triazine. *Appl. Environ. Microbiol.* **1993**, *59*, 4342–4346. [CrossRef] [PubMed]
27. Vonberg, D.; Vanderborght, J.; Cremer, N.; Pütz, T.; Herbst, M.; Vereecken, H. 20 years of long-term atrazine monitoring in a shallow aquifer in western Germany. *Water Res.* **2014**, *50*, 294–306. [CrossRef]
28. Qu, R.J.; Liu, H.X.; Feng, M.B.; Yang, X.; Wang, Z.Y. Investigation on intramolecular hydrogen bond and some thermodynamic properties of polyhydroxylated anthraquinones. *J. Chem. Eng. Data* **2012**, *57*, 2442–2455. [CrossRef]
29. Wang, X.L.; Gu, W.E.; Guo, E.; Cui, C.Y.; Li, Y. Assessment of long-range transport potential of polychlorinated Naphthalenes based on three-dimensional QSAR models. *Environ. Sci. Pollut. Res. Int.* **2017**, *24*, 14802–14818. [CrossRef]
30. Veerasamy, R.; Rajak, H.; Jain, A.; Sivadasan, S.; Varghese, C.P.; Agrawal, R.K. Validation of QSAR models—Strategies and importance. *Int. J. Drug Des. Discov.* **2011**, *2*, 511–519.
31. Karns, J.S. Gene sequence and properties of an s-triazine ring-cleavage enzyme from *Pseudomonas* sp. strain NRRLB-12227. *Appl. Environ. Microbiol.* **1999**, *65*, 3512–3517. [CrossRef]
32. Asgher, M.; Bhatti, H.N.; Ashraf, M.; Legge, R.L. Recent developments in biodegradation of industrial pollutants by white rot fungi and their enzyme system. *Biodegradation* **2008**, *19*, 771–783. [CrossRef]
33. Protein Data Bank. Available online: https://www.rcsb.org (accessed on 15 July 2022).
34. UniProt. Available online: https://www.uniprot.org (accessed on 6 August 2022).
35. Sun, Y.H. Researches on the Oxidation Mechanisms of Typical Organic Pollutants in the Atmosphere. Ph.D. Thesis, Shandong University, Jinan, China, 2015.
36. Sun, P.X.; Zhao, W.J. Strategies to control human health risks arising from antibiotics in the environment: Molecular modification of QNs for enhanced plant-microbial synergistic degradation. *Int. J. Environ. Res. Public Health* **2021**, *18*, 10610. [CrossRef]
37. Shen, Y.Y. Study on the Behavior Characteristics of Petroleum Containment in the Soil and Rhizoremediation. Ph.D. Thesis, Chang'an University, Xi'an, China, 2013.
38. Lipson, D.A.; Raab, T.K.; Schmidt, S.K.; Monson, R.K. An empirical model of amino acid transformations in an alpine soil. *Soil. Biol. Biochem.* **2001**, *33*, 189–198. [CrossRef]
39. Chapin, F.S.; Matson, P.A.; Mooney, H.A. Terrestrial nutrient cycling. In *Principles of Terrestrial Ecosystem Ecology*, 1st ed.; Springer: New York, NY, USA, 2002; pp. 97–122. [CrossRef]

40. Li, J.B.; Luo, C.L.; Zhang, D.Y.; Cai, X.X.; Jiang, L.F.; Zhao, X.; Zhang, G. Diversity of the active phenanthrene degraders in PAH-polluted soil is shaped by ryegrass rhizosphere and root exudates. *Soil. Biol. Biochem.* **2019**, *128*, 100–110. [CrossRef]
41. Fu, R.; Li, X.; Zhao, Y.; Pu, Q.; Li, Y.; Gu, W. Efficient and synergistic degradation of fluoroquinolones by bacteria and microalgae: Design of environmentally friendly substitutes, risk regulation and mechanism analysis. *J. Hazard. Mater.* **2022**, *437*, 129384. [CrossRef]
42. Li, X.; Hou, Y.; Li, Q.; Gu, W.; Li, Y. Molecular design of high-efficacy and high drug safety Fluoroquinolones suitable for a variety of aerobic biodegradation bacteria. *J. Environ. Manag.* **2021**, *299*, 113628. [CrossRef] [PubMed]
43. Xue, J.Q.; Chen, X.Y.; Zhao, Y.Y.; Li, Y. Exposure to high-performance benzotriazole ultraviolet stabilizers: Advance in toxicological effects, environmental behaviors and remediation mechanism using in-silica methods. *Chemosphere* **2023**, *315*, 137699. [CrossRef] [PubMed]
44. Gu, W.W.; Zhao, Y.Y.; Li, Q.; Li, Y. Environmentally friendly polychlorinated naphthalenes (PCNs) derivatives designed using 3D-QSAR and screened using molecular docking, density functional theory and health-based risk assessment. *J. Hazard. Mater.* **2019**, *363*, 316–327. [CrossRef] [PubMed]
45. Liao, D.; Qian, J.; Chisholm, D.A.; Jordan, D.B.; Diner, B.A. Crystal structures of the photosystem II D1 C-terminal processing protease. *Nat. Struct. Mol. Biol.* **2000**, *7*, 749–753. [CrossRef]
46. Jackson, C.J.; Coppin, C.W.; Carr, P.D.; Aleksandrov, A.; Wilding, M.; Sugrue, E.; Ubels, J.; Paks, M.; Newman, J.; Peat, T.S.; et al. 300-fold increase in production of the Zn^{2+}-dependent dechlorinase TrzN in soluble form via apoenzyme stabilization. *Appl. Environ. Microbiol.* **2014**, *80*, 4003–4011. [CrossRef]
47. Sharma, D.; De Falco, L.; Padavattan, S.; Rao, C.; Geifman-Shochat, S.; Liu, C.F.; Davey, C.A. PARP1 exhibits enhanced association and catalytic efficiency with γH2A.X-nucleosome. *Nat. Commun.* **2019**, *10*, 5751. [CrossRef]
48. Pan, J.L.; Xu, Q.J.; Liu, H.X. Typhoon disaster risk assessment based on AHP-TOPSIS optimal combination empowerment. *J. Nanning Norm. Univ. (Nat. Sci. Ed.)* **2021**, *38*, 60–67. [CrossRef]
49. Zeng, X.L.; Qu, R.J.; Feng, M.B.; Chen, J.; Wang, L.S.; Wang, Z.Y. Photodegradation of Polyfluorinated Dibenzo-p-Dioxins in Organic Solvents: Experimental and Theoretical Studies. *Environ. Sci. Technol.* **2016**, *50*, 8128–8134. [CrossRef] [PubMed]
50. Hou, Y.L.; Zhao, Y.Y.; Li, Y. Environmentally friendly fluoroquinolone derivatives with lower plasma protein binding rate designed using 3D-QSAR, molecular docking and molecular dynamics simulation. *Int. J. Environ. Res. Public Health* **2020**, *17*, 6626. [CrossRef]
51. Yang, L.Z.; Liu, M. A double-activity (green algae toxicity and bacterial genotoxicity) 3D-QSAR model based on the comprehensive index method and its application in fluoroquinolones' modification. *Int. J. Environ. Res. Public Health* **2020**, *17*, 942. [CrossRef]
52. Morris, T.F.; Murrell, T.S.; Beegle, D.B.; Camberato, J.J.; Ferguson, R.B.; Grove, J.; Ketterings, Q.; Kyveryga, P.M.; Laboski, C.A.M.; McGrath, J.M.; et al. Strengths and limitations of nitrogen rate recommendations for corn and opportunities for improvement. *Agron. J.* **2018**, *110*, 1–37. [CrossRef]
53. Liu, X.W.; Wang, K.Q.; Zhao, Y.Y.; Duan, X.; Zhang, Y. Characteristics of maize root exudates at seedling stage and their response to rhizosphere soil enzyme activities under contour reverse-slope terrace. *J. Sichuan Agric. Univ.* **2021**, *39*, 477–485. [CrossRef]
54. Zhu, H.F.; Zhang, L.; Li, S.S.; Wang, Y.J.; Sun, S.R.; Chen, J.; Kou, S.M.; Jin, Q.; Xiao, M. The rhizosphere and root exudates of maize seedlings drive plasmid mobilization in soil. *Appl. Soil Ecol.* **2018**, *124*, 194–202. [CrossRef]
55. Majumdar, A.; Ghosh, D. Genetic algorithm parameter optimization using Taguchi robust design for multi-response optimization of experimental and historical data. *Int. J. Comput. Appl.* **2015**, *127*, 26–32. [CrossRef]
56. Childers, M.C.; Daggett, V. Insights from molecular dynamics simulations for computational protein design. *Mol. Syst. Des. Eng.* **2017**, *2*, 9–33. [CrossRef] [PubMed]
57. Kendrick, M.R.; Benstead, J.P. Temperature and nutrient availability interact to mediate growth and body stoichiometry in a detritivorous stream insect. *Freshwater Biol.* **2013**, *58*, 1820–1830. [CrossRef]
58. Shapir, N.; Cheng, G.; Sadowsky, M.J.; Wackett, L.P. Purification and characterization of TrzF: Biuret hydrolysis by allophanate hydrolase supports growth. *Appl. Environ. Microbiol.* **2006**, *72*, 2491–2495. [CrossRef]
59. Salahinejad, M.; Ghasemi, J.B. 3D-QSAR studies on the toxicity of substituted benzenes to Tetrahymena pyriformis: CoMFA, CoMSIA and VolSurf approaches. *Ecotoxicol. Environ. Saf.* **2014**, *105*, 128–134. [CrossRef] [PubMed]
60. US EPA. *Estimation Programs Interface Suite™ for Microsoft®Windows, v 4.11 or Insert Version Used*; United States Environmental Protection Agency: Washington, DC, USA, 2012.

toxics

MDPI

Article

Development of an Improved Sulfur-Oxidizing Bacteria-Based Ecotoxicity Test for Simple and Rapid On-Site Application

Heonseop Eom

Department of Civil Engineering, Keimyung University, 1095 Dalgubeol-daero, Dalseo-gu, Daegu 42601, Republic of Korea; heom@kmu.ac.kr; Tel.: +82-53-580-5706

Abstract: Microbial toxicity tests are considered efficient screening tools for the assessment of water contamination. The objective of this study was to develop a sulfur-oxidizing bacteria (SOB)-based ecotoxicity test with high sensitivity and reproducibility for simple and rapid on-site application. To attain this goal, we developed a 25 mL vial-based toxicity kit and improved our earlier SOB toxicity test technique. The current study applied a suspended form of SOB and shortened the processing time to 30 min. Moreover, we optimized the test conditions of the SOB toxicity kit in terms of initial cell density, incubating temperature, and mixing intensity during incubation. We determined that 2×10^5 cells/mL initial cell density, 32 °C incubating temperature, and 120 rpm mixing intensity are the optimal test conditions. Using these test conditions, we performed SOB toxicity tests for heavy metals and petrochemicals, and obtained better detection sensitivity and test reproducibility, compared to earlier SOB tests. Our SOB toxicity kit tests have numerous advantages, including a straightforward test protocol, no requirement of sophisticated laboratory equipment, and no distortion of test results from false readings of end-points and properties of test samples, making it suitable for simple and rapid on-site application.

Keywords: ecotoxicity; sulfur-oxidizing bacteria; electrical conductivity

Citation: Eom, H. Development of an Improved Sulfur-Oxidizing Bacteria-Based Ecotoxicity Test for Simple and Rapid On-Site Application. *Toxics* **2023**, *11*, 352. https://doi.org/10.3390/toxics11040352

Academic Editors: Esref Demir and Sam Kacew

Received: 25 February 2023
Revised: 4 April 2023
Accepted: 6 April 2023
Published: 8 April 2023

1. Introduction

There are approximately 150,000 different chemicals in commercial use and their number and applications continue to grow [1,2]. Chemicals used in homes and diverse industries are released directly or indirectly into water systems [1,2]. Although chemicals can show unknown adverse effects on the environment and ecosystems, the presence of chemicals in the water environment does not necessarily represent a risk [1]. Chemicals exceeding levels of concern cause water contamination and pose threats to aquatic ecosystems and public health [1,3].

Conventionally, water contamination has been assessed based on physicochemical quantitative analyses of water quality parameters, including dissolved oxygen, solids, biochemical or chemical oxygen demand, various nutrients, and selected contaminants [4–7]. These physicochemical analyses are useful to understanding the fundamental properties of water and obtaining detailed quantitative information of specific contaminants [8–10]. However, such water quality evaluation is unable to reveal the biochemical effects of contamination on living organisms and the environment [9,11]. Moreover, physicochemical quantification usually requires advanced analytic equipment, skilled personnel, lengthy processing time, and high experimental expense, making it unsuitable for on-site simple and rapid toxicity screening of contaminated water [2,12].

As supplements or alternative to physicochemical quantitative analyses, biological tests that employ organisms such as invertebrates, fish, daphnia, and microorganisms, have been widely used in toxicity assessment of contaminated water [4,13–18]. Because biological tests, generally named ecotoxicity tests, evaluate toxicity based on changes in the response of organisms to contaminants, they can directly demonstrate the impacts of contaminants on living organisms and the environment [8,15,19–21]. Among the diverse trophic levels of organisms, microorganism-based tests are considered particularly efficient tools for routine toxicity evaluation because they provide easy test protocols, relatively short test time, cost effectiveness, and less ethical responsibility compared to other organism-based tests [21–23]. Moreover, as microorganisms have diverse ecological functions, microorganism-based tests can provide important toxicological information on ecosystems. Numerous microorganisms, including bioluminescent bacteria, nitrifying bacteria, oligotrophic bacteria, *Escherichia coli*, and microalgae, have been employed in ecotoxicity tests with measurement of luminescence, growth and respiration rate, turbidity, and photosynthesis, showing favorable performance in assessing toxicity of various inorganic and organic contaminants in water [24–28].

In our earlier studies, we demonstrated the application of sulfur-oxidizing bacteria (SOB), specifically *Acidithiobacillus caldus*, to toxicity tests of contaminated water [2,21,29–35]. SOB are chemolithoautotrophic and acidophilic bacteria ubiquitous in diverse environments, including the hydrosphere [35,36]. SOB gain energy from aerobic oxidation of sulfur and produce sulfate and hydrogen ions as by-products (Equation (1)) [21,37].

$$S^0 + H_2O + 1.5O_2 \rightarrow SO_4^{2-} + 2H^+, \Delta G^{\circ'} = -587 \text{ kJ/reaction} \quad (1)$$

Because electrical conductivity represents the ability to carry a current and is proportional to the concentration of ions, electrical conductivity is able to serve as a proxy for the microbial activity of SOB [21,35,38]. In the presence of contaminants, SOB activity is inhibited, resulting in less generation of sulfate and hydrogen ions. Hence, SOB tests evaluate the toxicity of contaminated water by comparing increases in electrical conductivity between test samples and the control (where no contaminants exist). Our earlier studies confirmed that this SOB test is a reliable toxicity-screening technique [21,29–35]. SOB tests showed favorable results from toxicity assessment of heavy metals, endocrine-disrupting compounds, inorganic nitrogen, and petrochemicals in water [2,21,29–35]. Moreover, SOB tests have the advantages of simple test methodology and low cost, and they do not require sophisticated instruments to measure microbial activity.

The objective of this study was to develop improved SOB toxicity tests for simple and rapid on-site application. To attain this goal, we used a suspended form of SOB in this study. We expected that this approach would be more advantageous to the application of identical amounts of SOB and making direct contact with contaminants. Moreover, for better field application, the present study developed a kit-type SOB test and decreased the processing time from several hours to 30 min. In addition, we optimized test conditions such as initial cell density, incubating temperature, and mixing intensity, yielding enhanced detection sensitivity and test reproducibility. As a result, our current SOB test represents an improvement in simple and rapid on-site toxicity assessment.

2. Materials and Methods

2.1. SOB Strain and Cultivation

In the current study, a specific SOB strain, *Acidithiobacillus caldus*, was employed as the test organism. The SOB were obtained from Kangwon National University (Chuncheon, Republic of Korea) and cultivated in a liquid medium in a 500 mL conical glass flask. The medium for SOB was prepared according to Johnson et al. (1987) [39] and Duquesne et al. (2003) [40]. The medium had 0.5 g $MgSO_4 \cdot 7H_2O$, 3 g $(NH_4)_2SO_4$, 0.5 g $K_2HPO_4 \cdot 3H_2O$, 0.1 g KCl, and 0.01 g $Ca(NO_3)_2$ per liter of distilled water. The pH of the medium was adjusted to 3 with 10% sulfuric acid. The medium was autoclaved for 1 h at 120 °C and subsequently cooled at room temperature before use. A filter-sterilized trace-element solution (10 mL) and sulfur powder (1 g), an energy source for SOB, were added to 100 mL medium. The trace-element solution included 11 mg $FeCl_3 \cdot 6H_2O$, 0.5 mg $CuSO_4 \cdot 5H_2O$, 2.0 mg H_3BO_4, 2.0 mg $MnSO_4 \cdot H_2O$, 0.8 mg $NaMoO_4 \cdot 2H_2O$, 0.6 mg $CoCl_2 \cdot 6H_2O$, and 0.9 mg $ZnSO_4 \cdot 7H_2O$ in 10 mL distilled water. Oxygen was continuously sparged to the medium to provide an electron acceptor for SOB. Cultivation was performed in a shaking incubator (JSSI-070, JSR, Gongju, Republic of Korea) at 37 °C with 50 rpm mixing intensity. SOB were cultivated for 3–4 d. To provide SOB with comparable activity in subsequent toxicity tests, we evaluated the activity of SOB by measuring changes in electrical conductivity for 1 h before being employed. Only SOB showing an increase in electrical conductivity of 0.10–0.12 mS/cm were used for toxicity tests.

2.2. SOB Toxicity Test and Optimization of Test Conditions

The SOB toxicity test kit consisted of a 25 mL flat-bottom glass vial with a rubber stopper and a plastic cap (Figure 1).

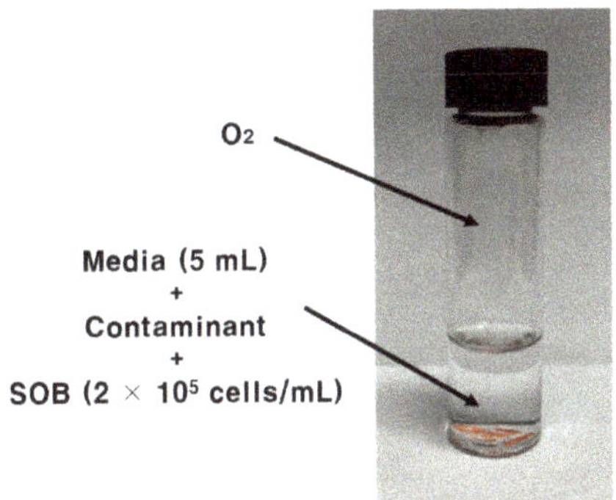

Figure 1. A SOB toxicity test kit. This 25 mL glass-vial-based SOB kit consists of 2×10^5 cells/mL of initial cell density and 5 mL of media. Headspace and media are purged with oxygen. Incubation is conducted in a shaking incubator with 120 rpm mixing intensity at 32 °C for 30 min.

SOB toxicity tests were conducted as follows. First, a certain amount of SOB and 5 mL of contaminant-spiked medium were added to the kit. Oxygen was sparged to the headspace of the kit and medium for around 1 min. Then, initial electrical conductivity (in the solution) was measured using an electrical conductivity meter (InLab 737, Mettler Toledo, Columbus, OH, USA). The kit was closed with a cap and rubber stopper and incubated with the vial lying on its side in a shaking incubator for 30 min. After incubation, electrical conductivity was determined again. All SOB toxicity tests were performed in triplicate.

Optimization for test conditions (initial cell density, incubating temperature, and mixing intensity during incubation) of the SOB toxicity kit was performed. We tested 10^5, 2×10^5, 5×10^5, and 10^6 cells/mL initial cell densities; 27, 32, 37, and 42 °C incubating temperatures; and 70, 100, 120, and 150 rpm mixing intensities. Hence, a total of 64 combinations of test conditions were evaluated with mercury (0.01, 0.05, 0.1, 0.2, and 0.5 mg/L Hg^{2+}). We assessed the detection sensitivity and reproducibility of SOB tests with (30 min) half-effective concentration (EC_{50}) and coefficient of variation (CV) for EC_{50} from triplicate kit tests, respectively.

Using the conditions obtained from the above optimization tests, SOB toxicity tests were conducted using heavy metals (Ag^{2+}, As^{3+}, CN^-, Cr^{6+}, Cu^{2+}, Hg^{2+}, and Zn^{2+}) and petrochemicals (benzene (B), toluene (T), ethylbenzene (E), and *p*-xylene (X), collectively referred to as BTEX) to evaluate improvement in the current technique compared to earlier SOB tests [31,33–35].

2.3. Chemicals and Laboratory Analyses

All chemicals and sulfur powder used in the present study were ACS grade and had at least 99.9% purity. They were all purchased from Sigma-Aldrich (St. Louis, MO, USA) and employed without further purification. Contaminants (heavy metals and petrochemicals (BTEX)) tested in the current study were prepared according to our earlier studies [31–33,35].

Concentrations of heavy metals were calculated based on ions. BTEX were prepared as follows. Each crude liquids of B, T, E, and X was diluted in a nutrient mineral buffer [21] with 0.1% dimethyl sulfoxide twice to create target concentrations of BTEX used in toxicity tests. Amounts of BTEX in the liquid phase of the test kits were determined before and after toxicity testing using high-performance liquid chromatography (HPLC, Water Corporation, Milford, MA, USA). The detailed methodology for using HPLC was presented by Eom et al. (2023) [35].

The toxicity of each contaminant was evaluated by SOB inhibition (%) using Equation (2). As described in this equation, toxic responses of SOB (SOB inhibition) to contaminants were determined by comparisons of increases in electrical conductivity between the controls (the test kits where no contaminant was spiked) and test samples (where specific concentrations of contaminant were spiked).

$$Inhibition(\%) = \left(1 - \frac{Increase\ in\ electrical\ conductivity\ in\ sample\ for\ 30\ min\ incubation}{Increase\ in\ electric\ conductivity\ in\ control\ for\ 30\ min\ incubation}\right) \times 100 \quad (2)$$

EC_{50} values for contaminants were determined by the Hillslope equation (Equation (3)). In all dose–response relations, the lowest and highest effects were set to 0% and 100%, respectively.

$$Y = Bottom + \left(\frac{Top - Bottom}{(1+10)^{((logEC_{50}-X)\times Hillslope)}}\right) \quad (3)$$

where X is the dose of contaminant, Y is the toxic response of SOB (SOB inhibition), *Top* is the maximum toxic response, and *Bottom* is the minimum response.

For filter sterilization of the trace-element solution, Nalgene bottle-top sterile filter (0.2 μm) was employed. SOB cell density was determined using a hemocytometer (Paul Marienfeld GmbH & Co. KG, Lauda-Königshofen, Germany). We first measured cell density in the liquid medium and then diluted it to target cell densities. To evaluate statistical significance among data (SOB inhibition and EC_{50} values), ANOVA analysis was performed. A p-value of less than 0.05 was seen as statistical significant.

3. Results and Discussion

3.1. Optimization for Test Conditions of SOB Toxicity Kit

The results (30 min EC_{50} for Hg^{2+} and CV values) from the optimization tests are summarized in Table 1.

Overall, the employment of smaller initial cell densities (10^5 and 2×10^5 cells/mL) was lower (30 min) EC_{50} values (for Hg^{2+}) than the application of larger initial cell densities (5×10^5 and 10^6 cells/mL) under identical incubating temperatures and mixing intensities. This result indicates that smaller initial cell concentrations yielded more favorable detection sensitivity than larger initial cell concentrations. In microbial toxicity tests, initial cell concentration is a vital factor in determining detection sensitivity because toxicant availability per cell depends on initial cell density [3,41]. Hence, it is expected that employment of smaller initial cell density can lead to improved toxicity detection sensitivity. Lin et al. (2005), Singh and Shrivastave (2015), and Eom et al. (2021) confirmed this advantage (better detection sensitivity) with the employment of smaller initial cell density [3,41,42]. However, it was reported that less initial cell concentration can negatively impact test reproducibility (Lin et al., 2005; Eom et al., 2021) [3,42]. Our data also support this disadvantage (poor test reproducibility) with the application of smaller initial cell density. Particularly, employment of 10^5 cells/mL of initial cell density resulted in fairly greater CV values than application of 2×10^5, 5×10^5, and 10^6 cells/mL of initial cell densities. For example, CV values from the tests in which 10^5, 2×10^5, 5×10^5, and 10^6 cells/mL of initial cell densities were employed ranged 8.3–12.5, 2.1–4.7, 2.3–5.0, and 1.2–5.0%, respectively. These data show that 10^5 cells/mL of initial cell density led to poor test reproducibility compared to the other initial cell concentrations. Among the four tested initial cell densities, we chose 2×10^5 cells/mL as the optimal initial cell density achieving favorable performance in both sensitivity and reproducibility. As discussed above, employment of 10^5 cells/mL of initial cell density resulted in poor test reproducibility, while application of 5×10^5 and 10^6 cells/mL of initial cell densities caused relatively inferior detection sensitivity (higher EC_{50} values) than 2×10^5 cells/mL of initial cell density.

Table 1. Results from optimization for test conditions of SOB toxicity kit (30 min EC_{50} for Hg^{2+} and CV values, depending on test conditions).

Initial Cell Density (Cells/mL)	Incubating Temperature (°C)	Mixing Intensity (rpm)	EC_{50} (μg/L)	CV (%)	Initial Cell Density (Cells/mL)	Incubating Temperature (°C)	Mixing Intensity (rpm)	EC_{50} (μg/L)	CV (%)
10^5	27	70	90.7	12.1	5×10^5	27	70	115.7	4.3
		100	84.7	12.5			100	99.3	5.0
		120	44.7	12.3			120	56.4	4.5
		150	48.3	11.4			150	64.7	4.7
	32	70	92.0	9.7		32	70	124.0	2.4
		100	85.3	8.3			100	114.1	2.6
		120	44.0	9.1			120	58.0	3.0
		150	44.7	8.5			150	62.3	4.0
	37	70	143.0	10.3		37	70	155.0	3.2
		100	115.3	10.3			100	133.3	3.0
		120	57.3	11.3			120	72.3	3.5
		150	60.0	10.0			150	69.7	4.4
	42	70	154.3	10.1		42	70	174.7	2.3
		100	124.7	10.2			100	147.0	2.4
		120	75.1	11.9			120	86.3	5.7
		150	72.3	10.2			150	88.0	5.0
2×10^5	27	70	93.3	4.3	10^6	27	70	142.3	3.3
		100	84.0	3.1			100	126.7	2.5
		120	44.7	4.6			120	67.7	2.3
		150	46.1	3.8			150	74.7	5.1
	32	70	96.3	2.2		32	70	146.0	3.1
		100	88.3	2.8			100	130.3	1.9
		120	38.0	2.6			120	70.7	2.9
		150	44.3	2.6			150	72.3	2.1
	37	70	144.7	3.8		37	70	160.0	1.7
		100	120.3	2.1			100	145.7	4.0
		120	60.3	2.5			120	82.0	3.7
		150	62.1	3.2			150	79.7	1.9
	42	70	165.3	2.4		42	70	187.0	2.8
		100	133.7	4.3			100	159.1	4.7
		120	78.3	2.7			120	92.7	1.2
		150	76.0	4.7			150	94.0	4.6

In terms of incubating temperature, lower temperatures (27 and 32 °C) resulted in more decreased EC_{50} values than higher temperatures (37 and 42 °C). For example, when incubating temperatures were 27, 32, 37, and 42 °C under 2×10^5 cells/mL initial cell density and 120 rpm mixing intensity, (30 min) EC_{50} values (for Hg^{2+}) were 44.7, 38.0, 60.3, and 78.3 μg/L, respectively, demonstrating that lower temperatures yielded better detection sensitivity than higher temperatures. We speculate that incubating temperature is associated with microbial activity of SOB. SOB are mesophilic bacteria. There are numerous studies reporting that SOB show active microbial activity up to 40–42 °C [43–45]. Our tests also found that as incubating temperatures rose from 27 °C to 42 °C, EC increased in the control tests (where no contaminant was spiked) also escalated, indicating that SOB were more active as the incubating temperatures increased. However, high activity of SOB did not necessarily lead to favorable sensitivity in toxicity detection. SOB with high microbial activity can be less inhibited by the toxicity of contaminants. Therefore, relatively low incubating temperatures (27 and 32 °C), where SOB showed less activity, achieved better sensitivity than higher incubating temperatures (37 and 42 °C). Test reproducibility, on the other hand, was not significantly affected by incubating temperatures. CV values from the tests incubated under 27, 32, 37, and 42 °C were not very different if identical initial cell density and mixing intensity were applied. Among the four tested temperatures, we chose 32 °C as the optimal incubating temperature. EC_{50} values from 27 °C and 32 °C incubating temperatures did not show statistical significance (p-value = 0.17). The CV values from 32 °C incubating temperature were slightly lower than those from 27 °C incubating temperature when initial cell densities and mixing intensities were same.

Concerning mixing intensity during incubation, 120 and 150 rpm showed better detection sensitivity than 70 and 100 rpm. For example, (30 min) EC_{50} values (for Hg^{2+}) from the tests with 70, 100, 120, and 150 rpm under 2×10^5 cells/mL initial cell density and 32 °C incubating temperature were analyzed at 96.3, 88.3, 38.0, and 44.3 µg/L, respectively. We conjecture that mixing intensity provides SOB increased opportunities for contact with oxygen, which is an e-acceptor of SOB, in the headspace of test kits and with nutrients in the medium. In our tested range, as mixing intensities raised (from 70 rpm to 150 rpm), EC increases in the control tests also escalated. This finding suggests that increased mixing intensity boosts the activity of SOB, which can have a negative impact on detection sensitivity as discussed above. However, mixing intensity can also give SOB the opportunity to interact with contaminants in water, which is a factor contributing to favorable sensitivity. Accordingly, increasing mixing intensity creates both positive and negative impacts on detection sensitivity. Considering EC_{50} data, the positive impact seems to be stronger than the negative impact in our optimization tests. Increased mixing intensity to 120 rpm led to better detection sensitivity. Test reproducibility was not substantially influenced by mixing intensity; four tested different mixing intensities resulted in largely similar CV values if identical initial cell densities and incubating temperatures were employed. Because 120 rpm yielded the lowest EC_{50} values and achieved comparable CV values compared to the others, we considered 120 rpm the optimal mixing intensity.

In summary, from the optimization tests, we determined that 2×10^5 cells/mL initial cell density, 32 °C incubating temperature, and 120 rpm mixing intensity were the optimal test conditions, allowing favorable detection sensitivity and reproducibility in our SOB kit tests.

3.2. Comparisons of SOB Toxicity Test Results between the Current Optimal and Earlier Techniques

Using the above optimal test conditions, we conducted toxicity tests for heavy metals (Ag^{2+}, As^{3+}, CN^{-}, Cr^{6+}, Cu^{2+}, Hg^{2+}, and Zn^{2+}) and petrochemicals (benzene, toluene, ethylbenzene, and p-xylene), and compared the results with data from our previous SOB tests (Table 2).

As shown in Table 2, the current tests resulted in generally lower EC_{50} values compared to earlier ones. For example, our earlier tests [31–34] obtained 1.76–3.62, 0.20, 4.90, 1.17–2.70, 5.00, 0.21–0.92, and 1.55 mg/L of (2 h) EC_{50} values for Ag^{2+}, As^{3+}, CN^{-}, Cr^{6+}, Cu^{2+}, Hg^{2+}, and Zn^{2+}, respectively; however, the current (30 min) EC_{50} values were significantly lower than the earlier data (currently, 0.195, 0.042, 0.673, 0.456, 0.859, 0.038, and 0.692 mg/L were obtained for Ag^{2+}, As^{3+}, CN^{-}, Cr^{6+}, Cu^{2+}, Hg^{2+}, and Zn^{2+}, respectively). Furthermore, we previously had 166.1, 94.4, 38.9, and 34.3 mg/L of (24 h) EC_{50} values for benzene, toluene, ethylbenzene, and p-xylene, respectively [35]; currently, 35.7, 20.5, 4.0, and 3.7 mg/L of (30 min) EC_{50} values were obtained for benzene, toluene, ethylbenzene, and p-xylene, respectively. (After 30 min incubation, BTEX concentrations in the liquid phase of test kits were remained above 94% of initial values, which is in the range of the OECD guidance. Because our test kits were tightly closed and completely sealed with parafilm, it was expected that no BTEX escaped from the test kits. Hence, volatilization of BTEX seems to make no significant impact on EC_{50} values.) These comparisons of EC_{50} values indicate that the current optimized test technique improved sensitivities for toxicity detection compared to our earlier technique. In addition, it is worth noting is that we obtained this improved sensitivity result even with a shorter incubating time (30 min).

Table 2. Comparisons of results from SOB toxicity tests between the current and earlier studies.

Contaminant		Earlier SOB Tests				Current SOB Tests		
		Processing Time (h)	EC_{50} (mg/L)	CV (%)	Reference	Processing Time (h)	EC_{50} (mg/L)	CV (%)
Heavy metal	Ag^{2+}	2	1.76–3.62	-	Gurung et al. (2015) [31]; Ahmed et al. (2018) [33]	0.5	0.195	3.1
	As^{3+}	2	0.2	11.5	Eom et al. (2019) [34]	0.5	0.047	4.5
	CN^{-}	2	4.9	12.7	Eom et al. (2019) [34]	0.5	0.676	3.3
	Cr^{6+}	2	1.17–2.7	10.5	Qambrani et al. (2016) [32]; Ahmed et al. (2018) [33]; Eom et al. (2019) [34]	0.5	0.456	3.0
	Cu^{2+}	2	5	-	Ahmed et al. (2018) [33]	0.5	0.860	2.4
	Hg^{2+}	2	0.21–0.92	8.7	Ahmed et al. (2018) [33]; Eom et al. (2019) [34]	0.5	0.038	2.6
	Zn^{2+}	2	1.55	-	Ahmed et al. (2018) [33]	0.5	0.692	3.4
Petrochemical	Benzene	24	166.1	9.8	Eom et al. (2023) [35]	0.5	35.849	4.6
	Toluene	24	94.4	9.5	Eom et al. (2023) [35]	0.5	20.575	3.8
	Ethylbenzene	24	38.9	9.7	Eom et al. (2023) [35]	0.5	4.038	4.1
	p-Xylenes	24	34.3	8.6	Eom et al. (2023) [35]	0.5	3.803	2.4

The current SOB tests also resulted in favorable test reproducibility. In our earlier SOB tests [34,35], CV values ranged from 8.6% to 12.7%. However, the CV values obtained from the present study were fairly lower (2.4–4.6%) than these earlier values. In addition, Van Coillie et al. (1982) and Blaise et al. (1986) reported that conventional algal toxicity tests accounted for 20–32% of test variability in terms of reproducibility [46,47]. Considering these data, one may conclude that the current SOB technique shows favorable test reproducibility.

3.3. *Advantages of SOB Toxicity Tests*

The present study developed an improved SOB-based toxicity test kit for simple and rapid on-site application. To achieve this, we applied a form of suspended SOB, rather than SOB attached on sulfur particles, decreased the processing time to 30 min, and optimized test conditions in terms of initial cell density, incubating temperature, and mixing intensity. In our earlier tests [2,21,29–35], SOB attached to the surface of insoluble sulfur particles were used as the test organism. This methodology made it difficult to apply identical amounts of SOB in the toxicity tests. We estimated the amount of SOB by the amount of sulfur particles. Furthermore, in our earlier tests [2,21,29–35], contact between SOB and contaminants was relatively indirect because SOB resided as a form of biofilm (attached on the surface of insoluble sulfur particles). However, the suspended form of SOB in this study yielded identical amounts of SOB and SOB made direct contact with contaminants.

As previously discussed, initial cell density determines toxicant availability per cell; incubating temperature and mixing intensity are relevant to microbial activity of SOB. By optimizing these conditions, the current SOB tests showed better detection sensitivity and test reproducibility than our previous technique. Furthermore, our SOB tests demonstrate more favorable detection sensitivity even when compared to other existing microbial ecotoxicity tests. For example, Dalzell et al. (2002) reported that toxicity tests using nitrifying bacteria, *Vibrio fischeri*, *E. coli*, and activated sludges resulted in 22.5–37.5 mg/L, 3.7–41.5 mg/L, 0.87–67.5 mg/L of EC_{50} values for Cr^{6+}, Cu^{2+}, and Zn^{2+}, respectively [48]. Cho et al. (2004) showed that *Vibrio fischeri*-based tests resulted in 0.8–1.6 mg/L, 17.2–18.9 mg/L, 12.6–25.2 mg/L of EC_{50} values for Hg^{2+}, Cr^{6+}, As^{3+}, respectively [15], which are much higher than our data. Kudlak et al. (2011) also reported that *Daphnia magna*-based toxicity tests obtained significantly greater EC_{50} values for Zn^{2+} (11.56 mg/L) and Hg^{2+} (9.6 mg/L) than our SOB tests [49].

Our SOB toxicity tests have numerous advantages compared to other microbial ecotoxicity tests. SOB are chemolithoautotrophic bacteria, indicating that SOB do not use organic matters as carbon and electron sources. Moreover, according to our earlier studies [9,34], SOB showed relatively constant activity even under severe conditions. These properties of SOB suggest that SOB are less affected by organic substances and characteristics of test samples. Hence, it is expected that SOB-based tests can be employed in diverse natural environments. In addition, the species of SOB used in the current study is acidophilic bacteria. Generally, contaminated water (by heavy metals or various organic contaminants) has low pH. However, most microbial ecotoxicity tests operate under neutral pH [23,38]. To use these tests, one must adjust pH to neutral. Because pH is a significant factor determining toxicity, such necessary adjustments of pH can inadvertently alter the initial toxicity of the test samples [38,50]. On the other hand, the current SOB tests do not require this pH adjustment; hence, the initial toxicity of test samples is not distorted. Employment of electric conductivity as an end-point measurement is also a merit of our SOB tests. In many microbial toxicity tests (such as bioluminescence bacteria or *Escherichia coli*-based tests), light absorbance is usually used as the end-point measurement [38,51,52]. This parameter is highly affected by the turbidity and color of the test samples, potentially causing false readings of the end-point. However, electric conductivity is independent of these characteristics of the test samples. Consequently, our SOB-based toxicity technique can lead to more accurate test results.

In the present study, we also focused on on-site application. In general, microbial ecotoxicity tests are laboratory-based tests. To measure microbial activity, laboratory analytical instruments are required. Despite the relatively shorter processing time of microbial toxicity tests compared to high trophic level organism-based ecotoxicity tests, several hours of processing of microbial toxicity tests are unsuitable for on-site applications. In contrast, our SOB tests require a simple portable electric-conductivity meter. The necessary processing time of the current SOB tests is only 30 min. Moreover, we developed portable kit-type toxicity tests for better mobility. These advantages make the current SOB toxicity test technique suitable for on-site applications. However, in spite of these numerous advantages, our SOB tests still need a portable incubating system for field applications. We are now developing this system.

Our future research includes further verification of the reliability of the current SOB toxicity tests with more diverse contaminants. In addition, we will employ this SOB test to evaluate biochemical and mixture toxicity of contaminants in various environments, including the hydrosphere, geosphere, and atmosphere.

4. Conclusions

The current study aimed to improve an existing SOB-based toxicity test technique for simple and rapid on-site application. To achieve this goal, we developed a 25 mL glass-vial-based toxicity kit test. We employed a form of suspended SOB and decreased the processing time to 30 min. To yield favorable detection sensitivity and test reproducibility, we determined the optimal test conditions of the SOB toxicity kit to be 2×10^5 cells/mL initial cell density, 32 °C incubating temperature, and 120 rpm mixing intensity. Compared to the test results from our earlier SOB technique, the current technique resulted in lower EC_{50} values from toxicity tests of diverse contaminants, such as heavy metals, inorganic nitrogen, endocrine-disrupting compounds, and petrochemicals. It also showed favorable CV values. Our SOB kit test has a number of advantages, such as no need for advanced analytic instruments and no distortion of test results from characteristics of test samples, making it suitable for simple and rapid on-site application.

Funding: This study was supported by the Sejong Science Fellowship (NRF-2021R1C1C2004788) funded by the National Research Foundation (NRF) of Republic of Korea.

Institutional Review Board Statement: Not applicable.

Informed Consent Statement: Not applicable.

Data Availability Statement: The data presented in this study are available upon request from the corresponding author.

Acknowledgments: We express our gratitude to the environmental biotechnology laboratory at Kangwon National University for distribution of SOB.

Conflicts of Interest: The author declares no conflict of interest.

References

1. Environmental Agency (UK). Chemicals: Challenges for the Water Environment. 2021. Available online: https://www.gov.uk/government/publications/chemicals-challenges-for-the-water-environment (accessed on 22 February 2023).
2. Van Ginkel, S.W.; Hassan, S.H.A.; Oh, S.-E. Detecting endocrine disrupting compounds in water using sulfur-oxidizing bacteria. *Chemosphere* **2010**, *81*, 294–297. [CrossRef] [PubMed]
3. Eom, H.; Park, M.; Jang, A.; Kim, S.; Oh, S.-E. A simple and rapid algal assay kit to assess toxicity of heavy metal-contaminated water. *Environ. Pollut.* **2021**, *269*, 116135. [CrossRef] [PubMed]
4. Niemirycz, E.; Nichthauser, J.; Staniszewska, M.; Nałęcz-Jawecki, G.; Bolałek, J. The Microtox® biological test: Application in toxicity evaluation of surface waters and sediments in Poland. *Oceanol. Hydrobiol. Stud.* **2007**, *36*, 151–163. [CrossRef]
5. Tyagi, S.; Sharma, B.; Singh, P.; Dobhal, R. Water Quality Assessment in Terms of Water Quality Index. *Am. J. Water Resour.* **2013**, *1*, 34–38. [CrossRef]
6. Bae, M.-J.; Park, Y.-S. Biological early warning system based on the responses of aquatic organisms to disturbances: A review. *Sci. Total. Environ.* **2014**, *466–467*, 635–649. [CrossRef]
7. Rahmanian, N.; Ali, S.H.B.; Homayoonfard, M.; Ali, N.J.; Rehan, M.; Sadef, Y.; Nizami, A.S. Analysis of Physiochemical Parameters to Evaluate the Drinking Water Quality in the State of Perak, Malaysia. *J. Chem.* **2015**, *2015*, 716125. [CrossRef]
8. Brayner, R.; Couté, A.; Livage, J.; Perrette, C.; Sicard, C. Micro-algal biosensors. *Anal. Bioanal. Chem.* **2011**, *401*, 581–597. [CrossRef]
9. Hassan, S.H.A.; Van Ginkel, S.W.; Oh, S.-E. Effect of organics and alkalinity on the sulfur oxidizing bacteria (SOB) biosensor. *Chemosphere* **2013**, *90*, 965–970. [CrossRef]
10. Gunjan, D.; Gargi, B. Toxicity tests to check water quality. *Int. Res. J. Environ. Sci.* **2015**, *4*, 87–90.
11. Hernando, M.D.; Fernández-Alba, A.R.; Tauler, R.; Barceló, D. Toxicity assays applied to wastewater treatment. *Talanta* **2005**, *65*, 358–366. [CrossRef]
12. Van Wezel, A.; Mons, M.; van Delft, W. New methods to monitor emerging chemicals in the drinkingwater production chain. *J. Environ. Monit.* **2010**, *12*, 80–89. [CrossRef]
13. Mishra, S.; Barik, S.K.; Ayyappan, S.; Mohapatra, B. Fish bioassays for evaluation of raw and bioremediated dairy effluent. *Bioresour. Technol.* **2000**, *72*, 213–218. [CrossRef]
14. Sponza, D.T. Incorporation of Toxicity Tests into the Turkish Industrial Discharge Monitoring Systems. *Arch. Environ. Contam. Toxicol.* **2002**, *43*, 186–197. [CrossRef]
15. Cho, J.-C.; Park, K.-J.; Ihm, H.-S.; Park, J.-E.; Kim, S.-Y.; Kang, I.; Lee, K.-H.; Jahng, D.; Lee, D.-H.; Kim, S.-J. A novel continuous toxicity test system using a luminously modified freshwater bacterium. *Biosens. Bioelectron.* **2004**, *20*, 338–344. [CrossRef]
16. Moreira-Santos, M.; Soares, A.M.; Ribeiro, R. An in situ bioassay for freshwater environments with the microalga Pseudokirchneriella subcapitata. *Ecotoxicol. Environ. Saf.* **2004**, *59*, 164–173. [CrossRef]
17. Barata, C.; Alañon, P.; Gutierrez-Alonso, S.; Riva, M.C.; Fernández, C.; Tarazona, J.V. A *Daphnia magna* feeding bioassay as a cost effective and ecological relevant sublethal toxicity test for Environmental Risk Assessment of toxic effluents. *Sci. Total. Environ.* **2008**, *405*, 78–86. [CrossRef] [PubMed]
18. Palma, P.; Palma, V.L.; Matos, C.; Fernandes, R.M.; Bohn, A.; Soares, A.; Barbosa, I. Effects of atrazine and endosulfan sulphate on the ecdysteroid system of *Daphnia magna*. *Chemosphere* **2008**, *74*, 676–681. [CrossRef] [PubMed]
19. Kim, S.D. Acute toxicity of copper in flow-through system with suspended particles. *Environ. Eng. Res.* **2001**, *6*, 131–138.
20. Hassan, S.H.; Oh, S.E. Improved detection of toxic chemicals by *Photobacterium phosphoreum* using modified Boss medium. *J. Photochem. Photobiol. B Biol.* **2010**, *101*, 16–21. [CrossRef]
21. Oh, S.-E.; Hassan, S.H.A.; Van Ginkel, S.W. A novel biosensor for detecting toxicity in water using sulfur-oxidizing bacteria. *Sens. Actuators B Chem.* **2011**, *154*, 17–21. [CrossRef]
22. Catterall, K.; Robertson, D.; Hudson, S.; Teasdale, P.R.; Welsh, D.T.; John, R. A sensitive, rapid ferricyanide-mediated toxicity bioassay developed using *Escherichia coli*. *Talanta* **2010**, *82*, 751–757. [CrossRef]
23. Hassan, S.H.A.; Van Ginkel, S.W.; Hussein, M.A.M.; Abskharon, R.; Oh, S.-E. Toxicity assessment using different bioassays and microbial biosensors. *Environ. Int.* **2016**, *92–93*, 106–118. [CrossRef]
24. Tada, Y.; Kobata, T.; Nakaoka, C. A simple and easy method for the monitoring of environmental pollutants using oligotrophic bacteria. *Lett. Appl. Microbiol.* **2001**, *32*, 12–15. [CrossRef] [PubMed]
25. Kim, B.C.; Park, K.S.; Kim, S.D.; Gu, M.B. Evaluation of a high throughput toxicity biosensor and comparison with a *Daphnia magna* bioassay. *Biosens. Bioelectron.* **2003**, *18*, 821–826. [CrossRef] [PubMed]

26. Eilersen, A.M.; Arvin, E.; Henze, M. Monitoring toxicity of industrial wastewater and specific chemicals to a green alga, nitrifying bacteria and an aquatic bacterium. *Water Sci. Technol.* **2004**, *50*, 277–283. [CrossRef] [PubMed]
27. Tencaliec, A.M.; Laschi, S.; Magearu, V.; Mascini, M. A comparison study between a disposable electrochemical DNA biosensor and a *Vibrio fischeri*-based luminescent sensor for the detection of toxicants in water samples. *Talanta* **2006**, *69*, 365–369. [CrossRef]
28. Zlatev, R.; Magnin, J.-P.; Ozil, P.; Stoytcheva, M. Bacterial sensors based on *Acidithiobacillus ferrooxidans*: Part I. Fe^{2+} and $S_2O_3{}^{2-}$ determination. *Biosens. Bioelectron.* **2006**, *21*, 1493–1500. [CrossRef] [PubMed]
29. Hassan, S.H.; Van Ginkel, S.W.; Kim, S.-M.; Yoon, S.-H.; Joo, J.-H.; Shin, B.-S.; Jeon, B.-H.; Bae, W.; Oh, S.-E. Isolation and characterization of *Acidithiobacillus caldus* from a sulfur-oxidizing bacterial biosensor and its role in detection of toxic chemicals. *J. Microbiol. Methods* **2010**, *82*, 151–155. [CrossRef] [PubMed]
30. Van Ginkel, S.W.; Hassan, S.H.A.; Ok, Y.S.; Yang, J.E.; Kim, Y.-S.; Oh, S.-E. Detecting Oxidized Contaminants in Water Using Sulfur-Oxidizing Bacteria. *Environ. Sci. Technol.* **2011**, *45*, 3739–3745. [CrossRef]
31. Gurung, A.; Shin, B.-S.; Oh, S.-E. Effect of different air flow rate on operation of sulfur-oxidizing bacteria (SOB) biosensor. *Geosystem Eng.* **2015**, *18*, 245–250. [CrossRef]
32. Qambrani, N.A.; Hwang, J.-H.; Oh, S.-E. Comparison of chromium III and VI toxicities in water using sulfur-oxidizing bacterial bioassays. *Chemosphere* **2016**, *160*, 342–348. [CrossRef]
33. Ahmed, N.; Oh, S.-E. Toxicity assessment of selected heavy metals in water using a seven-chambered sulfur-oxidizing bacterial (SOB) bioassay reactor. *Sens. Actuators B Chem.* **2018**, *258*, 1008–1014. [CrossRef]
34. Eom, H.; Hwang, J.-H.; Hassan, S.H.; Joo, J.H.; Hur, J.H.; Chon, K.; Jeon, B.-H.; Song, Y.-C.; Chae, K.-J.; Oh, S.-E. Rapid detection of heavy metal-induced toxicity in water using a fed-batch sulfur-oxidizing bacteria (SOB) bioreactor. *J. Microbiol. Methods* **2019**, *161*, 35–42. [CrossRef] [PubMed]
35. Eom, H.; Kim, S.; Oh, S.-E. Evaluation of joint toxicity of BTEX mixtures using sulfur-oxidizing bacteria. *J. Environ. Manag.* **2023**, *325*, 116435. [CrossRef] [PubMed]
36. Rana, K.; Rana, N.; Singh, B. Chapter 10—Applications of sulfur oxidizing bacteria. *Physiol. Biotechnol. Asp. Extrem.* **2020**, *2020*, 131–136. [CrossRef]
37. Madigan, M.T.; Martinko, J.M.; Parker, J. *Brock Biology of Microorganisms*, 10th ed.; Prentice Hall/Pearson Education: Hoboken, NJ, USA, 2003.
38. Eom, H.S.; Ashun, E.; Toor, U.A.; Oh, S.-E. A solid-phase direct contact bioassay using sulfur-oxidizing bacteria (SOB) to evaluate toxicity of soil contaminated with heavy metals. *Sens. Actuators B Chem.* **2020**, *305*, 127510. [CrossRef]
39. Johnson, D.B.; Macvicar, J.H.; Rolfe, S. A new solid medium for the isolation and enumeration of *Thiobacillus ferrooxidans* and acidophilic heterotrophic bacteria. *J. Microbiol. Methods* **1987**, *7*, 9–18. [CrossRef]
40. Duquesne, K.; Lebrun, S.; Casiot, C.; Bruneel, O.; Personné, J.-C.; Leblanc, M.; Elbaz-Poulichet, F.; Morin, G.; Bonnefoy, V. Immobilization of Arsenite and Ferric Iron by *Acidithiobacillus ferrooxidans* and Its Relevance to Acid Mine Drainage. *Appl. Environ. Microbiol.* **2003**, *69*, 6165–6173. [CrossRef]
41. Singh, P.K.; Shrivastava, A.K. Role of initial cell density of algal bioassay of toxic chemicals. *J. Basic Microbiol.* **2015**, *56*, 812–819. [CrossRef]
42. Lin, J.-H.; Kao, W.-C.; Tsai, K.-P.; Chen, C.-Y. A novel algal toxicity testing technique for assessing the toxicity of both metallic and organic toxicants. *Water Res.* **2005**, *39*, 1869–1877. [CrossRef]
43. Jaggi, A.R.C.; Aulakh, M.S.; Sharma, R. Temperature effects on soil organic sulphur mineralization and elemental sulphur oxidation in subtropical soils of varying pH. *Nutr. Cycl. Agroecosyst.* **1999**, *54*, 175–182. [CrossRef]
44. Yang, Z.-H.; Stöven, K.; Haneklaus, S.; Singh, B.R.; Schnug, E. Elemental Sulfur Oxidation by *Thiobacillus* spp. and Aerobic Heterotrophic Sulfur-Oxidizing Bacteria. *Pedosphere* **2010**, *20*, 71–79. [CrossRef]
45. Hassan, S.H.A.; Van Ginkel, S.W.; Oh, S.-E. Detection of Cr^{6+} by the Sulfur Oxidizing Bacteria Biosensor: Effect of Different Physical Factors. *Environ. Sci. Technol.* **2012**, *46*, 7844–7848. [CrossRef] [PubMed]
46. Van Coillie, R.; Couture, P.; Schoenert, R.; Thellen, C. *Mise au Point d'une 'Evaluation Rapide de Latoxicit'e Originale des Effluents et de Leurs Composantes al'Aide d'Algues*; Environnement Canada, Service de laProtection de 1'Environnement: Quebec, PQ, Canada, 1982.
47. Blaise, C.; Legault, R.; Bermingham, N.; Van Coillie, R.; Vasseur, P. A simple microplate algal assay technique for aquatic toxicity assessment. *Environ. Toxicol.* **1986**, *1*, 261–281. [CrossRef]
48. Dalzell, D.J.B.; Alte, S.; Aspichueta, E.; de la Sota, A.; Etxebarria, J.; Gutierrez, M.; Hoffmann, C.C.; Sales, D.; Obst, U.; Christofi, N. A comparison of five rapid direct toxicity assessment methods to determine toxicity of pollutants to activated sludge. *Chemosphere* **2002**, *47*, 535–545. [CrossRef] [PubMed]
49. Kudłak, B.; Wolska, L.; Namieśnik, J. Determination of EC50 toxicity data of selected heavy metals toward *Heterocypris incongruens* and their comparison to "direct-contact" and microbiotests. *Environ. Monit. Assess.* **2011**, *174*, 509–516. [CrossRef]
50. Olaniran, A.O.; Balgobind, A.; Pillay, B. Bioavailability of Heavy Metals in Soil: Impact on Microbial Biodegradation of Organic Compounds and Possible Improvement Strategies. *Int. J. Mol. Sci.* **2013**, *14*, 10197–10228. [CrossRef]

51. Bitton, G.; Garland, E.; Kong, I.; Morel, J.L.; Koopman, B. A direct solid-phase assay specific for heavy metal toxicity. I. methodology. *J. Soil Contam.* **1996**, *5*, 385–394. [CrossRef]
52. *ISO 11348*; Water quality—Determination of the inhibitory effets of water samples on the light emission of *Vibrio fischeri*. (Luminscent bacteria tests). International Organization for Standardization: Geneva, Switzerland, 2007.

toxics

MDPI

Article

A Realistic Mixture of Persistent Organic Pollutants Affects Zebrafish Development, Behavior, and Specifically Eye Formation by Inhibiting the Condensin I Complex

Gustavo Guerrero-Limón [1], Renaud Nivelle [1], Nguyen Bich-Ngoc [2], Dinh Duy-Thanh [1] and Marc Muller [1,*]

1 Laboratory for Organogenesis and Regeneration, GIGA Institute, University of Liège, 4000 Liège, Belgium; g.guerrero.limon@pm.me (G.G.-L.); rnivelle4550@gmail.com (R.N.); duythanh84bio@gmail.com (D.D.-T.)
2 VNU School of Interdisciplinary Studies, Vietnam National University (VNU), Hanoi 10000, Vietnam; nbngoc.87@gmail.com
* Correspondence: m.muller@uliege.be; Tel.: +32-473-993074

Abstract: Persistent organic pollutants (POPs) are posing major environmental and health threats due to their stability, ubiquity, and bioaccumulation. Most of the numerous studies of these compounds deal with single chemicals, although real exposures always consist of mixtures. Thus, using different tests, we screened the effects on zebrafish larvae caused by exposure to an environmentally relevant POP mixture. Our mixture consisted of 29 chemicals as found in the blood of a Scandinavian human population. Larvae exposed to this POP mix at realistic concentrations, or sub-mixtures thereof, presented growth retardation, edemas, retarded swim bladder inflation, hyperactive swimming behavior, and other striking malformations such as microphthalmia. The most deleterious compounds in the mixture belong to the per- and polyfluorinated acids class, although chlorinated and brominated compounds modulated the effects. Analyzing the changes in transcriptome caused by POP exposure, we observed an increase of insulin signaling and identified genes involved in brain and eye development, leading us to propose that the impaired function of the condensin I complex caused the observed eye defect. Our findings contribute to the understanding of POP mixtures, their consequences, and potential threats to human and animal populations, indicating that more mechanistic, monitoring, and long-term studies are imperative.

Keywords: SVHC; persistent organic pollutants; POP; PFOS; zebrafish; development; behavior; condensin I

Citation: Guerrero-Limón, G.; Nivelle, R.; Bich-Ngoc, N.; Duy-Thanh, D.; Muller, M. A Realistic Mixture of Persistent Organic Pollutants Affects Zebrafish Development, Behavior, and Specifically Eye Formation by Inhibiting the Condensin I Complex. *Toxics* **2023**, *11*, 357. https://doi.org/10.3390/toxics11040357

Academic Editors: Esref Demir and Sam Kacew

Received: 11 March 2023
Revised: 7 April 2023
Accepted: 7 April 2023
Published: 9 April 2023

1. Introduction

Sixty years ago, Rachel Carson started raising awareness about persistent organic pollutants (POPs) in her book "Silent Spring". She documented the deleterious effects caused by the indiscriminate use of DDT. Ever since, research has proven and continues to prove her point; POPs have been listed in the "Stockholm Convention on Persistent Organic Pollutants" [1]. The European Union and the United Nations Environmental Program define the persistent organic pollutants (POPs) as "chemical substances that are hard to degrade, with a tendency to bioaccumulate, transfer through the food web rather easily, transport across international boundaries, and having long half-lives" [1,2]. They have been extensively linked to adverse health effects [3–7]. POPs, though very relevant for modern life, are normally studied in a reductionist approach, where a single compound is targeted and tested. Nevertheless, POPs are rarely found as stand-alone compounds in nature [8–10]. Mixtures are the rule [11], and their effects have not been widely described yet. To understand potential threats resulting from this exposure, studies have been carried out using chemical mixtures with different approaches ranging from molecular biology to transgenerational studies, using in vitro, in vivo, and in silico techniques. Studies recently focused on a constructed mixture of POPs that was designed based on the levels found in the blood of a Scandinavian human population [12]. Using cell reporter assays, this mixture was found to antagonize the androgen receptor transactivation and nuclear translocation [13], to inhibit the transactivation activity of the aryl hydrocarbon receptor [14], and to induce cytotoxicity while it enhances nerve-growth-factor-induced neurite outgrowth in PC12 cells at high concentrations [15]. Microscopic high content analysis (HCA) revealed that some sub-mixtures affected cell number, nuclear area, and mitochondrial membrane potential in A-498 human kidney cells [16].

In this study we have chosen zebrafish larvae due to their many technical and practical advantages. To name a few, the zebrafish (*Danio rerio*) shares a non-negligible amount of genetic pool with humans (up to 80%), small size, and ease of maintenance in captivity, high fecundity and short times till adulthood and reproduction, amongst many others [17]. In this context, research using zebrafish has shed some light on the toxicological features of chemical compounds. Therefore, the aim of this research was to employ zebrafish larvae as a model organism to describe the many adverse developmental effects caused by realistic doses of the POP mixture and of specific sub-mixtures. Furthermore, we performed RNA-seq analysis on the larvae exposed to the POP mixture, aiming at elucidating the mechanism of action for specific observations.

2. Materials and Methods

2.1. Zebrafish Husbandry and Ethical Considerations

Adult wild-type zebrafish of the AB strain and the transgenic line *Tg(kdrl-mls:GFP)* [18] were obtained from breeding facilities at the GIGA-Institute, Liege, Belgium. Fish maintenance, breeding conditions, and egg production were described in detail [19,20] and are in accordance with internationally accepted standards. Animal care and all experimentation were conducted in compliance with Belgian and European laws (Authorization: LA1610002 Ethical commission protocol ULg19-2134 and Ulg19-2135).

2.2. Chemicals and POP Mixture

Dimethyl sulfoxide (DMSO, >99.9%, CAS number 67-68-5) was purchased from Sigma-Aldrich (Merck KGaA, Darmstadt, Germany). The stock solutions for the total POP mixture and six sub-mixtures were designed and prepared by the Norwegian University of Life Sciences, Oslo, Norway [12] as indicated in Supplementary Table S1. Briefly, the total POP mixture was designed to represent a mixture of 29 compounds at 1,000,000-fold the mean concentrations found in the blood of a Scandinavian population, while the sub-mixtures consisted of either one single class of these compounds (PFAA, Br, Cl) or of two combined (PFAA + Br, PFAA + Cl, Br + Cl) classes. Stock solutions of POP mixtures and their sub-mixtures were prepared in DMSO and stored at −20 until the day of testing. For all treatments, we used the stock solution (1,000,000×) that was further diluted on the testing day in E3 zebrafish raising media [21]. Next, the concentration of DMSO was corrected to achieve 0.1% in all cases, including the control groups.

2.3. Exposure Tests

Exposure tests were performed in 6 well-plates, with 25 fertilized eggs per well in 4 mL of E3 medium supplemented or not with the test compounds. For each experiment, 150 fertilized eggs were selected, 50 as controls and 100 for the specific treatment, to ensure a sufficient number of treated individuals for the tests. Each treatment was repeated at least three times in independent experiments. To keep stable chemical concentrations, we used a static-renewal approach where at least 90% of the media was refreshed every 24 h. Exposure started between 0 to 6 h post fertilization (hpf); the larvae were treated for at least 96 h. Finally, following the guidelines of the OECD Test number: 236, we tested 8 different concentrations (1×, 5×, 25×, 75×, 125×, 250×, 500×, 1000×) to estimate the median lethal concentration LC_{50}.

2.4. Morphological Observations

A set of morphological features was recorded, including presence of edemas, inflation of the swim bladder, eye malformations, etc. Pictures of treated and untreated larvae at different stages were taken. All observations were made with a stereomicroscope Leica M165 FC (Leica Microsystems©, Leica, Wetzlar, Germany). Standard length was estimated in fish at 5- and 10-days post fertilization (dpf) using FIJI line tool for measurement (ImageJ2, v. 2.3.0/1.53f).

2.5. Mitochondrial Toxicity

Estimation of the mitochondrial integrity in the blood vessels was conducted using the transgenic zebrafish line *Tg(kdrl-mls:GFP)*, which expresses the fluorescent protein GFP fused to a peptide targeting it to the mitochondria (MLS) under the control of the endothelial cell-specific promoter sequence of the zebrafish *kdrl* gene. Heterozygous parents for the transgene were crossed, offspring carrying the transgene were selected based on fluorescence at 24 hpf and separated in control and treated groups. Then, the exposure test was carried on as described above. Fluorescence intensity was observed, and pictures were taken at 120 hpf using the epifluorescence stereomicroscope Leica M165 FC (Leica Microsystems©). Then, fluorescence was quantified using FIJI. Since the transgene is expressed in all blood vessels (head, heart, etc.), to avoid overestimation of the intensity, values were obtained from sectioning the body in a lateral view and using only the tail, from the opening of the anus to the caudal peduncle. Each intensity value was determined using the corrected total fluorescence (CTF) [22] and expressed as RFU (relative fluorescence units).

2.6. Heart Rate

Heartbeats were counted manually using an inverted Nikon Eclipse TS100 microscope and a counter for 15 s. To obtain the beats per minute (BPM), measurements were multiplied by 4. The heart rate was estimated on 96 hpf larvae that were acclimated to the lighting conditions for no less than 5 min prior to counting; the larvae were not immobilized by anesthetics or other means. Each larva was observed sequentially at least three times. Ten larvae were observed per treatment and each experiment was performed at least in triplicate.

2.7. Behavior

Behavioral tests were conducted on zebrafish larvae at 98~120 hpf and every test was performed between 10:00 and 13:00 to maintain a constant position in the circadian cycle. During the entire exposure period to the chemicals, special care was taken to avoid the interference of environmental factors. Exposed larvae were shielded from loud noises, changes in the temperature of the incubator (27–28 °C) and the raising media (~26° at the time of testing), changing light conditions and activities in the room by putting them in a specific enclosure harboring its own, constant dark/light regime. Prior to each behavioral test, the zebrafish larvae were inspected under a stereomicroscope to select and transfer to the testing plates only individuals devoid of any malformation that might interfere with mobility outcome (e.g., yolk sac or pericardial oedemas, spinal aberrations, aberrations in pigmentation, and/or loss of equilibrium, etc.). The larvae were placed individually in a well of a 96-well plate and observed using a ViewPoint® Zebrabox system and its tracking software (ViewPoint Life Sciences, Lyon, France). The light level was set to 20% on the ViewPoint software (7.45 klux, TES 1337 light meter), while infrared light (850 nm) was used to track larval activity. We applied a light–dark cycle that lasted for a total of 1 h and consisted of 20 min of light, allowing for the larvae to acclimate to the situation and discarded from the analysis, followed by 10 min of darkness, 10 min of light, 10 min of darkness and 10 min of light. The video and tracking software were used to screen larval locomotion behavior for 10 s intervals; the distance travelled, and the time spent active were determined and, from these parameters, the mean swimming speed was also calculated by dividing the cumulated distance travelled by the total time spent active.

2.8. Injection of Antisense Oligonucleotide Morpholino

As previously described [23,24], one cell-stage embryos were injected with a concentration of 100 µM of MO^{p53} (MO, Gene Tools Inc., Philomath, OR, USA). The morpholino was diluted in Danieau buffer and 0.5% tetramethylrhodamine dextran (Invitrogen, Merelbeke, Belgium). To assess the effects of morpholino injection, 150 individuals were microinjected in two independent experiments, followed by exposure to chemicals as described above. Sequence of the morpholino oligonucleotide:

$$MO^{p53}\text{: 5}'\text{-GACCTCCTCTCCACTAAACTACGAT-3}'$$

2.9. RNA Extraction

RNA was extracted from pools of 65 larvae at 5 dpf using the RNA mini extraction kit (Qiagen, Hilden, Germany). Samples were lysed in RLT+ buffer with β-mercaptoethanol (Sigma-Aldrich, St. Louis, MO, USA) and homogenized at least 10 times with a 26-gauge needle in a 1 mL syringe. An amount of 22 µL of RNAse free water was used to resuspend total RNA. RNA extract was treated with DNAseI (Qiagen, Hilden, Germany) to avoid DNA contamination. Quantity (ng/µL) and quality (260/280 and 260/230 ratios) of each extract was assessed by nanodrop spectrophotometer measurements. Poor quality (260/280 < 2; 260/230 < 2) samples were subsequently purified by lithium chloride precipitation, followed by 2 times pellet washing with 70% ethanol, and resuspended in 51 µL of RNAse-free water and stored at −80 °C. The integrity of total RNA extracts was assessed with BioAnalyzer analysis and provided RIN (RNA integrity number) scores for each sample (Agilent, Santa Clara, CA, USA).

2.10. RNAseq

cDNA libraries were generated from 100 to 500 ng of extracted total RNA using the Illumina Truseq mRNA stranded kit (Illumina, San Diego, CA, USA) according to the manufacturer's instructions. cDNA libraries were then sequenced on a NovaSeq sequencing system, in 1 ×100 bp (single end). Approximatively 20–25 M reads were sequenced per sample. The sequencing reads were processed through the Nf-core rnaseq pipeline 3.0 [25] with default parameters and using the zebrafish reference genome (GRCz11) and the annotation set from Ensembl release 103 (www.ensembl.org; accessed 1 May 2020). Differential gene expression analysis was performed using DESeq2 pipeline [26]. Pathway and biological function enrichment analysis was performed using the WEB-based "Gene SeT AnaLysis Toolkit" (http://www.webgestalt.org; accessed on 10 November 2022) based on the integrated GO (Gene Ontology), KEGG (Kyoto Encyclopedia of Genes and Genomes) [27,28], Panther, and WikiPathways databases (all accessed on 10 November 2022 via http://www.webgestalt.org). An additional database was constructed using the Gene-mutant/Phenotype database from zfin (zfin.org; accessed on 6 March 2023). The cut-off values were set for the false discovery rate (FDR) to "adjusted p value < 0.05" and the fold change > 1.5.

2.11. Data and Statistical Analysis

For the estimation of the lethal concentration (LC_{50}), data were transferred to R (4.0.2) [29] and the command "*dose.p*" used in the library "MASS" [30]. Morphological and fluorescence data were transferred to Prism 9.0.0 (v86) (Graphpad, San Diego, CA, USA). Each data set was tested for normality (e.g., using a visual cue (QQ plot), D'Agostino–Darling and Shapiro–Wilk tests) and equal variances (Bartlett's test). Thus, parametric or non-parametric tests were performed, as indicated in each case in each figure.

Raw behavioral data sets consisted of tables holding the positions of each larva in each video frame (30 frames/second). This table was first trimmed to eliminate very short, oscillating, and likely artefactual movements, and then aggregated into 10-s periods for further analysis. These data were transferred to R version 4.0.2 to analyze motility during the dark and light phases. To assess behavior, we used linear mixed effect (LME) models within the "nlme" package [31]. Three dependent variables were used, either the "mean time spent active" (seconds), the "mean distance travelled" (mm), or the "mean swimming speed" (calculated as the mean distance travelled/mean time spent active) within each 10 s period, with "compound" and "time" as the categorical and continuous independent variables, and "batch" as a random effect. The "*Anova*" command within the "*car*" library [32] was used to extract the results for the main effects whereas the "*lsmeans*" command [33] within the "*emmeans*" library was used as a post-hoc test to compare groups against one another while adjusting for the means of other factors within the model [34]. Type II sum of squares was used for the model. Two kinds of analyses were performed (see also below in results): the "startle" response including the 10 s prior to change of phase (light to dark, or dark to light) with a length of 50 s and the values obtained for 560 s after the spike (the remaining time of the phase). Confidence was assigned at $\alpha = 95\%$ and a p-value of ≤ 0.05 was considered as significant, $p \leq 0.05$ (*), ≤ 0.01 (**), ≤ 0.001 (***).

3. Results and Discussion

3.1. LC_{50}–Chronic Exposure to Total POP Mix Is Lethal at Relatively High Doses

In a preliminary range-finding experiment, we exposed AB zebrafish fertilized eggs to eight concentrations (1×, 5×, 25×, 75×, 125×, 250×, 500×, 1000× the mean human blood concentration) of the total POP mixture and we monitored survival at 24, 48, 72, and 96 hpf (Figure 1A) compared to untreated controls. The median lethal concentration (LC_{50}) for this mixture was calculated at 386-fold the human blood concentration (386×). Consequently, in the following experiments, we limited the concentration range to 75×, 125×, and 250× the mean human blood concentration. This may seem high considering a normal population; however, we must consider that the chorion is a protective layer, which can be crossed easily by molecules with a size below 4000 Da [35]. The POPs studied here are hydrophobic and of small size; thus, they are potentially able to cross the chorion and to exert their effects right after adding the solutions to the media. Previous experiments revealed that only about 10% of, e.g., PFOS could be found in zebrafish larvae exposed (continuously, i.e., without medium change) for 96 hrs to the compound [36], while between 0 and 16% of the nominal amounts were found in larvae exposed to the POP mixture [37]. Given their high bioconcentration values (BCF factor used to estimate the potential to bioaccumulate) and the persistent nature of these chemicals [38,39], the harmful concentrations used here may be, eventually, reached in individuals that are constantly exposed, exerting their effects in later stages in life while continuing to accumulate through the many pathways of exposure [40–42]. Thus, we can assume that, in our experiments, larvae are exposed to concentrations of the POP mix that may be reached in exposed populations [43].

3.2. The POP Mix Significantly Reduces the Standard Length of Zebrafish Larvae

We first evaluated the effect of the POP mixture on general growth by measuring the standard length of the larvae at 5 dpf after continuous exposure to the POP125× (125× human blood concentration) mix. In parallel, we also tested equivalent concentrations of the different sub-mixtures. At 5 dpf, the average size of the larvae was significantly affected, fish treated with the total POP125× mix were significantly smaller by about 10% (from 3.4 to 3.1 mm) (Figure 1C). Among the single sub-mixes, only the PFAA mix resulted in a significantly decreased standard length, similar to but slightly less than the POP125× mix. Cl and Br mixes alone did not cause a clear effect. In line with these observations, only the binary mixtures containing PFAA (PFAA + Cl, PFAA + Br) caused a similar effect on standard length comparable to PFAA alone. Cl + Br had no effect. Thus, only those mixtures containing PFAA affected the size of zebrafish larvae at 5 dpf.

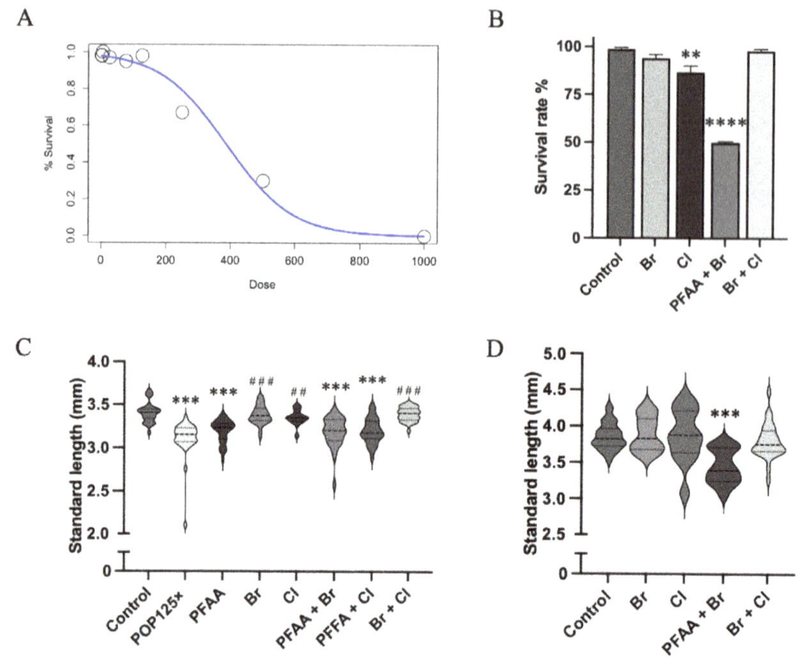

Figure 1. (**A**) Survival after 96 h of exposure to POP mixture. Survival decreases drastically at 386×. (**B**) Survival rate for POP mixtures and sub-mixtures at 10 dpf; ordinary one-way ANOVA and Dunnett's multiple comparisons test. (**C**) Standard length of fish at 5 dpf. Data presented as median with higher and lower quartiles for each treatment. Asterisks (*) indicate when significant differences were found compared to control, hash sign (#) when differences were found relative to PFAA alone. (**D**) Fish measured at 10 dpf. Missing groups due to high mortality rates were not included. Kruskal–Wallis and Dunn's multiple comparison tests, $p \leq 0.01$ (**), $p \leq 0.001$ (***), $p \leq 0.0001$ (****). In short, PFAA < Total Mix < Cl = Br = Control.

Previously, exposure to PFOA 4 ng/mL has been found to decrease the body length of 3 dpf zebrafish larvae, but not at 40 or 400 ng/mL [44]. In contrast, other PFAS (PFBA, PFHxA) did cause decrease of size at 40 and 400 ng/mL. Another study revealed that PFOS or PFOA decreased total body length at 200 or 2000 ng/mL, respectively [45]. Here, we used 217 ng/mL in the 125× POP mix for PFOA, in addition to PFOS and PFHxS, to analyze the effect on standard length at 5 dpf. Taken together, there is clear evidence that PFAS affects larval growth.

After halting the exposure (at 96 hpf) we kept the larvae to grow in normal E3 medium, free of POPs until 10 dpf. We observed a significant lethality during this period, which was not assessed in the preliminary experiment (Figure 1B). Among the surviving larvae, only those exposed to PFAA + Br were significantly smaller (Figure 1D), while all other treatments including PFAA, as well as the total mix, left no survivor. PCB and PBDE congeners were previously shown to impact survival of zebrafish larvae at concentrations around 1–5 µg/mL [46,47]; we did not observe a significant lethality induced by the Cl and Br mixtures here, suggesting that the congeners present in the POP mix are indeed less toxic.

3.3. Common Developmental Toxic Effects Such as Edemas and Non-Inflated Swim Bladder Were Commonly Found following Exposure to POP Mix

We also looked for other developmental defects induced by the different treatments, according to the recommendations for the Zebrafish Embryotoxicity test [48]. The most striking features observed were the presence of edemas and non-inflated swim bladder. Even at the lowest concentration of the total mix (POP 75×), a relatively large proportion of the population had edemas (~70%). Using the swim bladder as another phenotypical endpoint to assess developmental retardation, a high number of fish (close to 100%) presented developmental impairment of their swim bladders after 96 hpf compared to only 50% in the controls at this stage (Figure 2). While edemas are commonly observed in toxicity assays, they may have been caused in our experiments (whether yolk sac or pericardium) by the presence of PBDE47 [49]. The lack of inflation of the swim bladder is less common; it was previously reported as being affected by exposure to PFOA at 4.7 ng/mL [50] by interfering with thyroid hormone signaling. Other chemicals present in the mixtures may in addition contribute to disruption of thyroid hormone action [50]. Underdevelopment of the swim bladder would have important ecological consequences, impeding the normal swimming of the larvae at a crucial age.

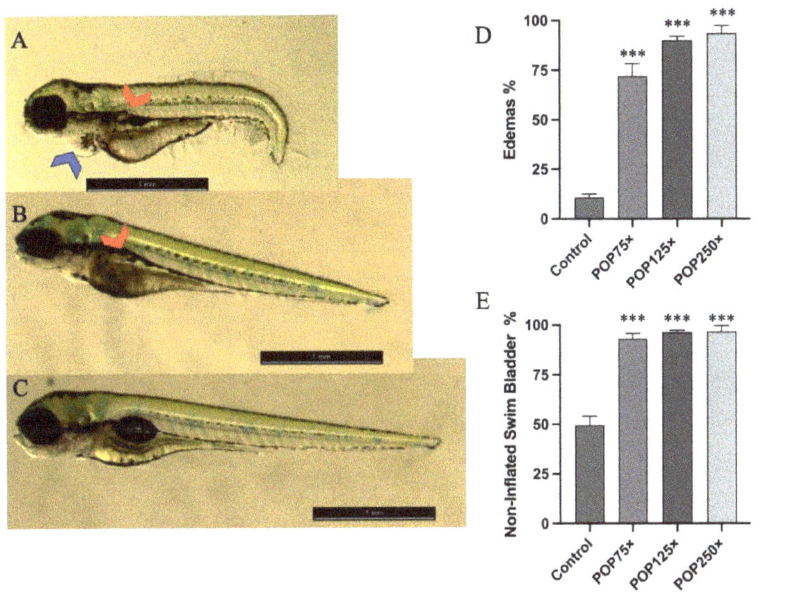

Figure 2. Examples of fish treated and untreated with the POP mix. Several malformations and size differences are striking. Pictures were cropped purposefully to enhance the differences found. (**A**) Fish treated with POP250x, red arrow pointing at the non-inflated swim bladder, blue arrow pointing at a pericardial edema. (**B**) Fish treated with POP125×. (**C**) Control, size bar = 1 mm. (**D**) Edema, as percentage of the population having this malformation. (**E**) Non-inflated swim bladder at 96 hpf, as percentage of population. Data presented as mean percentage of population having either malformation and standard deviation; ordinary one-way ANOVA and Tukey's multiple comparison test, $n = 100$, $p \leq 0.001$ (***).

The deleterious effects we describe here were only seen after 72 hpf. In a preliminary experiment (data not shown), we tested the capabilities of early developmental disruption of the POP mix and could not find a clear effect at stages earlier than 72 hpf, almost three full days of continuous exposure. A similar observation was made previously when testing pharmaceutical pollutants, some of which exerted their effects mainly on 72 and 96 hpf larvae [21]. We hypothesized that, though the chemicals would cross the chorion and be taken up by the embryos, the absence of some targeted molecules at these early stages would make them impervious to the POPs' effects. That would be the case of thyroid follicles that start developing after 96 hpf [51]; hence, some PBDEs would not be exerting their effects through this pathway until a later stage.

3.4. Striking Eye Malformation in Fish Treated with Any of the POP Mix and Its Sub-Mixes

One outstanding feature we observed was the pear-like shaped eyes, with dents in the polar regions on the eyeball of the fish treated with the POP mix (Figure 3A–C). Compared to the control group (~7%), treatment with the POP75× mix affected about 50% of affected larvae, while POP125× affected about 70% and POP250× close to 90% of the fish (Figure 3D). When we tested the sub-mixtures at 125× the mean human blood concentration, we observed that each of the single mixes caused a slightly lower fraction of affected individuals (Figure 3E) compared to the total POP125×. However, the dual combinations PFAA + Br or PFAA + Cl reached similar levels to POP125×; addition of either Br or Cl significantly increased the incidence of this malformation relative to PFAA alone, indicating that each sub-mixture contributed to various degrees to the effect caused by the POP mix.

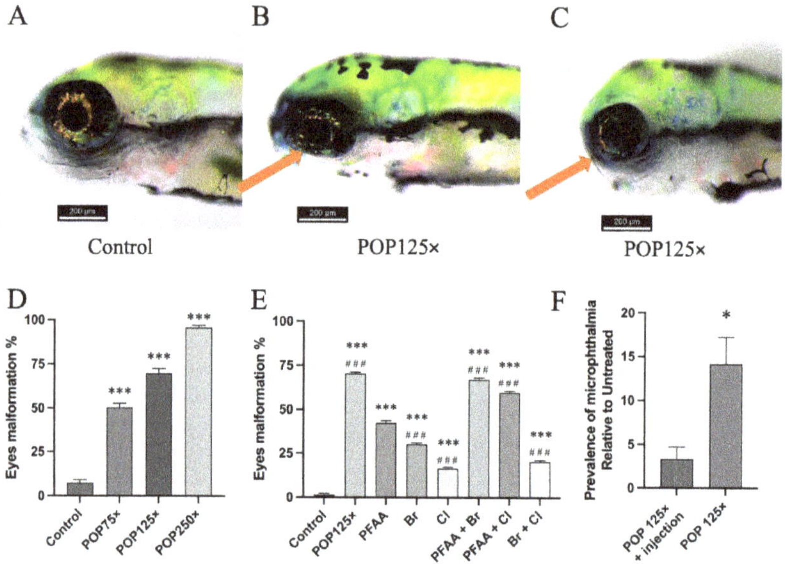

Figure 3. Example of eye malformation. (**A**) Untreated larva 4 dpf; (**B**) flattened eye with dents at both the upper and lower side of the eye and (**C**) eye hypoplasia of treated larvae with POP125×; (**D**) dose-based prevalence of eye malformation; (**E**) prevalence of eye malformation in larvae upon treatment with the different POP sub-mixtures. Data is presented as mean percentage of population having eye malformation and standard deviation; ordinary one-way ANOVA and Šidák's multiple comparison test, $n = 100$, $p \leq 0.001$ (***). Asterisks indicate when significant differences were found compared to control, hash sign (#) when differences were found compared to PFAA. (**F**) Prevalence of eye malformation present in fish exposed to POP125× relative to untreated individuals. Columns represent the effect of POP125 treatment in larvae previously injected with Mo^{p53} (POP125× + injection) or not (POP125×). A Fisher's exact test revealed the difference with a significance of $p = 0.055$ (*).

This malformation of the eyes was one of the most striking and unexpected effects. Previous studies have shown a link between eye malformation and certain compounds or the suppression of expression of certain genes. Two main eye malformations are described in the literature: either eyes were absent (anophthalmia), or their size was reduced (microphthalmia). The first is linked to the absence of genes such as *chokh/rx3* [52], while the second is linked to the expression of many different genes, such as *sox2* [53], *otx2* [54], *pax6a*, or *pax6b* [55]. Regarding chemical exposure, these two morphological aberrations have been described after treatments with a variety of chemicals, such as phenylthiourea [56], gold nanoparticles [57], di-butyl phthalate [58], and PCBs (Aroclor 1254) [59]. Retinal defects have been shown in workers exposed to solvents or heavy metal, and defects in photoreceptor cells were described in zebrafish exposed to PBDEs or PCBs [60]. However, to the best of our knowledge, this is the first time the pear-like shape and microphthalmia are described as a malformation caused by these kinds of mixed organic pollutants. We did not observe a correlation of this malformation with any of the other defects that we observed, indicating that a specific mechanism is involved.

One previously described zebrafish mutant, the *cap-g*^{s105} mutant, presents a reduction of retinal cell number and smaller eyes similar to what we observed here [61]. The *cap-g* gene codes for a component of the condensin I complex involved in the regulation of chromosome condensation and segregation during mitosis. The *cap-g*^{s105} mutation of this gene causes increased apoptosis in proliferating retinal stem cells, leading to a small eye phenotype, which could be partially rescued by interfering with the expression of the pro-apoptotic gene *p53*. We thus decided to test the effect of the POP125× mix on zebrafish larvae that had been previously micro-injected with antisense morpholino directed against the *p53* gene. Although injection of the MO^{p53} alone seems to generate some eye deformities on its own, the eye malformation induced by POP125× was significantly reduced in fish injected with MO^{p53} (Figure 3F), similar to what was observed for the *cap-g*^{105} zebrafish mutant [61]. Further support for this mechanism and the genes involved in this striking phenotype will be given in the transcriptome analysis section below.

3.5. Heart Rate Is Severely Affected after 96 h of Exposure, Especially When PFAAs Were Present

Next, we tested the effect of the total POP mix on the heart rate of the zebrafish larvae at 96 hpf. We witnessed a significant, dose-dependent increase in the heart rate upon treatment with the total mixture (Figure 4A). Testing the sub-mixtures at 125× concentration, the most pronounced effects were observed in those treatments where PFAA mix was present (Figure 4B). A weaker, but significant difference was observed as well with the Cl mix, while no significant effect was found using only the Br mix.

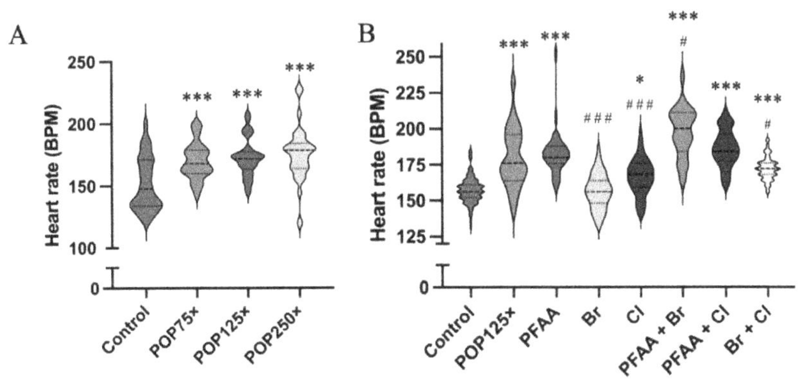

Figure 4. (**A**) Heart rate (BPM, beats per minute) of treated and untreated zebrafish using 3 different concentrations of the total POP mix. (**B**) Heart rate of all the fish exposed to the different treatments and measured at 4 dpf. Kruskal–Wallis test and Dunn's multiple comparison, $n = 30$, $p < 0.05$ (*), $p \leq 0.001$ (***). Asterisks indicate when significant differences were found compared to control, hash (#) sign when differences were found compared to PFAA.

All binary mixtures (PFAA + Br, PFAA + Cl, Br + Cl) significantly increased the heart rate; however, the PFAA mix exerted a dominant effect while the increase caused by Br + Cl was clearly lower compared to those due to any of the treatments where PFAA was present. The highest heart rate was recorded in the fish exposed to the binary mixture PFAA + Br (197 ± 17 BPM vs. Control = 157 ± 9 BPM).

PFOS and PFOA have been shown to increase the heart rate in 72 hpf zebrafish larvae at, respectively, 500 ng/mL and 75 μg/mL [62], indicating that PFOS may be the main agent here. Chemicals such as perfluorononanoic acid (PFNA) can alter gene expression linked to cardiac development by dysregulating genes such as *amhc, nppa, nkx2.5, edn1* and *tgfb2* [63], but no effect on heart rate was shown. Similarly, there are 12 dioxin-like PCBs (e.g., PCB 118) that have been associated with heart conditions such as hypertension and cardiac defects [64,65]. These effects have been linked to the activation of the aryl hydrocarbon (AhR). More interestingly, even small doses of 1,2,5,6-tetrabromocyclooctane (HBCD) cause arrythmia through dysregulating the function of sarcoplasmic/endoplasmic reticulum Ca^{2+} ATPase (SERCA2a) [66]. The latter is encoded by the *atp2a2a* and *atp2a2b* genes, of which only *atp2a2a* was significantly induced after exposure to POP125× (see below). Thus, the role of Atp2a2 in inducing arrythmia is very likely, although the precise mechanism remains unclear. Finally, the cardiotoxic effects of organochlorine pesticides have been clearly described previously [67]. The mechanisms may vary but most of the compounds within the different mixes undoubtedly have the potential to alter the heart rate, leading to cardiac conditions such as arrythmia, hypertension, and other cardiac defects.

3.6. *Fish Were Hyperactive and Responded Notably to Changes in Illumination*

To test for behavioral effects, indicative of potential neurological defects, we used a standard 10-min light–dark swimming activity protocol to assess the parameters "time spent active", "distance travelled", and "swimming speed" in 10 s intervals. Due to the high number of multiple malformations in the larvae in POP250x, this dose was excluded from the behavioral analyses.

3.6.1. Dark–Light Response

In Figure 5, we illustrate the changes in behavior observed after treatment with the total mix POP125× compared to the control, untreated larvae at 5 dpf. As expected, we observe an increase in all parameters in control larvae during the dark phase, compared to the light phase. We can also observe the initial increase in all parameters at the start of the dark phase, which decreases in time while the larvae acclimate to the new situation. This decrease is even much faster in the POP125×-treated larvae. In contrast, we observe a stronger spike in activity when switching to the light phase, indicating that the larvae do perceive the change in lighting conditions, but rapidly return to a slightly higher activity compared to the dark phase.

The effects recorded during the dark and light phases (excluding the spikes, see below) were different depending on the compounds used (Figure 6). For instance, POP125× was the only compound that increased the swimming speed (SWS) during the dark phase, whereas binary mixtures containing Cl (PFAA + Cl and Br + Cl) decreased the speed significantly compared to controls (Figure 6A). Cl alone decreased the speed significantly only when compared to PFAA, itself slightly, but not significantly, higher than control. Thus, it appears that Cl was mainly responsible for decreasing swimming speed during the dark phase. During the light phase, more striking effects on the swimming speed were seen. Overall, all treatments (except Br) caused a faster swimming speed; however, only those mixtures where PFAA was present caused a significant increase compared to the control group. Fish exposed to the POP125× mix were swimming the fastest, followed by the binary mixtures PFAA + Br and PFAA + Cl, and finally PFAA (Figure 6B). Thus, while only PFAA alone caused a significant increase in swimming speed, addition of Cl or Br in the binary mixtures further enhanced this effect.

Time spent active (TSA) was similarly significantly affected (Figure 6C,D). POP125× caused the highest increase in activity in both the dark and light phases. During the dark phase, the presence of PFAA and PFAA + Cl significantly decreased TSA values, similar to the Cl and Br + Cl groups. During the light phase, a dramatic increase in activity was observed in fish exposed to POP125×, while PFAA, PFAA + Br, and PFAA + Cl caused a significant, but weaker, increase. Br, Cl, and Br + Cl did not affect TSA relative to control, they were thus significantly different from PFAA.

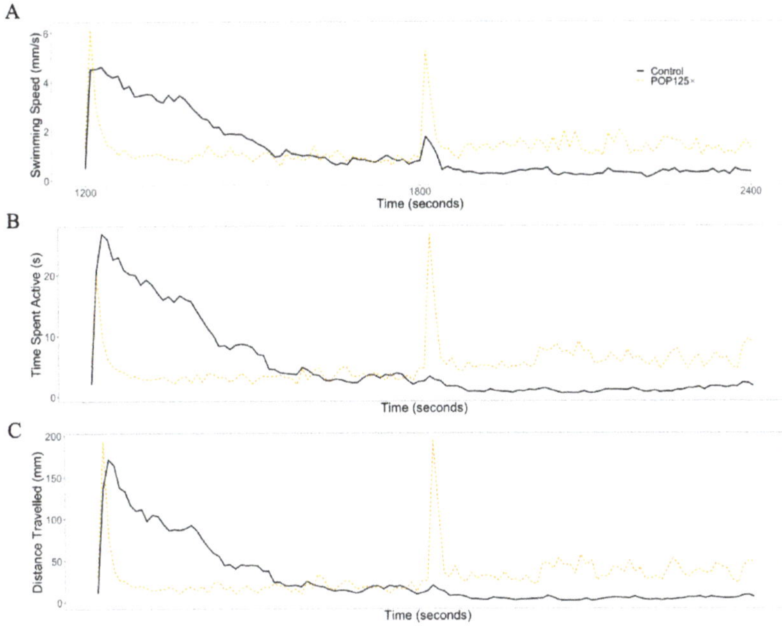

Figure 5. Actogram example of (**A**) swimming speed, (**B**) time spent active, and (**C**) distance travelled during 20 min of the tests, starting with 10 min dark phase followed by 10 min light. For comparison, control larvae are shown alongside POP125×-treated larvae. Two spikes can be appreciated when the fish entered a different phase.

The distance travelled (DT) during the dark phase was significantly higher in fish treated with POP125×, but lower in fish treated with PFAA + Cl and Br + Cl. During the light phase, a large increase in DT was seen upon POP125× treatment, while PFAA, PFAA + Br, and PFAA + Cl, caused weaker, but still significant, increases, similar to what was observed for the TSA (Figure 6E,F).

Behavior is a complex endpoint, hard to analyze, and where many variables could be playing a role and inducing changes. One of the first hypotheses we thought of to explain the altered behavior was related to compounds binding to brain aromatase or Cyp19a1b. This protein (or its isoforms) is present from 24 hpf [68] and several studies have reported changes in swimming behavior triggered by compounds such as fadrozole (a well-known aromatase inhibitor) or other endocrine disrupting chemicals like a PCB mixture (aroclor 1254), PBDE-47, or PFOA in various fish species [69–73]. However, involvement of classical endocrine disruptors was ruled out for environmental effects, either due to the high concentrations used (fadrozole), or to the observation that hormone antagonists did not revert the changes [73]. In our experiments, the most obvious alterations of behavior were caused by the PFAA sub-mix (with PFOA and PFOS at the highest concentrations in the mix). We also observed a significant effect of Br or Cl, also adding to the effect when used in combination with PFAA, although never reaching the extent of the full POP125× mix. Thus, we cannot discard eventual synergistic or additive effects caused by the presence of the other chemicals within the mix. The behavior altering properties of PFAAs have been described before, in various settings. At very low concentrations between 7–700 ng/L, PFOS or PFOA led to decreased activity in 5 dpf larvae when tested alone, but increased activity when both compounds were tested together [72]. Increased activity was also observed for PFOA at 400–4000 ng/mL [44] or at 300 to 2000 ng/mL considering only the dark phase [37], consistent with our findings. These authors also suggested that hyperactivity was linked to alterations in calcium signaling involving the ryanodine receptor Ryr and affecting muscular contractions. Also, sensitization of the RYR by PCBs (e.g., 28, 138 and 153) can cause a developmental neurotoxicity [74], hence affecting the photomotor response of the larvae.

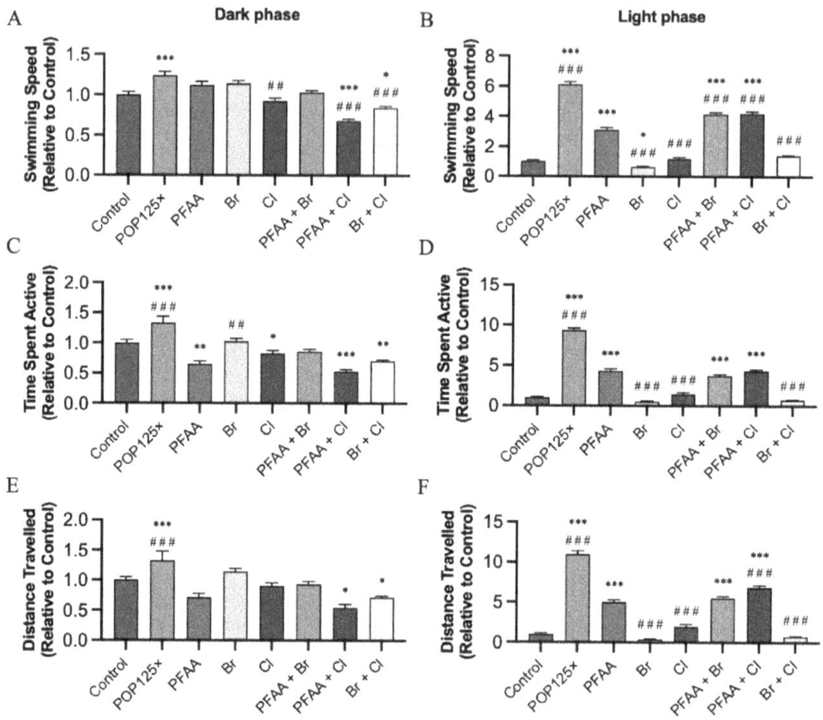

Figure 6. Behavior tests during the dark (**left**) and light (**right**) phases: swimming speed (**A**,**B**), time spent active (**C**,**D**), and distance travelled (**E**,**F**) for all treatments. All data were calculated excluding 50 s in the transition zone between light and dark phases, and the values were normalized relative to the corresponding controls for dark and light phases, $n = 72$, $p < 0.05$ (*), ≤ 0.01 (**), ≤ 0.001 (***). Asterisks (*) indicate when significant differences were found compared to control, hash (#) sign when differences were found against PFAA.

3.6.2. Startle Response

As mentioned above, we noticed that each time fish were recorded, we could observe a dramatic increase in their activity (spike) at the moment of drastic transition from light to dark or back to light. Thus, we decided to focus on this spike response by analyzing only the 50 s around the transition. For the light–dark transition, we observed an increased startle response for the POP125× mix, which was even higher for the PFAA sub-mix, but somehow attenuated by addition of Br in the PFAA + Br sub-mix and Cl in the PFAA + Cl sub-mix reaching significance only for swimming speed (Figure 7). Interestingly, Br alone had no effect, while Cl alone significantly increased all parameters relative to controls. In the dark–light transition, this increased response was also observed, this time higher for the POP125× mix compared to all other mixtures containing PFAA. Br and Cl alone or in combination had no effect at all compared to control, while only marginally modulating the effect of PFAA in binary mixtures.

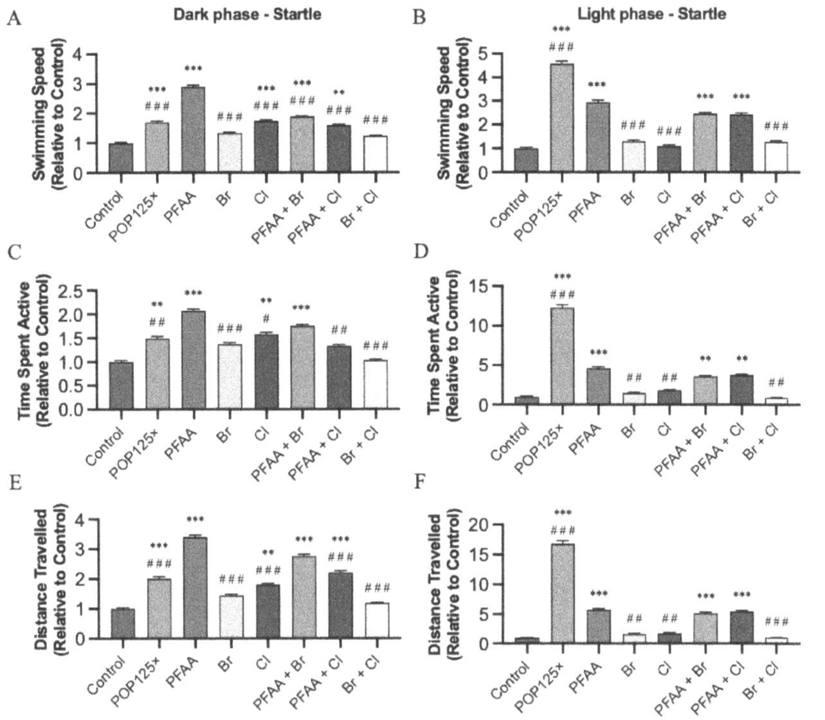

Figure 7. Swimming speed (**A**,**B**), time spent active (**C**,**D**), and distance traveled (**E**,**F**) for all treatments. Left column, parameters during the dark phase, right column, parameters during the light phase. All results were calculated using 50 s during the transition zone between light and dark phases, $n = 72$, $p \leq 0.01$ (**), ≤ 0.001 (***). Asterisks (*) indicate when significant differences were found compared to control, hash sign (#) when differences were found against PFAA.

Taken together, our results indicate that the immediate startle response, presumably corresponding to the larvae reacting to any change in environmental conditions, is amplified by the presence of POPs, especially PFAA. This is consistent with the overall higher activity, as described above; however, the Br and Cl components seem to modulate this startle response more strongly. PBDEs such as BDE-47, -99, -100, and -153 have been shown to alter behavior at concentrations as low as 2.5 μg/mL depending on the congener [46]. According to these authors, the drastic response of the larvae in lighting transition can be explained by alterations in the glutamatergic transmission and changes in electrical coupling in the presence of PBDEs or PCBs.

3.7. Mitochondria Responded Notably to POP Mixture

To test the effect of the POPs on mitochondria, we used the transgenic line *Tg(kdrl-mls:GFP)*, which expresses the green fluorescent protein GFP in the endothelial cells of the vasculature and directs it to the mitochondria via its Mls signal peptide [18]. Note that the expression of the endogenous *kdrl* gene is not significantly affected by POP treatment (see below and Table S2). Exposure of these embryos to POP mixes led to a significantly increased fluorescence with POP125×, while only combined sub-mixes PFAA + Br and Br + Cl caused a significant increase (Figure 8). The higher levels of activity in mitochondria seem to be linked to OCPs and PCBs, which are explained by a POP-induced imbalance in redox, hormone homeostasis, and mitochondrial dysfunction; the mechanisms are discussed in further detail in [75,76].

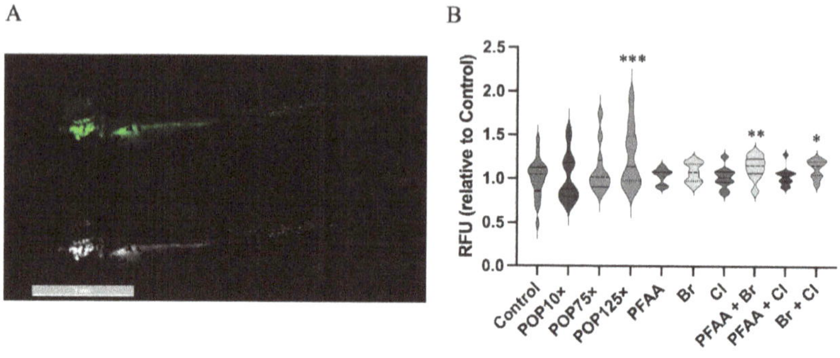

Figure 8. (**A**) Transgenic *Tg(kdrl-mls:GFP)* fish line treated with POP125×. (**B**) Plot with median values of the normalized fluorescence intensity of fish treated with different POP mixes at 96 hpf. There is a clear increasing trend. Kruskal–Wallis test and Dunn's multiple comparison, $n = 21$, $p < 0.05$ (*), $p < 0.01$ (**), $p < 0.001$ (***).

3.8. Gene Expression Is Severely Affected by Exposure to POPs

To gain further insight into the molecular mechanisms affected by POP exposure, we compared the whole genome transcriptome of control larvae to that of larvae treated with POP75× or POP125× by RNA-Seq analysis on whole larvae at 5 dpf. The number of differentially expressed genes (DEGs at $p_{adjust} < 0.05$) was 172 and 2466 for, respectively, POP75× and POP125× treatments, with 1531 genes that were upregulated and 935 that were downregulated by POP125×.

Interestingly, the huge majority (169/172) of DEGs affected at the lower concentration were also, and more strongly, affected at the higher concentration (Figure 9A, see also Table S2). As an example, the *fbxo32* (involved in muscle morphogenesis and homeostasis, and in larval behavior), *fosb* (transcription factor of the AP1 family), and *cdca7a* (regulation of hematopoietic stem cell differentiation, thymus development) genes were not significantly affected at POP75× (log(fold-change), respectively, 0.64, 0.57, and −0.65, $p_{adj} > 0.05$), but were strongly affected at POP125× concentration (log(fold-change) of, respectively, 2.45, 2.20, and −1.31, $p_{adj} << 0.05$) (Figure 9B). Another such gene is *calcoco1b*, coding for a calcium binding protein acting as a translation elongation factor.

The most highly, and most significantly, regulated genes are mostly upregulated genes. Among the most highly induced genes, we observe *fosl1a*, *fosb*, and *junba* which together form the AP1 transcription factor regulating cell proliferation, differentiation, and stress response [77]. Also, among these upregulated genes are the two paralogs *igfbp1a and igfbp1b*, coding for Igfbp proteins that interact with insulin-like growth factors (IGFs) to stabilize them and modulate their effects on growth and glucose metabolism. The *cyp24a1* gene, coding for a 1,25-dihydroxyvitamin D3 metabolizing enzyme, is significantly upregulated, possibly in relation to the growth retardation observed [78,79]. In contrast, *cyp2aa9* and *cyp2aa8*, coding for xenobiotic metabolizing enzymes [80–82], are among the most downregulated genes, indicating a response to inflammatory status [83]. In addition, many more genes (2297) were differentially expressed in POP125×-treated larvae relative to control, although we did not observe substantial lethality at this stage.

Pathway and gene ontology (Tables S3–S5) analysis points to a dysregulation of the cell cycle, but also of central nervous system development, motor activity, growth, response to stress, and metabolic processes, in particular insulin signaling and glucose metabolism (Tables S3 and S4). Using the list of genes involved in insulin signaling (Table S5) in GENEMANIA, we constructed a network of co-expressed genes encoding proteins with physical interaction (Figure 10). Strikingly, all these genes are significantly upregulated by POP125×, while also distributing in several modules.

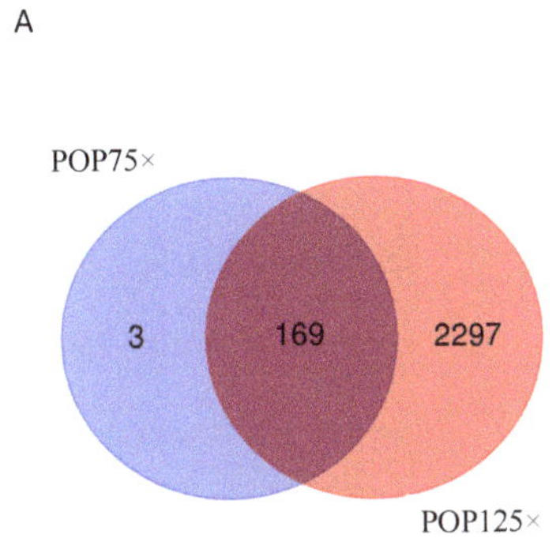

B

Zfin Name	POP75× log2(fold)	POP75× padj	POP125× log2(fold)	POP125× padj
fosl1a	1.35	3.52×10^{-04}	3.06	3.19×10^{-27}
psme4a	1.43	6.37×10^{-06}	2.81	1.69×10^{-36}
mmp9	1.60	7.46×10^{-06}	2.63	1.32×10^{-22}
pdk2b	1.45	6.74×10^{-06}	2.55	5.67×10^{-28}
calcoco1b	0.96	8.03×10^{-02}	2.53	2.91×10^{-16}
fbxo32	0.64	3.09×10^{-01}	2.45	1.48×10^{-17}
retreg1	1.18	4.88×10^{-03}	2.36	1.56×10^{-15}
igfbp1b	1.27	5.13×10^{-03}	2.34	2.50×10^{-15}
ucp3	1.31	7.46×10^{-06}	2.31	3.19×10^{-27}
slc16a9a	1.15	3.58×10^{-03}	2.27	6.25×10^{-18}
igfbp1a	1.12	3.21×10^{-03}	2.24	3.11×10^{-20}
fosb	0.57	4.75×10^{-01}	2.20	6.94×10^{-14}
lcat	1.19	2.12×10^{-03}	2.06	8.90×10^{-14}
cbx7a	1.16	3.41×10^{-04}	2.05	1.06×10^{-19}
oat	1.44	2.36×10^{-05}	2.05	2.50×10^{-15}
cyp2aa9	-0.74	1.72×10^{-01}	-1.23	1.45×10^{-04}
cdca7a	-0.65	2.40×10^{-01}	-1.31	6.29×10^{-05}
ca4b	-0.91	7.11×10^{-02}	-1.33	1.23×10^{-04}
krt1c19e	-0.81	4.47×10^{-02}	-1.42	4.95×10^{-07}
cyp2aa8	-1.11	5.06×10^{-03}	-2.18	4.06×10^{-14}

Figure 9. (**A**) Venn diagram comparing the lists of genes affected by POP1 or POP2 treatment. (**B**) Most highly and significantly regulated genes upon treatment with both POP total mix concentrations at 5 dpf. Log(fold change) and significance (adjusted *p*-value) are shown. Upregulated genes are in red, downregulated in green, while genes discussed in the text are highlighted in yellow.

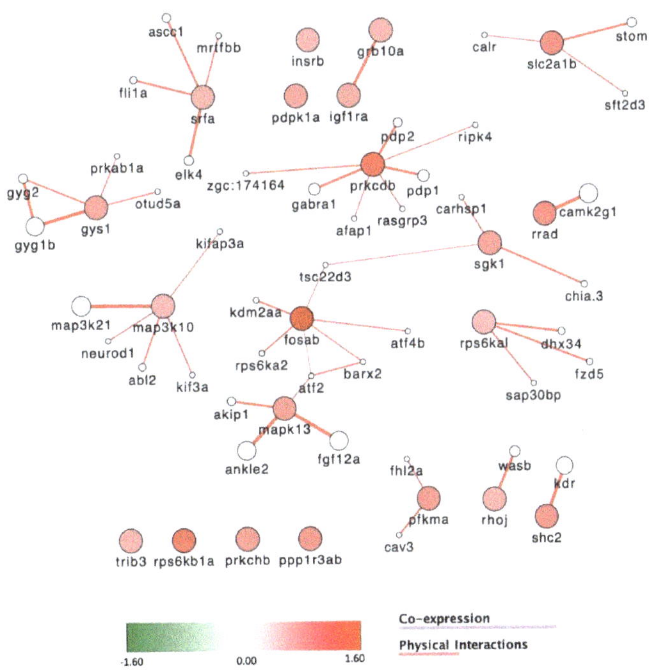

Figure 10. Differentially expressed genes involved in insulin signaling and response. These genes are all upregulated upon treatment with POP125× and distribute into specific co-expression and physically interacting modules.

One of these modules centers around *gys1*, indicating increased glycogen synthesis activity. Other modules involve the insulin receptor (*insrb*), the Glut1 glucose transporter (*slc2a1b*) [84], a transcriptional regulatory module centered on the serum response factor (*srfa*), or regulatory protein kinases such as the MAPkinases *map3k10*, *map3k21* and *mapk13* genes, the protein kinase C gene *prkcdb* or the serum/glucocorticoid regulated kinase 1 gene (*sgk1*). Together, these observations indicate an increase in metabolism in the POP-treated larvae.

As may be expected from exposure to environmental toxins, oxidoreductase molecular function was identified as significantly affected, with all the genes in the list interestingly downregulated. These genes code for detoxifying enzymes such as alcohol dehydrogenases (*adh5*, *adh8b*), aldehyde dehydrogenases (*aldh16a1*, *aldh9a1a.1*), or cytochromes (*cyb5r3*, *cyp2aa3*, *cyp2r1*, *cyp4v8*, *cyp51*).

Gene enrichment analysis based on the mutant descriptions at zfin (zfin.org; accessed on 6 March 2023)) revealed that mutations in a significant number of the affected genes cause defects in development of the nervous system and the eye, as well as in the mitotic cell cycle (Tables S3 and S4). Based on these results, we used the GENEMANIA database to construct networks for the genes involved in eye and brain development (Figure 11). In both networks, we notice that the genes are all downregulated and that they build a tight network of co-regulated genes. About half of the genes affecting either eye or brain development are common to both networks. When we focused on the genes whose products were shown to physically interact (Figure 11), some similarities and some differences were observed. One common module is formed by the *aurka* (aurora kinase A, a histone serine kinase), the histone deacetylase gene *hdac8*, the polo-like protein serine/threonine kinase gene *plk1*, and the *fbxo5* gene coding for a predicted ubiquitin ligase inhibitor. In the brain, this module is connected through the CDK–cyclin pair Cdk1-Ccnb1 to the gene products of *rpa1* and *smc4*, predicted to be involved, respectively, in DNA repair, replication, and chromosome organization. Two smaller modules are formed by the *birc5a*, *cdca8*, *nono*, and *sfpq* genes, all of which were shown to affect both brain and eye formation when mutated (zfin.org). In the eye, the *cdk1* gene is connected to several members of the condensin I complex, including Smc2, Smc4, Ncapg, Ncaph, and Ncapd2 (Figure 11). This observation is reminiscent of the putative involvement of the *cap-g* gene in the eye defect that we observed. Interestingly, not only mutation of the *cap-g* gene, but also morpholino knockdown of the *capd2* and *caph* genes (coding for two other components of the condensing I complex), led to reduction of retinal cell number and smaller eyes similar to what we observed here [61].

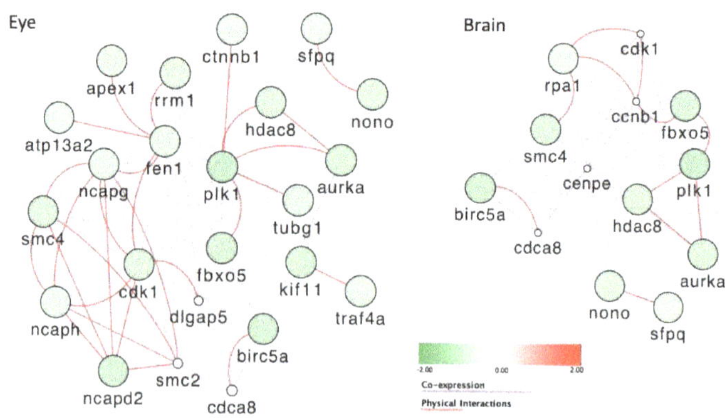

Figure 11. Differentially expressed genes that are involved in brain and eye development. These genes are all downregulated in zebrafish larvae treated with POP125× and form co-expression and physically interacting networks.

Note that the description of the *cap-g* mutant also mentions a behavioral effect, presumably based on oculomotor and optokinetic tests mainly revealing visual impairment [61]; thus, we cannot rule out that the eye defect we observed may impact the larval behavior.

Previous studies involving RNA-Seq analysis of whole larvae exposed to the POP mixture in a similar setting, albeit at lower concentrations, revealed interesting results. The PPAR pathway was shown to be affected [37], similar to our observation that the nuclear receptor genes *pparda* and *ppargc1a* were upregulated (log(fold) 0.89 and 0.64, respectively) along with those for other nuclear receptors such as *rxrab*. Conversely, our data set did not reveal calcium ion transport or signaling to be significantly affect, nor the genes involved in ryanodine receptor signaling such as *ryr1a*, *ryr1b*, *ryr2*, *myl7*, *actc1*, or *tnnc1*.

Another concern with environmental pollution by POPs is their effect on sexual maturation of growing embryos. While we did not perform fertility experiments in our larval development tests, data concerning the effects of sexual hormones on 2–4 dpf larvae are readily available [85,86]. Among the genes most highly upregulated by estradiol (E2) [85], only *vtg1* was moderately and non-significantly downregulated (Table S5), other vitellogenin genes and the aromatase gene *cyp19a1b* were not affected.

In contrast, some of the most regulated genes by 17-ß-testosterone [86] were also significantly affected upon POP125× treatment, such as *insig1*, *col10a1a*, *matn1*, *tmx3a*, *gnb3a*, or *gng13b*. Altogether, these results tend to argue for the presence of an androgen antagonistic activity in the POP mix, as was previously shown in cellular assays [13].

The effect on retina and eye development is particularly interesting. The observed defect phenocopies exactly that previously found in a *capg* mutant and in morphants for *capd2* and *caph* [61], all components of the condensin I complex specifically required for retina and eye formation. Microinjection of *p53* morpholino oligonucleotides at the one–two cell stage was able to rescue the *capg* phenotype [61], but also the eye defect in larvae exposed to POP125×. Whole larvae RNA-Seq revealed that neither the *p53* gene *tp53* expression was affected, nor did the term apoptosis appear in pathway analysis, further supporting the notion that decreasing condensin I action will specifically induce *p53*-dependent apoptosis in the developing retina.

4. Conclusions

This study sheds some new light on the effects of a realistic mixture of POPs on vertebrate physiology, with a focus on early developmental stages. Although informing on the effect on wildlife when present in the environment, our findings are also relevant for pregnant women relative to the health of the unborn child.

As discussed above, we believe that the concentrations tested here (POP75× and POP125×) represent realistic doses that may be reached in the environment, but also in human blood due to the stability and bioaccumulation characteristics of these compounds. Furthermore, we have been using a mixture, and sub-mixtures thereof, that represent the composition of that found in a Scandinavian population. Although not directly transposable to humans, our results obtained on zebrafish larvae allow us to identify potential risks to human fetuses and can inform about the molecular mechanisms that may be involved.

The effects that we observed, and discussed above, at non-lethal concentrations of the POP-mix are growth and developmental retardation as illustrated by the retarded inflation of the swim bladder, increased heart rate, increased metabolism, and mostly increased behavioral activity. Differential expression of genes upon exposure to POP125× was indeed consistent with an increase of metabolism (insulin signaling), but further revealed impacts on musculo-skeletal development and function, brain development, and several signaling pathways. Additional studies will be required in the future to investigate which component of the POP mix is responsible for any specific effect and through which molecular mechanism it acts.

The effects observed here are mainly due to the PFAA sub-mix; this single sub-mix, alone or in combination, caused effects close to those observed with the total POP mixture. This is of high relevance at the present time, as these substances (per- and polyfluoroalkyl acids) are intensely scrutinized as the most harmful chemical group for their deleterious effects on environment and human health. Initiatives and regulations have been introduced very recently by the European Chemicals Agency [87] and the US Environmental Protection

Agency [88] to investigate the origins and effects of these compounds, aiming at reducing or banning their use. Furthermore, the spotlight of public opinion has been drawn to the topic by recent movies such as "Dark waters" [89] inspired by a story in the New York Times [90] and a recent report in The Guardian [91].

Contributions of the Cl and Br sub-mixes were observed, enhancing the effect of the PFAA mix for eye malformations, behavior, mitochondrial effect, and weakly for heart rate. One interesting exception was observed for the startle response during the light to dark transition, where the PFAA sub-mix alone caused a stronger increase compared to the POP mix, which was attenuated upon addition of either the Cl or Br sub-mix in the binary mixtures.

In conclusion, our study highlights the need to study environmental pollution, not only on single compounds, but rather to consider the more realistic situation of exposure to mixtures. One striking result from our studies is that, although some of the compounds within our mixtures are described as endocrine disrupting chemicals, we did not observe significant estrogenic effects. This illustrates the fact that individual compounds in the mix may antagonize the activities of other chemicals. Our results comparing specific sub-mixtures to the total mixture further support this conclusion, with sub-mixes either enhancing or suppressing the effects of another. Finally, we also reveal a novel defect caused by POP contamination on eye development, which we propose to result from inhibition of the condensin I complex.

Supplementary Materials: The following supporting information can be downloaded at: https://www.mdpi.com/article/10.3390/toxics11040357/s1. Figure S1: Network of genes affected by POP125× treatment and involved in eye (A) and brain (B) development. (C) Venn diagram comparing number of genes involved in brain and eye development. Table S1: List of compounds and concentrations in the POP mix and in the PFAA, Br, and CL sub-mixtures. Table S2: List of differentially expressed genes in zebrafish larvae exposed to POP75× and POP125×: gene Ids and names, log2(FoldChange), *p*-value, and adjusted *p*-value. Table S3: Gene ontology analysis of the genes affected by POP125× treatment. Enrichment analysis of up- or downregulated genes in Genotype–Phenotype database, GO-terms, KEGG, Reactome, and Wikipathways databases. Table S4: Gene ontology analysis of the genes affected by POP125× treatment: GSEA analysis using the Genotype–Phenotype database, GO-terms, KEGG, Reactome, and Wikipathways databases. Table S5: List of genes differentially expressed upon POP75× or POP125× treatment, and involved in brain, eye development, insulin signaling, oxido-reductive regulation, estrogen, or androgen regulation.

Author Contributions: Conceptualization, M.M. and G.G.-L.; methodology, M.M., G.G.-L., R.N., N.B.-N. and D.D.-T.; software, G.G.-L., N.B.-N. and R.N.; validation, M.M. and G.G.-L.; formal analysis, M.M.; investigation, G.G.-L.; resources, M.M.; data curation, M.M.; writing—original draft preparation, G.G.-L.; writing—review and editing, G.G.-L. and M.M.; supervision, M.M.; project administration, M.M.; funding acquisition, M.M. All authors have read and agreed to the published version of the manuscript.

Funding: This research was funded by the European Union's Horizon 2020 research and innovation program under the Marie Skłodowska-Curie Innovative Training Network (ITN) program PROTECTED [Grant agreement No. 722634]. M.M. is a "Maître de Recherche" at "Fonds National de Recherche Scientifique (FNRS)".

Institutional Review Board Statement: Animal care and all experimentation were conducted in compliance with Belgian and European laws. They were approved by the "Ethical commission of Université de Liège (Authorization: LA1610002 Ethical commission protocol ULg19-2134 and Ulg19-2135)".

Informed Consent Statement: Not applicable.

Data Availability Statement: Raw and processed sequencing data have been deposited in NCBI's Gene Expression Omnibus [92] and are accessible through GEO Series accession number GSE208019 (https://www.ncbi.nlm.nih.gov/geo/query/acc.cgi?acc=GSE208019; accessed on 8 April 2023).

Acknowledgments: This project has received funding from the European Union's Horizon 2020 research and innovation program under the Marie Skłodowska-Curie Innovative Training Network (ITN) program PROTECTED, Grant agreement No. 722634. The authors would like to thank Jérémie Zappia for technical help, and the GIGA zebrafish platform (H. Pendeville-Samain) for taking care of and delivering the zebrafish larvae, the GIGA imaging platform for their help and support with microscopy, the GIGA genomic platform for sequencing, and the GIGA bioinformatics platform for data analysis.

Conflicts of Interest: The authors declare no conflict of interest. The funders had no role in the design of the study; in the collection, analyses, or interpretation of data; in the writing of the manuscript; or in the decision to publish the results.

References

1. United Nations. Stockholm Convention on Persistent Organic Pollutants. *Treaty Ser.* **2006**, *2256*, 119.
2. European Parliament and Council. Regulation (EC) No 850/2004 of the European Parliament and of the Council of 29 April 2004 on persistent organic pollutants and amending Directive 79/117/EEC. *Off. J. Eur. Union* **2004**, L158/157–149.
3. Guo, W.; Pan, B.; Sakkiah, S.; Yavas, G.; Ge, W.; Zou, W.; Tong, W.; Hong, H. Persistent Organic Pollutants in Food: Contamination Sources, Health Effects and Detection Methods. *Int. J. Environ. Res. Public Health* **2019**, *16*, 4361. [CrossRef]
4. Gregoraszczuk, E.L.; Ptak, A. Endocrine-Disrupting Chemicals: Some Actions of POPs on Female Reproduction. *Int. J. Endocrinol.* **2013**, *2013*, 828532. [CrossRef] [PubMed]
5. Brody, J.G.; Moysich, K.B.; Humblet, O.; Attfield, K.R.; Beehler, G.P.; Rudel, R.A. Environmental pollutants and breast cancer: Epidemiologic studies. *Cancer* **2007**, *109*, 2667–2711. [CrossRef]
6. Sanderson, J.T. The steroid hormone biosynthesis pathway as a target for endocrine-disrupting chemicals. *Toxicol. Sci.* **2006**, *94*, 3–21. [CrossRef] [PubMed]
7. Ljunggren, S.A.; Helmfrid, I.; Salihovic, S.; van Bavel, B.; Wingren, G.; Lindahl, M.; Karlsson, H. Persistent organic pollutants distribution in lipoprotein fractions in relation to cardiovascular disease and cancer. *Environ. Int.* **2014**, *65*, 93–99. [CrossRef]
8. Van Oostdam, J.C.; Dewailly, E.; Gilman, A.; Hansen, J.C.; Odland, J.O.; Chashchin, V.; Berner, J.; Butler-Walker, J.; Lagerkvist, B.J.; Olafsdottir, K.; et al. Circumpolar maternal blood contaminant survey, 1994–1997 organochlorine compounds. *Sci. Total Environ.* **2004**, *330*, 55–70. [CrossRef]
9. Knutsen, H.K.; Kvalem, H.E.; Thomsen, C.; Froshaug, M.; Haugen, M.; Becher, G.; Alexander, J.; Meltzer, H.M. Dietary exposure to brominated flame retardants correlates with male blood levels in a selected group of Norwegians with a wide range of seafood consumption. *Mol. Nutr. Food Res.* **2008**, *52*, 217–227. [CrossRef]
10. Polder, A.; Skaare, J.U.; Skjerve, E.; Loken, K.B.; Eggesbo, M. Levels of chlorinated pesticides and polychlorinated biphenyls in Norwegian breast milk (2002–2006), and factors that may predict the level of contamination. *Sci. Total Environ.* **2009**, *407*, 4584–4590. [CrossRef]
11. Haug, L.S.; Salihovic, S.; Jogsten, I.E.; Thomsen, C.; van Bavel, B.; Lindstrom, G.; Becher, G. Levels in food and beverages and daily intake of perfluorinated compounds in Norway. *Chemosphere* **2010**, *80*, 1137–1143. [CrossRef] [PubMed]
12. Berntsen, H.F.; Berg, V.; Thomsen, C.; Ropstad, E.; Zimmer, K.E. The design of an environmentally relevant mixture of persistent organic pollutants for use in in vivo and in vitro studies. *J. Toxicol. Environ. Health Part A* **2017**, *80*, 1002–1016. [CrossRef] [PubMed]
13. McComb, J.; Mills, I.G.; Muller, M.; Berntsen, H.F.; Zimmer, K.E.; Ropstad, E.; Verhaegen, S.; Connolly, L. Human blood-based exposure levels of persistent organic pollutant (POP) mixtures antagonise androgen receptor transactivation and translocation. *Environ. Int.* **2019**, *132*, 105083. [CrossRef] [PubMed]
14. Doan, T.Q.; Berntsen, H.F.; Verhaegen, S.; Ropstad, E.; Connolly, L.; Igout, A.; Muller, M.; Scippo, M.L. A mixture of persistent organic pollutants relevant for human exposure inhibits the transactivation activity of the aryl hydrocarbon receptor in vitro. *Environ. Pollut.* **2019**, *254*, 113098. [CrossRef] [PubMed]
15. Yadav, A.; Amber, M.; Zosen, D.; Labba, N.A.; Huiberts, E.H.W.; Samulin Erdem, J.; Haugen, F.; Berntsen, H.F.; Zienolddiny, S.; Paulsen, R.E.; et al. A human relevant mixture of persistent organic pollutants (POPs) and perfluorooctane sulfonic acid (PFOS) enhance nerve growth factor (NGF)-induced neurite outgrowth in PC12 cells. *Toxicol. Lett.* **2021**, *338*, 85–96. [CrossRef]
16. Amber, M.; Xie, Y.; Berntsen, H.F.; Zimmer, K.E.; Ropstad, E.; Verhaegen, S.; Connolly, L. Effects of Defined Mixtures of Persistent Organic Pollutants (POPs) on Pre-lethal Cytotoxicity in the Human A-498 Kidney Cell Line In Vitro. *Expo. Health* **2021**, *13*, 465–475. [CrossRef]
17. Segner, H. Zebrafish (Danio rerio) as a model organism for investigating endocrine disruption. *Comp. Biochem. Physiol. C Toxicol. Pharm.* **2009**, *149*, 187–195. [CrossRef]
18. Herkenne, S.; Ek, O.; Zamberlan, M.; Pellattiero, A.; Chergova, M.; Chivite, I.; Novotna, E.; Rigoni, G.; Fonseca, T.B.; Samardzic, D.; et al. Developmental and Tumor Angiogenesis Requires the Mitochondria-Shaping Protein Opa1. *Cell Metab.* **2020**, *31*, 987–1003.e1008. [CrossRef]

19. Lammer, E.; Carr, G.J.; Wendler, K.; Rawlings, J.M.; Belanger, S.E.; Braunbeck, T. Is the fish embryo toxicity test (FET) with the zebrafish (Danio rerio) a potential alternative for the fish acute toxicity test? *Comp. Biochem. Physiol. C Toxicol. Pharmacol.* **2009**, *149*, 196–209. [CrossRef]
20. Westerfield, M. A guide for the laboratory use of zebrafish (Danio rerio). In *The Zebrafish Book*, 5th ed.; University of Oregon Press: Eugene, OR, USA, 2007.
21. Pruvot, B.; Quiroz, Y.; Voncken, A.; Jeanray, N.; Piot, A.; Martial, J.A.; Muller, M. A panel of biological tests reveals developmental effects of pharmaceutical pollutants on late stage zebrafish embryos. *Reprod. Toxicol.* **2012**, *34*, 568–583. [CrossRef]
22. Philip, A.M.; Wang, Y.; Mauro, A.; El-Rass, S.; Marshall, J.C.; Lee, W.L.; Slutsky, A.S.; dosSantos, C.C.; Wen, X.Y. Development of a zebrafish sepsis model for high-throughput drug discovery. *Mol. Med.* **2017**, *23*, 134–148. [CrossRef] [PubMed]
23. Dalcq, J.; Pasque, V.; Ghaye, A.; Larbuisson, A.; Motte, P.; Martial, J.A.; Muller, M. Runx3, Egr1 and Sox9b form a regulatory cascade required to modulate BMP-signaling during cranial cartilage development in zebrafish. *PLoS ONE* **2012**, *7*, e50140. [CrossRef]
24. Larbuisson, A.; Dalcq, J.; Martial, J.A.; Muller, M. Fgf receptors Fgfr1a and Fgfr2 control the function of pharyngeal endoderm in late cranial cartilage development. *Differentiation* **2013**, *86*, 192–206. [CrossRef]
25. Ewels, P.A.; Peltzer, A.; Fillinger, S.; Patel, H.; Alneberg, J.; Wilm, A.; Garcia, M.U.; Di Tommaso, P.; Nahnsen, S. The nf-core framework for community-curated bioinformatics pipelines. *Nat. Biotechnol.* **2020**, *38*, 276–278. [CrossRef] [PubMed]
26. Love, M.I.; Huber, W.; Anders, S. Moderated estimation of fold change and dispersion for RNA-seq data with DESeq2. *Genome Biol.* **2014**, *15*, 550. [CrossRef]
27. Kanehisa, M.; Sato, Y.; Kawashima, M. KEGG mapping tools for uncovering hidden features in biological data. *Protein Sci.* **2022**, *31*, 47–53. [CrossRef]
28. Kanehisa, M.; Goto, S. KEGG: Kyoto encyclopedia of genes and genomes. *Nucleic Acids Res.* **2000**, *28*, 27–30. [CrossRef] [PubMed]
29. R Core Team. *R: A Language and Environment for Statistical Computing (4.0.2)*; R Foundation for Statistical Computing: Vienna, Austria, 2020; Available online: http://www.r-project.org/index.html (accessed on 9 March 2023).
30. Venables, W.N.; Ripley, B.D. *Modern Applied Statistics with S*, 4th ed.; Springer: New York, NY, USA, 2002; ISBN 0-387-95457-0.
31. Pinheiro, J.; Bates, D.M. *Mixed-Effects Models in S and S-PLUS*; Springer: New York, NY, USA, 2000. [CrossRef]
32. Fox, J.; Weisberg, S. *An {R} Companion to Applied Regression*, 3rd ed.; SAGE Publications Inc.: Thousand Oaks, CA, USA, 2019; Available online: https://socialsciences.mcmaster.ca/jfox/Books/Companion/ (accessed on 9 March 2023).
33. Lenth, R.V. Emmeans: Estimated Marginal Means, aka Least-Squares Means. In *R Package Version 1.7.5*; R Foundation for Statistical Computing: Vienna, Austria, 2022; Available online: https://CRAN.R-project.org/package=emmeans (accessed on 9 March 2023).
34. Lenth, R.V. Least-Squares Means: The R Package lsmeans. *J. Stat. Softw.* **2016**, *69*, 1–33. [CrossRef]
35. Pelka, K.E.; Henn, K.; Keck, A.; Sapel, B.; Braunbeck, T. Size does matter–Determination of the critical molecular size for the uptake of chemicals across the chorion of zebrafish (Danio rerio) embryos. *Aquat. Toxicol.* **2017**, *185*, 1–10. [CrossRef] [PubMed]
36. Khezri, A.; Fraser, T.W.; Nourizadeh-Lillabadi, R.; Kamstra, J.H.; Berg, V.; Zimmer, K.E.; Ropstad, E. A Mixture of Persistent Organic Pollutants and Perfluorooctanesulfonic Acid Induces Similar Behavioural Responses, but Different Gene Expression Profiles in Zebrafish Larvae. *Int. J. Mol. Sci.* **2017**, *18*, 291. [CrossRef] [PubMed]
37. Christou, M.; Fraser, T.W.K.; Berg, V.; Ropstad, E.; Kamstra, J.H. Calcium signaling as a possible mechanism behind increased locomotor response in zebrafish larvae exposed to a human relevant persistent organic pollutant mixture or PFOS. *Environ. Res.* **2020**, *187*, 109702. [CrossRef]
38. Fox, K.; Zauke, G.P.; Butte, W. Kinetics of bioconcentration and clearance of 28 polychlorinated biphenyl congeners in zebrafish (Brachydanio rerio). *Ecotoxicol. Env. Saf.* **1994**, *28*, 99–109. [CrossRef] [PubMed]
39. US-EPA. Emerging Contaminants Perfluorooctane Sulfonate (PFOS) and Perfluorooctanoic Acid (PFOA). In *Emerging Contaminants Fact Sheet; Report 505F14001*; 2014. Available online: https://nepis.epa.gov (accessed on 8 April 2023).
40. De Silva, A.O.; Armitage, J.M.; Bruton, T.A.; Dassuncao, C.; Heiger-Bernays, W.; Hu, X.C.; Karrman, A.; Kelly, B.; Ng, C.; Robuck, A.; et al. PFAS Exposure Pathways for Humans and Wildlife: A Synthesis of Current Knowledge and Key Gaps in Understanding. *Environ. Toxicol. Chem.* **2021**, *40*, 631–657. [CrossRef] [PubMed]
41. Sunderland, E.M.; Hu, X.C.; Dassuncao, C.; Tokranov, A.K.; Wagner, C.C.; Allen, J.G. A review of the pathways of human exposure to poly- and perfluoroalkyl substances (PFASs) and present understanding of health effects. *J. Expo. Sci. Environ. Epidemiol.* **2019**, *29*, 131–147. [CrossRef]
42. Sinclair, G.M.; Long, S.M.; Jones, O.A.H. What are the effects of PFAS exposure at environmentally relevant concentrations? *Chemosphere* **2020**, *258*, 127340. [CrossRef] [PubMed]
43. Kato, K.; Ye, X.; Calafat, A.M. PFASs in the General Population. In *Toxicological Effects of Perfluoroalkyl and Polyfluoroalkyl Substances*; De Witt, J.C., Ed.; Series Molecular and Integrative Toxicology; Nunes, C., Ed. Humana Press: Totowa, NJ, USA, 2015; pp. 51–76. [CrossRef]
44. Wasel, O.; Thompson, K.M.; Freeman, J.L. Assessment of unique behavioral, morphological, and molecular alterations in the comparative developmental toxicity profiles of PFOA, PFHxA, and PFBA using the zebrafish model system. *Environ. Int.* **2022**, *170*, 107642. [CrossRef]

45. Jantzen, C.E.; Annunziato, K.A.; Bugel, S.M.; Cooper, K.R. PFOS, PFNA, and PFOA sub-lethal exposure to embryonic zebrafish have different toxicity profiles in terms of morphometrics, behavior and gene expression. *Aquat. Toxicol.* **2016**, *175*, 160–170. [CrossRef]
46. Usenko, C.Y.; Robinson, E.M.; Usenko, S.; Brooks, B.W.; Bruce, E.D. PBDE developmental effects on embryonic zebrafish. *Environ. Toxicol. Chem.* **2011**, *30*, 1865–1872. [CrossRef]
47. Singleman, C.; Zimmerman, A.; Harrison, E.; Roy, N.K.; Wirgin, I.; Holtzman, N.G. Toxic Effects of Polychlorinated Biphenyl Congeners and Aroclors on Embryonic Growth and Development. *Environ. Toxicol. Chem.* **2021**, *40*, 187–201. [CrossRef]
48. *OECD Test No. 236: Fish Embryo Acute Toxicity (FET) Test*; OECD Publishing: Paris, France, 2013.
49. Abe, F.R.; de Oliveira, A.A.S.; Marino, R.V.; Rialto, T.C.R.; Oliveira, D.P.; Dorta, D.J. A comparison of developmental toxicity of brominated and halogen-free flame retardant on zebrafish. *Ecotoxicol. Environ. Saf.* **2021**, *208*, 111745. [CrossRef]
50. Godfrey, A.; Hooser, B.; Abdelmoneim, A.; Horzmann, K.A.; Freemanc, J.L.; Sepulveda, M.S. Thyroid disrupting effects of halogenated and next generation chemicals on the swim bladder development of zebrafish. *Aquat. Toxicol.* **2017**, *193*, 228–235. [CrossRef]
51. Raldua, D.; Babin, P.J. Simple, rapid zebrafish larva bioassay for assessing the potential of chemical pollutants and drugs to disrupt thyroid gland function. *Environ. Sci. Technol.* **2009**, *43*, 6844–6850. [CrossRef] [PubMed]
52. Loosli, F.; Staub, W.; Finger-Baier, K.C.; Ober, E.A.; Verkade, H.; Wittbrodt, J.; Baier, H. Loss of eyes in zebrafish caused by mutation of chokh/rx3. *EMBO Rep.* **2003**, *4*, 894–899. [CrossRef]
53. Pavlou, S.; Astell, K.; Kasioulis, I.; Gakovic, M.; Baldock, R.; van Heyningen, V.; Coutinho, P. Pleiotropic effects of Sox2 during the development of the zebrafish epithalamus. *PLoS ONE* **2014**, *9*, e87546. [CrossRef]
54. Chassaing, N.; Sorrentino, S.; Davis, E.E.; Martin-Coignard, D.; Iacovelli, A.; Paznekas, W.; Webb, B.D.; Faye-Petersen, O.; Encha-Razavi, F.; Lequeux, L.; et al. OTX2 mutations contribute to the otocephaly-dysgnathia complex. *J. Med. Genet.* **2012**, *49*, 373–379. [CrossRef] [PubMed]
55. Takamiya, M.; Stegmaier, J.; Kobitski, A.Y.; Schott, B.; Weger, B.D.; Margariti, D.; Cereceda Delgado, A.R.; Gourain, V.; Scherr, T.; Yang, L.; et al. Pax6 organizes the anterior eye segment by guiding two distinct neural crest waves. *PLoS Genet.* **2020**, *16*, e1008774. [CrossRef] [PubMed]
56. Li, Z.; Ptak, D.; Zhang, L.; Walls, E.K.; Zhong, W.; Leung, Y.F. Phenylthiourea specifically reduces zebrafish eye size. *PLoS ONE* **2012**, *7*, e40132. [CrossRef]
57. Kim, K.T.; Zaikova, T.; Hutchison, J.E.; Tanguay, R.L. Gold nanoparticles disrupt zebrafish eye development and pigmentation. *Toxicol. Sci.* **2013**, *133*, 275–288. [CrossRef]
58. Barbagallo, S.; Baldauf, C.; Orosco, E.; Roy, N.M. Di-butyl phthalate (DBP) induces defects during embryonic eye development in zebrafish. *Ecotoxicology* **2022**, *31*, 178–185. [CrossRef]
59. Wang, Y.P.; Hong, Q.; Qin, D.N.; Kou, C.Z.; Zhang, C.M.; Guo, M.; Guo, X.R.; Chi, X.; Tong, M.L. Effects of embryonic exposure to polychlorinated biphenyls on zebrafish (Danio rerio) retinal development. *J. Appl. Toxicol.* **2012**, *32*, 186–193. [CrossRef]
60. Li, M.; Yang, T.; Gao, L.; Xu, H. An inadvertent issue of human retina exposure to endocrine disrupting chemicals: A safety assessment. *Chemosphere* **2021**, *264*, 128484. [CrossRef] [PubMed]
61. Seipold, S.; Priller, F.C.; Goldsmith, P.; Harris, W.A.; Baier, H.; Abdelilah-Seyfried, S. Non-SMC condensin I complex proteins control chromosome segregation and survival of proliferating cells in the zebrafish neural retina. *BMC Dev. Biol.* **2009**, *9*, 40. [CrossRef] [PubMed]
62. Hagenaars, A.; Vergauwen, L.; De Coen, W.; Knapen, D. Structure-activity relationship assessment of four perfluorinated chemicals using a prolonged zebrafish early life stage test. *Chemosphere* **2011**, *82*, 764–772. [CrossRef]
63. Gong, H.; Du, J.; Xu, J.; Yang, Y.; Lu, H.; Xiao, H. Perfluorononanoate and Perfluorobutane Sulfonate Induce Cardiotoxic Effects in Zebrafish. *Environ. Toxicol. Chem.* **2022**, *41*, 2527–2536. [CrossRef] [PubMed]
64. Singleman, C.; Holtzman, N.G. PCB and TCDD derived embryonic cardiac defects result from a novel AhR pathway. *Aquat. Toxicol.* **2021**, *233*, 105794. [CrossRef]
65. Everett, C.J.; Mainous, A.G., 3rd; Frithsen, I.L.; Player, M.S.; Matheson, E.M. Association of polychlorinated biphenyls with hypertension in the 1999-2002 National Health and Nutrition Examination Survey. *Environ. Res.* **2008**, *108*, 94–97. [CrossRef] [PubMed]
66. Wu, M.; Zuo, Z.; Li, B.; Huang, L.; Chen, M.; Wang, C. Effects of low-level hexabromocyclododecane (HBCD) exposure on cardiac development in zebrafish embryos. *Ecotoxicology* **2013**, *22*, 1200–1207. [CrossRef]
67. El-Nahhal, Y.; El-Nahhal, I. Cardiotoxicity of some pesticides and their amelioration. *Environ. Sci. Pollut. Res. Int.* **2021**, *28*, 44726–44754. [CrossRef]
68. Loerracher, A.K.; Braunbeck, T. Cytochrome P450-dependent biotransformation capacities in embryonic, juvenile and adult stages of zebrafish (Danio rerio)-a state-of-the-art review. *Arch. Toxicol.* **2021**, *95*, 2299–2334. [CrossRef]
69. Schmidt, K.; Steinberg, C.E.W.; Pflugmacher, S.; Staaks, G.B.O. Xenobiotic substances such as PCB mixtures (Aroclor 1254) and TBT can influence swimming behavior and biotransformation activity (GST) of carp (Cyprinus carpio). *Environ. Toxicol.* **2004**, *19*, 460–470. [CrossRef]
70. Schmidt, K.; Staaks, G.B.O.; Pflugmacher, S.; Steinberg, C.E.W. Impact of PCB mixture (Aroclor 1254) and TBT and a mixture of both on swimming behavior, body growth and enzymatic biotransformation activities (GST) of young carp (Cyprinus carpio). *Aquat. Toxicol.* **2005**, *71*, 49–59. [CrossRef] [PubMed]

71. Chou, C.-T.; Hsiao, Y.-C.; Ko, F.-C.; Cheng, J.-O.; Cheng, Y.-M.; Chen, T.-H. Chronic exposure of 2,2′,4,4′-tetrabromodiphenyl ether (PBDE-47) alters locomotion behavior in juvenile zebrafish (Danio rerio). *Aquat. Toxicol.* **2010**, *98*, 388–395. [CrossRef]
72. Haimbaugh, A.; Wu, C.C.; Akemann, C.; Meyer, D.N.; Connell, M.; Abdi, M.; Khalaf, A.; Johnson, D.; Baker, T.R. Multi- and Transgenerational Effects of Developmental Exposure to Environmental Levels of PFAS and PFAS Mixture in Zebrafish (Danio rerio). *Toxics* **2022**, *10*, 334. [CrossRef] [PubMed]
73. Fraser, T.W.K.; Khezri, A.; Lewandowska-Sabat, A.M.; Henry, T.; Ropstad, E. Endocrine disruptors affect larval zebrafish behavior: Testing potential mechanisms and comparisons of behavioral sensitivity to alternative biomarkers. *Aquat. Toxicol.* **2017**, *193*, 128–135. [CrossRef] [PubMed]
74. Yaghoobi, B.; Miller, G.W.; Holland, E.B.; Li, X.; Harvey, D.; Li, S.; Lehmler, H.J.; Pessah, I.N.; Lein, P.J. Ryanodine receptor-active non-dioxin-like polychlorinated biphenyls cause neurobehavioral deficits in larval zebrafish. *Front. Toxicol.* **2022**, *4*, 947795. [CrossRef] [PubMed]
75. Lee, S.; Ko, E.; Lee, H.; Kim, K.T.; Choi, M.; Shin, S. Mixed Exposure of Persistent Organic Pollutants Alters Oxidative Stress Markers and Mitochondrial Function in the Tail of Zebrafish Depending on Sex. *Int. J. Environ. Res. Public Health* **2021**, *18*, 9539. [CrossRef]
76. Ko, E.; Kim, D.; Kim, K.; Choi, M.; Shin, S. The action of low doses of persistent organic pollutants (POPs) on mitochondrial function in zebrafish eyes and comparison with hyperglycemia to identify a link between POPs and diabetes. *Toxicol. Mech. Methods* **2020**, *30*, 275–283. [CrossRef]
77. Jochum, W.; Passegue, E.; Wagner, E.F. AP-1 in mouse development and tumorigenesis. *Oncogene* **2001**, *20*, 2401–2412. [CrossRef]
78. Knuth, M.M.; Mahapatra, D.; Jima, D.; Wan, D.; Hammock, B.D.; Law, M.; Kullman, S.W. Vitamin D deficiency serves as a precursor to stunted growth and central adiposity in zebrafish. *Sci. Rep.* **2020**, *10*, 16032. [CrossRef]
79. Aceto, J.; Nourizadeh-Lillabadi, R.; Maree, R.; Dardenne, N.; Jeanray, N.; Wehenkel, L.; Alestrom, P.; van Loon, J.J.; Muller, M. Zebrafish bone and general physiology are differently affected by hormones or changes in gravity. *PLoS ONE* **2015**, *10*, e0126928. [CrossRef]
80. Su, T.; Ding, X. Regulation of the cytochrome P450 2A genes. *Toxicol. Appl. Pharmacol.* **2004**, *199*, 285–294. [CrossRef] [PubMed]
81. Goldstone, J.V.; McArthur, A.G.; Kubota, A.; Zanette, J.; Parente, T.; Jonsson, M.E.; Nelson, D.R.; Stegeman, J.J. Identification and developmental expression of the full complement of Cytochrome P450 genes in Zebrafish. *BMC Genom.* **2010**, *11*, 643. [CrossRef] [PubMed]
82. Kubota, A.; Bainy, A.C.; Woodin, B.R.; Goldstone, J.V.; Stegeman, J.J. The cytochrome P450 2AA gene cluster in zebrafish (Danio rerio): Expression of CYP2AA1 and CYP2AA2 and response to phenobarbital-type inducers. *Toxicol. Appl. Pharm.* **2013**, *272*, 172–179. [CrossRef] [PubMed]
83. Hartig, E.I.; Zhu, S.; King, B.L.; Coffman, J.A. Cortisol-treated zebrafish embryos develop into pro-inflammatory adults with aberrant immune gene regulation. *Biol. Open* **2016**, *5*, 1134–1141. [CrossRef]
84. Delcourt, N.; Quevedo, C.; Nonne, C.; Fons, P.; O'Brien, D.; Loyaux, D.; Diez, M.; Autelitano, F.; Guillemot, J.C.; Ferrara, P.; et al. Targeted identification of sialoglycoproteins in hypoxic endothelial cells and validation in zebrafish reveal roles for proteins in angiogenesis. *J. Biol. Chem.* **2015**, *290*, 3405–3417. [CrossRef]
85. Hao, R.; Bondesson, M.; Singh, A.V.; Riu, A.; McCollum, C.W.; Knudsen, T.B.; Gorelick, D.A.; Gustafsson, J.A. Identification of estrogen target genes during zebrafish embryonic development through transcriptomic analysis. *PLoS ONE* **2013**, *8*, e79020. [CrossRef]
86. Haggard, D.E.; Noyes, P.D.; Waters, K.M.; Tanguay, R.L. Transcriptomic and phenotypic profiling in developing zebrafish exposed to thyroid hormone receptor agonists. *Reprod. Toxicol.* **2018**, *77*, 80–93. [CrossRef]
87. European Chemical Agency. Per- and Polyfluoroalkyl Substances (PFASs). 2023. Available online: https://echa.europa.eu/hot-topics/perfluoroalkyl-chemicals-pfas (accessed on 3 March 2023).
88. US-EPA. Per- and Polyfluoroalkyl Substances (PFAS). 2023. Available online: https://www.epa.gov/pfas (accessed on 6 March 2023).
89. Focus features. Dark Waters, YouTube 2019. Available online: https://www.youtube.com/watch?v=RvAOuhyunhY (accessed on 6 March 2023).
90. Rich, N. The Lawyer Who Became DuPont's Worst Nightmare. *The New York Times*; 2016. Available online: https://www.nytimes.com/2016/2001/2010/magazine/the-lawyer-who-became-duponts-worst-nightmare.html (accessed on 16 March 2023).
91. Salvidge, R.; Hosea, L. What Are PFAS, How Toxic Are They and How Do You Become Exposed? Everything You Need to Know about 'Forever Chemicals' Detected in Air, Water, Soils, Sediments and Rain. *The Guardian*; 2023. Available online: https://www.theguardian.com/environment/2023/feb/2023/what-are-pfas-forever-chemicals-how-toxic-are-they-and-how-do-you-become-exposed (accessed on 16 March 2023).
92. Edgar, R.; Domrachev, M.; Lash, A.E. Gene Expression Omnibus: NCBI gene expression and hybridization array data repository. *Nucleic Acids Res.* **2002**, *30*, 207–210. [CrossRef]

International Journal of *Molecular Sciences*

MDPI

Article

Protective Effects of Selenium Nanoparticles against Bisphenol A-Induced Toxicity in Porcine Intestinal Epithelial Cells

Zaozao Pan [1,†], Jiaqiang Huang [2,†], Ting Hu [1], Yonghong Zhang [1], Lingyu Zhang [1], Jiaxi Zhang [1], Defeng Cui [1], Lu Li [3], Jing Wang [3,*] and Qiong Wu [1,*]

1 Animal Science and Technology College, Beijing University of Agriculture, Beijing 100096, China
2 Beijing Advanced Innovation Center for Food Nutrition and Human Health, Department of Nutrition and Health, China Agricultural University, Beijing 100193, China
3 Institute of Animal Husbandry and Veterinary Medicine, Beijing Academy of Agriculture and Forestry Sciences, Beijing 100097, China
* Correspondence: wangjing76@baafs.net.cn (J.W.); 20188602@bua.edu.cn (Q.W.)
† These authors contributed equally to this work.

Abstract: Bisphenol A (BPA) is widely used to harden plastics and polycarbonates and causes serious toxic effects in multiple organs, including the intestines. Selenium, as an essential nutrient element for humans and animals, exhibits a predominant effect in various physiological processes. Selenium nanoparticles have attracted more and more attention due to their outstanding biological activity and biosafety. We prepared chitosan-coated selenium nanoparticles (SeNPs) and further compared the protective effects, and investigated the underlying mechanism of SeNPs and inorganic selenium (Na_2SeO_3) on BPA-induced toxicity in porcine intestinal epithelial cells (IPEC-J2). The particle size, zeta potential, and microstructure of SeNPs were detected by using a nano-selenium particle size meter and a transmission electron microscope. IPEC-J2 cells were exposed to BPA alone or simultaneously exposed to BPA and SeNPs or Na_2SeO_3. The CCK8 assay was performed to screen the optimal concentration of BPA exposure and the optimal concentration of SeNPs and Na_2SeO_3 treatment. The apoptosis rate was detected by flow cytometry. Real-time PCR and Western blot methods were used to analyze the mRNA and protein expression of factors related to tight junctions, apoptosis, inflammatory responses and endoplasmic reticulum stress. Increased death and morphological damage were observed after BPA exposure, and these increases were attenuated by SeNPs and Na_2SeO_3 treatment. BPA exposure disturbed the tight junction function involved with decreased expression of tight junction protein Zonula occludens 1 (ZO-1), occludin, and claudin-1 proteins. Proinflammatory response mediated by the transcription factor nuclear factor-k-gene binding (NF-κB), such as elevated levels of *interleukin-1β(IL-1β)*, *interleukin-6 (IL-6)*, *interferon-γ (IFN-γ)*, *interleukin-17 (IL-17)*, and tumor necrosis *factor-α* (*TNF-α*) expression was induced at 6 and 24 h after BPA exposure. BPA exposure also disturbed the oxidant/antioxidant status and led to oxidative stress. IPEC-J2 cell apoptosis was induced by BPA exposure, as indicated by increased BCL-2-associated X protein (Bax), caspase 3, caspase 8, and caspase 9 expression and decreased B-cell lymphoma-2 (Bcl-2) and Bcl-xl expression. BPA exposure activated the endoplasmic reticulum stress (ERS) mediated by the receptor protein kinase receptor-like endoplasmic reticulum kinase (PERK), Inositol requiring enzyme 1 (IRE1α), and activating transcription factor 6 (ATF6). We found that treatment with SeNPs and Na_2SeO_3 can alleviate the intestinal damage caused by BPA. SeNPs were superior to Na_2SeO_3 and counteracted BPA-induced tight junction function injury, proinflammatory response, oxidative stress, apoptosis, and ERS stress. Our findings suggest that SeNPs protect intestinal epithelial cells from BPA-induced damage, partly through inhibiting ER stress activation and subsequently attenuating proinflammatory responses and oxidative stress and suppressing apoptosis, thus enhancing the intestinal epithelial barrier function. Our data indicate that selenium nanoparticles may represent an effective and reliable tool for preventing BPA toxicity in animals and humans.

Citation: Pan, Z.; Huang, J.; Hu, T.; Zhang, Y.; Zhang, L.; Zhang, J.; Cui, D.; Li, L.; Wang, J.; Wu, Q. Protective Effects of Selenium Nanoparticles against Bisphenol A-Induced Toxicity in Porcine Intestinal Epithelial Cells. *Int. J. Mol. Sci.* **2023**, *24*, 7242. https://doi.org/10.3390/ijms24087242

Academic Editors: Esref Demir and Sam Kacew

Received: 23 February 2023
Revised: 5 April 2023
Accepted: 9 April 2023
Published: 14 April 2023

Keywords: Bisphenol A; selenium nanoparticles; tight junction; oxidative stress; apoptosis; endoplasmic reticulum stress

1. Introduction

The intestinal tract is the organ with the largest surface area in direct contact with the external environment of the animal body, which has critical biological functions such as digestion, absorption, metabolism, and immunity [1]. The intestinal barrier mainly consists of four aspects: intestinal commensal microorganisms, chemical barrier, physical barrier, and immune barrier [2]. Among these, intestinal barrier integrity largely relies on the intestinal epithelial cells (IECs). The IEC is the first cell boundary between the luminal environment and the body against outside hostile stimuli. The choice of the IEC's unique permeability can guarantee the body's absorption of nutrients and effectively suppress pathogenic microorganisms and harmful substances through the barrier to enter the blood to maintain the body's normal physiological function [3]. The IEC barrier's permeability is regulated by the junction complex formed between adjacent intestinal epithelial cells [4]. The intercellular junction complex is mainly composed of tight junctions, which mainly include claudins, occludin, and ZO families [5]. Maintaining the normal expression and distribution of tight junction proteins is essential for intestinal barrier function.

BPA is one of the most widely used industrial compounds in the world, mainly in the production of plastic containers, toys, tableware, medical devices, and polycarbonate bottles. As an endocrine-disrupting chemical, BPA has estrogen-like and anti-androgen properties, causing significant damage to human tissues and organs, such as the reproductive system, immune system, and neuroendocrine system [6]. Humans and animals are exposed to BPA mainly through dietary, transdermal, and inhalation, in which dietary is considered the main route [7]. Studies have demonstrated that after BPA enters the human body, it first destroys the intestinal epithelial barrier functions, intestinal immune systems, and reproductive systems, and thus leads to the occurrence of various metabolic diseases [8]. Mice dietary intake of BPA first destroys the morphological structure of intestinal epithelial cells by inhibiting the expression of tight junction proteins and increasing intestinal permeability [9]. Similarly, the oral administration of BPA can affect the intestinal barrier functions and then accumulate in the intestine, liver, and gonads of pigs, leading to reproductive toxicity [10]. Further studies also indicated that dietary exposure to BPA destroys the gastrointestinal mucosal layer of human colon cancer cells (HCT116), leading to increased apoptosis and slower progression associated with intestinal epithelial cell proliferation [11]. BPA treatment increases intestinal permeability and disrupts the barrier function by increasing the chemical marker content and tight junction expression in intestinal tissues and blood circulation [12]. It remains unclear how BPA damages the intestinal epithelial barrier functions.

The endoplasmic reticulum (ER) is composed of branched tubules and flat reticular capsules that are present in all eukaryotic cells and extend into the cytoplasm to surround the nuclear membrane. The endoplasmic reticulum can be divided into two types: the rough endoplasmic reticulum full of ribosomes and the smooth endoplasmic reticulum rich in lipid synthetase. The rough endoplasmic reticulum is mainly involved in the biosynthesis, folding, processing and modification of soluble and membrane proteins. The rough endoplasmic reticulum is of paramount importance for adaptive responses to biotic stresses due to increased unfolded or misfolded proteins. ERS is well known as the accumulation of unfolded proteins in ER lumen caused by the increased protein synthesis or multiple conditions, and ER responds to ER stress by activating the unfolded protein response (UPR), an adaptive intracellular signal to cope with ER stress and help to sustain cell survival and normal functions [13]. In response to ER stress, ER-resident chaperone binding immunoglobulin protein (Bip) dissociates from the luminal domains of the three protein sensors, leading to the activation of proteins PERK, IRE1, and ATF6 to restore intracellular homeostasis [14,15]. Notably, ER stress in intestinal epithelium evokes a range of adverse cellular responses, such as oxidative stress, inflammatory responses, and apoptosis, which are all involved in the pathogenesis of intestinal barrier dysfunction [16–18]. Dietary intake of BPA results in an abnormal increase in the amount of unfolded and misfolded proteins in the ER and then induces ER stress [19]. Studies have shown that BPA can reveal persistent ER stress by activating ATF6 and IRE1 pathways in mammals [20]. Moreover, BPA activates ER stress and stimulates oxidative stress, thereby promoting cell apoptosis and damaging intestinal barrier functions [21].

As an essential nutrient element for humans and animals, selenium has important biological functions for human health, such as anti-cancer, anti-inflammatory, and anti-oxidation. In recent years, selenium has been applied in the form of inorganic selenium, organic selenium, and nano-selenium. Studies demonstrated that the use of inorganic sodium selenite ameliorates BPA toxicity in the liver, testis, and lungs of mice [22]. However, because inorganic selenium is easy to combine with vitamins and has poor stability, low bioavailability, and high biological toxicity, it is difficult to control the safe use dose in the actual use process [23]. In recent years, selenium nanoparticles, as a new type of elemental selenium, have attracted much attention. Compared with inorganic selenium and organic selenium, selenium nanoparticles have higher bioavailability, stronger biological activity, lower toxicity, and better compatibility as therapeutic drug carriers, and are easy to synthesize and store [24,25]. Previous studies have shown that the intestinal villus circumference and height of fish fed with a nano-selenium diet were greater than those of fish fed with sodium selenite, suggesting that nano-selenium can be more effective in maintaining intestinal integrity [26]. Recent studies have shown that, compared with inorganic selenium and organic selenium, selenium nanoparticles have obvious biological beneficial effects on the alleviation of cadmium-induced inflammation via NF-κB/IκB pathway in the heart [27]. Nano-selenium is superior to inorganic selenium and organic selenium in attenuating the cardiotoxic effects of cadmium by activating the aryl hydrocarbon receptor pathway [28]. Studies also have shown that selenium nanoparticles attenuate BPA-induced testicular toxicity by inhibiting oxidative stress in male rats [29]. The molecular mechanism underlying the protective effects of selenium nanoparticles against BPA-induced toxicity needs to be further explored.

In the present study, we established the model of BPA-stimulated porcine IPEC-J2 cells and further analyzed the effects of prepared SeNPs on ER stress and the downstream apoptosis, oxidative stress, and inflammatory pathways.

2. Results

2.1. Preparation and Characterization of SeNPs

The size distribution, zeta potential, microstructure, and stability of particles of SeNPs are presented in Figure 1. SeNPs were prepared by in-situ reduction of selenite and vitamin C using chitosan as the carrier material, and the concentration of SeNPs was 76.424 mg/g. The characteristic peak of the absorption maximum for SeNPs was 149.7 nm, indicating the small particle size of SeNPs (Figure 1A). The zeta potential of SeNPs ranged from −45.8 to 31.4 mV, and a signature peak appeared at 27.7 mV (Figure 1B). The TEM images showed that the SeNPs were distributed as well-dispersed spherical particles (Figure 1C). The presence of pepsin and trypsin in the gastroenteric fluid in the digestive system affects the structure and particle size of SeNPs. The stability of SeNPs was measured after gastric and intestinal digestion. The absorption values of SeNPs did not change after 1–3 h of gastric digestion and 1–2 h of intestinal digestion but increased after 2–3 h of gastric digestion and 3 h of intestinal digestion (Figure 1D).

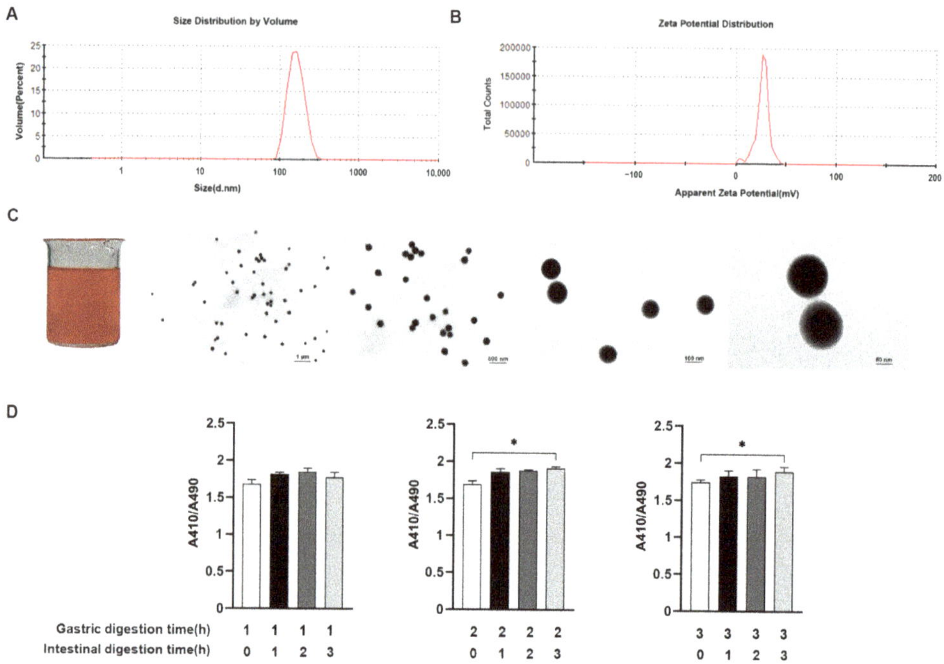

Figure 1. Determination of physicochemical properties and stability of SeNPs. The size distribution (**A**) and zeta potential (**B**) of prepared SeNPs were measured. Schematic diagram of macroscopic and microscopic morphology of SeNPs (**C**). Changes in particle size of SeNPs were measured after digestion in vitro (**D**). Data are presented as the means ± SEM of three independent experiments. * $p < 0.05$.

2.2. Establishment of the Model of BPA-Exposed and Na_2SeO_3 or SeNPs-Treated IPECs

The viability and morphological characteristics of IPECs were measured and observed to determine the optimal exposure dose of BPA and treatment dose of Na_2SeO_3 or SeNPs. Cell viability decreased when exposed to the increased BPA concentration, and cell viability decreased to about 60% at a BPA concentration of 50 μM, whether 6 or 24 h after BPA exposure (Figure 2A). Thus, we choose 50 μM as the BPA exposure concentration to allow for cell damage without disruption of the cell monolayer in subsequent experiments. After adding 5–50 μg/mL of SeNPs or Na_2SeO_3, a low dose of SeNPs or Na_2SeO_3 decreased the cell death induced by 50 μM of BPA, but a high dose of SeNPs or Na_2SeO_3 led to cell death. SeNPs or Na_2SeO_3 at the concentration of 15 μg/mL exhibited the best inhibitory effect on BPA-induced cell damage (Figure 2B). Under the optical microscope, at 24 h, untreated control cells appeared normal, and the cell monolayer was intact. At 24 h after BPA exposure, cells were broken, and the boundary was blurred. However, the addition of 15 μg/mL SeNPs or Na_2SeO_3 attenuated the degree of disruption of the cell monolayer induced by BPA. Compared with cells in the Na_2SeO_3 + BPA group, the number of vacuolar dead cells was less in the SeNPs + BPA group (Figure 2C).

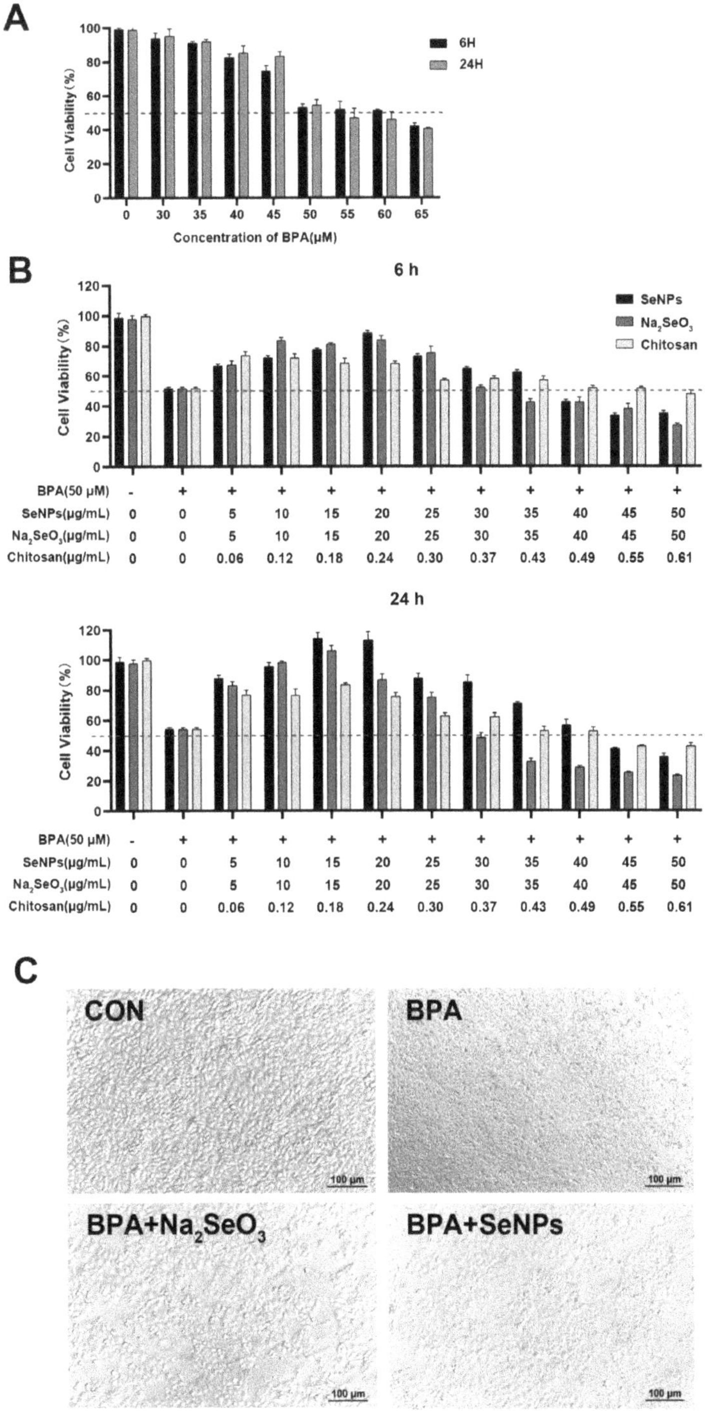

Figure 2. Effects of Na_2SeO_3 and SeNPs on viability and morphology of IPEC-J2 after BPA exposure. The dashed line represented the standard line of 50% cell viability. The optimal concentration of BPA

was determined by measuring the cell viability after BPA exposure (**A**). The optimal concentration of Na_2SeO_3 or SeNPs was determined by measuring the cell viability when cells were treated with Na_2SeO_3 or SeNPs, followed by BPA exposure (**B**). The morphological changes of IPEC-J2 were observed when cells were treated with 15 μg/mL Na_2SeO_3 or SeNPs followed by 50 μM of BPA exposure for 24 h (**C**). Data are presented as the means ± SEM of three independent experiments.

2.3. SeNPs Maintained the TJ Expression in BPA-Exposed IPECs

To investigate the effect of SeNPs on intestinal integrity in BPA-exposed cells, the expression of tight junction proteins was determined. Compared with untreated control IPECs, the expression of ZO-1 and occludin proteins was decreased at 6 and 24 h after BPA exposure in cells only exposed to BPA (Figure 3A,B). Compared with cells only exposed to BPA, the expression of ZO-1 and occludin proteins was decreased at 6 h in the Na_2SeO_3 + BPA group but increased in the SeNPs + BPA group. Compared with untreated control IPECs, BPA exposure led to the decreased expression of ZO-1 and occludin proteins at 24 h, whereas Na_2SeO_3 or SeNPs treatment enhanced the ZO-1 and occludin proteins expression. BPA exposure led to decreased expression of claudin-1 protein at 6 h, regardless of with or without SeNPs or Na_2SeO_3 treatment. Compared with untreated control IPECs, BPA exposure led to the decreased expression of claudin-1 protein at 24 h, whereas Na_2SeO_3 or SeNPs treatment enhanced the claudin-1 protein expression.

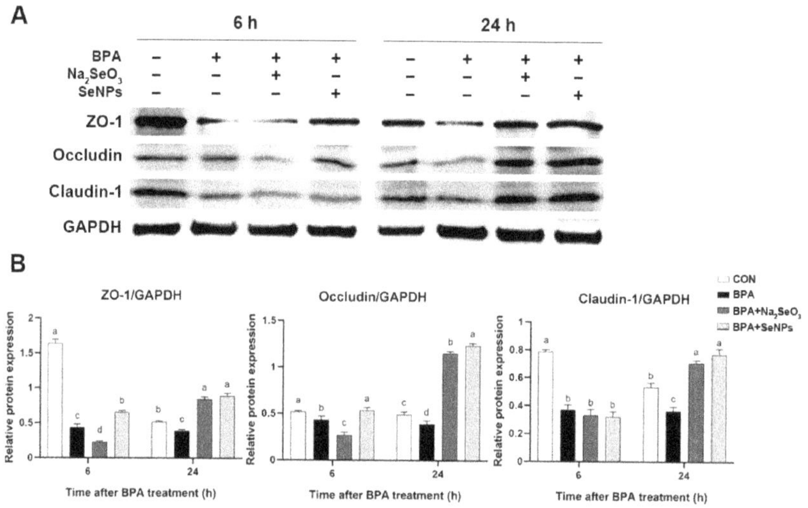

Figure 3. The effects of Na_2SeO_3 and SeNPs treatment on the TJ expression of IPEC-J2 cells after BPA exposure. Representative panels showing expression of ZO-1, Occludin, and Claudin-1 proteins by Na_2SeO_3 or SeNPs treatment followed by BPA exposure at 6 and 24 h (**A**). Expression of GAPDH was measured as an internal control. Results are presented as the ratio of ZO-1, Occludin, and Claudin-1 band intensity to that of GAPDH (**B**). Data are presented as the means ± SEM of three independent experiments. [a,b,c,d] Mean values within a row with different superscript letters were significantly different ($p < 0.05$).

2.4. SeNPs Attenuated the BPA-Induced IPEC Apoptosis

According to the flow cytometry results, IPECs exposed to BPA alone had a higher percentage of apoptosis at 24 h compared with untreated control cells (Figure 4A). Treatment with SeNPs or Na_2SeO_3 resulted in a decrease in the percentage of apoptosis during BPA exposure. Cells in the SeNPs + BPA group had a lower percentage of apoptosis than cells in the Na_2SeO_3 + BPA group. Western blot results showed that BPA exposure increased the expression of pro-apoptotic protein caspase-3 and Bax at 6 and 24 h compared with untreated control cells, whereas treatment with SeNPs or Na_2SeO_3 attenuated this increase (Figure 4B,C). The BPA-induced down-regulation of the expression of anti-apoptotic protein Bcl-2 and Bcl-xl was observed, and this down-regulation was attenuated by SeNPs treatment at 6 and 24 h but not by Na_2SeO_3 treatment at 6 h. Consistently, treatment with SeNPs and Na_2SeO_3 reversed the increase in the mRNA expression of *caspase-8*, *caspase-9*, and *Bax* and the decrease in the mRNA expression of *Bcl-2* induced by BPA exposure (Figure 4D).

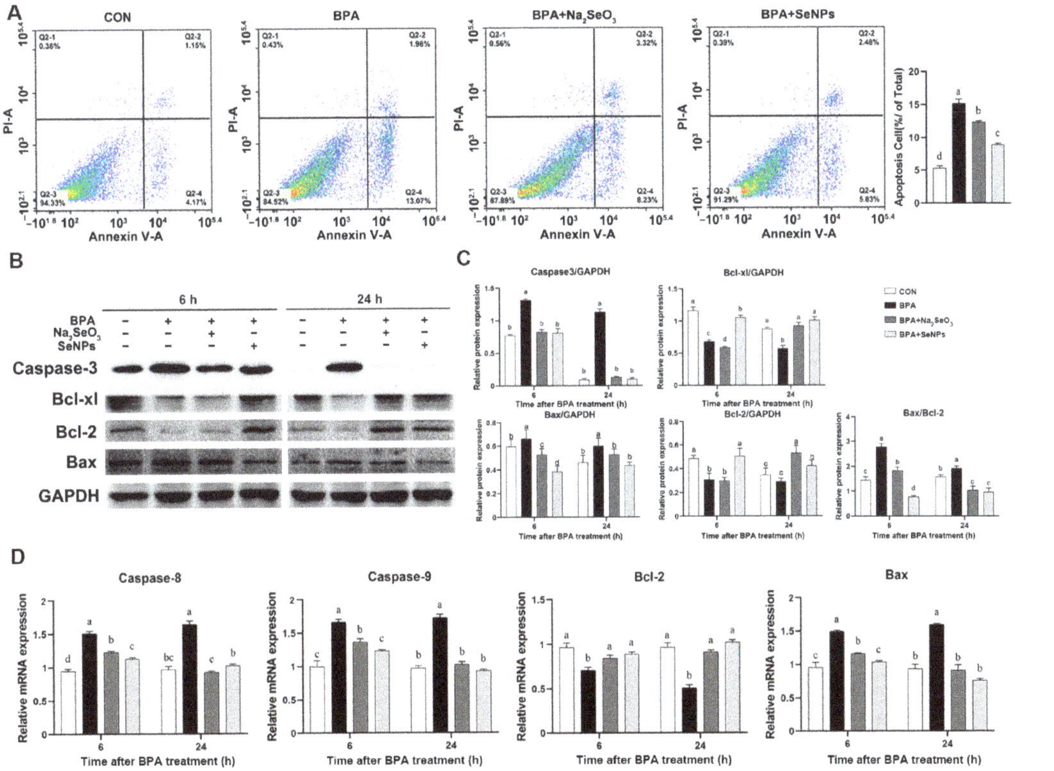

Figure 4. Effects of Na_2SeO_3 and SeNPs treatment on apoptosis of IPEC-J2 cells induced by BPA. Cell apoptosis was determined by flow cytometric analysis and analyzed (**A**). Representative panels showing expression of caspase 3, Bcl-xl, Bax, and Bcl-2 proteins by Western blotting at 6 and 24 h after BPA challenge (**B**). Expression of GAPDH was measured as an internal control. Results are presented as the ratio of caspase 3, Bcl-xl, Bax, and Bcl-2 protein band intensities to that of GAPDH, and the cellular protein Bcl-2 was used as an internal control for Bax (**C**). The relative expression of mRNAs for the *caspase-8*, *caspase-9*, *Bcl-2*, and *Bax* genes was analyzed by quantitative real-time PCR (**D**). Data are presented as the means ± SEM of three independent experiments. ^{a,b,c,d} Mean values within a row with different superscript letters were significantly different ($p < 0.05$).

2.5. The Effects of Na_2SeO_3 and SeNPs on Inflammatory Pathways in BPA-exposed IPECs

The Western blot results showed that cells exposed to BPA alone had a higher expression of p-NF-κB and p-IκB proteins compared with the untreated control cells at 6 and 24 h, but this increase was inhibited by SeNPs or Na_2SeO_3 treatment (Figure 5A,B). SeNPs exhibited a better inhibitory effect on the BPA-induced increase in the p-NF-κB expression at 24 h and in the p-IκB expression at 6 and 24 h. Compared with the untreated control group, BPA exposure increased the mRNA expression of pro-inflammatory cytokine *IL-1β*,

IL-6, *IFN-γ*, *IL-17*, and *TNF-α* at 6 and 24 h, but this increase was attenuated by SeNPs or Na_2SeO_3 treatment (Figure 5C). Compared with cells in the Na_2SeO_3 + BPA group, cells in the SeNPs + BPA group had a lower expression of *IL-1β*, *IL-6*, and *IL-17* during BPA exposure. Compared with the untreated control group, BPA exposure led to decreased mRNA expression of *IL-10*, but this decrease was reversed by SeNPs or Na_2SeO_3 treatment.

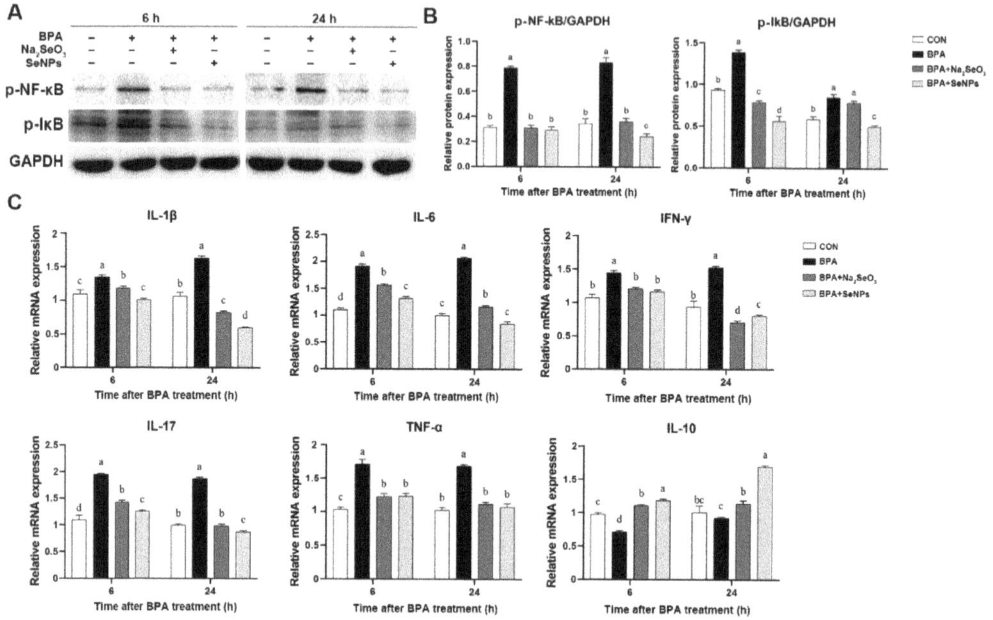

Figure 5. Effect of Na_2SeO_3 and SeNPs treatment on the inflammatory response of IPEC-J2 cells after BPA exposure. Representative panels showing expression of p-NF-κB and p-IκB by Western blotting at 6 and 24 h after BPA exposure (**A**). Expression of GAPDH was measured as an internal control. Results are presented as the ratio of p-NF-κB and p-IκB band intensity to that of GAPDH (**B**). The relative expression of mRNAs for the *IL-1β*, *IL-6*, *IFN-γ*, *IL-17*, *TNF-α*, and *IL-10* genes was analyzed by quantitative real-time PCR (**C**). Data are presented as the means ± SEM of three independent experiments. ^{a,b,c,d} Mean values within a row with different superscript letters were significantly different ($p < 0.05$).

2.6. SeNPs Maintained the Antioxidant Capacity of IPEC-J2 Cells during BPA Exposure

The antioxidant capacity of IPEC-J2 cells was assessed by measuring the abundance of malondialdehyde (MDA), total antioxidant capacity (T-AOC), superoxide dismutase (SOD), catalase (CAT), and glutathione peroxidase (GSH-Px). Compared with the untreated control group, BPA exposure increased the abundance of MDA at 6 and 24 h, and this increase was inhibited by SeNPs or Na_2SeO_3 treatment (Figure 6A). Compared with the untreated control group, BPA resulted in decreased abundance of T-AOC, SOD, CAT, and GSH-Px at 6 and 24 h (Figure 6B–E). SeNPs or Na_2SeO_3 treatment inhibited even reversed the BPA-induced decrease in the abundance of T-AOC, SOD, CAT, and GSH-Px at 24 h compared with BPA alone group. SeNPs, but not Na_2SeO_3 treatment, reversed the decreased abundance of SOD induced by BPA at 6 h.

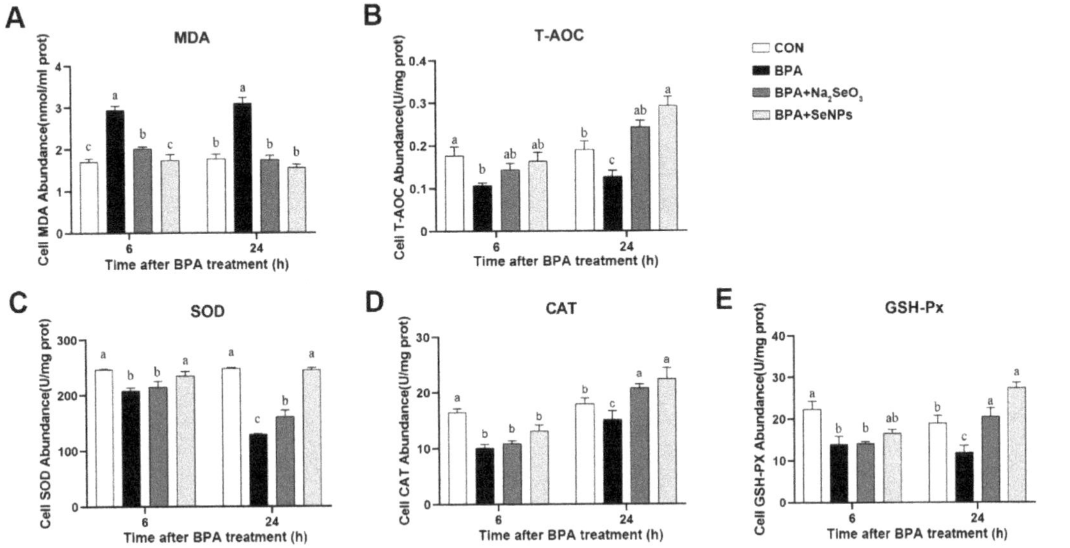

Figure 6. Effects of Na_2SeO_3 and SeNPs treatment on oxidative stress of IPEC-J2 cells after BPA exposure. Oxidative stress was analyzed by measuring the level of MDA (**A**), T-AOC (**B**), SOD (**C**), CAT (**D**), and GSH-Px (**E**). Data are presented as the means ± SEM of three independent experiments. [a,b,c,] Mean values within a row with different superscript letters were significantly different ($p < 0.05$).

2.7. SeNPs Ameliorated the BPA-Induced ERS in IPEC-J2 Cells

Compared with untreated control cells, BPA exposure led to increased expression of ERS marker protein Bip at 24 h, whereas this increase was attenuated by SeNPs or Na_2SeO_3 treatment (Figure 7A,B). For the PERK/ATF4 pathway, BPA exposure increased the expression of p-PERK, translational initiation factor 2α (eIf2α), and ATF4 proteins at 6 and 24 h, and this increase was inhibited by SeNPs or Na_2SeO_3 treatment. Compared with Na_2SeO_3 treatment, cells treated with SeNPs showed lower expression of p-PERK and eIf2α at 6 h and lower expression of ATF4 at 24 h (Figure 7A,B). For the IRE1 and ATF6 pathways, BPA exposure increased the expression of IRE1 and ATF6 compared with the untreated control group, and this increase was inhibited by SeNPs or Na_2SeO_3 treatment. SeNPs treatment exhibited a better inhibitory effect on the increased IRE1 protein expression induced by BPA at 6 h than Na_2SeO_3 treatment (Figure 7A,B). Compared with untreated control cells, BPA exposure led to increased expression of C/EBP homologous protein (CHOP) at 6 h, whereas this increase was attenuated by SeNPs or Na_2SeO_3 treatment.

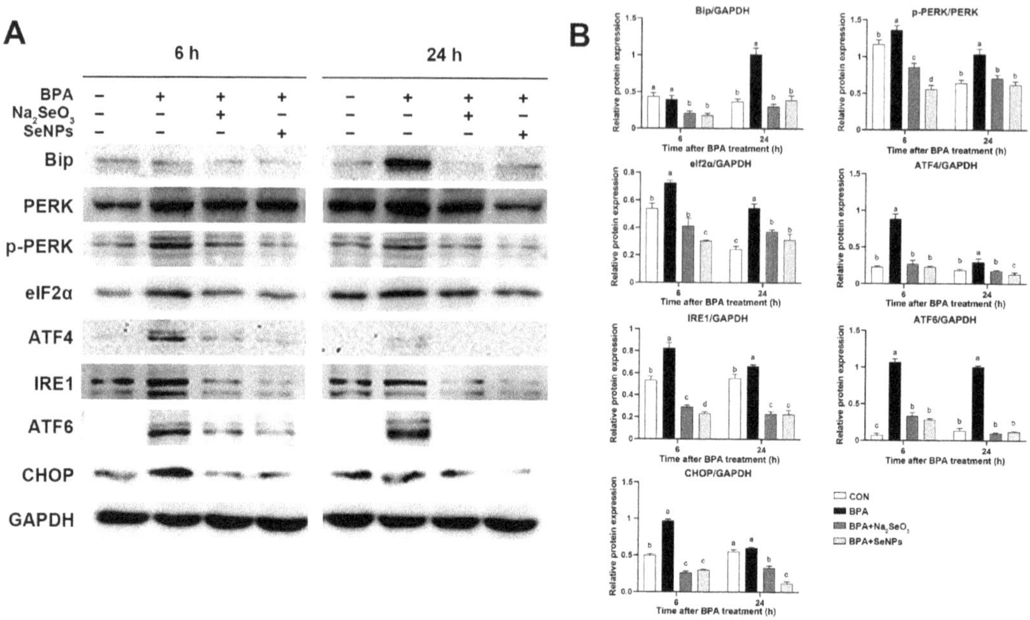

Figure 7. Effects of Na_2SeO_3 and SeNPs treatment on ER stress of IPEC-J2 cells after BPA exposure. Representative panels showing expression of Bip, PERK, p-PERK, ATF4, ATF6, eIf2α, p-IRE1, and CHOP proteins by Western blotting (**A**). Expression of GAPDH was measured as an internal control. The cellular protein PERK was used as an internal control for p-PERK, and GAPDH was used as an internal control for other proteins (**B**). Data are presented as the means ± SEM of three independent experiments. [a,b,c,d] Mean values within a row with different superscript letters were significantly different ($p < 0.05$).

3. Discussion

As a common environmental contaminant, BPA has been shown to cause potential damage to many tissues and organs (lung, liver, kidney, skin, and mucous membrane) of humans and animals [30,31]. The intestine is especially vulnerable to the adverse effects of BPA. Previous studies showed that selenium and its nano form have powerful protective effects against oxidative stress, DNA damage, and apoptosis in response to BPA cisplatin and ionizing radiation exposure in vivo [32,33]. In this study, we employed the IPEC-J2 cell line, widely used in mimicking intestinal epithelial cells in in vitro studies, to evaluate the adverse effects of BPA exposure on the intestinal epithelial barrier function and further revealed the protective effects and mechanism of SeNPs. Our results showed that SeNPs alleviated the development of BPA-induced toxicity and maintained the intestinal barrier functions by inhibiting ER stress and downstream intestinal epithelial cell apoptosis, inflammation, and oxidative stress.

The intestinal epithelial barrier acts as the first line of defense and plays a vital role in nutrition and immunoregulation. A layer of epithelial cells bounds together via intercellular junction proteins and maintains intestinal barrier integrity. Dietary BPA uptake destroys the morphology of the colonic epithelium and increases the pathology score by decreasing the expression of tight junction proteins (ZO-1, occludin, and claudin-1) in the colonic epithelium of mice [34]. The destruction of the intestinal barrier helps BPA to penetrate the intestine and then invade other organ systems. Dietary exposure of 50 μg/kg/day of BPA induces increased intestinal permeability and consequently leads to hepatic steatosis in CD-1 mice [35]. Studies found that long-term exposure to BPA for 22 weeks reduced the tight junctions in the colon of male C57BL/6J mice, resulting in the dysfunction of the gut barrier and impaired cognitive function [36]. Consistent with previous studies, our results showed that BPA exposure significantly decreased the expression of tight junction proteins both at 6 and 24 h. However, SeNPs treatment maintained the normal expression of tight junction proteins during BPA exposure at 24 h. We found that the protective effects of SeNPs were superior to Na_2SeO_3, as shown that Na_2SeO_3 treatment led to a lower expression of ZO-1 and Occludin proteins than BPA exposure alone. In a model of fluorine-induced chronic oxidative stress of broilers, selenomethionine exhibited a better protective effect on ameliorating tight junction network impairment than sodium selenite [37]. Nano-bio selenium can effectively improve the performance and intestinal integrity of broilers compared to the common organic and inorganic sources of selenium [38]. Our results suggest that SeNPs maintained the intestinal barrier function by increasing the expression of tight junction proteins to attenuate the adverse effects of BPA.

Apoptosis is programmed death of a self-protective nature after the host cell senses external risk factors [39]. There are two main pathways of apoptosis, the extrinsic apoptotic pathway mediated by caspase 8 and the endogenous pathway mediated by Bcl 2, caspase 3, caspase 7, and caspase 9 [40]. Studies have shown that cell apoptosis is the main signal pathway of BPA-induced tissue damage, including the brain [41], liver [42], and lung [43]. It has been shown that BPA-treated rat and human stem cells have a significant decrease in cell viability and substantial apoptosis as early as 10 min of exposure and at physiologically relevant low doses [44]. At the same time, it was found that BPA induced apoptosis of mouse interstitial cells through oxidative stress [45]. Treatment with Se significantly alleviated Cd-induced apoptosis in chicken livers, as evidenced by a reduction in the production of NO, iNOS activity, the number of apoptotic cells, and mRNA and protein expression levels of caspase-3 and Bax [46]. In the rat thyroid follicular cell model, selenium reduced the proportion of cell death. In addition, high doses of nano-selenium incubation may prevent tunicamycin-induced ER stress cell apoptosis through the maintenance of membrane integrity and the reduction of caspase 3/7 activity [47]. In the present study, BPA exposure decreased the expression of anti-apoptotic Bcl-2 and Bcl-xl, while conversely, BPA increased the Bax/Bcl-2 ratio and the expressions of pro-apoptotic Bax and effector protein caspase 3, caspase 8, and caspase 9, indicating that BPA had a strong pro-apoptotic effect on pig intestinal epithelial cells. A larger ratio of Bax/Bcl-2 directs cells toward apoptosis [48]. SeNPs or Na_2SeO_3 treatment alleviated the above apoptosis marker protein during BPA exposure at 6 and 24 h. Compared with the Na_2SeO_3 treatment, SeNPs treatment exhibited a better inhibitory effect on BPA-induced intestinal epithelial cell apoptosis, as shown by the lower expression of caspase-9 and Bax and the higher expression of Bcl-2 in the SeNPs group at 6 and 24 h.

The transcription factor NF-κB, as an important immunoregulatory factor, promotes the high expression of pro-inflammatory cytokine TNF-α and interleukin family, which have a significant disruptive effect on the expression and distribution of tight junction proteins and impair the intestinal epithelial barrier function [49,50]. BPA has been demonstrated to modulate the function of the immune system and increases the susceptibility to infections by virtue of acting as a pro-inflammatory molecule [51]. Long-term exposure to BPA-induced TLR4-dependent hypothalamic inflammation exacerbates diet-induced prediabetes [52]. Oral exposure to 4 μg/kg bw/d BPA exacerbates allergic inflammation in a mouse model of food allergy [53]. Consistent with previous reports, the current study found that BPA promoted the activation of NF-κB and inhibitor of NF-κB (IκB) and significantly elevated the expression of intestinal inflammation-related genes, including IL-6, IL-17, IL-1β, IFN-γ, and TNF-α. These results reflected that the activation of innate immune responses was an important mechanism of BPA-induced impairment of the intestinal epithelial barrier function. Selenium can control exaggerated immunological responses and persistent inflammation [54]. In this study, compared with Na_2SeO_3, SeNPs could reduce the excessive expression of inflammatory factors caused by BPA, indicating selenium nanoparticles exhibit considerable promise as a more effective and reliable tool in controlling and preventing intestinal inflammation during BPA exposure.

Oxidative stress refers to an imbalance between oxidants and antioxidants in the body produced by free radicals and is known to be a significant factor influencing disease occurrence [55]. BPA can increase MDA content in rat testicular tissue and thus leads to oxidative stress and functional damage [56]. Selenium modulates antioxidant system status by affecting the activity of antioxidant enzymes via selenoproteins and selenic nucleic acids [57,58]. Dietary intake of organic selenium can increase the activity of GSH-Px and significantly decrease the content of MDA in the serum and placenta of the sows [59]. In this study, BPA exposure led to a decrease in the content of T-AOC, SOD, CAT, and GSH-Px and an increase in the content of MDA, while SeNPs or Na_2SeO_3 treatment alleviated the effects of BPA on oxidative stress. We also found that cells pretreated with SeNPs had a higher abundance of T-AOC, SOD, CAT, and GSH-Px and a lower content of MDA compared with cells pretreated with Na_2SeO_3. These results were in line with the study reporting that dietary supplement of nano-Se exhibits better effects on elevating the levels of GSH-Px and T-AOC in laying hens compared with Na_2SeO_3 [60].

Under the stimulation of the endoplasmic reticulum by intracellular and extracellular stress factors, the number of unfolded and misfolded proteins in ER increases abnormally, which induces ER stress. After ER stress occurs, UPR is activated to clear unfolded and misfolded proteins. ER stress is considered an intermediate pathway to induce the other pathways and damage cells during BPA exposure. Single or combined exposure to BPA and mono(2-ethylhexyl) phthalate cause serious toxicological outcomes in the HepG2 cell line, including altered oxidant/antioxidant status, aggravated ER stress, and apoptosis [61]. BPA exposure induces neurotoxicity in rats involved with neuronal ER stress, apoptosis, and JAK1/STAT1 signaling pathway [41]. Long-term BPA exposure disturbed lipid metabolism and induced oxidative stress, ER stress, apoptosis, autophagy, and inflammatory response in the liver of common carp [42]. BPA triggers apoptosis by activating ER stress in human endometrial stromal cells [62]. ER stress might be related to the disturbance of the intestinal barrier in IPEC-J2 cells after heat exposure [63]. Pterostilbene restores the intestinal barrier function by inhibiting ER stress in piglets [64]. It was reported that selenium deficiency leads to apoptosis by inducing ER stress in the swine small intestine and IPEC-J2 cells, indicating that selenium has a protective effect on porcine intestinal health through modulating ER stress [65]. A previous study showed that ER stress and oxidative stress showed a high correlation, and the lack of selenium triggers ER stress and oxidative stress, and apoptosis in chicken hepatocytes [66]. In this study, we determined the status of three UPR pathways mediated by receptor proteins PERK, IRE1, and ATF6. BPA exposure significantly activated UPR, as shown by the increased expression of Bip, PERK, eIf2α, ATF4, IRE1α, ATF6, and CHOP proteins. Interestingly, we also found that the activated UPR was mitigated by SeNPs and Na_2SeO_3 treatment. Here, BPA exposure disrupted tight junction function and induced proinflammatory response, oxidative stress, apoptosis, and ER stress in intestinal epithelial cells, while SeNPs treatment could alleviate these effects.

In conclusion, our results suggested that, compared with inorganic selenium, selenium nanoparticles were superior in alleviating BPA-induced proinflammatory response, oxidative stress, and apoptosis in porcine intestinal epithelial cells through ameliorating ER stress, thus maintaining the intestinal epithelial barrier function. Our study provides a theoretical basis for the toxic mechanism of BPA in swine and further complements the risk assessment of BPA in domestic animals, and pushes forward a solution for BPA-induced damage by probing the role of selenium nanoparticles.

4. Materials and Methods

4.1. Preparation of SeNPs

The chitosan was dissolved in deionized water and stirred for 6 h. The chitosan (2 mg/mL) was mixed vigorously with the freshly prepared sodium selenite solution (20 mmol/L) at a ratio of 1:4 (*v/v*) for 60 min. Then, the Vc solution (80 mmol/L) was added to the sodium selenite solution with an equal volume, stirred at 35 °C for 12 h, and lyophilized to obtain SeNPs.

4.2. Dynamic Light Scattering (DLS)

DLS was employed to record the variation in the intensity of the scattered light on the microsecond time scale based on particles in gas or liquid being subjected to Brownian motions. To investigate the stability of colloidal particles, the Z-average size and polymer dispersity index (PDI) of SeNPs were measured by the Nanosizer ZS90 instrument (Malvern Instruments, Malvern, UK). The refractive indices of water and the SeNPs solution were taken as 1.333. In general, zeta potentials reflect the electrostatic repulsion between dispersed particles, and a high value of zeta potentials represents a good dispersion of colloid particles. In this work, zeta potentials were performed by using Nanosizer. All measurements were conducted at 25 $\pm$ 2 °C in triplicate.

4.3. Transmission Electron Microscopy (TEM)

SeNPs were fixed in 2.5% glutaraldehyde for 4 h and then in 1% osmium tetroxide for 1.5 h at 4 °C, after which they were dehydrated in gradient ethanol solutions and propylene oxide. The resin was then impregnated with acetone for 12 h and polymerized on a polymerized plate at 40 °C for 48 h. Ultrathin sections (1 mm) were cut with the Ultramicrotome Leica EM UC7 (Leica, Wetzlar, Germany) and then stained with uranyl acetate and lead citrate [67]. The sections were examined under a TEM (JEM-2100, JEOL, Tokyo, Japan).

4.4. Gastric and Intestinal Fluid Digestion Test

For the gastric fluid digestion test, 1 mg of SeNPs was mixed with 10 mL of gastric fluid (pepsin, $\geq$30 U/mL) and adjusted to pH 1.2 with 6 M HCl. Then, the mixture was incubated at 37 °C for 3 h in a thermostat water bath. The samples were taken at the digestion time of 0, 1, 2, and 3 h. For the intestinal fluid digestion test, after the gastric digestion for 4 h was centrifuged at 10,000$\times$ g for 30 min, the precipitation was collected and mixed with 10 mL of intestinal fluid (pancreatin, $\geq$40 U/mL). The mixture was then adjusted to pH 7.4 with 1 mol/L of $NaHCO_3$, followed by incubation at 37 °C for 3 h in the thermostat water bath. The samples were taken at the digestion time of 0, 1, 2, and 3 h.

4.5. Cell Culture and Treatment

The IPEC-J2 cell line is a porcine colonic epithelial cell line. Cryopreserved cells were extracted from liquid nitrogen and continuously cultured for three passages for subsequent experiments. The cells were cultured in DMEM-F12 containing 10% fetal bovine serum (Invitrogen, Carlsbad, CA, USA) at 37 °C in an atmosphere of 5% CO_2 and 95% air at 95% relative humidity. The medium was replaced every 2 days. The cells were treated under 1 of 4 conditions, as follows: (i) medium (CON group); (ii) 50 μmol of BPA (RHAWN, China) exposure (BPA group); (iii) pre-incubation with 32.85 μg/mL of sodium selenite following 50 μmol of BPA exposure (BPA + Na_2SeO_3 group); (iv) pre-incubation with 15 μg/mL of SeNPs following 50 μmol of BPA exposure (BPA + SeNPs group). Cell samples were harvested at 6 and 24 h after BPA exposure, respectively.

4.6. Cell Counting Kit-8 (CCK8) Assay

IPEC-J2 cells were seeded at a density of 1×10^4 per well in 96-well plates and cultured for 24 h. According to the BPA concentration range reported in recent research, cells were treated with 0, 10, 20, 40, 50, 60, and 80 μM of BPA for 6 and 24 h, respectively. The final DMSO concentration in all cell cultures was adjusted to 0.01% (*v/v*). A Cell Counting Kit-8 (Solarbio, Beijing, China) was employed to determine cell viability. The absorbances at 450 nm were determined by a microplate reader (INNO-M, TeCK, China). The highest concentration of BPA that did not considerably impact cell viability was chosen as the favorable concentration for the subsequent assays. All the experiments were carried out independently in triplicate.

4.7. Morphological Observation

Morphological Observation: The cells were seeded into a 6-well plate, and different BPA, Na_2SeO_3, or SeNPs treatments were carried out. At 24 h after BPA exposure, cell morphology was observed under a light microscope (Olympus Corporation, Tokyo, Japan).

4.8. Western Blotting

Total protein was extracted by lysis buffer for Western blotting with 100 mM of phenylmethanesulfonylfluoride, and 25 μg of total protein sample was subjected to SDS-polyacrylamide gel electrophoresis under reducing conditions. Separated proteins were transferred to nitrocellulose membranes in Tris-glycine buffer containing 20% methanol at 4 °C. The membranes were blocked with 5% skim milk for 2 h, following incubated overnight with diluted primary antibodies against Bcl-xl (1:5000, Abcam, Cambridge, UK), caspase 3 (1:2000, Cell signaling technology, Boston, MA, USA), Bip (1:2000, Cell signaling technology), PERK (1:5000, Abcam), p-PERK (1:1000, Bioss), ATF4 (1:2000, Proteintech, Chicago, IL, USA), ATF6 (1:2000, Proteintech), eIf2α (1:1000, Proteintech, Wuhan, China), p-IRE1(1:1000, Bioss), CHOP (1:1000, Bioss), NF-κB (1:1000, Bioss), IκB (1:2000, Cell signaling technology), ZO-1 (1:1000, Bioss, Beijing, China), Occludin (1:1000, Bioss), Claudin-1 (1:1000, Bioss), and GAPDH (1:1000, Proteintech, China) followed by goat anti-rabbit IgG (H+L; 1:1000, Proteintech, China). The gray values of protein bands were measured by ImageJ software version 6.1 (Bio-Rad Laboratories, Hercules, CA, USA). The cellular protein PERK was used as an internal control for p-PERK, and GAPDH was used as an internal control for other proteins.

4.9. Quantitative Real-Time PCR (qRT-PCR) Analysis

Total RNA samples were isolated by Trizol reagent (Invitrogen) according to the manufacturer's instructions. The RNA was reverse transcribed by a cDNA synthesis kit (Promega, Madison, WI, USA). Specific primers (*caspase-8*, *caspase-9*, *Bcl-2*, *Bax*, *Il-1β*, *Il-6*, *Il-17*, *Il-10*, *Tnf-*, *Ifn-γ*, and *Gapdh*) were designed by Primer-BLAST at the National Center for Biotechnology Information and were shown in Table 1. The dried RNA pellets were resuspended in 50 μL of diethylpyrocarbonate-treated water. The concentration and purity of the total RNA were determined using a spectrophotometer. cDNA was synthesized from 1 μg of the total RNA using oligo dT primers and Superscript II reverse transcriptase according to the manufacturer's instructions (Promega, Beijing, China), and cDNA was stored at 80 °C. Reactions were performed in a 20 μL of reaction mixture containing 10 μL of 2X SYBR Green I PCR Master Mix,1 μL of cDNA, 1 μL of each primer (10 μM), and 7 μL of PCR-grade water. The optimal conditions for PCR were 95 °C for 2 min, followed by 40 cycles of denaturation for 15 s at 95 °C, annealing for 1 min at 60 °C, and elongation for 50 s at 72 °C. The qRT-PCR was performed with StepOne™ 96 system (Roche, Basel, Switzerland). The relative abundance of mRNA of each gene was calculated according to the $2^{-\Delta\Delta Ct}$ metho d and was normalized to the mean expression of GAPDH.

Table 1. Sequences of oligonucleotide primers used for real-time PCR, length of the respective PCR product, and gene accession number.

Gene Product [a]	Primer Direction [b]	Sequence (5′ to 3′)	Product Size (bp)	Accession Number
GADPH	F	CCAGAACATCATCCCTGCTT	229	XM_021091114
	R	GTCCTCAGTGTAGCCCAGGA		
Caspase-8	F	TGGAGGACGTTTTCACAGGGC	133	XM_021074712
	R	AGTTGTAACCGGAGGCAAATCC		
Caspase-9	F	AACTTCTGCCATGAGTCGGG	135	XM_013998997
	R	GAGGTGGCTGGCCTTGG		
Bcl-2	F	AGCATGCGGCCTCTATTTGAT	107	XM_021099593
	R	CACTTATGGCCCAGATAGGCA		

Table 1. *Cont.*

Gene Product [a]	Primer Direction [b]	Sequence (5′ to 3′)	Product Size (bp)	Accession Number
Bax	F	AGCAGATCATGAAGACAGGGG	137	XM_003127290
	R	ACACTCGCTCAACTTCTTGGT		
IL-6	F	GGCTGTGCAGATTAGTACC	124	JQ_839263
	R	CTGTGACTGCAGCTTATCC		
IL-10	F	CTTGTTGCTGACCGGGTCTC	110	HQ_236499
	R	TCTCTGCCTTCGGCATTACG		
IL-17	F	GACGGCCCTCAGATTACTCC	125	KF_646141
	R	AGCATTGATACAGCCCGAGT		
IL-1β	F	GCCAACGTGCAGTCTATGGAGTG	91	XM_021085847
	R	GGTGGAGAGCCTTCAGCATGTG		
TNF-α	F	GCCCTTCCACCAACGTTTTC	97	JF_831365
	R	CAAGGGCTCTTGATGGCAGA		
Ifn-γ	F	CAGGCCATTCAAAGGAGCAT	150	MH_538101
	R	GAGTTCACTGATGGCTTTGCG		

[a] GAPDH = glyceraldehyde-3-phosphate dehydrogenase. [b] F = forward; R = reverse.

4.10. Antioxidant Determination

The activities of antioxidant GSH-PX (A005-1-1), catalase (CAT, A007-1-1), superoxide dismutase (SOD, A001-3-1), total antioxidant capacity (T-AOC, A015-2-1), and the concentration of malondialdehyde (MDA, A003-1-1) were analyzed using the corresponding commercial assay kit (Nanjing Institute of Jiancheng Biological Engineering, Nanjing, China) according to the manufacturer's instructions. The antioxidant activity of the above parameters was calculated based on the protein content of the cell samples, and the assay was performed using the method described by Bradford.

4.11. Detection of Apoptosis by Flow Cytometry

The percentage of apoptotic cells was determined using a commercial Annexin V-FITC/PI Apoptosis Detection Kit (Kangwei Biotech, Shenzhen, China). Following the manufacturer's instructions, the harvested cell pellets were washed twice with pre-cooled PBS and then resuspended in 0.5 mL of 1X binding buffer. Afterward, cells were incubated with FITC-labeled Annexin V for 15 min and were stained with 50 μg/mL of propidium iodide for 5 min before detection. Samples were analyzed by a flow cytometer (Beckman, Brea, CA, USA). The data analysis was performed using CytExpert software version 2.4 (Beckman Coulter, Brea, CA, USA), and the apoptosis percentage was referred to as the ratio of apoptosis cells to total cells.

4.12. Statistical Analysis

All of the experimental data are expressed as the means ± SEM. The differences between the groups were determined by a 1-way analysis of variance (ANOVA) test followed by Tukey's honestly significant difference post hoc test using GraphPad Prism 5 software (Graphpad Software Inc., San Diego, CA, USA). A p-value of <0.05 was considered indicative of statistical significance.

Author Contributions: Conceptualization, Q.W., J.H. and J.W.; Methodology, Z.P., T.H., L.Z., D.C., Y.Z., L.L. and J.Z.; Software, Z.P. and J.Z.; Data Analysis, Z.P.; Writing—Original Draft Preparation, Z.P.; Writing—Review & Editing, Q.W., J.H. and J.W.; Project Administration, Q.W. and J.W.; Funding Acquisition, Q.W. and J.W. All authors have read and agreed to the published version of the manuscript.

Funding: This work was supported financially by the National Natural Science Foundation of China (project no. 32202881) and the innovation and development project of the Institute of Animal Husbandry and Veterinary Medicine, BAAFS (XMS202303).

Institutional Review Board Statement: Not applicable.

Informed Consent Statement: Not applicable.

Data Availability Statement: All data are presented within the manuscript.

Conflicts of Interest: The authors declare no conflict of interest.

References

1. Pabst, R.; Russell, M.W.; Brandtzaeg, P. Tissue distribution of lymphocytes and plasma cells and the role of the gut. *Trends Immunol.* **2008**, *29*, 206–208. [CrossRef] [PubMed]
2. Gonzalez-Correa, C.A.; Mulett-Vásquez, E.; Miranda, D.A.; Gonzalez-Correa, C.H.; Gómez-Buitrago, P.A. The colon revisited or the key to wellness, health and disease. *Med. Hypotheses* **2017**, *108*, 133–143. [CrossRef] [PubMed]
3. Jacobi, S.K.; Odle, J. Nutritional factors influencing intestinal health of the neonate. *Adv. Nutr.* **2012**, *3*, 687–696. [CrossRef] [PubMed]
4. Ren, Z.; Guo, C.; Yu, S.; Zhu, L.; Wang, Y.; Hu, H.; Deng, J. Progress in mycotoxins affecting intestinal mucosal barrier function. *Int. J. Mol. Sci.* **2019**, *20*, 11. [CrossRef] [PubMed]
5. Schneeberger, E.E.; Lynch, R.D. The tight junction: A multifunctional complex. *Am. J. Physiol. Cell Physiol.* **2004**, *286*, C1213–C1228. [CrossRef]
6. Ma, Y.; Liu, H.; Wu, J.; Yuan, L.; Wang, Y.; Du, X.; Wang, R.; Marwa, P.W.; Petlulu, P.; Chen, X.; et al. The adverse health effects of bisphenol A and related toxicity mechanisms. *Environ. Res.* **2019**, *176*, 108575. [CrossRef]
7. Cimmino, I.; Fiory, F.; Perruolo, G.; Miele, C.; Beguinot, F.; Formisano, P.; Oriente, F. Potential mechanisms of bisphenol A (BPA) contributing to human disease. *Int. J. Mol. Sci.* **2020**, *21*, 16. [CrossRef]
8. Murata, M.; Kang, J.H. Bisphenol A (BPA) and cell signaling pathways. *Biotechnol. Adv.* **2018**, *36*, 311–327. [CrossRef]
9. Gonkowski, S. Bisphenol A (BPA)-Induced changes in the number of serotonin-positive cells in the mucosal layer of porcine small intestine-the preliminary studies. *Int. J. Mol. Sci.* **2020**, *21*, 3. [CrossRef]
10. Makowska, K.; Szymańska, K.; Całka, J.; Gonkowski, S. The influence of bisphenol A (BPA) on the occurrence of selected active substances in neuregulin 1 (NRG1)-positive enteric neurons in the porcine large intestine. *Int. J. Mol. Sci.* **2021**, *22*, 19. [CrossRef]
11. Qu, W.; Zhao, Z.; Chen, S.; Zhang, L.; Wu, D.; Chen, Z. Bisphenol A suppresses proliferation and induces apoptosis in colonic epithelial cells through mitochondrial and MAPK/AKT pathways. *Life Sci.* **2018**, *208*, 167–174. [CrossRef] [PubMed]
12. Wang, K.; Qiu, L.; Zhu, J.; Sun, Q.; Qu, W.; Yu, Y.; Zhao, Z.; Yu, Y.; Shao, G. Environmental contaminant BPA causes intestinal damage by disrupting cellular repair and injury homeostasis in vivo and in vitro. *Biomed. Pharmacother.* **2021**, *137*, 111270. [CrossRef] [PubMed]
13. Walter, P.; Ron, D. The unfolded protein response: From stress pathway to homeostatic regulation. *Science* **2011**, *334*, 1081–1086. [CrossRef] [PubMed]
14. Iurlaro, R.; Muñoz-Pinedo, C. Cell death induced by endoplasmic reticulum stress. *FEBS J.* **2016**, *283*, 2640–2652. [CrossRef]
15. Ma, X.; Dai, Z.; Sun, K.; Zhang, Y.; Chen, J.; Yang, Y.; Tso, P.; Wu, G.; Wu, Z. Intestinal epithelial cell endoplasmic reticulum stress and inflammatory bowel disease pathogenesis: An update review. *Front. Immunol.* **2017**, *8*, 1271. [CrossRef]
16. Chaudhari, N.; Talwar, P.; Parimisetty, A.; Lefebvre d'Hellencourt, C.; Ravanan, P. A molecular web: Endoplasmic reticulum stress, inflammation, and oxidative stress. *Front. Cell. Neurosci.* **2014**, *8*, 213. [CrossRef]
17. Holczer, M.; Márton, M.; Kurucz, A.; Bánhegyi, G.; Kapuy, O. A Comprehensive systems biological study of autophagy-apoptosis crosstalk during endoplasmic reticulum stress. *Biomed. Res. Int.* **2015**, *2015*, 319589. [CrossRef]
18. Vancamelbeke, M.; Vermeire, S. The intestinal barrier: A fundamental role in health and disease. *Expert Rev. Gastroenterol. Hepatol.* **2017**, *11*, 821–834. [CrossRef]
19. Asahi, J.; Kamo, H.; Baba, R.; Doi, Y.; Yamashita, A.; Murakami, D.; Hanada, A.; Hirano, T. Bisphenol A induces endoplasmic reticulum stress-associated apoptosis in mouse non-parenchymal hepatocytes. *Life Sci.* **2010**, *87*, 431–438. [CrossRef]
20. Tabas, I.; Ron, D. Integrating the mechanisms of apoptosis induced by endoplasmic reticulum stress. *Nat. Cell Biol.* **2011**, *13*, 184–190. [CrossRef]
21. Szegezdi, E.; Logue, S.E.; Gorman, A.M.; Samali, A. Mediators of endoplasmic reticulum stress-induced apoptosis. *EMBO Rep.* **2006**, *7*, 880–885. [CrossRef]
22. Fang, Y.; Zhou, Y.; Zhong, Y.; Gao, X.; Tan, T. Effect of vitamin E on reproductive functions and anti-oxidant activity of adolescent male mice exposed to bisphenol A. *Wei Sheng Yan Jiu* **2013**, *42*, 18–22. [PubMed]
23. Abedelhaffez, A.S.; El-Aziz, E.A.A.; Aziz, M.A.A.; Ahmed, A.M. Lung injury induced by Bisphenol A: A food contaminant, is ameliorated by selenium supplementation. *Pathophysiology* **2017**, *24*, 81–89. [CrossRef] [PubMed]
24. Shi, L.G.; Yang, R.J.; Yue, W.B.; Xun, W.J.; Zhang, C.X.; Ren, Y.S.; Shi, L.; Lei, F.L. Effect of elemental nano-selenium on semen quality, glutathione peroxidase activity, and testis ultrastructure in male Boer goats. *Anim. Reprod. Sci.* **2010**, *118*, 248–254. [CrossRef] [PubMed]
25. El-Deep, M.H.; Ijiri, D.; Ebeid, T.A.; Ohtsuka, A. Effects of dietary nano-selenium supplementation on growth performance, antioxidative status, and immunity in broiler chickens under thermoneutral and high ambient temperature conditions. *J. Poult. Sci.* **2016**, *53*, 274–283. [CrossRef]

26. Saffari, S.; Keyvanshokooh, S.; Zakeri, M.; Johari, S.A.; Pasha-Zanoosi, H.; Mozanzadeh, M.T. Effects of dietary organic, inorganic, and nanoparticulate selenium sources on growth, hemato-immunological, and serum biochemical parameters of common carp (*Cyprinus carpio*). *Fish Physiol. Biochem.* **2018**, *44*, 1087–1097. [CrossRef] [PubMed]
27. Ge, J.; Guo, K.; Zhang, C.; Talukder, M.; Lv, M.W.; Li, J.Y.; Li, J.L. Comparison of nanoparticle-selenium, selenium-enriched yeast and sodium selenite on the alleviation ofcadmium-induced inflammation via NF-kB/IκB pathway in heart. *Sci. Total Environ.* **2021**, *773*, 145442. [CrossRef]
28. Ge, J.; Guo, K.; Huang, Y.; Morse, P.D.; Zhang, C.; Lv, M.W.; Li, J.L. Comparison of antagonistic effects of nanoparticle-selenium, selenium-enriched yeast and sodium selenite against cadmium-induced cardiotoxicity via AHR/CAR/PXR/Nrf2 pathways activation. *Comp. Study* **2022**, *105*, 108992. [CrossRef]
29. Khalaf, A.A.; Ahmed, W.; Moselhy, W.A.; Abdel-Halim, B.R.; Ibrahim, M.A. Protective effects of selenium and nano-selenium on bisphenol-induced reproductive toxicity in male rats. *Hum. Exp. Toxicol.* **2019**, *38*, 398–408. [CrossRef]
30. Acaroz, U.; Ince, S.; Arslan-Acaroz, D.; Gurler, Z.; Demirel, H.H.; Kucukkurt, I.; Eryavuz, A.; Kara, R.; Varol, N.; Zhu, K. Bisphenol-A induced oxidative stress, inflammatory gene expression, and metabolic and histopathological changes in male Wistar albino rats: Protective role of boron. *Toxicol. Res.* **2019**, *8*, 262–269. [CrossRef]
31. Soundararajan, A.; Prabu, P.; Mohan, V.; Gibert, Y.; Balasubramanyam, M. Novel insights of elevated systemic levels of bisphenol-A (BPA) linked to poor glycemic control, accelerated cellular senescence and insulin resistance in patients with type 2 diabetes. *Mol. Cell. Biochem.* **2019**, *458*, 171–183. [CrossRef] [PubMed]
32. Abdel-Halim, B.; Khalaf, A.; Moselhy, W.; Ahmed, W.M. Protective effect of nano-selenium and ionized selenium against the testicular damage, endocrine disruptor and testicular ultrastructure of bisphenol A in albino male rats. *Asian J. Anim. Vet. Adv.* **2016**, *11*, 653–664. [CrossRef]
33. Fahmy, H.A.; Abd El Azim, A.S.; Gharib, O.A. Protective Effects of omega-3 fatty acids and/or nanoselenium on cisplatin and ionizing radiation induced liver toxicity in rats. *Indian J. Pharm. Educ. Res.* **2016**, *50*, 649–656. [CrossRef]
34. Feng, L.; Chen, S.; Zhang, L.; Qu, W.; Chen, Z. Bisphenol A increases intestinal permeability through disrupting intestinal barrier function in mice. *Environ. Pollut.* **2019**, *254*, 112960. [CrossRef] [PubMed]
35. Feng, D.; Zhang, H.; Jiang, X.; Zou, J.; Li, Q.; Mai, H.; Su, D.; Ling, W.; Feng, X. Bisphenol A exposure induces gut microbiota dysbiosis and consequent activation of gut-liver axis leading to hepatic steatosis in CD-1 mice. *Environ. Pollut.* **2020**, *265*, 114880. [CrossRef]
36. Ni, Y.; Hu, L.; Yang, S.; Ni, L.; Ma, L.; Zhao, Y.; Zheng, A.; Jin, Y.; Fu, Z. Bisphenol A impairs cognitive function and 5-HT metabolism in adult male mice by modulating the microbiota-gut-brain axis. *Chemosphere* **2021**, *282*, 130952. [CrossRef]
37. Chen, S.; Xue, Y.; Shen, Y.; Ju, H.; Zhang, X.; Liu, J.; Wang, Y. Effects of different selenium sources on duodenum and jejunum tight junction network and growth performance of broilers in a model of fluorine-induced chronic oxidative stress. *Poult. Sci.* **2022**, *101*, 101664. [CrossRef] [PubMed]
38. Ali, F.; Saeed, K.; Fatemeh, H. Nano-Bio selenium synthesized by bacillus subtilis modulates broiler performance, intestinal morphology and microbiota, and expression of tight junction's proteins. *Bio. Trace Elem. Res.* **2022**, *200*, 1811–1825. [CrossRef] [PubMed]
39. Elmore, S. Apoptosis: A review of programmed cell death. *Toxicol. Pathol.* **2007**, *35*, 495–516. [CrossRef]
40. Tummers, B.; Green, D.R. Caspase-8: Regulating life and death. *Immunol. Rev.* **2017**, *277*, 76–89. [CrossRef] [PubMed]
41. Caglayan, C.; Kandemir, F.M.; Ayna, A.; Gür, C.; Küçükler, S.; Darendelioğlu, E. Neuroprotective effects of 18β-glycyrrhetinic acid against bisphenol A-induced neurotoxicity in rats: Involvement of neuronal apoptosis, endoplasmic reticulum stress and JAK1/STAT1 signaling pathway. *Metab. Brain Dis.* **2022**, *37*, 1931–1940. [CrossRef] [PubMed]
42. Gu, Z.; Jia, R.; He, Q.; Cao, L.; Du, J.; Feng, W.; Jeney, G.; Xu, P.; Yin, G. Alteration of lipid metabolism, autophagy, apoptosis and immune response in the liver of common carp (*Cyprinus carpio*) after long-term exposure to bisphenol A. *Ecotoxicol. Environ. Saf.* **2021**, *211*, 111923. [CrossRef] [PubMed]
43. Liu, S.H.; Su, C.C.; Lee, K.I.; Chen, Y.W. Effects of bisphenol a metabolite 4-Methyl-2,4-bis(4-hydroxyphenyl)pent-1-ene on lung function and type 2 pulmonary alveolar epithelial cell growth. *Sci. Rep.* **2016**, *6*, 39254. [CrossRef] [PubMed]
44. Harnett, K.G.; Chin, A.; Schuh, S.M. BPA and BPA alternatives BPS, BPAF, and TMBPF, induce cytotoxicity and apoptosis in rat and human stem cells. *Ecotoxicol. Environ. Saf.* **2021**, *216*, 112210. [CrossRef] [PubMed]
45. Zhang, M.; Ma, B.; Yang, S.; Wang, J.; Chen, J. Bisphenol A (BPA) induces apoptosis of mouse Leydig cells via oxidative stress. *Environ. Toxicol.* **2023**, *38*, 312–321. [CrossRef]
46. Zhang, R.; Yi, R.; Bi, Y.; Xing, L.; Bao, J.; Li, J. The effect of selenium on the Cd-Induced apoptosis via NO-mediated mitochondrial apoptosis pathway in chicken liver. *Bio. Trace Elem. Res.* **2017**, *178*, 310–319. [CrossRef]
47. Nettore, I.C.; De Nisco, E.; Desiderio, S.; Passaro, C.; Maione, L.; Negri, M.; Albano, L.; Pivonello, R.; Pivonello, C.; Portella, G.; et al. Selenium supplementation modulates apoptotic processes in thyroid follicular cells. *BioFactors* **2017**, *43*, 415–423. [CrossRef]
48. Zaldivar, V.; Magri, M.L.; Zárate, S.; Jaita, G.; Eijo, G.; Radl, D.; Ferraris, J.; Pisera, D.; Seilicovich, A. Estradiol increases the Bax/Bcl-2 ratio and induces apoptosis in the anterior pituitary gland. *Neuroendocrinology* **2009**, *90*, 292–300. [CrossRef]
49. Kim, K.Y.; Oh, T.W.; Do, H.J.; Yang, J.H.; Yang, I.J.; Jeon, Y.H.; Go, Y.H.; Ahn, S.C.; Ma, J.Y.; Park, K.I. Acer palmatum thumb. Ethanol extract alleviates interleukin-6-induced barrier dysfunction and dextran sodium sulfate-induced colitis by improving intestinal barrier function and reducing inflammation. *J. Immunol. Res.* **2018**, *2018*, 5718396. [CrossRef]

50. Zhou, H.Y.; Zhu, H.; Yao, X.M.; Qian, J.P.; Yang, J.; Pan, X.D.; Chen, X.D. Metformin regulates tight junction of intestinal epithelial cells via MLCK-MLC signaling pathway. *Eur. Rev. Med. Pharmacol. Sci.* **2017**, *21*, 5239–5246.
51. Araiza, V.; Mendoza, M.S.; Castro, K.E.N.; Cruz, S.M.; Rueda, K.C.; de Leon, C.T.G.; Morales Montor, J. Bisphenol A, an endocrine-disruptor compund, that modulates the immune response to infections. *Front. Biosci.* **2021**, *26*, 346–362. [CrossRef]
52. Ma, Q.; Deng, P.; Lin, M.; Yang, L.; Li, L.; Guo, L.; Zhang, L.; He, M.; Lu, Y.; Pi, H.; et al. Long-term bisphenol A exposure exacerbates diet-induced prediabetes via TLR4-dependent hypothalamic inflammation. *J. Hazard. Mater.* **2021**, *402*, 123926. [CrossRef]
53. Misme-Aucouturier, B.; De Carvalho, M.; Delage, E.; Dijoux, E.; Klein, M.; Brosseau, C.; Bodinier, M.; Guzylack-Piriou, L.; Bouchaud, G. Oral exposure to bisphenol A exacerbates allergic inflammation in a mouse model of food allergy. *Toxicology* **2022**, *472*, 153188. [CrossRef]
54. Raza, A.; Johnson, H.; Singh, A.; Sharma, A.K. Impact of selenium nanoparticles in the regulation of inflammation. *Arch. Biochem. Biophys.* **2022**, *732*, 109466. [CrossRef] [PubMed]
55. Pan, T.; Liu, T.; Tan, S.; Wan, N.; Zhang, Y.; Li, S. Lower selenoprotein T expression and immune response in the immune organs of broilers with exudative diathesis due to selenium deficiency. *Biol. Trace Elem. Res.* **2018**, *182*, 364–372. [CrossRef]
56. Tiwari, D.; Vanage, G. Bisphenol A Induces Oxidative stress in bone marrow cells, lymphocytes, and reproductive organs of holtzman Rats. *Int. J. Toxicol.* **2017**, *36*, 142–152. [CrossRef] [PubMed]
57. Wu, X.; Dai, S.; Hua, J.; Hu, H.; Wang, S.; Wen, A. Influence of dietary copper methionine concentrations on growth performance, digestibility of nutrients, serum lipid profiles, and immune defenses in broilers. *Biol. Trace Elem. Res.* **2019**, *191*, 199–206. [CrossRef] [PubMed]
58. Huo, B.; He, J.; Shen, X. Effects of selenium-deprived habitat on the immune index and antioxidant capacity of przewalski's Gazelle. *Biol. Trace Elem. Res.* **2020**, *198*, 149–156. [CrossRef]
59. Mou, D.; Ding, D.; Yan, H.; Qin, B.; Dong, Y.; Li, Z.; Che, L.; Fang, Z.; Xu, S.; Lin, Y.; et al. Maternal supplementation of organic selenium during gestation improves sows and offspring antioxidant capacity and inflammatory status and promotes embryo survival. *Food Funct.* **2020**, *11*, 7748–7761. [CrossRef] [PubMed]
60. Meng, T.; Liu, Y.L.; Xie, C.Y.; Zhang, B.; Huang, Y.Q.; Zhang, Y.W.; Yao, Y.; Huang, R.; Wu, X. Effects of different selenium sources on laying performance, egg selenium concentration, and antioxidant capacity in laying hens. *Biol. Trace Elem. Res.* **2019**, *189*, 548–555. [CrossRef]
61. Ozkemahli, G.; Erkekoglu, P.; Ercan, A.; Zeybek, N.D.; Yersal, N.; Kocer-Gumusel, B. Effects of single or combined exposure to bisphenol A and mono(2-ethylhexyl) phthalate on oxidant/antioxidant status, endoplasmic reticulum stress, and apoptosis in HepG2 cell line. *Environ. Sci. Pollut. Res. Int.* **2023**, *30*, 12189–12206. [CrossRef] [PubMed]
62. Ferreira, R.; Amaral, C.; Correia-da-Silva, G.; Almada, M.; Borges, M.; Cunha, S.C.; Fernandes, J.O.; Teixeira, N. Bisphenols A, F, S and AF trigger apoptosis and/or endoplasmic reticulum stress in human endometrial stromal cells. *Toxicology* **2022**, *478*, 153282. [CrossRef] [PubMed]
63. Zhou, J.Y.; Huang, D.G.; Zhu, M.; Gao, C.Q.; Yan, H.C.; Li, X.G.; Wang, X.Q. Wnt/β-catenin-mediated heat exposure inhibits intestinal epithelial cell proliferation and stem cell expansion through endoplasmic reticulum stress. *J. Cell. Physiol.* **2020**, *235*, 5613–5627. [CrossRef]
64. Monceaux, K.; Gressette, M.; Karoui, A.; Pires Da Silva, J.; Piquereau, J.; Ventura-Clapier, R.; Garnier, A.; Mericskay, M.; Lemaire, C. Ferulic acid, pterostilbene, and tyrosol protect the heart from ER-stress-induced injury by activating SIRT1-dependent deacetylation of eIF2α. *Int. J. Mol. Sci.* **2022**, *23*, 12. [CrossRef] [PubMed]
65. Zheng, Y.; Zhang, B.; Guan, H.; Jiao, X.; Yang, J.; Cai, J.; Liu, Q.; Zhang, Z. Selenium deficiency causes apoptosis through endoplasmic reticulum stress in swine small intestine. *Biofactors* **2021**, *47*, 788–800. [CrossRef]
66. Yao, L.; Du, Q.; Yao, H.; Chen, X.; Zhang, Z.; Xu, S. Roles of oxidative stress and endoplasmic reticulum stress in selenium deficiency-induced apoptosis in chicken liver. *Biometals* **2015**, *28*, 255–265. [CrossRef]
67. Yang, Y.; Fan, X.; Ji, Y.; Li, J.; Dai, Z.; Wu, Z. Glycine represses endoplasmic reticulum stress-related apoptosis and improves intestinal barrier by activating mammalian target of rapamycin complex 1 signaling. *Anim. Nutr.* **2022**, *8*, 1–9. [CrossRef]

toxics MDPI

Review

Drosophila as a Robust Model System for Assessing Autophagy: A Review

Esref Demir [1,2,3,4,*] and Sam Kacew [5]

1 Massachusetts General Hospital, Harvard Medical School, Boston, MA 02129, USA
2 F.M. Kirby Neurobiology Center, Boston Children's Hospital, 300 Longwood Avenue, Boston, MA 02115, USA
3 Department of Neurobiology, Harvard Medical School, Boston, MA 02115, USA
4 Medical Laboratory Techniques Program, Department of Medical Services and Techniques, Vocational School of Health Services, Antalya Bilim University, 07190 Antalya, Turkey
5 R. Samuel McLaughllin Center for Population Health Risk Assessment, Institute of Population Health, University of Ottawa, 1 Stewart (320), Ottawa, ON K1N 6N5, Canada; sam.kacew@uottawa.ca
* Correspondence: esref.demir@antalya.edu.tr; Tel.: +90-242-245-00-88; Fax: +90-242-245-01-00

Abstract: Autophagy is the process through which a body breaks down and recycles its own cellular components, primarily inside lysosomes. It is a cellular response to starvation and stress, which plays decisive roles in various biological processes such as senescence, apoptosis, carcinoma, and immune response. Autophagy, which was first discovered as a survival mechanism during starvation in yeast, is now known to serve a wide range of functions in more advanced organisms. It plays a vital role in how cells respond to stress, starvation, and infection. While research on yeast has led to the identification of many key components of the autophagy process, more research into autophagy in more complex systems is still warranted. This review article focuses on the use of the fruit fly *Drosophila melanogaster* as a robust testing model in further research on autophagy. *Drosophila* provides an ideal environment for exploring autophagy in a living organism during its development. Additionally, *Drosophila* is a well-suited compact tool for genetic analysis in that it serves as an intermediate between yeast and mammals because evolution conserved the molecular machinery required for autophagy in this species. Experimental tractability of host–pathogen interactions in *Drosophila* also affords great convenience in modeling human diseases on analogous structures and tissues.

Keywords: *Drosophila melanogaster*; autophagy; in vivo animal model system; Atg; fat body; development

Citation: Demir, E.; Kacew, S. *Drosophila* as a Robust Model System for Assessing Autophagy: A Review. *Toxics* **2023**, *11*, 682. https://doi.org/10.3390/toxics11080682

Academic Editor: Guido Cavaletti

Received: 29 May 2023
Revised: 7 August 2023
Accepted: 7 August 2023
Published: 8 August 2023

1. Introduction

1.1. The Morphology of Autophagy

Eukaryotic cells naturally destroy and recycle damaged cellular components through a process known as autophagy, which is activated by a variety of environmental and developmental triggers [1]. In eukaryotic cells, the main protein degradation pathways are the proteasome and lysosomal breakdown. Autophagy is a lysosomal degradation process that can be classified into different routes based on how the intracellular material reaches the lysosome. Mammalian cells exhibit three different versions of autophagy, namely, microautophagy, macroautophagy, and chaperone-mediated autophagy, each distinguished by their morphological characteristics [2,3]. During macroautophagy, large portions of cytoplasm are engulfed by a membrane cistern called a phagophore or isolation membrane, which then forms a double-membrane autophagosome [3]. The anatomical steps of autophagosome formation are illustrated in Figure 1. This autophagosome then fuses with the endolysosomal components for degradation. In microautophagy, late endosomes (lysosomes) directly engulf tiny portions of the cytoplasm. Chaperone-mediated autophagy is another process that allows individual proteins to reach the lysosomal lumen by means of Hsc70 and the lysosome-associated membrane protein 2A (Lamp-2A) in mammalian cells [3,4]. It is worth noting that, as there is no homolog for the protein Lamp2A in *Drosophila melanogaster*, chaperone-mediated autophagy induced by Lamp-2A will not take place. In all forms of autophagy, the cargo material is degraded by acidic lysosomal hydrolases, and the resulting monomers are recycled back to the cytosol for use in energy production and biosynthesis [4]. Autophagy serves a critical function in maintaining cellular homeostasis by responding to stresses or disturbances such as nutrient starvation. When the biogenesis of autophagosomes is inhibited, it leads to the accumulation of selective autophagic cargo, including Ref(2)P/p62-positive aggregates of polyubiquitinated proteins. These aggregates can be cytotoxic and can contribute to cellular dysfunction [5]. It has been observed that autophagy is often impaired in cancer cells, leading to uncontrolled cell growth [4]. Similarly, autophagy dysfunction, or misregulation, has been linked to several neurodegenerative disorders [4,5]. This review focuses specifically on macroautophagy, which is a bulk degradation pathway conserved over the course of eukaryotic evolution and responsible for the clearance of whole organelles, long-lived cytosolic proteins, and aggregates within eukaryotic cells [6,7]. Neurodegenerative illnesses are becoming more prevalent in older populations, which has a devastating impact on both individuals and their communities. Khurana and Lindquist [8] showed that *Saccharomyces cerevisiae*, a developing yeast, plays a special role in the arsenal against neurodegeneration. *S. cerevisiae*, as a basic eukaryotic organism, providesinteraction-scale mechanistic insights into cell-autonomous neurodegenerative pathways. To recognize and describe these components, a number of PD models, including non-mammalian eukaryotic creatures, have been established. Surguchov [9] has discussed recent discoveries in three model organisms for Parkinson's disease, including yeast, *Drosophila*, and the nematode *Caenorhabditis elegans*, which revealed unique processes and pinpointed fresh causes of the condition. These non-mammalian models and human cells function similarly in many conserved molecular and cellular pathways.

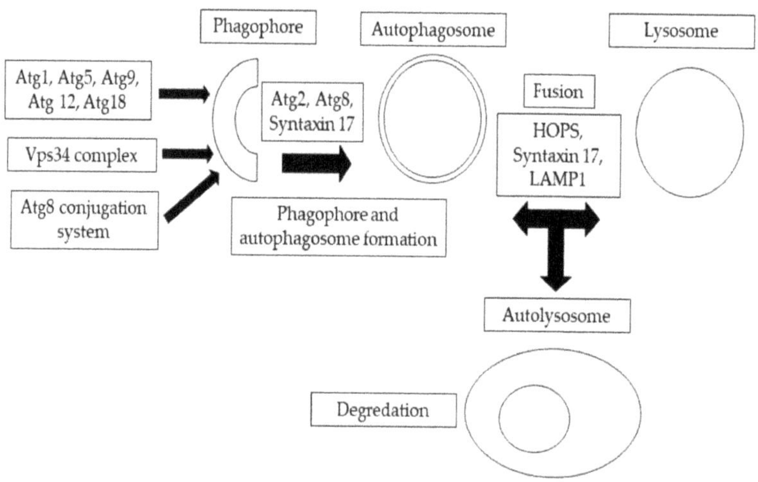

Figure 1. A summary of autophagic structures. The development of phagophores and double-membrane autophagosomes is mediated by the sequential and coordinated activity of Atg protein complexes.

1.2. Homologous Autophagy Proteins between Yeast/Drosophila/Mammals

The proteins responsible for the formation of autophagosomes were first discovered in the yeast *S. cerevisiae* and are referred to as Atg (Autophagy-related) proteins [10]. These proteins have counterparts in higher eukaryotic organisms, such as *D. melanogaster*, and their roles are highly similar [11]. The identification of the Atg proteins marked a significant step forward in understanding autophagy. Specifically, 18 Atg proteins in *Drosophila* make up five complexes that operate the autophagic process in this species [12,13]. Investigations in yeast have uncovered 33 genes, referred to as ATG, that play a role in autophagy. Much of this genetic material has been preserved in organisms with increased complexity [11,14]. The formation and expansion of the autophagosomal membrane, which encloses cellular components for degradation, is regulated by the ATG proteins that are conserved across eukaryotic organisms [15,16]. The fusion of autophagosomes with lysosomes, which occurs through the action of Rab7 and the binding of HOPS and SNARE proteins in yeast, is a crucial step in the autophagic process [17]. It is noteworthy that the processes of autophagosome–lysosome fusion and the subsequent degradation of autophagic cargo vary significantly between yeast and animal cells. However, the HOPS tethering complex and Rab7 are conserved across *Drosophila* and mammals [18–20]. Additionally, in animal cells, autophagosome degradation also requires the Rab2 protein [12,21], unlike in yeast where SNAREs are not homologous [22,23]. Furthermore, molecular motors such as dynein and kinesins play a role in the movement of autophagosomes to lysosomes for degradation [24,25].

The term "autophagy" describes a set of procedures used by eukaryotic cells to recycle and degrade cellular components. The Atg1 complex regulates autophagy's molecular mechanism, which has three main stages: initiation, the production of autophagosomes, and expansion and completion of the membrane [26]. Autophagy is activated by the Atg1 complex in all eukaryotic organisms. The formation of autophagosomal membranes is led by the Vps34 complex, and the autophagosome membrane expands depending on two distinct ubiquitin-like protein conjugation machineries involving Atg8 and Atg12–Atg5. After autophagosome completion, vesicles merge with lysosomes, creating autolysosomes. In this respect, Table 1 highlights the central autophagy genes from humans and baker's yeast (*S. cerevisiae*), and their counterparts in *D. melanogaster* [27]. Each model has its own advantages or limitations. *Drosophila* is an excellent genetic model but a poor one for biochemistry and physiology, while mammals present difficulties for large-scale genetic screening. Yeasts are single-celled and do not go through development.

Table 1. Genes that are conserved in autophagosome induction, nucleation, and expansion27.

Autophagy Process	Model Organisms/Gene			References
	Homo sapiens	*D. melanogaster*	*S. cerevisiae*	
Induction	ULK1, ULK2	Atg1	ATG1	[28–33]
	mTOR	dTOR	TOR	[29,34–37]
	HARBIl	Atg13	ATG13	[38–40]
	-	-	ATG17	[41,42]
	-	-	ATG29	[43]
	-	-	ATG31	[44]
	FIB200, RB1CC1	CG1347	-	[45]
	ATG101	CG7053	-	[45,46]
Nucleation	BECN1	Atg6	ATG6	[29,47–51]
	PIK3C3	Pi3K59F	VPS34	[47,49,52]
	PIK3R4	Ird1	VPS15	[49,53,54]
	ATG14 (barkor)	CG11877	ATG14	[43,49,55,56]
	UVRAG	CG616	-	[49,57]
	SH3GLB1	endoB	-	[58]
	BCL2	buffy	-	[59–61]
	AMBRA1	-	-	[62]
Expansion	ATG2A, ATG2B	Atg2	ATG2	[63–65]
	ATG3	Atg3	ATG3	[66–68]
	ATG4A,B,C,D	Atg4	ATG4	[69–71]
	ATG5	Atg5	ATG5	[32,72–75]
	ATG7	Atg7	ATG7	[68,69,72,73,76]
	ATG9A	Atg9	ATG9	[74,77]
	LC3, GABARAP, GABARAPL2	Atg8a, Atg8b	ATG8	[67,70,78–83]
	ATG10	CG12821	ATG10	[66,73]
	ATG12	Atg12	ATG12	[41,66,73]
	ATG16L1, ATG16L2	CG31033	ATG16	[84,85]
	ATG18	Atg18	ATG18	[77,86,86–88]

The Atg1 protein, a serine/threonine kinase found in various organisms, has a crucial role in the process of autophagy. Studies have shown that Atg1 is necessary for autophagy in mammalian cells [28–31] and *Drosophila* [29,89], nevertheless the composition and function of the Atg1 complex varies between species [41]. For instance, in yeast, the TOR pathway regulates the formation of the Atg1-Atg13-Atg17 [41] and controls autophagy induction through phosphorylation of Atg13. When nutrient levels are low, the dephosphorylation of Atg13 increases its affinity for Atg1-Atg17, triggering autophagy [41]. Neither Atg17 nor Atg29 and Atg31, which both interact with Atg17, are present in equivalent form in either *Drosophila* or humans [43,78]. In contrast to yeast, the Atg1 and Atg13 proteins in *Drosophila* and mammals constitute a more stable structure independent of the activity of the TOR pathway. The Atg1 ortholog in mammals also exhibits interactions with Atg13 independently of the nutrition or starvation status [34,46]. Furthermore, *Drosophila* orthologs of both Atg101 and FIP200, necessary for autophagosome generation in mammals, have been identified but not yet tested for their role in autophagy [34,45,46]. Not only do the *Drosophila* Atg1 and mammalian unc-51-like kinase 1 (ULK1) complexes have different functions but also the overexpression of *Drosophila* Atg1 triggers autophagy, whereas the overexpression of Ulk1 in mammals suppresses it [32,38]. The source of the disparity is uncertain; theories have been proposed that attribute it to the effect of extra regulatory proteins [90,91]. In contrast, Atg101 is found in the majority of eukaryotes apart from budding yeast. It has been suggested that the control of autophagy initiation by the Atg1 complex and its regulatory mode may have developed from yeast to animal cells [90]. Specifically, previous work indicates that Atg101 mutants have reduced lifespan, increased oxidative stress, and impaired mitochondrial function, and Atg101 is required for autophagy, a process that is known to be involved in age-related processes [92]. These findings suggest that Atg101 may be a key regulator of tissue homeostasis and aging in *Drosophila*.

After the Atg1 complex triggers autophagy in yeast, *Drosophila*, and mammalian cells, a PI3P-enriched structure may be seen where autophagosomes develop. PI3P is a phospholipid manufactured by the enzyme PI3K (phosphatidylinositol 3-kinase), whose role in the formation of autophagosomes (membrane-bound vesicles used to transport cellular components for degradation) is still unknown. However, it is thought that PI3P may recruit additional components to the autophagosome, as several yeast proteins that bind to PI3P and localize to the autophagosome have been shown to depend on PI3K activity [93]. Although there are several proteins involved in autophagy, only one of them, namely Atg18, has an ortholog that is found in both *Drosophila* and mammals. Additionally, it has only been confirmed that the ortholog found in *Drosophila* is essential for autophagy to occur [32,79].

The yeast PI3K complex, a team of vital proteins including Vps34, Vps15, Atg6, and Atg14 [94], are the masterminds behind the formation of autophagosomes. These proteins are not only present in yeast but also in *Drosophila* and mammals [95]. Like *Drosophila*, mammals too have three versions of PI3K, but the type III PI3K, Vps34, stands out by activating autophagy through its production of PIP3 [96]. Likewise, Vps34, a member of the class III PI3K family, has a role in initiating autophagy through the production of PIP3. However, things get a bit more complicated in mammals, as the Vps34-Vps15-Atg6 team can be found working alongside other proteins like Ambra1, Atg14, Rubicon, and UVRAG [95]. Even though the *Drosophila* body contains proteins comparable in function to UVRAG, Rubicon, and Atg14, which are found in the Vps34 complex in other organisms, it is not currently understood how these proteins function in the Vps34 complex in *Drosophila*. They may play a similar role as in other organisms or they may have unique functions, and further testing is needed for more evidence on their role in the Vps34 complex in *Drosophila*.

Vesicle expansion is a process that allows the vesicles (small, enclosed spaces within a cell) to grow in size, mediated by two groups of proteins known as ubiquitin-like groups: Atg5-Atg12-Atg16 and Atg8. These groups have remained largely unchanged throughout the evolution from yeast to mammalian organisms [97]. The Atg16 complex is a collection of proteins that localizes to the autophagosome and plays a crucial role in the formation of the autophagosome membrane. The autophagosome is a structure that encloses cellular components that are intended for destruction [84,98]. The Atg12 protein is covalently conjugated to Atg5 through a process that resembles the conjugation of ubiquitin to a target protein, which involves two other proteins called Atg10 and Atg7. The second of these corresponds to a protein called E1 enzyme that is responsible for activating ubiquitin involved in the ubiquitin conjugation process [72]. Atg10 is a protein contributing to the generation of the Atg5-Atg12 complex, which functions similarly to an E2 ubiquitin-conjugating enzyme, a protein that plays a role in the ubiquitin system, but it is not related to those found in the ubiquitin system [99]. Once the Atg5-Atg12 complex is formed, it is linked to Atg16 in a non-covalent manner to form the completed complex [84]. Nevertheless, despite the presence of orthologs in *Drosophila* for each of these proteins, further experiments are warranted to show whether the orthologs of Atg10 and Atg16 play a part in the autophagy pathway in *Drosophila*.

The other conjugation structure involved in autophagosome generation is a process that links Atg8 to a type of lipid called phosphatidylethanolamine (PE) through an amide bond [66]. This process helps in the formation of the autophagosome membrane, which is essential for the autophagosome to enclose the cellular materials that are targeted for degradation. The Atg8–PE conjugation is crucial in the formation and expansion of the autophagosome. Conjugation of Atg8 to PE begins with the action of the cysteine protease Atg4, which cleaves Atg8 [69]. Once Atg8 is cleaved, it is bound to Atg7. In the next phase, Atg8 is transported to Atg3, which is an enzyme comparable to E2 ubiquitin-conjugating enzymes, which then catalyzes the conjugation of Atg8 to PE. Humans have four proteins that are equivalent to yeast ATG8: ATG8L, GATE16, MAP1LC3, and GABARAP. These human orthologs of yeast ATG8 are also conjugated to PE in a comparable fashion to the way it occurs in yeast [67,81]. In *Drosophila,* there are two genes that code for Atg8 proteins (Atg8a and Atg8b) that are found in the autophagosomes [29,82,83]. Such proteins probably have some overlap in function, as deleting the Atg8a gene results in a less severe phenotype than what would be expected for a protein that has such a vital part in autophagosome formation [32,79]. *Drosophila* also has equivalents of Atg3, Atg4, and Atg7, which have been proven to play a part in the tightly regulated pathway of autophagy [29,68,100].

1.3. Drosophila as a Model

Comparative genetic research revealed a significant degree of genetic similarity between humans and a variety of other animals, including *D. melanogaster*. It has been estimated that roughly 60% of the fly genome is comparable to that of humans and that approximately 77% of known disease genes in *Homo sapiens*, including those involved in diabetes, autism, and carcinoma, have matching sequences in *D. melanogaster* [101], which highlights the value of this species in exploring human biology and relevant risk factors [102,103].

This means that the *Drosophila* fruit fly can be used in a variety of experiments because it is adaptable and can be used in combination with other testing methods, which renders it an ideal choice with optimal adaptability and versatility for larger-scale experiments, such as RNAi or mutant screens, requiring a large number of samples [104]. Genome sequencing has greatly facilitated identifying *Drosophila* genes that are almost identical to human genetic material associated with certain diseases. Previous cDNA (complementary DNA) analyses [105] detected 289 of these corresponding genes in fruit flies, and around three-quarters of the genes causing certain human diseases, such as diabetes, autism, and cancer, were found to have a functional corresponding gene in *Drosophila* [102,103]. In this regard, FlyBase, a database containing data on the genetics, genomics, and biology of *D. melanogaster*, has more than 800 reports that detail the links between specific human diseases and *Drosophila* genes, offering researchers valuable insight into the genetic basis of human diseases, which may lead to the development of new treatments and therapies (FB2018_03) [106]. Another important advantage of *Drosophila* studies is that they are exempt from the ethical limitations associated with the use of more complex organisms such as mammals [107]. In that vein, the principles of humane experimental technique known as the 3Rs (replacement, refinement, and reduction), which have been widely accepted as guidelines for humane animal treatment in toxicology experiments, can be readily put into practice with this species [108]. The fundamentals of cell biology, from gene expression and neuron synapse formation to cell signaling and differentiation, are shared between humans and *Drosophila* flies. Even immune signaling pathways, when exposed to cytokines, remain remarkably similar between the two species [109]. The GAL4/UAS system, in combination with CRISPR, has been effectively modeled by employing *Drosophila* as a dynamic tool [110,111]. CRISPR's tissue-specific genome editing can confine mutations to specific cells or tissue [110], making research on flies applicable to vertebrates, including humans. More recently, Trinca and Malik [112] have investigated the impact of gamma radiation on autophagy using *Drosophila* as a testing model, discovering that exposure to radiation resulted in elevated autophagy levels in two different cell types (gut and brain), which suggests that autophagy might have a function in the initial response to exposure to radiation.

While *Drosophila* offers a range of benefits as an experimental model organism for research on human biology, there are also several limitations to consider. One such limitation is that the difference in typical body temperatures of adult fruit flies and humans (18–27 °C versus 36–37 °C) may result in multiple variations in host–pathogen interactions. Additionally, the lack of certain factors in *Drosophila* can limit the types of pathogenesis studies that can be conducted using this organism. Furthermore, the lack of sialic acid as a main surface molecule on fly cells may prevent the study of certain viruses that depend on this molecule for entry into host cells [113]. Another key limitation to consider is that *Drosophila* has a less complex immune system compared to mammals. At the same time, in comparison to mammals, flies are much cheaper to maintain, can be mailed by standard post, and do not require authorization to transport. Thus, despite such drawbacks and restrictions, *Drosophila* remains a valuable model organism for researchers in gaining insights into basic biological processes before moving on to more complex testing models [114–116].

1.4. The Life Cycle of Drosophila

Drosophila, a holometabolous insect, follows a developmental process characterized by prepupal and pupal stages of immobility during which the entire organism undergoes a complete metamorphosis. *Drosophila* exhibits a rapid life cycle (9–10 days at 25 °C) (Figure 2). This process entails the histolysis of larval tissues, and diploid cells undergo proliferation and differentiation to generate adult organs, which are fully developed by the time of emergence as an imago from the pupal case. In the context of fruit fly development, three distinct larval stages, L1, L2, and L3, can be identified and are separated by highly regulated molting transitions [117]. At the mid-L3 stage, a slight elevation in levels of 20-hydroxyecdysone in combination with levels of juvenile hormone causes an alteration in larval behavior, wherein they exit their food source and commence searching for an environment to metamorphose, and developmental autophagy triggered by ecdysone occurs in the majority of tissues in the larvae that have multiple sets of chromosomes [83]. In autophagy, the cells in such tissues break down and recycle their own components for freeing up stored biological matter, which can be used as an energy source by diploid cells in the case of programmed cell death. Autophagy helps ensure that the cells have enough energy and nutrients to complete this metamorphosis process [83,118]. It is likely that autophagy plays a significant role during metamorphosis, a period of developmentally regulated starvation that spans five days. Also occurring during the remodeling of muscles in pupae, autophagy is thought to contribute to the DNA fragmentation in cells that provide nutrition stability to their neighboring cells (nurse cells) during egg cell formation [21,119].

Figure 2. Fruit flies go through four distinct life stages: embryo, larva, pupa, and adult. After 6 to 8 h, the eggs begin to hatch, producing first-instar larvae that develop into second- and third-instar larvae. The larvae turn into pupae around day five. On the ninth or tenth day, the fruit fly enters its adult form.

1.5. Research Tools Available for Drosophila

In *Drosophila*, the visualization of autophagy involves detecting the presence of Atg8, a protein that is found in autophagosomes in mammals, flies, and yeast, which suggests that autophagy is a conserved process across species and that Atg8 is a valuable marker for detecting autophagy in different organisms [73]. *Drosophila* Atg8 antibodies instrumental in the distribution, localization, and expression levels of proteins in cells and tissues are not readily available, but the Gal4/UAS system, a powerful tool for gene expression in *D. melanogaster*, allows for tissue-specific or inducible expression of target genes by using a Gal4 transcriptional activator. By using a GFP-tagged version of Atg8, researchers can study the distribution, localization, and expression levels of Atg8 in specific tissues or under specific conditions. The GFP tag makes it possible to visualize Atg8 in living cells, which helps researchers understand the dynamics of autophagy processes [29,47,83]. While GFP-tagged Atg8 transgenes have been widely used as indicators of autophagosome formation, we should remember that there are some concerns over the validity of this approach. One concern is that the GFP-tagging of Atg8 may interfere with its normal function in autophagy, potentially leading to inaccurate conclusions about autophagosome formation. Additionally, some studies have reported that GFP-tagging can alter the localization and distribution of Atg8 within cells, further complicating the interpretation of results. Therefore, it is important to consider these potential limitations when interpreting results obtained using GFP-tagged Atg8 reagents and to corroborate findings with other methods [120]. In order to overcome the limitations associated with Atg8-GFP-based autophagosome detection, many *Drosophila* researchers supplement it with additional assays, such as lysosomal markers (e.g., LysoTracker Red, LAMP1). By using multiple markers, researchers can get a more comprehensive and reliable understanding of autophagosome formation and autophagy activity. In addition to using multiple markers, another correlative measurement of autophagic activity involves measuring transcriptional upregulation of Atg genes [82,121,122]. These complementary approaches provide a more robust assessment of autophagy activity and help overcome the limitations associated with using Atg8-GFP as the sole indicator of autophagosome formation.

Electron microscopy (EM), despite being non-quantitative, is considered the gold standard for demonstrating the presence of autophagosomes, as it provides direct and high-resolution visualization of cellular structures; thus, it is still widely employed in autophagy research involving experiments on *Drosophila* tissues. Detecting autophagosomes in cells provides insight into their autophagy activity, but the interpretation of the results can sometimes be challenging. For instance, increased autophagosome numbers may point to a rise in autophagy or a failure of autophagosome–lysosome fusion that causes a decrease in autophagy functionality. This is particularly crucial when determining the connection between autophagy and disease development. Therefore, efforts are underway to evaluate autophagic flux, which monitors autophagosome–lysosomal degradation by tracking the movement of cytoplasm, organelles, and other cargo. For instance, researchers evaluated autophagic flux through Western blotting to analyze the time-course of ubiquitinated aggregates on samples from *Drosophila* brain [123]. They discovered that, as autophagic activity declines in the brain as it ages, while the levels of ubiquitinated proteins elevate and Atg8 gene mutations exacerbate the accumulation of ubiquitinated proteins, expression of Atg8 prevents aggregate accumulation. These findings demonstrate that ubiquitinated protein levels can be considered a reliable indicator of alterations in autophagic flux, providing valuable insight into the dynamics of autophagy in diseases and aging.

Transmission electron microscopy (TEM) may afford some valuable data on organelle ultrastructure but requires additional tests to analyze autophagy and flux, and may not be practical for high-throughput genetic screens involving analysis of large numbers of genes or genetic variations in a rapid and systematic manner, including RNA interference (RNAi) screens, CRISPR-Cas9 screens, and whole-genome sequencing studies. In this regard, confocal laser scanning microscopy (CLSM) stands out as a commonly employed alternative for analyzing autophagy in *Drosophila*, as it allows for relatively quick and efficient analysis of large numbers of cells or tissues, although it may not provide the same level of detail as TEM. Staining with vital dyes such as acridine orange, ethidium bromide, and propidium iodide are commonly adopted to detect the presence of autophagosomes and lysosomes, and can also be used to quantify the amount of autophagic flux [124]. Fluorescent reporters, such as GFP-LC3, can label autophagic vesicles and track the dynamics of autophagy in real time. Moreover, the reagents employed in research on autophagy in *Drosophila* samples are often identical or comparable to those used in higher vertebrates, facilitating comparison between different species [1].

A key factor in autophagy, Atg8a covalently conjugates to phosphatidylethanolamine in autophagosomes, which is considered a marker of autophagy. Atg8a is found in various organisms, such as fruit flies, yeast (Atg8), and mammals (LC3). Transgenic *Drosophila* lines with Atg8a expression tagged with GFP or mCherry can now be obtained from public stock centers, and such lines allow in vivo experiments on autophagy under both starvation and developmental conditions [29,83].

In addition to being a marker for autophagy, Atg8a also plays an essential role in autophagosome formation, during which cells create a membrane-bound vesicle to enclose cellular materials for degradation and the protein Atg8a covalently conjugates to phosphatidylethanolamine, a type of lipid. This binding is essential for the formation of the autophagosome and allows the enclosed materials to be degraded and recycled. An E1-like ubiquitin-activating enzyme, Atg7 activates ubiquitin-like protein conjugation systems, while Atg3, an E2-like enzyme, catalyzes the transfer of the ubiquitin-like protein from Atg7 to the substrate. The Atg12-Atg5-Atg16 complex is an E3-like enzyme that facilitates the transfer of the ubiquitin-like protein from Atg3 to the substrate. Atg8a's advantage over other Atg proteins appears to be a greater amount of lipidated Atg8a remaining tagged along with completed autophagosomes, which suggests that Atg8a may have a role in the maturation of autophagosomes and may be important for their stability [15]. Atg8a forms a membrane anchor for autophagosome formation and maturation, where it is attached to the outer and inner membranes. In later stages of autophagosome development, Atg4 separates Atg8 from the external membrane, while it still stays bound to the inner membrane. Two abnormalities were consistently seen in V-ATPase-depleted fat body cells: a substantial increase in vesicle size in response to 4 h of starvation and an inappropriate buildup of mCh-Atg8a-labelled vesicles in the cells of well-fed animals [125].

By tagging LAMPs (lysosome-associated membrane proteins participating in the transfer of molecules into and out of the lysosome) with a fluorescent marker, lysosomes can be visualized in living cells and tracked over time, which allows researchers to study the dynamics of lysosomal trafficking and the role of lysosomes in autophagy and endosomal–lysosomal degradation regardless of their acidification status [12,20,126]. The mCherry-Atg8a reporter is particularly effective in labeling autolysosomes because it is targeted to the autophagosome and the fluorescent signal remains stable when it is transported to the lysosome. This reporter is therefore widely employed to specify autophagic structures, such as autolysosomes and autophagosomes, and in cases where autophagy is impaired, autophagosome formation is also impaired, resulting in no punctate signal being detected [127]. This fluorescence reporter can also indicate changes in the expression of autophagy-related proteins, which can be indicative of defects in autophagosome fusion or maturation [128]. The presence of large and bright autolysosomes in cells signals that they are undergoing normal cellular degradation processes, while small, faint autophagosomes in starved mutant cells suggest an inability to degrade contents efficiently [12,18,22,127].

Recent studies have reported switching to the use of *Drosophila* Atg8a fused to triple mCherry and LAMP reporters and that such triple-mCherry-tagged fluorescent reporters facilitate visualization of autophagic vesicles and lysosomes in fat body experiments and other tissues from larval wing disc or midgut cells of adult *Drosophila* [12,129].

The selective receptor of ubiquitinated proteins, commonly known as Ref(2)P/p62, is a remarkable autophagic cargo that can be tracked through fluorescent or HA-tagged reporters [20,39,130]. Not only can Ref(2)P/p62 be detected but it can also be the focus of autophagic degradation [5,131]. This means that when autophagy is functioning correctly, p62 levels are low, as it is quickly degraded; however, if autophagy is impaired, p62 accumulates and forms aggregates, and such accumulation can be visualized through the CLSM, allowing the measurement of autophagic degradation in a cell [12,20,39,132].

Lysotracker and Magic Red dyes are commonly used to stain fat cells in larvae [12,20,29,47,133], and are membrane permeable, accumulating in acidic organelles. The appearance of Lysotracker-positive vesicles in starved cells in cases of autophagy induced by starvation indicates a dramatic increase in the compartment known as an autolysosome [29,47]. Commercially available Magic Red dye, which fluoresces red upon intracellular Cathepsin B protease activity, can reliably identify and characterize functional lysosomes/autolysosomes that contain active cathepsin [20,133]. However, it is important to note that macrophages and nephrocytes also carry large endolysosomes or phagolysosomes, components in the elimination of cellular waste, including foreign particles and bacteria, through phagocytosis, which test positive for Lysotracker, so they may be confused with autolysosomes [134]. Hence, these dyes cannot be solely relied upon to indicate autophagy in such cells.

The lack of antibodies that allow the detection of specific endogenous proteins through Western blotting (WB) and indirect immunofluorescence (IF) methods, as well as examination of interactions between proteins by immunoprecipitation assays, is a drawback of *Drosophila* as a model organism. However, the widespread availability of antibodies against Atg8a, the most common marker, makes it feasible to perform WB and IF microscopy experiments [12,71,77,135–137]. In addition, commercially available anti-GABARAP antibodies in humans are another alternative for research involving WB and IF techniques in *Drosophila* [127,138]. Monitoring and detecting endogenous Atg8a-positive vesicles is a reliable method for identifying autophagosomes. There is also an antibody against the protein SNARE Syntaxin17 (Syx17), which is necessary for autophagosome–lysosome fusion. However, despite its localization in autophagosomes [23], anti-Syx17 antibody alone is not a definitive autophagosome marker, since it is also present in other organelles like mitochondria and endoplasmic reticulum [22,127]. Some commercially available anti-Atg5 and anti-Atg12 antibodies for *Drosophila* have been employed in studies using the IF method to track autophagy initiation as phagophore markers [12,23,139,140].

Handling fly tissues requires no special equipment, so IF methods could be performed in a fashion comparable to those employed while working on higher vertebrates. Fat bodies can be stained by inverting carcasses, fixing, and staining in small containers and then dissected and mounted after staining [127]. There are two different anti-Ref(2)P/p62 antibodies that may yield reliable results in studies employing WB and IF techniques [12,131,132].

The involvement of other intracellular vesicular transport pathways in autophagy may also be studied through markers for endosomes, the Golgi apparatus, and the endoplasmic reticulum. To that end, a comprehensive antibody toolkit was developed in Sean Munro's lab [141]. In addition, fluorescently tagged reporters for organelles in transgenic *Drosophila* offer some potential to analyze autophagy progression and mutant phenotypes. When retromer depletion was tested on cells, lysosomal hydrolases were found to be loaded improperly, and enlarged acidic autolysosomes accumulated, which might have been mistaken for increased autophagy, although TEM studies revealed that the cytoplasmic material in these vesicles was still intact [142].

TEM is widely considered a standard method for characterizing vesicular transport events [12]. Although TEM is the gold standard for visualizing autophagic vesicles, it is imperative to validate its results using other methods such as the CLSM and biochemical assays [143], which provides additional information about autophagy and its impact on cellular processes. Autophagic structures in *Drosophila* cells are almost identical in appearance to those observed in mammalian cells, which makes it relatively straightforward for researchers with prior TEM experience to identify autophagic vesicles in *Drosophila*.

Immuno-electron microscopy (Immuno-EM), as described by Lőrincz et al. [127], is another approach to the study of autophagy that can be performed through standard methods. Researchers often use acrylic resins, such as LR White, instead of epoxy resins, as well as milder chemical fixation methods to preserve the antigens. These adjustments are important to ensure that the antigens are preserved and can be successfully visualized and analyzed [23]. Also, utilizing an embedding method with a progressive temperature lowering may enhance the preservation of antigens during the Immuno-EM analysis [144]. As well as progressively lowering the temperature, techniques such as cryo-ultra-sectioning and sucrose infiltration of fixed samples may prove beneficial in immunogold labeling during such analyses. Another convenient tool could be correlative light and electron microscopy (CLEM), which combines the advantages of EM and light microscopy to provide high-resolution images of cellular structures and processes. Although CLEM demands task-specific equipment and expertise, it is appropriate for autophagy studies in *Drosophila* and can afford invaluable information about such complex biological processes.

Another method is acid phosphatase enzyme cytochemistry, a technique widely adopted in the past to discover and analyze lysosomes in various organisms. It involves detecting the activity of acid phosphatases, which are commonly found in lysosomes and autolysosomes, through specialized histochemical staining procedures; thus offering a powerful way of visualizing the ultrastructural characteristics of lysosomes and autolysosomes, though it is now a largely forgotten technique. The practical method of acid phosphatase enzyme cytochemistry in *Drosophila* involves incubating fixed tissue samples with a substrate solution to visualize the presence and distribution of acid phosphatase in lysosomes within cells. The deposition of an electron-dense precipitate serves as a visual indicator of the lysosomal localization of the enzyme [12,127,144].

Western Blotting (WB) is typically conducted with samples obtained from the entire animal body or tissues dissected from the body. To begin, the samples are boiled in a Laemmli buffer containing SDS for three minutes, then homogenized. This boiling process is repeated to extract protein, and two centrifugation steps remove fat and other debris. Atg8a is the primary protein commonly detected in WB experiments [127]. It is comparable to the detection of mammalian LC3 in blots [1]. The method distinguishes between the autophagosome-associated version of Atg8a (Atg8a-II) and the non-lipidated form (Atg8a-I) by their migration during gel electrophoresis. Effective separation of the two bands using a 13% or higher polyacrylamide gel is used to evaluate autophagy. Increased levels of Atg8a-II protein as compared to a loading control suggest increased autophagosome numbers [12,18,20,23]. A decrease in the amount of Atg8a-II protein suggests a problem in the lipidation of Atg8a or induction of autophagy. At the same time, WB results require careful interpretation as high levels of lipidated Atg8a can accumulate in some Atg mutants and elevated levels could imply high autophagic activity in the cells [128]. To avoid misinterpretation, Atg8a immunoblots should always be evaluated along with flux and morphological assays.

WB research may also involve the detection of TOR (target of rapamycin) activation, which is known to inhibit autophagy. In such cases, researchers can determine phosphorylation levels of common TOR targets like 4EBP1 and S6K to estimate TOR kinase activity through readily available antibody kits like phospho-4E-BP1 and anti-phospho-S6K [12,125,138,145].

With uses comparable to mammalian RFP-GFP-LC3B reporters [1], tandem-tagged mCherry-GFP-Atg8a reporters are commonly employed to track autophagic flux in flies [12,23,119,146]. When autophagosomes fuse with lysosomes, the low lysosomal pH quickly reduces the GFP signal, thus autophagosomes are observed in the form of small dots that are positive for GFP and mCherry, while autolysosomes test positive for mCherry only. If internal material fails to degrade, enlarged yellow structures are observed under microscopy [18,23,128]. However, when a significant number of small autophagosomes are grouped together, it may complicate the visualization of each vesicle through the CLSM. In these instances, ultrastructural analysis can be beneficial in determining whether there is a fusion defect between autophagosomes and lysosomes or whether autolysosomal degradation is impaired. A common method for measuring autophagic flux involves examining the buildup of Ref(2)P/p62-positive protein aggregates and ubiquitinated proteins within cells [127,128,131,132]. Researchers have recently created a GFP-p62 reporter driven by the tubulin promoter to avoid issues with Gal4/UAS-mediated p62 overexpression and gene regulation. The reporter, expressed at a constant low level in larval tissues, primarily reflects autophagic degradation, making it a highly sensitive indicator of disrupted autophagic flux [12,20].

Ref(2)P/p62 antibodies are employed by researchers utilizing the WB method to monitor autophagic degradation where a rise in p62 levels often signifies a block in the autophagic intracellular degradation system [12,131,132]. Other techniques involving WB as the main tool for measuring autophagic flux exist, primarily relying on the transformation of tagged Atg8a reporters to free GFP or mCherry within lysosomes [127,128,147].

Drosophila has been used for medium-throughput drug experiments on a range of diseases such as heart dysfunction, neurodegeneration, and obesity induced by high-fat intake. *Drosophila* larvae and adult flies can be cultured in compounds like rapamycin to trigger autophagy in this model organism [12,29]. Spermidine, an autophagy-triggering compound, has been shown to increase lifespan in several models and protect against toxicity from pesticide paraquat, which has been linked to Parkinson's disease (PD) and alpha-synuclein (α-Syn), a key protein found to play a role in the pathology of PD [127,148,149]. Toxicity assays have revealed that compounds mimicking ecdysone hormone in insects (like RH 5849) may initiate autophagy machinery in the fat body [12,83]. Chloroquine (CQ) is known to block the acidification of autolysosomes and trigger myopathies when fed to fly larvae, and this method of feeding CQ to larvae has been found to be effective in initiating muscle toxicity [12,150]. Bafilomycin A1 can also be utilized to prevent the merging and acidification of autophagosome and lysosome; however, due to the potential alteration of TOR signaling, caution must be exercised when evaluating the outcomes [12,151]. Additionally, further testing on flies could help validate possible drug candidates that may induce or enhance autophagy, such as AUTEN-67, known as autophagy enhancer-67 [12,152].

Another approach is to incubate fly tissues with drugs ex vivo, allowing for the determination of direct effects on a specific cell type and avoiding the potential toxicity of administering drugs to whole animals. Bafilomycin A1 is an example of a drug that can be tested using this strategy [125].

Drosophila is a valuable model system in part due to the abundance of genetic methods accessible to researchers without much effort. There are loss-of-function mutations available for almost all core autophagy-related genes. However, for the *Drosophila* counterparts of Atg101 and FIP200, which are newly discovered parts of the Atg1-Atg13 complex, no mutant alleles currently exist. This means that researchers do not have the tools to manipulate the function of these genes in the same way they can with the other core autophagy genes. In that regard, a database for *Drosophila* genetics, FlyBase, contains all information related to the current autophagy alleles and transgenic constructs, serving as a comprehensive resource for researchers to access up-to-date information on currently available genetic tools for autophagy research on *Drosophila*.

In *Drosophila*, tissues that are highly impacted by autophagy are postmitotic, which makes it challenging to conduct clonal analysis. This complicates the process of assessing the tissue-specific functions of autophagy genes, as their function is often pleiotropic. To overcome this challenge, RNAi technology has proven crucial, as it enables the precise control of gene knockdown in terms of both space and time. In contrast, genome-wide RNAi screens in *Caenorhabditis elegans* are commonly performed through feeding methods, a widely used practice in the study of gene function [153]. In *Drosophila*, RNAi feeding cannot be implemented, so alternative methods have been developed to produce transgenes expressing snapback/hairpin constructs. One of the most popular methods involves the utilization of Gal4/UAS [154–156] to specifically express a double-stranded RNAi hairpin, which results in the post-transcriptional silencing of target genes. Due to achievements reached by transgenic RNAi, research efforts are geared towards creating a large collection of hairpin lines on a genome-wide scale [155]. A hairpin construct is a piece of genetic material that folds back on itself to form a loop, with the two ends of the DNA strand base-pairing to form a hairpin-like structure. In this respect, researchers created a system for the targeted integration of these constructs into specific locations in the genome, allowing for the study of gene function and regulation. This system allows for the precise insertion of the hairpin construct at a specific location in the genome, allowing for the study of its effects on gene expression and cellular behavior [154,157]. Researchers collaborated with the *Drosophila* RNAi Screening Center to create a "second generation library" and produced transgenic RNAi lines that target the majority of the core autophagy genes, which are accessible on FlyBase [158,159].

The exact potential of using *Drosophila* genes for research on autophagy has yet to be understood. One reason for this is that the tissues most commonly targeted in autophagy studies on *Drosophila* are postmitotic and polyploid, making standard mosaic analysis less effective. Additionally, it is difficult to design a screening method for autophagy that does not have a readily observable morphology. While the RNAi-based evaluation of autophagy in well-known *Drosophila* cell lines shows some promise, there are questions about the biological relevance of these studies. The following section will discuss various approaches used for autophagy screening in *Drosophila*.

In the past, researchers demonstrated that changes in the *Drosophila* blue cheese gene (bchs) cause adult life expectancy to shorten as well as age-dependent brain degeneration and cell death [160]. Bchs and its human equivalent Alfy play a crucial role in the removal of protein aggregates through autophagy, although the exact mechanism is still unknown [161]. Studies employing stocks with deficiencies and chosen mutant variations managed to discover that mutations in the genes responsible for lysosomal trafficking altered the strong bchs eye appearance. Significantly, Atg1, Atg6, and Atg18 were found to accentuate the phenotype, implying that this method could accurately spot genes related to autophagy. Another study by Arsham and Neufeld [162] combined mosaic analysis and live-cell/fixed-cell imaging to screen autophagy regulators. They used the Flp-FRT system to generate homozygous mutant clones and analyzed the lysosomal activity in the mutant cells compared to surrounding wild-type tissue, identifying 79 transposon insertions heightening lysosomal activity. These studies highlight the importance of a multidisciplinary approach to the study of autophagy, as they demonstrate the importance of combining classical genetics with cutting-edge technologies such as microarray analysis and mass spectrometry. The results of these studies provide new avenues for further investigation into the mechanisms of autophagy in *Drosophila*.

D. melanogaster is one of the best-studied models for numerous mutagenic screens, development, and aging. This fly proceeds through well-defined stages during its life cycle, including embryo, larva, pupa, and adult, accomplishing complete phenotypic metamorphosis. These changes are induced by carefully controlled gene expression at the transcriptional, epigenetic, and translational levels. Most developmental gene expression investigations using *Drosophila* are currently based upon RNA in situ hybridization and transcriptome analysis which employ large-scale microarray/RNA-seq data sets, or a combination ofboth. To that end, two contemporaneous studies explored genome-wide transcript analysis of salivary glands undergoing cell death related to autophagy [121,122]. By using microarrays, Lee et al. discovered several fly ATG genes that showed an increase in transcription, including Atg2, Atg4, Atg5, and Atg7. Further research from the same laboratory revealed that Dynein Light Chain 1 plays a crucial role in inducing autophagy during cell death in salivary glands [163]. The study by Gorski et al. [121] involved a serial analysis of gene expression (SAGE) approach to examine the transcript levels in salivary glands undergoing autophagic cell death. The authors reported that they identified over 732 differentially expressed genes with unknown functions. Juhasz et al. [68] used microarray technology to investigate the initiation of autophagy in the larval fat body and discovered that the downregulation of FK506-binding protein 39 kDa (FKBP39) occurred during autophagy. The study revealed that FKBP39 inhibited autophagy, probably by exerting a modifying or controlling effect on the Foxo transcription factor.

In addition to gene expression analysis, two research teams employed high-throughput screening by proteomic techniques to investigate proteins involved in autophagy in the salivary glands and fat body of *Drosophila*, allowing for comprehensive analyses of proteins expressed during autophagy, providing a deeper understanding of the molecular events underlying this cellular process [164,165]. The latter team employed a shotgun proteomics method to uncover the proteins involved in the autophagic cell death of larval salivary glands. Their findings aligned with earlier microarray and SAGE studies, but also revealed new players in the process—namely Warts, a kinase in the Hippo pathway that was found to be vital for the regulation of autophagy and the programmed cell fate of salivary glands [166]. The former team of researchers, Kohler et al. [164] utilized a mass spectrometry technique with an isotope-coded affinity tag to uncover the players in the starvation-prompted autophagic response. By contrasting proteins from starved and normal fat body in *Drosophila*, they managed to identify 110 proteins that varied in regulation. One noteworthy discovery was the upregulation of the lipid desaturase Desat1 in the starved sample, which was found to be a necessary component for starvation-triggered autophagy and to be localized to structures positive for Atg5 and Atg8.

Drosophila is a valuable testing model in research employing comprehensive genome-wide high-throughput RNAi screening thanks to its accessible cell culture lines that are capable of quickly absorbing long dsRNAs introduced to the medium, leading to effective target gene suppression [167]. Additionally, the availability of extensive dsRNA libraries enables large-scale screenings to systematically study the functions of all genes predicted from genomic sequencing. Despite this, there has been no published report of a genome-wide autophagy screening using this system, which is somewhat surprising. Nevertheless, it is evident that cell cultures from *Drosophila* can be utilized to explore autophagic function in immunity, cell fate, and nutrient deprivation [59,71,168,169]. In their screening study, the research team led by Chittaranjan aimed to validate the involvement of genes in autophagic cell death of the salivary gland. To achieve this, they selected 460 genes that had been previously identified through expression studies as potentially playing a role in this process. The screening effort was considered moderately sized, indicating a significant number of genes were being tested, but not necessarily covering the entire genome [170]. The researchers triggered cell death in a tumorous hemocyte cell line from *Drosophila* using ecdysone, known to initiate metamorphosis and autophagy-related cell death. Through dsRNA analyses, they discovered 25 genes that may affect survival. Further study showed that the knockdown of genes including Atg2, Atg3, Atg5, Atg6, Atg7, Atg8a, and Atg8b resulted in lower levels of cell survival [170]. The results of their study provided valuable insights into the genetic processes in salivary gland cells leading to death.

1.6. The Role of Autophagy in Organ System Function and Developmental Processes

Autophagy can also act in response to infection, and an autophagy protein called Beclin was demonstrated to provide some protection against Sindbis virus (SB)-induced encephalitis in mice [171]. Moreover, autophagosomes can directly engulf bacterial and viral pathogens [172–174]. Conversely, certain pathogens, including poliovirus and rhinoviruses, can use the autophagic machinery to replicate as well as escaping degradation by autophagy [175].

Although *Drosophila* lack the adaptive immune response present in higher vertebrates, this fly can still be a reliable, useful model organism for studying the function of autophagy in the immune system. In this regard, the Listeria infection highlights both benefits and drawbacks of bacteria-induced autophagy in adult *Drosophila*. Microorganisms responsible for listeriosis trigger antibacterial autophagy in fruit fly cells, relying on the peptidoglycan recognition protein LE [169]. However, Listeria can escape autophagy in mammalian macrophages through ActA in the cytoplasm and Listeriolysin O, creating non-degradable phagosomes [176–178]. Such phagosomes limit bacterial growth, forming from failed autophagosome or lysosome fusion attempts. Despite these varying autophagy responses to Listeria-induced infection, studying Listeria and autophagy in *Drosophila* cells can still provide insight into Listeria's ability to evade autophagy.

Previous work in the field [71,179] once again confirmed the value of *Drosophila* as an experimental model in viral autophagy research. These studies found that vesicular stomatitis virus (VSV) infected both cultured and in vivo cells and triggered autophagy, and suppression of core autophagy genes with RNAi led to higher severity of VSV infection. Modifying the insulin signaling pathway in experiments was found to have an effect on virus replication, reflecting PI3K/Tor pathway's contribution to autophagy. To date, this is the only study to examine virus-induced autophagy in fruit flies. Such a research model could attract further research due to *Drosophila*'s rapid genetics and the evolutionary distance between host and virus, making it easier to identify potential pathways involving interactions between pathogens and autophagy [180].

Autophagy is a vital process for cellular health, as it helps maintain a balance between the production and breakdown of cellular components as well as protecting cells from damage caused by various environmental factors. Autophagy also helps recycle nutrients and energy under starvation conditions, such as when oxygen levels are low, as is the case in the heart muscles of mammals, facilitating the survival of the cells [181]. Such a protective role could have considerable implications for tumor cells, because autophagy may either contribute to the survival of tumor cells [182] or inhibit their growth [183]. Conditions such as Parkinson's, Alzheimer's, and Huntington's, which are all neurodegenerative diseases, involve the accumulation of large amounts of mutated proteins. Of all the human diseases related to autophagy, those concerning the unusual accumulation of proteins have the most advanced model systems in fruit flies, and increased autophagosome generation seen in such diseases could have protective functions [184]. Rapamycin is an immunosuppressant drug that has been shown to induce autophagy in some cell types. Rapamycin, an mTOR inhibitor, induces autophagy and speeds up the removal of these harmful substrates. Accordingly, treatment with the rapamycin analog has been found to cause declines in huntingtin aggregate accumulation and neurodegeneration. This suggests that induction of autophagy may be beneficial in the treatment of neurodegenerative diseases [185]. *Drosophila* neurodegenerative disease models have been employed in several studies to investigate the role autophagy plays in the toxicity of TDP-43, polyQ repeat, and Amyloid beta 42 [25,186–188]. By conducting a genetic analysis of the *Drosophila* version of Alfy, a protein associated with autophagy, researchers were able to verify its function in the context of Huntington's disease. To further investigate how Alfy works, the researchers implemented Huntington's pathological changes in the retina to examine its role in eliminating ubiquitin-positive protein inclusions and halting the degeneration of neurons. The results of this analysis showed that the *Drosophila* ortholog of Alfy is essential for such elimination and suppression of deterioration of neurons in vivo, confirming its role as a key player in autophagy. This research has been invaluable in furthering our understanding of the pivotal part of autophagy in the progression of Huntington's and has provided insight into potential therapeutic strategies for this devastating neurological disorder [160,161,189].

1.7. How Data from Drosophila Informs Insights into the Role(s) of Autophagy in Mammalian Physiology and Pathogenesis

Autophagy is essential for keeping cells balanced and functioning normally, as well as helping them to cope with challenges like a lack of nutrients. An insect's fat body stores fat and helps store and use nutrients and carries out vital metabolic activities, which means that it can be thought of as an insect's equivalent of a liver [190]. The fat body, which serves as the primary storage location, acts rather quickly in cases of deprivation by releasing amino acids, carbohydrates, and lipids [190]. By exposing larvae to sucrose solution (20%) in the laboratory, autophagy in fat body cells can be triggered due to amino acid starvation. This method is advantageous in comparison to submerging larvae in water, as the solution's high density enables larvae to float on the surface. Within an hour and a half, autophagy starts and it peaks in around three to five hours [29]. Larval exposure to sucrose solution has been found to trigger the synthesis of glycogen to great extents in fat cells, which is an interesting finding [191].

The fat body forms from the embryonic mesoderm and is made up of two lobes of cells arranged in monolayers for simple microscopic inspection. These benefits make fat cells a popular tool for researching autophagy in *Drosophila*. In addition, organs like the compound eye are also studied to explore the part of autophagy in neurodegenerative disorders [134]. In addition, the salivary glands are dependent upon both autophagy and apoptosis for death, while the midgut depends exclusively upon autophagy and represents the most reliable example of autophagic cell death in any organism. Autophagy in these tissues has been studied in both larvae and adults, allowing researchers to investigate the differences in autophagic processes between different developmental stages as well as the shrinkage and death of cells [79,192]. The larval midgut is an ideal organ for studying intracellular trafficking if epithelial cell polarization is desired, as it responds well to autophagy triggered by starvation in comparison to the fat body. Additionally, the ovaries of adult female *Drosophila* have also been used to analyze this type of autophagy [135].

Research into the Atg7-deleted mutant, the first autophagy gene null animal, showed a distinctive effect where *D. melanogaster* suffered delayed development, but no visible morphological defects. However, they were found to be more vulnerable to oxidative stress and nutrient deprivation, with a shorter life [82]. These phenotypes have been discovered in the presence of null mutants of certain genes like Atg5, Atg16, necessary for lipidation in *Drosophila* Atg8a protein, which is considered a homolog of mammalian LC3 proteins [136,193]. Lipidation in Atg8a and its mutants are both possible, likely due to the presence of residual autophagic degradation, which has been demonstrated in mammalian cells [194]. Overall, the function autophagy plays in aging and stress tolerance can be effectively studied using *D. melanogaster*. Research on a variety of model organisms, including mice, has shown that increased autophagy can help maintain cellular balance, thereby increasing longevity. For example, the expression of Atg8a and Atg1 in *Drosophila* neurons has been observed to extend lifespan by up to 50% in comparison to control animals. Additionally, moderate Atg1 expression in the fat body, intestine, and Malpighian tubules of *Drosophila* has been found to extend lifespan by altering mitochondrial genes and enhancing proteostasis [195]. The elevation in the amount of Ref(2)P (p62 in mammals), a significant cargo receptor in selective autophagy, not only enhances proteostasis but also mitochondrial function and mitophagy [196]. This mitophagy activation then leads to improved mitochondrial health and homeostasis, ultimately resulting in an extended survival or lifespan extension thanks to the activity of prolongevity pathways [197]. When mTOR is inhibited, a mitophagy-dependent decrease in cyclin E in germline stem cells (GSCs) and human induced pluripotent stem cells (hiPSCs) delays the normal G1/S transition, driving the cells toward reversible quiescence (G0) [33].

In *Drosophila*, while the selective autophagy of ubiquitinated proteins is understood well, organelle degradation has received little attention. However, Vincow et al. [198] used a proteomics-based method to show that the main autophagy pathway targets mitochondria for degradation in lysosomes, a cellular organelle acting as the cell's recycling center. During this process, subunits of the respiratory chain, the series of protein complexes responsible for generating energy within the mitochondria, are selectively removed independently of the protein Atg7. The degradation of mitochondria in lysosomes can occur through the formation of mitochondria-derived vesicles. These vesicles are formed through a process that is dependent on the protein Syntaxin 17, although the precise details of this process have yet to be fully investigated in fruit flies. The formation of these vesicles and their subsequent fusion with lysosomes is thought to represent an important mechanism for the targeted degradation of mitochondria through autophagy [199,200].

Testing on complete animals instead of cultured cells in autophagy-related research offers a series of benefits; for instance, this approach allows us to observe the organism as a whole, such as through neuromuscular evaluations in negative geotaxis (climbing) assays, while it helps gain better insight into the intricate and tissue-specific regulation of autophagy by examining the information exchange between various tissues and cells through certain metabolites and hormones [129,138,201].

The ubiquitin–proteasome system and autophagy are two important mechanisms for maintaining protein quality control in cells. The ubiquitin–proteasome system acts like a "cleanup crew" for short-lived proteins, marking them for destruction with a chemical tag called ubiquitin and then breaking them down in a structure called the proteasome. Autophagy, conversely, can be thought of as a "recycling plant" for larger cellular structures and damaged organelles. Autophagy engulfs these structures in a membrane-bound sac, which then fuses with a degradation center called a lysosome to break down the contents and recycle the resulting molecules. In periods of starvation, autophagy serves as a homeostatic response to nutrient deprivation, whereas during periods of abundant nutrition, it is virtually undetectable in yeast; however, after about 30 min of nitrogen deprivation, autophagosome generation increases significantly. Thus, autophagy appears to emerge as a vital mechanism for maintaining cellular homeostasis in periods of nutrient deficiency [202]. The appearance of autophagosomes occurs within one hour of starvation in the fat body of the fruit fly (*D. melanogaster*), a multifunctional organ that performs functions similar to those of the liver in mammals in storing nutrients and providing a source of energy [29]. Similarly, research in mice has shown that under starvation conditions, autophagy increases in many organs as a way for cells to conserve energy and recycle cellular components for use as an energy source. This response is particularly pronounced in muscle tissue, where autophagy helps preserve muscle mass and prevent muscle wasting [203]. Additionally, it has been observed that autophagy is upregulated by a range of tissues in newborn mice, potentially as a means of adapting to the sudden lack of nutrients resulting from detachment from the placenta [84]. Thus, these findings demonstrate that the ability of autophagy to protect the organism in case of starvation appears to be a rudimentary function preserved during the evolution of eukaryotic organisms.

In yeast cells, the regulation of autophagy in response to starvation is primarily controlled by the Target of Rapamycin (TOR) pathway, which senses the availability of nutrients in the environment and adjusts cellular metabolism accordingly. Under conditions of high nutrient availability, the TOR pathway suppresses autophagy, allowing cells to grow and divide, whereas when nutrients become scarce, the TOR pathway is inhibited, leading to the activation of autophagy as a way for cells to conserve energy and recycle cellular components [96]. Similarly, higher eukaryotic organisms, such as *Drosophila*, also control and regulate autophagy by means of the PI3K pathway, which is upstream of the TOR pathway which senses changes in the levels of insulin and other growth factors and adjusts cellular metabolism [204]. Previous research confirmed that both the TOR and insulin pathways regulate autophagy in the fat body of *Drosophila* larvae [27]. The fat body cells of organisms have been observed to exhibit a rapid autophagic response when subjected to starvation, treatment with the drug rapamycin, or genetic inactivation of the TOR pathway [29,32,83]. This is evidenced by the strong induction of autophagy in this tissue following the loss of PI3K or insulin receptor function [29,32,83]. It has thus been demonstrated that the pathways that regulate autophagy under nutrient-scarce conditions are conserved in *Drosophila*. Furthermore, equivalent ATG genes contributing to autophagosome generation in the *Drosophila* fat body model implies that autophagy in this fruit fly species is dependent on preserved features of ATG machinery.

Previous research demonstrated TOR-mediated autophagy across *Drosophila* and mammals. In their study, Kim et al. [205] discovered that Rag GTPases activate TOR when amino acid signaling is present. Experiments on cell cultures from fruit flies and mammals revealed that a reduction in Rag gene expression weakens the impact of amino acids on the TOR pathway. To further verify the role of Rag in the TOR-mediated regulation of autophagy and cell size regulation, they conducted an in vivo testing in the fat body of fruit flies. In a subsequent study, Li et al. [206] explored the regulation of TOR activity and autophagy by Rab and Arf family GTPases. They discovered that these GTPases also regulate TOR, though not through direct interaction with it, unlike Rag.

2. Future Directions

Several studies utilizing the *Drosophila* model demonstrated that several signaling pathways and treatments play well-established roles in control of aging and cancer. *Drosophila* researchers have access to an extensive set of tools for investigating autophagy. The remarkable conservation of the autophagy machinery between this fruit fly species and humans has made cross-species research particularly valuable, making possible the discovery of up-to-now uncharted territories in the autophagy pathway. Of particular note, though, it is crucial to use multiple, complementary assays to accurately determine the status of autophagy in *Drosophila*, just as with various other species. Given that this testing model's cell biology and physiology exhibit striking similarities to those of humans and that there are well-established models for analyzing various pathologies, such as the progression of cancer and neurodegeneration, it has proven to be the ideal system for studying the fundamental mechanisms and regulation of autophagy. With the establishment of cultured cell models for autophagic cell death, bacterial autophagy, and viral autophagy, the amount of research involving RNAi screens for such phenomena should increase in the coming years. Based on recent knowledge gained from autophagy research, this review article will hopefully help us gain a better understanding of autophagy machinery at the molecular level in *Drosophila* and humans. There are, however, still many aspects of autophagy in flies that need to be explored, such as selective organelle degradation and selective autophagy, or xenophagy. Further advancements in the understanding of the function and regulatory pathways of autophagy in *Drosophila* should yield new insights for grasping the significance of this process.

Author Contributions: E.D.: Conception, and Writing—Original Manuscript. S.K.: Conception, and Writing—Original Manuscript. All authors have read and agreed to the published version of the manuscript.

Funding: The authors declare no funding.

Institutional Review Board Statement: Not applicable.

Informed Consent Statement: Not applicable.

Data Availability Statement: Not applicable.

Acknowledgments: Eşref Demir was supported by a Fulbright Scholarship from the Turkish Fulbright Commission-Fulbright Visiting Scholar Program (Ankara, Turkey) for the 2022–2023 academic year.

Conflicts of Interest: The authors declare no conflict of interest.

References

1. Klionsky, D.J.; Abdelmohsen, K.; Abe, A.; Abedin, M.J.; Abeliovich, H.; Acevedo Arozena, A.; Adachi, H.; Adams, C.M.; Adams, P.D.; Adeli, K.; et al. Guidelines for the use and interpretation of assays for monitoring autophagy (3rd edition). *Autophagy* **2016**, *12*, 1–222. [CrossRef] [PubMed]
2. Cuervo, A.M. Autophagy: Many paths to the same end. *Mol. Cell. Biochem.* **2004**, *263*, 55–72. [CrossRef] [PubMed]
3. McPhee, C.K.; Baehrecke, E.H. Autophagy in *Drosophila melanogaster*. *Biochim. Biophys. Acta* **2009**, *1793*, 1452–1460. [CrossRef]
4. Mizushima, N.; Levine, B.; Cuervo, A.M.; Klionsky, D.J. Autophagy fights disease through cellular self-digestion. *Nature* **2008**, *451*, 1069–1075. [CrossRef] [PubMed]
5. Lippai, M.; Szatmári, Z. Autophagy-from molecular mechanisms to clinical relevance. *Cell Biol. Toxicol.* **2017**, *33*, 145–168. [CrossRef] [PubMed]
6. Levine, B.; Klionsky, D.J. Development by self-digestion: Molecular mechanisms and biological functions of autophagy. *Dev. Cell* **2004**, *6*, 463–477. [CrossRef] [PubMed]
7. Mizushima, N.; Ohsumi, Y.; Yoshimori, T. Autophagosome formation in mammalian cells. *Cell. Struct. Funct.* **2002**, *27*, 421–429. [CrossRef]
8. Khurana, V.; Lindquist, S. Modelling neurodegeneration in *Saccharomyces cerevisiae*: Why cook with baker's yeast? *Nat. Rev. Neurosci.* **2010**, *11*, 436–449. [CrossRef]
9. Surguchov, A. Invertebrate models untangle the mechanism of neurodegeneration in Parkinson's disease. *Cells* **2021**, *10*, 407. [CrossRef]
10. Tsukada, M.; Ohsumi, Y. Isolation and characterization of autophagy-defective mutants of *Saccharomyces cerevisiae*. *FEBS Lett.* **1993**, *333*, 169–174. [CrossRef]
11. Klionsky, D.J.; Cregg, J.M.; Dunn, W.A., Jr.; Emr, S.D.; Sakai, Y.; Sandoval, I.V.; Sibirny, A.; Subramani, S.; Thumm, M.; Veenhuis, M.; et al. A unified nomenclature for yeast autophagy-related genes. *Dev. Cell* **2003**, *5*, 539–545. [CrossRef] [PubMed]
12. Lőrincz, P.; Tóth, S.; Benkő, P.; Lakatos, Z.; Boda, A.; Glatz, G.; Zobel, M.; Bisi, S.; Hegedűs, K.; Takáts, S.; et al. Rab2 promotes autophagic and endocytic lysosomal degradation. *J. Cell Biol.* **2017**, *216*, 1937–1947. [CrossRef] [PubMed]
13. Mulakkal, N.C.; Nagy, P.; Takats, S.; Tusco, R.; Juhász, G.; Nezis, I.P. Autophagy in *Drosophila*: From historical studies to current knowledge. *Biomed. Res. Int.* **2014**, *2014*, 273473. [CrossRef] [PubMed]
14. Yorimitsu, T.; Klionsky, D.J. Autophagy: Molecular machinery for self-eating. *Cell Death Differ.* **2005**, *12* (Suppl. 2), 1542–1552. [CrossRef] [PubMed]
15. Mizushima, N.; Yoshimori, T.; Ohsumi, Y. The role of Atg proteins in autophagosome formation. *Annu. Rev. Cell Dev. Biol.* **2011**, *27*, 107–132. [CrossRef] [PubMed]
16. Yang, Z.; Klionsky, D.J. An overview of the molecular mechanism of autophagy. *Curr. Top. Microbiol. Immunol.* **2009**, *335*, 1–32. [CrossRef]
17. Reggiori, F.; Ungermann, C. Autophagosome Maturation and Fusion. *J. Mol. Biol.* **2017**, *429*, 486–496. [CrossRef]
18. Takáts, S.; Pircs, K.; Nagy, P.; Varga, Á.; Kárpáti, M.; Hegedűs, K.; Kramer, H.; Kovács, A.L.; Sass, M.; Juhász, G. Interaction of the HOPS complex with Syntaxin 17 mediates autophagosome clearance in *Drosophila*. *Mol. Biol. Cell* **2014**, *25*, 1338–1354. [CrossRef] [PubMed]
19. Jiang, P.; Nishimura, T.; Sakamaki, Y.; Itakura, E.; Hatta, T.; Natsume, T.; Mizushima, N. The HOPS complex mediates autophagosome-lysosome fusion through interaction with syntaxin 17. *Mol. Biol. Cell* **2014**, *25*, 1327–1337. [CrossRef] [PubMed]
20. Hegedűs, K.; Takáts, S.; Boda, A.; Jipa, A.; Nagy, P.; Varga, K.; Kovács, A.L.; Juhász, G. The Ccz1-Mon1-Rab7 module and Rab5 control distinct steps of autophagy. *Mol. Biol. Cell* **2016**, *27*, 3132–3142. [CrossRef] [PubMed]

21. Fujita, N.; Huang, W.; Lin, T.H.; Groulx, J.F.; Jean, S.; Nguyen, J.; Kuchitsu, Y.; Koyama-Honda, I.; Mizushima, N.; Fukuda, M.; et al. Genetic screen in *Drosophila* muscle identifies autophagy-mediated T-tubule remodeling and a Rab2 role in autophagy. *eLife* **2017**, *6*, e23367. [CrossRef]
22. Itakura, E.; Kishi-Itakura, C.; Mizushima, N. The hairpin-type tail-anchored SNARE syntaxin 17 targets to autophagosomes for fusion with endosomes/lysosomes. *Cell* **2012**, *151*, 1256–1269. [CrossRef]
23. Takáts, S.; Nagy, P.; Varga, Á.; Pircs, K.; Kárpáti, M.; Varga, K.; Kovács, A.L.; Hegedűs, K.; Juhász, G. Autophagosomal Syntaxin17-dependent lysosomal degradation maintains neuronal function in *Drosophila*. *J. Cell Biol.* **2013**, *201*, 531–539. [CrossRef] [PubMed]
24. Mauvezin, C.; Neisch, A.L.; Ayala, C.I.; Kim, J.; Beltrame, A.; Braden, C.R.; Gardner, M.K.; Hays, T.S.; Neufeld, T.P. Coordination of autophagosome-lysosome fusion and transport by a Klp98A-Rab14 complex in *Drosophila*. *J. Cell Sci.* **2016**, *129*, 971–982. [CrossRef] [PubMed]
25. Ravikumar, B.; Acevedo-Arozena, A.; Imarisio, S.; Berger, Z.; Vacher, C.; O'Kane, C.J.; Brown, S.D.; Rubinsztein, D.C. Dynein mutations impair autophagic clearance of aggregate-prone proteins. *Nat. Genet.* **2005**, *37*, 771–776. [CrossRef] [PubMed]
26. Xie, Z.; Klionsky, D.J. Autophagosome formation: Core machinery and adaptations. *Nat. Cell Biol.* **2007**, *9*, 1102–1109. [CrossRef]
27. Zirin, J.; Perrimon, N. *Drosophila* as a model system to study autophagy. *Semin. Immunopathol.* **2010**, *32*, 363–372. [CrossRef]
28. Matsuura, A.; Tsukada, M.; Wada, Y.; Ohsumi, Y. Apg1p, a novel protein kinase required for the autophagic process in *Saccharomyces cerevisiae*. *Gene* **1997**, *192*, 245–250. [CrossRef]
29. Scott, R.C.; Schuldiner, O.; Neufeld, T.P. Role and regulation of starvation-induced autophagy in the *Drosophila* fat body. *Dev. Cell* **2004**, *7*, 167–178. [CrossRef]
30. Yan, J.; Kuroyanagi, H.; Kuroiwa, A.; Matsuda, Y.; Tokumitsu, H.; Tomoda, T.; Shirasawa, T.; Muramatsu, M. Identification of mouse ULK1, a novel protein kinase structurally related to *C. elegans* UNC-51. *Biochem. Biophys. Res. Commun.* **1998**, *246*, 222–227. [CrossRef]
31. Yan, J.; Kuroyanagi, H.; Tomemori, T.; Okazaki, N.; Asato, K.; Matsuda, Y.; Suzuki, Y.; Ohshima, Y.; Mitani, S.; Masuho, Y.; et al. Mouse ULK2, a novel member of the UNC-51-like protein kinases: Unique features of functional domains. *Oncogene* **1999**, *18*, 5850–5859. [CrossRef]
32. Scott, R.C.; Juhász, G.; Neufeld, T.P. Direct induction of autophagy by Atg1 inhibits cell growth and induces apoptotic cell death. *Curr. Biol.* **2007**, *17*, 1–11. [CrossRef] [PubMed]
33. Taslim, T.H.; Hussein, A.M.; Keshri, R.; Ishibashi, J.R.; Chan, T.C.; Nguyen, B.N.; Liu, S.; Brewer, D.; Harper, S.; Lyons, S.; et al. Stress-induced reversible cell-cycle arrest requires PRC2/PRC1-mediated control of mitophagy in *Drosophila* germline stem cells and human iPSCs. *Stem Cell Rep.* **2023**, *18*, 269–288. [CrossRef] [PubMed]
34. Hosokawa, N.; Sasaki, T.; Iemura, S.; Natsume, T.; Hara, T.; Mizushima, N. Atg101, a novel mammalian autophagy protein interacting with Atg13. *Autophagy* **2009**, *5*, 973–979. [CrossRef]
35. Jung, C.H.; Ro, S.H.; Cao, J.; Otto, N.M.; Kim, D.H. mTOR regulation of autophagy. *FEBS Lett.* **2010**, *584*, 1287–1295. [CrossRef]
36. Ganley, I.G.; Lam, D.H.; Wang, J.; Ding, X.; Chen, S.; Jiang, X. ULK1.ATG13.FIP200 complex mediates mTOR signaling and is essential for autophagy. *J. Biol. Chem.* **2009**, *284*, 12297–12305. [CrossRef] [PubMed]
37. Noda, T.; Ohsumi, Y. Tor, a phosphatidylinositol kinase homologue, controls autophagy in yeast. *J. Biol. Chem.* **1998**, *273*, 3963–3966. [CrossRef]
38. Chan, E.Y.; Kir, S.; Tooze, S.A. siRNA screening of the kinome identifies ULK1 as a multidomain modulator of autophagy. *J. Biol. Chem.* **2007**, *282*, 25464–25474. [CrossRef]
39. Chang, Y.Y.; Neufeld, T.P. An Atg1/Atg13 complex with multiple roles in TOR-mediated autophagy regulation. *Mol. Biol. Cell* **2009**, *20*, 2004–2014. [CrossRef]
40. Funakoshi, T.; Matsuura, A.; Noda, T.; Ohsumi, Y. Analyses of APG13 gene involved in autophagy in yeast, *Saccharomyces cerevisiae*. *Gene* **1997**, *192*, 207–213. [CrossRef] [PubMed]
41. Kamada, Y.; Funakoshi, T.; Shintani, T.; Nagano, K.; Ohsumi, M.; Ohsumi, Y. Tor-mediated induction of autophagy via an Apg1 protein kinase complex. *J. Cell Biol.* **2000**, *150*, 1507–1513. [CrossRef]
42. Kabeya, Y.; Kamada, Y.; Baba, M.; Takikawa, H.; Sasaki, M.; Ohsumi, Y. Atg17 functions in cooperation with Atg1 and Atg13 in yeast autophagy. *Mol. Biol. Cell* **2005**, *16*, 2544–2553. [CrossRef]
43. Kawamata, T.; Kamada, Y.; Suzuki, K.; Kuboshima, N.; Akimatsu, H.; Ota, S.; Ohsumi, M.; Ohsumi, Y. Characterization of a novel autophagy-specific gene, ATG29. *Biochem. Biophys. Res. Commun.* **2005**, *338*, 1884–1889. [CrossRef] [PubMed]
44. Kabeya, Y.; Kawamata, T.; Suzuki, K.; Ohsumi, Y. Cis1/Atg31 is required for autophagosome formation in *Saccharomyces cerevisiae*. *Biochem. Biophys. Res. Commun.* **2007**, *356*, 405–410. [CrossRef] [PubMed]
45. Hara, T.; Takamura, A.; Kishi, C.; Iemura, S.; Natsume, T.; Guan, J.L.; Mizushima, N. FIP200, a ULK-interacting protein, is required for autophagosome formation in mammalian cells. *J. Cell Biol.* **2008**, *181*, 497–510. [CrossRef] [PubMed]
46. Mercer, C.A.; Kaliappan, A.; Dennis, P.B. A novel, human Atg13 binding protein, Atg101, interacts with ULK1 and is essential for macroautophagy. *Autophagy* **2009**, *5*, 649–662. [CrossRef] [PubMed]
47. Juhász, G.; Hill, J.H.; Yan, Y.; Sass, M.; Baehrecke, E.H.; Backer, J.M.; Neufeld, T.P. The class III PI(3)K Vps34 promotes autophagy and endocytosis but not TOR signaling in *Drosophila*. *J. Cell Biol.* **2008**, *181*, 655–666. [CrossRef] [PubMed]
48. Qu, X.; Yu, J.; Bhagat, G.; Furuya, N.; Hibshoosh, H.; Troxel, A.; Rosen, J.; Eskelinen, E.L.; Mizushima, N.; Ohsumi, Y.; et al. Promotion of tumorigenesis by heterozygous disruption of the beclin 1 autophagy gene. *J. Clin. Investig.* **2003**, *112*, 1809–1820. [CrossRef] [PubMed]

49. Kihara, A.; Noda, T.; Ishihara, N.; Ohsumi, Y. Two distinct Vps34 phosphatidylinositol 3-kinase complexes function in autophagy and carboxypeptidase Y sorting in *Saccharomyces cerevisiae*. *J. Cell Biol.* **2001**, *152*, 519–530. [CrossRef] [PubMed]
50. Kametaka, S.; Okano, T.; Ohsumi, M.; Ohsumi, Y. Apg14p and Apg6/Vps30p form a protein complex essential for autophagy in the yeast, *Saccharomyces cerevisiae*. *J. Biol. Chem.* **1998**, *273*, 22284–22291. [CrossRef]
51. Yue, Z.; Jin, S.; Yang, C.; Levine, A.J.; Heintz, N. Beclin 1, an autophagy gene essential for early embryonic development, is a haploinsufficient tumor suppressor. *Proc. Natl. Acad. Sci. USA* **2003**, *100*, 15077–15082. [CrossRef]
52. Petiot, A.; Ogier-Denis, E.; Blommaart, E.F.; Meijer, A.J.; Codogno, P. Distinct classes of phosphatidylinositol 3′-kinases are involved in signaling pathways that control macroautophagy in HT-29 cells. *J. Biol. Chem.* **2000**, *275*, 992–998. [CrossRef]
53. Lindmo, K.; Brech, A.; Finley, K.D.; Gaumer, S.; Contamine, D.; Rusten, T.E.; Stenmark, H. The PI 3-kinase regulator Vps15 is required for autophagic clearance of protein aggregates. *Autophagy* **2008**, *4*, 500–506. [CrossRef]
54. Vanhaesebroeck, B.; Leevers, S.J.; Ahmadi, K.; Timms, J.; Katso, R.; Driscoll, P.C.; Woscholski, R.; Parker, P.J.; Waterfield, M.D. Synthesis and function of 3-phosphorylated inositol lipids. *Annu. Rev. Biochem.* **2001**, *70*, 535–602. [CrossRef]
55. Itakura, E.; Kishi, C.; Inoue, K.; Mizushima, N. Beclin 1 forms two distinct phosphatidylinositol 3-kinase complexes with mammalian Atg14 and UVRAG. *Mol. Biol. Cell* **2008**, *19*, 5360–5372. [CrossRef]
56. Sun, Q.; Fan, W.; Chen, K.; Ding, X.; Chen, S.; Zhong, Q. Identification of Barkor as a mammalian autophagy-specific factor for Beclin 1 and class III phosphatidylinositol 3-kinase. *Proc. Natl. Acad. Sci. USA* **2008**, *105*, 19211–19216. [CrossRef]
57. Liang, C.; Feng, P.; Ku, B.; Dotan, I.; Canaani, D.; Oh, B.H.; Jung, J.U. Autophagic and tumour suppressor activity of a novel Beclin1-binding protein UVRAG. *Nat. Cell Biol.* **2006**, *8*, 688–699. [CrossRef]
58. Takahashi, Y.; Coppola, D.; Matsushita, N.; Cualing, H.D.; Sun, M.; Sato, Y.; Liang, C.; Jung, J.U.; Cheng, J.Q.; Mulé, J.J.; et al. Bif-1 interacts with Beclin 1 through UVRAG and regulates autophagy and tumorigenesis. *Nat. Cell Biol.* **2007**, *9*, 1142–1151. [CrossRef]
59. Hou, Y.C.; Chittaranjan, S.; Barbosa, S.G.; McCall, K.; Gorski, S.M. Effector caspase Dcp-1 and IAP protein Bruce regulate starvation-induced autophagy during *Drosophila melanogaster* oogenesis. *J. Cell Biol.* **2008**, *182*, 1127–1139. [CrossRef]
60. Pattingre, S.; Tassa, A.; Qu, X.; Garuti, R.; Liang, X.H.; Mizushima, N.; Packer, M.; Schneider, M.D.; Levine, B. Bcl-2 antiapoptotic proteins inhibit Beclin 1-dependent autophagy. *Cell* **2005**, *122*, 927–939. [CrossRef]
61. Saeki, K.; Yuo, A.; Okuma, E.; Yazaki, Y.; Susin, S.A.; Kroemer, G.; Takaku, F. Bcl-2 down-regulation causes autophagy in a caspase-independent manner in human leukemic HL60 cells. *Cell Death Differ.* **2000**, *7*, 1263–1269. [CrossRef]
62. Fimia, G.M.; Stoykova, A.; Romagnoli, A.; Giunta, L.; Di Bartolomeo, S.; Nardacci, R.; Corazzari, M.; Fuoco, C.; Ucar, A.; Schwartz, P.; et al. Ambra1 regulates autophagy and development of the nervous system. *Nature* **2007**, *447*, 1121–1125. [CrossRef]
63. Laczkó-Dobos, H.; Maddali, A.K.; Jipa, A.; Bhattacharjee, A.; Végh, A.G.; Juhász, G. Lipid profiles of autophagic structures isolated from wild type and Atg2 mutant *Drosophila*. *Biochim. Biophys. Acta Mol. Cell Biol. Lipids* **2021**, *1866*, 158868. [CrossRef]
64. Kotani, T.; Kirisako, H.; Koizumi, M.; Ohsumi, Y.; Nakatogawa, H. The Atg2-Atg18 complex tethers pre-autophagosomal membranes to the endoplasmic reticulum for autophagosome formation. *Proc. Natl. Acad. Sci. USA* **2018**, *115*, 10363–10368. [CrossRef]
65. Osawa, T.; Ishii, Y.; Noda, N.N. Human ATG2B possesses a lipid transfer activity which is accelerated by negatively charged lipids and WIPI4. *Genes Cells* **2020**, *25*, 65–70. [CrossRef]
66. Ichimura, Y.; Kirisako, T.; Takao, T.; Satomi, Y.; Shimonishi, Y.; Ishihara, N.; Mizushima, N.; Tanida, I.; Kominami, E.; Ohsumi, M.; et al. A ubiquitin-like system mediates protein lipidation. *Nature* **2000**, *408*, 488–492. [CrossRef]
67. Tanida, I.; Sou, Y.S.; Minematsu-Ikeguchi, N.; Ueno, T.; Kominami, E. Atg8L/Apg8L is the fourth mammalian modifier of mammalian Atg8 conjugation mediated by human Atg4B, Atg7 and Atg3. *FEBS J.* **2006**, *273*, 2553–2562. [CrossRef]
68. Juhász, G.; Puskás, L.G.; Komonyi, O.; Erdi, B.; Maróy, P.; Neufeld, T.P.; Sass, M. Gene expression profiling identifies FKBP39 as an inhibitor of autophagy in larval *Drosophila* fat body. *Cell Death Differ.* **2007**, *14*, 1181–1190. [CrossRef]
69. Kirisako, T.; Ichimura, Y.; Okada, H.; Kabeya, Y.; Mizushima, N.; Yoshimori, T.; Ohsumi, M.; Takao, T.; Noda, T.; Ohsumi, Y. The reversible modification regulates the membrane-binding state of Apg8/Aut7 essential for autophagy and the cytoplasm to vacuole targeting pathway. *J. Cell Biol.* **2000**, *151*, 263–276. [CrossRef]
70. Hemelaar, J.; Lelyveld, V.S.; Kessler, B.M.; Ploegh, H.L. A single protease, Apg4B, is specific for the autophagy-related ubiquitin-like proteins GATE-16, MAP1-LC3, GABARAP, and Apg8L. *J. Biol. Chem.* **2003**, *278*, 51841–51850. [CrossRef]
71. Shelly, S.; Lukinova, N.; Bambina, S.; Berman, A.; Cherry, S. Autophagy is an essential component of *Drosophila* immunity against vesicular stomatitis virus. *Immunity* **2009**, *30*, 588–598. [CrossRef]
72. Mizushima, N.; Sugita, H.; Yoshimori, T.; Ohsumi, Y. A new protein conjugation system in human. The counterpart of the yeast Apg12p conjugation system essential for autophagy. *J. Biol. Chem.* **1998**, *273*, 33889–33892. [CrossRef]
73. Mizushima, N.; Noda, T.; Yoshimori, T.; Tanaka, Y.; Ishii, T.; George, M.D.; Klionsky, D.J.; Ohsumi, M.; Ohsumi, Y. A protein conjugation system essential for autophagy. *Nature* **1998**, *395*, 395–398. [CrossRef]
74. Peng, D.; Ruan, C.; Fu, S.; He, C.; Song, J.; Li, H.; Tu, Y.; Tang, D.; Yao, L.; Lin, S.; et al. Atg9-centered multi-omics integration reveals new autophagy regulators in *Saccharomyces cerevisiae*. *Autophagy* **2021**, *17*, 4453–4476. [CrossRef]
75. Kametaka, S.; Matsuura, A.; Wada, Y.; Ohsumi, Y. Structural and functional analyses of APG5, a gene involved in autophagy in yeast. *Gene* **1996**, *178*, 139–143. [CrossRef]
76. Tanida, I.; Mizushima, N.; Kiyooka, M.; Ohsumi, M.; Ueno, T.; Ohsumi, Y.; Kominami, E. Apg7p/Cvt2p: A novel protein-activating enzyme essential for autophagy. *Mol. Biol. Cell* **1999**, *10*, 1367–1379. [CrossRef]

77. Nagy, P.; Hegedűs, K.; Pircs, K.; Varga, Á.; Juhász, G. Different effects of Atg2 and Atg18 mutations on Atg8a and Atg9 trafficking during starvation in *Drosophila*. *FEBS Lett.* **2014**, *588*, 408–413. [CrossRef] [PubMed]
78. Kawamata, T.; Kamada, Y.; Kabeya, Y.; Sekito, T.; Ohsumi, Y. Organization of the pre-autophagosomal structure responsible for autophagosome formation. *Mol. Biol. Cell* **2008**, *19*, 2039–2050. [CrossRef]
79. Berry, D.L.; Baehrecke, E.H. Growth arrest and autophagy are required for salivary gland cell degradation in *Drosophila*. *Cell* **2007**, *131*, 1137–1148. [CrossRef]
80. Kabeya, Y.; Mizushima, N.; Yamamoto, A.; Oshitani-Okamoto, S.; Ohsumi, Y.; Yoshimori, T. LC3, GABARAP and GATE16 localize to autophagosomal membrane depending on form-II formation. *J. Cell Sci.* **2004**, *117*, 2805–2812. [CrossRef]
81. Tanida, I.; Tanida-Miyake, E.; Nishitani, T.; Komatsu, M.; Yamazaki, H.; Ueno, T.; Kominami, E. Murine Apg12p Has a Substrate Preference for Murine Apg7p over Three Apg8p Homologs. *Biochem. Biophys. Res. Commun.* **2002**, *292*, 256–262. [CrossRef]
82. Juhász, G.; Erdi, B.; Sass, M.; Neufeld, T.P. Atg7-dependent autophagy promotes neuronal health, stress tolerance, and longevity but is dispensable for metamorphosis in *Drosophila*. *Genes Dev.* **2007**, *21*, 3061–3066. [CrossRef]
83. Rusten, T.E.; Lindmo, K.; Juhász, G.; Sass, M.; Seglen, P.O.; Brech, A.; Stenmark, H. Programmed autophagy in the *Drosophila* fat body is induced by ecdysone through regulation of the PI3K pathway. *Dev. Cell* **2004**, *7*, 179–192. [CrossRef]
84. Kuma, A.; Mizushima, N.; Ishihara, N.; Ohsumi, Y. Formation of the approximately 350-kDa Apg12-Apg5.Apg16 multimeric complex, mediated by Apg16 oligomerization, is essential for autophagy in yeast. *J. Biol. Chem.* **2002**, *277*, 18619–18625. [CrossRef]
85. Mizushima, N.; Kuma, A.; Kobayashi, Y.; Yamamoto, A.; Matsubae, M.; Takao, T.; Natsume, T.; Ohsumi, Y.; Yoshimori, T. Mouse Apg16L, a novel WD-repeat protein, targets to the autophagic isolation membrane with the Apg12-Apg5 conjugate. *J. Cell Sci.* **2003**, *116*, 1679–1688. [CrossRef]
86. Ochaba, J.; Lukacsovich, T.; Csikos, G.; Zheng, S.; Margulis, J.; Salazar, L.; Mao, K.; Lau, A.L.; Yeung, S.Y.; Humbert, S.; et al. Potential function for the Huntingtin protein as a scaffold for selective autophagy. *Proc. Natl. Acad. Sci. USA* **2014**, *111*, 16889–16894. [CrossRef] [PubMed]
87. Polson, H.E.; de Lartigue, J.; Rigden, D.J.; Reedijk, M.; Urbé, S.; Clague, M.J.; Tooze, S.A. Mammalian Atg18 (WIPI2) localizes to omegasome-anchored phagophores and positively regulates LC3 lipidation. *Autophagy* **2010**, *6*, 506–522. [CrossRef] [PubMed]
88. Backues, S.K.; Orban, D.P.; Bernard, A.; Singh, K.; Cao, Y.; Klionsky, D.J. Atg23 and Atg27 act at the early stages of Atg9 trafficking in S. cerevisiae. *Traffic* **2015**, *16*, 172–190. [CrossRef]
89. Meléndez, A.; Tallóczy, Z.; Seaman, M.; Eskelinen, E.L.; Hall, D.H.; Levine, B. Autophagy genes are essential for dauer development and life-span extension in *C. elegans*. *Science* **2003**, *301*, 1387–1391. [CrossRef] [PubMed]
90. Mizushima, N. The role of the Atg1/ULK1 complex in autophagy regulation. *Curr. Opin. Cell Biol.* **2010**, *22*, 132–139. [CrossRef]
91. Chang, Y.Y.; Neufeld, T.P. Autophagy takes flight in *Drosophila*. *FEBS Lett.* **2010**, *584*, 1342–1349. [CrossRef] [PubMed]
92. Guo, T.; Nan, Z.; Miao, C.; Jin, X.; Yang, W.; Wang, Z.; Tu, Y.; Bao, H.; Lyu, J.; Zheng, H.; et al. The autophagy-related gene Atg101 in *Drosophila* regulates both neuron and midgut homeostasis. *J. Biol. Chem.* **2019**, *294*, 5666–5676. [CrossRef]
93. Strømhaug, P.E.; Reggiori, F.; Guan, J.; Wang, C.W.; Klionsky, D.J. Atg21 is a phosphoinositide binding protein required for efficient lipidation and localization of Atg8 during uptake of aminopeptidase I by selective autophagy. *Mol. Biol. Cell* **2004**, *15*, 3553–3566. [CrossRef] [PubMed]
94. He, C.; Klionsky, D.J. Regulation mechanisms and signaling pathways of autophagy. *Annu. Rev. Genet.* **2009**, *43*, 67–93. [CrossRef]
95. Simonsen, A.; Tooze, S.A. Coordination of membrane events during autophagy by multiple class III PI3-kinase complexes. *J. Cell Biol.* **2009**, *186*, 773–782. [CrossRef] [PubMed]
96. Jacinto, E.; Hall, M.N. Tor signalling in bugs, brain and brawn. *Nat. Rev. Mol. Cell Biol.* **2003**, *4*, 117–126. [CrossRef]
97. Ohsumi, Y. Molecular dissection of autophagy: Two ubiquitin-like systems. *Nat. Rev. Mol. Cell Biol.* **2001**, *2*, 211–216. [CrossRef]
98. Fujita, N.; Itoh, T.; Omori, H.; Fukuda, M.; Noda, T.; Yoshimori, T. The Atg16L complex specifies the site of LC3 lipidation for membrane biogenesis in autophagy. *Mol. Biol. Cell* **2008**, *19*, 2092–2100. [CrossRef] [PubMed]
99. Shintani, T.; Mizushima, N.; Ogawa, Y.; Matsuura, A.; Noda, T.; Ohsumi, Y. Apg10p, a novel protein-conjugating enzyme essential for autophagy in yeast. *Embo J.* **1999**, *18*, 5234–5241. [CrossRef] [PubMed]
100. Thumm, M.; Kadowaki, T. The loss of *Drosophila* APG4/AUT2 function modifies the phenotypes of cut and Notch signaling pathway mutants. *Mol. Genet. Genom.* **2001**, *266*, 657–663. [CrossRef] [PubMed]
101. Reiter, L.T.; Potocki, L.; Chien, S.; Gribskov, M.; Bier, E. A systematic analysis of human disease-associated gene sequences in *Drosophila melanogaster*. *Genome Res.* **2001**, *88*, 1114–1125. [CrossRef] [PubMed]
102. Kim, S.K.; Tsao, D.D.; Suh, G.S.B.; Miguel-Aliaga, I. Discovering signaling mechanisms governing metabolism and metabolic diseases with *Drosophila*. *Cell Metab.* **2021**, *33*, 1279–1292. [CrossRef] [PubMed]
103. Lloyd, T.E.; Taylor, J.P. Flightless flies: *Drosophila* models of neuromuscular disease. *Ann. N. Y. Acad. Sci.* **2010**, *1184*, e1–e20. [CrossRef]
104. Zhou, J.; Tong, C. Design and Methods of Large-Scale RNA Interference Screens in *Drosophila*. *Methods Mol. Biol.* **2016**, *1470*, 163–169. [CrossRef]
105. Rubin, G.M.; Yandell, M.D.; Wortman, J.R.; Gabor Miklos, G.L.; Nelson, C.R.; Hariharan, I.K.; Fortini, M.E.; Li, P.W.; Apweiler, R.; Fleischmann, W.; et al. Comparative genomics of the eukaryotes. *Science* **2000**, *287*, 2204–2215. [CrossRef] [PubMed]
106. Gramates, L.S.; Marygold, S.J.; Santos, G.d.; Urbano, J.-M.; Antonazzo, G.; Matthews, B.B.; Rey, A.J.; Tabone, C.J.; Crosby, M.A.; Emmert, D.B.; et al. FlyBase at 25: Looking to the future. *Nucleic Acids Res.* **2017**, *45*, D663–D671. [CrossRef]
107. Jennings, B. *Drosophila*—A versatile model in biology & medicine. *Mater. Today* **2011**, *14*, 190–195. [CrossRef]

108. Flecknell, P. Replacement, Reduction, Refinement. *ALTEX—Altern. Anim. Exp.* **2002**, *19*, 73–78.
109. Ng, R.; Salem, S.S.; Wu, S.T.; Wu, M.; Lin, H.H.; Shepherd, A.K.; Joiner, W.J.; Wang, J.W.; Su, C.Y. Amplification of *Drosophila* Olfactory Responses by a DEG/ENaC Channel. *Neuron* **2019**, *104*, 947–959.e5. [CrossRef]
110. Gratz, S.J.; Ukken, F.P.; Rubinstein, C.D.; Thiede, G.; Donohue, L.K.; Cummings, A.M.; O'Connor-Giles, K.M. Highly specific and efficient CRISPR/Cas9-catalyzed homology-directed repair in *Drosophila*. *Genetics* **2014**, *196*, 961–971. [CrossRef]
111. Gratz, S.J.; Rubinstein, C.D.; Harrison, M.M.; Wildonger, J.; O'Connor-Giles, K.M. CRISPR-Cas9 Genome Editing in *Drosophila*. *Curr. Protoc. Mol. Biol.* **2015**, *111*, 31.2.1–31.2.20. [CrossRef] [PubMed]
112. Trinca, T.; Malik, B. Brief Report: Ionising Radiation Induces Autophagy in the Adult *Drosophila* Brain and Intestine. *Res. Sq.* **2022**. [CrossRef]
113. Hughes, T.T.; Allen, A.L.; Bardin, J.E.; Christian, M.N.; Daimon, K.; Dozier, K.D.; Hansen, C.L.; Holcomb, L.M.; Ahlander, J. *Drosophila* as a genetic model for studying pathogenic human viruses. *Virology* **2012**, *423*, 1–5. [CrossRef] [PubMed]
114. Demir, E. Adverse biological effects of ingested polystyrene microplastics using *Drosophila melanogaster* as a model in vivo organism. *J. Toxicol. Environ. Health A* **2021**, *84*, 649–660. [CrossRef]
115. Demir, E. The potential use of *Drosophila* as an in vivo model organism for COVID-19-related research: A review. *Turk. J. Biol.* **2021**, *45*, 559–569. [CrossRef]
116. Demir, E. Mechanisms and biological impacts of graphene and multi-walled carbon nanotubes on *Drosophila melanogaster*: Oxidative stress, genotoxic damage, phenotypic variations, locomotor behavior, parasitoid resistance, and cellular immune response. *J. Appl. Toxicol.* **2022**, *42*, 450–474. [CrossRef]
117. Ashburner, M.; Bergman, C.M. *Drosophila melanogaster*: A case study of a model genomic sequence and its consequences. *Genome Res.* **2005**, *15*, 1661–1667. [CrossRef]
118. Lee, C.Y.; Cooksey, B.A.; Baehrecke, E.H. Steroid regulation of midgut cell death during *Drosophila* development. *Dev. Biol.* **2002**, *250*, 101–111. [CrossRef]
119. Nezis, I.P.; Shravage, B.V.; Sagona, A.P.; Lamark, T.; Bjørkøy, G.; Johansen, T.; Rusten, T.E.; Brech, A.; Baehrecke, E.H.; Stenmark, H. Autophagic degradation of dBruce controls DNA fragmentation in nurse cells during late *Drosophila melanogaster* oogenesis. *J. Cell Biol.* **2010**, *190*, 523–531. [CrossRef]
120. Kuma, A.; Matsui, M.; Mizushima, N. LC3, an autophagosome marker, can be incorporated into protein aggregates independent of autophagy: Caution in the interpretation of LC3 localization. *Autophagy* **2007**, *3*, 323–328. [CrossRef]
121. Gorski, S.M.; Chittaranjan, S.; Pleasance, E.D.; Freeman, J.D.; Anderson, C.L.; Varhol, R.J.; Coughlin, S.M.; Zuyderduyn, S.D.; Jones, S.J.; Marra, M.A. A SAGE approach to discovery of genes involved in autophagic cell death. *Curr. Biol.* **2003**, *13*, 358–363. [CrossRef]
122. Lee, C.Y.; Clough, E.A.; Yellon, P.; Teslovich, T.M.; Stephan, D.A.; Baehrecke, E.H. Genome-wide analyses of steroid- and radiation-triggered programmed cell death in *Drosophila*. *Curr. Biol.* **2003**, *13*, 350–357. [CrossRef]
123. Cumming, R.C.; Simonsen, A.; Finley, K.D. Quantitative analysis of autophagic activity in *Drosophila* neural tissues by measuring the turnover rates of pathway substrates. *Methods Enzymol.* **2008**, *451*, 639–651. [CrossRef] [PubMed]
124. Pandey, A.; Trigun, S.K. Fisetin induces apoptosis in colorectal cancer cells by suppressing autophagy and down-regulating nuclear factor erythroid 2-related factor 2 (Nrf2). *J. Cell Biochem.* **2023**. [CrossRef]
125. Mauvezin, C.; Nagy, P.; Juhász, G.; Neufeld, T.P. Autophagosome-lysosome fusion is independent of V-ATPase-mediated acidification. *Nat. Commun.* **2015**, *6*, 7007. [CrossRef]
126. Pulipparacharuvil, S.; Akbar, M.A.; Ray, S.; Sevrioukov, E.A.; Haberman, A.S.; Rohrer, J.; Krämer, H. *Drosophila* Vps16A is required for trafficking to lysosomes and biogenesis of pigment granules. *J. Cell Sci.* **2005**, *118*, 3663–3673. [CrossRef]
127. Lőrincz, P.; Mauvezin, C.; Juhász, G. Exploring Autophagy in *Drosophila*. *Cells* **2017**, *6*, 22. [CrossRef] [PubMed]
128. Nagy, P.; Varga, Á.; Kovács, A.L.; Takáts, S.; Juhász, G. How and why to study autophagy in *Drosophila*: It's more than just a garbage chute. *Methods* **2015**, *75*, 151–161. [CrossRef] [PubMed]
129. Katheder, N.S.; Khezri, R.; O'Farrell, F.; Schultz, S.W.; Jain, A.; Rahman, M.M.; Schink, K.O.; Theodossiou, T.A.; Johansen, T.; Juhász, G.; et al. Microenvironmental autophagy promotes tumour growth. *Nature* **2017**, *541*, 417–420. [CrossRef]
130. de Castro, I.P.; Costa, A.C.; Celardo, I.; Tufi, R.; Dinsdale, D.; Loh, S.H.; Martins, L.M. *Drosophila* ref(2)P is required for the parkin-mediated suppression of mitochondrial dysfunction in pink1 mutants. *Cell Death Dis.* **2013**, *4*, e873. [CrossRef]
131. Nezis, I.P.; Simonsen, A.; Sagona, A.P.; Finley, K.; Gaumer, S.; Contamine, D.; Rusten, T.E.; Stenmark, H.; Brech, A. Ref(2)P, the *Drosophila melanogaster* homologue of mammalian p62, is required for the formation of protein aggregates in adult brain. *J. Cell Biol.* **2008**, *180*, 1065–1071. [CrossRef]
132. Pircs, K.; Nagy, P.; Varga, A.; Venkei, Z.; Erdi, B.; Hegedus, K.; Juhasz, G. Advantages and limitations of different p62-based assays for estimating autophagic activity in *Drosophila*. *PLoS ONE* **2012**, *7*, e44214. [CrossRef]
133. Mauvezin, C.; Ayala, C.; Braden, C.R.; Kim, J.; Neufeld, T.P. Assays to monitor autophagy in *Drosophila*. *Methods* **2014**, *68*, 134–139. [CrossRef]
134. Lőrincz, P.; Takáts, S.; Kárpáti, M.; Juhász, G. iFly: The eye of the fruit fly as a model to study autophagy and related trafficking pathways. *Exp. Eye Res.* **2016**, *144*, 90–98. [CrossRef]
135. Barth, J.M.; Szabad, J.; Hafen, E.; Köhler, K. Autophagy in *Drosophila* ovaries is induced by starvation and is required for oogenesis. *Cell Death Differ.* **2011**, *18*, 915–924. [CrossRef] [PubMed]

136. Varga, K.; Nagy, P.; Arsikin Csordás, K.; Kovács, A.L.; Hegedűs, K.; Juhász, G. Loss of Atg16 delays the alcohol-induced sedation response via regulation of Corazonin neuropeptide production in *Drosophila*. *Sci. Rep.* **2016**, *6*, 34641. [CrossRef] [PubMed]
137. Nagy, P.; Kárpáti, M.; Varga, A.; Pircs, K.; Venkei, Z.; Takáts, S.; Varga, K.; Erdi, B.; Hegedűs, K.; Juhász, G. Atg17/FIP200 localizes to perilysosomal Ref(2)P aggregates and promotes autophagy by activation of Atg1 in *Drosophila*. *Autophagy* **2014**, *10*, 453–467. [CrossRef] [PubMed]
138. Kim, J.; Neufeld, T.P. Dietary sugar promotes systemic TOR activation in *Drosophila* through AKH-dependent selective secretion of Dilp3. *Nat. Commun.* **2015**, *6*, 6846. [CrossRef]
139. Rojas-Ríos, P.; Chartier, A.; Pierson, S.; Séverac, D.; Dantec, C.; Busseau, I.; Simonelig, M. Translational Control of Autophagy by Orb in the *Drosophila* Germline. *Dev. Cell* **2015**, *35*, 622–631. [CrossRef]
140. Hindle, S.; Afsari, F.; Stark, M.; Middleton, C.A.; Evans, G.J.; Sweeney, S.T.; Elliott, C.J. Dopaminergic expression of the Parkinsonian gene LRRK2-G2019S leads to non-autonomous visual neurodegeneration, accelerated by increased neural demands for energy. *Hum. Mol. Genet.* **2013**, *22*, 2129–2140. [CrossRef]
141. Riedel, F.; Gillingham, A.K.; Rosa-Ferreira, C.; Galindo, A.; Munro, S. An antibody toolkit for the study of membrane traffic in *Drosophila melanogaster*. *Biol. Open* **2016**, *5*, 987–992. [CrossRef] [PubMed]
142. Maruzs, T.; Lőrincz, P.; Szatmári, Z.; Széplaki, S.; Sándor, Z.; Lakatos, Z.; Puska, G.; Juhász, G.; Sass, M. Retromer Ensures the Degradation of Autophagic Cargo by Maintaining Lysosome Function in *Drosophila*. *Traffic* **2015**, *16*, 1088–1107. [CrossRef] [PubMed]
143. Eskelinen, E.L.; Reggiori, F.; Baba, M.; Kovács, A.L.; Seglen, P.O. Seeing is believing: The impact of electron microscopy on autophagy research. *Autophagy* **2011**, *7*, 935–956. [CrossRef]
144. Lőrincz, P.; Lakatos, Z.; Maruzs, T.; Szatmári, Z.; Kis, V.; Sass, M. Atg6/UVRAG/Vps34-containing lipid kinase complex is required for receptor downregulation through endolysosomal degradation and epithelial polarity during *Drosophila* wing development. *BioMed Res. Int.* **2014**, *2014*, 851349. [CrossRef]
145. Takáts, S.; Varga, Á.; Pircs, K.; Juhász, G. Loss of *Drosophila* Vps16A enhances autophagosome formation through reduced Tor activity. *Autophagy* **2015**, *11*, 1209–1215. [CrossRef] [PubMed]
146. Szatmári, Z.; Kis, V.; Lippai, M.; Hegedus, K.; Faragó, T.; Lorincz, P.; Tanaka, T.; Juhász, G.; Sass, M. Rab11 facilitates cross-talk between autophagy and endosomal pathway through regulation of Hook localization. *Mol. Biol. Cell* **2014**, *25*, 522–531. [CrossRef]
147. Nagy, P.; Varga, A.; Pircs, K.; Hegedűs, K.; Juhász, G. Myc-driven overgrowth requires unfolded protein response-mediated induction of autophagy and antioxidant responses in *Drosophila melanogaster*. *PLoS Genet.* **2013**, *9*, e1003664. [CrossRef]
148. Büttner, S.; Broeskamp, F.; Sommer, C.; Markaki, M.; Habernig, L.; Alavian-Ghavanini, A.; Carmona-Gutierrez, D.; Eisenberg, T.; Michael, E.; Kroemer, G.; et al. Spermidine protects against α-synuclein neurotoxicity. *Cell Cycle* **2014**, *13*, 3903–3908. [CrossRef]
149. Eisenberg, T.; Knauer, H.; Schauer, A.; Büttner, S.; Ruckenstuhl, C.; Carmona-Gutierrez, D.; Ring, J.; Schroeder, S.; Magnes, C.; Antonacci, L.; et al. Induction of autophagy by spermidine promotes longevity. *Nat. Cell Biol.* **2009**, *11*, 1305–1314. [CrossRef]
150. Zirin, J.; Nieuwenhuis, J.; Perrimon, N. Role of autophagy in glycogen breakdown and its relevance to chloroquine myopathy. *PLoS Biol.* **2013**, *11*, e1001708. [CrossRef] [PubMed]
151. Juhász, G. Interpretation of bafilomycin, pH neutralizing or protease inhibitor treatments in autophagic flux experiments: Novel considerations. *Autophagy* **2012**, *8*, 1875–1876. [CrossRef] [PubMed]
152. Papp, D.; Kovács, T.; Billes, V.; Varga, M.; Tarnóci, A.; Hackler, L., Jr.; Puskás, L.G.; Liliom, H.; Tárnok, K.; Schlett, K.; et al. AUTEN-67, an autophagy-enhancing drug candidate with potent antiaging and neuroprotective effects. *Autophagy* **2016**, *12*, 273–286. [CrossRef]
153. Kamath, R.S.; Fraser, A.G.; Dong, Y.; Poulin, G.; Durbin, R.; Gotta, M.; Kanapin, A.; Le Bot, N.; Moreno, S.; Sohrmann, M.; et al. Systematic functional analysis of the *Caenorhabditis elegans* genome using RNAi. *Nature* **2003**, *421*, 231–237. [CrossRef] [PubMed]
154. Okamura, K.; Chung, W.J.; Ruby, J.G.; Guo, H.; Bartel, D.P.; Lai, E.C. The *Drosophila* hairpin RNA pathway generates endogenous short interfering RNAs. *Nature* **2008**, *453*, 803–806. [CrossRef]
155. Dietzl, G.; Chen, D.; Schnorrer, F.; Su, K.C.; Barinova, Y.; Fellner, M.; Gasser, B.; Kinsey, K.; Oppel, S.; Scheiblauer, S.; et al. A genome-wide transgenic RNAi library for conditional gene inactivation in *Drosophila*. *Nature* **2007**, *448*, 151–156. [CrossRef] [PubMed]
156. Brand, A.H.; Perrimon, N. Targeted gene expression as a means of altering cell fates and generating dominant phenotypes. *Development* **1993**, *118*, 401–415. [CrossRef]
157. Perrimon, N.; Mathey-Prevot, B. Applications of high-throughput RNA interference screens to problems in cell and developmental biology. *Genetics* **2007**, *175*, 7–16. [CrossRef] [PubMed]
158. Ni, J.Q.; Liu, L.P.; Binari, R.; Hardy, R.; Shim, H.S.; Cavallaro, A.; Booker, M.; Pfeiffer, B.; Markstein, M.; Wang, H.; et al. A *Drosophila* resource of transgenic RNAi lines for neurogenetics. *Genetics* **2009**, *182*, 1089–1100. [CrossRef]
159. Ni, J.Q.; Markstein, M.; Binari, R.; Pfeiffer, B.; Liu, L.P.; Villalta, C.; Booker, M.; Perkins, L.; Perrimon, N. Vector and parameters for targeted transgenic RNA interference in *Drosophila melanogaster*. *Nat. Methods* **2008**, *5*, 49–51. [CrossRef] [PubMed]
160. Simonsen, A.; Cumming, R.C.; Finley, K.D. Linking lysosomal trafficking defects with changes in aging and stress response in *Drosophila*. *Autophagy* **2007**, *3*, 499–501. [CrossRef]
161. Filimonenko, M.; Isakson, P.; Finley, K.D.; Anderson, M.; Jeong, H.; Melia, T.J.; Bartlett, B.J.; Myers, K.M.; Birkeland, H.C.; Lamark, T.; et al. The selective macroautophagic degradation of aggregated proteins requires the PI3P-binding protein Alfy. *Mol. Cell* **2010**, *38*, 265–279. [CrossRef] [PubMed]

162. Arsham, A.M.; Neufeld, T.P. A genetic screen in *Drosophila* reveals novel cytoprotective functions of the autophagy-lysosome pathway. *PLoS ONE* **2009**, *4*, e6068. [CrossRef] [PubMed]
163. Batlevi, Y.; Martin, D.N.; Pandey, U.B.; Simon, C.R.; Powers, C.M.; Taylor, J.P.; Baehrecke, E.H. Dynein light chain 1 is required for autophagy, protein clearance, and cell death in *Drosophila*. *Proc. Natl. Acad. Sci. USA* **2010**, *107*, 742–747. [CrossRef]
164. Kohler, K.; Barth, J.M.I.; Hafen, E.; Brunner, E.; Mohanty, S.; Boucke, K.; Greber, U.F.; Xue, L.G.; Wenk, M.R. A combined proteomic and genetic analysis identifies a role for the lipid desaturase Desat1 in starvation-induced autophagy in *Drosophila*. *Autophagy* **2009**, *5*, 980–990. [CrossRef] [PubMed]
165. Martin, D.N.; Balgley, B.; Dutta, S.; Chen, J.; Rudnick, P.; Cranford, J.; Kantartzis, S.; DeVoe, D.L.; Lee, C.; Baehrecke, E.H. Proteomic analysis of steroid-triggered autophagic programmed cell death during *Drosophila* development. *Cell Death Differ.* **2007**, *14*, 916–923. [CrossRef]
166. Dutta, S.; Baehrecke, E.H. Warts is required for PI3K-regulated growth arrest, autophagy, and autophagic cell death in *Drosophila*. *Curr. Biol.* **2008**, *18*, 1466–1475. [CrossRef] [PubMed]
167. Clemens, J.C.; Worby, C.A.; Simonson-Leff, N.; Muda, M.; Maehama, T.; Hemmings, B.A.; Dixon, J.E. Use of double-stranded RNA interference in *Drosophila* cell lines to dissect signal transduction pathways. *Proc. Natl. Acad. Sci. USA* **2000**, *97*, 6499–6503. [CrossRef] [PubMed]
168. Kim, Y.-I.; Ryu, T.; Lee, J.; Heo, Y.-S.; Ahnn, J.; Lee, S.-J.; Yoo, O. A genetic screen for modifiers of *Drosophila* caspase Dcp-1 reveals caspase involvement in autophagy and novel caspase-related genes. *BMC Cell Biol.* **2010**, *11*, 9. [CrossRef]
169. Yano, T.; Mita, S.; Ohmori, H.; Oshima, Y.; Fujimoto, Y.; Ueda, R.; Takada, H.; Goldman, W.E.; Fukase, K.; Silverman, N.; et al. Autophagic control of listeria through intracellular innate immune recognition in drosophila. *Nat. Immunol.* **2008**, *9*, 908–916. [CrossRef] [PubMed]
170. Chittaranjan, S.; McConechy, M.; Hou, Y.C.; Freeman, J.D.; Devorkin, L.; Gorski, S.M. Steroid hormone control of cell death and cell survival: Molecular insights using RNAi. *PLoS Genet.* **2009**, *5*, e1000379. [CrossRef]
171. Liang, X.H.; Kleeman, L.K.; Jiang, H.H.; Gordon, G.; Goldman, J.E.; Berry, G.; Herman, B.; Levine, B. Protection against fatal Sindbis virus encephalitis by beclin, a novel Bcl-2-interacting protein. *J. Virol.* **1998**, *72*, 8586–8596. [CrossRef]
172. Gutierrez, M.G.; Master, S.S.; Singh, S.B.; Taylor, G.A.; Colombo, M.I.; Deretic, V. Autophagy is a defense mechanism inhibiting BCG and Mycobacterium tuberculosis survival in infected macrophages. *Cell* **2004**, *119*, 753–766. [CrossRef] [PubMed]
173. Nakagawa, I.; Amano, A.; Mizushima, N.; Yamamoto, A.; Yamaguchi, H.; Kamimoto, T.; Nara, A.; Funao, J.; Nakata, M.; Tsuda, K.; et al. Autophagy defends cells against invading group A Streptococcus. *Science* **2004**, *306*, 1037–1040. [CrossRef] [PubMed]
174. Singh, S.B.; Davis, A.S.; Taylor, G.A.; Deretic, V. Human IRGM induces autophagy to eliminate intracellular mycobacteria. *Science* **2006**, *313*, 1438–1441. [CrossRef]
175. Jackson, W.T.; Giddings, T.H., Jr.; Taylor, M.P.; Mulinyawe, S.; Rabinovitch, M.; Kopito, R.R.; Kirkegaard, K. Subversion of cellular autophagosomal machinery by RNA viruses. *PLoS Biol.* **2005**, *3*, e156. [CrossRef]
176. Birmingham, C.L.; Canadien, V.; Kaniuk, N.A.; Steinberg, B.E.; Higgins, D.E.; Brumell, J.H. Listeriolysin O allows Listeria monocytogenes replication in macrophage vacuoles. *Nature* **2008**, *451*, 350–354. [CrossRef]
177. Meyer-Morse, N.; Robbins, J.R.; Rae, C.S.; Mochegova, S.N.; Swanson, M.S.; Zhao, Z.; Virgin, H.W.; Portnoy, D. Listeriolysin O is necessary and sufficient to induce autophagy during Listeria monocytogenes infection. *PLoS ONE* **2010**, *5*, e8610. [CrossRef]
178. Yoshikawa, Y.; Ogawa, M.; Hain, T.; Yoshida, M.; Fukumatsu, M.; Kim, M.; Mimuro, H.; Nakagawa, I.; Yanagawa, T.; Ishii, T.; et al. Listeria monocytogenes ActA-mediated escape from autophagic recognition. *Nat. Cell Biol.* **2009**, *11*, 1233–1240. [CrossRef] [PubMed]
179. Cherry, S. VSV infection is sensed by *Drosophila*, attenuates nutrient signaling, and thereby activates antiviral autophagy. *Autophagy* **2009**, *5*, 1062–1063. [CrossRef] [PubMed]
180. Deretic, V.; Levine, B. Autophagy, immunity, and microbial adaptations. *Cell Host Microbe* **2009**, *5*, 527–549. [CrossRef]
181. Yan, L.; Sadoshima, J.; Vatner, D.E.; Vatner, S.F. Autophagy: A novel protective mechanism in chronic ischemia. *Cell Cycle* **2006**, *5*, 1175–1177. [CrossRef]
182. Degenhardt, K.; Mathew, R.; Beaudoin, B.; Bray, K.; Anderson, D.; Chen, G.; Mukherjee, C.; Shi, Y.; Gélinas, C.; Fan, Y.; et al. Autophagy promotes tumor cell survival and restricts necrosis, inflammation, and tumorigenesis. *Cancer Cell* **2006**, *10*, 51–64. [CrossRef]
183. Mathew, R.; Kongara, S.; Beaudoin, B.; Karp, C.M.; Bray, K.; Degenhardt, K.; Chen, G.; Jin, S.; White, E. Autophagy suppresses tumor progression by limiting chromosomal instability. *Genes. Dev.* **2007**, *21*, 1367–1381. [CrossRef] [PubMed]
184. Yuan, J.; Lipinski, M.; Degterev, A. Diversity in the mechanisms of neuronal cell death. *Neuron* **2003**, *40*, 401–413. [CrossRef] [PubMed]
185. Ravikumar, B.; Vacher, C.; Berger, Z.; Davies, J.E.; Luo, S.; Oroz, L.G.; Scaravilli, F.; Easton, D.F.; Duden, R.; O'Kane, C.J.; et al. Inhibition of mTOR induces autophagy and reduces toxicity of polyglutamine expansions in fly and mouse models of Huntington disease. *Nat. Genet.* **2004**, *36*, 585–595. [CrossRef]
186. Hanson, K.A.; Kim, S.H.; Wassarman, D.A.; Tibbetts, R.S. Ubiquilin modifies TDP-43 toxicity in a *Drosophila* model of amyotrophic lateral sclerosis (ALS). *J. Biol. Chem.* **2010**, *285*, 11068–11072. [CrossRef]
187. Ling, D.; Song, H.J.; Garza, D.; Neufeld, T.P.; Salvaterra, P.M. Abeta42-induced neurodegeneration via an age-dependent autophagic-lysosomal injury in *Drosophila*. *PLoS ONE* **2009**, *4*, e4201. [CrossRef] [PubMed]

188. Ravikumar, B.; Imarisio, S.; Sarkar, S.; O'Kane, C.J.; Rubinsztein, D.C. Rab5 modulates aggregation and toxicity of mutant huntingtin through macroautophagy in cell and fly models of Huntington disease. *J. Cell Sci.* **2008**, *121*, 1649–1660. [CrossRef]
189. Simonsen, A.; Cumming, R.C.; Brech, A.; Isakson, P.; Schubert, D.R.; Finley, K.D. Promoting basal levels of autophagy in the nervous system enhances longevity and oxidant resistance in adult *Drosophila*. *Autophagy* **2008**, *4*, 176–184. [CrossRef] [PubMed]
190. Arrese, E.L.; Soulages, J.L. Insect fat body: Energy, metabolism, and regulation. *Annu. Rev. Entomol.* **2010**, *55*, 207–225. [CrossRef] [PubMed]
191. Erdi, B.; Nagy, P.; Zvara, A.; Varga, A.; Pircs, K.; Ménesi, D.; Puskás, L.G.; Juhász, G. Loss of the starvation-induced gene Rack1 leads to glycogen deficiency and impaired autophagic responses in *Drosophila*. *Autophagy* **2012**, *8*, 1124–1135. [CrossRef] [PubMed]
192. Denton, D.; Shravage, B.; Simin, R.; Mills, K.; Berry, D.L.; Baehrecke, E.H.; Kumar, S. Autophagy, not apoptosis, is essential for midgut cell death in *Drosophila*. *Curr. Biol.* **2009**, *19*, 1741–1746. [CrossRef]
193. Kim, M.; Sandford, E.; Gatica, D.; Qiu, Y.; Liu, X.; Zheng, Y.; Schulman, B.A.; Xu, J.; Semple, I.; Ro, S.H.; et al. Mutation in ATG5 reduces autophagy and leads to ataxia with developmental delay. *eLife* **2016**, *5*, e12245. [CrossRef]
194. Tsuboyama, K.; Koyama-Honda, I.; Sakamaki, Y.; Koike, M.; Morishita, H.; Mizushima, N. The ATG conjugation systems are important for degradation of the inner autophagosomal membrane. *Science* **2016**, *354*, 1036–1041. [CrossRef]
195. Bjedov, I.; Cochemé, H.M.; Foley, A.; Wieser, D.; Woodling, N.S.; Castillo-Quan, J.I.; Norvaisas, P.; Lujan, C.; Regan, J.C.; Toivonen, J.M.; et al. Fine-tuning autophagy maximises lifespan and is associated with changes in mitochondrial gene expression in *Drosophila*. *PLOS Genet.* **2020**, *16*, e1009083. [CrossRef]
196. Aparicio, R.; Rana, A.; Walker, D.W. Upregulation of the Autophagy Adaptor p62/SQSTM1 Prolongs Health and Lifespan in Middle-Aged *Drosophila*. *Cell Rep.* **2019**, *28*, 1029–1040.e5. [CrossRef]
197. Rana, A.; Oliveira, M.P.; Khamoui, A.V.; Aparicio, R.; Rera, M.; Rossiter, H.B.; Walker, D.W. Promoting Drp1-mediated mitochondrial fission in midlife prolongs healthy lifespan of *Drosophila melanogaster*. *Nat. Commun.* **2017**, *8*, 448. [CrossRef]
198. Vincow, E.S.; Merrihew, G.; Thomas, R.E.; Shulman, N.J.; Beyer, R.P.; MacCoss, M.J.; Pallanck, L.J. The PINK1-Parkin pathway promotes both mitophagy and selective respiratory chain turnover in vivo. *Proc. Natl. Acad. Sci. USA* **2013**, *110*, 6400–6405. [CrossRef] [PubMed]
199. Juhász, G. A mitochondrial-derived vesicle HOPS to endolysosomes using Syntaxin-17. *J. Cell Biol.* **2016**, *214*, 241–243. [CrossRef]
200. McLelland, G.L.; Lee, S.A.; McBride, H.M.; Fon, E.A. Syntaxin-17 delivers PINK1/parkin-dependent mitochondrial vesicles to the endolysosomal system. *J. Cell Biol.* **2016**, *214*, 275–291. [CrossRef] [PubMed]
201. O'Farrell, F.; Wang, S.; Katheder, N.; Rusten, T.E.; Samakovlis, C. Two-tiered control of epithelial growth and autophagy by the insulin receptor and the ret-like receptor, stitcher. *PLoS Biol.* **2013**, *11*, e1001612. [CrossRef] [PubMed]
202. Takeshige, K.; Baba, M.; Tsuboi, S.; Noda, T.; Ohsumi, Y. Autophagy in yeast demonstrated with proteinase-deficient mutants and conditions for its induction. *J. Cell Biol.* **1992**, *119*, 301–311. [CrossRef] [PubMed]
203. Mizushima, N.; Yamamoto, A.; Matsui, M.; Yoshimori, T.; Ohsumi, Y. In vivo analysis of autophagy in response to nutrient starvation using transgenic mice expressing a fluorescent autophagosome marker. *Mol. Biol. Cell* **2004**, *15*, 1101–1111. [CrossRef]
204. Corradetti, M.N.; Guan, K.L. Upstream of the mammalian target of rapamycin: Do all roads pass through mTOR? *Oncogene* **2006**, *25*, 6347–6360. [CrossRef] [PubMed]
205. Kim, E.; Goraksha-Hicks, P.; Li, L.; Neufeld, T.P.; Guan, K.L. Regulation of TORC1 by Rag GTPases in nutrient response. *Nat. Cell Biol.* **2008**, *10*, 935–945. [CrossRef]
206. Li, L.; Kim, E.; Yuan, H.; Inoki, K.; Goraksha-Hicks, P.; Schiesher, R.L.; Neufeld, T.P.; Guan, K.L. Regulation of mTORC1 by the Rab and Arf GTPases. *J. Biol. Chem.* **2010**, *285*, 19705–19709. [CrossRef]

toxics

MDPI

Article

Uptake of Breathable Nano- and Micro-Sized Polystyrene Particles: Comparison of Virgin and Oxidised nPS/mPS in Human Alveolar Cells

Antonio Laganà [1,2], Giuseppa Visalli [1], Alessio Facciolà [1], Consuelo Celesti [3], Daniela Iannazzo [3] and Angela Di Pietro [1,*]

[1] Department of Biomedical and Dental Sciences and Morphofunctional Imaging, University of Messina, 98125 Messina, Italy; antonio.lagana1@studenti.unime.it (A.L.); giuseppa.visalli@unime.it (G.V.); alessio.facciola@unime.it (A.F.)
[2] Istituto Clinico Polispecialistico C.O.T., Cure Ortopediche Traumatologiche s.p.a., 98124 Messina, Italy
[3] Department of Electronic Engineering, Industrial Chemistry and Engineering, University of Messina, 98125 Messina, Italy; consuelo.celesti@unime.it (C.C.); daniela.iannazzo@unime.it (D.I.)
* Correspondence: angela.dipietro@unime.it

Abstract: Airborne micro- and nanoplastics are widely spread and pose a risk to human health. The third polymer plastic most commonly produced and present in atmospheric fallout is polystyrene (PS). For these reasons and for a more realistic assessment of biological effects, we examined in-home oxidised (ox-, simulating photoaging) nPS/mPS (0.1 and 1 μm), comparing the effects with virgin ones (v-). On human alveolar cells (A549), we quantified the cellular uptake, using FITC-functionalised nPS/mPS, while cytotoxicity, changes in the acidic compartment, ROS production, mitochondrial function, and DNA damage were assessed to study the effects of internalised v- and ox-nPS/mPS. The results showed that the uptake was dose-dependent and very fast (1 h), since, at the lowest dose (1.25 μg/well), it was 20.8% and 21.8% of nPS and mPS, respectively. Compared to v-, significant ROS increases, DNA damage, and mitochondrial impairment were observed after exposure to ox-nPS/mPS. The enhancement of effects due to environmental aging processes highlighted the true potential impact on human health of these airborne pollutants.

Keywords: environmental wear; uptake; cytotoxicity; ROS overproduction; mitochondrial dysfunction

Citation: Laganà, A.; Visalli, G.; Facciolà, A.; Celesti, C.; Iannazzo, D.; Di Pietro, A. Uptake of Breathable Nano- and Micro-Sized Polystyrene Particles: Comparison of Virgin and Oxidised nPS/mPS in Human Alveolar Cells. *Toxics* **2023**, *11*, 686. https://doi.org/10.3390/toxics11080686

Academic Editors: Esref Demir and Sam Kacew

Received: 7 July 2023
Revised: 3 August 2023
Accepted: 7 August 2023
Published: 10 August 2023

1. Introduction

Plastics, due to their malleability, high versatility and low cost, are currently widely used, and their production is only slightly lower than that of concrete and steel [1]. Despite the remarkable durability, around 50% of the total mass of currently manufactured plastics is disposable. This aspect has greatly increased their presence in different environmental matrices, where improperly disposed plastic waste undergoes a slow and partial abiotic and biotic degradation. Mechanical action of wind and wave motions and prolonged exposure to ultraviolet (UV) light (photo-oxidation), as well as degradative microbiological processes), cause the fragmentation of plastic waste, producing secondary microplastics (<5 mm) and, following further fragmentation, secondary nanoplastics with a diameter ≤0.1 μm [2]. Primary microplastics and nanoplastics, intentionally and directly produced on a micro- and nanoscale as constituents of specific products, also contribute to plastic pollution [3]. Micro- and nanoplastics are persistent in the environment and can interact very easily with biological systems [4,5]. Indeed, they have been found in sediment [6], soil [7], seawater [8], high mountain lake ecosystems [9] and air [10], but also in many foods and beverages like shellfish, cooking salt, drinking water and beer [11]. Due to their hydrophobicity, micro- and nanoplastics are highly bioavailable and are able to bioaccumulate along the trophic scale [12,13].

Humans are inevitably exposed to micro- and nanoplastics, mainly via ingestion and inhalation [14–19], as confirmed by the presence of microplastics in different body regions as well as in faecal excretion [20–24].

Airborne micro- and nanoplastics are derived from a variety of sources, including synthetic fibres, waste disposal products, incinerators, agricultural practices (such as PS peat and sewage sludge used as fertilizer), as well as road traffic [10,25–27]. In particular, tyre wear particles (TWPs) and brake wear particles (BWPs) are formed by complex mixtures of metal and mainly microplastics. It has been estimated that, in ambient air, around 4% and 11% of the respirable (fine) and inhalable (coarse) particulate matter (PM) are, respectively, formed by micro- and nanoplastics derived from only TWPs and BWPs [28].

Moreover, indoor exposure to airborne micro- and nanoplastics appears to be highly relevant, considering the lower dilution volumes, the time spent in indoor environments (on average, 70–90% of our lifetimes) and the several sources of these pollutants (synthetic textile fibres, upholstery and furnishing objects or building materials) [29].

We currently know a lot about the pathogenesis of airborne micro- and nano-PM (fine and ultrafine PM), as well as metal- and carbon-based engineered nanoparticles [30–32]. The $\leq$2.5 μm particles can overcome mucociliary clearance and reach the alveolar surface, wherein nanoparticles can bypass the phagocytic activity of macrophages (alveolar clearance). They easily cross the pulmonary epithelial barrier and enter the bloodstream, from where they are distributed to various anatomical regions [33]. For micro- and nanoparticles, the number of surface atoms per unit mass is increased by several orders of magnitude, greatly enhancing the surface area for chemical reactions, while charge, polarity and the presence of surface reactive groups are fundamental in regulating cellular uptake and biological effects [34]. The pathogenesis of respirable micro- and nanoplastics is poorly known, but it can be assumed that the trigger is determined by oxidative stress, which causes lipid peroxidation, protein and DNA damage, mitochondrial dysfunction and inflammation in response to tissue damage [35].

Since the largest share of airborne micro- and nanoplastics that humans inevitably inhale are the secondary ones, subjected in the environment to a variable photo-oxidation, the aim of this in vitro study was to gain a more realistic assessment of the hazard by studying the biological effects of aged micro- and nanoplastics. In human alveolar epithelial cells (A549 cell line), the effects of in-home oxidised polystyrene nanoplastics and microplastics (ox-nPS/mPS), with diameters of 0.1 and 1 μm, were compared to virgin ones (v- nPS/mPS). In particular, after quantifying the uptake, the cytotoxic effects (i.e., viability loss, changes in acidic compartment, ROS production, mitochondrial impairment and DNA damage) in v- and ox-nPS/mPS were compared. The presence on the particle surface of carboxyl, alkoxyl and hydroxyl groups, simulating the photoaging due to ultraviolet (UV) radiation, induced by an in-home oxidative process, enhanced the particle reactivity, increasing the risk for the exposed subjects.

2. Materials and Methods

2.1. Exposure Conditions and Cell Models

Virgin nPS (average size 100 nm) and mPS (average size 1 μm) were purchased from Sigma-Merck (Milan, Italy; code: BCC8557 and BCC9279). The choice of diameters was based on the evidence that both mPS and nPS are part of the respirable fraction of inhalable particulate matter ($\leq$2.5 μm).

nPS/mPS oxidation was performed by the method reported by Mielczarski et al. (2011) [36]. Briefly, to allow the presence of carboxyl, alkoxyl and hydroxyl groups on the particle surface, aliquots of stock suspensions in phosphate-buffered saline (PBS, 1:10 ratio) were treated at 80 °C for 2 h. The suspensions were characterised by Fourier-Transform Infrared (FT-IR) spectrometry, dynamic light-scattering (DLS), scanning electron microscopy (SEM) and UV-Vis spectrophotometry. As previously reported [37], dynamic light-scattering DLS analyses and SEM observations confirmed the same average size of the functionalised microplastics, suggesting that oxidation occurred only at the surface of the particles and that it did not cause their aggregation, either in PBS or in cell medium suspensions. UV-Vis and FTIR spectra highlighted the presence of oxygen functionalities, such as carbonyl and phenol groups. To quantify spectrofluorometrically the cellular uptake, these functionalised particles were covalently bound to fluorescein isothiocyanate (FITC). Specifically, the conjugation was performed using 1-ethyl-3-(3-dimethylaminopropyl) carbodiimide (EDC) and hydroxybenzotriazole (HOBt) as coupling reagents, and a polyethylene glycol (PEG) linker. This had two amino groups, one of which was previously bound to FITC, while the other was bound to functionalised particles (i.e., ox-mPS- and ox-nPS). The reaction steps, as well as the purification steps, were reported in detail in a previous study [37]. The complexes were then analysed by FTIR spectroscopy while the photoluminescence (PL) properties were investigated by dynamic light scattering (DLS) [37]. To assess the stability of the conjugates under the experimental conditions, a time course (1, 3, 6 and 24 h) was performed using mPS-FITC suspended in cell medium with 2% FBS and in saline solution at pH 6.5 (early endosomes) and 4.5 (mature endosomes). Briefly, after 1, 3, 6 and 24 h, the suspensions were centrifuged (12,000$\times$ g for 10 min) and a fluorometric reading was carried out (ex 485 nm, em 535 nm) in both supernatants and mPS resuspended in the same volume of fresh solutions. While the emission values were almost constant in FITC-mPS suspensions, no emissions were recorded in the supernatants, with the exception of the ones at pH 4.5 at 24 h. Compared with the respective suspensions, the emission values of the latter were within 10%.

The cellular uptake and the biological effects of virgin (v-) and oxidised (ox-) nPS/mPS suspensions were assessed in the human alveolar cell line A549 (ATCC-CCL-185 Tm), which is the model of choice for in vitro studies of airborne pollutants. The cells were cultured in F-12K medium (Gibco™ 21127022) supplemented with 2 mM of L-glutamine, 10% of inactivated foetal bovine serum (FBS), and 1% penicillin/streptomycin/amphotericin, at 37 °C in a 5% CO_2/95% air humidified atmosphere. For all experiments, the exposure treatment for times and doses established in the experimental protocol was performed in semiconfluent monolayers grown for 24–36 h and incubated with nPS/mPS suspensions that were set up in cell medium containing 2% FBS (maintenance cell medium). Although the presence of proteins could partially neutralise the effect of the particles due to the so-called "protein corona", we believe that this protocol best simulates what actually happens. The corona effect hinders the intake due to the particles' increased bulk and the loss of hydrophobicity, which is known to promote the interaction of particles with cell membranes [38]. Moreover, the protocol meant that exposure to xenobiotics occurred under physiological conditions and not in cells which, kept in suboptimal conditions, could lead to an overestimation of the effects. Cells treated with PBS without nPs/mPS were used as a negative control while cells treated with suitable compounds were used as positive controls.

2.2. Cellular Uptake of nPS/mPS

The stock suspensions of nPS-FITC/mPS-FITC (80 µg mL^{-1}) were diluted in culture medium and added (100 µL/well), in the range of 1.25–20 µg/well, to A549 cells that had been grown for 24 h in 96-well microplates (final density 4 × 10^4 cells/well). After 0.5, 1, 3 and 24 h, fluorometric readings were carried out at the excitation and emission wavelengths of 485 nm and 535 nm, respectively, by using a microplate reader (Tecan Italia, Milan, Italy). After recording the emission values in each well, the medium was removed, the monolayer was washed repeatedly with PBS, to remove uninternalised particles, and emission values were recorded to measure the percentage uptake (i.e., intracellular nPS-FITC/mPS-FITC). Moreover, A549 monolayers grown in chamber slides were examined with confocal laser scanning microscopy (CLSM) using the Leica TCS SP2 instrument (Leica Microsystems, Wetzlar, Germany), with Leica Confocal software (version 2.0) used to process the images, which were acquired in both fluorescence and phase contrast. A Leica DM IRB fluorescence microscope (Leica Microsystems) was used to select the optical fields.

2.3. Viability Assays

In A549 cells, v-and ox-nPS/mPS-induced cytotoxicity was evaluated by using the colorimetric MTT assay, based on the reduction of 3,(4,5-dimethiazol-2)-2,5-difeniltetrazolium bromide, catalysed by cellular NAD(P)H-dependent dehydrogenases. Briefly, after verifying the absence of particle interference in the spectrophotometric detection of cell viability, the assay was performed in cells cultured for 24 h in 96-well microplates, to which the appropriate volume of stock suspensions in PBS (10 mg mL^{-1}) was added in the medium. The tested concentrations ranged from 12.5 to 200 μg mL^{-1}. Dimethyl sulphoxide (DMSO, 10%) was used as a positive control, and eight different wells were treated for each dilution. After following our standardised protocol [37], the enzymatic activity was quantified by spectrophotometric measurement at 540 nm, using a microplate reader (Tecan Italia). The optical density (OD) values obtained for each sample were compared to the mean OD of the negative control, which was arbitrarily considered corresponding to 100% viability.

2.4. Assessment of the Cellular Acidic Compartment

Changes in the endocytic apparatus (late endosomes and lysosomes), due to the uptake of virgin and oxidised nPS/mPS, were examined by employing metachromatic fluorophore Acridine Orange (AO), which is captured by protons and collected in the acidic compartment. Here, the highly concentrated probe will emit red fluorescence, while it will release green fluorescence in the cytosol and nucleus, where AO scarcely accumulates. The loss of red fluorescence is indicative of acidic compartment damage [39]. The analyses were performed in semi-confluent A549 monolayers grown in chamber slides and treated for 3 and 24 h at 37 °C with nPS/mPS suspensions (100 μg mL^{-1}). After removing the medium and washing repeatedly with PBS, AO solution (5 μg mL^{-1}) was added and CLSM was used to assess the endocytic apparatus and other morphological changes which were nPS/mPS-induced. To quantify the acid compartment, the image-processing program Image J (imagej.nih.gov/ij/index.html, accessed on 15 September 2022) was used to calculate the cellular area which emitted red fluorescence. These values were expressed as %, referring to the total area of each cell, and at least 100 cells were analysed for each slide.

2.5. Evaluation of ROS Production

To test the pro-oxidant effect of v- and ox-nPS/mPS, ROS were measured by using the 2′,7′-dichlorofluorescein-diacetate (DCF-DA) probe (Merck Life Science S.r.l., Milan, Italy). After crossing cell membranes by passive diffusion, the reagent is hydrolysed rapidly by cellular esterases to 2′,7′-dichlorofluorescein (DCFH). This non-fluorescent compound is oxidised in the presence of ROS, forming the highly fluorescent molecule 2′,7′-dichlorofluorescein (DCF). Briefly, after repeated washing with PBS, sub-confluent A549 monolayers (80%) in 96-well microplates were loaded with the probe solution (1 μM) prepared in PBS containing 10 mM of D-glucose (pH 7.4) and were incubated at 37 °C for 30 min [40]. After washing in PBS to remove the non-internalised probe, cells were treated with v- and ox-nPS/mPS suspensions (in the range of 25–200 μg mL^{-1}). The fluorometric readings were carried out in the intervals of 0.5–24 h by using a microplate reader (Tecan Italia) at the excitation and emission wavelengths of 485 and 535 nm, respectively. ROS production was calculated as the percentage change (Δ%) compared with control cells.

2.6. Mitochondrial Transmebrane Potential

To assess mitochondrial impairment induced by v- and ox-nPS/mPS, we measured transmembrane potential by the incorporation of the fluorescent probe rhodamine 123 (R123) (Invitrogen Molecular Probes, Eugene, OR, USA). The chemical properties of the cationic fluorochrome R123 allow mitochondrial membrane crossing and storage in the matrix only in functional mitochondria that possess a transmembrane potential ($\Delta\Psi m$), which is indicative of an active proton gradient maintained during oxidative phosphorylation [41]. A549 monolayers, grown in 96-well plates, were treated with 100 μg mL^{-1}

of v- and ox-nPS/mPS suspensions for 24 h. After incubation at 37 °C and washing in PBS to remove the non-internalised particles, cells were treated with the probe solution (10 μM final concentration) and incubated for 10 min at 37 °C. Fluorimetric readings were carried out using a microplate reader (Tecan Italia, Milan, Italy) set to 535 and 595 nm as the excitation and emission wavelengths, respectively. In comparison to the control cells, the percentage changes of emission values were calculated for each sample.

2.7. Assessment of DNA Damage by the Comet Assay

A549 cells treated for 24 h with v- and ox-mPS/mPS suspensions (100 μg mL^{-1}), were assessed for DNA integrity by using the alkaline version of the comet assay [42]. Tests were performed in duplicate on about 2×10^4 cells for each spot, and the electrophoresis was carried out for 30 min at 300 mA and 25 V (0.86 V cm^{-1}). The slides, stained with ethidium bromide (20 μg mL^{-1}), were imaged using a DMIRB fluorescence microscope (Leica Microsystems), equipped with a digital camera (Power Shot S50; Canon, Milan, Italy), at 400× total magnification. Samples were run in duplicate, and images of 100 cells per slide were acquired randomly and analysed by using the Comet Assay Software Project (CASP) software (http://ww25.casplab.com/?subid1=20230810-1122-084e-8a7d-57f935e283f5 (accessed on 15 September 2022)). %TDNA (i.e., %DNA in the tail) was considered the parameter of DNA damage.

2.8. Statistical Analyses

All data are presented as the mean ± standard error (SE) based on at least three independent experiments. Analyses were performed using the Statistica programme (version 10). Lilliefors and Shapiro–Wilk normality tests were used to assess data distribution patterns. The relationships between different parameters were assessed by using the Pearson correlation coefficient, while the *t* Test was used to assess the differences between samples. Significance was accepted at $p < 0.05$.

3. Results

3.1. Cellular Uptakes of nPS/mPS

To spectrofluorimetrically quantify the uptake of nPS-FITC/mPS-FITC in A549 cells, we generated a time course of emission values. After ascertaining the absence of free FITC in the particle suspensions in cell medium in the interval 3–24 h, preliminary abiotic tests showed that the emission values (expressed in arbitrary fluorescence units [AFU]) under the experimental conditions were 97.13 and 72.45 for 1 μg of mPS and nPS, respectively. On this basis, in the range of 1.25–20 μg/well, we calculated the internalised amount (μg) at different times. After 1 h, at the lowest dose to which the cells were exposed, 20.8% and 21.8% of the amount of nPS and mPS, respectively, was internalised.

The results in Figure 1A are expressed as internalised amounts (μg/well) of nPS/mPS and clearly highlight the significant dose–effect correlation for both sizes of plastic particles (Pearson correlation coefficient [r] > 0.99). On the other hand, nPS and mPS showed distinct trends as a function of the exposure time (Figure 1B,C). The amounts of internalised mPS decreased as the exposure time increased for all tested doses; the percentage decrease (Δ%) was between 20% (at the lower doses of 1.25–2.5 μg/well) and 30% (at the higher doses of 5–20 μg/well). For nPS, this trend was observed only at the higher exposure doses (Δ% > 30), while the uptake at lower doses increased by an average of 20% during the entire exposure period.

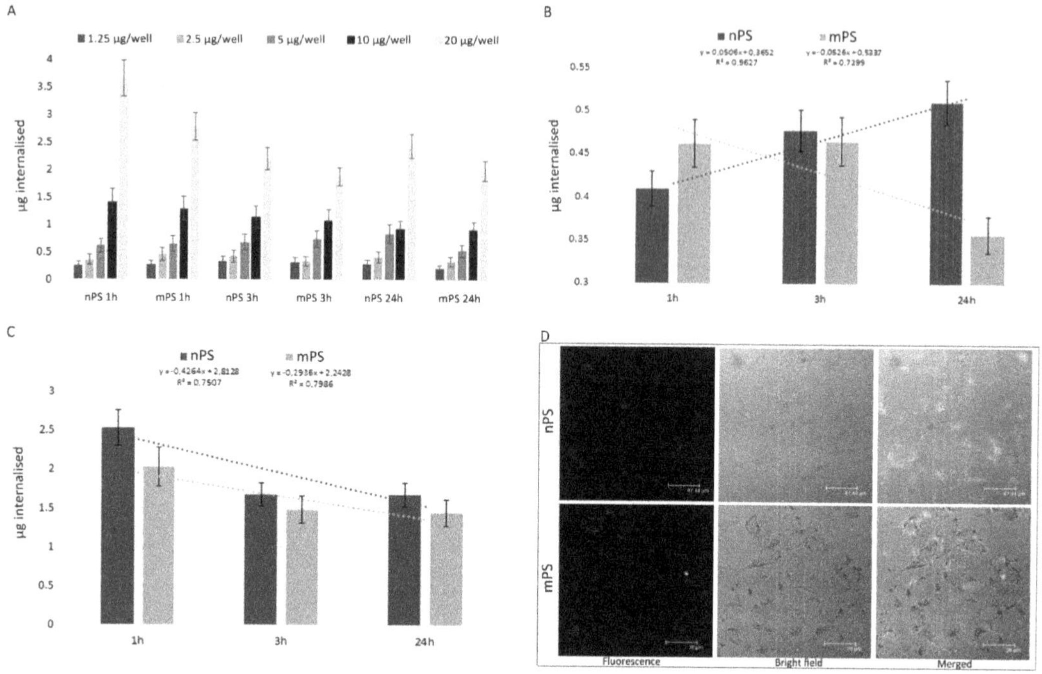

Figure 1. Cellular uptakes of PS/mPS. (**A**) Time course of spectrofluorimetric measurements of internalised μg of nPS-FITC/mPS-FITC in A549 cells treated in the range of 1.25–20 μg/well. Highly significant dose–effect correlation for both sizes of plastic particles was observed ($p < 0.01$ to Pearson test). (**B**,**C**) Trends of nPS-FITC/mPS-FITC uptake as a function of the exposure time in A549 cells exposed in the range of 1.25–2.5 μg/well (**B**) and in the range of 5–20 μg/well (**C**). In (**B**,**C**) each bar reports the average ± SE of internalised μg (based on FITC Emission value) recorded for 1.25 and 2.5 μg/well (lower doses) and 5, 10 and 20 μg/well of nPS-FITC/mPS-FITC, respectively. The graphs highlight the different trend over time of nPS-FITC/mPS-FITC uptake, showing a positive finding only for the lower concentrations of nPS. (**D**) CLSM images of nPS-FITC/mPS-FITC internalisation in A549 semiconfluent monolayers treated for 1 h at the dose corresponding to 10 μg/well. The cells exposed to nPS or mPS exhibited fair cytoplasmatic fluorescence, which was more intense in mPS-FITC-treated cells. In contrast microscopy image of these latter cells, intracytoplasmic aggregates of mPS were clearly visible.

Considering the PS density and the size of the particles, we calculated the total number of internalised particles: ~200 mPS and 200,000 nPS per μg internalised. The markedly higher number of internalised nPS highlights the greater surface area developed by the last ones. Because the experiments were performed in 96-well microplates with an average of 4×10^4 cells/well, regardless of the particle size, the uptake was in the range of 5.1–91.2 pg/cell.

CLSM confirmed the data and showed that the cells exposed to nPS/mPS exhibited fair cytoplasmatic fluorescence, which was clearly more intense in cells treated with the more fluorescent mPS (Figure 1D). In summary, the kinetics of nPS/mPS uptake indicated that internalisation was extremely fast, with a higher amount of internalised nPS.

3.2. *nPS/mPS-Induced Changes of the Acidic Compartment*

The evaluation of the acid compartment gives an insight about the cellular ability to neutralize foreign particles. Therefore, employing the metachromatic fluorophore AO, the microscopic analyses of the endocytic apparatus in cells treated with v-nPS/mPS for

3 h highlighted a very bulky acidic compartment made up of numerous red organelles, clearly showing the consistent internalisation of both particles and, at the same time, the integrity of mature endosomes that almost completely occupied the perinuclear cytosol (Figure 2A,B).

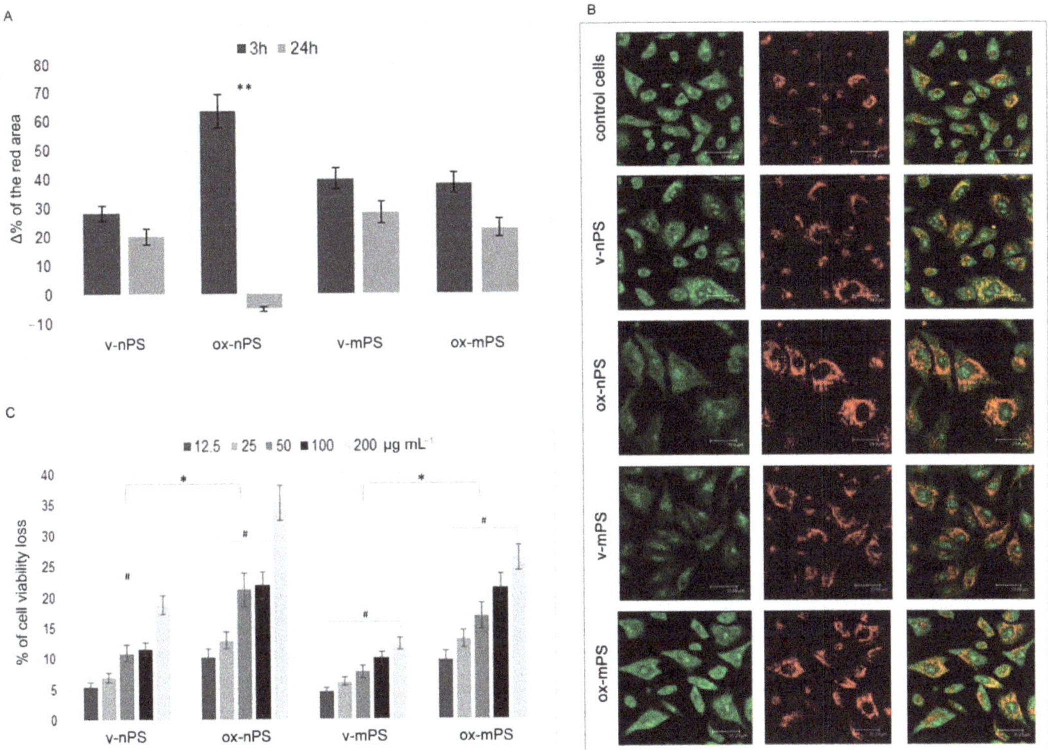

Figure 2. Changes in acidic compartment and cytotoxicity, v- and ox-nPS/mPS-induced. (**A**) The acidic compartment was assessed by employing metachromatic fluorophore AO which gives red colour when, due to the lower pH values, the dye builds up. The graph (**A**) reports Δ% of the area emitting red fluorescence in exposed cells in comparison to control cells. The experiments were performed in semi-confluent A549 monolayers grown in chamber slides and treated for 3 and 24 h at 37 °C with 100 μg mL^{-1} of v- and ox-nPS/mPS suspensions. A significant difference between 3 and 24 h was observed for ox-nPS ($p < 0.01$ to *t*-test). (**B**) Representative CLSM images in control and exposed cells (3 h). A549 cells exposed to v- and ox-nPS/mPS suspensions showed a very bulky acidic compartment made up of numerous red organelles that, particularly for ox-nPS, almost completely occupied the perinuclear cytosol. (**C**) Results of MTT assay in A549 treated for 24 h in the range of 12.5–200 μg mL^{-1}, corresponding to 2.3×10^2–3.68×10^3 particles/well and 2.3 10^5–3.68×10^6 particles/well for mPS and nPS, respectively. In comparison to the virgin nPS/mPS, the percentages of cell viability loss were significantly higher in ox-nPS and mPS (* $p < 0.05$; ** $p < 0.01$ to *t*-test). For all plastic particles, the decrease in cell viability was positively related to the exposure dose (# $p < 0.01$ to Pearson test). Compared to mPS, nPS showed a moderately higher cytotoxic effect.

Compared to control cells, in the v-nPS/mPS-treated cells the Δ% of the area which emitted red fluorescence was 28.3 and 40.1, respectively, on average. In contrast to mPS, for which the values of v- and ox- were superimposable, a significant increase in the phagosomal compartment was observed in cells treated with ox-nPS (Δ% 65.5). Further changes in the endocytic apparatus were observed after 24 h when, in the cells treated with both v-nPS and v- and ox-mPS, a moderate reduction in the area of the acidic compartment was observed. This, underlining a possible spill from the endocytic apparatus, was more evident in cells treated with ox-nPS in which acidic organelles were also less developed than the control cells (Δ% −5.1), and the differences compared to 3 h were highly significant ($p < 0.01$). The results obtained showed a greater alteration of the acid compartment in cells exposed to nPS than in mPS, which is more attributable to ox-nPS.

3.3. Cytotoxicity nPS/mPS-Induced

We evaluated cytotoxicity in our cell model by using the MTT assay. The tested doses ranged from 12.5 to 200 μg mL^{-1}, corresponding to 2.3×10^2–3.68×10^3 particles/well and 2.3×10^5–3.68×10^6 particles/well for mPS and nPS, respectively, with an exposure time of 24 h. Both v-nPS and mPS had a moderate cytotoxic effect. Up to 200 μg mL^{-1}, cell viability was >80% and, at the lowest exposure dose, cell viability was only about 5% lower than the control cells (Figure 2C). Unlike virgins, a more marked cytotoxicity was elicited by oxidised particles. In comparison to the virgin nPS/mPS, the percentage of cell viability loss was 1.9-fold and more than double in cells treated with ox-nPS and mPS, respectively ($p < 0.05$). For all plastic particles, the decrease in cell viability was positively related to the exposure dose. For v- and ox-nPS, the percentage of cell viability loss ranged between 5.4 and 18.8 and between 10.1 and 35.3, respectively ($p < 0.01$), while for mPS, these percentages ranged from 4.5 to 12.3 and from 9.7 to 26.3, respectively ($p < 0.01$). In conclusion, the assay underlined the moderately higher cytotoxic effect of nanosized particles, especially in the oxidised rather than in the virgin ones.

3.4. nPS/mPS Increased ROS Production

Figure 3A,B report the time course (0.5–24 h) of ROS production in A549 cells treated with v- and ox-nPS/mPS in the range of 25–200 μg mL^{-1}. Similar to the kinetic uptake data, ROS overproduction was already evident after 0.5 h, especially for ox-nPS/mPS. Over time, ROS levels progressively increased, and r coefficients to the Pearson test were always >0.95 and similar to those calculated for the positive control (H_2O_2 300 μM). In the interval of 0.5–24 h, DCF emission values increased on average by 15-fold for nPS and 17-fold for mPS, while no dose effect was observed for v-nPS/mPS. The pro-oxidant effect of v-mPS was significantly higher, with ROS values which were, on average, double compared to those of the v-nPS ($p < 0.05$) and only 20% lower than positive control. Compared to v-nPS/mPS, the increase was surprisingly smaller and equal to 40% for ox-nPS (P n.s). Conversely, ROS production induced by the ox-mPS was significantly increased, with emission values more than double ($p < 0.05$) in comparison to the v-mPS.

Moreover, ROS overproduction induced by ox-mPS was positively related to doses ($p < 0.01$), and the probe emission values were double compared to the positive control already at 50 μg mL^{-1} (Figure 3B).

CLSM observations of the semiconfluent monolayers of v- and ox-nPS/mPS-treated cells (100 μg mL^{-1}) for 3 h confirmed the results (Figure 3C), highlighting diffuse green cytosolic fluorescence, which was more intense in mPS-treated cells. Moreover, as shown by magnification of phase contrast micrographs, intracellular clusters of particles are clearly visible in the mPS-treated cells (Figure 3D). In summary, ROS overproduction was very fast, size-dependent and higher in cells treated with oxidised particles.

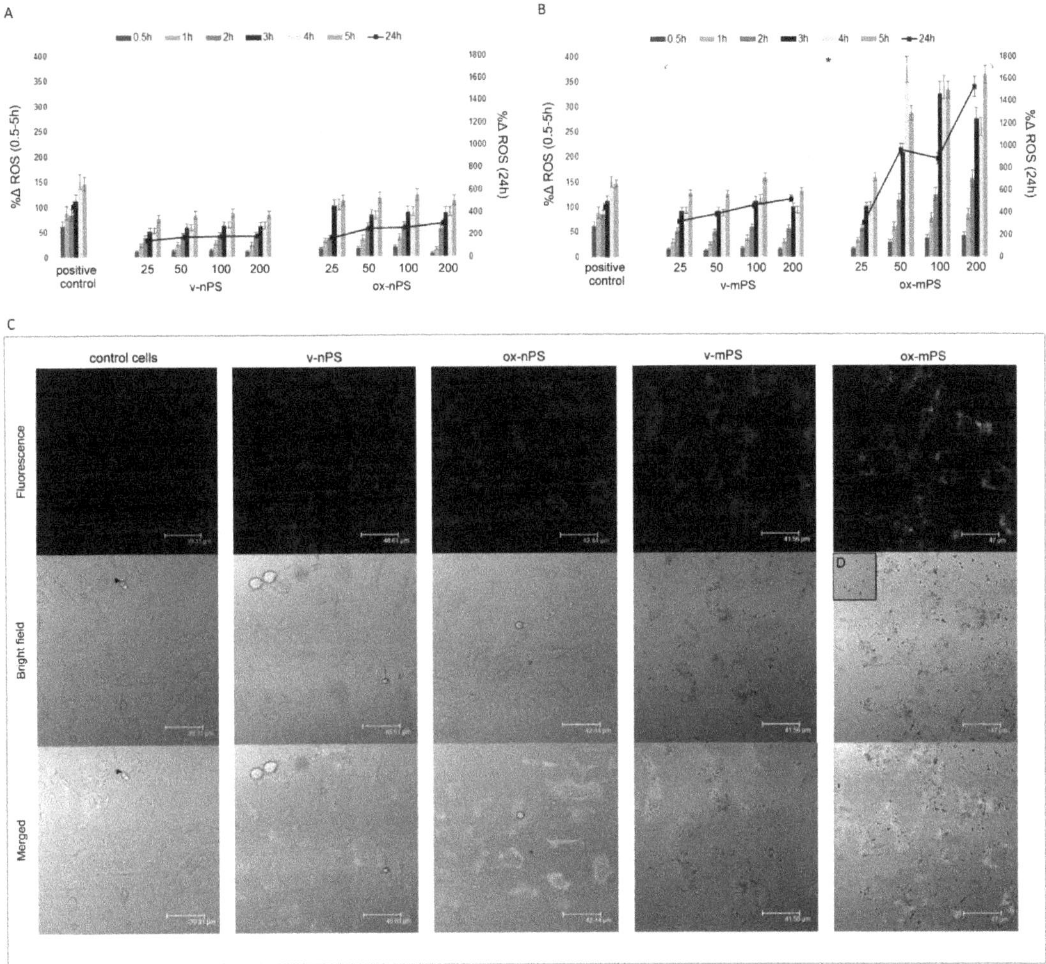

Figure 3. ROS overproduction induced by v- and ox-nPS/mPS. (**A**,**B**) Time courses (0.5–24 h) of ROS in A549 cells treated with nPS and mPS, respectively, in the range of 25–200 μg mL^{-1}. H_2O_2 (300 μM) was used as positive control and the values are reported as %Δ relative to control cells. The two scales on the y axis show the values recorded in the interval 0.5–5 h on the left and those recorded at 24 h on the right. For v- and ox-nPS/mPS, exposure time and ROS values were always significantly related ($r > 0.95$; $p < 0.01$ to Pearson test). No dose effect was observed for v- nPS/mPS and for ox-nPS; conversely, ROS overproduction was positively related to doses ($p < 0.01$) for ox-mPS. Compared to v-nPS, significantly higher ROS values were observed for v-mPS ($p < 0.05$ to *t*-test). No significant differences were observed between v- and ox-nPS, while the pro-oxidant effect of ox-mPS was significantly higher compared to v-mPS (* $p < 0.05$). (**C**) CLSM images of the semiconfluent monolayers exposed to v- and ox-nPS/mPS (100 μg mL^{-1}) for 3 h and treated with DCF-DA probe. A diffuse green cytosolic fluorescence, more intense in ox-mPS-treated cells, is shown. Moreover, as shown by (**D**) magnification of phase contrast micrograph, intracellular clusters of particles are clearly visible in the mPS-treated cells.

3.5. *nPS/mPS Induced Mitochondrial Dysfunction*

Transmembrane potential (ΔΨm) was detected to assess mitochondrial dysfunction induced by v- and ox-nPS/mPS. The experiments (Figure 4A) highlighted a moderate decrease in nPS-treated cells (%Δ 16 vs. control cells), without differences between v- and ox-nPS/mPS. Almost the same decrease was recorded in cells treated with ox-mPS. Instead, the virgin counterpart of these micro-sized particles did not alter mitochondrial function (%Δ −5 vs. control cells), clearly highlighting that ROS overproduction induced by v-mPS was directly attributable to the particles and not secondary to mitochondrial dysfunction. In summary, v-/ox-nPS and ox-mPS caused a modest mitochondrial dysfunction in treated cells.

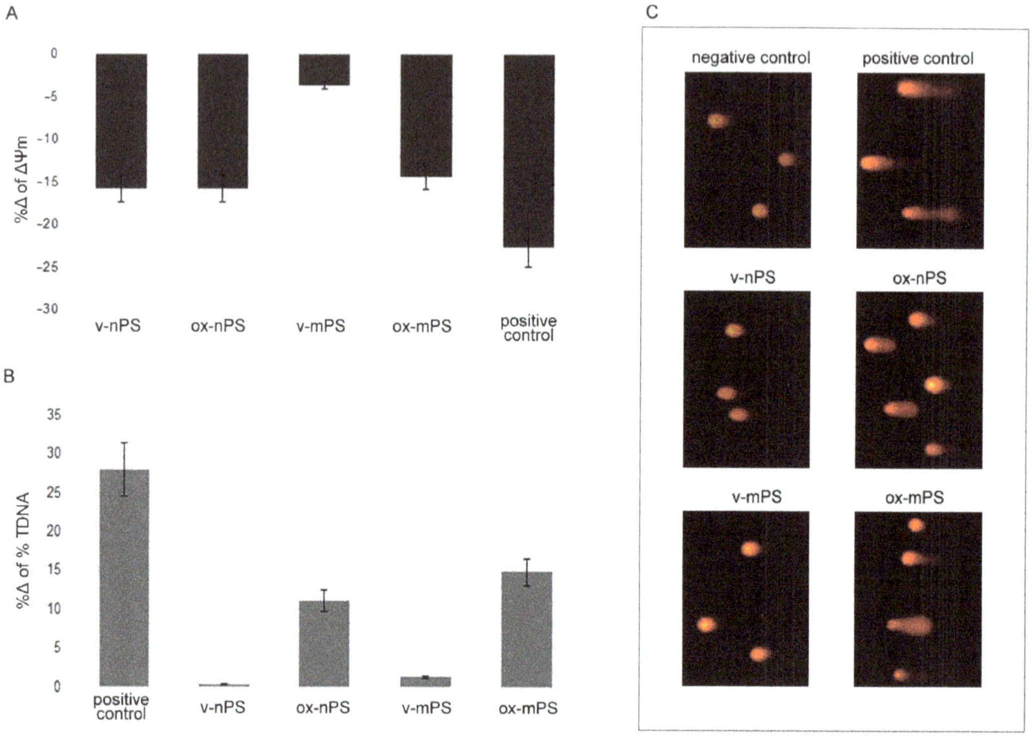

Figure 4. Mitochondrial impairment and DNA damage induced by v- and ox-nPS/mPS. (**A**) %Δ of transmembrane potential (ΔΨm) measured spectrophotometrically by the employment of R123. Conversely to v-mPS, both v- and ox-nPS and ox-mPS caused a moderate mitochondrial impairment (%Δ 16 vs. control cells). (**B**) Results expressed as %Δ of % TDNA to comet assay. The test was performed after overnight exposure to 100 μg mL^{-1} suspensions of v- and ox-nPS/mPS. H_2O_2 (300 μM) was used as positive control. Unlike v- nPS/mPS, the ones oxidised caused a moderate increase in DNA damage. (**C**) Representative images of comet assay that was replicated three times with similar results. Compared to v- nPS/mPS-treated cells, higher level of DNA damage is observed in ox-nPS/mPS-treated cells.

3.6. nPS/mPS-Induced DNA Damage

The genotoxicity of v- and ox-nPS/mPS was assessed by the comet assay performed after overnight exposure to 100 μg mL^{-1} suspensions. H_2O_2-treated cells (300 μM) served as the positive control. In our cell model, v-nPS/mPS did not cause DNA damage and the %TDNA values almost completely overlapped with the control cells (Figure 4B,C). Conversely, both the oxidised particles were genotoxic and, compared to positive control, the %TDNA values were only 15% (nPS) and 10% (mPS) lower.

The results were consistent with ROS overproduction and highlight the oxidative DNA damage induced by exposure to oxidised nPS/mPS.

4. Discussion

The potential adverse health effects of breathable micro- and nanoplastics in humans are still poorly studied, despite the increasing quantities of airborne micro- and nanoplastics that can be found in the ambient air and, above all, in indoor environments. Several factors contribute to make plastic air pollution a threat for human health. These include both the features of micro- and nanoplastics and the anatomical and physiological characteristics of humans.

As reported in the introduction, the progressive fragmentation of plastics and their low density favour their long stay in the air (fly particles). Other intrinsic features of the plastic particles, such as hydrophobicity, favour a closer interaction with cell membranes, inducing higher translocation rates [10].

Regarding humans, we must consider how the defence mechanisms of the respiratory system are undoubtedly less efficient against microparticles ($\leq$2.5 μm) and, above all, nanoparticles ($\leq$ 0.1 μm). In particular, while microparticles are phagocytosed by alveolar macrophages, a process which is much slower than mucociliary clearance and able to trigger the inflammatory cascade, nanoparticles may bypass macrophage clearance. To maximise gas exchange, the alveolar epithelium is extremely extended (about 140 m^2), and the alveolar–capillary barrier is particularly thin (<1 μm), which greatly favours the ability of nanoparticles to cross it and enter the bloodstream. Moreover, considering the volumes of air breathed daily (always >10 m^3 in adults), even a reduced presence of airborne micro- and nanoplastics (a single particle L^{-1} corresponds to 10,000 respired particles) would cause their accumulation both in the lungs and, via the bloodstream, in other organs, triggering pathogenic processes.

Despite some limitations, mainly due to the lack of characterization of nPS/mPS in the biological environment where proteins and other components can modify the sizes and the cell-particle interactions, our in vitro study improves our knowledge of the effects of inhaled airborne micro- and nanoplastics. We highlighted the enhanced damage attributable to the surface changes of the particles due to oxidative processes, undergone during their environmental stay. As previously reported [37], the "artificial aging" used by us significantly increased the presence of carboxyl, alkoxyl and hydroxyl groups on their surfaces, simulating the photoaging endured by the particles once released into the environment [43]. The increase in oxygen-containing groups after the oxidative process was confirmed by Fourier-Transform Infrared (FT-IR) spectrometry, dynamic light-scattering (DLS), scanning electron microscopy (SEM) and UV-Vis spectrophotometry [37], which coincided with that which was reported by Biale et al. (2021) [44], showing surface-limited formation of oxidised aromatic structures in PS particles without any involvement of the overall polymer mass. As assumed, the presence of oxygen-containing groups on the particle surface increased the reactivity, and our results were confirmed by several studies [45–48].

By using homemade FITC-loaded nPS/mPS, we quantified the uptake and highlighted the speed at which both nPS and mPS were internalised. Similar to that which has been demonstrated for other particles [30], hydrophobicity favours the close interaction between micro- and nanoparticles and cell membranes, justifying both the speed of the process and the yield, as confirmed by the internalised doses, which ranged from 5.1 to 91.19 pg/cell. For both nPS and mPS, the process was significantly dose-dependent, while only the internalisation of nPS at low doses was time-dependent. The observed internalisation rate, equal to approximately one fifth at low exposure doses, and the volumes of air breathed daily (always >10 m^3 in adults), clearly underscore the potential impact of these emergent pollutants on human health. Even a reduced presence of airborne micro- and nanoplastics (a single particle L^{-1} corresponds to 10,000 respired particles daily) would cause their accumulation both in lungs and, via the bloodstream, in other organs, triggering pathogenic processes. In humans, the defence mechanisms of the respiratory system are undoubtedly less efficient against microparticles ($\leq$2.5 μm) and, above all, nanoparticles ($\leq$0.1 μm). In particular, while microparticles are phagocytosed by alveolar macrophages, a process which is much slower than mucociliary clearance and able to trigger the inflammatory cascade, nanoparticles may bypass macrophage clearance. To maximise gas exchange, the alveolar epithelium is extremely extended (~140 m^2 in adults), and the alveolar–capillary barrier is particularly thin (<1 μm), which greatly favours the ability of nanoparticles to cross it and enter the bloodstream.

Despite endocytosis being the main pathway of particle internalisation in all cells, we cannot exclude the possibility that nanoparticles can cross cell membranes via the energy-independent diffusion process. Diffusion is gradient-dependent and, albeit partially, is counteracted by the frictional coefficient of the particle that is in turn related to both the medium viscosity and the interactions between particles and macromolecules diluted in the solvent [49,50].

Considering energy-dependent endocytosis, our cell models allowed us to verify internalisation in epithelial alveolar cells (pinocytosis) which involves actin polymerisation, as shown by Varma et al. [51]. Therefore, the process requires GTPase activity and can be receptor-mediated (clathrin-dependent endocytosis or caveola-mediated endocytosis) [50]. The role of the endocytosis pathway had been confirmed in two models of intestinal cells by using inhibitors of caveola- and clathrin-mediated endocytosis [51,52]. After membrane invagination, the particles are internalised in early endosomes, which merge with lysosomes to form endolysosomes (late endosomes); our results revealed the fast load (i.e., 3 h) of v- and ox-nPS/mPS in the acidic compartment, highlighted by the enlargement of endolysosomes. Extending observation times, the reduction of the acid compartment, particularly evident for ox-nPS, could be attributed to endolysosomal permeabilisation, producing irreversible cytoplasmic acidification, enzymolysis and apoptosis. As confirmed by the lower viability recorded by the MTT test, the endolysosomal permeabilisation induced by ox-nPS was more marked, while the moderate cytotoxicity leads us to believe that a limited number of cells were involved in this effect after exposure to the v-nPS/mPS and to ox-mPS.

Despite the massive seizure in the endocytic apparatus, a share of nPS/mPS is randomly localised in the cell cytoplasm, causing the observed oxidative damage. We have reported similar results in HT-29 cells exposed to 3 and 10 μm PS particles [17], and similar results were obtained in Caco-2 cells [53].

For both nPS and mPS, the cellular-induced redox imbalance was time-dependent and higher in cells treated with oxidised mPS, confirming their intracellular bioavailability and the most powerful pro-oxidant effect of aged particles. Surprisingly, the redox imbalance of aged particles was more evident for mPS, despite the higher surface/mass ratio of the nPS, which notably increases reactivity [32]. Conceivably, this result is imputable to the higher cytotoxicity elicited by ox-nPS. The detachment of a large number of cells with internalised nPS did not allow us to assess oxidative damage in its entirety.

The most powerful pro-oxidant effect of aged particles was highlighted by the results of the comet assay and mitochondrial transmembrane potential and, unlike the virgin counterpart, oxidised nPS/mPS was able to cause DNA damage and mitochondrial dysfunction.

Since oxidation is the most important degradation process which plastics undergo during their aging in the environment, our results highlight in a more realistic way the potential health risk of the general population, mainly exposed to aged nPS/mPS, downsizing the value of the first in vitro studies almost always performed using virgin micro- and nano-polystyrene particles. However, it should be emphasised that nPS impaired mitochondrial function, regardless of whether it was virgin or oxidised. The decreases in transmembrane potential after exposure to v- and ox-nPS confirmed the results of Wu et al. (2019) [53] who observed transmembrane depolarisation in Caco-2 cells exposed to virgin nanoplastics. In addition to nPS, ox-mPS also decreased transmembrane potential, confirming that mitochondrial impairment, by triggering a vicious circle, further contributes to ROS overproduction, which was significantly increased in cells exposed to ox-mPS. Moreover, considering the key role played by mitochondria in triggering apoptosis [54], the observed mitochondrial impairment is certainly not to be underestimated in outlining the pathogenetic mechanism of these emergent airborne pollutants.

5. Conclusions

Overall, our results highlight the potential negative effects of the respirable fraction of plastic particles on human health. These airborne particles, remaining in the environment for relatively prolonged times, undergo numerous degradation processes, including photochemical ones that cause oxidation, increasing their reactivity. Simulating the photoaging process and comparing the effects induced by oxidised nPS/mPS to those of virgins, we clearly demonstrated that the greatest damage is elicited by the former, underlining the importance of performing the risk assessment using environmentally aged particles. Although the effects induced by these airborne pollutants are much less powerful than those of other airborne particles (combustion by-products, metals, engineered nanoparticles, etc.), their potential impact on human health cannot be excluded, especially if urgent action is not taken to limit their presence in the environment, significantly increased also due to massive use of face masks during the SARS-CoV-2 pandemic.

Author Contributions: Conceptualization, A.D.P.; methodology, G.V., A.L. and C.C.; formal analysis, G.V., A.L. and A.F.; data curation, A.D.P. and D.I.; writing—original draft preparation, A.D.P. All authors have read and agreed to the published version of the manuscript.

Funding: This research received no external funding.

Institutional Review Board Statement: Not applicable.

Informed Consent Statement: Not applicable.

Data Availability Statement: Data is contained within the article.

Acknowledgments: The authors acknowledge support from the University of Messina through the APC initiative.

Conflicts of Interest: The authors declare no conflict of interest.

References

1. Li, W.C.; Tse, H.F.; Fok, L. Plastic waste in the marine environment: A review of sources, occurrence and effects. *Sci. Total Environ.* **2016**, *566–567*, 333–349.
2. Domínguez-Jaimes, L.P.; Cedillo-González, E.I.; Luévano-Hipólito, E.; Acuña-Bedoya, J.D.; Hernández-López, J.M. Degradation of primary nanoplastics by photocatalysis using different anodized TiO_2 structures. *J. Hazard. Mater.* **2021**, *413*, 125452. [CrossRef]
3. Hirt, N.; Body-Malapel, M. Immunotoxicity and intestinal effects of nano- and microplastics: A review of the literature. *Part. Fibre Toxicol.* **2020**, *17*, 57.
4. Prata, J.C.; Da Costa, J.P.; Lopes, I.; Duarte, A.C.; Rocha-Santos, T. Environmental exposure to microplastics: An overview on possible human health effects. *Sci. Total Environ.* **2020**, *702*, 134455.
5. Najahi, H.; Alessio, N.; Squillaro, T.; Conti, G.O.; Ferrante, M.; Di Bernardo, G.; Galderisi, U.; Messaoudi, I.; Minucci, S.; Banni, M. Environmental microplastics (EMPs) exposure alter the differentiation potential of mesenchymal stromal cells. *Environ. Res.* **2022**, *214*, 114088.
6. Li, Y.; Lu, Z.; Zheng, H.; Wang, J.; Chen, C. Microplastics in surface water and sediments of Chongming Island in the Yangtze Estuary, China. *Environ. Sci. Eur.* **2020**, *32*, 15. [CrossRef]
7. Yang, L.; Zhang, Y.; Kang, S.; Wang, Z.; Wu, C. Microplastics in soil: A review on methods, occurrence, sources, and potential risk. *Sci. Total Environ.* **2021**, *780*, 146546.
8. Alfaro-Núñez, A.; Astorga, D.; Cáceres-Farías, L.; Bastidas, L.; Soto Villegas, C.; Macay, K.C.; Christensen, J.H. Microplastic pollution in seawater and marine organisms across the Tropical Eastern Pacific and Galápagos. *Sci. Rep.* **2021**, *11*, 6424.
9. Pastorino, P.; Prearo, M.; Pizzul, E.; Elia, A.C.; Renzi, M.; Ginebreda, A.; Barceló, D. High-mountain lakes as indicators of microplastic pollution: Current and future perspectives. *Water Emerg. Contam. Nanoplast.* **2022**, *1*, 3.
10. Facciolà, A.; Visalli, G.; Pruiti Ciarello, M.; Di Pietro, A. Newly Emerging Airborne Pollutants: Current Knowledge of Health Impact of Micro and Nanoplastics. *Int. J. Environ. Res. Public Health* **2021**, *18*, 2997.
11. Lim, X. Microplastics are everywhere—But are they harmful? *Nature* **2021**, *593*, 22–25. [CrossRef]
12. Saley, A.M.; Smart, A.C.; Bezerra, M.F.; Burnham, T.L.U.; Capece, L.R.; Lima, L.F.O.; Carsh, A.C.; Williams, S.L.; Morgan, S.G. Microplastic accumulation and biomagnification in a coastal marine reserve situated in a sparsely populated area. *Mar. Pollut. Bull.* **2019**, *146*, 54–59.
13. Miller, M.E.; Hamann, M.; Kroon, F.J. Bioaccumulation and biomagnification of microplastics in marine organisms: A review and meta-analysis of current data. *PLoS ONE* **2020**, *15*, e0240792.

14. Dris, R.; Gasperi, J.; Mirande, C.; Mandin, C.; Guerrouache, M.; Langlois, V.; Tassin, B. A first overview of textile fibers, including microplastics, in indoor and outdoor environments. *Environ. Pollut.* **2017**, *221*, 453–458. [CrossRef]
15. Campanale, C.; Massarelli, C.; Savino, I.; Locaputo, V.; Uricchio, V.F. A Detailed Review Study on Potential Effects of Microplastics and Additives of Concern on Human Health. *Int. J. Environ. Res. Public Health* **2020**, *17*, 1212. [CrossRef]
16. Enyoh, C.E.; Shafea, L.; Verla, A.W.; Verla, E.N.; Qingyue, W.; Chowdhury, T.; Paredes, M. Microplastics Exposure Routes and Toxicity Studies to Ecosystems: An Overview. *Environ. Anal. Health Toxicol.* **2020**, *35*, e2020004.
17. Visalli, G.; Facciolà, A.; Pruiti Ciarello, M.; De Marco, G.; Maisano, M.; Di Pietro, A. Acute and Sub-Chronic Effects of Microplastics (3 and 10 μm) on the Human Intestinal Cells HT-29. *Int. J. Environ. Res. Public Health.* **2021**, *18*, 5833.
18. Zarus, G.M.; Muianga, C.; Hunter, C.M.; Pappas, R.S. A review of data for quantifying human exposures to micro and nanoplastics and potential health risks. *Sci. Total Environ.* **2021**, *756*, 144010.
19. Bredeck, G.; Halamoda-Kenzaoui, B.; Bogni, A.; Bogni, A.; Lipsa, D.; Bremer-Hoffmann, S. Tiered testing of micro- and nanoplastics using intestinal in vitro models to support hazard assessments. *Environ. Int.* **2022**, *158*, 106921.
20. Schwabl, P.; Köppel, S.; Königshofer, P.; Bucsics, T.; Trauner, M.; Reiberger, T.; Liebmann, B. Detection of Various Microplastics in Human Stool: A Prospective Case Series. *Ann. Intern. Med.* **2019**, *171*, 453–457.
21. Ibrahim, Y.S.; Tuan Anuar, S.; Azmi, A.A.; Wan Mohd Khalik, W.M.A.; Lehata, S.; Hamzah, S.R.; Ismail, D.; Ma, Z.F.; Dzulkarnaen, A.; Zakaria, Z.; et al. Detection of microplastics in human colectomy specimens. *JGH Open* **2020**, *5*, 116–121. [CrossRef]
22. Amato-Lourenço, L.F.; Carvalho-Oliveira, R.; Júnior, G.R.; Dos Santos Galvão, L.; Ando, R.A.; Mauad, T. Presence of airborne microplastics in human lung tissue. *J. Hazard Mater.* **2021**, *416*, 126124.
23. Braun, T.; Ehrlich, L.; Henrich, W.; Koeppel, S.; Lomako, I.; Schwabl, P.; Liebmann, B. Detection of Microplastic in Human Placenta and Meconium in a Clinical Setting. *Pharmaceutics* **2021**, *13*, 921. [CrossRef]
24. Ragusa, A.; Svelato, A.; Santacroce, C.; Catalano, P.; Notarstefano, V.; Carnevali, O.; Papa, F.; Rongioletti, M.C.A.; Baiocco, F.; Draghi, S.; et al. Plasticenta: First evidence of microplastics in human placenta. *Environ. Int.* **2021**, *146*, 106274. [CrossRef]
25. Prata, J.C. Airborne microplastics: Consequences to human health? *Environ. Pollut.* **2018**, *234*, 115–126.
26. Chen, G.; Feng, Q.; Wang, J. Mini-review of microplastics in the atmosphere and their risks to humans. *Sci. Total Environ.* **2020**, *703*, 135504. [CrossRef]
27. Liao, Z.; Ji, X.; Ma, Y.; Lv, B.; Huang, W.; Zhu, X.; Fang, M.; Wang, Q.; Wang, X.; Dahlgren, R.; et al. Airborne microplastics in indoor and outdoor environments of a coastal city in Eastern China. *J. Hazard. Mater.* **2021**, *417*, 126007. [CrossRef]
28. Evangeliou, N.; Grythe, H.; Klimont, Z.; Heyes, C.; Eckhardt, S.; Lopez-Aparicio, S.; Stohl, A. Atmospheric transport is a major pathway of microplastics to remote regions. *Nat. Commun.* **2020**, *11*, 3381. [CrossRef]
29. Xumiao, L.; Prata, J.C.; Alves, J.R.; Duarte, A.C.; Rocha-Santos, T.; Cerqueira, M. Airborne microplastics and fibers in indoor residential environments in Aveiro, Portugal. *Environ. Adv.* **2021**, *6*, 100134. [CrossRef]
30. Visalli, G.; Currò, M.; Iannazzo, D.; Pistone, A.; Pruiti Ciarello, M.; Acri, G.; Testagrossa, B.; Bertuccio, M.P.; Squeri, R.; Di Pietro, A. In vitro assessment of neurotoxicity and neuroinflammation of homemade MWCNTs. *Environ. Toxicol. Pharmacol.* **2017**, *56*, 121–128.
31. Facciolà, A.; Visalli, G.; La Maestra, S.; Ceccarelli, M.; D'Aleo, F.; Nunnari, G.; Pellicanò, G.F.; Di Pietro, A. Carbon nanotubes and central nervous system: Environmental risks, toxicological aspects and future perspectives. *Environ. Toxicol. Pharmacol.* **2019**, *65*, 23–30.
32. Visalli, G.; Facciolà, A.; Currò, M.; Laganà, P.; La Fauci, V.; Iannazzo, D.; Pistone, A.; Di Pietro, A. Mitochondrial Impairment Induced by Sub-Chronic Exposure to Multi-Walled Carbon Nanotubes. *Int. J. Environ. Res. Public Health* **2019**, *16*, 792.
33. Puisney, C.; Baeza-Squiban, A.; Boland, S. Mechanisms of Uptake and Translocation of Nanomaterials in the Lung. *Adv. Exp. Med. Biol.* **2018**, *1048*, 21–36.
34. Trovato, M.C.; Andronico, D.; Sciacchitano, S.; Ruggeri, R.M.; Picerno, I.; Di Pietro, A.; Visalli, G. Nanostructures: Between natural environment and medical practice. *Rev. Environ. Health* **2018**, *33*, 295–307.
35. Hu, M.; Palić, D. Micro- and nano-plastics activation of oxidative and inflammatory adverse outcome pathways. *Redox Biol.* **2020**, *37*, 101620. [CrossRef]
36. Mielczarski, J.A.; Jeyachandran, Y.L.; Mielczarski, E.; Rai, B. Modification of polystyrene surface in aqueous solutions. *J. Colloid Interface Sci.* **2011**, *362*, 532–539.
37. Visalli, G.; Laganà, A.; Facciolà, A.; Iaconis, A.; Curcio, J.; Pollino, S.; Celesti, C.; Scalese, S.; Libertino, S.; Iannazzo, D.; et al. Enhancement of biological effects of oxidised nano- and microplastics in human professional phagocytes. *Environ. Toxicol. Pharmacol.* **2023**, *99*, 104086. [CrossRef]
38. Ter Liu, N.; Tang, M.; Ding, J. The interaction between nanoparticles-protein corona complex and cells and its toxic effect on cells. *Chemosphere* **2020**, *245*, 125624.
39. Visalli, G.; Bertuccio, M.P.; Iannazzo, D.; Piperno, A.; Pistone, A.; Di Pietro, A. Toxicological assessment of multi-walled carbon nanotubes on A549 human lung epithelial cells. *Toxicol. In Vitro* **2015**, *29*, 352–362. [CrossRef]
40. Micale, R.T.; La Maestra, S.; Di Pietro, A.; Visalli, G.; Baluce, B.; Balansky, R.; Steele, V.E.; De Flora, S. Oxidative stress in the lung of mice exposed to cigarette smoke either early in life or in adulthood. *Arch. Toxicol.* **2013**, *87*, 915–918. [CrossRef]
41. Di Pietro, A.; Baluce, B.; Visalli, G.; La Maestra, S.; Micale, R.; Izzotti, A. Ex vivo study for the assessment of behavioral factor and gene polymorphisms in individual susceptibility to oxidative DNA damage metals-induced. *Int. J. Hygiene Environ. Health* **2011**, *214*, 210–218.

42. Visalli, G.; Baluce, B.; La Maestra, S.; Micale, R.T.; Cingano, L.; De Flora, S.; Di Pietro, A. Genotoxic damage in the oral mucosal cells of subjects carrying restorative dental fillings. *Arch. Toxicol.* **2013**, *87*, 2247–2248. [CrossRef]
43. Zhang, X.; Xia, M.; Zhao, J.; Cao, Z.; Zou, W.; Zhou, Q. Photoaging enhanced the adverse effects of polyamide microplastics on the growth, intestinal health, and lipid absorption in developing zebrafish. *Environ. Int.* **2022**, *158*, 106922. [CrossRef]
44. Biale, G.; La Nasa, J.; Mattonai, M.; Corti, A.; Vinciguerra, V.; Castelvetro, V.; Modugno, F.A. Systematic Study on the Degradation Products Generated from Artificially Aged Microplastics. *Polymers* **2021**, *13*, 1997.
45. Zhu, K.; Jia, H.; Sun, Y.; Dai, Y.; Zhang, C.; Guo, X.; Wang, T.; Zhu, L. Enhanced cytotoxicity of photoaged phenol-formaldehyde resins microplastics: Combined effects of environmentally persistent free radicals, reactive oxygen species, and conjugated carbonyls. *Environ. Int.* **2020**, *145*, 106137.
46. Shi, X.; Chen, Z.; Liu, X.; Wei, W.; Ni, B.J. The photochemical behaviors of microplastics through the lens of reactive oxygen species: Photolysis mechanisms and enhancing photo-transformation of pollutants. *Sci. Total Environ.* **2022**, *846*, 157498.
47. Völkl, M.; Jérôme, V.; Weig, A.; Jasinski, J.; Meides, N.; Strohriegl, P.; Scheibel, T.; Freitag, R. Pristine and artificially-aged polystyrene microplastic particles differ in regard to cellular response. *J. Hazard. Mater.* **2022**, *435*, 128955. [CrossRef]
48. Yu, X.; Lang, M.; Huang, D.; Yang, C.; Ouyang, Z.; Guo, X. Photo-transformation of microplastics and its toxicity to Caco-2 cells. *Sci. Total Environ.* **2022**, *806*, 150954.
49. Visalli, G.; Facciolà, A.; Iannazzo, D.; Piperno, A.; Pistone, A.; Di Pietro, A. The role of the iron catalyst in the toxicity of multi-walled carbon nanotubes (MWCNTs). *J. Trace Elem. Med. Biol.* **2017**, *43*, 153–160.
50. Katayama, K.; Nomura, H.; Ogataa, H.; Eitoku, T. Diffusion coefficients for nanoparticles under flow and stop-flow conditions. *Phys. Chem. Chem. Phys.* **2009**, *11*, 10494–10499.
51. Varma, S.; Dey, S.; Palanisamy, D. Cellular Uptake Pathways of Nanoparticles: Process of Endocytosis and Factors Affecting their Fate. *Curr. Pharm. Biotechnol.* **2022**, *23*, 679–706. [PubMed]
52. Bonanomi, M.; Salmistraro, N.; Porro, D.; Pinsino, A.; Colangelo, A.M.; Gaglio, D. Polystyrene micro and nano-particles induce metabolic rewiring in normal human colon cells: A risk factor for human health. *Chemosphere* **2022**, *303*, 134947. [PubMed]
53. Wu, B.; Wu, X.; Liu, S.; Wang, Z.; Chen, L. Size-dependent effects of polystyrene microplastics on cytotoxicity and efflux pump inhibition in human Caco-2 cells. *Chemosphere* **2019**, *221*, 333–341. [PubMed]
54. Visalli, G.; Baluce, B.; Bertuccio, M.; Picerno, I.; Di Pietro, A. Mitochondrial-mediated apoptosis pathway in alveolar epithelial cells exposed to the metals in combustion-generated particulate matter. *J. Toxicol. Environ. Health A* **2015**, *78*, 697–709. [CrossRef]

toxics

Review

Environmental Health and Toxicology: Immunomodulation Promoted by Endocrine-Disrupting Chemical Tributyltin

Ricardo Correia da Silva [1,2], Mariana Pires Teixeira [1,3], Luciana Souza de Paiva [4,5,*,†] and Leandro Miranda-Alves [1,2,3,6,*,†]

1 Laboratório de Endocrinologia Experimental-LEEx, Instituto de Ciências Biomédicas, Universidade Federal do Rio de Janeiro, Rio de Janeiro 21941-902, Brazil; rcorreia.ufrj@gmail.com (R.C.d.S.); mari.piress@gmail.com (M.P.T.)
2 Programa de Pós-Graduação em Ciências Morfológicas, Instituto de Ciências Biomédicas, Universidade Federal do Rio de Janeiro, Rio de Janeiro 21941-902, Brazil
3 Programa de Pós-Graduação em Endocrinologia, Faculdade de Medicina, Universidade Federal do Rio de Janeiro, Rio de Janeiro 21941-902, Brazil
4 Departamento de Imunobiologia, Instituto de Biologia, Universidade Federal Fluminense, Niterói 24210-201, Brazil
5 Programa de Pós-Graduação em Patologia, Faculdade de Medicina, Universidade Federal Fluminense, Niterói 24210-201, Brazil
6 Programa de Pós-Graduação em Farmacologia e Química Medicinal, Instituto de Ciências Biomédicas, Universidade Federal do Rio de Janeiro, Rio de Janeiro 21941-902, Brazil
* Correspondence: lupaiva@id.uff.br (L.S.d.P.); leandro.alves@icb.ufrj.br (L.M.-A.)
† These authors contributed equally to this work.

Abstract: Tributyltin (TBT) is an environmental contaminant present on all continents, including Antarctica, with a potent biocidal action. Its use began to be intensified during the 1960s. It was effectively banned in 2003 but remains in the environment to this day due to several factors that increase its half-life and its misuse despite the bans. In addition to the endocrine-disrupting effect of TBT, which may lead to imposex induction in some invertebrate species, there are several studies that demonstrate that TBT also has an immunotoxic effect. The immunotoxic effects that have been observed experimentally in vertebrates using in vitro and in vivo models involve different mechanisms; mainly, there are alterations in the expression and/or secretion of cytokines. In this review, we summarize and update the literature on the impacts of TBT on the immune system, and we discuss issues that still need to be explored to fill the knowledge gaps regarding the impact of this endocrine-disrupting chemical on immune system homeostasis.

Keywords: endocrine-disrupting chemical; tributyltin; immune system; organotin; cytokines; immunotoxicity

Citation: da Silva, R.C.; Teixeira, M.P.; de Paiva, L.S.; Miranda-Alves, L. Environmental Health and Toxicology: Immunomodulation Promoted by Endocrine-Disrupting Chemical Tributyltin. *Toxics* **2023**, *11*, 696. https://doi.org/10.3390/toxics11080696

Academic Editors: Esref Demir, Sam Kacew and Jamie DeWitt

Received: 27 June 2023
Revised: 24 July 2023
Accepted: 5 August 2023
Published: 12 August 2023

1. Introduction

1.1. Tributyltin

Organotin compounds are tin-based chemicals made up of hydrocarbons [1]. The use of these compounds ranges from industrial to agricultural biocidal agents such as antifungals, acaricides and molluscicides, and they are applied as wood preservatives and used in antifouling paints. Tributyltin (TBT), one of these organotin compounds, is a general name used to refer to a set of compounds distinguished by the presence of the (C4H9)3Sn group and low water solubility, with tributyltin oxide being a famous example [2,3]. TBT began to be widely used in antifouling paints in the mid-1960s, given its low cost and effectiveness in protecting ship hulls and underwater marine equipment against biofouling (TBT hinders the growth of algae, barnacles, mollusks and other organisms on ships' hulls) [4].

After years of extensive use, adverse effects of TBT were described as it spread into the marine environment, where it is harmful to aquatic organisms [5]. For example, in the invertebrate *Nucella lapillus*, a low concentration of 1 ng/L was able to induce imposex, the development of male sexual characteristics in females, as seen in populations of this invertebrate along the coast of the United Kingdom [6]. In the commercial oyster *Crassostrea gigas*, a concentration of 20 ng/L of TBT affected larval growth, and concentrations smaller than 2 ng/L inhibited its calcification, making it impossible to fish for oysters in Arcachon Bay, France [7]. The imposex effect has also been described in vertebrates. The first report of imposex in vertebrates showed that doses of 0.1 µg/g of TBT present in the diet of *Paralichthys olivaceus* were sufficient for the masculinization of females [8]. Up to 2011, the imposex effect of TBT has been described in more than 260 species of marine gastropods. Moreover, by this time, environmental damage and economic losses had already spread over several continents [9].

Several restrictions against TBT use in antifouling paints were first imposed in France, the United Kingdom and other developed countries during the 1980s [10]. Years later, the International Maritime Organization (IMO) passed a global ban on the application of antifouling paints containing TBT (from 1 January 2003) and a ban on their presence on the surface of ships' hulls (from January 2008) [11]. Even so, in 2004, it was estimated that 70–80% of the world naval fleet used TBT copolymer in its composition, given the economic benefits [10]. Also, in 2014, a United States company was fined for clandestinely producing and selling antifouling paints with TBT to various regions of the Caribbean. Nonetheless, the problem persists because, as recently as 2021, these paints could still be found for sale on the internet and be sent to different regions such as the Caribbean, Central America and Oceania [12].

Several factors can increase the half-life of TBT in the environment, such as its ability to be deposited in marine sediments (because TBT compounds exhibit significant lipid solubility and are preferentially absorbed by organic matter in soils or sediment) and to continue to be released into the environment for up to 100 years, according to mathematical models [4,13]. These factors have contributed to TBT values exceeding 7000 ng Sn/g in an environmental reserve in the Virgin Islands [14]. TBT also bioaccumulates in several marine species that are at the base of the food chain [15]. TBT can be found in these organisms even after 20 years of the initial contamination, maintained by their lipid solubility. Furthermore, the biomagnification of contamination occurs in the food chain, making it possible to find TBT residues in fish, seabirds and marine mammals [16].

Given the risk to human health due to the ingestion of products contaminated with TBT (e.g., seafood, water), the World Health Organization (WHO) has defined an acceptable daily intake value of 250 ng/Kg/day of TBT [17]. This value was extrapolated by a factor of 100 due to toxicity, kinetics and inter-individual differences tests performed in rats, in which the addition of TBT to the diet led to a reduction in the weight and function of the thymus of these animals [18]. Human exposure to TBT mainly occurs through the consumption of contaminated water and beverages. However, the consumption of marine food, in particular, has also been identified as a significant pathway for human exposure [19–21]. In Taiwan, TBT concentrations in oysters ranged from 320 to 1510 ng g^{-1} dry wt., depending on sampling locations. The highest TBT concentration (86–91% of total butyltin compounds) was 1510 ng g^{-1} dry wt., found in oysters from the Hsiangshan coastal area. Fishermen showed oyster consumption values of 94.1 and 250 g d^{-1} for typical and maximum exposure, respectively. The maximum intake of 250 g d^{-1} by fishermen was almost twice that of the general population (139 g d^{-1}), indicating potential health risks for those exposed to these contaminated oysters [21]. In Portugal, 32% of the 28 duplicate diet samples from members of the University of Aveiro showed the presence of organotin compounds. These compounds were detected at relatively low levels, with TBT being found in only two of the samples [22]. In "Isla Grande Atacama," northern Chile, the daily ingestion of 90 to 173 g of *Thaisella chocolata* (equivalent to four to eight organisms) from the most contaminated sites resulted in the consumption of TBT levels that exceeded the tolerable daily intake recommended by the European Food Safety Authority for tin (not exceeding 0.0015% in food composition and 100 ng Sn Kg^{-1}) [23]. Nevertheless, it is important to highlight that distinct diets may result in different levels of TBT contamination in human blood and tissues, but this is not the only determining factor for exposure to TBT [5].

Studies have shown that chronic exposure to TBT, even in values lower than the acceptable daily intake, may lead to different complications. Newborn C57BL/6J offspring exposed in utero to 0.05 or 0.5 mg/Kg of TBT (administered to mothers via intraperitoneal injection every 24 h from the 12th day of gestation) exhibited accelerated adipocyte differentiation in the liver, testis and mammary glands. In adulthood, these offspring displayed increased epididymal adipose mass [24]. Moreover, exposure of stem cells derived from 8-week-old C57BL/6J mice, which were previously exposed to TBT (0.1 mg/Kg) in utero starting at the 16th day of gestation, to 50 nM of TBT for 14 days resulted in the predominant induction of adipogenesis over osteogenesis. These findings suggest that prenatal TBT exposure alters the differentiation potential of stem cells, favoring adipogenic lineage commitment [25]. Female rats treated with 100 ng/Kg/day of TBT showed signs of dysfunction of the hypothalamic–pituitary–adrenal axis, including inflammation, oxidative stress and fibrosis [26]. Additionally, treatment of female rats with 100 or 500 ng/Kg/day of TBT via gavage resulted in abnormalities in renal function, including decreased glomerular filtration rate, elevated levels of proteinuria, inflammation, oxidative stress and kidney fibrosis [27]. Furthermore, an extensive amount of toxic effects due to TBT have already been described, mainly related to endocrine, metabolic and reproductive dysfunctions [28–33].

Finally, another important adverse effect of TBT is its immunotoxicity. In the present review, we aim to summarize and update the current literature on the impact of TBT on the immune system, opening the door for new questions about the impact of this endocrine disruptor.

1.2. The Immune System

In response to a stimulus, like a challenge by a pathogen, the immune system generates innate and adaptive immune responses. Briefly, the innate response is immediate, less specific and primarily involves the action of monocytes, neutrophils and natural killer (NK) cells. In contrast, the adaptive response is built gradually, being more specific and long-lasting. It is mainly orchestrated by B lymphocytes, the production of antibodies and by T lymphocytes that coordinate cell-mediated immune response [34,35].

Regarding T lymphocytes, the population of $T\alpha\beta$ lymphocytes is mainly divided into two main subpopulations, CD4+ and CD8+ T lymphocytes, named according to their expression of some surface protein, called clusters of differentiation (CD). These cells have different functions. CD8+ T lymphocytes can induce the death of tumor cells or cells infected by viral pathogens via class I major histocompatibility complex (MHC) recognition and produce inflammatory mediators (interleukin (IL)-2, interferon gamma (IFN-γ) and tumor necrosis factor alpha -TNF-α) [36,37]. CD4+ T lymphocytes are characterized by the recognition of class II MHC, expressed on the surface of antigen-presenting cells (APCs), and can differentiate into different subpopulations depending on the cytokines present in the microenvironment [36,38]. The main subpopulations of CD4+ T lymphocytes include: T helper (Th)1 cells, which are capable of promoting a cell-mediated effector response against viruses, intracellular bacteria and protozoa, and are characterized by the production of cytokines such as IFN-γ and IL-2 [39]; Th2 cells, which are involved in the coordination of the humoral response, mainly against parasites (for example, helminths), as well as in the pathophysiology of several allergies (including asthma and atopic dermatitis), and are characterized by the production of IL-4, IL-5 and IL-13 [36,39]; and Th17 cells, which are involved in the defense against pathogens in the mucosa and in autoimmune diseases, where there is a hyperactivation of Th17 (as in rheumatoid arthritis), and are characterized by the production of IL-17 and IL-22 [36,40,41]. Regulatory T cells are another important subtype of T lymphocytes. These have a primary function of suppressing the activation of the immune system, more specifically inhibiting the activation and expansion of $CD4^{+}$ and

CD8+ T lymphocytes and B cell activation, preventing the exacerbation of inflammation. Regulatory T cells can be characterized by the expression of CD4, CD25 and FOXP3 molecules and by the secretion of granzyme B, TGF-β and IL-10 [42].

The balance of synthesis and secretion (autocrine and paracrine signaling) of different cytokines by immune cells and other cell types, such as endothelial cells, fibroblasts and bone marrow cells, maintains the multicellular network of communication in the microenvironment [43]. In fact, cytokines are the basis of communication for the initiation, maintenance and termination of immune responses to infections, and any change in them can lead to extremes, such as immunosuppression or the establishment of chronic inflammation [44]. TNF-α, for example, modulates the function of effector immune cells, such as neutrophils, promoting an increase in their activity, adherence and chemotaxis. On the other hand, IFN-γ promotes an increase in the antitumor activity of natural killer cells, the phagocytic activity of macrophages and the production of nitric oxide, in addition to the increased expression of MHC class I and II molecules. Interleukin-1β (IL-1β) affects the maturation, proliferation and synthesis of immunoglobulins by B lymphocytes, in addition to stimulating the synthesis of acute-phase proteins at the beginning of the inflammatory process [43,45,46]. In contrast, other cytokines, such as IL-10, have an immunoregulatory function, inhibiting the synthesis of cytokines such as IFN-γ, TNF-α, IL-12, IL-2 and IL-1β, thus inhibiting the inflammatory process and promoting a change in the profile of macrophages from the activated state to the tissue-resident phenotype [43,47,48].

It is worth noting that the immune system does not act alone; it functions in close connection with other systems, such as the endocrine system [49]. The mechanism of communication between the immune and endocrine systems occurs in "two-way" process, as some immune cells are capable of producing hormones and are also sensitive to hormonal action [50–55]. A good example is that of estrogen, which can modulate the production of cytokines, as well as the differentiation, proliferation and even apoptosis of cells of the immune system such as T and B lymphocytes, mast cells, basophils and eosinophils via estrogen receptors (ERs) that are expressed by these cells [54]. Cell populations of the immune system may differ in the expression of ERs, such as CD4+ T lymphocytes that express greater amounts of ERα than B lymphocytes, which, in contrast, express greater amounts of ERβ than CD4+ T cells. In parallel, CD8+ T lymphocytes express both ERα and ERβ in low amounts, but at equal rates [56,57]. Interestingly, TBT has also been described as an antagonist of human ERs by inhibiting the transcriptional activation of the ER-dependent reporter gene and the interaction between the ligand-binding domain of the β isoform (ERβ LBD) and the steroid receptor coactivator-1 (SRC1) [58,59]. Additionally, TBT acts as an inhibitor of aromatase, the enzyme accountable for the conversion of testosterone to estrogen, as well as the estrogen receptor in zebrafish, thereby reducing the effects of ethinylestradiol [60]. Additionally, numerous studies have revealed that TBT disrupts estrogen signaling, affecting various tissues, as shown in Figure 1 [28,30,33,61–65].

It is also known that different cytokines and chemokines have targets in the hypothalamic–pituitary–thyroid (HPT) axis. Deregulation in the balance of these molecules can impact the HPT axis, mainly thyroid function [66]. Therefore, exposure to endocrine disruptors like TBT can impact innate and/or adaptive responses, more specifically interfering with the cellular and humoral immune responses, as well as the lifespan of immune cells [67].

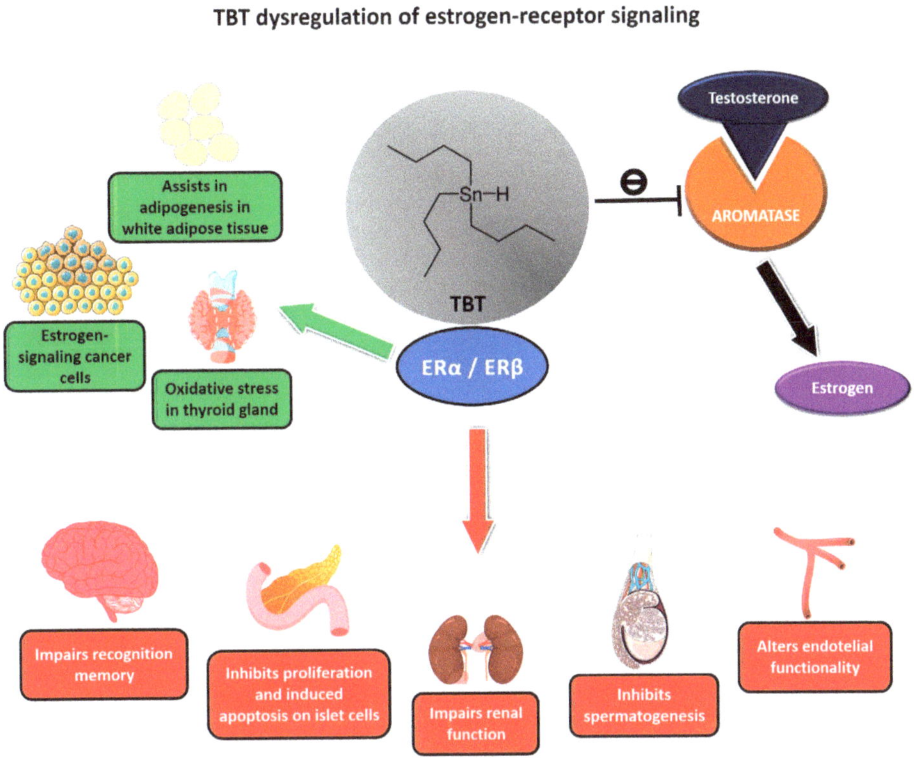

Figure 1. TBT impacts estrogen action and signaling. TBT may affect different systems by positively (green) or negatively (red) modulating ERα and ERβ, acting as an agonist or antagonist depending on the model. Additionally, TBT can block aromatase activity, preventing the conversion of testosterone to estrogen, leading to masculinization of female gastropods, for example.

2. TBT and the Immune System

The relationship between TBT contamination and immunotoxicity was first observed in animals that live exclusively or mainly in aquatic environments. In in vivo experiments with the fish *Oncorhynchus mykiss*, TBT led to thymic atrophy and reduced circulating lymphocyte population [68]. Moreover, bottlenose dolphins (*Tursiops truncatus*) and sea otters (*Enhydra lutris*) that lived in contaminated sites were found dead along the US coast and had high tissue concentrations of TBT. The mortality of both species has been linked to a lower ability of the immune system to fight infectious diseases [69,70]. In vitro experiments using seal cells (*Phoca vitulina*) showed that doses of between 50 and 200 nM of TBT and its metabolite dibutyltin (DBT) reduced the antitumor capacity of NK cells and decreased the proliferation of T cells, and that 100–200 nM of DBT decreased macrophage phagocytic activity [71]. Over the last decades, tests have been carried out using animal models (mainly mice) and human cells, which will be addressed throughout this article, as a way of inferring whether TBT's immunotoxic effects might also be seen in humans.

2.1. Mice Models

C57BL/6 mice that had TBT inserted in their diet for a period of two weeks showed a reduction in the number of lymphocytes in the spleen and lymph nodes, thymic atrophy and an increase in serum immunoglobulin M, but a decrease in immunoglobulin G [72]. In addition, BALB/c mice injected with 100 nM of TBT showed an increase in serum levels of the cytokines IFN-γ, TNF-α and IL-13 and the chemokines MIP-1β and RANTES. A decrease in the cytokine IL-2, which is essential for the maturation of B and T lymphocytes, was also observed [44]. In splenic cells of CBA/J mice stimulated in vitro with anti-mouse CD3 antibodies, exposure to 0.1 μM of TBT was sufficient to inhibit the secretion of IFN-γ and IL-4 by T CD4 cells [73]. ICR mice treated in vivo with doses of 4 and 20 mg/Kg of TBT showed a positive linear increase in thymocyte apoptosis and Fas expression, indicating that TBT may lead to cell apoptosis [74]. Furthermore, low doses of TBT (0.1, 1, 3 and 10 nM) were also capable of inducing apoptosis of Balb/c thymocytes in vitro via an increase in reactive oxygen species (ROS), a reduction in growth hormone (GH), depolarization of the mitochondrial membrane and the activation of caspase-3 [75]. In Balb/c and C57BL/6 T lymphocytes, it was seen that TBT (100 nM) in vitro induces the differentiation of T cells to the Th2 phenotype, which is characterized by the production of IL-10 and suppression of the production of IL-12, one of the main cytokines responsible for the differentiation of CD4+ T lymphocytes into the Th1 phenotype [76,77]. It was also observed that in C57BL/6 mice treated in vivo with TBT (6 μmol/Kg), there was greater ovalbumin-induced airway inflammation than in untreated mice. Furthermore, in the bronchoalveolar lavage fluid of these TBT-treated animals, there were increased numbers of eosinophils as well as increased IL-5 levels and IgE levels in serum, which was expected, given that Th2 cells are involved in the response to allergens [77]. Part of this effect was due to an increase in oxidative stress in secondary lymphoid organs of C57BL/6 animals promoted by TBT treatment, and, interestingly, Th17 lymphocytes had no role in the increased inflammation observed in the respiratory tract [78].

After oral treatment of ICR mice with 1, 10 and 20 mg/Kg of TBT, a decrease in the percentage of CD4+ and CD8+ T cell populations was observed in the thymus, but with an increase in the percentage of the CD4+CD8+ and CD4-CD8- T cell populations and decreased expression of IL-7 [79]. According to the same study, the spleens of ICR animals were also analyzed during TBT treatment, and a decrease in the percentage of naive CD44-CD62L+ and effector/memory CD44+ T cells was observed. Collectively, these data suggest that TBT impairs the development of T cells in the thymus and spleen [79]. Regarding B lymphocytes, in an C57BL/6 in vitro model, exposure to TBT promoted apoptosis of pro-B cells. In an C57BL/6 ex vivo model, TBT directly affected the differentiation of B cells, in addition to altering the bone marrow microenvironment [80]. In macrophages, exposure of the murine J774.1 cell line to TBT (1–1.5 μM) induced an increase in TNF-α expression and an increase in caspase-3 activity, leading to an increase in the apoptosis of these cells [81]. Additionally, treatment of the J774A.1 cell line with 0.4, 0.6, 1 and 1.2 μM of TBT promoted cell death possibly via receptor-interacting protein kinase 1 (RIP1) and receptor-interacting protein kinase 3 (RIP3), which are related to necroptosis. Furthermore, bone marrow-derived macrophages from TBT-treated C57BL/6 mice also displayed increased apoptosis [82]. Exposure to TBT also induced changes in RAW 264.7 murine macrophages, leading to activation of the inflammasome complex [83]. Finally, macrophages from C57BL/6 mice treated with doses of 250 and 500 μg/Kg of TBT showed activation of the peroxisome proliferator-activated receptor gamma (PPARγ) pathway and increased expression of genes related to lipogenesis and lipid metabolism, which could be related to a role of the innate immune system in the TBT-promoted obesogenic effect [84].

2.2. Human Models

In humans, TBT (1 μM) induced neutrophil apoptosis in vitro via a caspase-dependent mechanism [85]. Moreover, TBT (2.5–200 nM) in vitro decreased the viability of peripheral blood mononuclear cells (PBMCs) and was able to modulate the production of IL-1β and IFN-γ in a dose-dependent manner, with the TBT effect varying according to the concentration and duration of the exposure [86,87]. Also, in PBMCs and in the same concentrations of TBT (ranging from 2.5 to 200 nM) with varying in vitro exposures of 10 min, 1, 6 and 24 h, TBT induced the activation or increased expression levels of ribosomal protein S6 (S6), eukaryotic initiation factor 4B (eIF4B) and eIF4E in these cells. Surprisingly, this activation or elevation occurred at concentrations known to induce pro-inflammatory cytokine production, despite the absence of concomitant mRNA upregulation for these proteins [88]. Specifically, in monocyte-free human PBMCs, in addition to an increase in IL-1β synthesis and secretion, an increase in IL-6 synthesis and secretion was observed using the same concentrations of TBT [89,90]. These data were corroborated, given that the inhibition of Toll-like receptors (TLRs 4, 1/2 and 8) in PBMCs without monocytes treated with TBT leads to a significant decrease in the production of IL-1β and IL-6. This effect probably occurs because exposure to TBT causes the activation of these TLRs, promoting the activation of the MAPK pathway [91].

In human NKs, TBT also modulates IL-1β and TNF-α, promoting an increase in the secretion of these cytokines at low doses (5–50 nM for IL-1β and 5–100 nM for TNF-α), but a decrease at high doses (200 nM for both) [86,92]. Other in vitro studies also showed that NKs exposed to TBT (200–300 nM) had lower expression of perforins and granzyme B, as well as lower cytotoxic activity against tumors (TBT 25–500 nM) [93,94]. Additionally, within 10 min of TBT treatment, dosages ranging from 25 to 300 nM promoted the activation of MAP3K and its associated proteins, such as c-Raf and protein kinase C (PKC), in in vitro human NK cells. The activation of MAPK3 and apoptosis signal-regulating kinase 1 (ASK1) in human NK cells was another effect of TBT, but this occurred within an hour of exposure. In this approach, the innate potential of NK cells to effectively kill target cells may be disrupted by TBT, which could impair the activation of this pathway in a subsequent encounter with tumor cells or infected cells [95,96]. Specifically, in human B lymphocytes, in vitro treatment with TBT (100 nM) reduced the proliferation, survival and differentiation of mature B cells [97]. Finally, in human marrow cells, TBT (1 nM) led to a decrease in the percentage of CD19+CD22+ B cells in a mechanism independent of the PPARγ pathway [98,99].

When exposed to concentrations of 0.2 and 0.5 μM of TBT for various times (3, 6, 12 and 24 h) and subjected to whole-genome gene expression microarray analysis, the human T lymphocyte cell line Jurkat revealed that TBT treatment elicits immunotoxic effects by inducing endoplasmic reticulum (ER) stress, subsequently leading to an increase in intracellular Ca^{2+} levels. This elevation in Ca^{2+} levels triggers the activation of the nuclear factor of activated T cells (NFAT) and nuclear factor-kappa B (NF-κB), resulting in T cell activation, the induction of oxidative stress and, ultimately, cell apoptosis [100]. In Jurkat cells, doses ranging from 200 nM to 1 μM of TBT also induced apoptotic responses within 1 to 24 h of treatment, with the recruitment of caspase-8 and caspase-10 by TRAIL-R2. Interestingly, in Jurkat cells deficient in caspase-8, the apoptotic effects of TBT are only slightly reduced, whereas the inhibition of caspase-10 prevents all TBT-induced apoptotic effects [101]. Additionally, in Jurkat cells knockout for DNA fragmentation factor 40 (DFF40), exposure to TBT for 24 h at concentrations of 0.2, 0.4 and 0.6 μM did not induce DNA fragmentation, apoptosis and ROS production at the same rate as observed in wild-type cells. These findings indicate that DFF40 may play an important role in regulating cellular susceptibility to TBT and its contribution to the maintenance of DNA stability [102].

2.3. Other Models

In a zebrafish model (*Danio rerio*), chronic exposure for 8 weeks with different doses of TBT (1, 10 and 100 ng/L) led to a decrease in the activity of antioxidant enzymes (superoxide dismutase (SOD), catalase and glutathione peroxidase), intestinal lysozyme and immunoglobulin M (IgM) and to an increased expression of TNF-α, IL-1β, IL-6, NF-κB p65 and heat shock proteins HSP70 and HSP90 in the intestines, indicating that TBT induces oxidative stress and immunotoxicity in zebrafish [103]. In another protocol of chronic exposure to TBT with zebrafish for 6 weeks at doses of 10, 100 and 300 ng/L, a decrease in the amount of lysozyme and IgM was also observed, along with a dysregulation in the production of thyroid hormones [104].

Unsaturated fatty acid levels in muscle tissue were increased in *Gobiocypris rarus* fish after chronic exposure to environmentally relevant amounts of TBT (1, 10 and 100 ng/L) for 60 days. Pro-inflammatory cytokines TNF-a, IL-1 and the NF-κB transcription factor were also upregulated in the muscle tissue, as was antioxidant enzyme activity, suggesting potential TBT effects on the growth of fish and their nutritional value [105]. Additionally, after being exposed to TBT concentrations of 50 and 500 ng/L for 60 days, the lined seahorse (*Hippocampus erectus*) showed significant tin accumulation, liver damage, changes in antioxidant defenses (including increased SOD activity and decreased catalase activity) and upregulation of 20 genes linked to antioxidant defense, immune responses and inflammation [106].

In *Takifugu obscurus* fish, different water concentrations of TBT (1.962, 3.924 and 9.81 μg/L) for 96 h led to increased production of ROS and concomitant upregulation of CD28 (a known costimulatory receptor present in the surface of T cells) in the gills and liver of these animals, suggesting that CD28 plays a role in the response to TBT toxicity [107]. Also, in the *Takifugu obscurus* model, the same concentrations of TBT in water (1.962, 3.924 and 9.81 μg/L) for 96 h and chronic exposure with 900 ng/L for 30 days induced lower mRNA expression of TLRs 2 and 3 in the gills and higher mRNA expression of TLR18 and TLR22 in the liver and gills when compared to animals not exposed to TBT, confirming that these tissues are vital sites in the initial response to TBT exposure [108].

2.4. TBT and Other Endocrine-Disrupting Chemicals

Sensitization to allergens is one of the side effects caused by endocrine disruptors in the immune system, as they induce the breakdown of homeostasis through changes in the production of cytokines and chemokines, as summarized in this work in relation to exposure to TBT (Table 1). In the context of bisphenol S (BPS), water intake of 0.4 μg/Kg/day of BPS for 6 weeks concomitant with exposure to ovalbumin (OVA) every 2 weeks increased lung inflammation in C3H/HeJ mice, as well as anti-OVA IgE and IgG1 levels in serum and IL-5, IL-13, IL-33 and eotaxin levels in bronchoalveolar lavage fluid [109]. In the mediastinal lymph nodes of these same animals, there was an increase in the number of total cells and antigen-presenting cells such as dendritic cells. In addition, a restimulation of lymph node cells with OVA in vitro led to an increase in cell proliferation and cytokine production of the Th2 lymphocyte profile, such as IL-4, IL-5 and IL-13, indicating that BPS may lead to an increase in the number of Th2 cells and greater sensitization to allergens. Similar effects have already been observed for TBT in C57BL/6 mice models [76,77,109]. Furthermore, increased levels of IL-4 and decreased levels of IL-12 were observed in the umbilical cord blood of newborns whose mothers had higher amounts of monoethyl phthalate (MEP), a metabolite of di-ethyl phthalate (DEP), in blood and urine (during weeks 24 to 28 of gestation), indicating a possible polarization of newborn T cells towards the Th2 phenotype [110]. Bisphenol A (BPA) at doses of 10, 30 and 50 μM for 12 h led to a greater translocation of the transcription factor NF-κB p65 and to an increase in the production of cytokines IL-1β, IL-6 and TNF-α, together with nitric oxide and prostaglandin E2 (PGE2), in murine macrophages of the RAW264.7 lineage, suggesting that BPA can induce a pro-inflammatory response in these cells [111].

Table 1. Summary of changes in cytokine/chemokine synthesis and/or secretion caused by TBT treatment.

Cytokine/Chemokine	Experimental Model	Effect of TBT
IL-1β	Human/in vitro—zebrafish—*Gobiocypris rarus*	- Increased synthesis and secretion at low dose, inhibited at high dose [86,87,89]. - Increased expression in the intestine [103]. - Increased expression in muscle [105].
IL-2	BALB/c	- Decreased in serum [44].
IL-4	C57BL/6 and CBA/J mice spleen cells	- Indirect increase [78]. - Inhibited in vitro [73].
IL-5	C57BL/6	- Increased in serum [77].

Table 1. *Cont.*

Cytokine/Chemokine	Experimental Model	Effect of TBT
IL-6	Human/in vitro—zebrafish	- Increased synthesis and secretion [89,90]. - Increased expression in the intestine [103].
IL-7	ICR mice	- Decreased expression in the thymus [79].
IL-10	C57BL/6	- Indirect increase [76,77].
IL-12	C57BL/6	- Indirect decrease [76,77].
IL-13	BALB/c	- Increased in the serum [44].
IFN-γ	BALB/c, CBA/J mice spleen cells and human/in vitro	- Increased in serum [44]. - Increased synthesis and secretion at low dose, inhibition at high dose [86,87]. - Inhibited in vitro [73].
TNF-α	BALB/c/J774.1 cell line and human/in vitro—zebrafish and *Gobiocypris rarus*	- Increased in serum [44]. - Increased synthesis [81]. - Increased synthesis and secretion at low dose, inhibition at high dose [86,92]. - Increased expression in the intestine [103]. - Increased expression in muscle [105].
MIP-1β	BALB/c	- Increased in serum [44].
RANTES	BALB/c	- Increased in serum [44].

In the context of cancer, a single dose of BPA (250 ug/Kg) in newborn Balb/c mice is capable of inducing lung metastasis with increased intratumoral production of IL-1β, IL-6, IFN-γ, TNF-α and VEGF in a model of induced mammary tumorigenesis in which animals were injected in situ with 4T1 tumor cells when they reach sexual maturity [112]. Interestingly, 10^{-8} M of BPA was also able to increase the migration of breast ductal carcinoma in situ (DCIS) cells and RAW264.7 macrophages in an in vitro co-culture system. In in vivo experiments with Balb/c mice, exposure to 2.5 μg/L of BPA for 70 days promoted an increase in DCIS primary tumor growth rate and lymph node metastasis and a concomitant increase in protumorigenic M2 macrophages [113]. It has already been observed that treatment of C57BL/6 mice with 4 mg/Kg of di(2- ethylhexyl) phthalate (DEHP) for 21 days before the injection of B16F10 melanoma cells and for 7 days after the injection, reduced the polarization of macrophages into the M1 profile, but increased the polarization into the M2 profile, leading to tumor formation and growth [114]. This same polarization for the M2 profile is seen during in vitro exposure to benzophenone-3 (BP-3) in primary human macrophages [115]. Currently, there are no published data on how the exposure of macrophages to TBT would alter the immune response of these cells in the context of cancer, as previously described for other endocrine disruptors.

Similar to TBT, individual and co-exposure to the disruptors mancozeb (8000 mg/Kg/day) and fipronil (95 mg/Kg/day) for 29 days by oral gavage led to immunotoxicity in the spleen and thymus of Swiss albino mice (being more prominent in the treatment with both disruptors), as indicated by lower organ weight and cellularity, lower proliferation of splenocytes and thymocytes and higher rates of apoptosis and necrosis of these cells [116]. However, injection of Swiss mice with 50 μg/Kg of BPA for 6 weeks led to an increase in the number of lymphocytes and monocytes in the blood and an invasion of lymphocytes and eosinophils into the red pulp of the spleen [117]. In addition, doses of 100 μM of BPA, BPS, BPF and dimethyl terephthalate (DMTP) led to lower proliferation and viability of B lymphocytes isolated from the spleen of mice and stimulated with LPS, with BPA being the most toxic for B cells among these [118]. Interestingly, even though TBT is a PPARy agonist, it can induce a reduction in the mature population of B lymphocytes regardless of activation of the PPARy pathway by inducing changes in the bone marrow microenvironment that lead to adipogenesis in favor of lymphopoiesis [59,80,98,99]. It is important to mention that other endocrine disruptors that are agonists of the PPARγ pathway (such as phthalate metabolites) are well known to affect the differentiation of B lymphocytes in the bone marrow by inducing the apoptosis of B lymphocyte precursors [119–121].

3. Current Gaps in Literature

Many gaps need to be filled in the current knowledge about the effects of TBT on the immune system. For example, its effect on regulatory T cells is unclear; these cells are crucial for the maintenance of immune homeostasis through the synthesis and secretion of cytokines such as IL-10, TGF-β and IL-35 [122,123]. To date, there is only one study available in Japanese demonstrating that in the presence of TBT, regulatory T lymphocytes are more likely to enter apoptosis than Th2 lymphocytes. [124]. Moreover, TBT's impact on the functionality of γδ T cells is still unknown. These cells are widely present in peripheral tissues, where they promote tissue repair and immune surveillance in barrier tissues through the synthesis and secretion of various molecules such as IFN-γ, IL-17, IL-22, keratinocyte growth factor (KGF), insulin growth factor 1 (IGF1) and fibroblast growth factor 9 (FGF9) [125].

Unfortunately, there are also no studies on the effect of TBT on dendritic cells, even though these cells are very important for the effectiveness of the immune system, given that they bridge the gap between innate and adaptive immunity by capturing, processing and presenting antigens to T lymphocytes, mediating their polarization into effector cells [126]. In the case of other disruptors, it has already been seen that in vitro exposure to a mixture of BPA and BPF (10–50 μM) led to a decrease in the differentiation and maturation of human monocyte-derived dendritic cells, in addition to the loss of endocytic capacity and suppression of activation of NF-κB and ERK 1/2 pathways [127]. In human plasmacytoid dendritic cells exposed to DEHP, inhibition of the NF-κB and ERK pathways was also observed, along with lower expression of IFN-α and IFN-β. This led to changes in the cytokine secretion profile of CD4+ T cells activated by these dendritic cells, suppressing the production of IFN-γ, but increasing the production of IL-13 [128].

Another relevant point for this discussion is the lack of data about the effects of TBT on the polarization of macrophages into M1 and M2 profiles, as already described for other endocrine disruptors. This knowledge is important because M1 macrophages are mainly involved in the inflammatory response with the secretion of cytokines such as IL-1, IL-6, IFN-γ and TNF-α, and M2 macrophages are mostly involved in anti-inflammatory responses, producing cytokines such as IL-10 and TGF-β, so an imbalance between the two populations leads to the fatal loss of immune homeostasis [129].

There are only two studies on mast cells involving TBT. The first, from 30 years ago, shows that in vitro exposure of rat serosal mast cells (2.4×10^5 in 0.8 mL of medium) to 1 mM of TBT for 5 min leads to a strong inhibition of histamine secretion [130]. The second, a study of the impact of TBT on the coronary function, only reports an increase in mast cells in cardiac vessels of Wistar rats exposed to 100 ng/Kg of TBT per day by oral gavage [131]. As mast cells are cells present in all (but not only) mucosa of the body and synthesize and secrete various products such as IL-4, IL-6, TGF-β, biogenic amines, growth factors and proteases [132], the lack of knowledge regarding TBT's effect these cells is worrisome, since these cells may contribute to inflammation in the respiratory system. In fact, there are already enough data to demonstrate that exposure to TBT causes inflammation in the respiratory system [76–78].

Moreover, to the date of this publication, there are also no data comparing immunological alterations in males and females exposed to TBT. Evidence for divergence in responses between both sexes has already been shown for other endocrine disruptors, such as bisphenols [133] and even for TBT, but this was in the context of the nervous system [134].

4. Conclusions

In this article, we summarize the existing knowledge about the effects of TBT on immune system homeostasis. Studies using different animal and human models have shown that exposure to TBT is able to alter the function and viability of immune cells, which may impact immune responses. Changes in the homeostasis of cytokine and chemokine production (summarized in Table 1) were also described in human PBMCs and NK cells. Furthermore, murine models have shown similar alterations in cytokine profiles in the serum of TBT-exposed animals, as well as thymic atrophy and changes in cell populations of primary and secondary lymphoid organs. These changes, along with others directly or indirectly induced by TBT, lead to an increased sensitization of mice to allergens, as seen by the increase in the differentiation of CD4+ T lymphocytes to the Th2 profile, the decrease in the lymphocyte population present in secondary lymphoid organs and the apoptosis of murine thymocytes. Finally, in both human and murine models, TBT directly impacts B cell lymphopoiesis, decreasing the mature B cells (Figure 2).

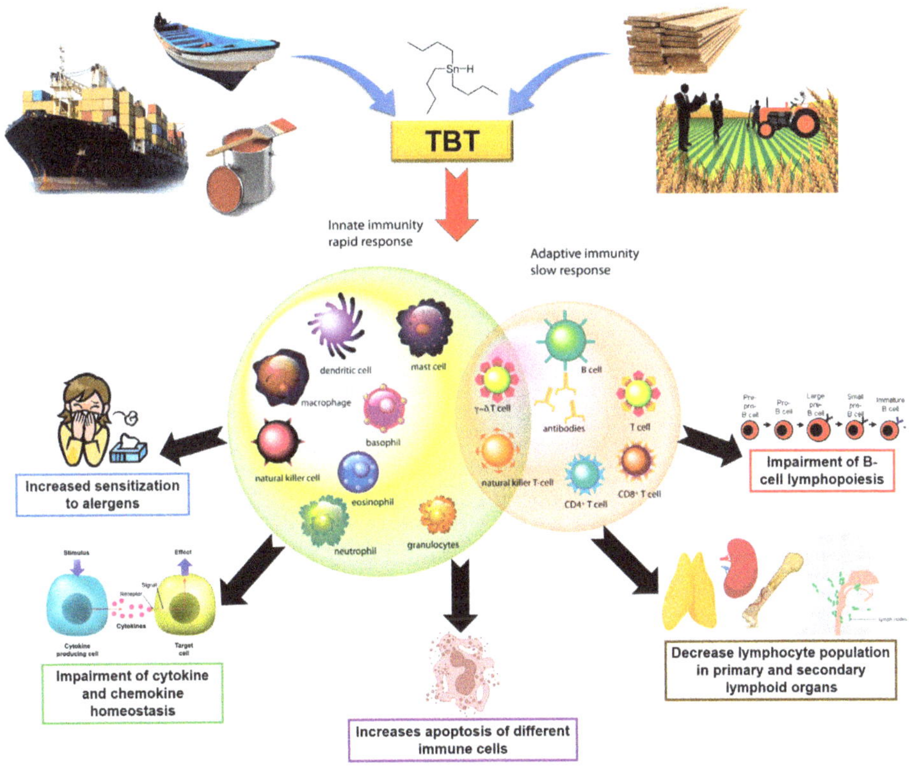

Figure 2. Summary of the main effects of TBT on the immune system.

Here, we have reviewed decades of observation and experimentation studies on the effects of TBT on the immune system. We hope that this work not only brings light to that which is already known about TBT, but that it also highlights the scarcity of our knowledge about this topic and the urgent need for additional studies. Moreover, we hope to call attention to the impacts of TBT on the environment and on human health, even in doses considered safe by government agencies worldwide.

Author Contributions: R.C.d.S.: Original Draft, Investigation and Writing. M.P.T.: Investigation, Supervision and Writing. L.S.d.P.: Writing—Review and Editing. L.M.-A.: Conceptualization, Investigation, Writing—Review and Editing, Supervision and Funding Acquisition. All authors have read and agreed to the published version of the manuscript.

Funding: This study was supported by Fundação Carlos Chagas Filho de Amparo à Pesquisa do Estado do Rio de Janeiro (CNE-FAPERJ E-26/200.798/2021; Apoio a Grupo Emergentes de Pesquisa do Estado do Rio de Janeiro, E-26/010.002198/2019; and Programa de Apoio a Projetos Temáticos FAPERJ, E-26/211.265/2021) and CNPq (PQ2 310654/2021-1). The funders had no role in the study design, data collection or analysis, in the decision to publish this manuscript or in its preparation.

Institutional Review Board Statement: Not applicable.

Informed Consent Statement: Not applicable.

Data Availability Statement: The images in this study were generated using the PowerPoint software and created from freely available digital image repositories (pngtree.com; pngwing.com accessed on 4 August 2023).

Acknowledgments: The authors would like to thank Fundação de Amparo à Pesquisa do Estado do Rio de Janeiro (FAPERJ), Coordenação de Aperfeiçoamento de Pessoal de Nível Superior (CAPES) and Conselho Nacional de Desenvolvimento Científico e Tecnológico (CNPq) for financial support.

Conflicts of Interest: The authors declare that there are no conflict of interest that could be perceived as prejudicing the impartiality of this manuscript.

References

1. Caseri, W. Initial Organotin Chemistry. *J. Organomet. Chem.* **2014**, *751*, 20–24. [CrossRef]
2. Maguire, R.J. Environmental aspects of tributyltin. *Appl. Organomet. Chem.* **1987**, *1*, 475–498. [CrossRef]
3. Davies, A.G. *Organotin Chemistry*, 2nd ed.; Wiley-VCH: Weinheim, Germany, 2004; ISBN 978-3-527-31023-4.
4. Kotrikla, A. Environmental management aspects for TBT antifouling wastes from the shipyards. *J. Environ. Manag.* **2009**, *90* (Suppl. 1), S77–S85. [CrossRef] [PubMed]
5. Antizar-Ladislao, B. Environmental levels, toxicity and human exposure to tributyltin (TBT)-contaminated marine environment. a review. *Environ. Int.* **2008**, *34*, 292–308. [CrossRef]
6. Gibbs, P.E.; Bryan, G.W.; Pascoe, P.L.; Burt, G.R. The use of the dog-whelk, *Nucella lapillus*, as an indicator of tributyltin (TBT) contamination. *J. Mar. Biol. Assoc. U. K.* **1987**, *67*, 507. [CrossRef]
7. Alzieu, C. Environmental impact of TBT: The French experience. *Sci. Total Environ.* **2000**, *258*, 99–102. [CrossRef]
8. Shimasaki, Y.; Kitano, T.; Oshima, Y.; Inoue, S.; Imada, N.; Honjo, T. Tributyltin causes masculinization in fish. *Environ. Toxicol. Chem.* **2003**, *22*, 141–144. [CrossRef] [PubMed]
9. Titley-O'Neal, C.P.; Munkittrick, K.R.; Macdonald, B.A. The effects of organotin on female gastropods. *J. Environ. Monit.* **2011**, *13*, 2360–2388. [CrossRef]
10. Yebra, D.M.; Kiil, S.; Dam-Johansen, K. Antifouling technology—Past, present and future steps towards efficient and environmentally friendly antifouling coatings. *Prog. Org. Coat.* **2004**, *50*, 75–104. [CrossRef]
11. International Marine Organization (IMO). Antifouling Systems. In *International Convention on the Control of Harmful Anti-Fouling Systems on Ships*; International Maritime Organization: London, UK, 2005.
12. Uc-Peraza, R.G.; Castro, Í.B.; Fillmann, G. An absurd scenario in 2021: Banned TBT-based antifouling products still available on the market. *Sci. Total Environ.* **2022**, *805*, 150377. [CrossRef]
13. Hamer, K.; Karius, V. Tributyltin release from harbour sediments--modelling the influence of sedimentation, bio-irrigation and diffusion using data from Bremerhaven. *Mar. Pollut. Bull.* **2005**, *50*, 980–992. [CrossRef] [PubMed]
14. Hartwell, S.I.; Apeti, D.A.; Mason, A.L.; Pait, A.S. An assessment of butyltins and metals in sediment cores from the St. Thomas East End Reserves, USVI. *Environ. Monit. Assess.* **2016**, *188*, 642. [CrossRef]
15. Takahashi, S.; Tanabe, S.; Takeuchi, I.; Miyazaki, N. Distribution and Specific Bioaccumulation of Butyltin Compounds in a Marine Ecosystem. *Arch. Environ. Contam. Toxicol.* **1999**, *37*, 50–61. [CrossRef] [PubMed]
16. Tanabe, S.; Prudente, M.; Mizuno, T.; Hasegawa, J.; Iwata, H.; Miyazaki, N. Butyltin contamination in marine mammals from North Pacific and Asian Coastal waters. *Environ. Sci. Technol.* **1998**, *32*, 193–198. [CrossRef]
17. World Health Organization; International Programme on Chemical Safety (WHO-IPCS). *Concise International Chemical Assessment Document 14. Tributyltin Oxide*; Office of Publications, World Health Organization: Geneva, Switzerland, 1999. Available online: https://apps.who.int/iris/bitstream/handle/10665/42186/9241530146.pdf (accessed on 25 February 2023).
18. Penninks, A.H. The evaluation of data-derived safety factors for bis (trin-butyltin) oxide. *Food Addit. Contam.* **1993**, *10*, 351–361. [CrossRef] [PubMed]
19. Forsyth, D.S.; Jay, B. Organotin leachates in drinking water from chlorinated poly(vinyl chloride) (CPVC) pipe. *Appl. Organomet. Chem* **1997**, *11*, 551–558. [CrossRef]

20. Azenha, M.; Vasconcelos, M.T. Butyltin compounds in Portuguese wines. *J. Agric. Food Chem.* **2002**, *50*, 2713–2716. [CrossRef]
21. Chien, L.C.; Hung, T.C.; Choang, K.Y.; Yeh, C.Y.; Meng, P.J.; Shieh, M.J.; Ha, B.C. Daily intake of TBT, Cu, Zn, Cd and As for fishermen in Taiwan. *Sci. Total Environ.* **2002**, *285*, 177–185. [CrossRef]
22. Sousa, A.C.A.; Coelho, S.D.; Pastorinho, M.R.; Taborda-Barata, L.; Nogueira, A.J.A.; Isobe, T.; Kunisue, T.; Takahashi, S.; Tanabe, S. Levels of TBT and other selected organotin compounds in duplicate diet samples. *Sci. Total Environ.* **2017**, *574*, 19–23. [CrossRef]
23. Mattos, Y.; Stotz, W.B.; Romero, M.S.; Bravo, M.; Fillmann, G.; Castro, Í.B. Butyltin contamination in Northern Chilean coast: Is there a potential risk for consumers? *Sci. Total Environ.* **2017**, *595*, 209–217. [CrossRef]
24. Grün, F.; Watanabe, H.; Zamanian, Z.; Maeda, L.; Arima, K.; Cubacha, R.; Gardiner, D.M.; Kanno, J.; Iguchi, T.; Blumberg, B. Endocrine-disrupting organotin compounds are potent inducers of adipogenesis in vertebrates. *Mol. Endocrinol.* **2006**, *20*, 2141–2155. [CrossRef]
25. Kirchner, S.; Kieu, T.; Chow, C.; Casey, S.; Blumberg, B. Prenatal exposure to the environmental obesogen tributyltin predisposes multipotent stem cells to become adipocytes. *Mol. Endocrinol.* **2010**, *24*, 526–539. [CrossRef] [PubMed]
26. Merlo, E.; Podratz, P.L.; Sena, G.C.; de Araújo, J.F.; Lima, L.C.; Alves, I.S.; Gama-de-Souza, L.N.; Pelição, R.; Rodrigues, L.C.; Brandão, P.A.; et al. The Environmental Pollutant Tributyltin Chloride Disrupts the Hypothalamic-Pituitary-Adrenal Axis at Different Levels in Female Rats. *Endocrinology* **2016**, *157*, 2978–2995. [CrossRef] [PubMed]
27. Coutinho, J.V.; Freitas-Lima, L.C.; Freitas, F.F.; Freitas, F.P.; Podratz, P.L.; Magnago, R.P.; Porto, M.L.; Meyrelles, S.S.; Vasquez, E.C.; Brandão, P.A.; et al. Tributyltin chloride induces renal dysfunction by inflammation and oxidative stress in female rats. *Toxicol. Lett.* **2016**, *260*, 52–69. [CrossRef]
28. Chen, Y.; Zuo, Z.; Chen, S.; Yan, F.; Chen, Y.; Yang, Z.; Wang, C. Reduction of spermatogenesis in mice after tributyltin administration. *Toxicology* **2008**, *251*, 21–27. [CrossRef]
29. Grün, F. The obesogen tributyltin. *Vitam. Horm.* **2014**, *94*, 277–325. [CrossRef]
30. Marques, V.B.; Faria, R.A.; Dos Santos, L. Overview of the Pathophysiological Implications of Organotins on the Endocrine System. *Front. Endocrinol.* **2018**, *9*, 101. [CrossRef]
31. Santos-Silva, A.P.; Andrade, M.N.; Pereira-Rodrigues, P.; Paiva-Melo, F.D.; Soares, P.; Graceli, J.B.; Dias, G.R.M.; Ferreira, A.C.F.; de Carvalho, D.P.; Miranda-Alves, L. Frontiers in endocrine disruption: Impacts of organotin on the hypothalamus-pituitary-thyroid axis. *Mol. Cell. Endocrinol.* **2018**, *460*, 246–257. [CrossRef] [PubMed]
32. Podratz, P.L.; Merlo, E.; de Araújo, J.F.P.; Ayub, J.G.M.; Pereira, A.F.Z.; Freitas-Lima, L.C.; da Costa, M.B.; Miranda-Alves, L.; Cassa, S.G.S.; Carneiro, M.T.W.D.; et al. Disruption of fertility, placenta, pregnancy outcome, and multigenerational inheritance of hepatic steatosis by organotin exposure from contaminated seafood in rats. *Sci. Total Environ.* **2020**, *723*, 138000. [CrossRef]
33. Andrade, M.N.; Melo-Paiva, F.D.; Teixeira, M.P.; Lima-Junior, N.C.; Soares, P.; Graceli, J.B.; Carvalho, D.P.; Morris, E.A.R.; Ferreira, A.C.F.; Miranda-Alves, L. Environmentally relevant dose of the endocrine disruptor tributyltin disturbs redox balance in female thyroid gland. *Mol. Cell. Endocrinol.* **2022**, *553*, 111689. [CrossRef]
34. Moss, R.B.; Moll, T.; El-Kalay, M.; Kohne, C.; Soo Hoo, W.; Encinas, J.; Carlo, D.J. Th1/Th2 cells in inflammatory disease states: Therapeutic implications. *Expert. Opin. Biol. Ther.* **2004**, *4*, 1887–1896. [CrossRef] [PubMed]
35. Nunes-Silva, A.; Dittz, D.; Santana, H.S.; Faria, R.A.; Freitas, K.M.; Coutinho, C.R.; de Melo Rodrigues, L.C.; Miranda-Alves, L.; Silva, I.V.; Graceli, J.B.; et al. The Pollutant Organotins Leads to Respiratory Disease by Inflammation: A Mini-Review. *Front. Endocrinol.* **2018**, *22*, 369. [CrossRef] [PubMed]
36. Koyasu, S.; Moro, K. Role of innate lymphocytes in infection and inflammation. *Front. Immunol.* **2012**, *3*, 101. [CrossRef]
37. Wang, Q.; Pan, W.; Liu, Y.; Luo, J.; Zhu, D.; Lu, Y.; Feng, X.; Yang, X.; Dittmer, U.; Lu, M.; et al. Hepatitis B virus-specific CD8+ T cells maintain functional exhaustion after antigen reexposure in an acute activation immune. *Environment. Front. Immunol.* **2018**, *9*, 219. [CrossRef]
38. Gutcher, I.; Becher, B. APC-derived cytokines and T cell polarization in autoimmune inflammation. *J. Clin. Investig.* **2007**, *117*, 1119–1127. [CrossRef] [PubMed]
39. Belizário, J.E.; Brandão, W.; Rossato, C.; Peron, J.P. Thymic and postthymic regulation of naïve $CD4^+$ T-Cell lineage fates in humans and mice models. *Mediat. Inflamm.* **2016**, *2016*, 9523628. [CrossRef] [PubMed]
40. Zambrano-Zaragoza, J.F.; Romo-Martínez, E.J.; Durán-Avelar, M.J.; García-Magallanes, N.; Vibanco-Pérez, N. Th17 cells in autoimmune and infectious diseases. *Int. J. Inflam.* **2014**, *2014*, 651503. [CrossRef]
41. Zayas, J.P.; Mamede, J.I. HIV infection and spread between Th17 cells. *Viruses* **2022**, *14*, 404. [CrossRef]
42. Rajendeeran, A.; Tenbrock, K. Regulatory T cell function in autoimmune disease. *J. Transl. Autoimmun.* **2021**, *4*, 100130. [CrossRef]
43. Borish, L.C.; Steinke, J.W. Cytokines and chemokines. *J. Allergy Clin. Immunol.* **2003**, *111* (Suppl. 2), S460–S475. [CrossRef]
44. Lawrence, S.; Pellom, S.T., Jr.; Shanker, A.; Whalem, M.M. Tributyltin exposure alters cytokine levels in mouse serum. *J. Immunotoxicol.* **2016**, *13*, 870–878. [CrossRef]
45. Gabay, C.; Lamacchia, C.; Palmer, G. IL-1 pathways in inflammation and human diseases. *Nat. Rev. Rheumatol.* **2010**, *6*, 232–241. [CrossRef] [PubMed]
46. Feng, P.; Jyotaki, M.; Kim, A.; Chai, J.; Simon, N.; Zhou, M.; Bachmanov, A.A.; Huang, L.; Wang, H. Regulation of bitter taste responses by tumor necrosis factor. *Brain Behav. Immun.* **2015**, *49*, 32–42. [CrossRef]
47. Galdiero, M.R.; Garlanda, C.; Jaillon, S.; Marone, G.; Mantovani, A. Tumor associated macrophages and neutrophils in tumor progression. *J. Cell. Physiol.* **2013**, *228*, 1404–1412. [CrossRef] [PubMed]

48. Nagata, K.; Nishiyama, C. IL-10 in mast cell-mediated immune responses: Anti-inflammatory and proinflammatory roles. *Int. J. Mol. Sci.* **2021**, *22*, 4972. [CrossRef] [PubMed]
49. Ponce-Regalado, M.D.; Pérez-Sánchez, G.; Rojas-Espinosa, O.; Arce-Paredes, P.; Girón-Peréz, M.I.; Pavón-Romero, L.; Becerril-Villanueva, E. NeuroImmunoEndocrinology: A brief historic narrative. *J. Leukoc. Biol.* **2022**, *112*, 97–114. [CrossRef]
50. Sakazaki, H.; Ueno, H.; Nakamuro, K. Estrogen receptor α in mouse splenic lymphocytes: Possible involvement in immunity. *Toxicol. Lett.* **2002**, *133*, 221–229. [CrossRef]
51. Lardone, P.J.; Guerrero, J.M.; Fernández-Santos, J.M.; Rubio, A.; Martín-Lacave, I.; Carrillo-Vico, A. Melatonin synthesized by T lymphocytes as a ligand of the retinoic acid-related orphan receptor. *J. Pineal Res.* **2011**, *51*, 454–466. [CrossRef]
52. Csaba, G. Immunoendocrinology: Faulty hormonal imprinting in the immune system. *Acta. Microbiol. Immunol. Hung.* **2014**, *61*, 89–106. [CrossRef]
53. Csaba, G. Effect of endocrine disruptor phytoestrogens on the immune system: Present and future. *Acta. Microbiol. Immunol. Hung.* **2018**, *65*, 1–14. [CrossRef]
54. Fan, Z.; Che, H.; Yang, S.; Chen, C. Estrogen and estrogen receptor signaling promotes allergic immune responses: Effects on immune cells, cytokines, and inflammatory factors involved in allergy. *Allergol. Immunopathol.* **2019**, *47*, 506–512. [CrossRef] [PubMed]
55. Tramontana, F.; Battisti, S.; Napoli, N.; Strollo, R. Immuno-Endocrinology of COVID-19: The Key Role of Sex Hormones. *Front. Endocrinol.* **2021**, *12*, 726696. [CrossRef] [PubMed]
56. Phiel, K.L.; Henderson, R.A.; Adelman, S.J.; Elloso, M.M. Differential estrogen receptor gene expression in human peripheral blood mononuclear cell populations. *Immunol. Lett.* **2005**, *97*, 107–113. [CrossRef] [PubMed]
57. Cunningham, M.; Gilkeson, G. Estrogen receptors in immunity and autoimmunity. *Clin. Rev. Allergy Immunol.* **2011**, *40*, 66–73. [CrossRef]
58. Cho, E.M.; Lee, H.S.; Moon, J.S.; Kim, I.S.; Sim, S.; Ohta, A. Organotin compounds act as inhibitor of transcriptional activation with human estrogen receptor. *J. Microbiol. Biotechnol.* **2012**, *22*, 378–384. [CrossRef]
59. Beg, M.A.; Zargar, U.R.; Sheikh, I.A.; Bajouh, O.S.; Abuzenadah, A.M.; Rehan, M. Organotin Antifouling Compounds and Sex-Steroid Nuclear Receptor Perturbation: Some Structural Insights. *Toxics* **2022**, *11*, 25. [CrossRef]
60. Berto-Júnior, C.; de Carvalho, D.P.; Soares, P.; Miranda-Alves, L. Tributyltin and Zebrafish: Swimming in Dangerous Water. *Front. Endocrinol.* **2018**, *9*, 152. [CrossRef] [PubMed]
61. Sharan, S.; Nikhil, K.; Roy, P. Effects of low dose treatment of tributyltin on the regulation of estrogen receptor functions in MCF-7 cells. *Toxicol. Appl. Pharmacol.* **2013**, *269*, 176–186. [CrossRef] [PubMed]
62. Zuo, Z.; Wu, T.; Lin, M.; Zhang, S.; Yan, F.; Yang, Z.; Wang, Y.; Wang, C. Chronic exposure to tributyltin chloride induces pancreatic islet cell apoptosis and disrupts glucose homeostasis in male mice. *Environ. Sci. Technol.* **2014**, *48*, 5179–5186. [CrossRef] [PubMed]
63. Bertuloso, B.D.; Podratz, P.L.; Merlo, E.; de Araújo, J.F.; Lima, L.C.; de Miguel, E.C.; de Souza, L.N.; Gava, A.L.; de Oliveira, M.; Miranda-Alves, L.; et al. Tributyltin chloride leads to adiposity and impairs metabolic functions in the rat liver and pancreas. *Toxicol. Lett.* **2015**, *235*, 45–59. [CrossRef]
64. Botelho, G.; Bernardini, C.; Zannoni, A.; Ventrella, V.; Bacci, M.L.; Forni, M. Effect of tributyltin on mammalian endothelial cell integrity. *Comp. Biochem. Physiol. C Toxicol. Pharmacol.* **2015**, *176–177*, 79–86. [CrossRef]
65. Ferraz da Silva, I.; Merlo, E.; Costa, C.S.; Graceli, J.B.; Rodrigues, L.C.M. Tributyltin Exposure Is Associated with Recognition Memory Impairments, Alterations in Estrogen Receptor α Protein Levels, and Oxidative Stress in the Brain of Female Mice. *Front. Toxicol.* **2021**, *3*, 654077. [CrossRef] [PubMed]
66. Croce, L.; Gangemi, D.; Ancona, G.; Liboà, F.; Bendotti, G.; Minelli, L.; Chiovato, L. The cytokine storm and thyroid hormone changes in COVID-19. *J. Endocrinol. Investig.* **2021**, *44*, 891–904. [CrossRef]
67. Nowak, K.; Jabłońska, E.; Ratajczak-Wrona, W. Immunomodulatory effects of synthetic endocrine disrupting chemicals on the development and functions of human immune cells. *Environ. Int.* **2019**, *125*, 350–364. [CrossRef] [PubMed]
68. Schwaiger, J.; Falk, H.F.; Bucher, F.; Orthuber, G.; Hoffmann, R.; Negele, R.D. A prolonged exposure of rainbow trout (*Oncorhynchus mykiss*) to sublethal concentrations of bis(tri-n-butyltin) oxide: Effects on leukocytes, lymphatic tissues and phagocytosis activity. In *Sublethal and Chronic Effects of Pollutants on Freshwater Fish*; Muller, R., Lloyd, R., Eds.; Blackwell Scientific: Cambridge, MA, USA, 1994; p. 113.
69. Kannan, K.; Senthilkumar, K.; Loganathan, B.G.; Takahashi, S.; Odell, D.K.; Tanabe, S. Evaluated levels of tributyltin and its breakdown products in Bottlenose Dolphins (*Tursiops truncatus*) found stranded along the US Atlantic and Gulf coasts. *Environ. Sci. Technol.* **1997**, *31*, 296–301. [CrossRef]
70. Murata, S.; Takahashi, S.; Agusa, T.; Thomas, N.J.; Kannan, K.; Tanabe, S. Contamination status and accumulation profiles of organotins in sea otters (*Enhydra lutris*) found dead along the coasts of California, Washington, Alaska (USA), and Kamchatka (Russia). *Mar. Pollut. Bull.* **2008**, *56*, 641–649. [CrossRef] [PubMed]
71. Frouin, H.; Lebeuf, M.; Saint-Louis, R.; Hammill, M.; Pelletier, É.; Fournier, M. Toxic effects of tributyltin and its metabolites on harbour seal (*Phoca vitulina*) immune cells in vitro. *Aquat. Toxicol.* **2008**, *90*, 243–251. [CrossRef]
72. Ohtaki, K.; Aihara, M.; Takahashi, H.; Fujita, H.; Takahashi, K.; Funabashi, T.; Hirasawa, T.; Ikezawa, Z. Effects of tributyltin on the emotional behavior of C57BL/6 mice and the development of atopic dermatitis-like lesions in DS-Nh mice. *J. Dermatol. Sci.* **2007**, *47*, 209–216. [CrossRef]

73. Yano, K.; Ohno, S.; Nakajima, Y.; Toyoshima, S.; Nakajin, S. Effects of various chemicals including endocrine disruptors and analogs on the secretion of Th1 and Th2 cytokines from anti CD3-stimulated mouse spleen cells. *J. Health Sci.* **2003**, *49*, 195–204. [CrossRef]
74. Chen, Q.; Zhang, Z.; Zhang, R.; Niu, Y.; Bian, X.; Zhang, Q. Tributyltin chloride-induced immunotoxicity and thymocyte apoptosis are related to abnormal Fas expression. *Int. J. Hyg. Environ. Health* **2011**, *214*, 145–150. [CrossRef]
75. Sharma, N.; Kumar, A. Mechanisms of immunotoxicological effects tributyltin chloride on murine thymocytes. *Cell Biol. Toxicol.* **2014**, *30*, 101–112. [CrossRef]
76. Kato, T.; Uchikawa, R.; Yamada, M.; Arizono, N.; Oikawa, S.; Kawanishi, S.; Nishio, A.; Nakase, H.; Kuribayashi, K. Environmental pollutant tributyltin promotes Th2 polarization and exacerbates airway inflammation. *Eur. J. Immunol.* **2004**, *34*, 1312–1321. [CrossRef]
77. Kato, T.; Tada-Oikawa, S.; Takahashi, K.; Saito, K.; Wang, L.; Nishio, A.; Hakamada-Tagushi, R.; Kawanishi, S.; Kuribayashi, K. Endocrine disruptors that deplete glutathione levels in APC promote Th2 polarization in mice leading to the exacerbation of airway inflammation. *Eur. J. Immunol.* **2006**, *36*, 1199–1209. [CrossRef] [PubMed]
78. Kato, T.; Tada-Oikawa, S.; Wang, L.; Murata, M.; Kuribayashi, K. Endocrine disruptors found in food contaminants enhance allergic sensitization through an oxidative stress that promotes the development of allergic airway inflammation. *Toxicol. Appl. Pharmacol.* **2013**, *273*, 10–18. [CrossRef] [PubMed]
79. Im, E.; Kim, H.; Kim, J.; Lee, H.; Yang, H. Tributyltin acetate-induced immunotoxicity is related to inhibition of T cell development in the mouse thymus. *Mol. Cell. Toxicol.* **2015**, *11*, 231–239. [CrossRef]
80. Baker, A.H.; Wu, T.H.; Bolt, A.M.; Gerstenfeld, L.C.; Mann, K.K.; Schlezinger, J.J. From the cover: Tributyltin alters the bone marrow microenvironment and suppresses B cell development. *Toxicol. Sci.* **2017**, *158*, 63–75. [CrossRef]
81. Nakano, K.; Tsunoda, M.; Konno, N. Tributyltin (TBT) increases TNFα mRNA expression and induces apoptosis in the murine macrophage cell line in vitro. *Environ. Health Prev. Med.* **2004**, *9*, 266–271. [CrossRef] [PubMed]
82. Ling, L.; Wen, J.; Tao, L.; Zhao, M.; Ge, W.; Wang, L.; Zhang, J.; Weng, D. RIP1 and RIP3 contribute to Tributyltin-induced toxicity in vitro and in vivo. *Chemosphere* **2019**, *218*, 589–598. [CrossRef]
83. Childers, G.M.; Perry, C.A.; Blachut, B.; Martin, N.; Bortner, C.D.; Sieber, S.; Li, J.L.; Fessler, M.B.; Harry, G.J. Assessing the Association of Mitochondrial Function and Inflammasome Activation in Murine Macrophages Exposed to Select Mitotoxic Tri-Organotin Compounds. *Environ. Health. Perspect.* **2021**, *129*, 47015. [CrossRef]
84. Jie, J.; Ling, L.; Yi, Y.; Tao, L.; Liao, X.; Gao, P.; Xu, Q.; Zhang, W.; Chen, Y.; Zhang, J.; et al. Tributyltin triggers lipogenesis in macrophages via modifying PPARγ pathway. *Environ. Pollut.* **2021**, *271*, 116331. [CrossRef] [PubMed]
85. Lavastre, V.; Girard, D. Tributyltin induces human neutrophil apoptosis and selective degradation of Cytoskeletal proteins by caspases. *J. Toxicol. Environ. Health A* **2002**, *65*, 1013–1024. [CrossRef] [PubMed]
86. Brown, S.; Whalem, M. Tributyltin alters secretion of interleukin1 beta from human immune cells. *J. Appl. Toxicol.* **2015**, *35*, 895–908. [CrossRef] [PubMed]
87. Lawrence, S.; Reid, J.; Whalem, M. Secretion of interferon-gamma (IFN-g) from human immune cells is altered by exposure to tributyltin (TBT) and dibutyltin (DBT). *Environ. Toxicol.* **2015**, *30*, 559–571. [CrossRef]
88. Ruff, A.; Lewis, M.; Whalen, M. Organotin and organochlorine toxicants activate key translational regulatory proteins in human immune cells. *Arch. Toxicol.* **2023**, *97*, 469–493. [CrossRef]
89. Brown, S.; Boules, M.; Hamza, N.; Wang, X.; Whalen, M. Synthesis of interleukin 1 beta and interleukin 6 in human lymphocytes is stimulated by tributyltin. *Arch. Toxicol.* **2018**, *92*, 2573–2586. [CrossRef] [PubMed]
90. Brown, S.; Wilburn, W.; Martin, T.; Whalen, M. Butyltin compounds alter secretion of interleukin 6 from human immune cells. *J. Appl. Toxicol.* **2018**, *38*, 201–218. [CrossRef]
91. Alcala, A.; Osborne, B.; Allen, B.; Seaton-Terry, A.; Kirkland, T.; Whalen, M. Toll-like receptors in the mechanism of tributyltin-induced production of pro-inflammatory cytokines, IL-1β and IL-6. *Toxicology* **2022**, *8*, 153177. [CrossRef]
92. Hurt, K.; Hurd-Brown, T.; Whalem, M. Tributyltin and dibutyltin alter secretion of tumor necrosis factor alpha from human natural killer cells and a mixture of T cells and natural killer cells. *J. Appl. Toxicol.* **2013**, *33*, 503–510. [CrossRef]
93. Thomas, L.D.; Shah, H.; Green, S.A.; Bankhurst, A.D.; Whalem, M.M. Tributyltin exposure causes decreased granzyme B and perforin levels in human natural killer cells. *Toxicology* **2003**, *200*, 221–233. [CrossRef]
94. Dudimah, F.D.; Odman-Ghazi, S.O.; Hatcher, F.; Whalem, M.M. Effect of tributyltin (TBT) on ATP levels in human natural killer (NK) cells: Relationship to TBT-induced decreases in NK function. *J. Appl. Toxicol.* **2007**, *27*, 86–94. [CrossRef]
95. Celada, L.J.; Whalen, M.M. Effects of butyltins on mitogen-activated-protein kinase kinase kinase and Ras activity in human natural killer cells. *J. Appl. Toxicol.* **2014**, *34*, 1002–1011. [CrossRef] [PubMed]
96. Rana, K.; Whalen, M. Activation of protein kinase C and protein kinase D in human natural killer cells: Effects of tributyltin, dibutyltin, and tetrabromobisphenol A. *Toxicol. Mech. Methods* **2015**, *25*, 680–688. [CrossRef] [PubMed]
97. De Santiago, A.; Aguilar-Santelises, M. Organotin compounds decrease in vitro survival, proliferation and differentiation of normal human B lymphocytes. *Hum. Exp. Toxicol.* **1999**, *118*, 619–624. [CrossRef]
98. Carfi', M.; Croera, C.; Ferrario, D.; Campi, V.; Bowe, G.; Pieters, R.; Gribaldo, L. TBTC induces adipocyte differentiation in human bone marrow long term culture. *Toxicology* **2008**, *249*, 11–18. [CrossRef]
99. Carfi', M.; Bowe, G.; Pieters, R.; Gribaldo, L. Selective inhibition of B lymphocytes in TBTC-treated human bone marrow long-term culture. *Toxicology* **2010**, *276*, 33–40. [CrossRef] [PubMed]

100. Katika, M.R.; Hendriksen, P.J.; van Loveren, H.; Peijnenburg, A. Exposure of Jurkat cells to bis (tri-n-butyltin) oxide (TBTO) induces transcriptomics changes indicative for ER- and oxidative stress, T cell activation and apoptosis. *Toxicol. Appl. Pharmacol.* **2011**, *254*, 311–322. [CrossRef]
101. Krug, H.F. Caspase-10 is the key initiator caspase involved in tributyltin-mediated apoptosis in human immune cells. *J. Toxicol.* **2012**, *2012*, 395482. [CrossRef]
102. Kulbay, M.; Johnson, B.; Bernier, J. DNA fragmentation factor 40 expression in T cells confers sensibility to tributyltin-induced apoptosis. *Toxicology.* **2019**, *426*, 152255. [CrossRef]
103. Zhang, C.N.; Zhang, J.L.; Ren, H.T.; Zhou, B.H.; Wu, Q.J.; Sun, P. Effect of tributyltin on antioxidant ability and immune responses of zebrafish (*Danio rerio*). *Ecotoxicol. Environ. Saf.* **2017**, *138*, 1–8. [CrossRef]
104. Li, Z.H.; Li, P. Effects of the tributyltin on the blood parameters, immune responses and thyroid hormone system in zebrafish. *Environ. Pollut.* **2021**, *268 Pt A*, 115707. [CrossRef]
105. Zhang, J.; Zhang, C.; Ma, D.; Liu, M.; Huang, S. Lipid accumulation, oxidative stress and immune-related molecules affected by tributyltin exposure in muscle tissues of rare minnow (*Gobiocypris rarus*). *Fish Shellfish Immunol.* **2017**, *71*, 10–18. [CrossRef]
106. Tang, L.; Zhang, Y.H.; Wang, X.; Zhang, C.C.; Qin, G.; Lin, Q. Effects of chronic exposure to environmental levels of tributyltin on the lined seahorse (*Hippocampus erectus*) liver: Analysis of bioaccumulation, antioxidant defense, and immune gene expression. *Sci. Total Environ.* **2021**, *801*, 149646. [CrossRef] [PubMed]
107. Fang, D.A.; Zhao, C.S.; Jiang, S.L.; Zhou, Y.F.; Xu, D.P. Toxic function of CD28 involving in the TLR/MyD88 signal pathway in the river pufferfish (*Takifugu obscurus*) after exposed to tributyltin chloride (TBT-Cl). *Gene* **2019**, *688*, 84–92. [CrossRef]
108. Zhao, C.S.; Fang, D.A.; Xu, D.P. Toll-like receptors (TLRs) respond to tributyltin chloride (TBT-Cl) exposure in the river pufferfish (*Takifugu obscurus*): Evidences for its toxic injury function. *Fish Shellfish Immunol.* **2020**, *99*, 526–534. [CrossRef]
109. Yanagisawa, R.; Koike, E.; Win-Shwe, T.T.; Takano, H. Effects of Oral Exposure to Low-Dose Bisphenol S on Allergic Asthma in Mice. *Int. J. Mol. Sci.* **2022**, *23*, 10790. [CrossRef] [PubMed]
110. Tsai, C.K.; Cheng, H.H.; Hsu, T.Y.; Wang, J.Y.; Hung, C.H.; Tsai, C.C.; Lai, Y.J.; Lin, Y.J.; Huang, H.C.; Chan, J.Y.H.; et al. Prenatal Exposure to Di-Ethyl Phthalate (DEP) Is Related to Increasing Neonatal IgE Levels and the Altering of the Immune Polarization of Helper-T Cells. *Int. J. Environ. Res. Public Health* **2021**, *18*, 6364. [CrossRef] [PubMed]
111. Huang, F.M.; Chang, Y.C.; Lee, S.S.; Yang, M.L.; Kuan, Y.H. Expression of pro-inflammatory cytokines and mediators induced by Bisphenol A via ERK-NFκB and JAK1/2-STAT3 pathways in macrophages. *Environ. Toxicol.* **2019**, *34*, 486–494. [CrossRef] [PubMed]
112. Palacios-Arreola, M.I.; Moreno-Mendoza, N.A.; Nava-Castro, K.E.; Segovia-Mendoza, M.; Perez-Torres, A.; Garay-Canales, C.A.; Morales-Montor, J. The Endocrine Disruptor Compound Bisphenol-A (BPA) Regulates the Intra-Tumoral Immune Microenvironment and Increases Lung Metastasis in an Experimental Model of Breast Cancer. *Int. J. Mol. Sci.* **2022**, *23*, 2523. [CrossRef]
113. Kim, H.; Kim, H.S.; Piao, Y.J.; Moon, W.K. Bisphenol A Promotes the Invasive and Metastatic Potential of Ductal Carcinoma In Situ and Protumorigenic Polarization of Macrophages. *Toxicol. Sci.* **2019**, *170*, 283–295. [CrossRef]
114. Lee, J.W.; Park, S.; Han, H.K.; Gye, M.C.; Moon, E.Y. Di-(2-ethylhexyl) phthalate enhances melanoma tumor growth via differential effect on M1-and M2-polarized macrophages in mouse model. *Environ. Pollut.* **2018**, *233*, 833–843. [CrossRef]
115. Gregory, K.J.; Morin, S.M.; Kubosiak, A.; Ser-Dolansky, J.; Schalet, B.J.; Jerry, D.J.; Schneider, S.S. The use of patient-derived breast tissue explants to study macrophage polarization and the effects of environmental chemical exposure. *Immunol. Cell Biol.* **2020**, *98*, 883–896. [CrossRef]
116. Bano, F.; Mohanty, B. Thyroxine modulation of immune toxicity induced by mixture pesticides mancozeb and fipronil in mice. *Life Sci.* **2020**, *240*, 117078. [CrossRef]
117. Shaibi, T.; Balug, H.N.; Ben-Othman, M.E.; Benjama, A.E.; Elhensheri, M.; Lwaleed, B.A.; Al-Griw, M.A. Exposure to low-dose bisphenol A induces spleen damage in a murine model: Potentially through oxidative stress? *Open Vet. J.* **2022**, *12*, 23–32. [CrossRef]
118. Jang, J.W.; Lee, J.W.; Yoon, Y.D.; Kang, J.S.; Moon, E.Y. Bisphenol A and its substitutes regulate human B cell survival via Nrf2 expression. *Environ. Pollut.* **2020**, *259*, 113907. [CrossRef] [PubMed]
119. Schlezinger, J.J.; Howard, G.J.; Hurst, C.H.; Emberley, J.K.; Waxman, D.J.; Webster, T.; Sherr, D.H. Environmental and endogenous peroxisome proliferator-activated receptor gamma agonists induce bone marrow B cell growth arrest and apoptosis: Interactions between mono(2-ethylhexyl)phthalate, 9-cis-retinoic acid, and 15-deoxy-Delta12,14-prostaglandin J2. *J. Immunol.* **2004**, *173*, 3165–3177. [CrossRef] [PubMed]
120. Bissonnette, S.L.; Teague, J.E.; Sherr, D.H.; Schlezinger, J.J. An endogenous prostaglandin enhances environmental phthalate-induced apoptosis in bone marrow B cells: Activation of distinct but overlapping pathways. *J. Immunol.* **2008**, *181*, 1728–1736. [CrossRef] [PubMed]
121. Giuliani, A.; Zuccarini, M.; Cichelli, A.; Khan, H.; Reale, M. Critical Review on the Presence of Phthalates in Food and Evidence of Their Biological Impact. *Int. J. Environ. Res. Public Health* **2020**, *17*, 5655. [CrossRef]
122. Dominguez-Villar, M.; Hafler, D.A. Regulatory T cells in autoimmune disease. *Nat. Immunol.* **2018**, *19*, 665–673. [CrossRef] [PubMed]
123. Shaw, D.M.; Merien, F.; Braakhuis, A.; Dulson, D. T-cells and their cytokine production: The anti-inflammatory and immunosuppressive effects of strenuous exercise. *Cytokine* **2018**, *104*, 136–142. [CrossRef]

124. Tada-Oikawa, S.; Murata, M.; Kato, T. [Preferential induction of apoptosis in regulatory T cells by tributyltin: Possible involvement in the exacerbation of allergic diseases]. *Nihon Eiseigaku Zasshi.* **2010**, *65*, 530–535. [CrossRef]
125. Ribot, J.C.; Lopes, N.; Silva-Santos, B. γδ T cells in tissue physiology and surveillance. *Nat. Rev. Immunol.* **2021**, *21*, 221–232. [CrossRef] [PubMed]
126. Shortman, K.; Liu, Y.J. Mouse and human dendritic cell subtypes. *Nat. Rev. Immunol.* **2002**, *2*, 151–161. [CrossRef] [PubMed]
127. Švajger, U.; Dolenc, M.S.; Jeras, M. In vitro impact of bisphenols BPA, BPF, BPAF and 17β-estradiol (E2) on human monocyte-derived dendritic cell generation, maturation and function. *Int. Immunopharmacol.* **2016**, *34*, 146–154. [CrossRef]
128. Kuo, C.H.; Hsieh, C.C.; Kuo, H.F.; Huang, M.Y.; Yang, S.N.; Chen, L.C.; Huang, S.K.; Hung, C.H. Phthalates suppress type I interferon in human plasmacytoid dendritic cells via epigenetic regulation. *Allergy* **2013**, *68*, 870–879. [CrossRef]
129. Yunna, C.; Mengru, H.; Lei, W.; Weidong, C. Macrophage M1/M2 polarization. *Eur. J. Pharmacol.* **2020**, *877*, 173090. [CrossRef] [PubMed]
130. Iwai, H.; Kurosawa, M.; Matsui, H.; Wada, O. Inhibitory effects of organotin compounds on histamine release from rat serosal mast cells. *Ind. Health.* **1992**, *30*, 77–84. [CrossRef]
131. dos Santos, R.L.; Podratz, P.L.; Sena, G.C.; Filho, V.S.; Lopes, P.F.; Gonçalves, W.L.; Alves, L.M.; Samoto, V.Y.; Takiya, C.M.; de Castro Miguel, E.; et al. Tributyltin impairs the coronary vasodilation induced by 17β-estradiol in isolated rat heart. *J. Toxicol. Environ. Health A* **2012**, *75*, 948–959. [CrossRef]
132. Wernersson, S.; Pejler, G. Mast cell secretory granules: Armed for battle. *Nat. Rev. Immunol.* **2014**, *14*, 478–494. [CrossRef]
133. Malaisé, Y.; Lencina, C.; Cartier, C.; Olier, M.; Ménar, S.; Guzylack-Piriou, L. Bisphenol A, S or F mother's dermal impregnation impairs offspring immune responses in a dose and sex-specific manner in mice. *Sci. Rep.* **2018**, *11*, 1650. [CrossRef]
134. Hernandez Scudder, M.E.; Young, R.L.; Thompson, L.M.; Kore, P.; Crews, D.; Hofmann, H.A.; Gore, A.C. EDCs Reorganize Brain-Behavior Phenotypic Relationships in Rats. *J. Endocr. Soc.* **2021**, *5*, bvab021. [CrossRef]

International Journal of
Molecular Sciences

MDPI

Review

Toxic Effects and Mechanisms of Polybrominated Diphenyl Ethers

Jinsong Xue *, Qingqing Xiao, Min Zhang, Dan Li and Xiaofei Wang *

School of Biology, Food and Environment, Hefei University, Hefei 230601, China; xiaoqq@hfuu.edu.cn (Q.X.); zhangmo@hfuu.edu.cn (M.Z.); lidan@hfuu.edu.cn (D.L.)
* Correspondence: xuejs@hfuu.edu.cn (J.X.); wangxf@hfuu.edu.cn (X.W.)

Abstract: Polybrominated diphenyl ethers (PBDEs) are a group of flame retardants used in plastics, textiles, polyurethane foam, and other materials. They contain two halogenated aromatic rings bonded by an ester bond and are classified according to the number and position of bromine atoms. Due to their widespread use, PBDEs have been detected in soil, air, water, dust, and animal tissues. Besides, PBDEs have been found in various tissues, including liver, kidney, adipose, brain, breast milk and plasma. The continued accumulation of PBDEs has raised concerns about their potential toxicity, including hepatotoxicity, kidney toxicity, gut toxicity, thyroid toxicity, embryotoxicity, reproductive toxicity, neurotoxicity, and immunotoxicity. Previous studies have suggested that there may be various mechanisms contributing to PBDEs toxicity. The present study aimed to outline PBDEs' toxic effects and mechanisms on different organ systems. Given PBDEs' bioaccumulation and adverse impacts on human health and other living organisms, we summarize PBDEs' effects and potential toxicity mechanisms and tend to broaden the horizons to facilitate the design of new prevention strategies for PBDEs-induced toxicity.

Keywords: polybrominated diphenyl ethers; exposures; toxic effects; toxic mechanisms

Citation: Xue, J.; Xiao, Q.; Zhang, M.; Li, D.; Wang, X. Toxic Effects and Mechanisms of Polybrominated Diphenyl Ethers. *Int. J. Mol. Sci.* **2023**, *24*, 13487. https://doi.org/10.3390/ijms241713487

Academic Editors: Esref Demir and Sam Kacew

Received: 31 July 2023
Revised: 25 August 2023
Accepted: 27 August 2023
Published: 30 August 2023

1. Introduction

Polybrominated diphenyl ethers (PBDEs) can inhibit combustion in organic material and suppress toxic fumes formation. They are thus found in various products, including electronics, vehicles, plastics, furnishings, polyurethane foams, building materials and textiles [1]. As halogenated organic compounds, PBDEs consist of two benzene rings connected by an oxygen atom. A total of 209 PBDE congeners named according to the number of bromine atoms and their position (i.e., ortho-, meta-, and para-substitution) are included in PBDEs [2,3]. Among these congeners, the major components available for commercial use are pentabromodiphenyl ether (penta-BDE), octabromodiphenyl ether (octa-BDE), and decabromodiphenyl ether (deca-BDE, PBDE-209) [4]. However, despite their efficiency in preventing fires, extensive use of PBDEs has engendered great safety concerns for the environment and public health [5]. Therefore, the present study aims to summarize the hazardous effects and potential mechanisms of PBDEs.

Since PBDEs are not chemically bonded to the polymer product, they are easily released into surroundings and become persistent organic pollutants, leading to contamination of the external environment [6]. PBDEs are distributed throughout the world and are frequently found in air, soil, water, and biota. For example, PBDEs enter the environment through atmospheric emissions from various sources, such as waste incineration, manufacturing, and recycling infrastructures [7]. The generated pollutants are transported long distances via airflow, resulting in deposition and accumulation in distant regions [8]. In addition, disposal of PBDE treatment materials in landfill sites and illegal sites may lead to the emission of leachate, thereby imposing major soil pollution issues [9]. Owning to the lipophilic and hydrophobic properties, PBDEs bind firmly to organic matter and remain in soils with reported half-lives of about 28 years [10]. The contaminated soils may also transfer PBDEs to suspended solids and sediments of aquatic environments via precipitation run-off [11]. Water is an important medium for PBDEs transmission [12]. Therefore, a persistent concern has been raised about the increasing levels of PBDEs in sewage treatment works [9,13]. PBDEs can enter the human body via ingestion of dust-bound PBDEs and inhalation of air-containing PBDEs [14]. In addition to the direct exposure, PBDEs are absorbed by the plants' roots and shoots that eventually enter the food chain [15]. PBDEs in fish, meat and livestock products may also result in a dietary risk and increase the PBDEs body burden in human beings [16,17].

People are unwittingly exposed to chemicals through their food, drinking water, the air they breathe, dust, and contact with consumer goods. Therefore, PBDEs may affect overall health by interacting with other substances. For instance, when microplastics and PBDEs were both present, the oxidative system was more severely disrupted than when either was present alone [18]. Interestingly, those who opt for the high-fat diet (HFD) are more at risk of BDE-209, exacerbating the advancement of non-alcoholic fatty liver disease (NAFLD) [19]. The combination of cadmium and PBDE-209 exposure resulted in more severe damage to hepatocytes [20]. Telomere length among newborns is linked to prenatal exposure to mixtures of per- and polyfuoroalkyl substances (PFAS) and PBDEs [21]. Besides, prenatal PBDEs can adversely affect child health, so the exposure of pregnant women to PBDEs cannot be ignored. For example, it was determined that prenatal exposures to PBDEs correlated with heightened liver injury risks, impaired cognitive performance, and fetal growth restriction in children [22–24]. In this review, we summarize the literature regarding the effects of combined and indirect exposures, which could further clarify PBDEs' detrimental effects.

PBDEs are ubiquitous toxicants frequently detected in human tissues [25]. As an essential organ for the metabolism of exogenous compounds, the liver serves as the main target for PBDEs [26,27]. For instance, the transplacental transfer of PBDEs from mother to fetus leads to tissue accumulation in the fetal liver [28]. Various PBDE congeners, such as PBDE-3, PBDE-7, PBDE-17, PBDE-99, PBDE-153, PBDE-197, and PBDE-209, have been reported to accumulate in blood, hair, and nails [29,30]. Hair and nail samples are non-invasive biomonitors. PBDE-47 and PBDE-99 were the predominant PBDEs detected in hair and nail samples [31]. Moreover, PBDEs have been found in human breast milk, cord blood and placentas. Therefore, it's inevitable that newborns are exposed to high levels of PBDEs during prenatal and postnatal periods [32]. Additionally, PBDEs were reported in adipose, kidney, lung, and semen [33].

Because of concerns regarding PBDEs' persistence, bioaccumulation, and potential toxicity, numerous studies have focused on delineating the underlying mechanisms. Herein, we review the literature addressing the effects on different tissues and mechanisms known to potentially contribute to PBDEs toxicity. The summarized information of this study may provide a clearer understanding of the impact of PBDEs on health.

2. Liver Toxicity

The liver is an essential organ for metabolic detoxification and is sensitive to environmental toxicants. Hence, the liver is susceptible to injury when exposed to xenobiotics [26,34,35]. For instance, significant liver weight increase and cell swelling, coupled with an elevated expression of cytochrome P450 (CYP1A2, CYP3A1, and CYP2B1) enzymes and genes and hepatocytic fatty degeneration, have been reported in PBDEs (the structures are shown in Figure 1) treated animals [36–41].

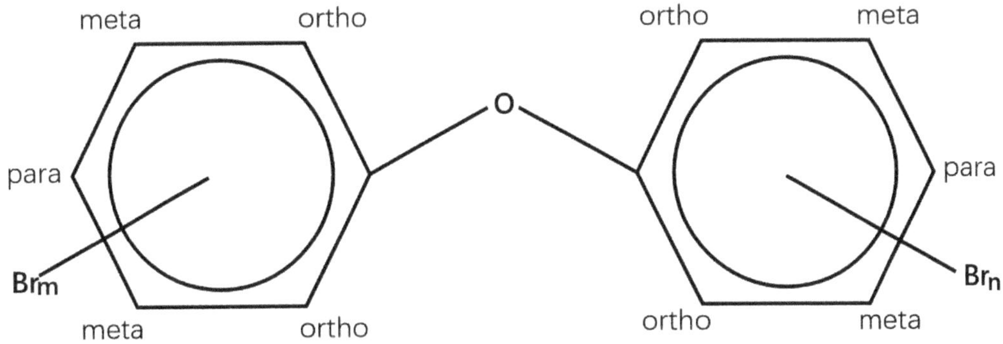

Figure 1. Structures of polybrominated diphenyl ethers (PBDEs). PBDEs consist of two benzene rings connected by an oxygen atom. A total of 209 PBDE congeners named according to the number of bromine atoms and their position (i.e., ortho-, meta-, and para-substitution) are included in PBDEs. m + n = 1–10.

2.1. Oxidative Damage and Apoptosis

A study on zebrafish has shown that PBDE-47 and PBDE-153 exposure markedly increased catalase (CAT) and superoxide dismutase (SOD) activities [26]. Additionally, the upregulation of apoptotic-regulated genes, including cysteine-aspartic acid protease-3 (*Caspase-3*) and tumor protein 53 (*P53*), as well as downregulation of anti-apoptotic genes, including B-cell lymphoma 2 (*Bcl2*) were observed in zebrafish treatment with PBDE-47 and PBDE-153, indicating PBDEs may regulate hepatic oxidative stress, DNA damage and apoptosis [26]. In addition, PBDE-47 and PBDE-32 reduced cell viability, generated reactive oxygen species (ROS) and triggered apoptosis in human hepatocellular carcinoma cell line HepG2 cells [42,43]. Shao et al. have analyzed the response of primary human fetal liver hematopoietic stem cells (HSCs) to PBDE-47 induction. They found higher concentrations of PBDE-47 may elicit overt ROS generation and lipid peroxidation, whereas N-acetylcysteine (NAC) can alleviate oxidative damage induced by PBDE-47 [44]. Analogously, trout liver cells exposed to PBDE-47 displayed a significant reduction in cell viability. The enhanced 6-carboxy-2′,7′-dichlorodihydrofluorescein diacetate (H2DCFDA) fluorescence in the presence of PBDE-47 indicated liver cells may be sensitive to PBDE-47 via a mechanism involving oxidative stress [45]. Zhang et al. investigated a rescue strategy using troxerutin to ameliorate PBDE-47-induced hepatocyte apoptosis. Perturbation of proteasome functions leads to endoplasmic reticulum (ER) stress, which is associated with apoptosis. They found that troxerutin efficaciously mitigates mice's liver apoptosis via modulating oxidative stress-mediated proteasome dysfunction. Furthermore, the downstream TNF receptor-associated factor 2 (TRAF2)/apoptosis signal-regulating kinase 1 (ASK1)/c-Jun N-terminal kinase (JNK) pathway was dramatically blocked by troxerutin in PBDE-47-treated mice livers [46]. Meanwhile, PBDE-47 promotes liver inflammation by inducing oxidative stress-triggered nicotinamide adenine dinucleotide (NAD^+) depletion. Troxerutin may abate oxidative stress, preventing the NAD^+-depletion-mediated loss of silent mating type information regulation 2 homolog 1 (Sirt1) and subsequent occurrence of inflammation [47]. In rat liver, PBDE-99 induced oxidative damage as evidenced by increased SOD activity and oxidized glutathione (GSSG) level, as well as decreased glutathione (GSH) level and CAT activity [48]. Likewise, PBDE-99 activated Caspases (i.e., Caspase-3 and Caspase-9) and generated toxic levels of ROS, thereby causing HepG2 cell apoptosis [49]. PBDE-209 and its quinone-type metabolite could induce an oxidative stress response, which activates ER stress and the autophagy-lysosomal system in hepatocytes [50,51]. Meanwhile, PBDE-209 disrupted calcium homeostasis, induced mitochondrial Ca^{2+} overload, and the subsequent cell apoptosis occurred [50,52]. Hu et al. have conducted several experiments to assess oxidative stress indicators. For example, increased ROS and lactate dehydrogenase (LDH) leakage have been observed in HepG2 cells dosed with PBDE-209 [53]. PBDE-209 could upregulate the activity of hepatic glutathione reductase (GR), and this elevation may compensate for cellular GSH depletion [54]. Interestingly, in 2013, samples of the kingfisher were collected from the e-waste recycling site and processed for biochemical analysis. The analysis showed that PBDEs, malondialdehyde (MDA) and ROS levels in kingfishers from e-waste sites were markedly increased compared with the normal group. Conversely, SOD and CAT activities in the liver from the exposed area were lower than in the reference group [55]. Transcriptional profiles of *O.melastigma* liver were analysed. The results discovered that PBDE-47 may activate phosphoinositide-3-kinase (PI3K) and mitogen-activated protein kinase (MAPK) pathway, which can modulate cell growth, proliferation, and survival [56]. The mechanisms are shown in Figures 2 and 3.

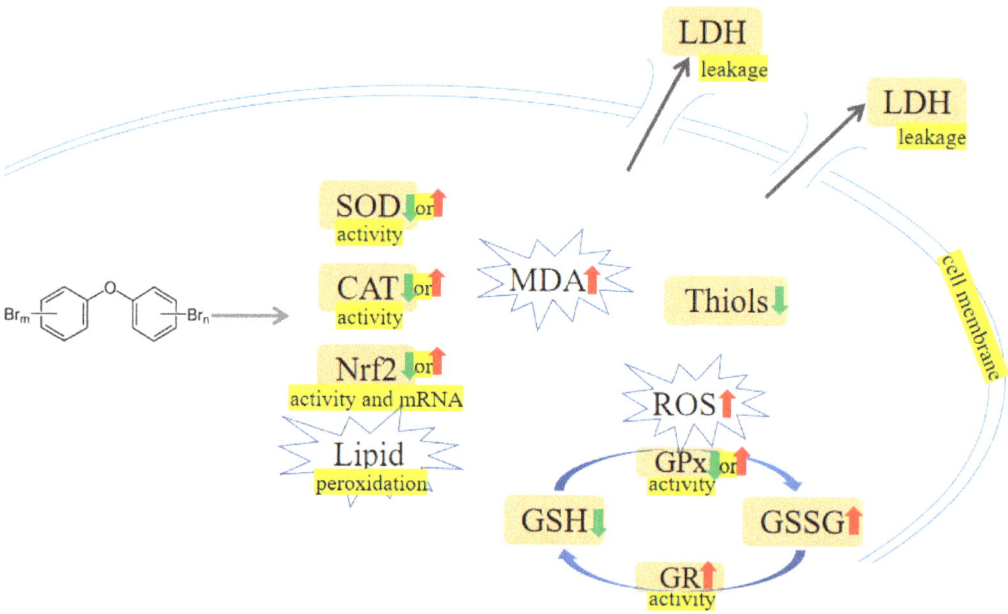

Figure 2. PBDEs-induced toxicity is associated with oxidative damage. PBDEs exposure can alter antioxidant enzyme activities, generate reactive oxygen species (ROS), increase malondialdehyde (MDA), and induce lactate dehydrogenase (LDH) leakage. Arrows indicate up (red colour), increased; down (green colour), decreased; up or down, increased or decreased (opposite research results exist). The yellow highlighted text is an explanation of the figure.

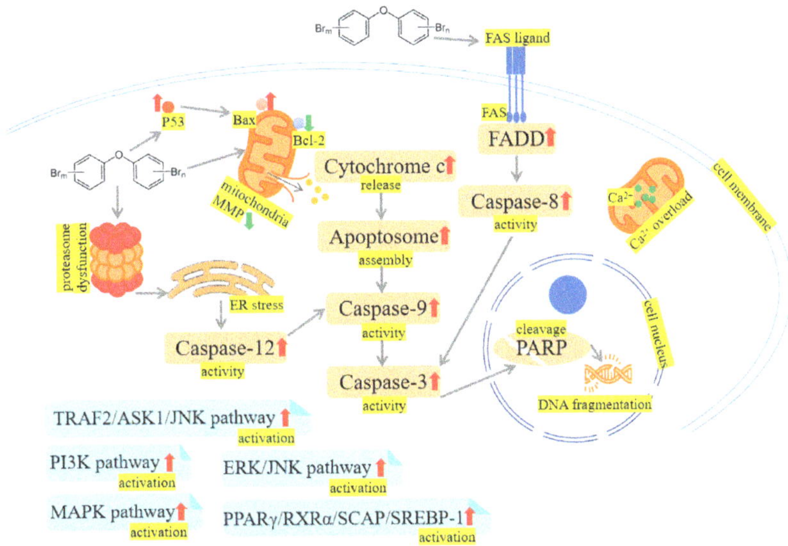

Figure 3. PBDEs-induced toxicity is associated with apoptosis. PBDEs exposure can reduce mitochondrial membrane potential (MMP), increase caspase activities, disrupt calcium homeostasis, induce endoplasmic reticulum (ER) stress and damage DNA. TNF receptor-associated factor 2 (TRAF2)/apoptosis signal-regulating kinase 1 (ASK1)/c-Jun N-terminal kinase (JNK) pathway, phosphoinositide-3-kinase (PI3K) pathway, extracellular signal-regulated kinase (ERK)/c-Jun N-terminal kinase (JNK) pathway, mitogen-activated protein kinase (MAPK) pathway and peroxisome proliferator-activated receptor γ (PPARγ)/retinoid X receptor α (RXRα)/sterol regulatory element-binding protein cleavage-activating protein (SCAP)/sterol regulatory element-binding protein-1 (SREBP-1) pathway are activated by PBDEs. Arrows indicate up (red colour), increased; down (green colour), decreased. The yellow highlighted text is an explanation of the figure.

2.2. Disturbance of Glucose and Lipid Metabolism

A growing body of evidence supports the idea that exposure to PBDEs is associated with metabolic dysfunction, with findings suggesting that these toxins may interfere with glucose and lipid metabolism. PBDE-47 and PBDE-153 have been reported to alter the blood-liver balance of lipids and disturb glucose metabolism in mice [1,57]. Moreover, to test if the aberrant metabolic phenotype is associated with altered liver epigenome, adult rats were exposed to PBDE-47, and functional analysis displayed that genes related to differentially methylated regions and differentially expressed miRNA were involved in lipid metabolism [58]. PBDE-71 has been found to reduce the activity of phosphoenolpyruvate carboxykinase, a key metabolic enzyme in hepatic glucose and lipid metabolism, and change the glucose: insulin ratio [59,60]. C57BL/6 mice that received PBDE-71 exhibited glucose intolerance, fasting hyperglycemia, retarded glucose clearance, and diminished thermogenic brown adipose tissue mass [61]. Zhu et al. reported PBDE-209 altered protein kinase A (PKA), phospho-PKA (p-PKA), adenosine 5′-monophosphate-activated protein kinase (AMPK), phospho-AMPK (p-AMPK), acetyl-CoA carboxylase (ACC), and fatty acid synthase (FAS) expression in rats' liver and LO2 cells (human normal liver cells). Besides, protein kinase cyclic adenosine monophosphate (cAMP)-activated catalytic subunit α (*PRKACA-1*) hypermethylation induced by PBDE-209 was observed in LO2 cells. Further study revealed that hypermethylation may contribute to disturbance of glycolipid metabolism [62]. Casella et al. exposed HepG2 cells to PBDEs (i.e., PBDE-47, PBDE-99 and PBDE-209) at 1 nM. The following Kyoto Encyclopedia of Genes and Genomes (KEGG) pathways and Gene Set Enrichment Analysis (GSEA) analyses were carried out, and the results indicated that PBDE-47 perturbed the glucose metabolism and hypoxia pathway; the ternary mixtures containing PBDE-47, PBDE-99 and PBDE-209 influenced lipid metabolism and PI3K/protein kinase B (AKT)/mammalian target of rapamycin (mTOR) signaling pathway. Meanwhile, PBDE-209 was reported to cause increased estrogen receptor α (*ERα*) and peroxisome proliferator-activated receptor α (*PPARα*) gene expression. These mechanism-based findings may reveal the potential relation between PBDEs and glycolipid metabolism [27]. PPARγ is an important nuclear receptor crucial in regulating lipid metabolism and glucose homeostasis [63]. Of interest, PBDE-47, a potential PPARγ ligand, could activate PPARγ [64,65]. PPARγ may push the adipocyte differentiation process forward by forming a positive-feedback loop with liver X receptor α (LXRα) [66]. It was reported that PPARγ activated by PBDE-47 may increase the expression of adipocyte-specific genes such as fatty acid binding protein 4 (*Fabp4*), lipoprotein lipase (*Lpl*), glucose transporter type 4 (*Slc2a4*), and adiponectin (*Adipoq*) [67]. Zhu et al. have found PBDE-209 led to histological impairment and lipid deposition, which were characterized by reduced glycogen and high-density lipoprotein (HDL) levels and increased low-density lipoprotein (LDL), glucose, triglyceride (TG) levels, and total cholesterol (CHOL) in mice livers. And besides they also found that LO2 cells' survival declined after PBDE-209 treatment. Further exploration revealed that PBDE-209 impaired glucose homeostasis via preventing PI3K/AKT/Glucose transporter type 4 (GLUT4) signaling pathway and induced lipid metabolic abnormality by triggering mTOR/PPARγ/retinoid X receptor α (RXRα) signaling pathways [68,69]. The mTOR pathway activated by PBDE-209 is responsible for the induction of PPARγ expression. Subsequently, PPARγ increases lipogenesis by combining with RXRα to form dimers [68]. Intriguingly, PPARγ inhibitor antagonized the alterations to the expression of p-mTOR, PPARγ, and RXRα and hindered TG accumulation provoked by PBDE-209, suggesting PPARγ may participate in modulating glucolipid metabolism [68]. Rats orally administered with PBDE-209 have shown hyperglycemia as compared to control rats. The reduced GSH and SOD implied that oxidative damage may contribute to PBDE-209-induced hyperglycemia and the onset of diabetes [70]. PBDE-209 has been reported to hinder glucose absorption, increase the levels of total cholesterol (TC), TG, aspartate transaminase (AST), alanine aminotransferase (ALT), and MDA through insulin receptor substrate-1 (IRS-1)/GLUT4 and IRS-1/PI3K/AKT/Glycogen synthase kinase 3β (GSK-3β) pathways, eventually interfering with glucolipid metabolism in buffalo rat liver cells with insulin resistance (IR-BRL) [71]. The mechanisms are shown in Figure 4.

2.3. Mitochondrial Damage

PBDE-47 increased *miR-34a-5p* level to trigger NAD^+ insufficiency via targeting nicotinamide phosphoribosyltransferase (NAMPT) expression. Subsequently, Sirtuin 3 (Sirt3)/forkhead box O-3 α (FOXO3α)/PTEN-induced putative kinase1 (PINK1) pathway-associated mitophagy was inhibited, which results in mitochondrial dysfunction and oxidative damages in mouse livers [72]. Fetal liver HSCs with PBDE-47 treatment showed a loss of mitochondrial membrane potential (MMP) [44]. DNA damage and mitochondrial impairment were detectable in cells after exposure to PBDE-47 and PBDE-32 [42]. Pazin et al. have found that PBDE-47 or PBDE-99 can influence membrane potential, mitochondrial inner membrane, oxygen consumption, mitochondrial swelling, and calcium release, which results in adenosine triphosphate (ATP) exhaustion [73]. As the energy-producing organelles inside cells, mitochondria are essential in maintaining energy supplies. In isolated liver mitochondria, Pereira et al. observed that PBDE-153 can interact with the mitochondrial membrane and disrupt MMP, thus causing ATP deficiency [74]. Meanwhile, they have also investigated the effects of PBDE-209 on rat liver mitochondria. The results showed PBDE-209-induced matrix swelling and ATP depletion. This process may contribute to reduced HepG2 cell viability [75]. The mechanisms are shown in Figure 5.

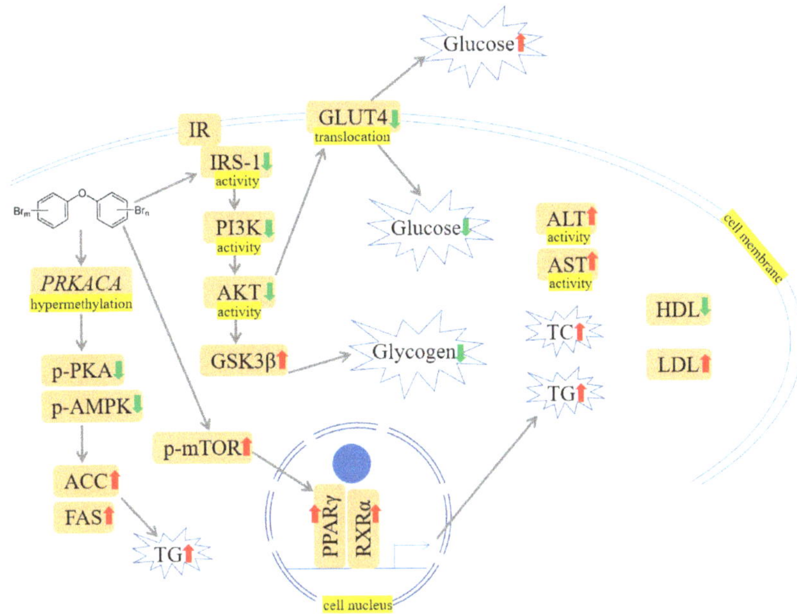

Figure 4. PBDEs-induced toxicity is associated with disturbances in glucose and lipid metabolism. PBDEs could increase glucose, total cholesterol (TC), triglyceride (TG), aspartate transaminase (ALT) activity, and aspartate transaminase (AST) activity. The PI3K/protein kinase B (AKT)/Glucose transporter type 4 (GLUT4) pathway is inhibited, and the mammalian target of rapamycin (mTOR)/PPARγ/RXRα pathway is elevated. Arrows indicate up (red colour), increased; down (green colour), decreased. The yellow highlighted text is an explanation of the figure.

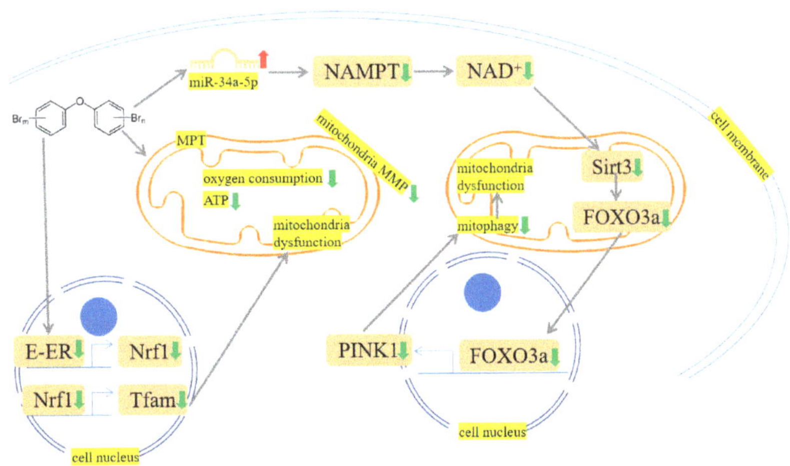

Figure 5. PBDEs-induced toxicity is associated with mitochondria damage. PBDEs exposure caused adenosine triphosphate (ATP) depletion, mitochondrial permeability transition (MPT) induction and mitochondria dysfunction. Sirtuin 3 (Sirt3)/forkhead box O-3 α (FOXO3α)/PINK1 pathway is suppressed by PBDE-209. Arrows indicate up (red colour), increased; down (green colour), decreased. The yellow highlighted text is an explanation of the figure.

2.4. Indirect Exposures

Indirect exposures occurred perinatally. For example, Dunnick et al. reported that PBDE-47 induced centrilobular hypertrophy and fatty change in pup livers on postnatal day (PND) 22. Liver transcriptomic changes were also measured, and the results showed that cytochrome p450 transcripts, nuclear factor erythroid 2-related factor 2 (Nrf2) antioxidant pathway transcripts and ATP-binding cassette (ABC) membrane transport transcripts were upregulated. These alterations elicited lipids, oncogenes, and epigenetic changes, which can lead to liver damage and tumorigenesis [76]. Perinatal exposure to PBDE-99 can disrupt the nongenomic actions of thyroid hormone (TH), thereby reducing the activity of the PI3K/AKT pathway in rat pup livers and affecting cell survival [77].

2.5. Combined Exposures

Combined exposures produced a series of public health issues. It has been reported that PBDEs are tightly associated with the occurrence of obesity and NAFLD. Further exploration revealed that the combination of PBDE-47 and HFD treatment reduced carnitine palmitoyltransferase 1α (*CPT1α*) gene expression, inhibiting fatty acid oxidation. Besides, the expression of microsomal TG transfer protein was inhibited by PBDE-47, which led to dysfunction of TG metabolism [78]. Co-exposure of nanoplastics and PBDE-47 leads to changed liver colour and atrophied liver in zebrafish larvae. The liver degeneration or necrosis may be associated with reduced antioxidant glutathione peroxidase 1 (*gpx1a*) gene and increased *CYP1A1* [79]. Using high-throughput sequencing approaches, Li et al. have proved that combined exposure of microplastics and PBDE-47 upregulated PPAR-related genes and reduced IL-17-associated genes [18]. Chen et al. have found that combined exposure to PBDE-209 and high fat resulted in elevated TG, MDA, and ROS levels in HepG2 cells, suggesting an increased lipid accumulation and oxidative stress. Similar to the in vitro results, mice receiving PBDE-209 and high fat showed elevated levels of sterol regulatory element-binding protein 1 (SREBP-1), stearoyl-CoA desaturase 1, and fatty acid synthase, thus promoting lipid deposition and NAFLD progression [19].

2.6. Others

Aside from those mentioned above, other effects and mechanisms of PBDEs on the liver are also worth mentioning. Crump et al. used an in vitro experiment to study the effects of PBDEs on cultured hepatocytes derived from embryonic chickens. They have found PBDE-71 diminished transthyretin (*TTR*), thyroid hormone–responsive spot 14-α (*THRSP14-α*), and liver fatty acid–binding protein (*FABP*) genes expression [80]. PBDE-71 can also induce hypomethylation at the T-Box Transcription Factor 3 (*Tbx3*) locus. As a transcription factor important in liver tumorigenesis, *Tbx3* hypomethylation in mouse liver cells indicated that PBDE-71 may engage in liver carcinoma development [81]. To gain knowledge about the toxicological mechanisms of PBDEs, primary Atlantic salmon hepatocytes were exposed to these congeners alone or in combination (PBDE-47, PBDE-153 and PBDE-154). Levels of endoplasmic reticulum-responsive genes vitellogenin (*VTG*) and zona pellucida 3 (*ZP3*) become elevated [82]. Early life exposure to PBDE-99 can induce hepatic inflammation and increase acetate and succinate levels [83]. To elucidate the PBDEs-gut microbiome interactions in modulating hepatic long noncoding RNAs (lncRNAs) and protein-coding genes (PCGs), conventional and germ-free mice were orally dosed with PBDE-47 or PBDE-99. LncRNAs increased the translational efficiency of PCGs, and this process might be a compensatory mechanism in response to PBDEs exposure. Pathway analysis of PCGs paired with lncRNAs displayed that PBDE-47 regulated nucleic acid, retinol metabolism and circadian rhythm, whereas PBDE-99 regulated fatty acid metabolism in conventional mice. Likewise, in germ-free mice, glutathione conjugation and transcriptional regulation were regulated by PBDE-47. In addition, the xenobiotic-metabolizing *CYP3A* genes and the fatty acid-metabolizing *CYP4* genes were modulated by PBDE-99 [84]. In Sueyoshi et al.'s study, human primary hepatocytes exposed to PBDE-47 exhibited upregulated CYP2B6 expression at both gene and protein levels. Because

CYP2B6 is a constitutive androstane receptor (CAR) target gene, the changed expression pattern suggested a cause-and-effect relationship between PBDE-47 and CAR pathway [85]. It has been reported PBDEs modulated several processes linked to pregnane X receptor (PXR) and CAR (i.e., protein ubiqutination, PPARα-RXRα activation) [86]. A further study exploring potential underlying mechanisms revealed that PBDE-209 could incur liver morphological alteration, cause oxidative stress, and subsequently reduce *PXR*, *CAR*, and CYP3A expression [87]. The effects and mechanisms of liver toxicity induced by PBDEs are shown in Table 1.

Table 1. Effects and mechanisms of liver toxicity induced by PBDEs.

Treatments	Effects and Mechanisms	References
PBDE-47 or -153, zebrafish	CAT activity↑, SOD activity↑, *Caspase-3*↑, *P53*↑, *Bcl-2*↓	[26]
PBDE-47 or -32, HepG2 cells, trout liver cells	Cell viability↓, ROS↑, apoptosis, DNA damage, mitochondrial impairment	[42,43,45]
PBDE-47, HSCs	ROS↑, lipid peroxidation, MMP↓	[44]
PBDE-47, -99, -209, HepG2 cells	*ERα*↑, *PPARα*↑, intracellular lipid accumulation	[27]
PBDE-47, CAR and PXR null mice	*CYP2B6*↑, CYP2B6↑	[85]
PBDE-47 or -99, isolated Wistar rat liver mitochondria	oxygen consumption↓, mitochondrial swelling, calcium release, ATP↓	[73]
PBDE-47, CD-1 mice, ICR mice, C57 BL/6 mice	Proteasome dysfunction, TRAF2/ASK1/JNK pathway↑, NAD^+ depletion, Sirt1↓, inflammation↑, abnormal insulin secretion, *miR-34a-5p*↑, Sirt3/FOXO3α/PINK1 pathway↓, mitochondrial dysfunction	[1,46,47,57,72]
PBDE-99, SD rats, HepG2 cells	SOD activity↑, CAT activity↓, GSSG↑, GSH↓, Caspase-3 activity↑, Caspase-9 activity↑, apoptosis	[48,49]
PBE-99, C57BL/6 mice	Inflammation, acetate↑, succinate↑	[83]
PBDE-209, C57BL/6 mice, ICR mice, LO2 cells	ER stress↑, mitochondrial Ca^{2+} overload, apoptosis, ROS↑, PI3K/AKT/GLUT4 pathway↓, mTOR/PPARγ/RXRα pathway↑, Glucose↑, TG↑, HDL↓, liver and adipose structures damage	[50,68,69]
PBDE-209, SD rats	Hyperglycemia, GSH↓, SOD activity↓, liver weight↑, liver/body weight ratio↑, serum total bilirubin and indirect bilirubin↑, oxidative stress, *PXR*↓, *CAR*↓, CYP3A↓	[70,87]
PBDE-209, IR-BRL cells	TC↑, TG↑, AST activity↑, ALT activity↑, MDA↑, IRS-1/PI3K/AKT/GSK-3β pathway↓, IRS-1/GLUT4↓	[71]
PBDE quinone, LO2 cells	ER stress↑, autophagy-lysosomal system↑, ROS↑	[51]
PBDE-209, SD rats, LO2 cells	*PRKACA-1* hypermethylation, TG↑, Glucose↑, PI3K/AKT/GLUT4 pathway↓, mTOR/PPARγ/RXRα pathway↑	[62,68]
PBDE-209, HepG2 cells, isolated mitochondria	Mitochondrial Ca^{2+} overload, apoptosis, ROS↑, LDH leakage, matrix swelling, ATP↓, cell viability↓	[52,53,75]
PBDE-209, *Carassius auratus*	GR activity↑, GSH↓	[54]
PBDEs in e-waste site, kingfisher (*Alcedo atthis*)	MDA↑, ROS↑, CAT activity↓, SOD activity↓	[55]
PBDE-47, marine medaka (*Oryzias melastigma*)	PI3K pathway activity↑, MAPK pathway activity↑	[56]
PBDE-71, Wistar rats	Glucose:insulin ratio↑	[59,60]
PBDE-71, C57BL/6 mice	Glucose intolerance, fasting hyperglycemia, retarded glucose clearance, diminished thermogenic brown adipose tissue mass	[61]
PBDE-153, isolated rat liver mitochondria	MMP↓, ATP↓, ROS↑	[74]
PBDE-47, Wistar Han rats, indirect exposure	Centrilobular hypertrophy, fatty change, *cytochrome p450*↑, *Nrf2*↑, lipid↑, oncogenes change, epigenetic change	[76]
PBDE-99, SD rats, indirect exposure	PIP3K/AKT pathway↓	[77]
PBDE-47 and high-fat diet, HepG2 cells, C57BL/6J mice, combined exposure	*CPT1α*↓, fatty acid oxidation↓, microsomal triglyceride transfer protein↓, sterol regulatory element-binding protein 1↑, stearoyl-CoA desaturase 1↑, fatty acid synthase↑, lipid deposition, NAFLD, MDA↑, ROS↑, lipid accumulation	[19,78]

Table 1. *Cont.*

Treatments	Effects and Mechanisms	References
PBDE-47 and nanoplastics, zebrafish, combined exposure	Darker/browner liver colour, atrophied liver, liver degeneration or necrosis, *gpx1a*↓, CYP1A1↑, mortality↑, voluntary movements↑, hatching rate↑, heart rate↓	[79]
PBDE-47 and microplastics, grouper (*Epinephelus moara*), combined exposure	PPAR-related genes↑, IL-17-related genes↓	[18]
PBDE-71, hepatocytes derived from embryonic chickens	*TTR*↓, *THRSP14-α*↓, *FABP*↓	[80]
PBDE-71, B6C3F1/N mice	*Tbx3* hypomethylation	[81]
PBDE-47, PBDE-153 and PBDE-154 (alone or in combination), primary Atlantic salmon hepatocytes	*VTG*↑, *ZP3*↑	[82]

↑ represents upregulation, ↓ represents downregulation.

3. Kidney Toxicity

3.1. Oxidative Damage and Apoptosis

To illuminate the effects of PBDEs on the kidney, adult male rats received a gavage dose of 1.2 mg/kg PBDE-99 for the study duration of 45 days. Decreased CAT activity and increased GSSG/GSH ratio were detected after PBDE-99 exposure [48]. Human embryonic kidney cells (HEK293) were exposed to PBDE-47, and a range of bioassays were performed. For instance, PBDE-47 could change Bcl-2 family-encoding gene expression, including Bcl-2-associated death promoter (*Bad*), Harakiri (*Hrk*) and *Bcl-2*. Besides, energy metabolism disturbances characterized by altered ethanol, GSH, creatine, aspartate, uridine diphosphate glucose (UDP)-glucose and NAD^+ were observed in PBDE-47 administration [88]. Ctenopharyngodon idellus kidney (CIK) cells treated with 100 μM PBDE-47 showed a drop in antioxidant enzymes, such as CAT, SOD, GPx, and total antioxidant capacity (T-AOC). A significant elevation in Bcl-2-associated X protein (Bax), Cytochrome c, and Caspase-3 was observed in PBDE-47 exposure compared to the normal group [89]. Consistently, a pharmacological study has shown troxerutin prevented cytochrome c release, Caspase activation, and poly ADP ribose polymerase (PARP) cleavage, raised antioxidative enzymes and Nrf2 activities, thus relieving the toxic effects of PBDE-47 on kidney [90]. Analogously, male broilers were exposed to PBDE-209 for 42 days. Swelling and granular degeneration of the renal tubular epithelium and atrophy and necrosis of glomeruli were observed. Additionally, oxidative stress indicators (MDA, GPx, GSH, SOD) were changed in the kidney [91]. Furthermore, PBDE-209 could upregulate apoptosis-related protein expression, including Bax/Bcl-2 ratio, p-extracellular signal-regulated kinase 1/2 (ERK1/2), p-JNK1/2, Bax, Cytochrome c and Caspase-3 [91]. It was reported that PBDE-209 does not affect kidney weight, while PBDE-209 supplement showed greater GSH and thiobarbituric acid reactive substances (TBARS) and reduced total -SH groups, with consequent exacerbation of nephrotoxicity [92]. The mechanisms are shown in Figures 2 and 3.

3.2. Combined Exposures

The combined exposure of PBDE-47 and cadmium (Cd) displayed cell rounding and swelling, eventually resulting in renal tubular epithelial cell damage. Using human kidney cells (HKC), Zhang et al. reported that intracellular LDH release, nucleotide-binding oligomerization domain-like receptor protein 3 containing pyrin domain (NLRP3), cleaved Caspase-1 and cleaved gasdermin D (GSDMD) were increased by co-exposure. Further, it has been found that co-exposure to PBDE-47 and Cd could give rise to mitochondrial dysfunction NLRP3 inflammasome and GSDMD-dependent pyroptosis. Interestingly, NAC, a ROS scavenger, could mitigate the percentage of apoptotic and necrotic cells inflicted by PBDE-47 and Cd [93].

3.3. Others

PBDE-47 can inhibit mitochondrial fusion and fission, causing MMP decreases, ROS overgeneration, ATP depletion, and cellular disintegration in porcine kidney-15 (PK15) cell [94]. Deeper cells investigation revealed that underlying AMPK-Sirtuin 1 (Sirt1)-Peroxisome proliferator-activated receptor γ coactivator 1-α (PGC-1α) signaling pathway that might be driving the toxic changes in CIK cells subjected to PBDE-47 [89]. By using the CIK cell line, Li et al. have found PBDE-47 can enhance cytoplasmic Ca^{2+} concentration, reduce *miR-140-5p* miRNA level, increase Toll-like Receptor 4 (TLR4) and nuclear factor-κB (NF-κB) mediated inflammatory factors release. Intriguingly, melatonin could protect against PBDE-47-triggered necroptosis via targeting miR-140-5p/TLR4/NF-κB pathway [95]. Similarly, another study showed that PBDE-47-treated mice had elevated ROS, NF-κB, urine albumin-to-creatinine ratio (ACR) and NLRP3 inflammasome level, while troxerutin effectively improved kidney injury elicited by PBDE-47 through inhibiting C-X-C chemokine ligand 12 receptor 4 (CXCR4)-TXNIP-NLRP3 inflammasome [96]. The effects and mechanisms of kidney toxicity induced by PBDEs are shown in Table 2.

Table 2. Effects and mechanisms of kidney toxicity induced by PBDEs.

Treatments	Effects and Mechanisms	References
PBDE-99, SD rats	CAT activity↓, GSSG/GSH↑	[48]
PBDE-47, HEK293 cells	Cell apoptosis, ROS↑, *Bax*↑, *Bad*↑, *Bcl-2*↑, *Hrk*↑, ethanol↑, GSH↓, creatine↓, aspartate↓, UDP-glucose↓, NAD^+↓	[88]
PBDE-47, CIK cells	CAT activity↓, SOD activity↓, GPx activity↓, T-AOC↓, Bax↑, Cytochrome C↑, Caspase-3↑	[89]
PBDE-47, C57BL/6 mice	Cytochrome c release, caspase activation, PARP cleavage, CAT activity↓, SOD activity↓, GPx activity↓, Nrf2 activity↓, ROS↑, NF-κB↑, ACR↑, NLRP3↑, CXCR4/TXNIP/NLRP3↑	[90,96]
PBDE-209, broilers	Swelling and granular degeneration of the renal tubular epithelium, atrophy and necrosis of glomeruli, MDA↑, GSH-Px↓, GSH↓, SOD↓, Bax/Bcl-2 ratio↑, p-ERK1/2↑, p-JNK1/2↑, Bax↑, Cytochrome c↑, Caspase-3↑	[91]
PBDE-209, Wistar rats	GSH↑, TBARS↑, -SH↓	[92]
PBDE-47 and Cd, HKC cells, combined exposure	Cell rounding, cell swelling, renal tubular epithelial cell damage, LDH release, NLRP3↑, cleaved Caspase-1↑, cleaved GSDMD↑, mitochondrial dysfunction, pyroptosis	[93]
PBDE-47, PK 15 cells	Mitochondrial fusion and fission↓, MMP↓, ROS↑, ATP↓, cellular disintegration	[94]
PBDE-47, CIK cells	AMPK-Sirt1-PGC-1α pathway↓, cytoplasmic Ca^{2+}↑, *miR-140-5p* miRNA↓, TLR4↑, NF-κB↑	[89,95]

↑ represents upregulation, ↓ represents downregulation.

4. Gut Toxicity

4.1. Oxidative Damage and Apoptosis

It was reported that the metabolic activities of bacteria in the guts were impacted by PBDE-71, including functions related to energy metabolism, virulence, respiration, cell division, cell signaling, and stress response. For instance, the disruption of epithelial barrier integrity, inflammatory response and anti-oxidant capacity were observed in male intestines after PBDE-71 exposure [97]. Li et al. chose the carcinoma colon-2 (Caco-2) cells model to study the toxicity mechanism of PBDE-209. The mRNA expression of the antioxidative defense factor, *Nrf2*, was suppressed by PBDE-209. Besides, Caco-2 cells exposed to PBDE-209 exhibited a rise in Fas cell surface death receptor (*FAS*) and *CYP1A1* mRNA expression levels [98]. The mechanisms are shown in Figures 2 and 3.

4.2. Intestinal Microbiome Disturbance

Maternal exposed to the PBDE-47 exhibited a distinctive profile in the microbiome of the gut, compared with the control dam, as shown by a decrease in genera *AF12* and

Oscillospira and an increase in the *Actinobacteria* phylum and genera *Blautia, Gemella* and *Phascolarctobacterium*. Serum metabolites strongly associated with the altered gut microbiota in response to PBDE-47 are likely involved in amino acid, lipid, carbohydrate, and energy metabolism [99]. The gut microbiome plays a crucial role in toxicological responses. The intestinal microbiome was required for PBDEs to reduce 3-indolepropionic acid. A tryptophan metabolite has been demonstrated to have protective properties against inflammation and diabetes [100]. PBDE-47 continuously increased the fecal and liver bile acids in the 12α hydroxylation pathway, corresponding to an up-regulation with the hepatic bile acid-synthetic enzyme *CYP7A1* and reduced farnesoid X receptor (FXR) signaling [101]. Exposure of ICR mice to PBDE-47 in-utero and during lactation in its early life may significantly cause a drop in microbial diversity and compositional alterations, and when combined with a HFD, may further exacerbate the progression of obesity and other metabolic illnesses [102,103]. Through the action of the gut microbiome, primary bile acids are converted into more lipophilic secondary bile acids that may be taken up by the host and interact with certain receptors. Both PBDE-47 and PBDE-99 decreased the proteins of sodium taurocholate cotransporting polypeptide (Ntcp) and organic anion transporting polypeptide 1b2 (Oatp1b2) in a gut microbiota-dependent manner [104]. Neonatal contact with PBDE-99 caused a lasting rise in *Akkermansia muciniphila* throughout the intestine, along with augmented hepatic levels of acetate and succinate, the expected byproducts of *A. muciniphila* [83]. The mechanisms are shown in Figure 6. The effects and mechanisms of gut toxicity induced by PBDEs are shown in Table 3.

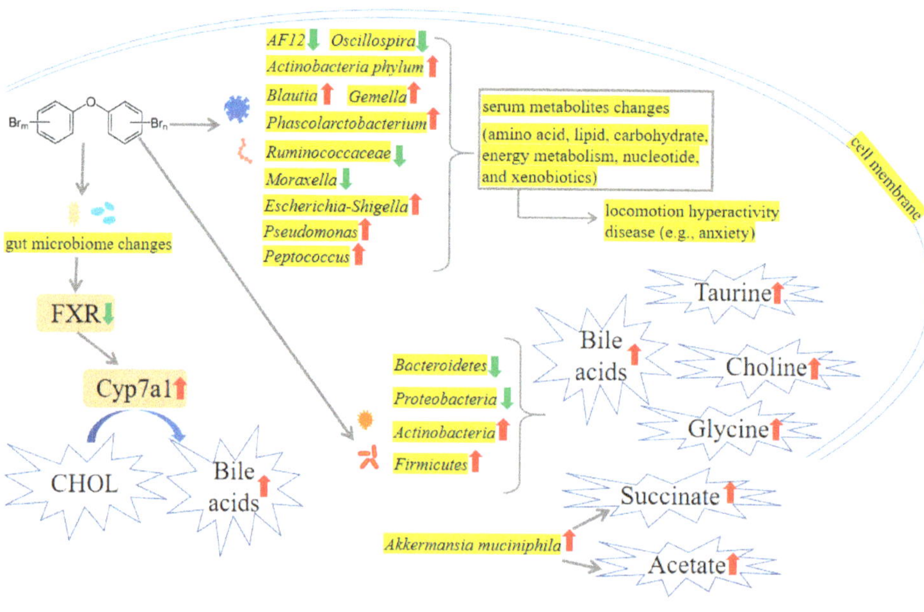

Figure 6. PBDEs-induced toxicity is associated with intestinal microbiome disturbance. PBDEs exposure increased *Actinobacteria phylum, Blautia, Gemella, Phascolarctobacterium, Escherichia-Shigella, Pseudomonas, Peptococcus, Actinobacteria, Firmicutes*, and *Akkermansia muciniphila*. In addition, *AF12, Oscillospira, Ruminococcaceae, Moraxella, Bacteroidetes*, and *Proteobacteria* are decreased by PBDEs treatment. These intestinal microbiome changes may upregulate bile acids, taurine, choline, glycine, succinate, and acetate levels. Arrows indicate up (red colour), increased; down (green colour), decreased. The yellow highlighted text is an explanation of the figure.

Table 3. Effects and mechanisms of gut toxicity induced by PBDEs.

Treatments	Effects and Mechanisms	References
PBDE-71, zebrafish	Disruption of epithelial barrier integrity, inflammatory response, and anti-oxidant capacity	[97]
PBDE-209, Caco-2 cells	*Nrf2*↓, *FAS*↑, *CYP1A1*↑	[98]
PBDE-47, SD rats	*AF12*↓, *Oscillospira*↓, *Actinobacteria phylum*↑, *Blautia*↑, *Gemella*↑, *Phascolarctobacterium*↑	[99]
PBDE-47, CD-1 mice, ICR mice	Fecal and liver bile acids↑, *CYP7A1*↑, FXR signaling↓, microbial diversity↓, microbial compositional alterations, worsen HFD-induced obesity, hepatic steatosis, and injury	[101–103]
PBDE-47 and -99, C57BL/6 mice	Unconjugated bile acids↑, *Akkermansia muciniphila*↑, acetate↑, succinate↑, Ntcp↓, Oatp1b2↓	[83,104]

↑ represents upregulation, ↓ represents downregulation.

5. Thyroid Toxicity

5.1. Hormonal Interferences

Huang et al. have demonstrated that even low concentrations of PBDEs could potentially affect THs in the general population [105]. Research has sought to fill the void by establishing a human PXR-overexpressing HepG2 cell model and a dual luciferase reporter assay system to examine the involvement of hPXR in the modifications of thyroid receptor (TR) expression caused by PBDE-47 in HepG2 cells. TR isoforms (TRα1 and TRβ1) were both observed to decrease in both mRNA and protein expression when the concentration of PBDE-47 was increased in HepG2-pCI-hPXR-neo cells, which may provide more evidence for the toxicological mechanisms of the disruption of the TH in the presence of PBDE-47 [106]. Consistently, Macaulay et al. have demonstrated that PBDE-47 negatively affected the early development of the zebrafish by reducing the TR [107]. Intriguingly, in larvae, PBDEs (PBDE-47 and PBDE-209) significantly stimulated several genes, which included *TRα* and *TRβ*, thyroglobulin (*TG*), thyroid peroxidase, *TTR*, corticotrophin-releasing hormone (*CRH*), sodium iodide symporter (*NIS*), thyronine deiodinase (*Dio1* and *Dio2*), uridinediphosphate-glucuronosyl-transferase (*UGT1ab*) and thyroid stimulating hormone (*TSH*) [108,109]. Lower plasma T4 and liver vitamin A concentrations were linked to PBDE-71 exposure [110]. Concentrations of total and free total thyroxine (FT4) were significantly decreased by PBDE-71 in a dose-dependent manner [111]. A positive link between TSH and almost all PBDE congeners was identified, while an inverse relationship between PBDE-154 and free triiodothyronine (FT3) and FT4 was found [112]. Conversely, lower TSH levels have been linked to exposure to PBDEs in pregnant women [2,113]. Interestingly, there was a significant positive relationship between serum PBDE-209 levels and total thyroxine (TT4), as well as a marginal positive relationship with total triiodothyronine (TT3), in all occupational workers after accounting for gender, age, body mass index (BMI), and duration of occupational exposure [114]. Exposure to PBDE-209 changed the thyroid gland's structure [115]. Hydroxylated PBDEs (OH-PBDEs) have a close structural affinity to TH, and have been demonstrated to interact antagonistically with human TTR, a T4 transport protein [116]. Consistently, changes in plasma FT4 levels in rainbow trout plasma, potentially caused by the metabolic activity of PBDE 209, might be due to the competition of OH-PBDEs for binding sites on TTR [117]. Analogously, Ibhazehiebo et al. have reported that PBDEs diminished TR-mediated gene expression by partially separating TR from TH response element (TRE) through the DNA binding domain (DBD) [118]. Fish exposed to PBDE-209 exhibited a drop in circulating TT4 and TT3 compared to controls [119]. CAR/PXR pathways may be the underlying cause of the decrease in TT4 following PBDE-47 exposure, which is evident in the elevated UGT activity and inducibility of genes in the CAR/PXR pathway, namely *CYP2B10*, *Ugt1a1*, *Ugt1a7*, *Ugt2b5* and multidrug resistance protein-associated protein (*Mrp3*) [120,121]. The mechanisms are shown in Figure 7.

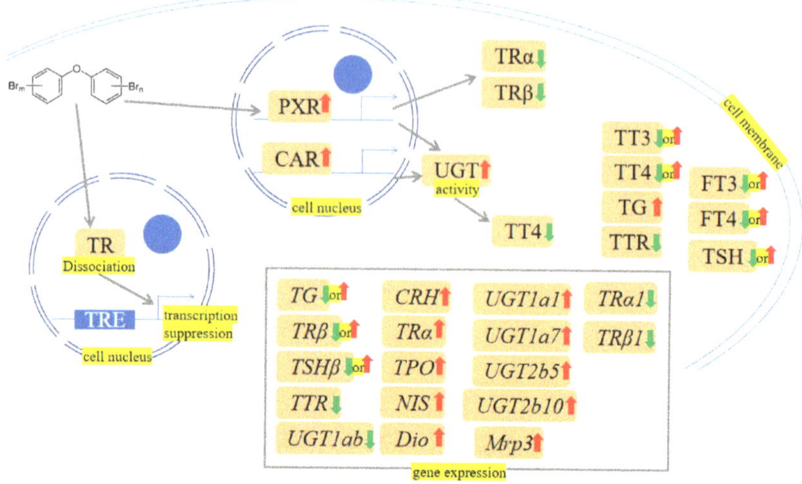

Figure 7. PBDEs-induced toxicity is associated with thyroid dysfunction. PBDEs exposure causes thyroid receptor (TR) dissociation from the TH response element (TRE), and the subsequently related gene expression may be affected. Constitutive androstane receptor (CAR)/pregnane X receptor (PXR) pathway may also be activated by PBDEs. Arrows indicate up (red colour), increased; down (green colour), decreased; up or down, increased or decreased (opposite research results exist). The yellow highlighted text is an explanation of the figure.

5.2. Oxidative Damage and Apoptosis

PBDE-47 augmented apoptosis in thyroid tissue, as revealed by Caspase-3 activation, PARP cleavage and DNA fragmentation. Additionally, the increased glucose-regulated protein 78 (GRP78), activating transcription factor 4 (ATF4), Caspase-12, C/EBP homologous protein (CHOP) and sequestome 1 (p62) accumulation were observed. These results indicate that excessive ER stress, defective autophagy and the resultant apoptosis are thought to be involved in maternal thyroid harm after perigestational PBDE-47 exposure [122–124]. Besides, further studies showed that oxidative damage and hypothalamic-pituitary-thyroid (HPT) axis-related gene alterations may be the underlying mechanisms involved in the toxicity of PBDEs (PBDE-47, PBDE-71, PBDE-99, PBDE-209) [125–134]. The mechanisms are shown in Figures 2 and 3.

5.3. Indirect Exposures

There was a correlation between elevated levels of maternal serum PBDEs 28 and 47 and increased maternal serum concentrations of T4 and T3 during the early second trimester of pregnancy [135]. Conversely, several studies revealed a negative association between hormones (T4 and T3) and PBDEs [136–140]. Positive associations between PBDEs and THs were found in the high-exposed population, while negative associations were reported in the low-exposed populations [141,142]. Discrepancies in the direction of correlations have been noted, and the potential explanations could include the low-dose and nonmonotonic effects of endocrine-disrupting chemicals [143]. The treatment of mothers with PBDE-71 affects the first filial generation (F1) females, as shown by a decrease in body weight and elevated serum T3 and T4 levels. In addition, thyroid gland weight and ovarian osteopontin mRNA were increased at five months of age [144]. Among infants delivered naturally and unassisted via vaginal delivery, prenatal polychlorinated biphenyls (PCBs) and PBDE exposures were associated with lower TT4 and FT4 levels [145].

5.4. Combined Exposures

Results showed that the co-exposition of polystyrene nanoplastics (PS-NPs) and the PBDE-47 aggravated the deformity in pericardial edema, yolk sac edema and curvature of the tail in the larvae of zebrafish. Interestingly, an investigation of the HPT axis-related genes showed that the expressions of *TSHβ*, *TG*, *Dio2* and *TRβ* were increased more prominently when both PBDE-47 and NPs were present, compared to PBDE-47 single exposure [146].

5.5. Others

Expression of Dio3 in placental cells is disrupted by low-dose PBDE-209, resulting in interference of TH. Modifications in the miRNA expression pattern of the miR379/656 cluster in the delta-like homolog 1 (Dlk1)-Dio3 imprinting domain, particularly of miR409-3p and miR668-3p, and/or changes in the DNA methylation of the cells, specifically the intergenic-differentially methylated regions (IG-DMR) and maternally expressed gene 3-DMR (MEG3-DMR) in the Dlk1-Dio3 imprinting region, may be responsible for the disturbance in Dio3 expression brought about by PBDE-209 [147]. The effects and mechanisms of thyroid toxicity induced by PBDEs are shown in Table 4.

Table 4. Effects and mechanisms of thyroid toxicity induced by PBDEs.

Treatments	Effects and Mechanisms	References
PBDE-47, HepG2 cells	*TRα1*↓, TRα1↓, *TRβ1*↓, TRβ1↓	[106]
PBDE-47, zebrafish	TR↓, head trunk angle↓, otic vesicle length↑, eye pigmentation↓, developmental delays	[107]
PBDE-47, zebrafish	*TTR*↓, *TG*↓, *TRβ*↓, *TSHβ*↓, *NIS*↑, *TPO*↑, *TRα*↑	[108]
PBDE-209, zebrafish	*CRH*↑, *TSHβ*↑, *NIS*↑, *TG*↑, *Dio1*↑, *Dio2*↑, *TRα*↑, *TRβ*↑, *TTR*↓, *UGT1ab*↓	[109]
PBDE-71, SD rats	Plasma T4↓, liver vitamin A↓, body weight↓, T3 (F1 Female)↑, T4 (F1 Female)↑, thyroid gland weight↑, *osteopontin*↑	[110,144]
PBDE-71, C57BL/6 mice	TT4↓, FT4↓	[111]
PBDE-209, workers	Positive relationship between serum PBDE-209 levels and total TH	[114]
PBDE-209, rainbow trout	OH-BDE metabolites negatively correlated with the plasma FT4 levels	[117]
PBDE-209, fathead minnows	TT4↓, TT3↓	[119]
PBDE-47, C57 BL/6 mice	TT4↓, *Ugt1a1*↑, *Ugt1a7*↑, *Ugt2b5*↑, *CYP2B10*↑, *Mrp3*↑	[121]
PBDE-47, SD rats	Apoptosis, Caspase-3 activation, PARP cleavage, DNA fragmentation↑, GRP78↑, ATF4↑, CHOP↑, p62 accumulation, ER stress, defective autophagy	[122–124]
PBDE-28 or -47, human	TT4↑, FT4↑, TT3↑, FT3↑	[135]
PBDEs, human	Placental T4↓ (PBDE-99, or -100)	[136]
PBDEs, human	TT4↓ (PBDE-99, -154 or -196), TT3↓ (PBDE-47, -99, -100, -197, -203 or -207)	[137]
PBDEs, human	TT3↑ (PBDE-47, -66 or 85), TT4↑ (PBDE-66, -85, 153 or -154), TT4↓ (PBDE-209)	[138]
PBDEs, human	TT4↓ (PBDE-28, -47, -99, -100 or -153)	[140]
PBDEs, human	TT3↓ (PBDE-154), FT3↓ (PBDE-153, -183), T4/T3 ratio↑ (PBDE-100)	[142]
PBDE-47 and PS-NPs, zebrafish, combined exposure	Deformity in pericardial edema, yolk sac edema, the curvature of the tail, *TSHβ*↑, *TG*↑, *Dio2*↑, *TRβ*↑	[146]
PBDE-209, JEG-3 cells	Dio3↓, *Dio3*↓, *has-miR-668-3p*↓, *has-miR409-3p*↓	[147]

↑ represents upregulation, ↓ represents downregulation.

6. Embryotoxicity

6.1. Oxidative Damage and Apoptosis

In vivo and in vitro models have shown that PBDE-47 could activate the MAPK signaling pathway, thus changing impaired placental physiological function [148]. Embryonic development abnormalities in zebrafish exposed to PBDE-47 could be improved by ROS scavenging and JNK inhibition. Therefore, deficiencies in mitochondrial biogenesis and mitochondrial dynamics, which may lead to ROS/JNK activation, could be the reason for

PBDE-47-induced developmental toxicity [149,150]. Meanwhile, supplementation with the antioxidant NAC could partly reverse the ROS increase and octamer-binding transcription factor 4 (OCT4) downregulation caused by PBDE-209 exposure [151]. It was reported that PBDE exposure caused a decrease in the expression of pluripotent genes such as *OCT4*, SRY-box transcription factor 2 (*SOX2*) and Nanog homeobox (*NANOG*) and prompted apoptosis in embryonic stem cells (ESCs) [151,152]. The mechanisms are shown in Figures 2 and 3.

6.2. Combined Exposures

Exposure to multiple chemicals is a common occurrence in the environment. For example, embryos temporarily postpone hatching when encountering PBDE-209 and $nSiO_2$ at 60 h post fertilization. PBDE-209-nano-SiO_2 co-exposure showed a decreased heartbeat and increased arrhythmia in zebrafish larvae compared to individual PBDE-209 treatments [153].

6.3. Others

Embryonic developmental abnormalities of SD rats (e.g., soft tissue syndactyly or malposition of the distal phalanges and decreased ossification of the sixth sternebra), zebrafish (e.g., embryo yolk sac, pericardial edema, spine deformation, neurobehavioral abnormalities and blood vessels damage) and kestrels (e.g., shorter humerus length and reduced total thyroid weight) arise as a result of being exposed to PBDEs [154–158]. The Dio3 is of great significance in maintaining the fetal thyroid equilibrium. PBDE-47 correlated with an increase in placental *Dio3* methylation, whereas this effect was only observed in female infants [159]. Epidemiological studies have indicated that PBDE-47 can bring about adverse pregnancy results. PBDE-47-treated mice displayed decreased vascular endothelial growth factor-A (VEGF-A) protein expression, indicating that PBDE-47 may disrupt placental angiogenesis [160]. Transcriptomic analysis of the PBDE-47 effect suggested concentration-dependent changes in the expression of genes, including stress pathways such as inflammation and metabolism of lipids/cholesterol, as well a process underlying the fate of trophocytes, such as differentiation, migration, and morphology of the vascular system [161]. Mediation analyses revealed that PBDEs exposure could potentially impact fetal growth through insulin-like growth factor 2 (IGF2) methylation [162]. Chi et al.'s multivariate statistical analysis revealed a marked change in the metabolic profile resulting from PBDE-209 embryotoxicity in maternal serum. Administration of PBDE-209 at a dosage of 2500 mg/kg caused considerable disturbances to thyroid hormone metabolism, the tricarboxylic acid cycle (TCA) cycle, and lipid metabolism in maternal mice, leading to substantial inhibition of fetal growth and development [163]. Consistently, Du et al.'s study demonstrated that prenatal PBDE-209 induces upregulation of endothelin-1 (ET-1) and inducible nitric oxide synthase (iNOS) and downregulation of endothelial nitric oxide synthase (eNOS) in the placenta, which in turn is associated with reduced birth weight of the newborns [164]. Gestational exposure to PBDE-209 can reduce placental weight, impede placental vascular growth, and cause cell death. Mechanically, gestational exposure to PBDE-209 augmented the expression of GRP78 and activated pancreatic endoplasmic reticulum kinase (PERK) signaling [165]. The exposure to PBDE-47, PBDE-99 and PBDE-209 at realistic concentrations led to lethal and sublethal changes and impaired gene expression involved in TH homeostasis, including *TSHβ*, *TTR*, thyroxine-binding globulin (*Tbg*) and *Dio1*, leading to an abnormal development of zebrafish embryos [166]. The effects and mechanisms of embryotoxicity induced by PBDEs are shown in Table 5.

Table 5. Effects and mechanisms of embryotoxicity induced by PBDEs.

Treatments	Effects and Mechanisms	References
PBDE-47, ICR mice	MAPK signaling↑, changed placental function, low birth weight, stillbirth rate↑, plasma testosterone↓, progesterone↓, growth hormone↓, compromised fetal development	[148]
PBDE-47, zebrafish	Embryonic development abnormalities, ROS↑, JNK activity↑	[149,150]
PBDE-209 or -47, hESCs, mESCs	ROS↑, OCT4↓, apoptosis, *OCT4*↓, *SOX2*↓, *NANOG*↓	[151,152]
PBDE-209 and $nSiO_2$, zebrafish, combined exposure	Postpone hatching, heartbeat↓, arrhythmia↑, malformation↑	[153]
PBDEs, SD rats, zebrafish, *common terns, kestrels*	Soft tissue syndactyly or malposition of the distal phalanges and decreased ossification of the sixth sternebra (rats), embryo yolk sac, pericardial edema, spine deformation, neurobehavioral abnormalities, and blood vessels damage (zebrafish), shorter humerus length and reduced total thyroid weight (kestrels)	[154–158]
PBDE-47, human	Placental Dio3 methylation (female infants)↑	[159]
PBDE-47, ICR mice	Adverse pregnancy results, VEGF-A↓, placental angiogenesis↓	[160]
PBDE-47, CTB	Cell viability↓, Global CpG methylation↑	[161]
PBDEs, human	Fetal growth retardation (PBDE-206, PBDE-17-190, PBDE-196-209), aberrant methylation of *HSD11B2* and *IGF2*	[162]
PBDE-209, C57 mice	Fetal growth and development↓, TCA cycle↓, accelerated lipolysis, TH↓	[163]
PBDE-209, SD rats	ET-1↑, iNOS↑, eNOS↓, birth weight of the newborns↓	[164]
PBDE-209, C57BL/6 mice	Placental vascular growth↓, placental cell death, GRP78↑, PERK signaling↑	[165]
PBDEs, zebrafish	Yolk and pericardial edema, tail, and head malformation, reduced and extremely reduced heartbeat rate, blood stasis and spinal curvature, cardiac edema, damage of eye structure and hydrocephaly, liver vacuolization (PBDE-47, -99, -209), *TSHβ*↑, *TTR*↑, *Tbg*↑, *Dio1*↑ (PBDE-47, -99), *Dio1*↑ (PBDE-209)	[166]

↑ represents upregulation, ↓ represents downregulation.

7. Reproductive Toxicity

7.1. Oxidative Damage and Apoptosis

Oxidative stress and reduced testosterone levels induced by PBDE-209 can cause DNA damage and activate the ataxia-telangiectasia-mutated (ATM)/checkpoint kinases 2 (Chk2), ataxia telangiectasia and Rad3-related (ATR)/checkpoint kinases 1 (Chk1) and DNA-dependent protein kinase catalytic subunit (DNA-PKcs)/X-ray repair cross-complementing protein 4 (XRCC4)/DNA ligase IV pathways [167,168]. Furthermore, PBDE-209 may damage the mitochondrial function by reducing the length of the telomeres, decreasing the activity of telomerase, and activating PPARγ/RXRα/sterol regulatory element-binding protein cleavage-activating protein (SCAP)/SREBP-1 pathway, resulting in cell apoptosis [169,170]. Following maternal exposure to PBDE-209 during lactation, prepubertal mice offspring exhibited impaired germ cell proliferation, affected steroidogenesis and increased germ cell apoptosis, alongside modifications in the expression of various cell survival and apoptotic markers and a reduction in the expression of gap junction connexin 43 (cx43) and cyclin-dependent kinase inhibitor 1B (p27Kip1) [171–175]. Similarly, a substantial drop in the net reproductive rate and intrinsic increase rate transpired when the concentration of PBDE-47 was high. PBDE-47 had a strongly detrimental effect on the ultrastructure of the ovary. Further studies showed that PBDE-47's reproductive toxicity was attributed to the ROS-mediated pathway [176,177]. PBDE-47 induced mitochondrial disruption (e.g., aberrant distribution and diminished MMP), which can induce apoptosis of early leptotene spermatocytes and affect the maturation of oocytes [178,179]. GC2 (immortalized mouse spermatocyte) exposed to PBDE-47 have reduced cell viability, and condensation of nuclear and vacuolated mitochondria. Meanwhile, the deletion of ATP synthase subunit β (Atp5b) or ubiquinol-cytochrome-c reductase complex core protein 1 (Uqcrc1) led to a decrease in MMP and triggered apoptosis in GC2. This may be part of the underlying mechanism of the association between PBDE-47 and spermatocytes [180]. PBDE-99 could induce Leydig cell apoptosis via increasing ROS, triggering the ERK1/2 pathway, and inhibiting the ubiquitination degradation pathway [181]. PBDEs (PBDE-47, PBDE-99 and PBDE-100) are capable of initiating apoptosis, both through the extrinsic and intrinsic pathways, upon extended exposure periods; this can cause early malfunction of the corpus luteum [182]. PBDE-3 decreased serum testosterone levels and Leydig cell size by decreasing extracellular signal-regulated kinase 1/2 (ERK1/2), AKT, and AMPK phosphorylation and elevating ROS production [183,184]. Caenorhabditis elegans (worm) that underwent PBDEs (PBDE-3 and PBDE-47) treatment showed reduced life spans, impeded fecundity, and delayed egg-laying. Most interestingly, mutants of C. elegans p53-like protein (*cep-1*), DNA damage checkpoint proteins (*hus-1*) and mitogen-activated protein kinase (*mek-1* and *sek-1*) rescinded the germ cell apoptosis induced by PBDEs [185,186]. The mechanisms are shown in Figures 2 and 3.

7.2. Epigenetic Inheritance

Moreover, the effects on reproductive health may even be transgenerational. For instance, PBDE-47 treatment displayed that protamine and transition protein genes were, on average, reduced by four-fold in the testes, suggesting that histone-protamine exchange could be disrupted during spermatogenesis, triggering an aberrant sperm epigenome [187]. According to Suvorov et al., sperm samples collected on PND 65 and PND 120 yielded 21 DMRs and 9 DMRs, respectively [188]. Following exposure to PBDE-209, there was a decrease in anogenital distance and the percentage of sperm with normal morphology. Further exploration revealed that, in contrast to the control group, the PBDE-209 group had 215 genes exhibiting hyper-methylation and 83 genes displaying hypo-methylation [189]. The GC-2spd mouse spermatocyte line was utilized to analyze the poisonous effects of PBDE-209 on methylation and spermatogenesis. The results indicated that PBDE-209-induced spermatogenesis damage was due to its disturbance of SET domain-containing protein 8 (SETD8)/Histone H4K20 monomethylation (H4K20me1)-linked histone methylation, inhibition of meiosis initiation and cell cycle progression, which in turn caused long-term male reproductive toxicity [190,191].

7.3. Mitochondrial Damage

PBDEs showed a negative relationship with semen mobility and sperm quality [192–195]. For example, PBDE-47 likely interferes with the ER-Nrf1-mitochondrial transcription factor A (Tfam)-mitochondria pathway, thereby reducing mitochondrial function, impairing spermatogenesis, and damaging germ cells [196]. Shan et al.'s findings unveiled that PBDE-47 impaired mitochondrial function and cholesterol transport, ultimately leading to a reduction in progesterone synthesis [197]. After administering a small amount of PBDE-99, Talsness et al. noticed adverse ultrastructural changes in the mitochondria of the F1 female offspring's ovary [198]. The mechanisms are shown in Figure 5.

7.4. Combined Exposures

In male rats treated with PBDE 47, HFD exacerbated the damage to the seminiferous epithelia and decreased testosterone levels, which decreased the number of spermatozoa. Further mechanistic exploration revealed HFD triggered PBDE47-induced dosage-sensitive sex reversal adrenal hypoplasia congenital critical region on X chromosome gene 1 (DAX-1) expression and lowered steroidogenic acute regulatory protein (StAR) and 3β-hydroxysteroid dehydrogenase (3β-HSD) levels in rat testicular interstitials [199].

7.5. Others

The effect of PBDE-209 on blood-testis barrier ultrastructure was destructive, with the destruction of tight junctions, ectoplasmic specialization structures with broken tight junctions, and a lack of actin microfilaments [200]. Moreover, the elevation of estrogen receptor signaling caused by PBDE-209 leads to disruption of the blood testis barrier in male mice of the F1 generation [201]. PBDEs disrupted gonadal development and had a reduced fecundity [202–206]. In adult mice, PBDE-209 treatment caused a diminished sperm quality and arrested meiotic prophase I at the early-pachytene stage during spermatogenesis [207–209]. Postnatal exposure to PBDE-209 at a low dose from day one to day five results in lower testosterone and androgen receptor (Ar) and TRα transcripts in Sertoli cells, along with an imbalance in the TRα splicing variants ratio, causing a decreased testicular size and hindered spermatogenesis [210]. Analysis of 42 differentially expressed proteins in the testis revealed that downregulating histone variants and parvalbumins associated with PBDE-47 may impede spermatogenesis and lead to infertility in fishes. The increase in VTGs and apolipoprotein A–I suggested that PBDE-47 acts like a mimicker of estrogen and causes reproductive dysfunction [211,212]. At higher concentrations, PBDE-47 caused prolonged hyperactivation of autophagy, which ultimately caused ovary damage [213]. In addition, fish exposed to PBDE-47 during early life stages had reduced clutch size and lower fecundity than the control group [214]. Interestingly, PBDE-71 significantly increased malformation and the percentage of males in the F1 generation and reduced frequencies of male courtship behaviors [215,216]. Arowolo et al. indicated that PBDEs and their metabolites, when present at environmental levels, may impact male reproductive health through AR antagonism, testosterone signaling, cAMP production, mechanistic target of rapamycin complex one (mTORC1) signaling and TH transport [217]. In addition, it appears that PBDE-47 has the potential to enhance the sensitivity of adult Leydig cells to cAMP when synthesizing androgen [218]. The effects and mechanisms of reproductive toxicity induced by PBDEs are shown in Table 6.

Table 6. Effects and mechanisms of reproductive toxicity induced by PBDEs.

Treatments	Effects and Mechanisms	References
PBDE-209, ICR mice, CD-1 mice, Parkes strain mice, Balb/c mice, Sertoli cells	Oxidative stress, testosterone↓, DNA damage, ATM/Chk2↑, ATR/Chk1↑, DNA-PKcs/XRCC4/DNA ligase IV pathways↑, impaired germ cell proliferation, germ cell apoptosis↑, cx43↓, p27Kip1↓, ER signaling↑, impaired blood-testis barrier, sperm quality↓, arrested meiotic prophase I, testicular size↓, spermatogenesis↓	[167,168,171–175,201,207–210]
PBDE-209, SD rats	Mitochondrial function↓, telomeres length↓, telomerase activity↓, PPARγ/RXRα/SCAP/SREBP-1↑, cell apoptosis, anogenital distance↓, abnormal sperm morphology, blood-testis barrier ultrastructure damage, tight junctions damage, ectoplasmic specialization structures with broken tight junctions, actin microfilaments↓	[169,170,189,200]
PBDE-47, *Brachionus plicatilis*, SD rats	Reproductive rate↓, intrinsic increase rate↓, impaired ultrastructure of the ovary, ROS↑	[176,177]
PBDE-47, marine medaka (*Oryzias melastigma*), manila clam *Ruditapes philippinarum*	Histone variants↓, parvalbumins↓, spermatogenesis↓, infertility, vitellogenins↑, apolipoprotein A-I↑, reproductive dysfunction	[211,212]
PBDE-47, mice, SD rats	Mitochondrial disruption, aberrant distribution, MMP↓, apoptosis	[178,179]
PBDE-47, Wistar rats, SD rats	Histone-protamine exchange↓, aberrant sperm epigenome, DMRs↑, autophagy↑, ovary damage, testosterone signaling disruption, AR antagonism, mTORC1 signaling↑, replacement of thyroid hormone from transporting proteins, cAMP↑	[187,188,213,217]
PBDE-47, GC2 cells, ICR mice	Cell viability↓, condensation of nuclear, vacuolated mitochondria, Atp5b↓, Uqcrc↓, MMP↓, apoptosis, ER-Nrf1-Tfam-mitochondria pathway disturbance, mitochondria function↓, spermatogenesis↓, germ cells damage	[180,196]

Table 6. *Cont.*

Treatments	Effects and Mechanisms	References
PBDE-47, *Fathead Minnows (Pimephales promelas)*	Clutch size↓, fecundity↓	[214]
PBDE-47, BeWo Cells	Mitochondria function↓, cholesterol transport↓, progesterone synthesis↓	[197]
PBDE-47, ICR mice, GC-2 cells	Spermatogenesis damage, SETD8/H4K20me1-linked histone methylation disturbance, meiosis initiation↓, cell cycle progression↓, male reproductive toxicity	[190,191]
PBDE-71, zebrafish, male American kestrels (*Falco sparverius*)	Malformation, percentage of male↑, male courtship behaviors↓	[215,216]
PBDE-99, Leydig cells	ROS↑, ERK1/2 pathway↑, ubiquitination degradation pathway↓, apoptosis	[181]
PBDE-99, Wistar rats	Adverse ultrastructural changes of mitochondria	[198]
PBDEs, Luteal cells	Malfunction of the corpus luteum, initiating apoptosis (PBDE-47, -99, -100)	[182]
PBDE-3, SD rats	ROS↑, serum testosterone↓, Leydig cell size↓, p-ERK1/2↓, p-AKT↓, p-AMPK↓	[183,184]
PBDE-3 or -47, *Caenorhabditis elegans*	Life spans↓, fecundity↓, delayed egg-laying, ROS↑, DNA damage	[185,186]
PBDE-47 and high-fat diet, SD rats, combined exposure	Exacerbated the damage to the seminiferous epithelia, testosterone↓, spermatozoa↓, DAX-1↑, StAR↓, 3β-HSD↓	[199]

↑ represents upregulation, ↓ represents downregulation.

8. Neurotoxicity

8.1. Apoptosis

A concentration-dependent increase in the protein expression of Fas and Fas-associated death domain (FADD), as well as activation of Caspases (Caspase-8 and Caspase-3), was detected, implying involvement of the death-receptor pathway in the PBDE-209-induced Neuro-2a cell apoptosis [219]. By increasingthe expression of phosphodiesterases (*PDEs*) that modify intracellular cyclic guanosine monophosphate (cGMP) levels and reducing the Bcl-2/Bax ratio, apoptosis induction was induced by PBDE-209 [220]. The autophagy proteins, such as microtubule-associated protein-I light chain 3 (LC3)-II and Beclin-1, and apoptosis proteins, including Bcl-2, Caspase-3 and PARP, were changed after PBDE-209 administration. This modulation could reduce the learning and memory capabilities of the offspring [221]. Adult rats exposed to PBDE-153 exhibited impaired learning ability, reduced spontaneous activity and neuron apoptosis [222]. Costa et al. have found that antagonists of glutamate ionotropic receptors reduced the toxicity of PBDE-47 in mouse cerebellar neurons, suggesting PBDE-47 may heighten extracellular glutamate, which then stimulates ionotropic glutamate receptors and brings about increased calcium levels, oxidative stress, and finally, cell death [223,224]. PBDE-47 has increased Caspase-3, Caspase-12 and cytochrome c levels in the rats' hippocampus [225]. Cytotoxic evaluation has indicated that PBDE-99 exhibits cytotoxicity against rat cerebellar granule neurons (CGNs). Besides, a decrease in the expression of brain-derived neurotrophic factor (BDNF) and Bcl-2 was also detected after PBDE-99 treatment [226]. The mechanisms are shown in Figure 3.

8.2. Disease Induction

Consistent with the cytotoxicity reported above, perinatal exposure to PBDE-99 through gestation and ingestion of maternal breast milk may lead to learning difficulties, BDNF downregulation and free radicals' production in the offspring of rats [227,228]. Similarly, it was reported that PBDE-47 could reduce BDNF production and increase the risk of post-partum depression [229]. Besides, PBDEs (PBDE-209, PBDE-206 and PBDE-203) were shown to reduce BDNF concentration and increase calcium/calmodulin-dependent protein kinase II (CaMKII) levels in mice hippocampus [230,231]. There is a significant correlation between PBDE-47 and PBDE-99 exposures and depression symptoms among the pregnant cohort [232]. Perinatal exposure to PBDE-47 has been found to decrease the length of the dendrites, the complexity of the branching patterns, and the density of the spines in the prefrontal cortex of offspring. These effects may contribute to autistic behavior [233]. Wang et al. have found that PBDE-71 leads to a significant decrease in serotonin levels and levels of tyrosine hydroxylase and dopamine transporter proteins [234,235]. 6-OH-PBDE-47 is a highly metabolized form of PBDE-47 in vivo. 6-OH-PBDE-47 administration could induce motor defect by impairing the dopaminergic system and promote α-synuclein aggregation by inhibiting ubiquitination and autophagy, indicating that the presence of 6-OH-PBDE-47 in the brain could pose a risk for developing Parkinson's disease (PD) [236].

8.3. Intestinal Microbiome Disturbance

PBDE-47 exposure during gestational and lactational periods displayed hyperactivity and anxiety-like behavior. Furthermore, 16S rRNA sequencing of fecal samples revealed a distinctive community composition of gut microbes after exposure to PBDE-47, which manifestes as a decrease in genera *Ruminococcaceae* and *Moraxella*, and an increase in genera *Escherichia-Shigella*, *Pseudomonas* and *Peptococcus*. Qiu et al. have discovered that the changes in the intestinal flora are involved in the alterations in serum metabolite levels, and both are correlated with locomotion hyperactivity and anxiety [237]. The mechanisms are shown in Figure 6.

8.4. Combined Exposures

When exposed together, PS-NPs and PBDE-47 coalesced into bigger particles. Neurodevelopmental toxicity (e.g., accelerated voluntary movements) in zebrafish larvae was heightened with concurrent exposure to PS-NPs and PBDE-47. Besides, the expression of the acetylcholinesterase (*ache*) and the cholinergic receptor nicotinic 7 α subunit (*chrn7α*) genes, which are associated with the development of neurocentral cells, was significantly decreased, mainly in the co-exposure group [79]. Analogously, the chemical mixtures (PBDE-47, 6-OH-PBDE-47 and 6-MeO-PBDE-47) caused a decrease in AChE activity, implying the potential neurological responses of such treatment [238]. Generally, the brain obtains free fatty acids from the systemic circulation and further alters them into structural and signaling lipids to guarantee proper neurotransmission [239]. A mixture of PCBs and PBDEs (MIX) caused neurobehavioral defects, and further studies observed impaired mitochondrial function and lipid metabolism regulation [239,240]. Pregnant C57BL/6J female mice were exposed to PBDE-209/Pb mixture. The male offspring have increased pro-inflammation cytokines, such as tumor necrosis factor α (TNFα), interferon γ (IFNγ), interleukin 4 (IL-4), interleukin 6 (IL-6), interleukin 10 (IL-10) and interleukin 17 A (IL-17 A) in the serum. Moreover, the male offspring displayed decreased neuronal cells in the cornu ammonis 1 (CA1) and CA3 subregions of the hippocampus and impaired learning behavior [241]. Meanwhile, ROS scavenger NAC can reduce locomotor dysfunction induced by co-exposure (PBDE-209 and Pb), suggesting ROS may be a major factor in eliciting developmental neurotoxicity [242]. Analysis of chemicals revealed that PBDE-209 was taken up and processed by zebrafish larvae, and the presence of titanium dioxide nanoparticles (nano-TiO_2) increased the rate of PBDE-209 absorption. The joint presence of nano-TiO_2 and PBDE-209 decreased locomotion activity and downregulation of specific genes and proteins related to the central nervous system of developing zebrafish larvae [243]. PBDE-99, in combination with methylmercury, augmented developmental neurotoxic effects, including impaired negative geotaxis reflexes and motor coordination [244].

8.5. Others

Chen et al. indicated common PBDE congeners might be toxic agents in neural precursors, which cause functional changes and induce transcriptome changes in pathways that regulate neurodevelopment, hormone signaling, and the response to stress in the environment [245]. PBDEs (PBDE-99 and PBDE-47) exposure in the neonatal period disrupts the normal development of the brain and causes a disturbance in spontaneous behavior [246–248]. In the cortex, a high level of growth-associated protein-43 (Gap 43), a neuronal growth-related protein, was observed [246,247]. PBDE-47 increased the spontaneous coil activity in the embryos of zebrafish under high-intensity light and decreased the locomotion in the larvae of zebrafish. These locomotion effects were negatively correlated with tissue PBDE-47 levels and might be related to pathways for early neurogenesis, the central nervous system and development of the axes [249,250]. PBDE-47 could interfere with neurogenic locus notch homolog protein (NOCTH), GSK3β and T3 signaling, which may affect neurogenesis [251]. Azar et al. have reported that prenatal PBDE exposure is correlated with a decrease in cognitive ability in preschool-age boys, but no such association was seen in girls at the concentrations of exposure in Canada [252]. It has been established that primary sensory neurons are susceptible to the neurotoxic effects of PBDE-209 [253]. C57BL/6J mice were given an oral dose of 20 mg/kg PBDE-209 from day 1 to 21. A drop in TH and/or glial cell activity could impede hippocampal growth, resulting in behavioral difficulties [254]. PBDE-99 was provided to CD-1 Swiss females orally daily from gestational day (GD) 6 to PND 21. On PND 60, the treated mice exhibited an altered thigmotaxis, devoting more time to the centre of the arena than the control mice [255]. PBDE-99 inhibits the differentiation of a mouse and human neural progenitor cell (NPC) lineage into a lineage of oligodendroglial based on species-specific actions [256]. Startle reactions to acoustic stimuli were intensified by PBDE-71 at PND 90, displaying the delay of sensory responsiveness [257]. In zebrafish larvae, hyperactivity was seen when PBDE-71 was present in low concentrations, whilst higher concentrations led to decreased activity during the dark period [258]. Disruption of calcium balance can be caused by PBDE-71, resulting in decreased cholinergic function and locomotor activity [259]. Perinatal exposure to PBDE-71 induced transcriptional alterations, including neurofilaments and cell adhesion molecules (i.e., N-cadherin and CAMKII, and cytokines) [260]. PBDE-47 exposure drastically affects spontaneous movements, decreases the response to touch and speed of swimming, and alters larvae's swimming behavior due to light stimulation. The inhibition of the axonal growth of primary and secondary motor neurons was found, which may contribute to these motor deficits [261]. Hedgehog signaling, a pathway involved in the development of embryos and neurogenesis, was suppressed due to PBDE-47 [262]. PBDE-47 and PBDE-99 could cause short- and long-term behavioral damage at low exposure levels [263]. Additionally, PBDEs (PBDE-47 and PBDE-49) delayed neuronal polarization, leading to a substantial decrease in axonal outgrowth within the first few days in vitro. Reduced ryanodine receptor (RyR) activity could block these axon inhibitory effects, indicating that a potential RyR-dependent mechanism is involved in PBDEs neurotoxicity [264]. The expression of two G1-phase-related regulatory factors genes, *p53* and cyclin-dependent kinase inhibitor 1 (*p21*), was significantly increased by PBDE-47 treatment. On the other hand, reduced gene expression of the *cyclin D1* and the cyclin-dependent kinase 2 (*CDK2*) occurred after PBDE-47 exposure [265]. Besides, PBDE-47 was reported to reduce the MMP and increase the release of cytochrome c to the cytoplasm [266]. Therefore, these processes might cause the reduced Neuro-2a cell proliferation [219,265,266]. PBDE-209 exposure at different developmental stages (i.e., pregnancy, lactation, intragastric administration, after weaning and prenatal to life) could alter the synaptic plasticity in adult rats [267]. PBDE-209 significantly affected dendritic branch number, synaptic protein levels and doublecortin in neurons [268,269]. PBDE-209 exposure to pregnant and lactating mice can disrupt the serum THs of the offspring, as it alters the expression of the Dio, thus resulting in neural impairment [270]. Calcium overload plays a vital role in neuronal function. After being subjected to PBDE-209, the concentration of Ca^{2+} in the hippocampus of the offspring was increased and impaired learning and memory occurred [271]. Roberts et al. have reported that PBDEs (PBDE-99 and PBDE-47) could reduce Dio2 activity in primary astrocyte cells and H4 glioma cells, which consequently caused neurodevelopmental deficits [272]. Using the Gesell Developmental Schedules (motor, adaptive, language, and social domains), researchers have found that prenatal PBDE exposure was linked to lower developmental quotients (DQs) in young kids [273]. The effects and mechanisms of neurotoxicity induced by PBDEs are shown in Table 7.

Table 7. Effects and mechanisms of neurotoxicity induced by PBDEs.

Treatments	Effects and Mechanisms	References
PBDE-209, Neuro-2a cells, HT-22 cells	FAS↑, FADD↑, Caspase-8↑, Caspase-3↑, apoptosis, *PDEs*↑, Bcl-2/Bax↓	[219,220]
PBDE-47 or -209, Neuro-2a cells	*P53*↑, *P21*↑, *cycline D1*↓, *CDK2*↓, *Nrf2*↑, MMP↓, Cytochrome c release↑, Caspase-9↑, Caspase-3↑, ROS↑, MDA↑, GSSG/GSH ratio↑	[265,266]
PBDE-209, SD rats, Wistar rats	LC3-III↑, Beclin-1↑, P62↓, cleaved caspase-3↑, cleaved PARP↑, Bcl-2↓, neurons death, synaptic plasticity↓	[221,267]
PBDE-153, SD rats	Learning ability↓, spontaneous activity↓, neuron apoptosis	[222]
PBDE-47, C57BL/6 mice cerebellar granule neurons, hNPCs	Extracellular glutamate↑, ionotropic glutamate receptors↑, calcium↑, oxidative stress, cell death	[223,224]
PBDE-47, SD rats	Caspase3↑, Caspase12↑, Cytochrome C↑, *Caspase3*↑, *Caspase12*↑, *Cytochrome C*↑, dendrites length↓, spines density↓, the behavior of autism, motor defect, impaired dopaminergic system, α-synuclein aggregation, ubiquitination↓, autophagy↓, PD risk↑, hyperactivity and anxiety-like behavior, *Ruminococcaceae* and *Moraxella*↓, *Escherichia-Shigella*↑, *Pseudomonas* and *Peptococcus*↑	[225,233,236,237]
PBDE-99, cerebellar granular neurons, SD rats	*BDNF*↓, Bcl-2↓, learning difficulties, free radicals↑	[226–228]
PBDE-99, CD-1 Swiss mice	Altered thigmotaxis, time in the centre of the arena↑	[255]
PBDE-47, human	BDNF↓, risk of post-partum depression↑	[229]
PBDEs, NMRI mice	BDNF↓, Ca/CaMKII↑ (PBDE-209, -206, -203)	[230,231]
PBDEs, human	Depression symptoms (PBDE-47, -99)	[232]
PBDE-71, zebrafish	Serotonin↓, TH↓, dopamine transporter protein↓, hyperactivity (low concentrations), activity during the dark period↓ (high concentrations), calcium balance disruption, cholinergic function↓, locomotor activity↓	[234,235,258,259]
PBDE-47 and PS-NPs, zebrafish, combinde exposure	Accelerated voluntary movements, mortality↑, darker/browner liver colour, atrophied liver, *ache*↓, *chrn7*↓	[79]
PBDE-71, SD rats	Delayed effects on sensory reactivity↓, startle reactions↑	[257]
PBDEs, *Daphnia magna*	AChe activity↓ (PBDE-47, 6-OH-PBDE-47 and 6-MeO-PBDE-47)	[238]
PBDEs and PCB, zebrafish, combinde exposure	Neurobehavioral defects, mitochondrial function↓, lipid metabolism regulation↓	[239,240]
PBDE-209 and Pb, C57BL/6 mice, combined exposure	TNFα↑, IFNγ↑, IL-4↑, IL-6↑, IL-10↑, IL-17 A↑, neuronal cells↓, impaired learning behavior	[241]
PBDE-209, C57BL/6 mice, ICR mice	Glial cell activity↓, hippocampal growth↓, behavioral difficulties, dendritic branches↓, synaptic proteins↓, doublecortin↑, weight gain↓, litter size of maternal mice↓, TT3↑, TT4↑, FT3↑, FT4↑, *dio1*↓ (livers), *dio2*↓ (livers), *dio3*↓ (livers), *dio1*↑ (brains), *dio3*↓ (brains), dio3↓, calcium overload, impaired learning and memory	[254,268–271]
PBDE-209 and Pb, zebrafish, combined exposure	Locomotor dysfunction, ROS↑, lipid peroxidation, DNA damage, antioxidant system↓	[242]
PBDE-209 and Pb, zebrafish, combined exposure	Locomotor dysfunction, ROS↑, lipid peroxidation, DNA damage, antioxidant system↓	[242]
PBDE-209 and nano-TiO_2, zebrafish, combined exposure	Locomotion activity↓, *mbp*↓, *a1-tubulin*↓, *gap-43*↓	[243]

Table 7. *Cont.*

Treatments	Effects and Mechanisms	References
PBDE-99 and MeHg, SD rats, combined exposure	Developmental neurotoxic effects, impaired negative geotaxis reflexes, impaired motor coordination	[244]
PBDE-99 or PBDE-47, NMRI mice, cerebral cortex cells, Wistar rats, human glial cells	Development of the brain↓, spontaneous behavior disturbance, Gap 43↑, Dio2 activity↓, neurodevelopmental deficits	[246–248,272]
PBDE-47, zebrafish	Spontaneous coil activity↑, locomotion↓, touch response↓, swimming speed↓, axonal growth↓, motor deficits, Hedgehog signaling↓	[249,250,261,262]
PBDE-47 or -209, hNSC	NOTCH, GSK3β and T3 signaling interference	[251]
PBDE-47 or -49, SD rats, primary hippocampal cell cultures	Neuronal polarization delay, axonal outgrowth↓, RyR activity↑	[264]

↑ represents upregulation, ↓ represents downregulation.

9. Immunotoxicity

9.1. Oxidative Damage and Apoptosis

PBDE-47 has been shown to significantly induce the formation of neutrophil extracellular traps (NETs), a central player in innate immune responses, and the mechanism may be linked to ROS [274]. Zhou et al. have shown that PBDE-47 could diminish the phagocytic ability and the bacteriolytic activity of *R. philippinarum* and *blue mussel mytilus edulis*. Further exploration revealed these changes may be related to the ROS imbalance, the MAPKs pathways, and the lysosomal membrane damage [275,276]. PBDE-47 and PBDE-209 have been reported to enhance ROS production and decrease GSH levels [277]. Exposure of harbour seal granulocytes to PBDEs (PBDE-47, PBDE-99 and PBDE-153) leads to oxidative stress by reducing thiols levels and increasing ROS production [278]. The mechanisms are shown in Figures 2 and 3.

9.2. Inflammatory Response

PBDE-47 can modulate the expression of an array of intracellular miRNAs, which are primarily involved in the regulation of the innate immunity response [279]. Persistent exposure to PBDE-47 can impair innate immunity in mammary tissue [280]. PBDE-47 can disrupt the secretion of proinflammatory cytokines (IL-6 and TNF-α) and interfere with basophil activation [281–284]. Peripheral blood mononuclear cells (PBMC) from subjects who had autism spectrum disorders (ASD) showed a higher response to lipopolysaccharide (LPS) when pretreated with PBDE-47 compared with the control group [285]. Consistently, resistance to the pathogen was compromised in minnow and rainbow trout after exposure to PBDE-47 [286,287]. Mice exposed to PBDE-209 were deprived of proliferative effects and the production of cytokines (IL-2, IFN-γ and TNF-α) in clusters of differentiation 4 (CD4) T cells and CD8 T cells [288,289]. PBDE-209 could induce immunotoxicity, which is characterized by atrophying immune organs, altering humoral and cellular immunity and gene expression [290,291]. Broiler chicks were supplied with PBDE-209, and the subsequent histopathologic examinations showed damaged and necrotic lymphocytes in the spleen and bursa and losses of lymphoid cells in the thymic gland. Interestingly, KEGG database analysis revealed that the cytokine-cytokine receptor interaction signal pathway was most significantly enriched [292]. Mated female C57BL/6J mice were orally administered with PBDEs, and a considerably reduced number of splenocytes and thymocytes were observed in offspring, suggesting PBDEs transferred from the dam affect the offspring's immune system [293]. Exposure to PBDEs (PBDE-47, PBDE-100 and PBDE-209) led to increased inflammatory cytokine expression, prostaglandin E2 (PGE2) levels, cAMP concentration and cyclooxygenase 2 (*COX-2*) gene expression, which illuminated that PBDEs-induced immune response might be attributed to PGE2-prostaglandin E receptors (EPs)-cAMP-cytokines signaling [294].

9.3. Combined Exposures

Mixtures of compounds, including carbamazepine (CBZ), $CdCl_2$, and PBDE-47, have a significant impact on head-kidney (HK) leucocyte cellular activities and the humoral response in the serum or skin mucus, as well as the dysregulation of pro-inflammatory factors [295]. The effects and mechanisms of immunotoxicity induced by PBDEs are shown in Table 8.

Table 8. Effects and mechanisms of immunotoxicity induced by PBDEs.

Treatments	Effects and Mechanisms	References
PBDE-47, SD rats	NETs formation, ROS↑	[274]
PBDE-47, *Ruditapes philippinarum* hemocytes, *Mytilus edulis*	Phagocytic ability↓, bacteriolytic activity↓, ROS↑, alteration of MAPKs pathways, lysosomal membrane damage	[275,276]
PBDE-47 or- 209, Kunming mice	ROS↑, GSH↓, macrophage accessory cell function↓	[277]
PBDEs, harbour seal immune cells	ROS↑, thiols↓ (PBDE-47, -99, -153)	[278]
PBDE-47, THP-1 macrophage-like cells, spleen-derived lymphocytes, BALB/c mice, human	Perturb the innate immune response, disrupt the secretion of proinflammatory cytokines (IL-6 and TNF-α) and interfere with basophil activation	[279,281–284]
PBDE-47, human PBMC	Higher response to LPS	[285]
PBDE-47, fathead minnows (*Pimephales promelas*), rainbow trout (*Oncorhynchus mykiss*)	Resistance to the pathogen↓, survival rate↓, head kidney impairment, immune factors↓, respiratory burst activity↑, immune-related genes↓, impaired immune organs	[286,287]
PBDE-209, C57BL/6 mice, Balb/c mice, SD rats	Proliferative effects↓, production of cytokines↓, atrophying immune organs, humoral and cellular immunity changes	[288–291]
PBDE-209, Broiler chicks	Damaged and necrotic lymphocytes, lymphoid cells in the thymic gland↓	[292]
PBDEs, pantropical spotted dolphin	Inflammatory cytokine↑, PGE2↑, cAMP↑, *COX-2*↑ (PBDE-47, -100, -209)	[294]
PBDE-47, CBZ and $CdCl_2$, gilthead sea bream (*Sparus aurata* L.), combined exposure	Dysregulation of pro-inflammatory factors and humoral response in the serum or skin mucus	[295]

↑ represents upregulation, ↓ represents downregulation.

10. Others

10.1. Diabetes Induction

An elevated prevalence of diabetes was observed in individuals who had been environmentally exposed to PBDE-47 [296]. The prevalence of gestational diabetes (GMD) is increasing rapidly in the world. It has been reported that PBDEs (PBDE-153, PBDE-154, etc.) could disturb maternal glucose homeostasis, augmenting the risk of gestational diabetes mellitus [297]. Interestingly, PBDE-209 has been shown to elevate blood glucose concentration and reduce insulin receptor (*InsR*) mRNA in HFD-fed mice [298].

10.2. Heart Toxicity

For 28 days, rats were supplied with corn oil containing PBDE-209 (5, 50, 500 mg/kg/day) orally. It was determined that PBDE-209 caused damage to the morphology and ultrastructure of the heart and abdominal aorta, increases in serum creatine kinase and LDH values, and changes in antioxidant enzyme activity [299]. PBDE-209 could stimulate the generation of ROS. Subsequently, ROS activated ER stress-related inositol-requiring enzyme 1α (IRE1α)/AKT/mTOR signaling pathway and elicited vascular endothelial cells apoptosis [300].

10.3. Eye Toxicity

To investigate the mechanism of PBDEs' effects on visual behavior, zebrafish embryos were subjected to PBDE-71. The histological analysis uncovered a diminished area of the inner plexiform layer, an enlarged inner nuclear layer, and a reduced density of ganglion cells in the retina of the PBDE-71 group. Behavioral tests unveiled that zebrafish larvae displayed hyperactive responses (i.e., enhanced saccadic eye movements and phototactic response) to

PBDE-71. Further biochemical analysis demonstrated that the patterns of mRNA and protein expression of opsin differed between normal and PBDE-71 groups [301]. It was reported that the thickness and area of the neural retina of human embryonic stem cell-derived retinal organoids (hESC-ROs) were diminished dose- and time-dependent due to exposure to PBDE-47. PBDE-47 exposure was associated with decreased cell proliferation, augmented cell apoptosis, aberrant differentiation and changed eye morphogenesis [302,303]. Moreover, metabolomics research of hESC-ROs demonstrated pronounced fluctuations in the metabolism of purine and glutathione after five weeks of PBDE-47 exposure [302]. It was reported that PBDE-71 increased the retinal and retinyl ester content in the eye of zebrafish [304]. Consistently, there has been a notable rise in the transcription of retinal dehydrogenase (*raldh2*) and a marked reduction in the transcription of retinol dehydrogenase (*rdh1*), cellular retinoic acid binding protein (*crabp1α* and *crabp2α*), retinol-binding protein (*rbp1α*) and retinoic acid receptor subunit (*raraa*) in zebrafish larvae, suggesting PBDE-71 exposure caused a disturbance of retinoid signaling [305].

10.4. Lung Toxicity

Inhaling PBDEs found in the air or dust can have detrimental effects on human health, particularly the respiratory system. For example, results indicated that PBDE-209 exposure can increase the leakage of LDH, inhibit the viability of cells, and increase the transcripts and protein levels of the inflammatory markers (IL-6 and IL-8) in human lung epithelial cells [306]. PBDEs inhalations may impair the health of the lungs through inflammation, oxidation stress, damage to the barrier integrity, uncontrolled production of mucous, and alterations in the physics and biochemical properties of airway fluids [307,308]. Normal human bronchial epithelial cells exposed to PBDEs (i.e., PBDE-47, PBDE-99 and PBDE-209) have been shown to cause oxidative stress, such as NADPH oxidase-4 (NOX-4) expression and ROS. A notable rise was noticed in the activation of DNA damage and repair processes (i.e., increased comet tail length and elevated histone family member X phosphorylation at ser139 site) [309]. The presence of PBDEs (PBDE-28, -47, -99, -100, -153, -154, -183, and -209) triggered membrane disruption and a release of LDH and also caused oxidative stress in cells, which was marked by a decrease in MMP and an emergence of ROS [310]. The effects and mechanisms of toxicity (diabetes, heart, eye, and lung) induced by PBDEs are shown in Table 9.

Table 9. Effects and mechanisms of toxicity (diabetes, heart, eye and lung) induced by PBDEs.

Treatments	Effects and Mechanisms	References
PBDE-47, SD rats	Risk of diabetes prevalence↑, hyperglycemia, scattered microvesicular steatosis	[296]
PBDEs, human	Glucose homeostasis disturbance, gestational diabetes mellitus↑	[297]
PBDE-209 and high-fat diet, C57BL/6 mice	Blood glucose↑, insulin signaling pathway↑, *GLUT4*↓, *TRα*↓, *AR*↓, *Insr*↓	[298]
PBDE-209, SD rats	Impaired morphology and ultrastructure of the heart and abdominal aorta, serum creatine kinase↑, LDH↑, IL-1β↑, IL-6↑, IL-10↑, TNFα↑, endothelial dysfunction, cardiovascular injury	[299]
PBDE-209, human umbilical vein endothelial cells	ROS↑, IRE1α/AKT/mTOR signaling pathway↑, autophagy↑, apoptosis	[300]

Table 9. *Cont.*

Treatments	Effects and Mechanisms	References
PBDE-71, zebrafish	Area of inner plexiform layer↓, inner nuclear layer↑, density of ganglion cells↓, hyperactive responses, retinal and retinyl ester content↑, *raldh2*↑, *rdh1*↓, *crabp1a*↓, *crabp2a*↓, *raraa*↓ *zfrho*↑, *zfuv*↑, *zfred*↑, *zfblue*↑, *and zfgr1*↑	[301,304,305]
PBDE-47, hESC-ROs, zebrafish	Thickness and area of the neural retina↓, cell proliferation↓, cell apoptosis, aberrant differentiation, abnormal eye morphogenesis	[302,303]
PBDE-209, human lung epithelial cell	LDH leakage↑, cell viability↓, IL-6↑, IL-8↑, *IL-6*↑, *IL-8*↑	[306]
PBDE-47, -99 or -209, A549 cells, pNHBE cells, human bronchial epithelial cells	Inflammation, oxidation stress, barrier integrity↓, uncontrolled production of mucous, alterations in physics and biochemical properties of airway fluids, NOX-4↑, ROS↑, DNA damage and repair processes↑	[307–309]
PBDEs, human lung epithelial cells	Membrane disruption, LDH leakage↑, oxidation stress, MMP↓, ROS↑	[310]

↑ represents upregulation, ↓ represents downregulation.

11. Discussion and Conclusions

As a class of typical persistent organic pollutants, PBDEs are among the most important brominated flame retardants classes, often used in plastic, textile, and electronic components and circuits, which are found in the environment through different paths and processes [311]. Food processing and packaging can lead to the contamination of food with PBDEs, thereby entering the food chain [312]. PBDEs are dispersed throughout different environmental media (water, soil, air, and dust) and pose serious threats to human health via multiple routes of exposure (ingestion, dermal, and inhalation) [313,314]. PBDEs and their metabolites have been proven to be hazardous to humans. The debromination of PBDEs occurs in organisms, resulting in a notable increase of congeners with fewer bromines [315]. PBDEs undergoing phase I metabolism give rise to OH-PBDEs metabolites in animals, which can cause more severe biological repercussions [33]. The OH-group's transformation into a MeO-group (i.e., MeO-PBDEs) is a phase II reaction. Nevertheless, metabolic processes do not yield significant changes in the elimination of PBDEs. Both OH- and MeO-PBDEs exhibit a relatively hydrophobic character, causing them to remain within the organism [316].

Considering the hazardous effects of PBDEs, it's necessary to take measures to prevent their release into the environment. An effective approach to minimize e-waste containing PBDEs in landfills is to classify and disassemble electronic devices for treatment using microbial technologies that can absorb and decompose toxic substances. Besides, anti-seepage systems and the continuous monitoring of PBDEs in their leachates are needed [317]. Interestingly, mosses can be utilized to track spatial patterns and temporal trends of atmospheric concentrations or deposition of PBDEs. This may offer a low-cost, feasible approach for facilitating the timely control and sustainable management of PBDEs pollution [318]. By using nanoscale zerovalent iron (nZVI) and palladized nZVI (nZVI/Pd), it's possible to debrominate lower-brominated PBDEs into diphenyl ether (DE), which is the fully debrominated form [319]. PBDEs mainly exist in the particulate phase of wastewater. Thus, sedimentation technology applied in wastewater treatment plants is also a vital method to remove PBDEs [320].

Various PBDE congeners and their metabolites may exert various biological effects by acting on different molecular targets. Despite being arranged hierarchically in Figures 2–7, mechanisms, and pathways of PBDEs toxicity are intertwined. Some studies have shown the opposite results. The potential explanations could include the low-dose effects, non-monotonic effects, compensate mechanisms and exposure route. The chemical formula $C_{12}H_{(9-0)}Br_{(1-10)}O$ has the sum of H and Br atoms fixed at 10. It is theorized that PBDEs comprise 209 congeners separated into ten homolog groups (ranging from mono- to decabromodiphenyl ethers). They belong to a similar structural group, i.e., they contain two aromatic bromine-containing rings (connected by an oxygen atom). Therefore, PBDEs are structurally similar and have analogous toxicological properties. Besides, different tissues may exhibit different toxic susceptibilities to the same compound. Given the complexity and diversity of signaling mechanisms responsible for PBDEs toxicity, it is difficult to consolidate them into particular classifications. Therefore, we summarize other mechanisms in the subtitle: "Others" and illustrate them in Tables 1–9.

Overall, we review the toxic effects of PBDEs on health, particularly focusing on liver toxicity, kidney toxicity, gut toxicity, thyroid toxicity, embryotoxicity, reproductive toxicity, neurotoxicity, immunotoxicity, heart toxicity and eye toxicity. We described the general mechanism of action, such as oxidative damage, apoptosis, disturbance of glucose and lipid metabolism, mitochondrial damage, intestinal microbiome disturbance, and hormonal interferences. This review highlights that PBDEs have a broad variety of toxic effects and mechanisms. Further studies regarding sub-cellular localization, actual environmental exposure levels, target tissue doses, tissue specificity, dose-effect relationship, and nuanced mechanisms are required. There is an ongoing need to further elucidate the specific mechanisms, assimilate new research outcomes into the existing framework of PBDE toxicity and expand existing measures to mitigate potential health effects associated with PBDE exposures.

12. Future Directions

Our review offers a summary of the toxic effects and mechanisms of PBDEs. Special attention should be paid to the persistence, bioaccumulation, and hazards of PBDEs. More work is needed to further explore the molecular mechanisms that explain the association between health outcomes and PBDEs. Understanding the adverse health effects and potential mechanisms of PBDEs helps in developing strategies to minimize exposure and mitigate potential health risks.

Author Contributions: Writing—original draft preparation, J.X., Q.X. and X.W.; Writing—review and editing, J.X., M.Z., D.L. and X.W.; Project Administration, J.X. and X.W.; Funding acquisition, J.X. and X.W. All authors have read and agreed to the published version of the manuscript.

Funding: This research was funded by the Key Project of Natural Science Research of Anhui High Education Institutions (2022AH051801), the Talent Research Foundation of Hefei University (20RC49), the Graduate Quality Engineering Project of Anhui High Education Institutions (2022hhsfkc049), the University Synergy Innovation Program of Anhui Province (GXXT-2022-001), the "Biology and Medicine" key subject of Hefei University (2023xk05). The APC was funded by the Key Project of Natural Science Research of Anhui High Education Institutions (2022AH051801).

Institutional Review Board Statement: Not applicable.

Informed Consent Statement: Not applicable.

Data Availability Statement: Not applicable.

Acknowledgments: We thank Hefei University for providing the working conditions to carry out this work.

Conflicts of Interest: The authors declare no conflict of interest.

References

1. Khalil, A.; Cevik, S.E.; Hung, S.; Kolla, S.; Roy, M.A.; Suvorov, A. Developmental Exposure to 2,2′,4,4′-Tetrabromodiphenyl Ether Permanently Alters Blood-Liver Balance of Lipids in Male Mice. *Front. Endocrinol.* **2018**, *9*, 548. [CrossRef]
2. Chevrier, J.; Harley, K.G.; Bradman, A.; Gharbi, M.; Sjödin, A.; Eskenazi, B. Polybrominated diphenyl ether (PBDE) flame retardants and thyroid hormone during pregnancy. *Environ. Health Perspect.* **2010**, *118*, 1444–1449. [CrossRef] [PubMed]

3. Yue, C.; Li, L.Y. Filling the gap: Estimating physicochemical properties of the full array of polybrominated diphenyl ethers (PBDEs). *Environ. Pollut.* **2013**, *180*, 312–323. [CrossRef] [PubMed]
4. Liu, Y.; Liu, Z.; Gong, A.; Qiu, L.; Zhang, W.; Li, J.; Li, F.; Bai, Y.; Li, J.; Gao, G. Cell changes and differential proteomic analysis during biodegradation of decabromodiphenyl ether (BDE-209) by Pseudomonas aeruginosa. *RSC Adv.* **2019**, *9*, 25048–25055. [CrossRef] [PubMed]
5. Wang, L.; Zou, W.; Zhong, Y.; An, J.; Zhang, X.; Wu, M.; Yu, Z. The hormesis effect of BDE-47 in HepG2 cells and the potential molecular mechanism. *Toxicol. Lett.* **2012**, *209*, 193–201. [CrossRef]
6. Zhang, Y.; Wang, W.; Song, J.; Ren, Z.; Yuan, H.; Yan, H.; Zhang, J.; Pei, Z.; He, Z. Environmental Characteristics of Polybrominated Diphenyl Ethers in Marine System, with Emphasis on Marine Organisms and Sediments. *BioMed Res. Int.* **2016**, *2016*, 1317232. [CrossRef]
7. McGrath, T.J.; Ball, A.S.; Clarke, B.O. Critical review of soil contamination by polybrominated diphenyl ethers (PBDEs) and novel brominated flame retardants (NBFRs); concentrations, sources and congener profiles. *Environ. Pollut.* **2017**, *230*, 741–757. [CrossRef]
8. Hassan, A.A.; Rylander, C.; Brustad, M.; Sandanger, T.M. Persistent organic pollutants in meat, liver, tallow and bone marrow from semi-domesticated reindeer (*Rangifer tarandus tarandus* L.) in Northern Norway. *Acta Vet. Scand.* **2013**, *55*, 57. [CrossRef]
9. O'Driscoll, K.; Robinson, J.; Chiang, W.S.; Chen, Y.Y.; Kao, R.C.; Doherty, R. The environmental fate of polybrominated diphenyl ethers (PBDEs) in western Taiwan and coastal waters: Evaluation with a fugacity-based model. *Environ. Sci. Pollut. Res. Int.* **2016**, *23*, 13222–13234. [CrossRef]
10. Andrade, N.A.; McConnell, L.L.; Torrents, A.; Ramirez, M. Persistence of polybrominated diphenyl ethers in agricultural soils after biosolids applications. *J. Agric. Food Chem.* **2010**, *58*, 3077–3084. [CrossRef]
11. Muresan, B.; Lorgeoux, C.; Gasperi, J.; Moilleron, R. Fate and spatial variations of polybrominated diphenyl ethers in the deposition within a heavily urbanized area: Case of Paris (France). *Water Sci. Technol. A J. Int. Assoc. Water Pollut. Res.* **2010**, *62*, 822–828. [CrossRef] [PubMed]
12. Wang, Y.; Zhang, Q.; Chen, S.; Cheng, L.; Jing, X.; Wang, X.; Guan, S.; Song, W.; Rao, Q. Determination of Polybrominated Diphenyl Ethers in Water Samples Using Effervescent-Assisted Dispersive Liquid-Liquid Icroextraction with Solidification of the Aqueous Phase. *Molecules* **2021**, *26*, 1376. [CrossRef]
13. Chow, K.L.; Man, Y.B.; Tam, N.F.Y.; Liang, Y.; Wong, M.H. Removal of decabromodiphenyl ether (BDE-209) using a combined system involving TiO(2) photocatalysis and wetland plants. *J. Hazard. Mater.* **2017**, *322 Pt A*, 263–269. [CrossRef]
14. Kim, J.S.; Klösener, J.; Flor, S.; Peters, T.M.; Ludewig, G.; Thorne, P.S.; Robertson, L.W.; Luthe, G. Toxicity assessment of air-delivered particle-bound polybrominated diphenyl ethers. *Toxicology* **2014**, *317*, 31–39. [CrossRef] [PubMed]
15. Wang, S.; Zhang, S.; Huang, H.; Niu, Z.; Han, W. Characterization of polybrominated diphenyl ethers (PBDEs) and hydroxylated and methoxylated PBDEs in soils and plants from an e-waste area, China. *Environ. Pollut.* **2014**, *184*, 405–413. [CrossRef] [PubMed]
16. Imm, P.; Knobeloch, L.; Buelow, C.; Anderson, H.A. Household exposures to polybrominated diphenyl ethers (PBDEs) in a Wisconsin Cohort. *Environ. Health Perspect.* **2009**, *117*, 1890–1895. [CrossRef] [PubMed]
17. Sethi, S.; Chen, X.; Kass, P.H.; Puschner, B. Polychlorinated biphenyl and polybrominated diphenyl ether profiles in serum from cattle, sheep, and goats across California. *Chemosphere* **2017**, *181*, 63–73. [CrossRef]
18. Li, H.; Li, Y.; Maryam, B.; Ji, Z.; Sun, J.; Liu, X. Polybrominated diphenyl ethers as hitchhikers on microplastics: Sorption behaviors and combined toxicities to Epinephelus moara. *Aquat. Toxicol.* **2022**, *252*, 106317. [CrossRef]
19. Chen, S.; Che, S.; Li, S.; Ruan, Z. The combined impact of decabromodiphenyl ether and high fat exposure on non-alcoholic fatty liver disease in vivo and in vitro. *Toxicology* **2021**, *464*, 153015. [CrossRef]
20. Wang, L.; Zheng, M.; Gao, Y.; Cui, J. In vitro study on the joint hepatoxicity upon combined exposure of cadmium and BDE-209. *Environ. Toxicol. Pharmacol.* **2018**, *57*, 62–69. [CrossRef]
21. Eick, S.M.; Goin, D.E.; Cushing, L.; DeMicco, E.; Park, J.S.; Wang, Y.; Smith, S.; Padula, A.M.; Woodruff, T.J.; Morello-Frosch, R. Mixture effects of prenatal exposure to per- and polyfluoroalkyl substances and polybrominated diphenyl ethers on maternal and newborn telomere length. *Environ. Health A Glob. Access Sci. Source* **2021**, *20*, 76. [CrossRef]
22. Midya, V.; Colicino, E.; Conti, D.V.; Berhane, K.; Garcia, E.; Stratakis, N.; Andrusaityte, S.; Basagaña, X.; Casas, M.; Fossati, S.; et al. Association of Prenatal Exposure to Endocrine-Disrupting Chemicals With Liver Injury in Children. *JAMA Netw. Open* **2022**, *5*, e2220176. [CrossRef] [PubMed]
23. Cowell, W.J.; Margolis, A.; Rauh, V.A.; Sjödin, A.; Jones, R.; Wang, Y.; Garcia, W.; Perera, F.; Wang, S.; Herbstman, J.B. Associations between prenatal and childhood PBDE exposure and early adolescent visual, verbal and working memory. *Environ. Int.* **2018**, *118*, 9–16. [CrossRef]
24. Ouidir, M.; Buck Louis, G.M.; Kanner, J.; Grantz, K.L.; Zhang, C.; Sundaram, R.; Rahman, M.L.; Lee, S.; Kannan, K.; Tekola-Ayele, F.; et al. Association of Maternal Exposure to Persistent Organic Pollutants in Early Pregnancy With Fetal Growth. *JAMA Pediatr.* **2020**, *174*, 149–161. [CrossRef]
25. Stapleton, H.M.; Eagle, S.; Anthopolos, R.; Wolkin, A.; Miranda, M.L. Associations between polybrominated diphenyl ether (PBDE) flame retardants, phenolic metabolites, and thyroid hormones during pregnancy. *Environ. Health Perspect.* **2011**, *119*, 1454–1459. [CrossRef]

26. Meng, S.; Chen, X.; Gyimah, E.; Xu, H.; Chen, J. Hepatic oxidative stress, DNA damage and apoptosis in adult zebrafish following sub-chronic exposure to BDE-47 and BDE-153. *Environ. Toxicol.* **2020**, *35*, 1202–1211. [CrossRef] [PubMed]
27. Casella, M.; Lori, G.; Coppola, L.; La Rocca, C.; Tait, S. BDE-47, -99, -209 and Their Ternary Mixture Disrupt Glucose and Lipid Metabolism of Hepg2 Cells at Dietary Relevant Concentrations: Mechanistic Insight through Integrated Transcriptomics and Proteomics Analysis. *Int. J. Mol. Sci.* **2022**, *23*, 14465. [CrossRef] [PubMed]
28. Schecter, A.; Johnson-Welch, S.; Tung, K.C.; Harris, T.R.; Päpke, O.; Rosen, R. Polybrominated diphenyl ether (PBDE) levels in livers of U.S. human fetuses and newborns. *J. Toxicol. Environ. Health. Part A* **2007**, *70*, 1–6. [CrossRef]
29. Kou, J.; Li, X.; Zhang, M.; Wang, L.; Hu, L.; Liu, X.; Mei, S.; Xu, G. Accumulative levels, temporal and spatial distribution of common chemical pollutants in the blood of Chinese adults. *Environ. Pollut.* **2022**, *311*, 119980. [CrossRef]
30. Zhao, X.; Chen, T.; Wang, D.; Du, Y.; Wang, Y.; Zhu, W.; Bekir, M.; Yu, D.; Shi, Z. Polybrominated diphenyl ethers and decabromodiphenyl ethane in paired hair/serum and nail/serum from corresponding chemical manufacturing workers and their correlations to thyroid hormones, liver and kidney injury markers. *Sci. Total Environ.* **2020**, *729*, 139049. [CrossRef]
31. Liu, L.Y.; Salamova, A.; He, K.; Hites, R.A. Analysis of polybrominated diphenyl ethers and emerging halogenated and organophosphate flame retardants in human hair and nails. *J. Chromatogr. A* **2015**, *1406*, 251–257. [CrossRef]
32. Tang, J.; Zhai, J.X. Distribution of polybrominated diphenyl ethers in breast milk, cord blood and placentas: A systematic review. *Environ. Sci. Pollut. Res. Int.* **2017**, *24*, 21548–21573. [CrossRef] [PubMed]
33. Cai, Z.; Hu, W.; Wu, R.; Zheng, S.; Wu, K. Bioinformatic analyses of hydroxylated polybrominated diphenyl ethers toxicities on impairment of adrenocortical secretory function. *Environ. Health Prev. Med.* **2022**, *27*, 38. [CrossRef] [PubMed]
34. Cheng, D.; Cao, K.; Wang, T.; Zhang, X.; Feng, M.; Liu, H. Evaluation of the oxidative stress in liver of crucian carp (Carassius auratus) exposed to 3,4,4′-tri-CDE, 2-MeO-3′,4,4′-tri-CDE, and 2-HO-3′,4,4′-tri-CDE. *Environ. Sci. Pollut. Res. Int.* **2019**, *26*, 5164–5175. [CrossRef] [PubMed]
35. Chaudhuri, D.; Ghate, N.B.; Panja, S.; Basu, T.; Shendge, A.K.; Mandal, N. Glycoside rich fraction from Spondias pinnata bark ameliorate iron overload induced oxidative stress and hepatic damage in Swiss albino mice. *BMC Complement. Altern. Med.* **2016**, *16*, 262. [CrossRef]
36. Lee, E.; Kim, T.H.; Choi, J.S.; Nabanata, P.; Kim, N.Y.; Ahn, M.Y.; Jung, K.K.; Kang, I.H.; Kim, T.S.; Kwack, S.J.; et al. Evaluation of liver and thyroid toxicity in Sprague-Dawley rats after exposure to polybrominated diphenyl ether BDE-209. *J. Toxicol. Sci.* **2010**, *35*, 535–545. [CrossRef]
37. Mercado-Feliciano, M.; Bigsby, R.M. The polybrominated diphenyl ether mixture DE-71 is mildly estrogenic. *Environ. Health Perspect.* **2008**, *116*, 605–611. [CrossRef] [PubMed]
38. Zhang, X.; Feng, M.; Liu, F.; Qin, L.; Qu, R.; Li, D.; Wang, Z. Subacute oral toxicity of BDE-15, CDE-15, and HODE-15 in ICR male mice: Assessing effects on hepatic oxidative stress and metals status and ascertaining the protective role of vitamin E. *Environ. Sci. Pollut. Res. Int.* **2014**, *21*, 1924–1935. [CrossRef]
39. Sanders, J.M.; Burka, L.T.; Smith, C.S.; Black, W.; James, R.; Cunningham, M.L. Differential expression of CYP1A, 2B, and 3A genes in the F344 rat following exposure to a polybrominated diphenyl ether mixture or individual components. *Toxicol. Sci. Off. J. Soc. Toxicol.* **2005**, *88*, 127–133. [CrossRef]
40. Sun, Z.; Zhang, Z.; Ji, M.; Yang, H.; Cromie, M.; Gu, J.; Wang, C.; Yang, L.; Yu, Y.; Gao, W.; et al. BDE47 induces rat CYP3A1 by targeting the transcriptional regulation of miR-23b. *Sci. Rep.* **2016**, *6*, 31958. [CrossRef]
41. Yang, J.; Zhu, J.; Chan, K.M. BDE-99, but not BDE-47, is a transient aryl hydrocarbon receptor agonist in zebrafish liver cells. *Toxicol. Appl. Pharmacol.* **2016**, *305*, 203–215. [CrossRef]
42. Saquib, Q.; Siddiqui, M.A.; Ahmed, J.; Al-Salim, A.; Ansari, S.M.; Faisal, M.; Al-Khedhairy, A.A.; Musarrat, J.; AlWathnani, H.A.; Alatar, A.A.; et al. Hazards of low dose flame-retardants (BDE-47 and BDE-32): Influence on transcriptome regulation and cell death in human liver cells. *J. Hazard. Mater.* **2016**, *308*, 37–49. [CrossRef]
43. Tang, S.; Liu, H.; Yin, H.; Liu, X.; Peng, H.; Lu, G.; Dang, Z.; He, C. Effect of 2, 2′, 4, 4′-tetrabromodiphenyl ether (BDE-47) and its metabolites on cell viability, oxidative stress, and apoptosis of HepG2. *Chemosphere* **2018**, *193*, 978–988. [CrossRef] [PubMed]
44. Shao, J.; White, C.C.; Dabrowski, M.J.; Kavanagh, T.J.; Eckert, M.L.; Gallagher, E.P. The role of mitochondrial and oxidative injury in BDE 47 toxicity to human fetal liver hematopoietic stem cells. *Toxicol. Sci. Off. J. Soc. Toxicol.* **2008**, *101*, 81–90. [CrossRef] [PubMed]
45. Shao, J.; Eckert, M.L.; Lee, L.E.; Gallagher, E.P. Comparative oxygen radical formation and toxicity of BDE 47 in rainbow trout cell lines. *Mar. Environ. Res.* **2008**, *66*, 7–8. [CrossRef] [PubMed]
46. Zhang, Z.F.; Shan, Q.; Zhuang, J.; Zhang, Y.Q.; Wang, X.; Fan, S.H.; Lu, J.; Wu, D.M.; Hu, B.; Zheng, Y.L. Troxerutin inhibits 2,2′,4,4′-tetrabromodiphenyl ether (BDE-47)-induced hepatocyte apoptosis by restoring proteasome function. *Toxicol. Lett.* **2015**, *233*, 246–257. [CrossRef] [PubMed]
47. Zhang, Z.F.; Zhang, Y.Q.; Fan, S.H.; Zhuang, J.; Zheng, Y.L.; Lu, J.; Wu, D.M.; Shan, Q.; Hu, B. Troxerutin protects against 2,2′,4,4′-tetrabromodiphenyl ether (BDE-47)-induced liver inflammation by attenuating oxidative stress-mediated NAD^+-depletion. *J. Hazard. Mater.* **2015**, *283*, 98–109. [CrossRef]
48. Albina, M.L.; Alonso, V.; Linares, V.; Bellés, M.; Sirvent, J.J.; Domingo, J.L.; Sánchez, D.J. Effects of exposure to BDE-99 on oxidative status of liver and kidney in adult rats. *Toxicology* **2010**, *271*, 51–56. [CrossRef]
49. Souza, A.O.; Pereira, L.C.; Oliveira, D.P.; Dorta, D.J. BDE-99 congener induces cell death by apoptosis of human hepatoblastoma cell line-HepG2. *Toxicol. Vitr. Int. J. Publ. Assoc. BIBRA* **2013**, *27*, 580–587. [CrossRef]

50. Che, S.; Chen, S.; Li, S.; Ruan, Z. Decabromodiphenyl ether initiates mitochondria-dependent apoptosis by disrupting calcium homeostasis in mice livers. *Chemosphere* **2022**, *291 Pt 1*, 132767. [CrossRef]
51. Wang, Y.; Xu, L.; Peng, L.; Fang, C.; Qin, Q.; Lv, X.; Liu, Z.; Yang, B.; Song, E.; Song, Y. Polybrominated diphenyl ethers quinone-induced intracellular protein oxidative damage triggers ubiquitin-proteasome and autophagy-lysosomal system activation in LO2 cells. *Chemosphere* **2021**, *275*, 130034. [CrossRef] [PubMed]
52. Yuan, J.; Sun, X.; Che, S.; Zhang, L.; Ruan, Z.; Li, X.; Yang, J. AhR-mediated CYP1A1 and ROS overexpression are involved in hepatotoxicity of decabromodiphenyl ether (BDE-209). *Toxicol. Lett.* **2021**, *352*, 26–33. [CrossRef] [PubMed]
53. Hu, X.Z.; Xu, Y.; Hu, D.C.; Hui, Y.; Yang, F.X. Apoptosis induction on human hepatoma cells Hep G2 of decabrominated diphenyl ether (PBDE-209). *Toxicol. Lett.* **2007**, *171*, 19–28. [CrossRef] [PubMed]
54. Zhao, A.; Liu, H.; Zhang, A.; Wang, X.; Zhang, H.; Wang, H. Effect of BDE-209 on glutathione system in Carassius auratus. *Environ. Toxicol. Pharmacol.* **2011**, *32*, 35–39. [CrossRef] [PubMed]
55. Wu, J.P.; Peng, Y.; Zhi, H.; Wu, S.K.; Chen, X.Y.; Zeng, Y.H.; Luo, X.J.; Mai, B.X. Contaminant-related oxidative distress in common kingfisher (Alcedo atthis) breeding at an e-waste site in South China. *Environ. Res.* **2020**, *182*, 109079. [CrossRef]
56. Yu, W.K.; Shi, Y.F.; Fong, C.C.; Chen, Y.; van de Merwe, J.P.; Chan, A.K.; Wei, F.; Bo, J.; Ye, R.; Au, D.W.; et al. Gender-specific transcriptional profiling of marine medaka (Oryzias melastigma) liver upon BDE-47 exposure. *Comp. Biochem. Physiol. Part D Genom. Proteom.* **2013**, *8*, 255–262. [CrossRef]
57. Liu, Z.L.; Jiang, S.R.; Fan, Y.; Wang, J.S.; Wang, M.L.; Li, M.Y. 2,2′,4,4′,5,5′-Hexabromophenyl ether (BDE-153) causes abnormal insulin secretion and disorders of glucose and lipid metabolism in mice. *J. Chin. Med. Assoc. JCMA* **2023**, *86*, 388–398. [CrossRef]
58. Suvorov, A.; Naumov, V.; Shtratnikova, V.; Logacheva, M.; Shershebnev, A.; Wu, H.; Gerasimov, E.; Zheludkevich, A.; Pilsner, J.R.; Sergeyev, O. Rat liver epigenome programing by perinatal exposure to 2,2′,4′4′-tetrabromodiphenyl ether. *Epigenomics* **2020**, *12*, 235–249. [CrossRef]
59. Nash, J.T.; Szabo, D.T.; Carey, G.B. Polybrominated diphenyl ethers alter hepatic phosphoenolpyruvate carboxykinase enzyme kinetics in male Wistar rats: Implications for lipid and glucose metabolism. *J. Toxicol. Environ. Health. Part A* **2013**, *76*, 142–156. [CrossRef]
60. Cowens, K.R.; Simpson, S.; Thomas, W.K.; Carey, G.B. Polybrominated Diphenyl Ether (PBDE)-Induced Suppression of Phosphoenolpyruvate Carboxykinase (PEPCK) Decreases Hepatic Glyceroneogenesis and Disrupts Hepatic Lipid Homeostasis. *J. Toxicol. Environ. Health. Part A* **2015**, *78*, 1437–1449. [CrossRef]
61. Kozlova, E.V.; Chinthirla, B.D.; Pérez, P.A.; DiPatrizio, N.V.; Argueta, D.A.; Phillips, A.L.; Stapleton, H.M.; González, G.M.; Krum, J.M.; Carrillo, V.; et al. Maternal transfer of environmentally relevant polybrominated diphenyl ethers (PBDEs) produces a diabetic phenotype and disrupts glucoregulatory hormones and hepatic endocannabinoids in adult mouse female offspring. *Sci. Rep.* **2020**, *10*, 18102. [CrossRef]
62. Zhu, Y.; Jing, L.; Li, X.; Zhou, G.; Zhang, Y.; Sang, Y.; Gao, L.; Liu, S.; Shi, Z.; Sun, Z.; et al. Decabromodiphenyl ether-induced PRKACA hypermethylation contributed to glycolipid metabolism disorder via regulating PKA/AMPK pathway in rat and L-02 cells. *Environ. Toxicol. Pharmacol.* **2022**, *90*, 103808. [CrossRef] [PubMed]
63. Gross, B.; Pawlak, M.; Lefebvre, P.; Staels, B. PPARs in obesity-induced T2DM, dyslipidaemia and NAFLD. *Nat. Rev. Endocrinol.* **2017**, *13*, 36–49. [CrossRef] [PubMed]
64. Abrha, A.; Suvorov, A. Transcriptomic Analysis of Gonadal Adipose Tissue in Male Mice Exposed Perinatally to 2,2′,4,4′-Tetrabromodiphenyl Ether (BDE-47). *Toxics* **2018**, *6*, 21. [CrossRef] [PubMed]
65. Fang, M.; Webster, T.F.; Ferguson, P.L.; Stapleton, H.M. Characterizing the peroxisome proliferator-activated receptor (PPARγ) ligand binding potential of several major flame retardants, their metabolites, and chemical mixtures in house dust. *Environ. Health Perspect.* **2015**, *123*, 166–172. [CrossRef]
66. Tung, E.W.; Boudreau, A.; Wade, M.G.; Atlas, E. Induction of adipocyte differentiation by polybrominated diphenyl ethers (PBDEs) in 3T3-L1 cells. *PLoS ONE* **2014**, *9*, e94583. [CrossRef] [PubMed]
67. Kamstra, J.H.; Hruba, E.; Blumberg, B.; Janesick, A.; Mandrup, S.; Hamers, T.; Legler, J. Transcriptional and epigenetic mechanisms underlying enhanced in vitro adipocyte differentiation by the brominated flame retardant BDE-47. *Environ. Sci. Technol.* **2014**, *48*, 4110–4119. [CrossRef]
68. Zhu, Y.; Jing, L.; Li, X.; Zheng, D.; Zhou, G.; Zhang, Y.; Sang, Y.; Shi, Z.; Sun, Z.; Zhou, X. Decabromodiphenyl ether disturbs hepatic glycolipid metabolism by regulating the PI3K/AKT/GLUT4 and mTOR/PPARγ/RXRα pathway in mice and L02 cells. *Sci. Total Environ.* **2021**, *763*, 142936. [CrossRef]
69. Zhu, Y.; Li, X.; Liu, J.; Zhou, G.; Yu, Y.; Jing, L.; Shi, Z.; Zhou, X.; Sun, Z. The effects of decabromodiphenyl ether on glycolipid metabolism and related signaling pathways in mice. *Chemosphere* **2019**, *222*, 849–855. [CrossRef]
70. Zhang, Z.; Sun, Z.Z.; Xiao, X.; Zhou, S.; Wang, X.C.; Gu, J.; Qiu, L.L.; Zhang, X.H.; Xu, Q.; Zhen, B.; et al. Mechanism of BDE209-induced impaired glucose homeostasis based on gene microarray analysis of adult rat liver. *Arch. Toxicol.* **2013**, *87*, 1557–1567. [CrossRef]
71. Mao, G.; Tang, J.; Liao, T.; Shi, X.; Dong, F.; Feng, W.; Chen, Y.; Zhao, T.; Wu, X.; Yang, L. Metabolism toxicity and susceptibility of decabromodiphenyl ether (BDE-209) exposure on BRL cells with insulin resistance. *Environ. Sci. Pollut. Res. Int.* **2022**, *29*, 91306–91324. [CrossRef] [PubMed]

72. Chen, F.; Feng, L.; Zheng, Y.L.; Lu, J.; Fan, S.H.; Shan, Q.; Zheng, G.H.; Wang, Y.J.; Wu, D.M.; Li, M.Q.; et al. 2, 2′, 4, 4′-tetrabromodiphenyl ether (BDE-47) induces mitochondrial dysfunction and related liver injury via eliciting miR-34a-5p-mediated mitophagy impairment. *Environ. Pollut.* **2020**, *258*, 113693. [CrossRef] [PubMed]
73. Pazin, M.; Pereira, L.C.; Dorta, D.J. Toxicity of brominated flame retardants, BDE-47 and BDE-99 stems from impaired mitochondrial bioenergetics. *Toxicol. Mech. Methods* **2015**, *25*, 34–41. [CrossRef]
74. Pereira, L.C.; Cabral Miranda, L.F.; Franco-Bernardes, M.F.; Tasso, M.J.; Duarte, F.V.; Inácio Varela, A.T.; Rolo, A.P.; Marques Palmeira, C.M.; Dorta, D.J. Mitochondrial damage and apoptosis: Key features in BDE-153-induced hepatotoxicity. *Chem.-Biol. Interact.* **2018**, *291*, 192–201. [CrossRef] [PubMed]
75. Pereira, L.C.; Souza, A.O.; Tasso, M.J.; Oliveira, A.M.C.; Duarte, F.V.; Palmeira, C.M.; Dorta, D.J. Exposure to decabromodiphenyl ether (BDE-209) produces mitochondrial dysfunction in rat liver and cell death. *J. Toxicol. Environ. Health. Part A* **2017**, *80*, 1129–1144. [CrossRef]
76. Dunnick, J.K.; Shockley, K.R.; Pandiri, A.R.; Kissling, G.E.; Gerrish, K.E.; Ton, T.V.; Wilson, R.E.; Brar, S.S.; Brix, A.E.; Waidyanatha, S.; et al. PBDE-47 and PBDE mixture (DE-71) toxicities and liver transcriptomic changes at PND 22 after in utero/postnatal exposure in the rat. *Arch. Toxicol.* **2018**, *92*, 3415–3433. [CrossRef]
77. Blanco, J.; Mulero, M.; Domingo, J.L.; Sanchez, D.J. Perinatal exposure to BDE-99 causes decreased protein levels of cyclin D1 via GSK3β activation and increased ROS production in rat pup livers. *Toxicol. Sci. Off. J. Soc. Toxicol.* **2014**, *137*, 491–498. [CrossRef] [PubMed]
78. Yang, C.; Zhu, L.; Kang, Q.; Lee, H.K.; Li, D.; Chung, A.C.K.; Cai, Z. Chronic exposure to tetrabromodiphenyl ether (BDE-47) aggravates hepatic steatosis and liver fibrosis in diet-induced obese mice. *J. Hazard. Mater.* **2019**, *378*, 120766. [CrossRef]
79. Wang, Q.; Chen, G.; Tian, L.; Kong, C.; Gao, D.; Chen, Y.; Junaid, M.; Wang, J. Neuro- and hepato-toxicity of polystyrene nanoplastics and polybrominated diphenyl ethers on early life stages of zebrafish. *Sci. Total Environ.* **2023**, *857 Pt 2*, 159567. [CrossRef] [PubMed]
80. Crump, D.; Chiu, S.; Egloff, C.; Kennedy, S.W. Effects of hexabromocyclododecane and polybrominated diphenyl ethers on mRNA expression in chicken (Gallus domesticus) hepatocytes. *Toxicol. Sci. Off. J. Soc. Toxicol.* **2008**, *106*, 479–487. [CrossRef]
81. Shimbo, T.; Dunnick, J.K.; Brix, A.; Mav, D.; Shah, R.; Roberts, J.D.; Wade, P.A. DNA Methylation Changes in Tbx3 in a Mouse Model Exposed to Polybrominated Diphenyl Ethers. *Int. J. Toxicol.* **2017**, *36*, 229–238. [CrossRef] [PubMed]
82. Søfteland, L.; Petersen, K.; Stavrum, A.K.; Wu, T.; Olsvik, P.A. Hepatic in vitro toxicity assessment of PBDE congeners BDE47, BDE153 and BDE154 in Atlantic salmon (*Salmo salar* L.). *Aquat. Toxicol.* **2011**, *105*, 246–263. [CrossRef] [PubMed]
83. Lim, J.J.; Dutta, M.; Dempsey, J.L.; Lehmler, H.J.; MacDonald, J.; Bammler, T.; Walker, C.; Kavanagh, T.J.; Gu, H.; Mani, S.; et al. Neonatal Exposure to BPA, BDE-99, and PCB Produces Persistent Changes in Hepatic Transcriptome Associated With Gut Dysbiosis in Adult Mouse Livers. *Toxicol. Sci. Off. J. Soc. Toxicol.* **2021**, *184*, 83–103. [CrossRef] [PubMed]
84. Li, C.Y.; Cui, J.Y. Regulation of protein-coding gene and long noncoding RNA pairs in liver of conventional and germ-free mice following oral PBDE exposure. *PLoS ONE* **2018**, *13*, e0201387. [CrossRef]
85. Sueyoshi, T.; Li, L.; Wang, H.; Moore, R.; Kodavanti, P.R.; Lehmler, H.J.; Negishi, M.; Birnbaum, L.S. Flame retardant BDE-47 effectively activates nuclear receptor CAR in human primary hepatocytes. *Toxicol. Sci. Off. J. Soc. Toxicol.* **2014**, *137*, 292–302. [CrossRef]
86. Zhang, A.; Li, C.Y.; Kelly, E.J.; Sheppard, L.; Cui, J.Y. Transcriptomic profiling of PBDE-exposed HepaRG cells unveils critical lncRNA- PCG pairs involved in intermediary metabolism. *PLoS ONE* **2020**, *15*, e0224644. [CrossRef]
87. Sun, Y.; Wang, Y.; Liang, B.; Chen, T.; Zheng, D.; Zhao, X.; Jing, L.; Zhou, X.; Sun, Z.; Shi, Z. Hepatotoxicity of decabromodiphenyl ethane (DBDPE) and decabromodiphenyl ether (BDE-209) in 28-day exposed Sprague-Dawley rats. *Sci. Total Environ.* **2020**, *705*, 135783. [CrossRef]
88. Wu, H.; Cao, L.; Li, F.; Lian, P.; Zhao, J. Multiple biomarkers of the cytotoxicity induced by BDE-47 in human embryonic kidney cells. *Chemosphere* **2015**, *126*, 32–39. [CrossRef]
89. Luan, P.; Zhang, H.; Chen, X.; Zhu, Y.; Hu, G.; Cai, J.; Zhang, Z. Melatonin relieves 2,2,4,4-tetrabromodiphenyl ether (BDE-47)-induced apoptosis and mitochondrial dysfunction through the AMPK-Sirt1-PGC-1α axis in fish kidney cells (CIK). *Ecotoxicol. Environ. Saf.* **2022**, *232*, 113276. [CrossRef]
90. Shan, Q.; Zhuang, J.; Zheng, G.; Zhang, Z.; Zhang, Y.; Lu, J.; Zheng, Y. Troxerutin Reduces Kidney Damage against BDE-47-Induced Apoptosis via Inhibiting NOX2 Activity and Increasing Nrf2 Activity. *Oxidative Med. Cell. Longev.* **2017**, *2017*, 6034692. [CrossRef]
91. Sun, S.; Jin, Y.; Yang, J.; Zhao, Z.; Rao, Q. Nephrotoxicity and possible mechanisms of decabrominated diphenyl ethers (BDE-209) exposure to kidney in broilers. *Ecotoxicol. Environ. Saf.* **2021**, *208*, 111638. [CrossRef] [PubMed]
92. Milovanovic, V.; Buha, A.; Matovic, V.; Curcic, M.; Vucinic, S.; Nakano, T.; Antonijevic, B. Oxidative stress and renal toxicity after subacute exposure to decabrominated diphenyl ether in Wistar rats. *Environ. Sci. Pollut. Res. Int.* **2018**, *25*, 7223–7230. [CrossRef] [PubMed]
93. Zhang, Y.; Hu, B.; Qian, X.; Xu, G.; Jin, X.; Chen, D.; Tang, J.; Xu, L. Transcriptomics-based analysis of co-exposure of cadmium (Cd) and 2,2′,4,4′-tetrabromodiphenyl ether (BDE-47) indicates mitochondrial dysfunction induces NLRP3 inflammasome and inflammatory cell death in renal tubular epithelial cells. *Ecotoxicol. Environ. Saf.* **2022**, *241*, 113790. [CrossRef] [PubMed]
94. Sun, S.; Zhao, Z.; Rao, Q.; Li, X.; Ruan, Z.; Yang, J. BDE-47 induces nephrotoxicity through ROS-dependent pathways of mitochondrial dynamics in PK15 cells. *Ecotoxicol. Environ. Saf.* **2021**, *222*, 112549. [CrossRef]

95. Li, X.; Zhang, H.; Qiao, S.; Ma, W.; Cai, J.; Zhang, X.; Zhang, Z. Melatonin administration alleviates 2,2,4,4-tetra-brominated diphenyl ether (PBDE-47)-induced necroptosis and secretion of inflammatory factors via miR-140-5p/TLR4/NF-κB axis in fish kidney cells. *Fish Shellfish Immunol.* **2022**, *128*, 228–237. [CrossRef]
96. Shan, Q.; Zheng, G.H.; Han, X.R.; Wen, X.; Wang, S.; Li, M.Q.; Zhuang, J.; Zhang, Z.F.; Hu, B.; Zhang, Y.; et al. Troxerutin Protects Kidney Tissue against BDE-47-Induced Inflammatory Damage through CXCR4-TXNIP/NLRP3 Signaling. *Oxidative Med. Cell. Longev.* **2018**, *2018*, 9865495. [CrossRef]
97. Chen, L.; Hu, C.; Lok-Shun Lai, N.; Zhang, W.; Hua, J.; Lam, P.K.S.; Lam, J.C.W.; Zhou, B. Acute exposure to PBDEs at an environmentally realistic concentration causes abrupt changes in the gut microbiota and host health of zebrafish. *Environ. Pollut.* **2018**, *240*, 17–26. [CrossRef]
98. Li, S.; Yuan, J.; Che, S.; Zhang, L.; Ruan, Z.; Sun, X. Decabromodiphenyl ether induces ROS-mediated intestinal toxicity through the Keap1-Nrf2 pathway. *J. Biochem. Mol. Toxicol.* **2022**, *36*, e22995. [CrossRef]
99. Gao, H.; Wan, X.; Xiao, B.; Yang, K.; Wang, Y.; Zhang, C.; Li, P.; Liu, L.; Xia, T.; Wang, A.; et al. Impacts of PBDE-47 exposure before, during and after pregnancy on the maternal gut microbiome and its association with host metabolism. *Ecotoxicol. Environ. Saf.* **2021**, *222*, 112530. [CrossRef]
100. Scoville, D.K.; Li, C.Y.; Wang, D.; Dempsey, J.L.; Raftery, D.; Mani, S.; Gu, H.; Cui, J.Y. Polybrominated Diphenyl Ethers and Gut Microbiome Modulate Metabolic Syndrome-Related Aqueous Metabolites in Mice. *Drug Metab. Dispos. Biol. Fate Chem.* **2019**, *47*, 928–940. [CrossRef]
101. Gomez, M.V.; Dutta, M.; Suvorov, A.; Shi, X.; Gu, H.; Mani, S.; Yue Cui, J. Early Life Exposure to Environmental Contaminants (BDE-47, TBBPA, and BPS) Produced Persistent Alterations in Fecal Microbiome in Adult Male Mice. *Toxicol. Sci. Off. J. Soc. Toxicol.* **2021**, *179*, 14–30. [CrossRef] [PubMed]
102. Popli, S.; Badgujar, P.C.; Agarwal, T.; Bhushan, B.; Mishra, V. Persistent organic pollutants in foods, their interplay with gut microbiota and resultant toxicity. *Sci. Total Environ.* **2022**, *832*, 155084. [CrossRef] [PubMed]
103. Wang, D.; Yan, J.; Teng, M.; Yan, S.; Zhou, Z.; Zhu, W. In utero and lactational exposure to BDE-47 promotes obesity development in mouse offspring fed a high-fat diet: Impaired lipid metabolism and intestinal dysbiosis. *Arch. Toxicol.* **2018**, *92*, 1847–1860. [CrossRef] [PubMed]
104. Li, C.Y.; Dempsey, J.L.; Wang, D.; Lee, S.; Weigel, K.M.; Fei, Q.; Bhatt, D.K.; Prasad, B.; Raftery, D.; Gu, H.; et al. PBDEs Altered Gut Microbiome and Bile Acid Homeostasis in Male C57BL/6 Mice. *Drug Metab. Dispos. Biol. Fate Chem.* **2018**, *46*, 1226–1240. [CrossRef]
105. Huang, F.; Wen, S.; Li, J.; Zhong, Y.; Zhao, Y.; Wu, Y. The human body burden of polybrominated diphenyl ethers and their relationships with thyroid hormones in the general population in Northern China. *Sci. Total Environ.* **2014**, *466–467*, 609–615. [CrossRef]
106. Hu, X.; Zhang, J.; Jiang, Y.; Lei, Y.; Lu, L.; Zhou, J.; Huang, H.; Fang, D.; Tao, G. Effect on metabolic enzymes and thyroid receptors induced by BDE-47 by activation the pregnane X receptor in HepG2, a human hepatoma cell line. *Toxicol. Vitr. Int. J. Publ. Assoc. BIBRA* **2014**, *28*, 1377–1385. [CrossRef]
107. Macaulay, L.J.; Chen, A.; Rock, K.D.; Dishaw, L.V.; Dong, W.; Hinton, D.E.; Stapleton, H.M. Developmental toxicity of the PBDE metabolite 6-OH-BDE-47 in zebrafish and the potential role of thyroid receptor β. *Aquat. Toxicol.* **2015**, *168*, 38–47. [CrossRef]
108. Chan, W.K.; Chan, K.M. Disruption of the hypothalamic-pituitary-thyroid axis in zebrafish embryo-larvae following waterborne exposure to BDE-47, TBBPA and BPA. *Aquat. Toxicol.* **2012**, *108*, 106–111. [CrossRef]
109. Chen, Q.; Yu, L.; Yang, L.; Zhou, B. Bioconcentration and metabolism of decabromodiphenyl ether (BDE-209) result in thyroid endocrine disruption in zebrafish larvae. *Aquat. Toxicol.* **2012**, *110–111*, 141–148. [CrossRef]
110. Ellis-Hutchings, R.G.; Cherr, G.N.; Hanna, L.A.; Keen, C.L. Polybrominated diphenyl ether (PBDE)-induced alterations in vitamin A and thyroid hormone concentrations in the rat during lactation and early postnatal development. *Toxicol. Appl. Pharmacol.* **2006**, *215*, 135–145. [CrossRef]
111. Fowles, J.R.; Fairbrother, A.; Baecher-Steppan, L.; Kerkvliet, N.I. Immunologic and endocrine effects of the flame-retardant pentabromodiphenyl ether (DE-71) in C57BL/6J mice. *Toxicology* **1994**, *86*, 49–61. [CrossRef] [PubMed]
112. Xu, X.; Liu, J.; Zeng, X.; Lu, F.; Chen, A.; Huo, X. Elevated serum polybrominated diphenyl ethers and alteration of thyroid hormones in children from Guiyu, China. *PLoS ONE* **2014**, *9*, e113699. [CrossRef] [PubMed]
113. Cowell, W.J.; Sjödin, A.; Jones, R.; Wang, Y.; Wang, S.; Whyatt, R.M.; Factor-Litvak, P.; Bradwin, G.; Hassoun, A.; Oberfield, S.; et al. Pre- and Postnatal Polybrominated Diphenyl Ether Concentrations in Relation to Thyroid Parameters Measured During Early Childhood. *Thyroid Off. J. Am. Thyroid Assoc.* **2019**, *29*, 631–641. [CrossRef] [PubMed]
114. Chen, T.; Niu, P.; Kong, F.; Wang, Y.; Bai, Y.; Yu, D.; Jia, J.; Yang, L.; Fu, Z.; Li, R.; et al. Disruption of thyroid hormone levels by decabrominated diphenyl ethers (BDE-209) in occupational workers from a deca-BDE manufacturing plant. *Environ. Int.* **2018**, *120*, 505–515. [CrossRef]
115. Cheng, L.; Yang, J.; Rao, Q.; Liu, Z.; Song, W.; Guan, S.; Zhao, Z.; Song, W. Toxic effects of Decabromodiphenyl ether (BDE-209) on thyroid of broiler chicks by transcriptome profile analysis. *Ecotoxicol. Environ. Saf.* **2021**, *219*, 112305. [CrossRef]
116. Yang, J.; Chan, K.M. Evaluation of the toxic effects of brominated compounds (BDE-47, 99, 209, TBBPA) and bisphenol A (BPA) using a zebrafish liver cell line, ZFL. *Aquat. Toxicol.* **2015**, *159*, 138–147. [CrossRef]
117. Feng, C.; Xu, Y.; Zhao, G.; Zha, J.; Wu, F.; Wang, Z. Relationship between BDE 209 metabolites and thyroid hormone levels in rainbow trout (Oncorhynchus mykiss). *Aquat. Toxicol.* **2012**, *122–123*, 28–35. [CrossRef]

118. Ibhazehiebo, K.; Iwasaki, T.; Kimura-Kuroda, J.; Miyazaki, W.; Shimokawa, N.; Koibuchi, N. Disruption of thyroid hormone receptor-mediated transcription and thyroid hormone-induced Purkinje cell dendrite arborization by polybrominated diphenyl ethers. *Environ. Health Perspect.* **2011**, *119*, 168–175. [CrossRef]
119. Noyes, P.D.; Lema, S.C.; Macaulay, L.J.; Douglas, N.K.; Stapleton, H.M. Low level exposure to the flame retardant BDE-209 reduces thyroid hormone levels and disrupts thyroid signaling in fathead minnows. *Environ. Sci. Technol.* **2013**, *47*, 10012–10021. [CrossRef]
120. Zhou, T.; Ross, D.G.; DeVito, M.J.; Crofton, K.M. Effects of short-term in vivo exposure to polybrominated diphenyl ethers on thyroid hormones and hepatic enzyme activities in weanling rats. *Toxicol. Sci. Off. J. Soc. Toxicol.* **2001**, *61*, 76–82. [CrossRef]
121. Richardson, V.M.; Staskal, D.F.; Ross, D.G.; Diliberto, J.J.; DeVito, M.J.; Birnbaum, L.S. Possible mechanisms of thyroid hormone disruption in mice by BDE 47, a major polybrominated diphenyl ether congener. *Toxicol. Appl. Pharmacol.* **2008**, *226*, 244–250. [CrossRef] [PubMed]
122. Li, P.; Liu, L.; Zhou, G.; Tian, Z.; Luo, C.; Xia, T.; Chen, J.; Niu, Q.; Dong, L.; Zhao, Q.; et al. Perigestational exposure to low doses of PBDE-47 induces excessive ER stress, defective autophagy and the resultant apoptosis contributing to maternal thyroid toxicity. *Sci. Total Environ.* **2018**, *645*, 363–371. [CrossRef] [PubMed]
123. Li, P.; Gao, H.; Dong, L.; Liu, L.; Zhou, G.; Luo, C.; Tian, Z.; Xia, T.; Wang, A.; Zhang, S. Perinatal low-dose PBDE-47 exposure hampered thyroglobulin turnover and induced thyroid cell apoptosis by triggering ER stress and lysosomal destabilization contributing to thyroid toxicity in adult female rats. *J. Hazard. Mater.* **2020**, *392*, 122265. [CrossRef]
124. Ma, R.; Li, B.; Zhang, C.; Lei, R.; Zhang, X.; Wang, C.; Zhang, S.; Wang, A. Roles of endoplasmic reticulum stress in 2,2′,4,4′-tetrabromodiphenylether-induced thyroid cell apoptosis and autophagy. *Ecotoxicol. Environ. Saf.* **2021**, *214*, 112077. [CrossRef]
125. Wang, Y.; Chen, T.; Sun, Y.; Zhao, X.; Zheng, D.; Jing, L.; Zhou, X.; Sun, Z.; Shi, Z. A comparison of the thyroid disruption induced by decabrominated diphenyl ethers (BDE-209) and decabromodiphenyl ethane (DBDPE) in rats. *Ecotoxicol. Environ. Saf.* **2019**, *174*, 224–235. [CrossRef] [PubMed]
126. Técher, R.; Houde, M.; Verreault, J. Associations between organohalogen concentrations and transcription of thyroid-related genes in a highly contaminated gull population. *Sci. Total Environ.* **2016**, *545–546*, 289–298. [CrossRef]
127. Arkoosh, M.R.; Van Gaest, A.L.; Strickland, S.A.; Hutchinson, G.P.; Krupkin, A.B.; Dietrich, J.P. Alteration of thyroid hormone concentrations in juvenile Chinook salmon (Oncorhynchus tshawytscha) exposed to polybrominated diphenyl ethers, BDE-47 and BDE-99. *Chemosphere* **2017**, *171*, 1–8. [CrossRef]
128. Lema, S.C.; Dickey, J.T.; Schultz, I.R.; Swanson, P. Dietary exposure to 2,2′,4,4′-tetrabromodiphenyl ether (PBDE-47) alters thyroid status and thyroid hormone-regulated gene transcription in the pituitary and brain. *Environ. Health Perspect.* **2008**, *116*, 1694–1699. [CrossRef]
129. Han, Z.; Li, Y.; Zhang, S.; Song, N.; Xu, H.; Dang, Y.; Liu, C.; Giesy, J.P.; Yu, H. Prenatal transfer of decabromodiphenyl ether (BDE-209) results in disruption of the thyroid system and developmental toxicity in zebrafish offspring. *Aquat. Toxicol.* **2017**, *190*, 46–52. [CrossRef]
130. Mortensen, Å.K.; Verreault, J.; François, A.; Houde, M.; Giraudo, M.; Dam, M.; Jenssen, B.M. Flame retardants and their associations with thyroid hormone-related variables in northern fulmars from the Faroe Islands. *Sci. Total Environ.* **2022**, *806 Pt 2*, 150506. [CrossRef]
131. Kang, H.M.; Lee, Y.H.; Kim, B.M.; Kim, I.C.; Jeong, C.B.; Lee, J.S. Adverse effects of BDE-47 on in vivo developmental parameters, thyroid hormones, and expression of hypothalamus-pituitary-thyroid (HPT) axis genes in larvae of the self-fertilizing fish Kryptolebias marmoratus. *Chemosphere* **2017**, *176*, 39–46. [CrossRef] [PubMed]
132. Wu, L.; Li, Y.; Ru, H.; Xie, H.; Yao, F.; Ni, Z.; Zhong, L. Parental exposure to 2,2′,4,4′ -pentain polybrominated diphenyl ethers (BDE-99) causes thyroid disruption and developmental toxicity in zebrafish. *Toxicol. Appl. Pharmacol.* **2019**, *372*, 11–18. [CrossRef] [PubMed]
133. Yu, L.; Deng, J.; Shi, X.; Liu, C.; Yu, K.; Zhou, B. Exposure to DE-71 alters thyroid hormone levels and gene transcription in the hypothalamic-pituitary-thyroid axis of zebrafish larvae. *Aquat. Toxicol.* **2010**, *97*, 226–233. [CrossRef] [PubMed]
134. Zheng, X.; Zhu, Y.; Liu, C.; Liu, H.; Giesy, J.P.; Hecker, M.; Lam, M.H.; Yu, H. Accumulation and biotransformation of BDE-47 by zebrafish larvae and teratogenicity and expression of genes along the hypothalamus-pituitary-thyroid axis. *Environ. Sci. Technol.* **2012**, *46*, 12943–12951. [CrossRef] [PubMed]
135. Vuong, A.M.; Webster, G.M.; Romano, M.E.; Braun, J.M.; Zoeller, R.T.; Hoofnagle, A.N.; Sjödin, A.; Yolton, K.; Lanphear, B.P.; Chen, A. Maternal Polybrominated Diphenyl Ether (PBDE) Exposure and Thyroid Hormones in Maternal and Cord Sera: The HOME Study, Cincinnati, USA. *Environ. Health Perspect.* **2015**, *123*, 1079–1085. [CrossRef]
136. Li, Z.M.; Hernandez-Moreno, D.; Main, K.M.; Skakkebæk, N.E.; Kiviranta, H.; Toppari, J.; Feldt-Rasmussen, U.; Shen, H.; Schramm, K.W.; De Angelis, M. Association of In Utero Persistent Organic Pollutant Exposure With Placental Thyroid Hormones. *Endocrinology* **2018**, *159*, 3473–3481. [CrossRef]
137. Li, Z.M.; Albrecht, M.; Fromme, H.; Schramm, K.W.; De Angelis, M. Persistent Organic Pollutants in Human Breast Milk and Associations with Maternal Thyroid Hormone Homeostasis. *Environ. Sci. Technol.* **2020**, *54*, 1111–1119. [CrossRef] [PubMed]
138. Zheng, J.; He, C.T.; Chen, S.J.; Yan, X.; Guo, M.N.; Wang, M.H.; Yu, Y.J.; Yang, Z.Y.; Mai, B.X. Disruption of thyroid hormone (TH) levels and TH-regulated gene expression by polybrominated diphenyl ethers (PBDEs), polychlorinated biphenyls (PCBs), and hydroxylated PCBs in e-waste recycling workers. *Environ. Int.* **2017**, *102*, 138–144. [CrossRef]

139. Ernest, S.R.; Wade, M.G.; Lalancette, C.; Ma, Y.Q.; Berger, R.G.; Robaire, B.; Hales, B.F. Effects of chronic exposure to an environmentally relevant mixture of brominated flame retardants on the reproductive and thyroid system in adult male rats. *Toxicol. Sci. Off. J. Soc. Toxicol.* **2012**, *127*, 496–507. [CrossRef]
140. Makey, C.M.; McClean, M.D.; Braverman, L.E.; Pearce, E.N.; He, X.M.; Sjödin, A.; Weinberg, J.M.; Webster, T.F. Polybrominated Diphenyl Ether Exposure and Thyroid Function Tests in North American Adults. *Environ. Health Perspect.* **2016**, *124*, 420–425. [CrossRef]
141. Turyk, M.E.; Persky, V.W.; Imm, P.; Knobeloch, L.; Chatterton, R.; Anderson, H.A. Hormone disruption by PBDEs in adult male sport fish consumers. *Environ. Health Perspect.* **2008**, *116*, 1635–1641. [CrossRef] [PubMed]
142. Lin, S.M.; Chen, F.A.; Huang, Y.F.; Hsing, L.L.; Chen, L.L.; Wu, L.S.; Liu, T.S.; Chang-Chien, G.P.; Chen, K.C.; Chao, H.R. Negative associations between PBDE levels and thyroid hormones in cord blood. *Int. J. Hyg. Environ. Health* **2011**, *214*, 115–120. [CrossRef] [PubMed]
143. Vandenberg, L.N.; Colborn, T.; Hayes, T.B.; Heindel, J.J.; Jacobs, D.R., Jr.; Lee, D.H.; Shioda, T.; Soto, A.M.; vom Saal, F.S.; Welshons, W.V.; et al. Hormones and endocrine-disrupting chemicals: Low-dose effects and nonmonotonic dose responses. *Endocr. Rev.* **2012**, *33*, 378–455. [CrossRef]
144. Blake, C.A.; McCoy, G.L.; Hui, Y.Y.; LaVoie, H.A. Perinatal exposure to low-dose DE-71 increases serum thyroid hormones and gonadal osteopontin gene expression. *Exp. Biol. Med.* **2011**, *236*, 445–455. [CrossRef]
145. Herbstman, J.B.; Sjödin, A.; Apelberg, B.J.; Witter, F.R.; Halden, R.U.; Patterson, D.G.; Panny, S.R.; Needham, L.L.; Goldman, L.R. Birth delivery mode modifies the associations between prenatal polychlorinated biphenyl (PCB) and polybrominated diphenyl ether (PBDE) and neonatal thyroid hormone levels. *Environ. Health Perspect.* **2008**, *116*, 1376–1382. [CrossRef] [PubMed]
146. Wang, Q.; Li, Y.; Chen, Y.; Tian, L.; Gao, D.; Liao, H.; Kong, C.; Chen, X.; Junaid, M.; Wang, J. Toxic effects of polystyrene nanoplastics and polybrominated diphenyl ethers to zebrafish (Danio rerio). *Fish Shellfish Immunol.* **2022**, *126*, 21–33. [CrossRef] [PubMed]
147. Qin, C.; Cheng, Y.; Shang, X.; Wang, R.; Wang, R.; Hao, X.; Li, S.; Wang, Y.; Li, Y.; Liu, X.; et al. Disturbance of the Dlk1-Dio3 imprinted domain may underlie placental Dio3 suppression and extracellular thyroid hormone disturbance in placenta-derived JEG-3 cells following decabromodiphenyl ether (BDE209) exposure. *Toxicology* **2021**, *458*, 152837. [CrossRef]
148. Zhu, Y.; Tan, Y.Q.; Leung, L.K. Exposure to 2,2′,4,4′-tetrabromodiphenyl ether at late gestation modulates placental signaling molecules in the mouse model. *Chemosphere* **2017**, *181*, 289–295. [CrossRef]
149. Wang, C.; Yang, L.; Hu, Y.; Zhu, J.; Xia, R.; Yu, Y.; Shen, J.; Zhang, Z.; Wang, S.L. Isoliquiritigenin as an antioxidant phytochemical ameliorates the developmental anomalies of zebrafish induced by 2,2′,4,4′-tetrabromodiphenyl ether. *Sci. Total Environ.* **2019**, *666*, 390–398. [CrossRef]
150. Zhuang, J.; Pan, Z.J.; Mengqiu, L.; Hong, F.S.; Zhu, C.K.; Wu, N.; Chang, G.; Wang, H.; Zhao, X.X. BDE-47 induced apoptosis in zebrafish embryos through mitochondrial ROS-mediated JNK signaling. *Chemosphere* **2020**, *258*, 127385. [CrossRef]
151. Du, L.; Sun, W.; Zhang, H.; Chen, D. BDE-209 inhibits pluripotent genes expression and induces apoptosis in human embryonic stem cells. *J. Appl. Toxicol. JAT* **2016**, *36*, 659–668. [CrossRef] [PubMed]
152. Xu, Q.; Yu, M.; Zhou, Y.; Huang, Z.; Huang, X.; Xu, B.; Zhou, K.; Chen, X.; Xia, Y.; Wang, X.; et al. Effects of 2,2′,4,4′-tetrabromodiphenyl ether on the development of mouse embryonic stem cells. *Reprod. Toxicol.* **2021**, *106*, 18–24. [CrossRef] [PubMed]
153. Chao, S.J.; Huang, C.P.; Chen, P.C.; Huang, C. Teratogenic responses of zebrafish embryos to decabromodiphenyl ether (BDE-209) in the presence of nano-SiO(2) particles. *Chemosphere* **2017**, *178*, 449–457. [CrossRef]
154. Berger, R.G.; Lefèvre, P.L.; Ernest, S.R.; Wade, M.G.; Ma, Y.Q.; Rawn, D.F.; Gaertner, D.W.; Robaire, B.; Hales, B.F. Exposure to an environmentally relevant mixture of brominated flame retardants affects fetal development in Sprague-Dawley rats. *Toxicology* **2014**, *320*, 56–66. [CrossRef]
155. Qin, L.; Liu, F.; Liu, H.; Wei, Z.; Sun, P.; Wang, Z. Evaluation of HODE-15, FDE-15, CDE-15, and BDE-15 toxicity on adult and embryonic zebrafish (Danio rerio). *Environ. Sci. Pollut. Res. Int.* **2014**, *21*, 14047–14057. [CrossRef] [PubMed]
156. Rattner, B.A.; Lazarus, R.S.; Heinz, G.H.; Karouna-Renier, N.K.; Schultz, S.L.; Hale, R.C. Comparative embryotoxicity of a pentabrominated diphenyl ether mixture to common terns (Sterna hirundo) and American kestrels (Falco sparverius). *Chemosphere* **2013**, *93*, 441–447. [CrossRef]
157. Tanaka, Y.; Fujiwara, M.; Shindo, A.; Yin, G.; Kitazawa, T.; Teraoka, H. Aroclor 1254 and BDE-47 inhibit dopaminergic function manifesting as changes in locomotion behaviors in zebrafish embryos. *Chemosphere* **2018**, *193*, 1207–1215. [CrossRef]
158. Xing, X.; Kang, J.; Qiu, J.; Zhong, X.; Shi, X.; Zhou, B.; Wei, Y. Waterborne exposure to low concentrations of BDE-47 impedes early vascular development in zebrafish embryos/larvae. *Aquat. Toxicol.* **2018**, *203*, 19–27. [CrossRef]
159. Kim, S.; Cho, Y.H.; Won, S.; Ku, J.L.; Moon, H.B.; Park, J.; Choi, G.; Kim, S.; Choi, K. Maternal exposures to persistent organic pollutants are associated with DNA methylation of thyroid hormone-related genes in placenta differently by infant sex. *Environ. Int.* **2019**, *130*, 104956. [CrossRef]
160. Liu, W.; Li, S.; Zhou, Q.; Fu, Z.; Liu, P.; Cao, X.; Xi, S. 2, 2′, 4, 4′-tetrabromodiphenyl ether induces placental toxicity via activation of p38 MAPK signaling pathway in vivo and in vitro. *Ecotoxicol. Environ. Saf.* **2022**, *244*, 114034. [CrossRef]
161. Robinson, J.F.; Kapidzic, M.; Hamilton, E.G.; Chen, H.; Puckett, K.W.; Zhou, Y.; Ona, K.; Parry, E.; Wang, Y.; Park, J.S.; et al. Genomic Profiling of BDE-47 Effects on Human Placental Cytotrophoblasts. *Toxicol. Sci. Off. J. Soc. Toxicol.* **2019**, *167*, 211–226. [CrossRef] [PubMed]

162. Zhao, Y.; Song, Q.; Ge, W.; Jin, Y.; Chen, S.; Zhao, Y.; Xiao, X.; Zhang, Y. Associations between in utero exposure to polybrominated diphenyl ethers, pathophysiological state of fetal growth and placental DNA methylation changes. *Environ. Int.* **2019**, *133 Pt B*, 105255. [CrossRef]
163. Chi, Y.; Xia, H.; Su, M.; Song, P.; Qi, X.; Cui, Y.; Cao, Y.; Chen, T.; Qiu, Y.; Zhao, A.; et al. Metabonomic phenotyping reveals an embryotoxicity of deca-brominated diphenyl ether in mice. *Chem. Res. Toxicol.* **2011**, *24*, 1976–1983. [CrossRef]
164. Du, P.; Li, Z.; Du, L.; Zhang, H.; Zhou, Y.; Sun, W.; Xiao, X.; He, Y.; Sun, B.; Yu, Y.; et al. The effects of PBDE-209 exposure during pregnancy on placental ET-1 and eNOS expression and the birth weight of offspring. *Int. J. Dev. Neurosci. Off. J. Int. Soc. Dev. Neurosci.* **2015**, *43*, 86–91. [CrossRef] [PubMed]
165. Zhao, Y.; Fan, K.; Zhu, Y.; Zhao, Y.; Cai, J.; Jin, L. Gestational exposure to BDE-209 induces placental injury via the endoplasmic reticulum stress-mediated PERK/ATF4/CHOP signaling pathway. *Ecotoxicol. Environ. Saf.* **2022**, *233*, 113307. [CrossRef]
166. Zezza, D.; Tait, S.; Della Salda, L.; Amorena, M.; Merola, C.; Perugini, M. Toxicological, gene expression and histopathological evaluations of environmentally realistic concentrations of polybrominated diphenyl ethers PBDE- 47, PBDE-99 and PBDE-209 on zebrafish embryos. *Ecotoxicol. Environ. Saf.* **2019**, *183*, 109566. [CrossRef] [PubMed]
167. Li, X.; Zhu, Y.; Zhang, C.; Liu, J.; Zhou, G.; Jing, L.; Shi, Z.; Sun, Z.; Zhou, X. BDE-209 induces male reproductive toxicity via cell cycle arrest and apoptosis mediated by DNA damage response signaling pathways. *Environ. Pollut.* **2019**, *255 Pt 1*, 113097. [CrossRef]
168. Tseng, L.H.; Hsu, P.C.; Lee, C.W.; Tsai, S.S.; Pan, M.H.; Li, M.H. Developmental exposure to decabrominated diphenyl ether (BDE-209): Effects on sperm oxidative stress and chromatin DNA damage in mouse offspring. *Environ. Toxicol.* **2013**, *28*, 380–389. [CrossRef]
169. Li, X.; Liu, J.; Zhou, G.; Sang, Y.; Zhang, Y.; Jing, L.; Shi, Z.; Zhou, X.; Sun, Z. BDE-209 and DBDPE induce male reproductive toxicity through telomere-related cell senescence and apoptosis in SD rat. *Environ. Int.* **2021**, *146*, 106307. [CrossRef]
170. Zhang, Y.; Li, X.; Jing, L.; Zhou, G.; Sang, Y.; Gao, L.; Jiang, S.; Shi, Z.; Ge, W.; Sun, Z.; et al. Decabromodiphenyl ether induces male reproductive toxicity by activating mitochondrial apoptotic pathway through glycolipid metabolism dysbiosis. *Chemosphere* **2021**, *285*, 131512. [CrossRef]
171. Sarkar, D.; Singh, S.K. Maternal exposure to polybrominated diphenyl ether (BDE-209) during lactation affects germ cell survival with altered testicular glucose homeostasis and oxidative status through down-regulation of Cx43 and p27Kip1 in prepubertal mice offspring. *Toxicology* **2017**, *386*, 103–119. [CrossRef] [PubMed]
172. Sarkar, D.; Singh, S.K. Inhibition of testicular steroidogenesis and impaired differentiation of Sertoli cells in peripubertal mice offspring following maternal exposure to BDE-209 during lactation suppress germ cell proliferation. *Toxicol. Lett.* **2018**, *290*, 83–96. [CrossRef] [PubMed]
173. Sarkar, D.; Singh, V.K.; Singh, S.K. Maternal BDE-209 exposure during lactation perturbs steroidogenesis, germ cell kinetics and THRα1 expression in testes of prepubertal mice offspring. *Food Chem. Toxicol. Int. J. Publ. Br. Ind. Biol. Res. Assoc.* **2018**, *122*, 104–119. [CrossRef]
174. Sarkar, D.; Joshi, D.; Singh, S.K. Maternal BDE-209 exposure during lactation causes testicular and epididymal toxicity through increased oxidative stress in peripubertal mice offspring. *Toxicol. Lett.* **2019**, *311*, 66–79. [CrossRef] [PubMed]
175. Sarkar, D.; Singh, S.K. Decabromodiphenyl ether (BDE-209) exposure to lactating mice perturbs steroidogenesis and spermatogenesis in adult male offspring. *Ecotoxicol. Environ. Saf.* **2021**, *209*, 111783. [CrossRef]
176. Wang, H.; Tang, X.; Sha, J.; Chen, H.; Sun, T.; Wang, Y. The reproductive toxicity on the rotifer Brachionus plicatilis induced by BDE-47 and studies on the effective mechanism based on antioxidant defense system changes. *Chemosphere* **2015**, *135*, 129–137. [CrossRef]
177. Zhang, Z.; Zhang, X.; Sun, Z.; Dong, H.; Qiu, L.; Gu, J.; Zhou, J.; Wang, X.; Wang, S.L. Cytochrome P450 3A1 mediates 2,2′,4,4′-tetrabromodiphenyl ether-induced reduction of spermatogenesis in adult rats. *PLoS ONE* **2013**, *8*, e66301. [CrossRef]
178. Sun, M.H.; Li, X.H.; Xu, Y.; Xu, Y.; Sun, S.C. Exposure to PBDE47 affects mouse oocyte quality via mitochondria dysfunction-induced oxidative stress and apoptosis. *Ecotoxicol. Environ. Saf.* **2020**, *198*, 110662. [CrossRef]
179. Huang, S.; Cui, Y.; Guo, X.; Wang, L.; Li, S.; Lu, Y.; Bi, Y.; Huang, X.; Lin, M.; Xia, Y.; et al. 2,2′,4,4′-Tetrabromodiphenyl ether disrupts spermatogenesis, impairs mitochondrial function and induces apoptosis of early leptotene spermatocytes in rats. *Reprod. Toxicol.* **2015**, *51*, 114–124. [CrossRef]
180. Huang, S.; Wang, J.; Cui, Y. 2,2′,4,4′-Tetrabromodiphenyl ether injures cell viability and mitochondrial function of mouse spermatocytes by decreasing mitochondrial proteins Atp5b and Uqcrc1. *Environ. Toxicol. Pharmacol.* **2016**, *46*, 301–310. [CrossRef]
181. Zhao, T.; Tang, X.; Li, D.; Zhao, J.; Zhou, R.; Shu, F.; Jia, W.; Fu, W.; Xia, H.; Liu, G. Prenatal exposure to environmentally relevant levels of PBDE-99 leads to testicular dysgenesis with steroidogenesis disorders. *J. Hazard. Mater.* **2022**, *424 Pt B*, 127547. [CrossRef]
182. Gregoraszczuk, E.L.; Siembida, M.; Grzyb, D.; Rak-Mardyła, A. Polybrominated diphenylethers (PBDEs) act as apoptotic factors in the corpus luteum in addition to having a short-term stimulatory effect on progesterone secretion by luteal cells. *Toxicol. Mech. Methods* **2012**, *22*, 131–138. [CrossRef] [PubMed]
183. Chen, X.; Dong, Y.; Tian, E.; Xie, L.; Wang, G.; Li, X.; Chen, X.; Chen, Y.; Lv, Y.; Ni, C.; et al. 4-Bromodiphenyl ether delays pubertal Leydig cell development in rats. *Chemosphere* **2018**, *211*, 986–997. [CrossRef]
184. Li, Y.; Ma, F.; Li, Z.; Yu, Y.; Yan, H.; Tahir, A.; Zheng, W.; Li, X.; Huang, T.; Ge, R.S. Exposure to 4-bromodiphenyl ether during pregnancy blocks testis development in male rat fetuses. *Toxicol. Lett.* **2021**, *342*, 38–49. [CrossRef] [PubMed]

185. You, X.; Xi, J.; Cao, Y.; Zhang, J.; Luan, Y. 4-Bromodiphenyl Ether Induces Germ Cell Apoptosis by Induction of ROS and DNA Damage in Caenorhabditis elegans. *Toxicol. Sci. Off. J. Soc. Toxicol.* **2017**, *157*, 510–518. [CrossRef] [PubMed]
186. You, X.; Xi, J.; Liu, W.; Cao, Y.; Tang, W.; Zhang, X.; Yu, Y.; Luan, Y. 2,2′,4,4′-tetrabromodiphenyl ether induces germ cell apoptosis through oxidative stress by a MAPK-mediated p53-independent pathway. *Environ. Pollut.* **2018**, *242 Pt A*, 887–893. [CrossRef]
187. Khalil, A.; Parker, M.; Brown, S.E.; Cevik, S.E.; Guo, L.W.; Jensen, J.; Olmsted, A.; Portman, D.; Wu, H.; Suvorov, A. Perinatal exposure to 2,2′,4′4′ -Tetrabromodiphenyl ether induces testicular toxicity in adult rats. *Toxicology* **2017**, *389*, 21–30. [CrossRef] [PubMed]
188. Suvorov, A.; Shershebnev, A.; Wu, H.; Medvedeva, Y.; Sergeyev, O.; Pilsner, J.R. Perinatal exposure to low dose 2,2′,4,4′-tetrabromodiphenyl ether (BDE-47) alters sperm DNA methylation in adult rats. *Reprod. Toxicol.* **2018**, *75*, 136–143. [CrossRef]
189. Hsu, P.C.; Li, Z.K.; Lai, C.S.; Tseng, L.H.; Lee, C.W.; Cheng, F.J.; Chang, C.Y.; Chen, J.R. Transgenerational effects of BDE-209 on male reproduction in F3 offspring rats. *Chemosphere* **2021**, *272*, 129829. [CrossRef]
190. Li, X.; Zhang, Y.; Gao, L.; Yang, X.; Zhou, G.; Sang, Y.; Xue, J.; Shi, Z.; Sun, Z.; Zhou, X. BDE-209 induced spermatogenesis disorder by inhibiting SETD8/H4K20me1 related histone methylation in mice. *Sci. Total Environ.* **2023**, *864*, 161162. [CrossRef]
191. Li, X.; Zhang, Y.; Dong, X.; Zhou, G.; Sang, Y.; Gao, L.; Zhou, X.; Sun, Z. DNA methylation changes induced by BDE-209 are related to DNA damage response and germ cell development in GC-2spd. *J. Environ. Sci.* **2021**, *109*, 161–170. [CrossRef] [PubMed]
192. Abdelouahab, N.; Ainmelk, Y.; Takser, L. Polybrominated diphenyl ethers and sperm quality. *Reprod. Toxicol.* **2011**, *31*, 546–550. [CrossRef]
193. Albert, O.; Huang, J.Y.; Aleksa, K.; Hales, B.F.; Goodyer, C.G.; Robaire, B.; Chevrier, J.; Chan, P. Exposure to polybrominated diphenyl ethers and phthalates in healthy men living in the greater Montreal area: A study of hormonal balance and semen quality. *Environ. Int.* **2018**, *116*, 165–175. [CrossRef] [PubMed]
194. Jenkins, J.A.; Olivier, H.M.; Draugelis-Dale, R.O.; Eilts, B.E.; Torres, L.; Patiño, R.; Nilsen, E.; Goodbred, S.L. Assessing reproductive and endocrine parameters in male largescale suckers (Catostomus macrocheilus) along a contaminant gradient in the lower Columbia River, USA. *Sci. Total Environ.* **2014**, *484*, 365–378. [CrossRef] [PubMed]
195. Li, S.; Che, S.; Chen, S.; Ruan, Z.; Zhang, L. Hesperidin partly ameliorates the decabromodiphenyl ether-induced reproductive toxicity in pubertal mice. *Environ. Sci. Pollut. Res. Int.* **2022**, *29*, 90391–90403. [CrossRef] [PubMed]
196. Huang, S.; Xia, J.; Zhang, X.; Zhou, T.; Wang, J.; Liu, T.; Xu, S.; Liang, G. 2,2′,4,4′-Tetrabromodiphenyl ether disrupts spermatogenesis in mice by interfering with the ER-Nrf1-Tfam-mitochondria pathway. *Toxicol. Ind. Health* **2022**, *38*, 182–191. [CrossRef]
197. Shan, A.; Li, M.; Li, X.; Li, Y.; Yan, M.; Xian, P.; Chang, Y.; Chen, X.; Tang, N.J. BDE-47 Decreases Progesterone Levels in BeWo Cells by Interfering with Mitochondrial Functions and Genes Related to Cholesterol Transport. *Chem. Res. Toxicol.* **2019**, *32*, 621–628. [CrossRef]
198. Talsness, C.E.; Shakibaei, M.; Kuriyama, S.N.; Grande, S.W.; Sterner-Kock, A.; Schnitker, P.; de Souza, C.; Grote, K.; Chahoud, I. Ultrastructural changes observed in rat ovaries following in utero and lactational exposure to low doses of a polybrominated flame retardant. *Toxicol. Lett.* **2005**, *157*, 189–202. [CrossRef]
199. Zhang, Z.; Yu, Y.; Xu, H.; Wang, C.; Ji, M.; Gu, J.; Yang, L.; Zhu, J.; Dong, H.; Wang, S.L. High-fat diet aggravates 2,2′,4,4′-tetrabromodiphenyl ether-inhibited testosterone production via DAX-1 in Leydig cells in rats. *Toxicol. Appl. Pharmacol.* **2017**, *323*, 1–8. [CrossRef]
200. Geng, X.; Wei, Y.; Geng, W.; Zhang, T.; Ding, T.; Xu, J.; He, H.; Gao, X.; Zhai, J. BDE-209 disrupted the blood-testis barrier integrity by inhibiting estrogen receptor α signaling pathway in Sprague-Dawley rats. *Environ. Sci. Pollut. Res. Int.* **2023**, *30*, 47349–47365. [CrossRef]
201. Zhai, J.; Geng, X.; Ding, T.; Li, J.; Tang, J.; Chen, D.; Cui, L.; Wang, Q. An increase of estrogen receptor α protein level regulates BDE-209-mediated blood-testis barrier disruption during spermatogenesis in F1 mice. *Environ. Sci. Pollut. Res. Int.* **2019**, *26*, 4801–4820. [CrossRef] [PubMed]
202. Li, J.B.; Li, Y.Y.; Shen, Y.P.; Zhu, M.; Li, X.H.; Qin, Z.F. 2,2′,4,4′-tetrabromodipheny ether (BDE-47) disrupts gonadal development of the Africa clawed frog (Xenopus laevis). *Aquat. Toxicol.* **2020**, *221*, 105441. [CrossRef]
203. Han, J.; Won, E.J.; Lee, M.C.; Seo, J.S.; Lee, S.J.; Lee, J.S. Developmental retardation, reduced fecundity, and modulated expression of the defensome in the intertidal copepod Tigriopus japonicus exposed to BDE-47 and PFOS. *Aquat. Toxicol.* **2015**, *165*, 136–143. [CrossRef] [PubMed]
204. He, J.; Yang, D.; Wang, C.; Liu, W.; Liao, J.; Xu, T.; Bai, C.; Chen, J.; Lin, K.; Huang, C.; et al. Chronic zebrafish low dose decabrominated diphenyl ether (BDE-209) exposure affected parental gonad development and locomotion in F1 offspring. *Ecotoxicology* **2011**, *20*, 1813–1822. [CrossRef] [PubMed]
205. Huang, Y.; Zhu, G.; Peng, L.; Ni, W.; Wang, X.; Zhang, J.; Wu, K. Effect of 2,2′,4,4′-tetrabromodiphenyl ether (BDE-47) on sexual behaviors and reproductive function in male zebrafish (Danio rerio). *Ecotoxicol. Environ. Saf.* **2015**, *111*, 102–108. [CrossRef] [PubMed]
206. Lefèvre, P.L.C.; Nardelli, T.C.; Son, W.Y.; Sadler, A.R.; Rawn, D.F.K.; Goodyer, C.; Robaire, B.; Hales, B.F. Polybrominated Diphenyl Ethers in Human Follicular Fluid Dysregulate Mural and Cumulus Granulosa Cell Gene Expression. *Endocrinology* **2021**, *162*, bqab003. [CrossRef]
207. Zhai, J.; Geng, W.; Zhang, T.; Wei, Y.; He, H.; Chen, W. BDE-209 induce spermatocytes arrest at early-pachytene stage during meiotic prophase I in mice. *Toxicology* **2022**, *467*, 153061. [CrossRef]

208. Tseng, L.H.; Lee, C.W.; Pan, M.H.; Tsai, S.S.; Li, M.H.; Chen, J.R.; Lay, J.J.; Hsu, P.C. Postnatal exposure of the male mouse to 2,2′,3,3′,4,4′,5,5′,6,6′-decabrominated diphenyl ether: Decreased epididymal sperm functions without alterations in DNA content and histology in testis. *Toxicology* **2006**, *224*, 33–43. [CrossRef]
209. Zhang, T.; He, H.; Wei, Y.; Geng, W.; Zhai, J. Vitamin C supplementation rescued meiotic arrest of spermatocytes in Balb/c mice exposed to BDE-209. *Ecotoxicol. Environ. Saf.* **2022**, *242*, 113846. [CrossRef]
210. Miyaso, H.; Nakamura, N.; Naito, M.; Hirai, S.; Matsuno, Y.; Itoh, M.; Mori, C. Early postnatal exposure to a low dose of decabromodiphenyl ether affects expression of androgen and thyroid hormone receptor-alpha and its splicing variants in mouse Sertoli cells. *PLoS ONE* **2014**, *9*, e114487. [CrossRef]
211. Fong, C.C.; Shi, Y.F.; Yu, W.K.; Wei, F.; van de Merwe, J.P.; Chan, A.K.; Ye, R.; Au, D.W.; Wu, R.S.; Yang, M.S. iTRAQ-based proteomic profiling of the marine medaka (Oryzias melastigma) gonad exposed to BDE-47. *Mar. Pollut. Bull.* **2014**, *85*, 471–478. [CrossRef]
212. Liu, P.; Miao, J.; Song, Y.; Pan, L.; Yin, P. Effects of 2,2′,4,4′-tetrabromodipheny ether (BDE-47) on gonadogenesis of the manila clam Ruditapes philippinarum. *Aquat. Toxicol.* **2017**, *193*, 178–186. [CrossRef] [PubMed]
213. Wang, C.; Zhang, S.; Ma, R.; Zhang, X.; Zhang, C.; Li, B.; Niu, Q.; Chen, J.; Xia, T.; Li, P.; et al. Roles of endoplasmic reticulum stress, apoptosis and autophagy in 2,2′,4,4′-tetrabromodiphenyl ether-induced rat ovarian injury. *Reprod. Toxicol.* **2016**, *65*, 187–193. [CrossRef]
214. Thornton, L.M.; Path, E.M.; Nystrom, G.S.; Venables, B.J.; Sellin Jeffries, M.K. Early Life Stage Exposure to BDE-47 Causes Adverse Effects on Reproductive Success and Sexual Differentiation in Fathead Minnows (Pimephales promelas). *Environ. Sci. Technol.* **2016**, *50*, 7834–7841. [CrossRef] [PubMed]
215. Han, X.B.; Yuen, K.W.; Wu, R.S. Polybrominated diphenyl ethers affect the reproduction and development, and alter the sex ratio of zebrafish (Danio rerio). *Environ. Pollut.* **2013**, *182*, 120–126. [CrossRef] [PubMed]
216. Marteinson, S.C.; Bird, D.M.; Shutt, J.L.; Letcher, R.J.; Ritchie, I.J.; Fernie, K.J. Multi-generational effects of polybrominated diphenylethers exposure: Embryonic exposure of male American kestrels (Falco sparverius) to DE-71 alters reproductive success and behaviors. *Environ. Toxicol. Chem.* **2010**, *29*, 1740–1747. [CrossRef]
217. Arowolo, O.; Pilsner, J.R.; Sergeyev, O.; Suvorov, A. Mechanisms of Male Reproductive Toxicity of Polybrominated Diphenyl Ethers. *Int. J. Mol. Sci.* **2022**, *23*, 14229. [CrossRef]
218. Zhao, Y.; Ao, H.; Chen, L.; Sottas, C.M.; Ge, R.S.; Zhang, Y. Effect of brominated flame retardant BDE-47 on androgen production of adult rat Leydig cells. *Toxicol. Lett.* **2011**, *205*, 209–214. [CrossRef]
219. Chen, H.; Tang, X.; Zhou, B.; Xu, N.; Wang, Y. Mechanism of Deca-BDE-induced apoptosis in Neuro-2a cells: Role of death-receptor pathway and reactive oxygen species-mediated mitochondrial pathway. *J. Environ. Sci.* **2016**, *46*, 241–251. [CrossRef]
220. Chen, J.; Li, X.; Li, X.; Chen, D. The environmental pollutant BDE-209 regulates NO/cGMP signaling through activation of NMDA receptors in neurons. *Environ. Sci. Pollut. Res. Int.* **2018**, *25*, 3397–3407. [CrossRef]
221. Sun, W.; Du, L.; Tang, W.; Kuang, L.; Du, P.; Chen, J.; Chen, D. PBDE-209 exposure damages learning and memory ability in rats potentially through increased autophagy and apoptosis in the hippocampus neuron. *Environ. Toxicol. Pharmacol.* **2017**, *50*, 151–158. [CrossRef] [PubMed]
222. Zhang, H.; Li, X.; Nie, J.; Niu, Q. Lactation exposure to BDE-153 damages learning and memory, disrupts spontaneous behavior and induces hippocampus neuron death in adult rats. *Brain Res.* **2013**, *1517*, 44–56. [CrossRef] [PubMed]
223. Costa, L.G.; Tagliaferri, S.; Roqué, P.J.; Pellacani, C. Role of glutamate receptors in tetrabrominated diphenyl ether (BDE-47) neurotoxicity in mouse cerebellar granule neurons. *Toxicol. Lett.* **2016**, *241*, 159–166. [CrossRef]
224. Gassmann, K.; Schreiber, T.; Dingemans, M.M.; Krause, G.; Roderigo, C.; Giersiefer, S.; Schuwald, J.; Moors, M.; Unfried, K.; Bergman, Å.; et al. BDE-47 and 6-OH-BDE-47 modulate calcium homeostasis in primary fetal human neural progenitor cells via ryanodine receptor-independent mechanisms. *Arch. Toxicol.* **2014**, *88*, 1537–1548. [CrossRef] [PubMed]
225. He, P.; Wang, A.G.; Xia, T.; Gao, P.; Niu, Q.; Guo, L.J.; Chen, X.M. Mechanisms underlying the developmental neurotoxic effect of PBDE-47 and the enhanced toxicity associated with its combination with PCB153 in rats. *Neurotoxicology* **2009**, *30*, 1088–1095. [CrossRef]
226. Blanco, J.; Mulero, M.; López, M.; Domingo, J.L.; Sánchez, D.J. BDE-99 deregulates BDNF, Bcl-2 and the mRNA expression of thyroid receptor isoforms in rat cerebellar granular neurons. *Toxicology* **2011**, *290*, 305–311. [CrossRef]
227. Blanco, J.; Mulero, M.; Heredia, L.; Pujol, A.; Domingo, J.L.; Sánchez, D.J. Perinatal exposure to BDE-99 causes learning disorders and decreases serum thyroid hormone levels and BDNF gene expression in hippocampus in rat offspring. *Toxicology* **2013**, *308*, 122–128. [CrossRef]
228. Cheng, J.; Gu, J.; Ma, J.; Chen, X.; Zhang, M.; Wang, W. Neurobehavioural effects, redox responses and tissue distribution in rat offspring developmental exposure to BDE-99. *Chemosphere* **2009**, *75*, 963–968. [CrossRef]
229. Peltier, M.R.; Fassett, M.J.; Arita, Y.; Chiu, V.Y.; Takhar, H.S.; Getahun, D. Exposure to polybrominated diphenyl ether-47 increases the risk of post-partum depression. *J. Matern.-Fetal Neonatal Med.* **2022**, *35*, 8350–8354. [CrossRef]
230. Viberg, H.; Mundy, W.; Eriksson, P. Neonatal exposure to decabrominated diphenyl ether (PBDE 209) results in changes in BDNF, CaMKII and GAP-43, biochemical substrates of neuronal survival, growth, and synaptogenesis. *Neurotoxicology* **2008**, *29*, 152–159. [CrossRef]
231. Viberg, H. Exposure to polybrominated diphenyl ethers 203 and 206 during the neonatal brain growth spurt affects proteins important for normal neurodevelopment in mice. *Toxicol. Sci. Off. J. Soc. Toxicol.* **2009**, *109*, 306–311. [CrossRef]

232. Mutic, A.D.; Barr, D.B.; Hertzberg, V.S.; Brennan, P.A.; Dunlop, A.L.; McCauley, L.A. Polybrominated Diphenyl Ether Serum Concentrations and Depressive Symptomatology in Pregnant African American Women. *Int. J. Environ. Res. Public Health* **2021**, *18*, 3614. [CrossRef]
233. Li, Z.; You, M.; Che, X.; Dai, Y.; Xu, Y.; Wang, Y. Perinatal exposure to BDE-47 exacerbated autistic-like behaviors and impairments of dendritic development in a valproic acid-induced rat model of autism. *Ecotoxicol. Environ. Saf.* **2021**, *212*, 112000. [CrossRef] [PubMed]
234. Wang, X.; Yang, L.; Wu, Y.; Huang, C.; Wang, Q.; Han, J.; Guo, Y.; Shi, X.; Zhou, B. The developmental neurotoxicity of polybrominated diphenyl ethers: Effect of DE-71 on dopamine in zebrafish larvae. *Environ. Toxicol. Chem.* **2015**, *34*, 1119–1126. [CrossRef] [PubMed]
235. Wang, X.; Yang, L.; Wang, Q.; Guo, Y.; Li, N.; Ma, M.; Zhou, B. The neurotoxicity of DE-71: Effects on neural development and impairment of serotonergic signaling in zebrafish larvae. *J. Appl. Toxicol. JAT* **2016**, *36*, 1605–1613. [CrossRef]
236. Ji, F.; Zhu, Z.; Zhang, M.; Zhang, H.; Zhu, L.; Cai, X.; Liu, W.; Song, J.; Li, M.; Cai, Z. 6-OH-BDE-47 exposure-induced Parkinson's disease pathology in Sprague Dawley rat. *Sci. Total Environ.* **2020**, *711*, 135184. [CrossRef] [PubMed]
237. Qiu, H.; Gao, H.; Yu, F.; Xiao, B.; Li, X.; Cai, B.; Ge, L.; Lu, Y.; Wan, Z.; Wang, Y.; et al. Perinatal exposure to low-level PBDE-47 programs gut microbiota, host metabolism and neurobehavior in adult rats: An integrated analysis. *Sci. Total Environ.* **2022**, *825*, 154150. [CrossRef] [PubMed]
238. Liu, Y.; Guo, R.; Tang, S.; Zhu, F.; Zhang, S.; Yan, Z.; Chen, J. Single and mixture toxicities of BDE-47, 6-OH-BDE-47 and 6-MeO-BDE-47 on the feeding activity of Daphnia magna: From behavior assessment to neurotoxicity. *Chemosphere* **2018**, *195*, 542–550. [CrossRef]
239. Blanc, M.; Alfonso, S.; Bégout, M.L.; Barrachina, C.; Hyötyläinen, T.; Keiter, S.H.; Cousin, X. An environmentally relevant mixture of polychlorinated biphenyls (PCBs) and polybrominated diphenylethers (PBDEs) disrupts mitochondrial function, lipid metabolism and neurotransmission in the brain of exposed zebrafish and their unexposed F2 offspring. *Sci. Total Environ.* **2021**, *754*, 142097. [CrossRef]
240. Alfonso, S.; Blanc, M.; Joassard, L.; Keiter, S.H.; Munschy, C.; Loizeau, V.; Bégout, M.L.; Cousin, X. Examining multi- and transgenerational behavioral and molecular alterations resulting from parental exposure to an environmental PCB and PBDE mixture. *Aquat. Toxicol.* **2019**, *208*, 29–38. [CrossRef]
241. Chen, Y.; Liu, S.; Xu, H.; Zheng, H.; Bai, C.; Pan, W.; Zhou, H.; Liao, M.; Huang, C.; Dong, Q. Maternal exposure to low dose BDE209 and Pb mixture induced neurobehavioral anomalies in C57BL/6 male offspring. *Toxicology* **2019**, *418*, 70–80. [CrossRef]
242. Zhu, B.; Wang, Q.; Shi, X.; Guo, Y.; Xu, T.; Zhou, B. Effect of combined exposure to lead and decabromodiphenyl ether on neurodevelopment of zebrafish larvae. *Chemosphere* **2016**, *144*, 1646–1654. [CrossRef]
243. Wang, Q.; Chen, Q.; Zhou, P.; Li, W.; Wang, J.; Huang, C.; Wang, X.; Lin, K.; Zhou, B. Bioconcentration and metabolism of BDE-209 in the presence of titanium dioxide nanoparticles and impact on the thyroid endocrine system and neuronal development in zebrafish larvae. *Nanotoxicology* **2014**, *8* (Suppl. 1), 196–207. [CrossRef]
244. Zhao, W.; Cheng, J.; Gu, J.; Liu, Y.; Fujimura, M.; Wang, W. Assessment of neurotoxic effects and brain region distribution in rat offspring prenatally co-exposed to low doses of BDE-99 and methylmercury. *Chemosphere* **2014**, *112*, 170–176. [CrossRef]
245. Chen, H.; Seifikar, H.; Larocque, N.; Kim, Y.; Khatib, I.; Fernandez, C.J.; Abello, N.; Robinson, J.F. Using a Multi-Stage hESC Model to Characterize BDE-47 Toxicity During Neurogenesis. *Toxicol. Sci. Off. J. Soc. Toxicol.* **2019**, *171*, 221–234. [CrossRef]
246. Alm, H.; Kultima, K.; Scholz, B.; Nilsson, A.; Andrén, P.E.; Fex-Svenningsen, A.; Dencker, L.; Stigson, M. Exposure to brominated flame retardant PBDE-99 affects cytoskeletal protein expression in the neonatal mouse cerebral cortex. *Neurotoxicology* **2008**, *29*, 628–637. [CrossRef]
247. Alm, H.; Scholz, B.; Kultima, K.; Nilsson, A.; Andrén, P.E.; Savitski, M.M.; Bergman, A.; Stigson, M.; Fex-Svenningsen, A.; Dencker, L. In vitro neurotoxicity of PBDE-99: Immediate and concentration-dependent effects on protein expression in cerebral cortex cells. *J. Proteome Res.* **2010**, *9*, 1226–1235. [CrossRef]
248. Suvorov, A.; Girard, S.; Lachapelle, S.; Abdelouahab, N.; Sebire, G.; Takser, L. Perinatal exposure to low-dose BDE-47, an emergent environmental contaminant, causes hyperactivity in rat offspring. *Neonatology* **2009**, *95*, 203–209. [CrossRef]
249. Zheng, S.; Zhang, Q.; Wu, R.; Shi, X.; Peng, J.; Tan, W.; Huang, W.; Wu, K.; Liu, C. Behavioral changes and transcriptomic effects at embryonic and post-embryonic stages reveal the toxic effects of 2,2′,4,4′-tetrabromodiphenyl ether on neurodevelopment in zebrafish (Danio rerio). *Ecotoxicol. Environ. Saf.* **2022**, *248*, 114310. [CrossRef]
250. Chou, C.T.; Hsiao, Y.C.; Ko, F.C.; Cheng, J.O.; Cheng, Y.M.; Chen, T.H. Chronic exposure of 2,2′,4,4′-tetrabromodiphenyl ether (PBDE-47) alters locomotion behavior in juvenile zebrafish (Danio rerio). *Aquat. Toxicol.* **2010**, *98*, 388–395. [CrossRef]
251. Liang, S.; Liang, S.; Zhou, H.; Yin, N.; Faiola, F. Typical halogenated flame retardants affect human neural stem cell gene expression during proliferation and differentiation via glycogen synthase kinase 3 beta and T3 signaling. *Ecotoxicol. Environ. Saf.* **2019**, *183*, 109498. [CrossRef]
252. Azar, N.; Booij, L.; Muckle, G.; Arbuckle, T.E.; Séguin, J.R.; Asztalos, E.; Fraser, W.D.; Lanphear, B.P.; Bouchard, M.F. Prenatal exposure to polybrominated diphenyl ethers (PBDEs) and cognitive ability in early childhood. *Environ. Int.* **2021**, *146*, 106296. [CrossRef]
253. Bai, X.; Chen, T.; Gao, Y.; Li, H.; Li, Z.; Liu, Z. The protective effects of insulin-like growth factor-1 on neurochemical phenotypes of dorsal root ganglion neurons with BDE-209-induced neurotoxicity in vitro. *Toxicol. Ind. Health* **2017**, *33*, 250–264. [CrossRef]

254. Edwards, C.M.; Small, D.; Bell, T.; David-Drori, J.; Hansen, C.; Morris-Schaffer, K.; Canale, C.; Ng, J.; Markowski, V.P. Early postnatal decabromodiphenyl ether exposure reduces thyroid hormone and astrocyte density in the juvenile mouse dentate gyrus. *Physiol. Behav.* **2020**, *216*, 112798. [CrossRef]
255. Branchi, I.; Alleva, E.; Costa, L.G. Effects of perinatal exposure to a polybrominated diphenyl ether (PBDE 99) on mouse neurobehavioural development. *Neurotoxicology* **2002**, *23*, 375–384. [CrossRef]
256. Dach, K.; Bendt, F.; Huebenthal, U.; Giersiefer, S.; Lein, P.J.; Heuer, H.; Fritsche, E. BDE-99 impairs differentiation of human and mouse NPCs into the oligodendroglial lineage by species-specific modes of action. *Sci. Rep.* **2017**, *7*, 44861. [CrossRef]
257. Bowers, W.J.; Wall, P.M.; Nakai, J.S.; Yagminas, A.; Wade, M.; Li, N. Behavioral and thyroid effects of in utero and lactational exposure of Sprague-Dawley rats to the polybrominated diphenyl ether mixture DE71. *Neurotoxicol. Teratol.* **2015**, *52 Pt B*, 127–142. [CrossRef]
258. Chen, L.; Huang, C.; Hu, C.; Yu, K.; Yang, L.; Zhou, B. Acute exposure to DE-71: Effects on locomotor behavior and developmental neurotoxicity in zebrafish larvae. *Environ. Toxicol. Chem.* **2012**, *31*, 2338–2344. [CrossRef]
259. Wang, X.; Zhao, L.; Shi, Q.; Guo, Y.; Hua, J.; Han, J.; Yang, L. DE-71 affected the cholinergic system and locomotor activity via disrupting calcium homeostasis in zebrafish larvae. *Aquat. Toxicol.* **2022**, *250*, 106237. [CrossRef]
260. Gill, S.; Hou, Y.; Li, N.; Pulido, O.; Bowers, W. Developmental neurotoxicity of polybrominated diphenyl ethers mixture de71 in Sprague-Dawley rats. *J. Toxicol. Environ. Health. Part A* **2016**, *79*, 482–493. [CrossRef]
261. Chen, X.; Huang, C.; Wang, X.; Chen, J.; Bai, C.; Chen, Y.; Chen, X.; Dong, Q.; Yang, D. BDE-47 disrupts axonal growth and motor behavior in developing zebrafish. *Aquat. Toxicol.* **2012**, *120–121*, 35–44. [CrossRef]
262. Zheng, S.; Huang, W.; Liu, C.; Xiao, J.; Wu, R.; Wang, X.; Cai, Z.; Wu, K. Behavioral change and transcriptomics reveal the effects of 2, 2′, 4, 4′-tetrabromodiphenyl ether exposure on neurodevelopmental toxicity to zebrafish (Danio rerio) in early life stage. *Sci. Total Environ.* **2021**, *752*, 141783. [CrossRef]
263. Glazer, L.; Wells, C.N.; Drastal, M.; Odamah, K.A.; Galat, R.E.; Behl, M.; Levin, E.D. Developmental exposure to low concentrations of two brominated flame retardants, BDE-47 and BDE-99, causes life-long behavioral alterations in zebrafish. *Neurotoxicology* **2018**, *66*, 221–232. [CrossRef]
264. Chen, H.; Streifel, K.M.; Singh, V.; Yang, D.; Mangini, L.; Wulff, H.; Lein, P.J. From the Cover: BDE-47 and BDE-49 Inhibit Axonal Growth in Primary Rat Hippocampal Neuron-Glia Co-Cultures via Ryanodine Receptor-Dependent Mechanisms. *Toxicol. Sci. Off. J. Soc. Toxicol.* **2017**, *156*, 375–386. [CrossRef]
265. Chen, H.; Tang, X.; Zhou, B.; Xu, N.; Zhou, Z.; Fang, K.; Wang, Y. BDE-47 and BDE-209 inhibit proliferation of Neuro-2a cells via inducing G1-phase arrest. *Environ. Toxicol. Pharmacol.* **2017**, *50*, 76–82. [CrossRef] [PubMed]
266. Chen, H.; Tang, X.; Zhou, B.; Zhou, Z.; Xu, N.; Wang, Y. A ROS-mediated mitochondrial pathway and Nrf2 pathway activation are involved in BDE-47 induced apoptosis in Neuro-2a cells. *Chemosphere* **2017**, *184*, 679–686. [CrossRef] [PubMed]
267. Xing, T.; Chen, L.; Tao, Y.; Wang, M.; Chen, J.; Ruan, D.Y. Effects of decabrominated diphenyl ether (PBDE 209) exposure at different developmental periods on synaptic plasticity in the dentate gyrus of adult rats In vivo. *Toxicol. Sci. Off. J. Soc. Toxicol.* **2009**, *110*, 401–410. [CrossRef]
268. Mariani, A.; Fanelli, R.; Re Depaolini, A.; De Paola, M. Decabrominated diphenyl ether and methylmercury impair fetal nervous system development in mice at documented human exposure levels. *Dev. Neurobiol.* **2015**, *75*, 23–38. [CrossRef] [PubMed]
269. Xu, M.; Huang, Y.; Li, K.; Cheng, X.; Li, G.; Liu, M.; Nie, Y.; Geng, S.; Zhao, S. Developmental exposure of decabromodiphenyl ether impairs subventricular zone neurogenesis and morphology of granule cells in mouse olfactory bulb. *Arch. Toxicol.* **2018**, *92*, 529–539. [CrossRef]
270. Qian, B.; Wang, C.; Zhao, C.; Jiang, R.; Song, J. Effects of maternal exposure to BDE209 on neuronal development and transcription of iodothyronine deiodinase in offspring mice. *Toxicol. Mech. Methods* **2019**, *29*, 569–579. [CrossRef]
271. Qian, B.; Zen, Z.; Zheng, Z.; Wang, C.; Song, J. A preliminary study on the mechanism of the neurosteroid-mediated ionotropic receptor dysfunction in neurodevelopmental toxicity induced by decabromodiphenyl ether. *Ecotoxicol. Environ. Saf.* **2021**, *217*, 112198. [CrossRef] [PubMed]
272. Roberts, S.C.; Bianco, A.C.; Stapleton, H.M. Disruption of type 2 iodothyronine deiodinase activity in cultured human glial cells by polybrominated diphenyl ethers. *Chem. Res. Toxicol.* **2015**, *28*, 1265–1274. [CrossRef] [PubMed]
273. Ding, G.; Yu, J.; Cui, C.; Chen, L.; Gao, Y.; Wang, C.; Zhou, Y.; Tian, Y. Association between prenatal exposure to polybrominated diphenyl ethers and young children's neurodevelopment in China. *Environ. Res.* **2015**, *142*, 104–111. [CrossRef] [PubMed]
274. Ye, S.; Li, S.; Ma, Y.; Hu, D.; Xiao, F. Curcumin hinders PBDE-47-induced neutrophil extracellular traps release via Nrf2-associated ROS inhibition. *Ecotoxicol. Environ. Saf.* **2021**, *225*, 112779. [CrossRef]
275. Zhou, S.; Liu, J. In vitro immunotoxicity and possible mechanisms of 2,2′,4,4′-tetrabromodiphenyl ether (BDE-47) on Ruditapes philippinarum hemocytes. *Fish Shellfish Immunol.* **2022**, *127*, 386–395. [CrossRef]
276. Jiang, Y.; Tang, X.; Sun, T.; Wang, Y. BDE-47 exposure changed the immune function of haemocytes in Mytilus edulis: An explanation based on ROS-mediated pathway. *Aquat. Toxicol.* **2017**, *182*, 58–66. [CrossRef]
277. Lv, Q.Y.; Wan, B.; Guo, L.H.; Zhao, L.; Yang, Y. In vitro immune toxicity of polybrominated diphenyl ethers on murine peritoneal macrophages: Apoptosis and immune cell dysfunction. *Chemosphere* **2015**, *120*, 621–630. [CrossRef]
278. Frouin, H.; Lebeuf, M.; Hammill, M.; Masson, S.; Fournier, M. Effects of individual polybrominated diphenyl ether (PBDE) congeners on harbour seal immune cells in vitro. *Mar. Pollut. Bull.* **2010**, *60*, 291–298. [CrossRef]

279. Longo, V.; Longo, A.; Adamo, G.; Fiannaca, A.; Picciotto, S.; La Paglia, L.; Romancino, D.; La Rosa, M.; Urso, A.; Cibella, F.; et al. 2,2′4,4′-Tetrabromodiphenyl Ether (PBDE-47) Modulates the Intracellular miRNA Profile, sEV Biogenesis and Their miRNA Cargo Exacerbating the LPS-Induced Pro-Inflammatory Response in THP-1 Macrophages. *Front. Immunol.* **2021**, *12*, 664534. [CrossRef]
280. Lamkin, D.M.; Chen, S.; Bradshaw, K.P.; Xu, S.; Faull, K.F.; Sloan, E.K.; Cole, S.W. Low-dose exposure to PBDE disrupts genomic integrity and innate immunity in mammary tissue. *Front. Genet.* **2022**, *13*, 904607. [CrossRef]
281. Longo, V.; Aloi, N.; Lo Presti, E.; Fiannaca, A.; Longo, A.; Adamo, G.; Urso, A.; Meraviglia, S.; Bongiovanni, A.; Cibella, F.; et al. Impact of the flame retardant 2,2′4,4′-tetrabromodiphenyl ether (PBDE-47) in THP-1 macrophage-like cell function via small extracellular vesicles. *Front. Immunol.* **2022**, *13*, 1069207. [CrossRef] [PubMed]
282. Longo, V.; Longo, A.; Di Sano, C.; Cigna, D.; Cibella, F.; Di Felice, G.; Colombo, P. In vitro exposure to 2,2′,4,4′-tetrabromodiphenyl ether (PBDE-47) impairs innate inflammatory response. *Chemosphere* **2019**, *219*, 845–854. [CrossRef] [PubMed]
283. Xue, D.; Wei, J.; Lu, W.; Xia, B.; Li, S.; Liu, D.; Liu, N.; Wang, X.; Lin, G. BDE-47 disturbs the immune response of lymphocytes to LPS by downregulating NF-κB pathway. *Chemosphere* **2022**, *308 Pt 3*, 136562. [CrossRef]
284. Zota, A.R.; Geller, R.J.; Romano, L.E.; Coleman-Phox, K.; Adler, N.E.; Parry, E.; Wang, M.; Park, J.S.; Elmi, A.F.; Laraia, B.A.; et al. Association between persistent endocrine-disrupting chemicals (PBDEs, OH-PBDEs, PCBs, and PFASs) and biomarkers of inflammation and cellular aging during pregnancy and postpartum. *Environ. Int.* **2018**, *115*, 9–20. [CrossRef]
285. Ashwood, P.; Schauer, J.; Pessah, I.N.; Van de Water, J. Preliminary evidence of the in vitro effects of BDE-47 on innate immune responses in children with autism spectrum disorders. *J. Neuroimmunol.* **2009**, *208*, 130–135. [CrossRef] [PubMed]
286. Thornton, L.M.; Path, E.M.; Nystrom, G.S.; Venables, B.J.; Sellin Jeffries, M.K. Embryo-larval BDE-47 exposure causes decreased pathogen resistance in adult male fathead minnows (*Pimephales promelas*). *Fish Shellfish Immunol.* **2018**, *80*, 80–87. [CrossRef]
287. Zhou, Z.; Jian, X.; Zhou, B.; Lu, K.; Wang, Y. Changes in the immune function of rainbow trout (Oncorhynchus mykiss) provide insights into strategies against BDE-47 stress. *J. Hazard. Mater.* **2020**, *392*, 122212. [CrossRef]
288. Feng, Y.; Zeng, W.; Wang, Y.; Shen, H.; Wang, Y. Long-term exposure to high levels of decabrominated diphenyl ether inhibits CD4 T-cell functions in C57Bl/6 mice. *J. Appl. Toxicol. JAT* **2016**, *36*, 1112–1119. [CrossRef]
289. Zeng, W.; Wang, Y.; Liu, Z.; Khanniche, A.; Hu, Q.; Feng, Y.; Ye, W.; Yang, J.; Wang, S.; Zhou, L.; et al. Long-term exposure to decabrominated diphenyl ether impairs CD8 T-cell function in adult mice. *Cell. Mol. Immunol.* **2014**, *11*, 367–376. [CrossRef]
290. Liao, T.; Li, B.; Zhang, Z.; Feng, W.; Chen, Y.; Ding, Y.; Chen, H.; Zhao, T.; Mao, G.; Wu, X.; et al. Short-term exposure of decabromodiphenyl ether in female adult Balb/c mice: Immune toxicity and self-recovery. *Toxicol. Lett.* **2021**, *342*, 26–37. [CrossRef]
291. Liu, X.; Zhan, H.; Zeng, X.; Zhang, C.; Chen, D. The PBDE-209 exposure during pregnancy and lactation impairs immune function in rats. *Mediat. Inflamm.* **2012**, *2012*, 692467. [CrossRef] [PubMed]
292. Cheng, L.; Rao, Q.; Zhang, Q.; Song, W.; Guan, S.; Jiang, Z.; Wu, T.; Zhao, Z.; Song, W. The immunotoxicity of decabromodiphenyl ether (BDE-209) on broiler chicks by transcriptome profiling analysis. *Ecotoxicol. Environ. Saf.* **2022**, *232*, 113284. [CrossRef] [PubMed]
293. Hong, S.K.; Sohn, K.H.; Kim, I.Y.; Lee, J.K.; Ju, J.H.; Kim, J.H.; Lim, C.H.; Han, B.S.; Jung, H.C.; Lee, J.Y.; et al. Polybrominated Diphenyl Ethers Orally Administration to Mice Were Tansferred to Offspring during Gestation and Lactation with Disruptions on the Immune System. *Immune Netw.* **2010**, *10*, 64–74. [CrossRef] [PubMed]
294. Huang, Y.; Rajput, I.R.; Sanganyado, E.; Yajing, S.; Yu, F.; Liang, B.; Liu, W. Immune stimulation effect of PBDEs via prostaglandin pathway in pantropical spotted dolphin: An in vitro study. *Chemosphere* **2020**, *254*, 126717. [CrossRef] [PubMed]
295. Espinosa-Ruiz, C.; Manuguerra, S.; Morghese, M.; García-Beltrán, J.M.; Esteban, M.; Giuga, M.; Messina, C.M.; Santulli, A. Immunity and inflammatory responses in gilthead sea bream (*Sparus aurata* L.) exposed to sub-lethal mixture of carbamazepine, cadmium chloride and polybrominated diphenyl ether. *Fish Shellfish Immunol.* **2021**, *111*, 25–35. [CrossRef]
296. Zhang, Z.; Li, S.; Liu, L.; Wang, L.; Xiao, X.; Sun, Z.; Wang, X.; Wang, C.; Wang, M.; Li, L.; et al. Environmental exposure to BDE47 is associated with increased diabetes prevalence: Evidence from community-based case-control studies and an animal experiment. *Sci. Rep.* **2016**, *6*, 27854. [CrossRef]
297. Liu, X.; Zhang, L.; Li, J.; Meng, G.; Chi, M.; Li, T.; Zhao, Y.; Wu, Y. A nested case-control study of the association between exposure to polybrominated diphenyl ethers and the risk of gestational diabetes mellitus. *Environ. Int.* **2018**, *119*, 232–238. [CrossRef]
298. Yanagisawa, R.; Koike, E.; Win-Shwe, T.T.; Takano, H. Decabromodiphenyl ether exacerbates hyperglycemia in diet-induced obese mice. *Toxicology* **2019**, *412*, 12–18. [CrossRef]
299. Jing, L.; Sun, Y.; Wang, Y.; Liang, B.; Chen, T.; Zheng, D.; Zhao, X.; Zhou, X.; Sun, Z.; Shi, Z. Cardiovascular toxicity of decabrominated diphenyl ethers (BDE-209) and decabromodiphenyl ethane (DBDPE) in rats. *Chemosphere* **2019**, *223*, 675–685. [CrossRef]
300. Hou, Y.; Fu, J.; Sun, S.; Jin, Y.; Wang, X.; Zhang, L. BDE-209 induces autophagy and apoptosis via IRE1α/Akt/mTOR signaling pathway in human umbilical vein endothelial cells. *Environ. Pollut.* **2019**, *253*, 429–438. [CrossRef]
301. Chen, L.; Huang, Y.; Huang, C.; Hu, B.; Hu, C.; Zhou, B. Acute exposure to DE-71 causes alterations in visual behavior in zebrafish larvae. *Environ. Toxicol. Chem.* **2013**, *32*, 1370–1375. [CrossRef]
302. Li, M.; Zeng, Y.; Ge, L.; Gong, J.; Weng, C.; Yang, C.; Yang, J.; Fang, Y.; Li, Q.; Zou, T.; et al. Evaluation of the influences of low dose polybrominated diphenyl ethers exposure on human early retinal development. *Environ. Int.* **2022**, *163*, 107187. [CrossRef]

303. Xu, T.; Zhao, J.; Yin, D.; Zhao, Q.; Dong, B. High-throughput RNA sequencing reveals the effects of 2,2′,4,4′-tetrabromodiphenyl ether on retina and bone development of zebrafish larvae. *BMC Genom.* **2015**, *16*, 23. [CrossRef]
304. Chen, L.; Hu, C.; Huang, C.; Wang, Q.; Wang, X.; Yang, L.; Zhou, B. Alterations in retinoid status after long-term exposure to PBDEs in zebrafish (Danio rerio). *Aquat. Toxicol.* **2012**, *120–121*, 11–18. [CrossRef]
305. Xu, T.; Chen, L.; Hu, C.; Zhou, B. Effects of acute exposure to polybrominated diphenyl ethers on retinoid signaling in zebrafish larvae. *Environ. Toxicol. Pharmacol.* **2013**, *35*, 13–20. [CrossRef]
306. Zhang, Y.; Mao, P.; Li, G.; Hu, J.; Yu, Y.; An, T. Delineation of 3D dose-time-toxicity in human pulmonary epithelial Beas-2B cells induced by decabromodiphenyl ether (BDE209). *Environ. Pollut.* **2018**, *243 Pt A*, 661–669. [CrossRef]
307. Albano, G.D.; Moscato, M.; Montalbano, A.M.; Anzalone, G.; Gagliardo, R.; Bonanno, A.; Giacomazza, D.; Barone, R.; Drago, G.; Cibella, F.; et al. Can PBDEs affect the pathophysiologic complex of epithelium in lung diseases? *Chemosphere* **2020**, *241*, 125087. [CrossRef]
308. Anzalone, G.; Moscato, M.; Montalbano, A.M.; Albano, G.D.; Gagliardo, R.; Marchese, R.; Fucarino, A.; Nigro, C.L.; Drago, G.; Profita, M. PBDEs affect inflammatory and oncosuppressive mechanisms via the EZH2 methyltransferase in airway epithelial cells. *Life Sci.* **2021**, *282*, 119827. [CrossRef]
309. Montalbano, A.M.; Albano, G.D.; Anzalone, G.; Moscato, M.; Gagliardo, R.; Di Sano, C.; Bonanno, A.; Ruggieri, S.; Cibella, F.; Profita, M. Cytotoxic and genotoxic effects of the flame retardants (PBDE-47, PBDE-99 and PBDE-209) in human bronchial epithelial cells. *Chemosphere* **2020**, *245*, 125600. [CrossRef]
310. Zandona, A.; Jagić, K.; Dvoršćak, M.; Madunić, J.; Klinčić, D.; Katalinić, M. PBDEs Found in House Dust Impact Human Lung Epithelial Cell Homeostasis. *Toxics* **2022**, *10*, 97. [CrossRef]
311. Cruz, R.; Cunha, S.C.; Casal, S. Brominated flame retardants and seafood safety: A review. *Environ. Int.* **2015**, *77*, 116–131. [CrossRef]
312. Fraser, A.J.; Webster, T.F.; McClean, M.D. Diet contributes significantly to the body burden of PBDEs in the general U.S. population. *Environ. Health Perspect.* **2009**, *117*, 1520–1525. [CrossRef]
313. Wang, J.; Yan, Z.; Zheng, X.; Wang, S.; Fan, J.; Sun, Q.; Xu, J.; Men, S. Health risk assessment and development of human health ambient water quality criteria for PBDEs in China. *Sci. Total Environ.* **2021**, *799*, 149353. [CrossRef]
314. Jin, Y.; Li, J.; Deng, X.; Xia, B.; Song, Q.; Zhao, Y.; He, X.; Li, Y.; Xu, Z.; Xie, A.; et al. Association between fetal growth restriction and maternal exposure to polybrominated diphenyl ethers. *Ecotoxicol. Environ. Saf.* **2020**, *198*, 110623. [CrossRef]
315. Stapleton, H.M.; Letcher, R.J.; Baker, J.E. Debromination of polybrominated diphenyl ether congeners BDE 99 and BDE 183 in the intestinal tract of the common carp (Cyprinus carpio). *Environ. Sci. Technol.* **2004**, *38*, 1054–1061. [CrossRef]
316. Karpeta, A.; Warzecha, K.; Jerzak, J.; Ptak, A.; Gregoraszczuk, E.L. Activation of the enzymes of phase I (CYP2B1/2) and phase II (SULT1A and COMT) metabolism by 2,2′,4,4′-tetrabromodiphenyl ether (BDE47) in the pig ovary. *Reprod. Toxicol.* **2012**, *34*, 436–442. [CrossRef]
317. Zhang, Y.; Xi, B.; Tan, W. Release, transformation, and risk factors of polybrominated diphenyl ethers from landfills to the surrounding environments: A review. *Environ. Int.* **2021**, *157*, 106780. [CrossRef]
318. Harmens, H.; Foan, L.; Simon, V.; Mills, G. Terrestrial mosses as biomonitors of atmospheric POPs pollution: A review. *Environ. Pollut.* **2013**, *173*, 245–254. [CrossRef]
319. Zhuang, Y.; Jin, L.; Luthy, R.G. Kinetics and pathways for the debromination of polybrominated diphenyl ethers by bimetallic and nanoscale zerovalent iron: Effects of particle properties and catalyst. *Chemosphere* **2012**, *89*, 426–432. [CrossRef]
320. Xiang, N.; Zhao, X.; Meng, X.Z.; Chen, L. Polybrominated diphenyl ethers (PBDEs) in a conventional wastewater treatment plant (WWTP) from Shanghai, the Yangtze River Delta: Implication for input source and mass loading. *Sci. Total Environ.* **2013**, *461–462*, 391–396. [CrossRef]

toxics

MDPI

Article

Maternal Serum Concentrations of Per- and Polyfluoroalkyl Substances in Early Pregnancy and Small for Gestational Age in Southern Sweden

Ellen Malm [1,*], Andreas Vilhelmsson [1], Hannah Högfeldt [1], Isabelle Deshayes [1], Karin Källén [2], Stefan R. Hansson [3], Christian H. Lindh [1] and Lars Rylander [1]

1 Division of Occupational and Environmental Medicine, Department of Laboratory Medicine, Lund University, MV (Building 402a), 223 81 Lund, Sweden; andreas.vilhelmsson@med.lu.se (A.V.); ha5566ho-s@student.lu.se (H.H.); isabelle.deshayes@outlook.com (I.D.); christian.lindh@med.lu.se (C.H.L.); lars.rylander@med.lu.se (L.R.)

2 Tornblad Institute, Lund University, Biskopsgatan 7, 223 62 Lund, Sweden; karin.kallen@med.lu.se

3 Division of Obstetrics and Gynecology, Department of Clinical Sciences Lund, Lund University, 221 85 Lund, Sweden; stefan.hansson@med.lu.se

* Correspondence: ellen.malm@med.lu.se

Abstract: Small for gestational age (SGA) is considered an adverse birth outcome. Per- and polyfluoroalkyl substances (PFAS) have become increasingly investigated as contributing environmental factors, thus far with inconclusive results. The current study aimed to investigate the hypothesized association between increased maternal PFAS levels in early pregnancy and an increased risk for SGA birth. This population-based study used data from a sample of children born in Scania, Southern Sweden, between 1995 and 2009. Two groups were compared: cases born with SGA (n = 298) and non-SGA controls (n = 580). The cases consisted of two subgroups: one included women whose children's growth in late pregnancy was in the lowest quartile, and another included women from the remaining growth quartiles. Corresponding maternal serum samples were collected from a biobank and analyzed for concentrations of four types of PFAS: perfluorooctanoic acid (PFOA), perfluorononanoic acid (PFNA), perfluorohexane sulfonic acid (PFHxS), and perfluorooctane sulfonic acid (PFOS) using liquid chromatography–tandem mass spectrometry (LC/MS/MS). The results were combined with information from birth registers and analyzed using Mann–Whitney U-tests and logistic regression—unadjusted as well as adjusted for potential confounders. In conclusion, elevated maternal concentrations of PFAS were not associated with an increased risk of SGA birth. However, significant ORs were observed in a subgroup analysis restricted to women of Nordic origin (unadjusted OR 3.2 and adjusted OR 2.4) for PFHxS.

Keywords: SGA; PFAS; endocrine disrupting chemical; case control; biobank; fetal growth

Citation: Malm, E.; Vilhelmsson, A.; Högfeldt, H.; Deshayes, I.; Källén, K.; Hansson, S.R.; Lindh, C.H.; Rylander, L. Maternal Serum Concentrations of Per- and Polyfluoroalkyl Substances in Early Pregnancy and Small for Gestational Age in Southern Sweden. *Toxics* **2023**, *11*, 750. https://doi.org/10.3390/toxics11090750

Academic Editors: Esref Demir and Sam Kacew

Received: 3 July 2023
Revised: 28 August 2023
Accepted: 29 August 2023
Published: 4 September 2023

1. Introduction

In recent years, per- and polyfluorinated alkyl substances (PFAS), a class of widely used, highly persistent, industrially produced compounds, have become a cause for environmental concern and are associated with adverse health outcomes [1].

PFAS are a vast and complex group of manmade substances classified as organic pollutants with half-lives of several years [2,3]. Their common properties include chemical and thermal stability, hydrophobia and lipophobia, the ability to reduce aqueous surface tension, and the ability to create stable foams. As a result of their unique chemical structures, PFAS are highly persistent and suitable for industrial use, for example, as surfactants. In addition to their extensive industrial applications, PFAS can be found in commercial products, e.g., non-stick cookware, oil-, stain-, and water-resistant coatings in clothing, carpets, personal protective equipment, fire-extinguishing foams, and cosmetic products.

PFAS are released and distributed to the environment through direct emissions during the production, use, and disposal of products in which PFAS are present [4]. The environmental impact is further increased by indirect emissions from the transformation of precursor substances—PFAS created through the degradation of compounds in other substances [5]. Human exposure occurs primarily via diet and drinking water due to surface and groundwater contamination [6]. Although the effects of PFAS have become increasingly investigated, the extent of their environmental and biological impact has yet to be determined. During gestation, a fetus is exposed to maternal PFAS via placental transportation and passage [7]. Strong correlations between maternal and cord serum PFAS concentrations have been observed, enabling maternal concentrations to be used as a surrogate marker for fetal exposure during pregnancy [8]. PFAS have been described as endocrine-disrupting chemicals (EDC), and exposure may affect fetal growth and development [9,10].

Birth weight is an established indicator of fetal development and healthy birth outcomes [11]. Small for gestational age (SGA) is defined as a baby born smaller than the average for a specific week of pregnancy. Morbidity and mortality are significantly increased in infants born SGA when compared with appropriate-for-gestational-age infants (AGA), even from term, low-risk pregnancies [12]. While causes for SGA are diverse, the most common etiologies include maternal medical conditions, infections, substance use, and a variety of placental factors [13]. Numerous studies have investigated the association between fetal intrauterine PFAS exposure and adverse birth outcomes, such as reduced birth weight and SGA, but findings are inconsistent.

The most recent opinion by the European Food Safety Authority (EFSA) concluded that "there may be a casual association between PFOS and PFOA and birth weight" [14]. In addition, a systematic review and meta-analysis observed a statistically significant association between PFAS and reduced birth weight for four of the most widespread and frequently investigated PFAS: perfluorooctanoic acid (PFOA), perfluorononanoic acid (PFNA), perfluorohexane sulfonic acid (PFHxS), and perfluorooctane sulfonic acid (PFOS), as well as other types [15]. However, the authors highlighted the need to investigate the effect of specific types and doses of PFAS on birth outcomes, expanding our understanding of causality and dose–response curves.

The present study aimed to investigate the hypothesized association between maternal levels of specific PFAS in early pregnancy and the risk of SGA birth by combining medical birth register data with serum samples from a biobank in Scania, the southernmost county in Sweden.

2. Materials and Methods

2.1. Study Design and Sources

This study was approved by the Ethical Committee (Reg. No. 2014/696) at Lund University, Sweden, and was conducted as a register- and biobank-based case-control study where the following sources were used:

2.1.1. Biobank Serum Samples

This study was based in Scania, where the Southern Sweden Maternal Biobank (SSMB) has collected >250,000 maternal blood samples from infection screening in early pregnancy (weeks 10–14) since 1989. The coverage in the biobank has been shown to be >90% [16].

2.1.2. Register Data

Various maternal and obstetric registers and databases were utilized: the Swedish Medical Birth Register (MBR) and the Perinatal Revision South (PRS), a local Scanian birth register. As part of the Scanian regional ultrasound routines, examinations in early (around gestational week 18) and late (around gestational week 32) pregnancy were performed and registered in a separate database [17,18].

2.2. Cases and Controls

An *a priori* statistical power analysis was performed to determine the study population size. Based on the aim of including 300 SGA cases and 600 controls, detecting an odds ratio of 1.6 or higher with statistical significance ($p < 0.05$) was deemed possible while not exceeding the project budget.

Figure 1 illustrates the study selection process. The study population consisted of all singleton births in Scania between 1995 and 2009, identified through the registers mentioned above. Mothers diagnosed with preeclampsia were excluded due to an ongoing study on the association between PFAS and preeclampsia [16].

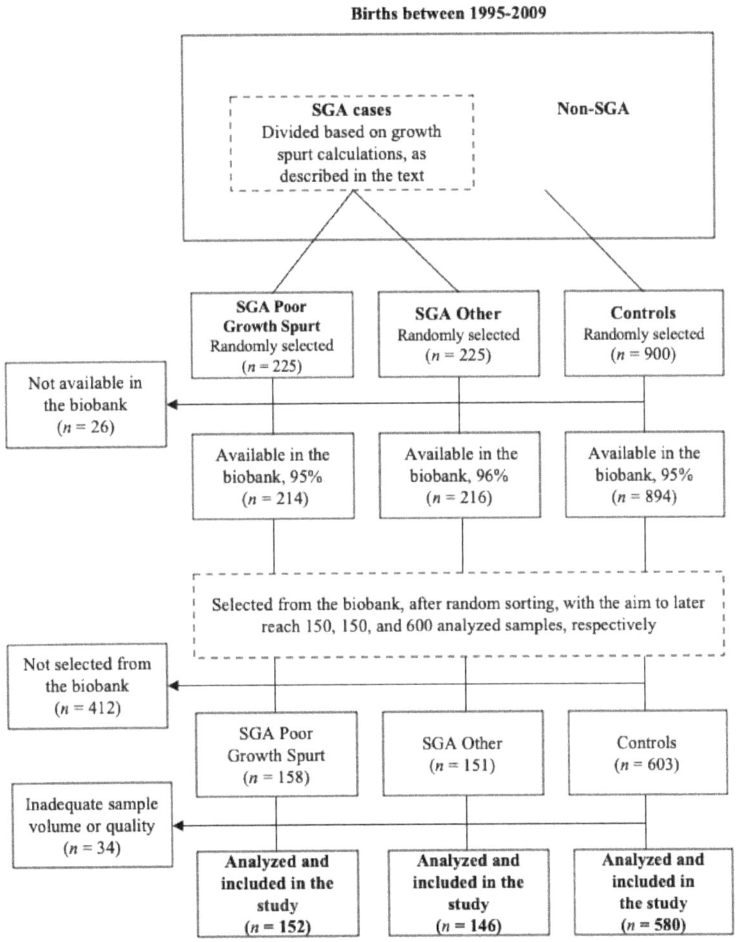

Figure 1. Flow diagram of the study selection process.

The SGA diagnosis was defined as birth weight below two standard deviations from the expected, according to the national intrauterine growth curve, which takes gestational week and gender into account [19]. Fetal growth in late pregnancy, i.e., between the second ultrasound and birth, was calculated for all SGA children. Among the women whose children's growth in late pregnancy was in the lowest quartile, 225 mothers were randomly selected. This case group was defined as "SGA with poor growth spurt". Likewise, a group of 225 women was randomly selected from the remaining upper three quartiles, defined as "SGA other". From the study population, 900 children with regular birth weights were randomly selected as controls.

Among the 225 SGA with a poor growth spurt, the 225 SGA other, and the 900 controls, 95%, 96%, and 95% serum samples were available in the biobank, respectively. The biobank was instructed to select samples with the aim of later reaching 150, 150, and 600 analyzed samples, respectively. After randomly sorting the individuals within the respective groups, the first 158 SGA with a poor growth spurt, 151 SGA other, and 603 controls were therefore selected for PFAS analysis—accounting for potential loss. The remaining samples were excluded from this study due to budgetary reasons. Although available in the biobank, not all samples met the criteria regarding sample volume to be analyzed for concentrations of PFAS. Six SGA with a poor growth spurt, five SGA other, and twenty-three controls could not be analyzed, and the final numbers of participants were 152 SGA with a poor growth spurt, 146 SGA other, and 580 controls.

2.3. Analysis of PFAS

The maternal serum samples were analyzed using liquid chromatography–tandem mass spectrometry (LC/MS/MS; QTRAP 5500; AB Sciex, Framingham, MA, USA) using a method by Norén et al. at the laboratory of Occupational and Environmental Medicine at Lund University [20]. In summary, internal standards for all compounds were added to aliquots of serum. The proteins were precipitated with acetonitrile, vigorously shaken for 30 min, and centrifuged and analyzed regarding concentrations of perfluorooctane sulfonic acid (PFOS), perfluorooctanoic acid (PFOA), perfluornonanoic acid (PFNA), and perfluorohexane sulfonic acid (PFHxS). All native and isotopically labeled standards were purchased from Wellington Laboratories (Guelph, ON, Canada). Acetonitrile, ammonium acetate, and methanol were from Merck (Darmstadt, Germany), and water was from a Milli-Q Integral 5 system (Millipore, Billerica, MA, USA). Homemade quality control (QC) samples were prepared by pooling serum samples, and MQ water was used for chemically blank samples. Every analyzed batch included calibration standards, QC samples, and chemically blank samples. The limit of detection (LOD) was determined from the blank samples.

The analyses were conducted in a duplicate and randomized order with LODs of 0.12 ng/mL for PFOS, 0.04 ng/mL for PFOA, and 0.03 ng/mL for PFNA and PFHxS. The coefficient of variation (CV) of the QC samples (n = 32) was 8% for PFHxS at 2 ng/mL and 10% at 3 ng/mL, for PFOA, 12% at 3 ng/mL and 9% at 4 ng/mL; for PFNA 10% at 2 ng/mL and 9% at 4 ng/mL; and for PFOS 7% at 12 ng/mL and 8% at 13 ng/mL. The laboratory participated successfully in the HBM4EU QA/QC program and participates in G-EQUAS for PFOS and PFOA analysis coordinated by the University of Erlangen-Nuremberg, Germany (Supplemental File S1).

2.4. Other Pregnancy Information

Additional information was obtained from the MBR, which includes data on practically all deliveries in Sweden with information acquired from medical records on prenatal, delivery, and neonatal care [21]. The following variables were obtained: maternal age, calendar year of birth, body mass index (BMI) kg/m^2, parity (primipara and multipara), smoking habits (non-smoker, 1–9 cigarettes/day, and >9 cigarettes a day), sex of the child, diabetes (yes/no), gestational diabetes (yes/no), involuntary childlessness for at least a year (yes/no), gestational week at partus, and maternal country of origin (eight categories).

2.5. Statistics

Background characteristics and PFAS concentrations for the cases and controls are presented as mean/median with minimum and maximum values for continuous variables or numbers and percentages for categorical variables. As a first step, comparisons of PFAS concentrations as continuous variables between cases and controls were analyzed with the Mann–Whitney U-test. In addition to managing the cases as a single group, the cases were divided into two subgroups (SGA with poor growth spurt and SGA other) and analyzed separately. As a second step, odds ratios (ORs) and 95% confidence intervals (CIs) were calculated using logistic regression models. In these analyses, the concentrations of the PFAS variables were categorized into quartiles based on their distribution among the controls. The lowest exposure quartiles were used as reference categories. The PFAS variables were analyzed individually. Crude analyses were completed with *a priori* defined models adjusted for maternal age (4 categories: <25, 26–30, 31–35, and >35 years), BMI (4 categories: <20, 20-<25, 25-<30, ≥30 kg/m^2), smoking (2 categories: yes and no), and parity (primipara and multipara). Based on prior research, women of Nordic origin have among the highest concentrations of PFAS; therefore, we conducted separate analyses including only women of Nordic origin [22]. In addition, it is reasonable to assume that these women are more homogeneous as a group compared to the entire study population. Furthermore, separate analyses were performed for infant sex, as previous studies have found sex-specific effects of PFAS on adverse birth outcomes [23,24]. Due to limited statistical power, separate analyses were not performed for the different case groups. Statistical significance was defined as p-values below 0.05. All statistical analyses were conducted in SPSS (Version 27.0) (IBM Corp., Armonk, NY, USA).

3. Results

Background characteristics and demographics of the study participants are presented in Table 1. The most pronounced differences between the cases and controls were the higher proportion of primiparas (66.8% vs. 47.1%) and smokers (23% vs. 9.6%), and the lower proportion of women originating from a Nordic country.

Table 1. Background characteristics among all cases (n = 298) and controls (n = 580) as well as for the two case groups.

Characteristics	Mean/Median (min, max)			
	SGA Poor Growth Spurt n = 152	**SGA Other** n = 146	**Cases (Total)** n = 298	**Controls** n = 580
Maternal birth year	1973/1973 (1956, 1988)	1974/1974 (1958, 1992)	1974/1973 (1956, 1992)	1973/1973 (1955, 1993)
Maternal age at childbirth	30/30 (19, 40)	28/29 (17, 41)	29/29 (17, 41)	29/29 (15, 42)
Calendar-year of childbirth	2003/2004 (1995, 2009)	2003/2004 (1995, 2009)	2003/2004 (1995, 2009)	2003/2004 (1995, 2009)
	n (%)			
BMI in early pregnancy (kg/m^2)				
<20	11 (8.3)	24 (19.2)	35 (13.6)	60 (11.9)
20-<25	75 (56.8)	72 (57.6)	147 (57.2)	286 (56.9)
25-<30	25 (18.9)	21 (16.8)	46 (17.9)	109 (21.7)
≥30	21 (15.9)	8 (6.4)	29 11.3)	48 (9.5)
Missing	20	21	41	77
Parity				
1	103 (67.8)	96 (65.8)	199 (66.8)	273 (47.1)
≥2	49 (32.2)	50 (34.2)	99 (33.2)	307 (52.9)
Smoking in early pregnancy				
Non-smokers	106 (77.9)	105 (76.1)	211 (77.0)	490 (90.4)
1–9 cig/day	23 (16.9)	22 (15.9)	45 (16.4)	39 (7.2)
>9 cig/day	7 (5.1)	11 (8.0)	18 (6.6)	13 (2.4)
Missing	16	8	24	38
Gender of child				
Male	65 (42.8)	76 (52)	141 (47.3)	299 (51.6)
Female	87 (57.2)	70 (47.9)	157 (52.7)	281 (48.4)
Diabetes (yes)	0 (0.0)	1 (0.7)	1 (0.3)	2 (0.3)
Gestational diabetes (yes)	2 (1.3)	4 (2.7)	6 (2.0)	11 (1.9)

Table 1. *Cont.*

Characteristics	Mean/Median (min, max)			
	SGA Poor Growth Spurt $n = 152$	SGA Other $n = 146$	Cases (Total) $n = 298$	Controls $n = 580$
Gestational week at partus				
<34	1 (0.7)	0 (0.0)	1 (0.3)	2 (0.03)
34-<37	25 (16.4)	11 (7.5)	36 (12.1)	13 (2.2)
≥37	126 (82.9)	135 (92.5)	261 (87.6)	565 (97.4)
Maternal country of origin				
Sweden	97 (64.8)	70 (50.3)	167 (59.2)	369 (65.8)
Other Nordic Countries	6 (4.2)	<5	8 (2.8)	20 (3.6)
Western Europe, USA, Australia, New Zealand	<5	<5	<5	6 (1.2)
Former Eastern Europe	8 (5.6)	13 (9.3)	21 (7.4)	48 (8.6)
Sub-Saharan Africa	<5	<5	6 (2.1)	13 (2.3)
The Middle East, North Africa	23 (16.1)	37 (26.6)	60 (20.3)	76 (13.5)
East Asia	<5	12 (8.6)	16 (5.7)	23 (4.1)
South/Central America	<5	<5	<5	6 (1.2)
Missing	9	7	16	19

The highest maternal serum concentrations were observed for PFOS in all study groups, followed by PFOA (Table 2). There were no significant differences between all cases and the controls when the PFAS were analyzed as continuous variables. The cases with poor growth spurt had significantly higher concentrations of PFOA and PFHxS as compared to the controls, whereas the case group SGA other had lower concentrations of PFOS compared to the controls.

Table 2. Maternal serum concentrations of the different PFAS compounds (ng/mL) among the cases and controls. For comparison, only statistically significant ($p < 0.05$) differences between groups are marked.

PFAS (ng/mL)	SGA Poor Growth Spurt $n = 152$	SGA Other $n = 146$	Cases (Total) $n = 298$	Controls $n = 580$
	Mean/Median (min, max)			
PFOA	2.9 */2.9 (0.3, 10.1)	2.5/2.1 (0.4, 8.5)	2.7/2.6 (0.3, 10.1)	2.5/2.1 (0.4, 9.4)
PFOS	14.4/12.0 (1.1, 71.1)	12.6 */9.8 (0.6, 126.9)	13.5/10.8 (0.6, 126.9)	13.2/10.7 (0.5, 54.4
PFNA	0.5/0.4 (0.04, 2.8)	0.5/0.4 (0.04, 2.2)	0.5/0.4 (0.04, 2.8)	0.5/0.4 (0.03, 3.5)
PFHxS	0.8 */0.7 (<LOD, 9.4)	0.6/0.5 (<LOD, 9.7)	0.7/0.6 (<LOD, 9.7)	0.6/0.5 (<LOD, 4.5)

PFOA: perfluorooctanoic acid; PFOS: perfluorooctane sulfonate; PFNA: perfluorononanoic acid; and PFHxS: perfluorohexane sulfonate. Limit of detection (LOD): PFOA: 0.04, PFOS: 0.12, PFNA: 0.03, and PFHxS: 0.03. * Significant differences ($p < 0.05$) between cases and controls using Mann–Whitney U-test.

When the different PFAS were analyzed as categorical variables and when the highest vs. lowest exposure categories were compared, the only statistically significant associations observed were ORs below one in the adjusted analysis for PFOS when all cases were included (OR 0.60, 95% CI 0.38–0.96, Table 3). This observation was driven by the group SGA other (unadjusted OR 0.49, 95% CI 0.28–0.85; adjusted OR 0.40, 95% CI 0.21–0.75). For PFHxS, the unadjusted analyses gave a significantly increased OR of 2.00 (95% CI 1.21–3.30) for SGA poor growth spurt, which decreased to 1.51 (95% CI 0.83–2.74) in the adjusted analyses and was no longer statistically significant.

Table 3. The associations between concentrations of PFAS in maternal serum samples taken in early pregnancy and small for gestational age (SGA). Odds ratios (OR) and 95% confidence intervals (CI) were obtained from logistic regressions.

PFAS (ng/mL)	Cases and Controls (*n*)				Cases (Total)				SGA Poor Growth Spurt				SGA Other			
					Unadjusted		Adjusted [a]		Unadjusted		Adjusted [a]		Unadjusted		Adjusted [a]	
	SGA Poor Growth Spurt	SGA Other	SGA (Total)	Controls	OR	95% CI	OR	95% CI	OR	95% CI	OR	95% CI	OR	95% CI	OR	95% CI
PFOA																
≤12.03	35	43	78	145	1.00	ref [b]	1.00	ref [b]	1.00	ref [b]	1.00	ref [b]	1.00	ref [b]	1.00	ref [b]
>12.03–19.62	16	27	43	145	**0.55**	**0.36–0.85**	**0.44**	**0.27–0.72**	**0.46**	**0.24–0.86**	**0.34**	**0.17–0.70**	0.63	0.37–1.07	**0.49**	**0.26–0.91**
>19.62–26.78	45	43	88	145	1.13	0.77–1.65	0.82	0.52–1.29	1.29	0.78–2.12	0.86	0.48–1.54	1.00	0.62–1.62	0.73	0.41–1.31
>26.78	56	33	89	145	1.14	0.78–1.67	0.75	0.47–1.19	1.60	0.99–2.59	0.93	0.52–1.65	0.77	0.46–1.28	0.55	0.30–1.01
PFOS																
≤6.66	40	47	87	145	1.00	ref [b]	1.00	ref [b]	1.00	ref [b]	1.00	ref [b]	1.00	ref [b]	1.00	ref [b]
>6.66–10.73	27	35	62	145	0.71	0.48–1.06	0.70	0.45–1.09	0.68	0.39–1.16	0.68	0.37–1.23	0.75	0.45–1.22	0.75	0.43–1.30
>10.73–18.09	47	41	88	145	1.01	0.70–1.47	0.87	0.57–1.35	1.18	0.73–1.90	0.99	0.57–1.73	0.87	0.54–1.41	0.78	0.45–1.35
>18.09	38	23	61	145	0.70	0.47–1.05	**0.60**	**0.38–0.96**	0.95	0.58–1.57	0.84	0.48–1.48	**0.49**	**0.28–0.85**	**0.40**	**0.21–0.75**
PFNA																
≤0.28	38	46	84	145	1.00	ref [b]	1.00	ref [b]	1.00	ref [b]	1.00	ref [b]	1.00	ref [b]	1.00	ref [b]
>0.28–0.41	27	34	61	145	0.73	0.49–1.09	0.65	0.41–1.03	0.71	0.41–1.23	0.61	0.33–1.11	0.74	0.45–1.22	0.67	0.37–1.19
>0.41–0.60	45	30	75	145	0.89	0.61–1.32	0.74	0.47–1.17	1.18	0.73–1.93	0.93	0.53–1.64	0.65	0.39–1.09	0.61	0.34–1.11
>0.60	42	36	78	145	0.93	0.63–1.36	0.82	0.52–1.29	1.11	0.67–1.81	0.88	0.49–1.58	0.78	0.48–1.28	0.78	0.44–1.40
PFHxS																
≤0.31	29	42	71	145	1.99	ref [b]	1.00	ref [b]	1.00	ref [b]	1.00	ref [b]	1.00	ref [b]	1.00	ref [b]
>0.31–0.53	32	37	69	145	0.97	0.65–1.46	0.97	0.62–1.52	1.10	0.64–1.92	1.13	0.61–2.07	0.88	0.54–1.45	0.81	0.46–1.43
>0.53–0.78	33	24	57	145	0.80	0.53–1.22	0.68	0.42–1.10	1.14	0.66–1.97	0.85	0.45–1.60	0.57	0.33–0.99	0.54	0.29–1.02
>0.78	58	43	101	145	1.42	0.97–2.08	1.16	0.74–1.84	**2.00**	**1.21–3.30**	1.51	0.83–2.74	1.02	0.63–1.66	0.92	0.51–1.64

PFOA: perfluorooctanoic acid; PFOS: perfluorooctane sulfonate; PFNA: perfluorononanoic acid; and PFHxS: perfluorohexane sulfonate. The PFAS variables were included in the models one at a time. [a] Adjusted for BMI (4 categories: <20, 20-<25, 25-<30, >30), parity (2 categories: 1 and ≥2), maternal smoking habits in early pregnancy (2 categories: yes or no), and age (4 categories: <20, 20-<25, 25-<30, 30). [b] Reference category.

In the analyses including only mothers of Nordic origin, significantly increased ORs were observed when the highest exposure category was compared to the lowest exposure category for PFOA (OR 2.85 95% CI 1.43–5.67), PFOS (OR 1.77 95% CI 1.02–3.10), as well as for PFHxS (OR 3.19 95% CI 1.60–6.34) (Table 4). In the adjusted analyses, PFHxS was the only one still statistically significant (OR 2.44, 95% CI 1.11–5.38). When separate analyses were performed for infant sex, two significant associations were observed. A decreased OR among boys for PFOS (adjusted OR 0.38 95% CI 0.20–0.74) was observed when the highest exposure category was compared to the lowest exposure category, and a significantly increased OR among girls was observed for PFHxS (adjusted OR 2.08 95% CI 1.08–4.02).

Table 4. The associations between concentrations of PFAS in maternal serum samples taken in early pregnancy and small for gestational age (SGA), including only women of Nordic origin in the analysis. Odds ratios (OR) and 95% confidence intervals (CI) were obtained from logistic regressions.

PFAS (ng/mL)	Women of Nordic Origin			
	Unadjusted		Adjusted [a]	
	OR	95% CI	OR	95% CI
PFOA				
≤12.03	1.00	ref [b]	1.00	ref [b]
>12.03–19.62	0.68	0.30–1.52	0.48	0.19–1.17
>19.62–26.78	**2.89**	**1.44–5.80**	1.87	0.85–4.12
>26.78	**2.85**	**1.43–5.67**	1.62	0.74–3.57
PFOS				
≤6.66	1.00	ref [b]	1.00	ref [b]
>6.66–10.73	1.00	0.54–1.84	0.92	0.46–1.85
>10.73–18.09	**1.77**	**1.02–3.10**	1.55	0.83–2.90
>18.09	1.33	0.76–2.31	1.11	0.59–2.09
PFNA				
≤0.28	1.00	ref [b]	1.00	ref [b]
>0.28–0.41	0.99	0.56–1.75	0.87	0.44–1.71
>0.41–0.60	1.17	0.67–2.05	1.14	0.59–2.20
>0.60	1.39	0.80–2.41	1.34	0.69–2.60
PFHxS				
≤0.31	1.00	ref[b]	1.00	ref [b]
>0.31–0.53	1.69	0.82–3.48	1.68	0.75–3.77
>0.53–0.78	1.67	0.82–3.43	1.46	0.65–3.28
>0.78	**3.19**	**1.60–6.34**	**2.44**	**1.11–5.38**

PFOA: perfluorooctanoic acid; PFOS: perfluorooctane sulfonate; PFNA: perfluorononanoic acid; and PFHxS: perfluorohexane sulfonate. The PFAS variables were included in the models one at a time. [a] Adjusted for BMI (4 categories: <20, 20-<25, 25-<30, >30), parity (2 categories: 1 and ≥2), maternal smoking habits in early pregnancy (2 categories: yes or no), and age (4 categories: <20, 20-<25, 25-<30, 30). [b] Reference category.

4. Discussion

The present study showed no statistically significant increased risk of SGA birth with higher maternal serum concentrations of PFAS in the general population in Scania, Southern Sweden. Results indicating a reduced risk of SGA birth with higher levels of PFOA and PFOS were observed in some instances—however sporadic and inconsistent across exposure quartiles. In contrast, we found significant associations between exposure to PFHxS and an increased risk of SGA birth in the analysis restricted to women of Nordic origin.

4.1. Previous Studies

Our main results are in agreement with recent EFSA opinions lacking strong support for the hypothesized association between increased PFAS exposure and SGA birth [14]. The most recent opinion reports a tendency among the examined studies towards an inverse correlation between both PFOS and PFOA and birth weight, while acknowledging the lack of strong indications that this translates into an increased risk of SGA birth.

The inconsistent and occasionally conflicting evidence from epidemiological studies on PFAS exposure and reduced birth weight is an ever-recurring predicament, perhaps further complicated by the physiological changes related to pregnancy. A recently updated meta- and bias-analysis on serum PFOA and birth weight found little or no association when exclusively including studies where blood was sampled from mothers in early pregnancy [15]. However, an association was observed when restricting the analysis to samples drawn later in pregnancy. The authors discuss a possible and repeatedly suggested explanation for the heterogeneity of the results: the timing of maternal blood sampling. More specifically, the differences may be attributed to a pregnancy-related expansion of plasma and blood volume and the parallel increase in glomerular filtration rate (GFR), which would be expected to result in reduced concentrations of PFOA later in pregnancy [25,26]. The authors argue that studies based on samples drawn in late pregnancy are more susceptible to confounding by low GFR [27]. An insufficient GFR during pregnancy leads to impaired fetal growth as well as higher maternal serum PFOA, which naturally would be more of a concern later in pregnancy as opposed to earlier.

4.2. Restricted Analyses

The analyses restricted to male infants and women of Nordic origin were specifically selected based on results from previous studies. Numerous research articles on the subject have presented sex-specific effects of PFAS, with male fetuses being more affected regarding birth weight [23,24]. In our study, higher PFOS concentrations resulted in a decreased risk of SGA among boys, whereas higher PFHxS concentrations resulted in an increased risk of SGA among girls. The reason for this pattern remains unknown. Our study found significant differences when restricting the analysis to women of Nordic origin. It is reasonable to assume that Nordic women are more homogenous as a group compared to the entire study population, which might reduce the risk of residual confounding. In addition, epidemiological studies have found PFAS serum levels to be among the highest in these countries [22]. The highest maternal serum concentrations were found in women of Swedish and Danish origin, compared to study participants born elsewhere. Unfortunately, no record of resident duration was available, but the discrepancy was speculated to be a result of differences in background exposure, lifestyle, diet, and perhaps even genetic differences in susceptibility to exposure. Regarding the discrepancy between the ORs in the unadjusted and adjusted models, it was found to be driven by parity when adjusting for all confounders separately. This observation agrees with prior research, as parity has been found to be a significant predictor for PFAS concentrations, decreasing with parity increase [28].

4.3. Strengths and Weaknesses

The strengths of our study include the use of the Southern Sweden Maternal Biobank, extensive birth registers, and a population-based design, which increases generalizability. The high-quality registers enabled us to adjust for important potential confounders, such as smoking habits and parity. By selecting subjects from a regional screening program, the risk of selection bias was minimized, and ultrasound data allowed us to take infant growth in the last trimester into account. Also, all samples were taken at the same gestational age and analyzed using state-of-the-art laboratory equipment. As previously mentioned, PFAS are stable molecules with half-lives of several years, meaning the concentration in a single sample can characterize exposure in epidemiological studies with relative accuracy.

Several limitations should be acknowledged. We could not account for the mother's educational level or socioeconomic status, which have been identified as key confounders. Furthermore, mothers born SGA are more likely to give birth to SGA children themselves, and maternal birth weight was not accounted for in our study. Pathophysiological mechanisms of intrauterine growth restriction (IUGR), namely infections, various maternal diseases, chromosomal aberration, immunization, and placenta-associated complications such as preeclampsia, were not accounted for. In addition, growth estimates based on ultrasound data are susceptible to operator-dependent differences as well as inherent random errors. This could potentially lead to a nondifferential misclassification and bias estimates towards the null. As pointed out in the introduction, thousands of specific molecules constitute the group of PFAS, and the current study chose to investigate only four of them. Although we selected the most widespread and frequently studied substances, this may limit the external validity and applicability of the results.

5. Conclusions

In conclusion, the present study found no association between higher maternal serum levels of PFAS in early pregnancy and an increased risk of SGA birth. However, an increased risk was observed for PFHxS when restricting the analysis to women of Nordic origin. This was a population- and biobank-based study using early-pregnancy sampling while also taking the fetal growth spurt in late pregnancy into account, which is, to our knowledge, a new approach. Future investigations of the potential fetotoxic effects of PFAS are needed to further our understanding of their biological impact on humans.

Supplementary Materials: The following supporting information can be downloaded at: https://www.mdpi.com/article/10.3390/toxics11090750/s1, Intercomparison programme 57, 2016 for toxicological analyses in biological materials.

Author Contributions: The study was designed by L.R., K.K. and S.R.H. Responsible for the chemical analyses was C.H.L. and L.R. performed the statistical analyses. E.M. drafted the manuscript. L.R., A.V., H.H., I.D., K.K., S.R.H. and C.H.L. contributed to drafting and editing the manuscript and approved the final version of the manuscript. All authors have read and agreed to the published version of the manuscript.

Funding: The work was supported by the Swedish Research Council for Sustainable Development (FORMAS, Dnr 2013-01205 and 2017-00699), the ReproUnion, which was co-financed by the European Union, Interreg VÖKS, and ALF government funding (Dnr 2018-0075), and Lund University.

Institutional Review Board Statement: The study was approved by the Ethical Committee (Reg. No. 2014/696) at Lund University, Sweden.

Informed Consent Statement: Patient consent was waived as the Swedish Biobanks in Medical Care Act (SFS 2002:297) allows human material collected for healthcare purposes to be used in research as the donor's consent is collected during sampling. A withdrawal of consent for saved samples can be submitted at any time.

Data Availability Statement: Due to the personal data collected and potentially identifying information contained within the data, the data are available upon request. Ethical approval by the Swedish Ethical Review Authority may be necessary, and requests for data may be sent to LUPOP—Lund University Population Research Platform via email (lupop@ed.lu.se).

Acknowledgments: The authors would like to thank the laboratory technicians Åsa Amilon, Marie Bengtsson, Agneta Kristensson, and Margareta Maxe for analyzing the samples.

Conflicts of Interest: The authors declare no conflict of interest. The funders had no role in the design of the study; in the collection, analyses, or interpretation of data; in the writing of the manuscript; or in the decision to publish the result.

References

1. Sunderland, E.M.; Hu, X.C.; Dassuncao, C.; Tokranov, A.K.; Wagner, C.C.; Allen, J.G. A review of the pathways of human exposure to poly- and perfluoroalkyl substances (PFASs) and present understanding of health effects. *J. Expo. Sci. Environ. Epidemiol.* **2019**, *29*, 131–147. [CrossRef] [PubMed]
2. Olsen, G.W.; Burris, J.M.; Ehresman, D.J.; Froehlich, J.W.; Seacat, A.M.; Butenhoff, J.L.; Zobel, L.R. Half-Life of Serum Elimination of Perfluorooctanesulfonate, Perfluorohexanesulfonate, and Perfluorooctanoate in Retired Fluorochemical Production Workers. *Environ. Health Perspect.* **2007**, *115*, 1298–1305. [CrossRef] [PubMed]

3. Li, Y.; Fletcher, T.; Mucs, D.; Scott, K.; Lindh, C.; Tallving, P.; Jakobsson, K. Half-lives of PFOS, PFHxS and PFOA after end of exposure to contaminated drinking water. *Occup. Environ. Med.* **2018**, *75*, 46–51. [CrossRef]
4. Schroeder, T.; Bond, D.; Foley, J. PFAS soil and groundwater contamination *via* industrial airborne emission and land deposition in SW Vermont and Eastern New York State, USA. *Environ. Sci. Process. Impacts* **2021**, *23*, 291–301. [CrossRef] [PubMed]
5. Perfluoroalkyl and Polyfluoroalkyl Substances in the Environment: Terminology, Classification, and Origins-Buck-2011-Integrated Environmental Assessment and Management-Wiley Online Library. Available online: https://setac.onlinelibrary.wiley.com/doi/10.1002/ieam.258 (accessed on 16 June 2022).
6. Vestergren, R.; Cousins, I.T.; Trudel, D.; Wormuth, M.; Scheringer, M. Estimating the contribution of precursor compounds in con-sumer exposure to PFOS and PFOA. *Chemosphere* **2008**, *73*, 1617–1624. [CrossRef] [PubMed]
7. Fei, C.; McLaughlin, J.K.; Tarone, R.E.; Olsen, J. Perfluorinated Chemicals and Fetal Growth: A Study within the Danish National Birth Cohort. *Environ. Health Perspect.* **2007**, *115*, 1677–1682. [CrossRef] [PubMed]
8. Manzano-Salgado, C.B.; Casas, M.; Lopez-Espinosa, M.-J.; Ballester, F.; Basterrechea, M.; Grimalt, J.O.; Jiménez, A.-M.; Kraus, T.; Schettgen, T.; Sunyer, J.; et al. Transfer of perfluoroalkyl substances from mother to fetus in a Spanish birth cohort. *Environ. Res.* **2015**, *142*, 471–478. [CrossRef] [PubMed]
9. Braun, J.M. Early-life exposure to EDCs: Role in childhood obesity and neurodevelopment. *Nat. Rev. Endocrinol.* **2017**, *13*, 161–173. [CrossRef]
10. Yang, M.; Park, M.S.; Lee, H.S. Endocrine Disrupting Chemicals: Human Exposure and Health Risks. *J. Environ. Sci. Health Part C Environ. Carcinog. Ecotoxicol. Rev.* **2006**, *24*, 183–224. [CrossRef]
11. Hendryx, M.; Chojenta, C.; Byles, J.E. Latent Class Analysis of Low Birth Weight and Preterm Delivery among Australian Women. *J. Pediatr.* **2020**, *218*, 42–48.e1. [CrossRef]
12. Madden, J.V.; Flatley, C.J.; Kumar, S. Term small-for-gestational-age infants from low-risk women are at significantly greater risk of adverse neonatal outcomes. *Am. J. Obstet. Gynecol.* **2018**, *218*, 525.e1–525.e9. [CrossRef]
13. Osuchukwu, O.O.; Reed, D.J. Small for Gestational Age. In *StatPearls*; StatPearls Publishing: Treasure Island, FL, USA, 2022. Available online: http://www.ncbi.nlm.nih.gov/books/NBK563247/ (accessed on 16 June 2022).
14. Risk to Human Health Related to the Presence of Perfluorooctane Sulfonic Acid and Perfluorooctanoic Acid in Food. Available online: https://www.efsa.europa.eu/en/efsajournal/pub/5194 (accessed on 28 October 2022).
15. Gui, S.-Y.; Chen, Y.-N.; Wu, K.-J.; Liu, W.; Wang, W.-J.; Liang, H.-R.; Jiang, Z.-X.; Li, Z.-L.; Hu, C.-Y. Association between Exposure to Per- and Polyfluoroalkyl Sub-stances and Birth Outcomes: A Systematic Review and Meta-Analysis. *Front. Public Health* **2022**, *10*, 855348. [CrossRef]
16. Rylander, L.; Lindh, C.H.; Hansson, S.R.; Broberg, K.; Källén, K. Per- and Polyfluoroalkyl Substances in Early Pregnancy and Risk for Preeclampsia: A Case-Control Study in Southern Sweden. *Toxics* **2020**, *8*, 43. [CrossRef]
17. Malmqvist, E.; Liew, Z.; Källén, K.; Rignell-Hydbom, A.; Rittner, R.; Rylander, L.; Ritz, B. Fetal growth and air pollution—A study on ultrasound and birth measures. *Environ. Res.* **2017**, *152*, 73–80. [CrossRef]
18. Lindell, G.; Maršál, K.; Källén, K. Impact of maternal characteristics on fetal growth in the third trimester: A population-based study. *Ultrasound Obstet. Gynecol.* **2012**, *40*, 680–687. [CrossRef]
19. Maršál, K.; Persson, P.-H.; Larsen, T.; Lilja, H.; Selbing, A.; Sultan, B. Intrauterine growth curves based on ultrasonically estimated foetal weights. *Acta Paediatr.* **1996**, *85*, 843–848. [CrossRef]
20. Norén, E.; Lindh, C.; Glynn, A.; Rylander, L.; Pineda, D.; Nielsen, C. Temporal trends, 2000–2017, of perfluoroalkyl acid (PFAA) concentrations in serum of Swedish adolescents. *Environ. Int.* **2021**, *155*, 106716. [CrossRef] [PubMed]
21. Cnattingius, S.; Källén, K.; Sandström, A.; Rydberg, H.; Månsson, H.; Stephansson, O.; Frisell, T.; Ludvigsson, J.F. The Swedish medical birth register during five decades: Documentation of the content and quality of the register. *Eur. J. Epidemiol.* **2023**, *38*, 109–120. [CrossRef]
22. Ode, A.; Rylander, L.; Lindh, C.H.; Källén, K.; Jönsson, B.A.G.; Gustafsson, P.; Olofsson, P.; Ivarsson, S.A.; Rignell-Hydbom, A. Determinants of maternal and fetal exposure and temporal trends of perfluorinated compounds. *Environ. Sci. Pollut. Res. Int.* **2013**, *20*, 7970–7978. [CrossRef]
23. Engström, K.; Axmon, A.; Nielsen, C.; Rignell-Hydbom, A. High in Utero Exposure to Perfluoroalkyl Substances from Drinking Water and Birth Weight: A Cohort Study among Infants in Ronneby, Sweden. *Int. J. Environ. Res. Public Health* **2022**, *19*, 2385. [CrossRef] [PubMed]
24. Lauritzen, H.B.; Larose, T.L.; Øien, T.; Sandanger, T.M.; Odland, J.; van de Bor, M.; Jacobsen, G.W. Maternal serum levels of perfluoroalkyl substances and organochlorines and indices of fetal growth: A Scandinavian case–cohort study. *Pediatr. Res.* **2017**, *81*, 33–42. [CrossRef] [PubMed]
25. Salas, S.P.; Marshall, G.; Gutiérrez, B.L.; Rosso, P. Time Course of Maternal Plasma Volume and Hormonal Changes in Women With Preeclampsia or Fetal Growth Restriction. *Hypertension* **2006**, *47*, 203–208. [CrossRef] [PubMed]
26. Cheung, K.L.; Lafayette, R.A. Renal Physiology of Pregnancy. *Adv. Chronic Kidney Dis.* **2013**, *20*, 209–214. [CrossRef] [PubMed]

27. Verner, M.-A.; Loccisano, A.E.; Morken, N.-H.; Yoon, M.; Wu, H.; McDougall, R.; Maisonet, M.; Marcus, M.; Kishi, R.; Miyashita, C.; et al. Associations of Perfluoroalkyl Substances (PFAS) with Lower Birth Weight: An Evaluation of Potential Confounding by Glomerular Filtration Rate Using a Physiologically Based Pharmacokinetic Model (PBPK). *Environ. Health Perspect.* **2015**, *123*, 1317–1324. [CrossRef]
28. Berg, V.; Nøst, T.H.; Huber, S.; Rylander, C.; Hansen, S.; Veyhe, A.S.; Fuskevåg, O.M.; Odland, J.; Sandanger, T.M. Maternal serum concentrations of per- and polyfluoroalkyl substances and their predictors in years with reduced production and use. *Environ. Int.* **2014**, *69*, 58–66. [CrossRef]

International Journal of
Environmental Research and Public Health

MDPI

Article

Transfer of Bisphenol A and Trace Metals from Plastic Packaging to Mineral Water in Ouagadougou, Burkina Faso

Boukary Sawadogo [1,*], Francis Ousmane Konaté [2], Yacouba Konaté [1], Ousmane Traoré [2], Seyram Kossi Sossou [1], Eric Sawadogo [1], Pane Bernadette Sourabié Ouattara [2] and Harouna Karambiri [1]

1 Laboratoire Eaux Hydro-Systèmes et Agriculture (LEHSA), Institut International D'Ingénierie de l'Eau et de l'Environnement (2iE), 1 Rue de la Science, Ouagadougou 01 BP 594, Burkina Faso; yacouba.konate@2ie-edu.org (Y.K.); seyram.sossou@2ie-edu.org (S.K.S.); eric.sawadogo2ie@gmail.com (E.S.); harouna.karambiri@2ie-edu.org (H.K.)
2 Agence Nationale Pour la Sécurité Sanitaire de L'Environnement, de L'Alimentation, du Travail et des Produits de Santé (ANSSEAT), Boulevard des Tensoba, Ouagadougou 09 BP 24, Burkina Faso; konatefrancis@yahoo.fr (F.O.K.); us_traore@yahoo.fr (O.T.); b_sourabie@yahoo.fr (P.B.S.O.)
* Correspondence: boukary.sawadogo@2ie-edu.org

Abstract: The consumption of packaged water is growing rapidly in both urban and rural centres in Burkina Faso. Bisphenol A (BPA) and trace metals are among the compounds used in the manufacture of plastic packaging, and their presence in water can pose a health risk to consumers due to their alleged toxicity. Therefore, this study explores the transfer of these compounds from plastic packaging to mineral water in Sudano-Sahelian climatic conditions. Ten samples of packaged sachet water commercialised in Ouagadougou were studied. An absence of BPA in the borehole water used to produce packaged water has been shown. The transfer of BPA into mineral water increases with storage temperature. The BPA that appears in packaged water degrades over time. BPA concentrations ranged from 0 to 0.38 mg/L after two weeks of storage, 0 to 0.8 mg/L after four weeks of storage and 0 to 0.35 mg/L after 8 weeks of storage. Analysis of the trace metals showed steadily increasing concentrations from the second to the sixth weeks, with concentrations ranging from 0 to 9.7 μg/L for cadmium and from 0 to 0.13 mg/L for iron in the sachet water samples.

Keywords: Bisphenol A; drinking water quality; plastic containers; sachets of mineral water; Sudano-Sahelian climatic conditions; trace metals; water micropollutants

Citation: Sawadogo, B.; Konaté, F.O.; Konaté, Y.; Traoré, O.; Sossou, S.K.; Sawadogo, E.; Sourabié Ouattara, P.B.; Karambiri, H. Transfer of Bisphenol A and Trace Metals from Plastic Packaging to Mineral Water in Ouagadougou, Burkina Faso. *Int. J. Environ. Res. Public Health* **2023**, *20*, 6908. https://doi.org/10.3390/ijerph20206908

Academic Editors: Esref Demir and Sam Kacew

Received: 7 July 2023
Revised: 8 August 2023
Accepted: 22 August 2023
Published: 11 October 2023

1. Introduction

Africa is currently characterised by rapid demographic growth, with one of the highest birth rates in the world [1]. The increasingly numerous population is become urbanised, with an average urbanisation rate of 4% [2]. The urban population growth dynamic has risen over the years from 31.0% in 1990 to 54.4% in 2020, while in West Africa, the growth rate has increased from 7.5% to over 34.0% from 1960 to 2020. This translates to more than 123 million inhabitants in West African cities in 2020 [3]. This new environment is bringing about profound changes in lifestyle and consumption patterns. Feeding and maintaining this urban population calls for intensification of production. Therefore, the situation places significant pressure on available resources, especially on the limited water resources. Burkina Faso, just like other countries in West Africa, faces major challenges in access to water and sanitation amidst the galloping population growth. Urbanisation in Ouagadougou remains heterogeneous [4]. Against this backdrop of urbanisation and the soaring water demand, it has been observed that people are increasingly using packaged water as drinking water, both at home and in the workplace [5]. Packaged water is produced by local industrial units. The hygienic appearance of these waters appeases consumer confidence. The situation has thus promoted the creation of numerous packaged water production units.

However, the rush to create packaged water production units must not compromise consumer protection, as failure to comply with specific requirements can lead to products that do not adhere to the potability standards in force for drinking water, especially as water is a mass consumption product. This is why quality control, through regular monitoring of the water produced, is an important part of the process.

Most water quality monitoring protocols focus on common parameters that can immediately impact consumer health. However, it is recognised that water can be contaminated in numerous ways [6–9]. There is natural contamination, i.e., depending on the composition of the rocks making up the aquifer [10–13], and/or anthropic contamination due to human activities [14]. The latter is complex because it varies according to human activities and different uses of the resource [15–17]. Some water pollutants can be toxic through a cumulative effect if ingested repeatedly. This is particularly true of certain micropollutants such as trace metals and bisphenol A (BPA) [18,19]. BPA, whose chemical name is 2,2-bis(4-hydroxyphenol) propane, is a synthetic chemical mainly used to manufacture plastics and resins. It is obtained through a reaction between two phenol equivalents and one acetone equivalent, with its molecule containing two phenol functional groups [20,21]. Mineral water is generally packaged in plastic containers produced with BPA [22]. Under normal conditions, BPA is a chemically stable substance, but it breaks down slowly at high temperatures and is released into the environment [23,24]. BPA is considered to be an endocrine disruptor [22,25]. It is also considered by several authors to be the cause of numerous diseases [19,26,27]. In recent years, BPA has been the subject of numerous studies worldwide. Authors have been particularly interested in its presence in water and its toxicity for humans and the environment [28,29]. Other studies have often focused on its transfer from plastic bottles to water or from cans to food [30,31]. Trace metals are also associated with various risks for humans in the event of high levels in drinking water [32–34]. The novelty of this study deals with the transfer of BPA and trace metals from plastic packaging to packaged mineral water; we assess their levels in packaged water being produced and commercialised in an urban centre in Africa. The external and internal factors involved in their migration from plastics to packaged water were studied.

2. Materials and Methods

2.1. Study Area

The study was carried out in Ouagadougou, the capital city of Burkina Faso. It is the country's largest economic and cultural centre, with an estimated population of two million six hundred and eighty-four thousand and fifty-two (2,684,052) inhabitants in 2020 [35]. The city is growing steadily, with an annual growth rate of 7.6% [5]. Ouagadougou is the capital of the Centre region, with twelve (12) districts subdivided into fifty-two (52) sectors. The city has a surface area of 2857.124 km^2 and is located between parallels 12°30′ and 12°25′ north latitude and meridians 1°27 and 1°35 west longitude. Figure 1 shows a map of the city of Ouagadougou.

2.2. Sampling and Preliminary Tests

The National Agency for Environmental, Food, Labour, and Health Product Safety in Burkina Faso (Ansseat) monitors water consumed in the city of Ouagadougou. Units that produce mineral water sign an assistance contract with this agency, allowing it to monitor their water quality. A sample of ten brands of mineral water packaged in sachets were selected from among the units that had signed a service contract with Ansseat for the study. In order to gain an overview of the transfer of these substances, the samples were selected based on the physical nature of the containers. Packaging from different water production units differs according to the presence or absence of a writing tab, as well as the roughness and thickness of the sachet. Other criteria used to select the samples were the city's geographical distribution of the production units. Table 1 presents the physical characteristics of the sachets while the geographical location of the samples in the city is shown in Figure 2. The final criterion for sample selection was biodegradability; the test was carried out using an XRF Tester as described by Hagiwara et al. [36]. The plastic containers were scanned to determine their iron, manganese, and cobalt concentrations and their biodegradability. Many metals such as iron, lead, cadmium, cobalt, manganese, antimony, and many other trace elements are used in the composition of plastics to improve their quality and accelerate their biodegradability. This therefore informed the selection of containers with high levels of these metals to determine whether they are transferred to the packaged water. Table 2 shows the iron, manganese, and cobalt contents given by the XRF tester, used as criteria for choosing our samples. Considering all the selectivity criteria, ten (10) samples were selected for the rest of the study (Table 1). For each production unit, samples were taken from a pack of forty sachets of packaged water (W1 to W10). The borehole water used to produce the sachet water (W1 to W10) are named F1 to F10, respectively. The quality of the borehole water was analysed.

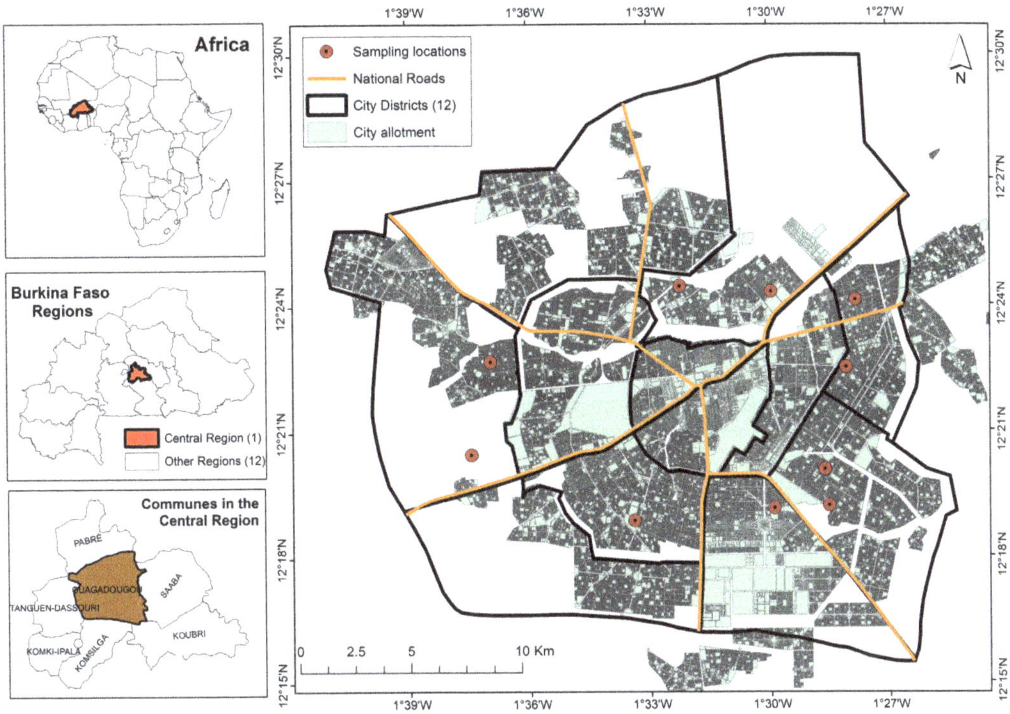

Figure 1. Location of the city of Ouagadougou.

Table 1. Physical characteristics of the water sachet and geographical distribution of the samples.

Sample Code	Roughness	Presence of Tongue
W1	Smooth	Yes
W2	Smooth	No
W3	Rough	Yes
W4	Smooth	Yes
W5	Rough	No
W6	Rough	No
W7	Smooth	No
W8	Smooth	No
W9	Rough	No
W10	Smooth	No

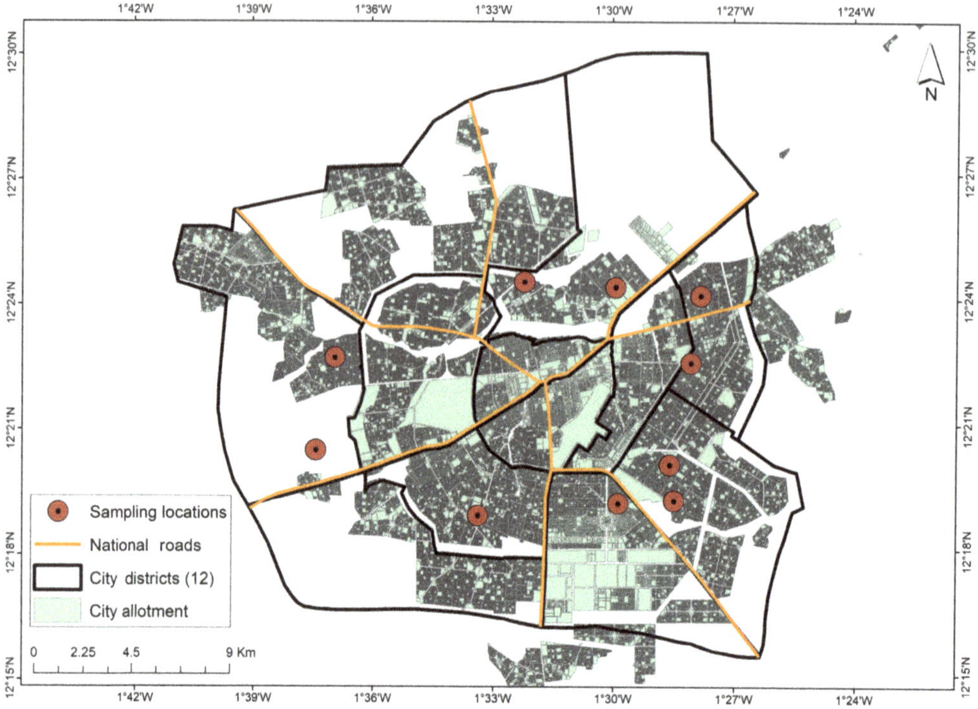

Figure 2. Geographical distribution of samples in the city of Ouagadougou.

2.3. *Analyses and Calibration*

BPA was analysed using an Agilent gas chromatograph-mass spectrometer (GC-MS) [31,37–39]. The analyses were carried out after an internal validation of the method. Before carrying out the analysis of BPA in the various samples, the method was verified. The purpose of this verification was to determine the effectiveness or efficiency of the method prior to its use in this study. A calibration range was prepared using standard solutions. Eight solutions of different concentrations (12.50 mg/L, 6.25 mg/L, 3.13 mg/L, 1.56 mg/L, 0.95 mg/L, 0.39 mg/L, 0.10 mg/L, and 0.05 mg/L) of BPA were prepared for GC-MS calibration using acetonitrile and ultrapure water as a solvent. The sample was spiked to verify the method's performance using the method described by Errico et al. [40]. Dichloromethane is the solvent used for PBA extraction. Table 3 shows the recovery rates for Bisphenol A using GC/MS as a function of the concentrations previously spiked. The internal procedure recommends a recovery rate of between 75% and 120%.

Once the method had been successfully tested, all the samples were treated in accordance with the method using the GC-MS. For each sample, the groundwater was analysed at ambient temperature. To assess the effect of temperature, the packaged water samples were placed in ovens set at 18 ± 2 °C, 29 ± 2 °C, and 39 ± 2 °C. The duration of conditioning varied from two, four, and eight weeks to study the combined effect of temperature and storage time. This was performed to monitor the behaviour of BPA and trace metals contained in the plastic container. These three sample conditioning temperatures were given by the meteorological surveys of the city of Ouagadougou:

18 °C as the average of the lowest annual temperatures;
29 °C as the average annual temperature;
39 °C as the average of the highest annual temperatures.

Samples collected from the boreholes were immediately acidified with nitric acid at a concentration around 5% v/v. Trace elements were measured with an Avio Perkin Elmer inductively coupled plasma atomic emission spectrophotometer (ICP-OES). The direct method was used to assay the samples. The calibration range was carried out using a multi-element standard.

Table 2. Iron, manganese, and cobalt content given by the XRF tester.

Sample Code	Parameters (mg/L)			Standard Results	Results P/F/X
	Fe	Mn	Co		
W1	33	38	2	Oxo Biodegradable	Validated
W2	24	48	2	Oxo Biodegradable	Validated
W3	52	16	3	Oxo Biodegradable	Undetermined
W4	38	25	3	Oxo Biodegradable	Validated
W5	58	12	2	Oxo Biodegradable	Undetermined
W6	73	13	2	Oxo Biodegradable	Undetermined
W7	81	17	2	Oxo Biodegradable	Validated
W8	75	32	3	Oxo Biodegradable	Validated
W9	45	47	4	Oxo Biodegradable	Validated
W10	70	33	4	Oxo Biodegradable	Validated

Table 3. Bisphenol A recovery rates using GC/MS.

Concentrations of Spiked Solutions (mg/L)	Concentrations Detected by GC/MS (mg/L)	Recovery Rate (%)
1.00	1.01	101
5.00	5.20	104
10.00	10.50	105
Average rate		103

3. Results and Discussion

3.1. Production, Transport and Storage of Packaged Water

The production and marketing of packaged water is an activity involving a chain of operations. A better interpretation of the results requires an in-depth examination of the various elements in the chain. Similarly, as several parameters can influence the fate of BPA and trace metals, the impact of transport and storage conditions on transfer to the product was highlighted.

Although all the production units use borehole water, the water circuit from the borehole to the bagging unit differs from one unit to another. The most complex systems consist of pre-treatment and demineralisation devices with units based on dense membranes and tanks for adding mineral substances, a contrast to simple circuits with water distribution from a storage tank. Based on the analysis, the semi-automated units seem to offer a better guarantee of the final product's quality and its compliance with good production standards than products from the smaller units.

After production, the finished product must be transported to wholesale distributors' warehouses or retailers' shelves, i.e., distribution and/or consumer centres. This stage of the production chain involves a variety of means of transport, including lorries, commercial vehicles, and tricycles. Over 60% of products are transported by tricycles because of their relatively low acquisition, servicing, and maintenance costs compared to commercial vehicles and lorries (insurance, technical inspection, motor vehicle tax). These vehicles offer greater flexibility of use and placement in a city where traffic is dense at certain times of the day (driving licence, ability to fill small orders quickly). In addition, more than 65% of means of transport are not covered, which exposes the conditioned water to dust and sunlight.

Storage occurs at three levels: after production, at the sale point, and at the consumer's home. Before transport, storage is carried out within the water production units. The observations show that the water samples produced are assembled on the floor or on pallets. The findings show that all the production units surveyed had suitable storage conditions. Storage is also carried out at the distribution point. At this level, wholesalers have warehouses in which products are stored. At these points, water packs are generally stored in metal grids provided by the producers. Even when fitted out, these racks often leave the products exposed to the sun, rain, and dust. Therefore, there is a clear risk of contamination and deterioration of water quality at the retail level. At the household level and in retail outlets, storage is relatively better. Consumers use water for direct consumption or storage in coolers, refrigerators, or equipped areas of the house.

Water is therefore stored under different conditions depending on the stage in the consumption chain. Storage at wholesalers, therefore, appears to be the weakest link in the chain, with conditions that do not guarantee product quality. More than half of the packaged water samples were stored in inadequate conditions. Exposure to the sun, wind, dust, and other adverse weather conditions poses the risk of deterioration in product quality. Previous studies have reported that photo-oxidation may be one of the degradation pathways for BPA [41]. It has also been shown that storage conditions do not depend on the brand of the product. It is more generally linked to the distribution company.

3.2. *Bisphenol A in Sources and Packaged Water in Ouagadougou*

The presence of bisphenol A in drinking water can have several origins, the greatest risk of which appears to be transfer from the container to the contents. To a lesser extent, the environmental conditions at the source and a lack of hygiene in handling may also be involved.

The presence of BPA in borehole water was assessed for all ten sources used to produce packaged water. The results of the analyses showed the absence of BPA in all the sampled groundwater used for the production, thus highlighting the fact that the level of BPA pollution in the city of Ouagadougou is low and cannot lead to contamination of the water table. This is justified by the Environment/Health report [42], which shows that the movement of BPA in the water and/or soil matrix is influenced by the physico-chemistry of the receiving soil and the water flowing through it, in particular the pH and the properties of the organic matter in the water. On the other hand, other results reported by Rudel et al. [43] showing concentrations of 1.41 μg/L in groundwater samples taken between a wastewater treatment plant and a municipal landfill, justify that the environmental state of the site on which the production unit is to be located must be closely scrutinised during the health inspection prior to the authorisation to open.

3.3. *BPA in Packaged Water*

3.3.1. Effect of Temperature on BPA Migration

Surveys and observations revealed poor storage conditions for packaged water, particularly regarding exposure to sunlight. As certain studies had revealed the possible effects of temperature on the stability of packaged water, investigations were carried out on the samples taken. The ten samples were therefore subjected to three different temperatures: 18 °C, 29 °C, and 39 °C. Samples were conditioned in ovens at these three temperatures, and the results of the analyses are shown in Figure 3. The results obtained show that although BPA was not present in the water leaving the production lines, BPA levels of up to 0.780 mg/L were obtained in the same packaged water exposed to high temperatures. This substance, previously absent in borehole water, was found in packaged water, suggesting that the containers were the source of contamination. This shows that the sachets used for packaging are made from polycarbonates and/or the resin is made from bisphenol A. These results are in line with those of Cadi et al. [44] who maintain that most plastics are made from polycarbonate or resin, and therefore contain bisphenol A.

It also appears that, at low temperatures, BPA is virtually absent from all the packaged waters studied after 14 days' storage. The results also show the presence of Bisphenol A in all the samples studied at ambient and high temperatures in the city of Ouagadougou.

Analysis of the results shows that the concentration of BPA in packaged water increases with temperature. Thus, a rise in temperature leads to an increase in BPA concentrations in the samples. Temperature therefore accelerates the fragmentation and depolymerization of polycarbonates and peroxide resins, leading to the release of monomers, including BPA [44]. These were also the results of the study by Brede et al. [45], who reported that the use of plastic containers leads to the migration of BPA in water, especially when the temperature is raised. For other authors, such as Calafat et al. [46], the concentration of BPA in water contained in plastic can increase up to 55 times compared to low temperatures. In this 4-week conditioning analysis, BPA concentrations ranged from a minimum of 0.032 mg/L to a maximum of 0.800 mg/L (Figure 3). Taking this maximum concentration into account, subjects weighing 10 kg (average weight of children aged 0–3 years) and 63 kg (average weight of adults) drinking 0.75 L and 1.5 L of sachet water per day, respectively, would have accumulated concentrations of 0.06 and 0.02 mg/kg body weight/day of BPA. Under these conditions of exposure, adults absorb a low dose compared to the acceptable daily intake (ADI), which is 0.05 mg/kg of body weight per day. If packaged water is their only source of BPA contamination, they would be safe. A child weighing 10 kg absorbs 0.06 mg/kg/day, which is higher than the ADI. Although it may not the only source of BPA contamination, the consumption of packaged water leads infants to absorb more than the TDI (0.05 mg/kg of body weight per day). This shows that children are the most exposed to BPA absorption, confirming the thesis of previous authors who maintain that young people in general, and foetuses and newborns in particular, are vulnerable to BPA [47].

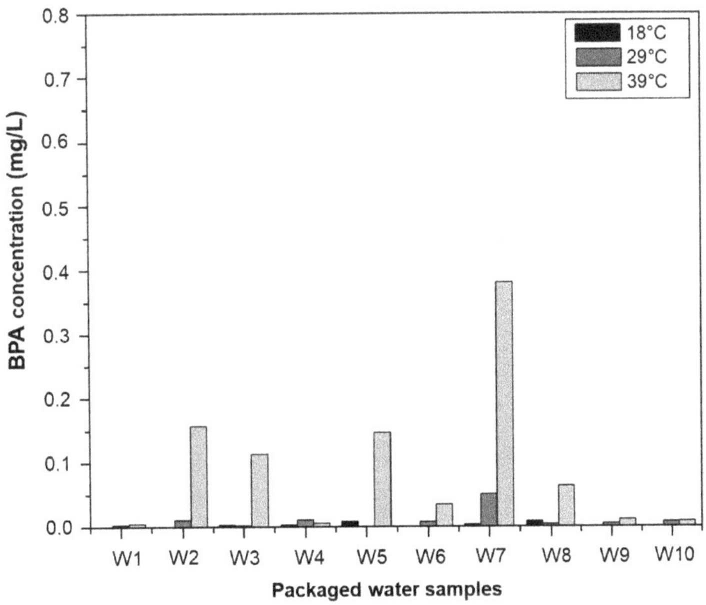

(a)

Figure 3. *Cont.*

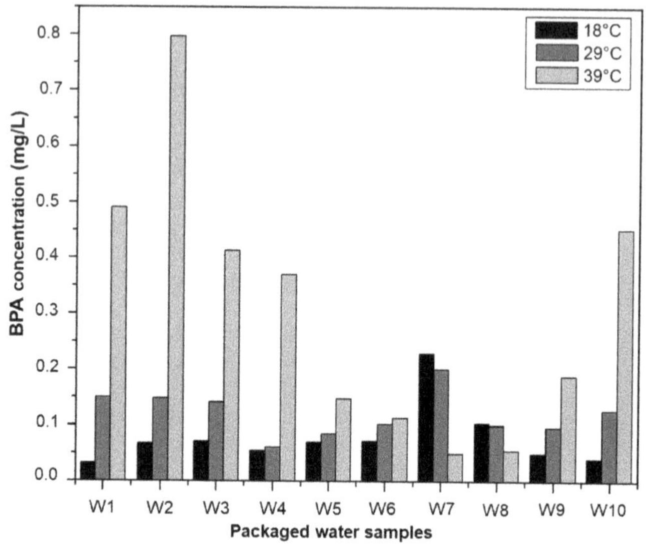

(b)

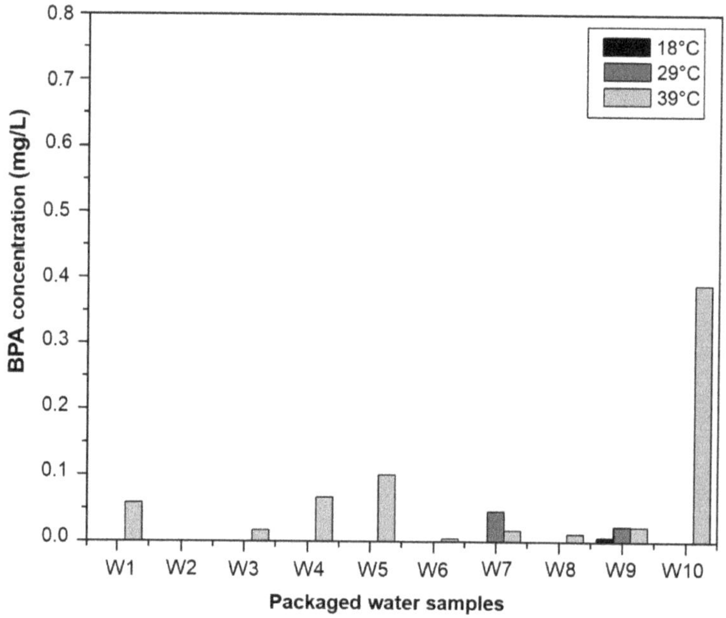

(c)

Figure 3. BPA in conditioned water as a function of conditioning time and temperature. (**a**): 2 weeks; (**b**): 4 weeks; (**c**): 8 weeks.

3.3.2. Effect of Storage Time on BPA Migration

Figure 3 also shows changes in BPA levels as a function of time. Concentrations at two weeks (Figure 3a) are lower than those at four weeks (Figure 3b). Therefore, the transfer of BPA into the water studied is influenced by the storage time. BPA concentrations in the same sample increased as storage time increased. This indicates that depolymerization does not occur abruptly, but rather as a function of time. This result is in line with those of Sajiki and Yonekubo [48], who showed that the concentration of BPA increases up until day 35. Yoshida et al. [49] reported that storage time is an important factor in BPA migration in products packaged or preserved in plastic containers. A comparison of the results at the two storage times shows different changes in the BPA content in the packaged water samples studied. The transfer of BPA from the sachet to the packaged water is not linear over time. This could be due to the simultaneous degradation of the sachet by physico-chemical and microbiological factors, which contribute to the release of BPA [50–52]. On the other hand, the tab of the sachet appears to have no significant impact on the transfer of BPA into the water, whatever the temperature and/or packaging time.

3.3.3. The Degradation of BPA in Packaged Water

BPA is not constantly increasing in packaged water. It degrades under certain conditions, and its content decreases with time. Figure 4 shows the regression of BPA concentrations at two, four, and eight weeks of storage. Between four and eight weeks, concentrations decrease as a function of time. This situation can be explained by a degradation or fragmentation of BPA as the exposure time increases. BPA in contact with oxygen oxidises and degrades, especially in the presence of water. Dissolved oxygen is a major player in the degradation of BPA, and is among the many factors that can also act in this way in packaged water. This is supported by Staples [53], who states that aerobic degradation is the dominant process in the decrease or disappearance of bisphenol A in an aquatic environment. In four weeks, degradation was complete in samples W2 and W6, almost complete in samples W1, W3, W4, W7, W8, and W9, and low in samples W5 and W10. This degradation time is close to that of West et al. [54], who found that under aerobic conditions, BPA degrades for 28 days. Other authors, such as Kang et Kondo [55], who carried out experiments in aerobic and anaerobic conditions with river water spiked with BPA, have concluded that BPA degrades rapidly in the presence of oxygen and that oxidation and/or the effect of anaerobic bacteria have little or virtually no capacity to degrade BPA. This thesis is supported by most authors, who confirm that BPA degrades in the presence of oxygen. Its half-life cycle in the environment is 3 to 5 days in the presence of oxygen. However, it is highly resistant in anoxic aquatic environments, with an assumed half-life of over a year [56]. Microbiology is also a parameter responsible for the elimination of BPA. Microbial analyses revealed the presence of the total germs in all the samples studied. These germs have an optimum growth temperature between 28 and 35 °C, which is close to 39 °C, the highest temperature in this study. The results of these analyses are shown in Table 4. Among these total germs, some are capable of degrading BPA to the point of total or partial mineralisation. This is also the opinion of a European Commission study which, in describing the conditions of the biological degradation of BPA, states that 60% would be mineralised into CO_2, 20% would constitute the carbon of bacterial cells, and 20% of soluble organic carbon would remain in the media [57].

3.3.4. Chloride and pH Effect on BPA Behaviour

Some authors such as Sakai et al. [58] and Yang et al. [59] have also proposed forms of BPA degradation by specific bacteria, showing that other physico-chemical reactions may be responsible for BPA degradation. This is the case in reactions with chlorides. Thus, by monitoring the presence of chloride ions in the samples for the three study temperatures, it appears that a drop in BPA concentrations is followed by a reduction in chloride concentration in the packaged water. There is then a possible reaction between BPA and chlorides to form new chlorinated substances that no longer have the same physico-chemical characteristics as the initial BPA. This analysis is supported by previous studies which have shown that chlorides react with BPA to give monochloroBPA, dichloroBPA, trichloroBPA, and tétrachloroBPA [60]. Table 5 shows changes in chloride concentrations as a function of time and temperature.

Table 4. Total germs in packaged water samples.

Samples	W1	W2	W3	W4	W5	W6	W7	W8	W9	W10
Total germs at 37 °C (CFU/L)	440	572	799	404	76	64	900	784	484	816

Table 5. Chloride content of samples after 60 days of conditioning.

Exposure Temperature	Chloride Concentration (mg/L)									
	W1	W2	W3	W4	W5	W6	W7	W8	W9	W10
18 °C	1.1	0.8	0.8	0.9	0.8	1.0	0.9	0.8	0.9	0.8
29 °C	0.9	0.8	0.8	0.9	0.7	0.9	0.8	0.8	0.9	0.7
39 °C	0.5	0.6	0.5	0.7	0.6	0.6	0.7	0.7	0.6	0.6

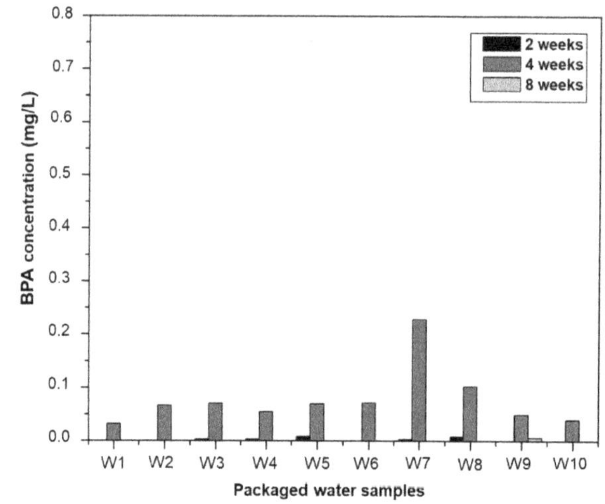

(**a**)

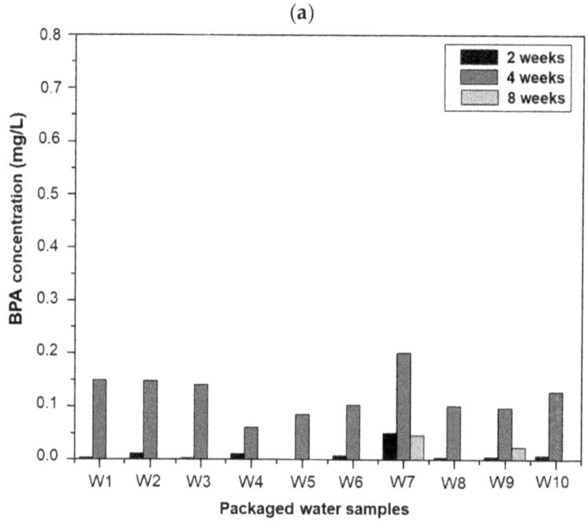

(**b**)

Figure 4. *Cont.*

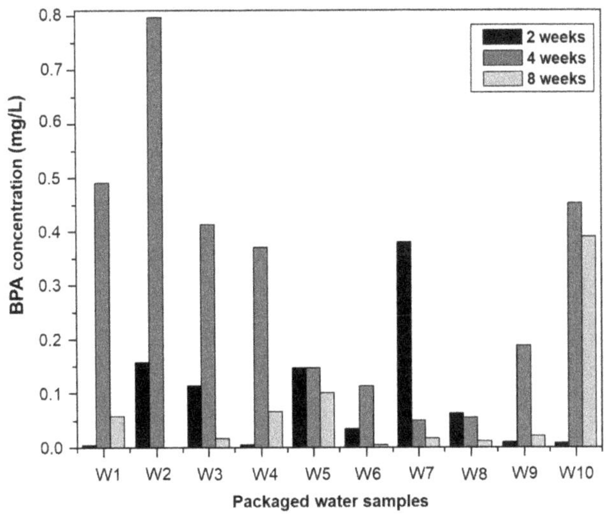

(c)

Figure 4. Evolution of BPA in conditioned water as a function of conditioning time and temperature. (**a**): 18 °C; (**b**): 19 °C; (**c**): 39 °C.

After eight weeks of conditioning, the results show an increased degradation of BPA in all samples except sample W10 at a temperature of 39 °C. This seems to be closely linked to pH, as observed for only the sample at 39 °C with a pH above 8, as shown in Table 6. This suggests that the degradation of BPA is not very efficient at high pH. Chauveheid et al. [60] reported in their study of exposure to BPA and its chlorinated derivatives that degradation of this substance occurs at around pH 7.7 and is clearly ineffective at pH 8.0 and above.

Table 6. pH of samples after 8 weeks of storage.

Exposure Temperature	pH Value									
	W1	W2	W3	W4	W5	W6	W7	W8	W9	W10
18 °C	7.3	7.6	6.9	7.2	7.3	7.2	7.2	7.0	7.2	7.5
29 °C	7.5	7.7	7.6	7.3	7.4	7.2	7.6	7.7	7.3	7.8
39 °C	7.6	7.8	7.7	7.4	7.5	7.8	7.7	7.9	7.4	8.1

3.4. Trace Metals in Packaged Water

All the packaged water carry an insignia printed in ink. The ink products can therefore migrate inside the product. These include trace metals. Cadmium is a micropollutant that can be transferred from the sachet to the water. Cadmium was absent in all samples of the borehole water sources used for the production. Concentrations varied according to the temperature and exposure time of the samples. Table 7 shows cadmium levels in packaged water at different temperatures and as a function of time. Concentrations ranged from 0.0 to 9.7 μg/L. These results show an increase in cadmium concentration in the conditioned water as a function of temperature and duration of storage. These concentrations increase as the exposure time increases and/or the temperature rises.

Table 7. Cadmium content in conditioned water samples.

Temperature	Cd Concentration (μg/L) *											Exposure Time
	W1	W2	W3	W4	W5	W6	W7	W8	W9	W10	Average	
18 °C	<DL	<DL	<DL	<DL	<DL	<DL	<DL	<DL	<DL	<DL	<DL	3 weeks
29 °C	<DL	<DL	<DL	<DL	<DL	<DL	<DL	0.9	<DL	<DL	0.1	
39 °C	2.0	1.0	2.5	<DL	<DL	<DL	<DL	5.9	<DL	<DL	1.1	
18 °C	1.0	<DL	<DL	<DL	<DL	<DL	<DL	<DL	<DL	<DL	0.1	6 weeks
29 °C	1.0	<DL	<DL	<DL	<DL	1.0	3.5	5.1	<DL	0.1	1.1	
39 °C	2.0	2.5	6.9	1.0	9.7	2.0	6.6	6.2	1.1	7.2	5.9	

* The admissible limit for Cd in drinking water in accordance with WHO guidelines is 3 μg/L.

After six weeks of storage, cadmium was found in all the packaged water studied at 39 °C. On the other hand, the trend in Figure 5 shows that the increase in temperature speeds up the transfer of cadmium to the drinking water. Plastic containers would therefore be responsible for releasing cadmium into packaged water. This migration of cadmium from the sachet to the packaged water is influenced by temperature and contact time. As some sachets have ink marks on the tabs and not on the sachets, we can conclude that the cadmium is not only in the ink, but also in the composition of the sachet.

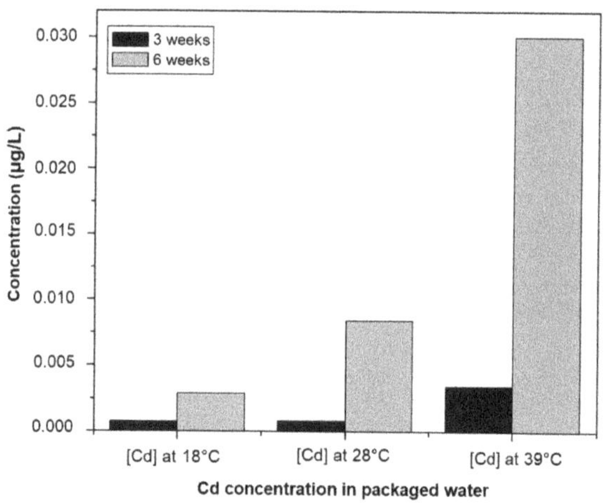

Figure 5. Cd evolution in sachet water as a function of time and temperature.

WHO guidelines for drinking water indicate a limit of 3.0 μg/L cadmium in drinking water. The results of analyses of samples stored for six weeks show that at 18 °C and 29 °C, concentrations vary from 0.0 to 5.1 μg/L, with the exception of W8 at 29 °C which recorded a concentration of 5.1 μg/L, above the recommended standard. At 39 °C, samples W1, W2, W4, W6, and W9, with concentrations of less than 3.0 μg/L, complied with the standard, while samples W3, W5, W7, W8, and W10 had concentrations ranging from 6.6 to 9.7 μg/L, which were two or three times higher than the limit value. Accordingly, depending on the temperature and storage time, some packaged waters may become unfit for consumption because they do not comply with the drinking water standards in force in Burkina Faso.

Table 8 shows the results of iron concentration in the borehole water used as a source for producing the packaged water. Iron was found in all the samples at concentrations ranging from 0.10 to 0.45 mg/L. The concentration for samples F1, F2, F3, F4, F5, F7, and F10 comply with the WHO allowable limit of 0.3 mg/L in drinking water. Samples F6 and F9, with concentrations of 0.45 and 0.40 mg/L, respectively, exceeded the standard limit. This excessive amount of iron is thought to be due to dissolution of the rocks followed by leaching of the soil or cracking of the parent rock [61].

Table 8. Iron content in borehole water samples.

Water Source	F1	F2	F3	F4	F5	F6	F7	F8	F9	F10
Iron concentration (mg/L)	0.25	0.20	0.15	0.30	0.10	0.45	0.25	0.25	0.40	0.10

The impact of environmental conditions such as temperature and storage time on iron transfer from the container to the water consumed was studied (Figure 6). The iron present in the borehole water was found to be completely absent in the packaged water samples W1, W2, W3, W4, W5, and W6, and in small quantities in the packaged water samples W7, W8, W9, and W10 after three weeks and at 18 °C. This can be explained by the treatment method adopted by the packaged water production units. These units sometimes use pre-treatment devices that can retain certain minerals such as iron. As the temperature was increased over time, iron was found in virtually all the samples at increasingly higher concentrations. Iron, like the cadmium present in the sachets, migrates into the packaged water as a function of temperature and time (Table 9). This is why iron concentrations are constantly increasing in packaged water samples. However, according to some authors, iron reacts with BPA in solution, and its concentration must therefore fall in the water [62]. This could be due to the fact that the quantity of iron released is greater than the quantity of iron that reacts with bisphenol A. This always gives an increasing trend that does not reveal this reaction. On the other hand, the absence of or very weak reactions with bisphenol A can be explained by the fact that many substances (oxygen, chloride, etc.) and many bacteria already react with BPA. This gives iron little or no opportunity to react with BPA. In light of Tables 7 and 9, the migration kinetics of the trace metals is not only a function of the content but also of the alloys between the iron and other trace metals. The degree to which substances migrate into the contents depends not only on the concentrations and chemical characteristics of the compounds, but also on the chemical composition of the plastic [63]. In other words, the presence of certain chemical compounds is likely to create alliances and prevent the migration of other compounds.

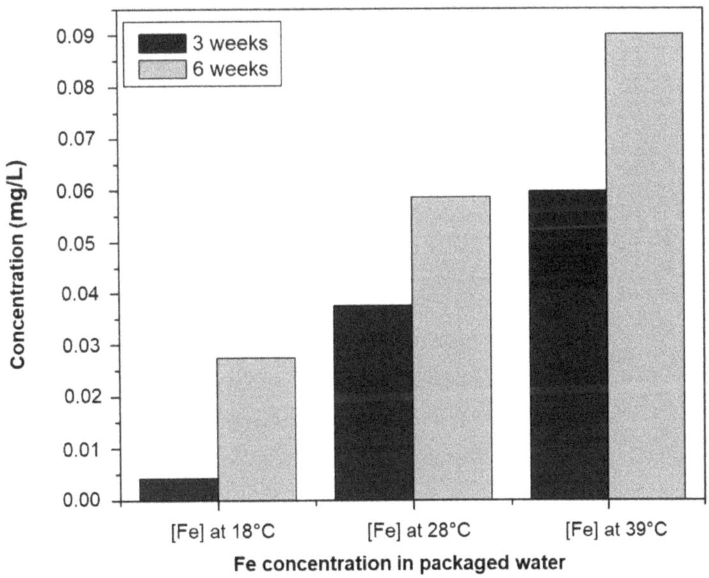

Figure 6. Fe evolution in sachet water as a function of time and temperature.

Table 9. Iron content in conditioned water samples.

Temperature	Fe Concentration (mg/L)											Exposure Time
	W1	W2	W3	W4	W5	W6	W7	W8	W9	W10	Average	
18 °C	<DL	<DL	<DL	<DL	<DL	<DL	0.015	0.017	0.001	0.009	0.004	3 weeks
29 °C	0.030	0.036	0.038	0.003	0.043	0.009	0.069	0.064	0.031	0.054	0.037	
39 °C	0.052	0.039	0.054	0.003	0.082	0.095	0.069	0.081	0.067	0.054	0.059	
18 °C	0.010	0012	<DL	0.008	0.036	0.032	0.059	0.039	0.044	0.034	0.027	6 weeks
29 °C	0.032	0.055	0.069	0.016	0.046	0.051	0.080	0.067	0.083	0.086	0.059	
39 °C	0.076	0.054	0.081	0.100	0.132	0.098	0.086	0.084	0.088	0.102	0.090	

4. Conclusions

This study assessed the influence of plastic on the quality of packaged water. The survey revealed differences in production conditions depending on the size of the packaged water production structure. Poor storage conditions were found in the places where packaged water is marketed. All the water sources analysed were BPA-free; some contained trace metals, but without exceeding the WHO guidelines for drinking water. The sachets are designed on the basis of BPA to ensure their resistance and transparency, and on the basis of metallic trace elements to accelerate their biodegradability. Under certain physico-chemical and biological conditions, these substances can migrate to the contents (packaged water). During the first thirty days, there was a massive migration of BPA into the water due to the biodegradation of the plastic, which could, in this case, be considered as a source of contamination of substances. After thirty days, BPA gradually degraded and began to disappear after sixty days in some samples. Bottled water contains less BPA when stored at low temperatures at the start of production (before two weeks) and late after production (after two months). BPA is toxic depending on the quantity absorbed and the weight of the consumer. The migration of chemical elements does not depend on the physical nature of the sachets, but rather on the type of chemical compound and storage conditions. Iron and cadmium are also used in the manufacture of sachets and constantly migrate into packaged water due to biodegradation and physico-chemical processes. The geographical location and the presence or absence of a tab on the sachet did not influence the results obtained.

Author Contributions: Conceptualization, B.S., F.O.K. and Y.K.; methodology, B.S., F.O.K., E.S., O.T. and Y.K.; software, B.S. and S.K.S.; validation, Y.K. and H.K.; formal analysis, E.S., B.S., F.O.K. and O.T.; investigation, E.S., B.S. and F.O.K.; resources, Y.K., H.K. and P.B.S.O.; data curation, E.S., B.S.; writing—original draft preparation, B.S. and E.S.; writing—review and editing, B.S., S.K.S. and Y.K.; visualization, F.O.K. and O.T.; supervision, Y.K.; project administration, Y.K. and H.K.; funding acquisition, H.K. All authors have read and agreed to the published version of the manuscript.

Funding: The APC was funded by World Bank Group under the Africa Centers of Excellence for Development Impact (ACE Impact) Project.

Institutional Review Board Statement: Not applicable.

Informed Consent Statement: Not applicable.

Data Availability Statement: Not applicable.

Acknowledgments: The authors acknowledge the Institut International d'ingénierie de l'Eau et de l'Environnement (2iE) and Agence nationale pour la sécurité sanitaire de l'environnement, de l'alimentation, du travail et des produits de santé (ANSSEAT) for their support. The authors also acknowledge the World Bank Group under the Africa Centers of Excellence for Development Impact (ACE Impact) Project for its support. The authors would also like to express their gratitude to Boukary Kabore, Ousmane Roland Yonaba and Agyeman Richard Osei for their support during data collection and manuscript preparation.

Conflicts of Interest: The authors declare no conflict of interest.

References

1. Africa: Population Growth Rate 2000–2030. Available online: https://www.statista.com/statistics/1224179/population-growth-in-africa/ (accessed on 5 July 2023).
2. ONU-Habitat Plan Stratégique du Programme des Nations Unies pour les Établissements Humains Pour la Période 2020–2023. 2020. Available online: https://unhabitat.org/sites/default/files/2019/12/strategic_plan_fr.pdf (accessed on 5 July 2023).
3. Denis, E. La Croissance Urbaine en Afrique de l'Ouest De l'Exploitation à la Prolifération. 2009. Available online: https://shs.hal.science/halshs-00371263 (accessed on 5 July 2023).
4. Zoma, V.; Kiemdé, A.; Sawadogo, Y. Principaux Défis de La Croissance Urbaine de Ouagadougou. *Regardsuds* **2022**, *2*, 77–92.
5. INSD Institut National de La Statistique et La Demographie, Ouagadougou (Projections Communales 2020). Available online: http://www.insd.bf/ (accessed on 21 November 2022).
6. Ahmed, J.; Wong, L.P.; Channa, N.; Ahmed, W.; Chua, Y.P.; Shaikh, M.Z. Arsenic Contamination and Potential Health Risk to Primary School Children through Drinking Water Sources. *Hum. Ecol. Risk Assess. Int. J.* **2022**, *29*, 369–389. [CrossRef]
7. Alarcón-Herrera, M.T.; Martin-Alarcon, D.A.; Gutiérrez, M.; Reynoso-Cuevas, L.; Martín-Domínguez, A.; Olmos-Márquez, M.A.; Bundschuh, J. Co-Occurrence, Possible Origin, and Health-Risk Assessment of Arsenic and Fluoride in Drinking Water Sources in Mexico: Geographical Data Visualization. *Sci. Total Environ.* **2020**, *698*, 134168. [CrossRef] [PubMed]
8. Saha, R.; Dey, N.C.; Rahman, M.; Bhattacharya, P.; Rabbani, G.H. Geogenic Arsenic and Microbial Contamination in Drinking Water Sources: Exposure Risks to the Coastal Population in Bangladesh. *Front. Environ. Sci.* **2019**, *7*, 57. [CrossRef]
9. Obianyo, J.I. Effect of Storage Containers on Water Quality. *Trop. J. Sci. Technol.* **2020**, *1*, 66.
10. Adeloju, S.B.; Khan, S.; Patti, A.F. Arsenic Contamination of Groundwater and Its Implications for Drinking Water Quality and Human Health in Under-Developed Countries and Remote Communities—A Review. *Appl. Sci.* **2021**, *11*, 1926. [CrossRef]
11. Ahmad, A.; Bhattacharya, P. Arsenic in Drinking Water: Is 10 Mg/L a Safe Limit? *Curr. Pollut. Rep.* **2019**, *5*, 1–3. [CrossRef]
12. Sawadogo, B.; Konaté, Y.; Sossou, S.K.; Ado Saidou, N.F.; Nouhou Moussa, A.W.; Karambiri, H. Removal of Sulphate Ions from Borehole Water Using Nanofiltration and Reverse Osmosis. *Water* **2022**, *14*, 3422. [CrossRef]
13. Uppal, J.S.; Zheng, Q.; Le, X.C. Arsenic in Drinking Water—Recent Examples and Updates from Southeast Asia. *Curr. Opin. Environ. Sci. Health* **2019**, *7*, 126–135. [CrossRef]
14. Ahoussi, E.; Soro, N.; Kouassi, A.; Soro, G.; Koffi, Y.; Zade, S. Application des méthodes d'analyses statistiques multivariées à l'étude de l'origine des métaux lourds (Cu^{2+}, Mn^{2+}, Zn^{2+} et Pb^{2+}) dans les eaux des nappes phréatiques de la ville d'Abidjan. *Int. J. Biol. Chem. Sci.* **2010**, *4*, 1753–1765. [CrossRef]
15. Kumar, S.; Islam, A.R.M.T.; Islam, H.M.T.; Hasanuzzaman, M.; Ongoma, V.; Khan, R.; Mallick, J. Water Resources Pollution Associated with Risks of Heavy Metals from Vatukoula Goldmine Region, Fiji. *J. Environ. Manag.* **2021**, *293*, 112868. [CrossRef]
16. Naik, M.R.; Barik, M.; Prasad, K.V.; Kumar, A.; Verma, A.K.; Sahoo, S.K.; Jha, V.; Sahoo, N.K. Hydro-Geochemical Analysis Based on Entropy and Geostatistics Model for Delineation of Anthropogenic Ground Water Pollution for Health Risks Assessment of Dhenkanal District, India. *Ecotoxicology* **2022**, *31*, 549–564. [CrossRef]
17. Tokatlı, C.; Varol, M. Variations, Health Risks, Pollution Status and Possible Sources of Dissolved Toxic Metal(Loid)s in Stagnant Water Bodies Located in an Intensive Agricultural Region of Turkey. *Environ. Res.* **2021**, *201*, 111571. [CrossRef]
18. Cantoni, B.; Cappello Riguzzi, A.; Turolla, A.; Antonelli, M. Bisphenol A Leaching from Epoxy Resins in the Drinking Water Distribution Networks as Human Health Risk Determinant. *Sci. Total Environ.* **2021**, *783*, 146908. [CrossRef]
19. Schlienger, J.-L. Impact métabolique de l'exposition au bisphénol A: État des lieux. *Médecine Mal. Metab.* **2014**, *8*, 340–345. [CrossRef]
20. AFSSA Avis de l'agence Française de Sécurité Santinaire Des Aliments Relatif Aux Bisphénol A Dans Les Biberons En Polycarbonate Susceptibles d'etre Chauffés Au Four à Micro-Ondes. 2008. Available online: https://www.anses.fr/fr/system/files/MCDA2008sa0141.pdf (accessed on 5 July 2023).
21. Vandenberg, L.N.; Hauser, R.; Marcus, M.; Olea, N.; Welshons, W.V. Human Exposure to Bisphenol A (BPA). *Reprod. Toxicol.* **2007**, *24*, 139–177. [CrossRef]
22. Monneret, C. Faut-il interdire le bisphénol A? *Ann. Pharm. Fr.* **2010**, *68*, 99–103. [CrossRef]
23. Schäfer, A.I.; Nghiem, L.D.; Oschmann, N. Bisphenol A Retention in the Direct Ultrafiltration of Greywater. *J. Membr. Sci.* **2006**, *283*, 233–243. [CrossRef]
24. Cooper, J.E.; Kendig, E.L.; Belcher, S.M. Assessment of Bisphenol A Released from Reusable Plastic, Aluminium and Stainless Steel Water Bottles. *Chemosphere* **2011**, *85*, 943–947. [CrossRef] [PubMed]
25. LEC Methode LEC Pour l'analyse Du Bisphénol A Dans Les Boissons, Fiche Technique, Laboratoire Expertises et Conseils Accredite Pour Les Analyses Œnologiques et Pour Les Analyses de Traces Sur Bois de Tonnellerie, Boissons Spiritueuses & Vins. Agrement Pour Certificat Exports. 2016. Available online: https://www.lec-cognac.com/wa_files/FT_Bisphenol_20A.pdf (accessed on 5 July 2023).
26. Nalbone, G.; Cicolella, A.; Laot-Cabon, S. L'exposition humaine au bisphénol A: Un facteur de risque environnemental des maladies métaboliques et cardiovasculaires. *Med. Longévité* **2011**, *3*, 42–51. [CrossRef]
27. Jiang, W.; Ding, K.; Huang, W.; Xu, F.; Lei, M.; Yue, R. Potential Effects of Bisphenol A on Diabetes Mellitus and Its Chronic Complications: A Narrative Review. *Heliyon* **2023**, *9*, e16340. [CrossRef] [PubMed]

28. Zafra, A.; del Olmo, M.; Suárez, B.; Hontoria, E.; Navalón, A.; Vílchez, J.L. Gas Chromatographic–Mass Spectrometric Method for the Determination of Bisphenol A and Its Chlorinated Derivatives in Urban Wastewater. *Water Res.* **2003**, *37*, 735–742. [CrossRef] [PubMed]
29. Prueitt, R.L.; Hixon, M.L.; Fan, T.; Olgun, N.S.; Piatos, P.; Zhou, J.; Goodman, J.E. Systematic Review of the Potential Carcinogenicity of Bisphenol A in Humans. *Regul. Toxicol. Pharmacol.* **2023**, *142*, 105414. [CrossRef]
30. Bhardwaj, L.K.; Sharma, A. Estimation of Physico-Chemical, Trace Metals, Microbiological and Phthalate in PET Bottled Water. *Chem. Afr.* **2021**, *4*, 981–991. [CrossRef]
31. Noureddine El Moussawi, S.; Ouaini, R.; Matta, J.; Chébib, H.; Cladière, M.; Camel, V. Simultaneous Migration of Bisphenol Compounds and Trace Metals in Canned Vegetable Food. *Food Chem.* **2019**, *288*, 228–238. [CrossRef]
32. Rowell, C.; Kuiper, N.; Preud'Homme, H. Is Container Type the Biggest Predictor of Trace Element and BPA Leaching from Drinking Water Bottles? *Food Chem.* **2016**, *202*, 88–93. [CrossRef]
33. Lu, T.; Peng, H.; Yao, F.; Nadine Ferrer, A.S.; Xiong, S.; Niu, G.; Wu, Z. Trace Elements in Public Drinking Water in Chinese Cities: Insights from Their Health Risks and Mineral Nutrition Assessments. *J. Environ. Manag.* **2022**, *318*, 115540. [CrossRef]
34. Vasseghian, Y.; Almomani, F.; Dragoi, E.-N. Health Risk Assessment Induced by Trace Toxic Metals in Tap Drinking Water: Condorcet Principle Development. *Chemosphere* **2022**, *286*, 131821. [CrossRef]
35. RGPH 2019: Le Burkina Faso Compte 20 487 979 Habitants. Available online: https://www.finances.gov.bf/forum/detail-actualites?tx_news_pi1%5Baction%5D=detail&tx_news_pi1%5Bcontroller%5D=News&tx_news_pi1%5Bnews%5D=210&cHash=d4f20bcadd89c7ad2ad4f420b4300dfa (accessed on 5 July 2023).
36. Hagiwara, K.; Inui, T.; Koike, Y.; Aizawa, M.; Nakamura, T. Speciation of Inorganic Arsenic in Drinking Water by Wavelength-Dispersive X-ray Fluorescence Spectrometry after in situ Preconcentration with Miniature Solid-Phase Extraction Disks. *Talanta* **2015**, *134*, 739–744. [CrossRef]
37. Cunha, S.C.; Fernandes, J.O. Assessment of Bisphenol A and Bisphenol B in Canned Vegetables and Fruits by Gas Chromatography–Mass Spectrometry after QuEChERS and Dispersive Liquid–Liquid Microextraction. *Food Control* **2013**, *33*, 549–555. [CrossRef]
38. De Toni, L.; Tisato, F.; Seraglia, R.; Roverso, M.; Gandin, V.; Marzano, C.; Padrini, R.; Foresta, C. Phthalates and Heavy Metals as Endocrine Disruptors in Food: A Study on Pre-Packed Coffee Products. *Toxicol. Rep.* **2017**, *4*, 234–239. [CrossRef] [PubMed]
39. Kyrila, G.; Katsoulas, A.; Schoretsaniti, V.; Rigopoulos, A.; Rizou, E.; Doulgeridou, S.; Sarli, V.; Samanidou, V.; Touraki, M. Bisphenol A Removal and Degradation Pathways in Microorganisms with Probiotic Properties. *J. Hazard. Mater.* **2021**, *413*, 125363. [CrossRef] [PubMed]
40. Errico, S.; Bianco, M.; Mita, L.; Migliaccio, M.; Rossi, S.; Nicolucci, C.; Menale, C.; Portaccio, M.; Gallo, P.; Mita, D.G.; et al. Migration of Bisphenol A into Canned Tomatoes Produced in Italy: Dependence on Temperature and Storage Conditions. *Food Chem.* **2014**, *160*, 157–164. [CrossRef]
41. Xu, J.; Zhao, C.; Wang, T.; Yang, S.; Liu, Z. Photo-Oxidation of Bisphenol A in Aqueous Solutions at Near Neutral PH by a Fe(III)-Carboxylate Complex with Oxalacetic Acid as a Benign Molecule. *Molecules* **2018**, *23*, 1319. [CrossRef] [PubMed]
42. Environnement/Santé, Évaluation Préalable Finale Pour Le Défi Concernant Le Phénol, 4,4′-(1-Méthyléthylidène)Bis (Bisphénol-A),Numéro de Registre Du Chemical Abstracts Service 80-05-7la Loi Canadienne Sur La Protection de l'environnement (1999). 2008. Available online: https://www.ec.gc.ca/ese-ees/default.asp?lang=Fr&n=3C756383-1 (accessed on 5 July 2023).
43. Rudel, R.A.; Melly, S.J.; Geno, P.W.; Sun, G.; Brody, J.G. Identification of Alkylphenols and Other Estrogenic Phenolic Compounds in Wastewater, Septage, and Groundwater on Cape Cod, Massachusetts. *Environ. Sci. Technol.* **1998**, *32*, 861–869. [CrossRef]
44. El Cadi, M.A.; Bouslimane, Y.; El Jaoudi, R.; Bouklouze, A.; Cherrah, Y. Bisphénol A: À nouveau risque, nouveau défi! *Med. Reprod.* **2012**, *14*, 151–155.
45. Brede, C.; Fjeldal, P.; Skjevrak, I.; Herikstad, H. Increased Migration Levels of Bisphenol A from Polycarbonate Baby Bottles after Dishwashing, Boiling and Brushing. *Food Addit. Contam.* **2003**, *20*, 684–689. [CrossRef]
46. Calafat, A.M.; Kuklenyik, Z.; Reidy, J.A.; Caudill, S.P.; Ekong, J.; Needham, L.L. Urinary Concentrations of Bisphenol A and 4-Nonylphenol in a Human Reference Population. *Environ. Health Perspect.* **2005**, *113*, 391–395. [CrossRef]
47. ARS Pollution Intérieure: Risques Sanitaires pour la Femme Enceinte et le Jeune Enfant. 2017. Available online: https://www.nouvelle-aquitaine.ars.sante.fr/media/9145 (accessed on 6 July 2023).
48. Sajiki, J.; Yonekubo, J. Degradation of Bisphenol-A (BPA) in the Presence of Reactive Oxygen Species and Its Acceleration by Lipids and Sodium Chloride. *Chemosphere* **2002**, *46*, 345–354. [CrossRef]
49. Yoshida, T.; Horie, M.; Hoshino, Y.; Nakazawa, H.; Horie, M.; Nakazawa, H. Determination of Bisphenol A in Canned Vegetables and Fruit by High Performance Liquid Chromatography. *Food Addit. Contam.* **2001**, *18*, 69–75. [CrossRef]
50. Tian, K.; Yu, Y.; Qiu, Q.; Sun, X.; Meng, F.; Bi, Y.; Gu, J.; Wang, Y.; Zhang, F.; Huo, H. Mechanisms of BPA Degradation and Toxicity Resistance in Rhodococcus Equi. *Microorganisms* **2023**, *11*, 67. [CrossRef] [PubMed]
51. Michałowicz, J. Bisphenol A—Sources, Toxicity and Biotransformation. *Environ. Toxicol. Pharmacol.* **2014**, *37*, 738–758. [CrossRef] [PubMed]
52. Liu, S.-H.; Su, C.-C.; Lee, K.-I.; Chen, Y.-W. Effects of Bisphenol A Metabolite 4-Methyl-2,4-Bis(4-Hydroxyphenyl)Pent-1-Ene on Lung Function and Type 2 Pulmonary Alveolar Epithelial Cell Growth. *Sci. Rep.* **2016**, *6*, 39254. [CrossRef] [PubMed]
53. Staples, C.A.; Dome, P.B.; Klecka, G.M.; Oblock, S.T.; Harris, L.R. A Review of the Environmental Fate, Effects, and Exposures of Bisphenol A. *Chemosphere* **1998**, *36*, 2149–2173. [CrossRef] [PubMed]

54. West, R.J.; Goodwin, P.A.; Klecka, G.M. Assessment of the Ready Biodegradability of Bisphenol A. *Bull. Environ. Contam. Toxicol.* **2001**, *67*, 106–112. [CrossRef]
55. Kang, J.-H.; Kondo, F. Bisphenol A Degradation by Bacteria Isolated from River Water. *Arch. Environ. Contam. Toxicol.* **2002**, *43*, 0265–0269. [CrossRef] [PubMed]
56. INERIS Données Technico-Économiques sur les Substances Chimiques en France: Bisphénol A, INERIS-DRC-10-102861-01251B. 2010. Available online: https://substitution-bp.ineris.fr/fr/document/donnees-technico-economiques-substances-chimiques-en-france-bisphenol (accessed on 7 July 2023).
57. Efsa Toxicokinetics of Bisphenol A—Scientific Opinion of the Panel on Food Additives, Flavourings, Processing Aids and Materials in Contact with Food (AFC) | EFSA. Available online: https://www.efsa.europa.eu/en/efsajournal/pub/759 (accessed on 21 November 2022).
58. Sakai, K.; Yamanaka, H.; Moriyoshi, K.; Ohmoto, T.; Ohe, T. Biodegradation of Bisphenol A and Related Compounds by *Sphingomonas* sp. Strain BP-7 Isolated from Seawater. *Biosci. Biotechnol. Biochem.* **2007**, *71*, 51–57. [CrossRef]
59. Yang, Y.; Wang, Z.; Xie, S. Aerobic Biodegradation of Bisphenol A in River Sediment and Associated Bacterial Community Change. *Sci. Total Environ.* **2014**, *470–471*, 1184–1188. [CrossRef]
60. Chauveheid, E.; Chorfi, S.; Scholdis, S.; Paquet, M.-H. L'exposition au bisphenol-a et a certains derives chlores par l'eau potable chloree. In Proceedings of the Conférence JIE, Bruxelles, Belgique, 23–24 May 2016; Volume 11.
61. Ouandaogo-Yameogo, S. Ressources En Eau Souterraine Du Centre Urbain de Ouagadougou Au Burkina Faso, Qualité et Vulnérabilité. These de Doctorat, Avignon. 2008. Available online: http://www.lha.univ-avignon.fr/Theses/Suzanne%20Yameogo/2008-11-28%20Suzanne%20YAMEOGO.pdf (accessed on 7 July 2023).
62. Han, Q.; Wang, H.; Dong, W.; Liu, T.; Yin, Y.; Fan, H. Degradation of Bisphenol A by Ferrate(VI) Oxidation: Kinetics, Products and Toxicity Assessment. *Chem. Eng. J.* **2015**, *262*, 34–40. [CrossRef]
63. Liao, C.; Kannan, K. A Survey of Bisphenol A and Other Bisphenol Analogues in Foodstuffs from Nine Cities in China. *Food Addit. Contam. Part A* **2014**, *31*, 319–329. [CrossRef]

International Journal of *Molecular Sciences*

MDPI

Review

Environmental Endocrinology: Parabens Hazardous Effects on Hypothalamic–Pituitary–Thyroid Axis

Damáris Barcelos Cunha Azeredo [1,2], Denilson de Sousa Anselmo [1,2], Paula Soares [3,4], Jones Bernardes Graceli [5], D'Angelo Carlo Magliano [1,2,6,†] and Leandro Miranda-Alves [1,2,3,7,8,*,†]

1 Laboratory of Experimental Endocrinology-LEEx, Institute of Biomedical Sciences, Federal University of Rio de Janeiro, Rio de Janeiro 21941-902, Brazil; damaris.barceloscunha@gmail.com (D.B.C.A.); denilson_anselmo@hotmail.com (D.d.S.A.); dc.magliano@gmail.com (D.C.M.)
2 Postgraduate Program in Endocrinology, Faculty of Medicine, Federal University of Rio de Janeiro, Rio de Janeiro 21941-902, Brazil
3 Cellular Signaling and Metabolism Group, i3S—Institute for Research and Innovation in Health, University of Porto, 420-135 Porto, Portugal; psoares@ipatimup.pt
4 Department of Pathology, Faculty of Medicine, University of Porto, 4200-139 Porto, Portugal
5 Laboratory of Cellular Toxicology and Endocrinology, Department of Morphology, Federal University of Espírito Santo, Vitória 29047-105, Brazil; jbgraceli@gmail.com
6 Morphology and Metabolism Group, Federal University of Fluminense, Niteroi 24020-150, Brazil
7 Postgraduate Program in Pharmacology and Medicinal Chemistry, Institute of Biomedical Sciences, Federal University of Rio de Janeiro, Rio de Janeiro 21941-902, Brazil
8 Postgraduate Program in Morphological Sciences, Institute of Biomedical Sciences, Federal University of Rio de Janeiro, Rio de Janeiro 21941-902, Brazil
* Correspondence: leandro.alves@icb.ufrj.br
† These authors contributed equally to this work.

Abstract: Parabens are classified as endocrine-disrupting chemicals (EDCs) capable of interfering with the normal functioning of the thyroid, affecting the proper regulation of the biosynthesis of thyroid hormones (THs), which is controlled by the hypothalamic–pituitary–thyroid axis (HPT). Given the crucial role of these hormones in health and the growing evidence of diseases related to thyroid dysfunction, this review looks at the effects of paraben exposure on the thyroid. In this study, we considered research carried out in vitro and in vivo and epidemiological studies published between 1951 and 2023, which demonstrated an association between exposure to parabens and dysfunctions of the HPT axis. In humans, exposure to parabens increases thyroid-stimulating hormone (TSH) levels, while exposure decreases TSH levels in rodents. The effects on THs levels are also poorly described, as well as peripheral metabolism. Regardless, recent studies have shown different actions between different subtypes of parabens on the HPT axis, which allows us to speculate that the mechanism of action of these parabens is different. Furthermore, studies of exposure to parabens are more evident in women than in men. Therefore, future studies are needed to clarify the effects of exposure to parabens and their mechanisms of action on this axis.

Keywords: parabens; toxicity; thyroid; endocrine disruptor; hypothalamus–pituitary–thyroid axis

Citation: Azeredo, D.B.C.; de Sousa Anselmo, D.; Soares, P.; Graceli, J.B.; Magliano, D.C.; Miranda-Alves, L. Environmental Endocrinology: Parabens Hazardous Effects on Hypothalamic–Pituitary–Thyroid Axis. *Int. J. Mol. Sci.* **2023**, *24*, 15246. https://doi.org/10.3390/ijms242015246

Academic Editor: Malgorzata Gabriela Wasniewska

Received: 10 August 2023
Revised: 28 September 2023
Accepted: 8 October 2023
Published: 17 October 2023

1. Introduction

The incidence of thyroid dysfunction (TD) has increased worldwide recently, particularly among women. Data from the literature show an increase of TD in women during reproductive age in a ratio of 4:1 compared to men. Furthermore, thyroid cancer has a higher incidence in women than in men [1–4], which suggests some influence of hormonal regulation on the development and establishment of TD. Nevertheless, men can also be affected by TD.

The HPT axis is a major modulator of synthesis and regulation of the prohormone tetraiodothyronine (T4) and the active hormone triiodothyronine (T3), both of which are crucial for the body's normal growth, development, and homeostasis. In the HPT axis, the thyrotropin-releasing hormone (TRH) is synthesized and secreted by the hypothalamus. TRH acts on its receptor in the pituitary, stimulating thyroid-stimulating hormone (TSH), which acts in the thyroid gland through its receptors, stimulating the production of the THs (T3 and T4). The THs are responsible for a negative feedback loop that regulates the secretion of TRH and TSH, reducing THs production and promoting hormone balance in the body [5–8].

The mechanism of THs synthesis and secretion is very complex and highly regulated and is influenced by endogenous and exogenous factors. Within this context, environmental factors, such as nutrients, viruses, and radiation, are determinants for the appearance of TD, but also the environmental pollution caused by EDCs needs to be considered. Many studies have reported a relation between EDCs exposure and alterations in the HPT axis function [9–16]. EDC compounds have been considered one of the major factors associated with functional disruption of the thyroid [17–22].

According to the Environmental Protection Agency (EPA) U.S., EDCs are any chemicals that can interfere with the normal functions of the endocrine system and lead to problems with reproduction (e.g., egg and sperm production) and development (e.g., healthy fetal growth) in both humans and wildlife [23]. Conversely, according to the World Health Organization (WHO), EDCs are exogenous substances or mixtures that alter the function(s) of the endocrine system and consequently cause adverse health effects in an intact organism, its progeny, or (sub)populations [24].

The EDCs can act by different mechanisms, namely: (1) interacting with or activating hormone receptors; (2) antagonizing hormone receptors; (3) altering hormone receptor gene and protein expression; (4) altering signal transduction in hormone-responsive cells; (5) inducing epigenetic modifications in hormone-producing or hormone-responsive cells; (6) altering hormone synthesis; (7) altering hormone transport across cell membranes; (8) altering hormone distribution or circulating levels of hormones; (9) altering hormone metabolism or clearance; and (10) altering the fate of hormone-producing or hormone-responsive cells [25]. However, the mechanisms of action of EDCs depend on specific actions at the cellular and tissue levels, as well as on circadian rhythms, seasonal changes, life stage, and sex [25].

Finally, EDCs can be classified according to their origin (natural or synthetic) and grouped according to their chemical composition. Compounds that are excreted by living beings are considered natural, such as phytoestrogens, flavonoids, and natural estrogens. Synthetic compounds can be of industrial or domestic origin, including products such as bisphenols (e.g., plastics), phthalates (e.g., plasticizers), heavy metals, pesticides, retardants (e.g., computers), and parabens (e.g., cosmetics). The main routes of exposure to these compounds are dermal, diet, or inhalation [26–29].

An increasing number of studies have been published on the association between exposure to parabens and endocrine-related diseases, especially in susceptible people, such as pregnant women and children. However, the physiological mechanisms involved in exposure to these compounds are still not fully understood, as there are few studies in the scientific literature that demonstrate the impacts of exposure to these compounds on the health of organisms. Therefore, we propose to review the evidence in the literature correlating exposure to parabens and the development of TD with a focus on human and animal models.

2. Thyroid Morphophysiology: An Overview

In mammals, the thyroid gland is composed of two lobes (right and left) situated anterolaterally to the trachea. The thyroid tissue is composed of several follicles and a large amount of blood vessels. The blood vessels in the thyroid are responsible for transporting oxygen and nutrients to thyroid cells, allowing them to perform their metabolic functions and produce thyroid hormones in adequate quantities. Additionally, these blood vessels also help remove waste products and metabolites from thyroid tissue. Finally, the hormones produced in the gland are released into the bloodstream to be distributed throughout the body. In summary, the blood vessels in the thyroid play a vital role in supporting the thyroid gland's function, ensuring the proper distribution and production of thyroid hormones. To a lesser extent, the parafollicular C cells produce calcitonin, which, together with the parathyroid hormone produced in the parathyroid, acts on calcium metabolism [30–32]. The main functional part of this gland is the thyroid follicle, which is widely distributed along the thyroid and is supported by loose connective tissue. These follicles are oval-shaped structures with a three-dimensional configuration, and their lumen is composed of colloid, a gelatinous substance composed of iodinated and non-iodinated thyroglobulin (TG), diiodothyronine (DIT), monoiodothyronine (MIT), T3, and T4. The lining of the follicle is composed of epithelial cells called follicular cells and thyrocytes, which can be cuboidal or squamous depending on the pituitary stimulus to produce the THs [33–36].

The main compound of the colloid is TG, a glycoprotein synthesized by follicular cells with tyrosine residues in its composition. TSH, by binding to the TSH receptor (TSHr) located in the basal domain, stimulates the production of several proteins involved in the synthesis of THs, such as TG. This precursor protein of THs has a sequence dominated by several cysteine-rich domains, a molecular weight equivalent to 600 kDa, and remarkable stability and solubility due to many disulfide bridges per monomer and about seventeen glycosylation sites. After its synthesis in thyrocytes, TG is secreted into the colloid, where it is stored. The synthesis and secretion of THs are dependent on iodine, and thus, TSH stimulates iodine uptake against the concentration gradient, increasing ion concentrations in the cell cytoplasm and in the follicle lumen. Thus, in the apical membrane of follicular cells, tyrosine residues are iodinated to iodotyrosine by thyroid peroxidase (TPO), an enzyme that produces THs in a reaction dependent on hydrogen peroxide, produced by the enzyme DUOX (Dual oxidase), which is also present in this region of the cell. After this step, a portion of the colloid undergoes endocytosis by thyrocytes and digestion by the action of cytoplasmic lysosomes. The TG is proteolyzed and releases free TH into the cytoplasm which will later be directed to the bloodstream. The remaining iodide from this reaction is recycled by the action of thyroid dehalogenase (Dehal1) and used again for hormone biosynthesis [37–42].

Hydrogen peroxide production is an essential step for iodide oxidation and thyroglobulin iodination for THs biosynthesis. In the thyroid, the oxidases Dual oxidase 1 (DUOX1) and Dual oxidase 2 (DUOX2) stand out, which are members of the NADPH oxidase (NOX) family of oxidoreductase enzymes, which are dependent on calcium to generate hydrogen peroxide. Under normal conditions, dual oxidases are highly expressed in the thyroid and other tissues (salivary gland, gastrointestinal tract). However, DUOX1 and DUOX2 are expressed only under physiological conditions in the thyroid [43]. There is 83% sequence similarity between DUOX1 and DUOX2, but they are differently regulated via direct phosphorylation: DUOX1 is activated by protein kinase A and DUOX2 is activated by protein kinase C. Both pathways are activated by calcium [44–46].

It is well established in the literature that TPO uses the hydrogen peroxide produced by DUOX to promote the oxidation of dietary iodine, which is then captured with the aid of the sodium/iodide symporter (NIS). This oxidized iodine is coupled to the tyrosine residues present in thyroglobulin, thus promoting the synthesis of THs. Although hydrogen peroxide is crucial for the biosynthesis of THs, when found in high concentrations of reactive oxygen species (ROS) in the body, it can have adverse effects on health. ROS include the superoxide anion (O_2^-), hydrogen peroxide, and hydroxyl radicals (OH), among others [47,48].

ROS formation can occur through enzymatic or non-enzymatic reactions when there is an imbalance between prooxidant factors and their elimination. This can be related to the action of endogenous or exogenous factors and can lead to a range of molecular damage [49–55]. The maintenance of redox homeostasis is promoted by molecules with antioxidant potential, which act in the regulation of ROS production and elimination. For this reason, the thyroid has a highly complex antioxidant system to protect its integrity, as it is continuously exposed to ROS for normal function and THs biosynthesis [56]. Some enzymes reduce ROS by minimizing or delaying the effects caused by these reactive species, providing a primary antioxidant defense, such as catalase, superoxide dismutase, glutathione peroxidase, and glutathione reductase [54–59].

3. Parabens

Parabens are chemical compounds classified as alkyl esters of parahydroxybenzoic acid (PHBA). Common parabens include benzylparaben (BeP), butylparaben (BuP), ethylparaben (EP), methylparaben (MP), and propylparaben (PP), which have structural differences between them (Table 1). These compounds show antifungal and antimicrobial potential and are widely used as preservatives in food, beverages, drugs, papers, and personal care products [60,61]. The main source of exposure is dermal absorption, but ingestion of products containing parabens is also an important pathway of exposure in the general population [62–64]. After ingestion, parabens are absorbed in the gastrointestinal tract and hydrolyzed by intestinal and liver esterases. The main metabolite is parahydroxybenzoic acid (PHBA), which is excreted as p-hydroxyhippuric acid (PHHA) in urine, bile, and feces within 24 to 48 h, making urinary paraben concentrations used as long-term urinary biomarkers in studies investigating human exposure levels [65–71].

Table 1. Chemical characteristics of parabens.

Class	No. CAS	Molecular Weight (g/mol)	Chemical Formula	Chemical Structure
Benzylparaben	94-18-8	228.2433	$C_{14}H_{12}O_3$	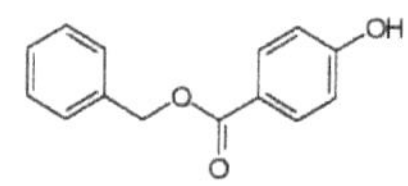
Butylparaben	94-26-8	194.2271	$C_{11}H_{14}O_3$	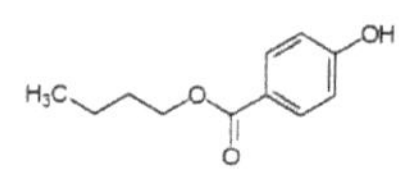
Ethylparaben	120-47-8	166.1739	$C_9H_{10}O_3$	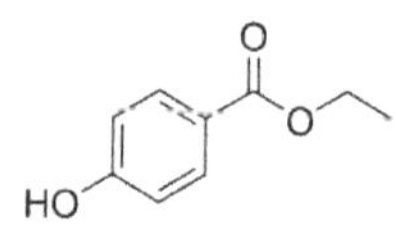
Methylparaben	99-76-3	152.1473	$C_8H_8O_3$	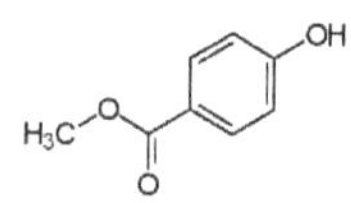
Propylparaben	94-13-3	180.2005	$C_{10}H_{12}O_3$	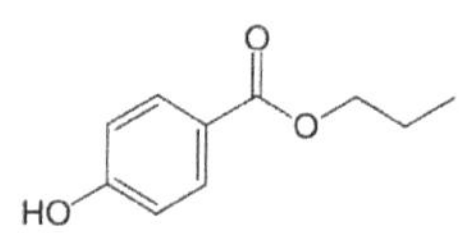

The indiscriminate use of these compounds in various products has aroused scientific interest in the effects of exposure to parabens. Since then, it has been shown that parabens have a potential estrogenic action capable of interfering with the body's homeostasis and thus could be classified as an EDC [72–74]. To regulate the use of parabens in Brazil, the Agência Nacional de Vigilância Sanitaria (ANVISA, Brazil) allows the use of up to 0.4% of individual parabens and up to 0.8% of conjugated parabens in personal care products, but there are no restrictions on the use of parabens in food products, except for propylparaben, which was banned [75]. In the European Union (EU), the maximum permissible use concentration is 0.4% for MP or EP and 0.19% for BuP or PP. As for its use in food, the European Food Safety Authority (EFSA) determines that the acceptable daily intake concentration is up to 10 mg/kg/day for MP or BuP [76,77].

In 2004, the EFSA review panel determined the No Observed Adverse Effect Level (NOAEL) for MP and EP to be 1000 mg/kg/day but considered that more data were needed to determine a specific NOAEL value for propylparaben [76]. Later, in 2008, the Cosmetic Ingredient Review (CIR) Expert Panel reviewed the safety assessment of MP, EP, PP, IPP, BuP, IBP, and BeP in cosmetic products, where it was determined that the NOAEL was 1000 mg /kg/day based on the results of Hoberman et al. (2008), which was considered the "most statistically powerful and well-conducted study on the effects of butylparaben on the male reproductive system" [77–79].

Despite this, several studies have demonstrated the harmful effects of exposure to parabens on the general population's health, even at concentrations considered safe by ANVISA (Brazilian Health Regulatory Agency), raising a series of concerns. Several parabens have also been found in human biological samples [80–85]. In samples of blood and breast milk, concentrations of 0.62 ng/mL of MP, 1.03 ng/mL of EP, 0.18 ng/mL of PP, and 0.05 ng/mL of BuP were found [82]. These compounds were also found in placental tissue samples, with concentrations of up to 11.77 ng/g of MP, EP, and PP [83]. In another study, concentrations between 0.14 and 0.50 µg/L of PP were also found in amniotic fluid samples [84]. Furthermore, EP concentrations between 0.13 and 0.16 µg/L were found in umbilical cord blood samples, and PP between 0.21 and 0.43 µg/L and BuP between 0.04 and 0.05 µg/L were found in men and women [85]. This suggests that parabens can cross the blood–placental barrier and affect fetal development during pregnancy. In addition, there is much evidence that exposure to parabens can interfere with the homeostasis of the thyroid gland, affecting the levels of synthesis and secretion of THs in different experimental models [85–87].

In the following sections, we will address the main effects of parabens on the proper functioning of THs based on articles published between 1951 and 2023. The searches were carried out on the PubMed platform using the terms "thyroid" and "paraben", and the first article directly related to the parabens was published in 1881. Tables 2–4 are organized according to the different types of parabens and their effects on the HPT axis.

Table 2. Effects of butylparaben on thyroid function.

Model	Exposure/Dose/Analyses	Main Results	References
Human (men)	Serum hormone analysis of Inhibin, FSH, LH, testosterone, estradiol, TSH, T3, and T4.	BuP was associated with ↑ TSH, T4, fT4 after 96 h of exposure.	[66]
Pregnant women (12–14 weeks)	Urine collection at 3 different gestational moments (16–20, 20–24, 24–28 gestation weeks) and hormone analyses.	BuP was associated with ↑ estradiol and progesterone with ↓ fT3 and fT4 at visit 3. tT3 and TSH levels did not change between visits.	[88]
Pregnant women—Boston (>15 weeks)	Collection of urine and blood at 4 different gestational moments (9, 17, 26, and 35 weeks of gestation).	BuP was associated with ↓ T3, ↓ T3/T4 ratio, and ↑ TSH.	[89]
Pregnant women—Puerto Rico	Urine collection at 3 different gestational moments (16–20, 20–24, 24–28 weeks of gestation).	Exposure to BuP has been associated with ↓ SHBG.	[90]
Pregnant women	Collection of maternal urine on the day of delivery and collection of umbilical cord blood for hormone measurement.	BuP is associated with ↑ boys' body weight at birth.	[72]
Male Wistar rats	Oral exposure BuP (10 mg/kg/day), BuP (50 mg/kg/day), and BuP+TCS (triclosan) (50 mg+10 mg/kg/day) for 60 days.	BuP (50 mg/kg/d) was associated with ↑ TSH and ↓ T3 and T4.	[91]

Table 2. *Cont.*

Model	Exposure/Dose/Analyses	Main Results	References
Female Wistar rats	Subcutaneous administration of BuP at doses of 1, 5, and 10 mg/kg/day for 7 and 21 days.	↑ TSH in BuP1 at 7 and 21 days; ↓ fT4 and tT4 at all concentrations (7 and 21 days); ↑ fT3, tT3, and TPO in SP1 and SP5 at 7 and 21 days.	[28]
Zebrafish larvae	Larvae were exposed to the following concentrations: 0, 2, 5, and 10 μM of BuP.	↓ T4 levels at most concentrations tested (BuP 5 and 10 μM) and ↓ T3 levels at all concentrations tested. That exposure also led to an increase in TSH gene expression at all concentrations of BuP.	[92]

Table 3. Effects of benzylparaben, isopropylparaben, isobutylparaben, and propylparaben on thyroid function.

Model	Exposure/Dose/Analyses	Main Results	References
Female Sprague Dawley rats	Oral exposure to MP, EP, PP, isopropylparaben (IPP), BuP, and isobutylparaben (IBP) (62.5; 250 and 1000 mg/kg/day) from the 21st to the 40th postnatal day.	PP and IPP were associated with ↓ T4 and estradiol and changes in thyroid weight.	[93]
Male and female Sprague Dawley rats	Injections of IPP, IBP, or mixture of IPP and IBP at 50, 100, 300, and 600 mg/kg bw dissolved in 100 mL of ethanol (99%), 5 days per week for 28 days.	The mixture of IPP and IBP induces a decrease of TSH in exposed individuals at an exposure of 600 mg/kg bw.	[94]
Pregnant women (PROTECT)	Blood collection at two different gestational moments for measurement of SHBG, TSH, fT3, fT4, and progesterone/estradiol ratio; urine collection for detection of phenols and parabens by high-performance liquid chromatography (HPLC).	↑ estradiol and progesterone at the last visit; ↓ fT3 and fT4 at the last visit with no changes in TSH levels.	[88]
Pregnant women—Puerto Rico	Urine and blood collection at 4 time points during pregnancy. Parabens were detected in urine by chromatography. In the blood, tT4, fT4, TSH, and T3 were measured.	PP was inversely associated with fT4.	[89]
Pregnant women—California	Urine and blood collection in the second gestational trimester and blood collection from neonates for measurement of tT4 and TSH.	PP was inversely associated with TSH levels with no changes in tT4 levels.	[95]
Pregnant women—Puerto Rico	Urine collection at 3 different gestational moments (16–20, 20–24, 24–28 weeks of gestation).	Exposure to PP was associated with ↓ SHBG and T3/T4 ratio.	[90]
Human (population of Wuhan, China)	Urine collection and detection of MP, EP, and PP.	PP has been associated with an increased risk of thyroid cancer.	[96]
Newborn human	Newborn blood spots were collected as part of the neonatal screening program, TSH and tT4 were assessed using immunofluorescence.	BuP increased TSH and decreased T4 hormone levels have been demonstrated in newborns and women with less than 150 μg/L of iodine.	[97]

Table 3. *Cont.*

Model	Exposure/Dose/Analyses	Main Results	References
Amphibian tadpoles	Oral exposure to PP (0.05; 0.5 and 5 mg/L) for 14 days.	An increase in PP concentrations in water has been associated with an acute toxic effect.	[87]
Zebrafish larvae	Larvae were exposed to the following concentrations: 0, 5, 10, and 20 μM of PP.	Serum T3 and T4 concentrations decreased at all concentrations tested. In 10 and 20 μM groups, PP increases TSH gene expression.	[92]

Table 4. Effects of ethylparaben and methylparaben on thyroid function.

Model	Exposure/Dose/Analyses	Main Results	References
Human	Urine samples were collected from patients at Wuhan Central Hospital who had thyroid disease and required surgery. Some types of parabens were detected in these samples, such as MP, EP, and PP.	MP and EP were found in urine samples in 99.06%, 95.29%, and 92%, respectively. There was a ↑ concentration of all parabens in the urine of both the nodule and cancer groups. MP and EP were associated with a benign nodule, especially when in higher concentrations. All three parabens studied were associated with an increased risk of thyroid cancer, with EP having the greatest association.	[96]
Mother—children	Urine samples from mothers of newborns were collected on the day of delivery. The concentrations of 5 parabens were determined by chromatography. Umbilical cord blood was collected immediately after birth, in which tT3, tT4, fT3, fT4, TSH, anti-TPO, and anti-TG were measured.	MP and EP were detected in the urine of the evaluated mothers. EP was positively related to increased tT3 in the umbilical cord and to anti-TPO. EP was correlated with increased birth weight in boys, but not in girls.	[72]
Human—Korea	Population study with 1254 people from Korea. Urine samples from this population were collected for analysis of the presence of EDC. Blood serum samples were also collected for measurement of tT4 and fT4, tT3 and fT3, TSH, anti-TPO, anti-thyroglobulin, thyroxine-binding globulin (TBG), and iodothyronine deiodinase (DIO) activity.	Parabens were found in most of the studied population (more than 90%). MP showed a positive association with altered levels of tT3. The increase in MP and EP parabens was correlated with an increase in TBG.	[59]
Pregnant women—Puerto Rico	Urine collection at 3 different gestational moments (16–20, 20–24, 24–28 weeks of gestation).	MP was associated with a decrease in SHBG. MP leads to a significant decrease in TSH and a decrease in the T3/T4 ratio particularly at weeks 24–28 of gestation.	[90]
Pregnant women—California	Urine and blood collection in the second gestational trimester and blood collection from neonates for measurement of tT4 and TSH.	MP was inversely associated with TSH levels with no changes in tT4 levels.	[95]
Pregnant women—Puerto Rico	Urine and blood collection at 4 time points during pregnancy. Parabens were detected in urine by chromatography. In the blood, tT4, fT4, TSH, and T3 were measured.	Urine samples that tested positive for the presence of MP were associated with increased T3 and negatively associated with fT4 at gestational age less than 21 weeks.	[89]

Table 4. *Cont.*

Model	Exposure/Dose/Analyses	Main Results	References
Pregnant women (12–14 weeks)	Urine collection at 3 different gestational moments (16–20, 20–24, 24–28 gestation weeks) and hormone analyses.	MP (293 ng/mL) was associated with a 7.70% increase in SHBG.	[98]
Human	Urine samples from a representative portion of the US population to assess urinary concentrations of triclosan and parabens.	Inverse associations have been found between parabens and circulating levels of thyroid hormones in adults, where women appear to be more vulnerable to exposure.	[99]
Mother–children	Maternal blood was collected during the first prenatal care visit for TSH measurement. MP was detected in meconium samples from newborns.	MP exposure leads to a decrease in gestational age, a significant change in newborn weight, and a decrease in maternal TSH levels. In addition, MP in meconium was associated with about a 16% decrease in tT3 and a decrease in fT4. MP may influence maternal thyroid physiology during pregnancy, and this may lead to the development of ADHD.	[81]
Mother–twin pairs	MP was extracted from urine samples of pregnant women using liquid-liquid extraction. Neonatal TSH levels were abstracted from medical records in China.	MP exposure in early pregnancy was associated with an increased intra-twin TSH difference.	[100]
Wistar rats	Oral exposure for 90 days to BPA (50 mg/kg) or BPA+MP (250 mg/kg).	A minimal thyroid receptor antagonistic effect was only observed after treatment with BPA+MP. MP demonstrated antioxidant properties by reducing lipid peroxidation and generation of hydroxyl radicals induced by exposure to BPA.	[101]
Zebrafish larvae	Larvae were exposed to the following concentrations: 0, 20, 50, and 100 μM of EP and 0, 20, 100, and 200 μM of MP.	Serum T3 concentrations decreased at most concentrations tested (EP at 50, 100 μM and MP at 20, 100, and 200 μM) and T4 concentrations decreased at all concentrations tested.	[92]

3.1. Parabens and TSH

The pituitary gland is an endocrine gland responsible for commanding and regulating various functions in the body. Located in the midline region of the brain within the *Sella turcica*, it consists of two portions of distinct embryological origin: the adenohypophysis and the neurohypophysis. The adenohypophysis comprises the thyrotrophs, the cells responsible for the synthesis and secretion of the TSH, which act in the thyroid gland [88].

Organisms exposed to EDCs appear to have the pituitary gland as a potential target of these compounds, which can lead to growth-related disturbances, metabolic dysfunctions, and alterations in reproduction and homeostasis. Although the action of EDCs on the HPT axis is not entirely clear, it is already known that these compounds can interfere with and change the HPT axis functions [102,103].

3.1.1. Parabens and TSH in Human

Hu et al. (2023) conducted a study in Wuhan, China, and found a positive association between exposure to MP in early pregnancy and a significant increase in TSH concentration in female twins [100]. An important function of TSH is to regulate the input of iodine into thyroid follicular cells. Coiffier et al. (2023) evaluated mother–child pairs from the French cohort and found an increase in TSH levels in boys and girls exposed to BuP [97]. Berger et al. (2018) demonstrated that the serum of pregnant women living in agricultural regions in Northern California presented reductions in TSH levels in association with concentrations of MP and PP in the urine samples [95]. Aker et al. (2019) found, in pregnancy samples originating from Puerto Rico, a general decrease in TSH in association with parabens detection, particularly MP, between 16–20 weeks of gestation [90]. Baker et al. (2020) studied a prospective cohort in Canada with meconium samples and found a positive correlation with the presence of MP in newborns who were later diagnosed with attention-deficit hyperactivity disorder (ADHD) [81]. The study showed that MP exposure can lead to a decrease in gestational age, a significant change in newborn weight, and a decrease in maternal TSH levels that can lead to thyroid complications in children [81]. The homeostasis and correct functioning of TSH and its target gland are important factors for the development and growth of a healthy organism, and it seems that exposure to parabens can lead to a decrease in TSH levels. There are still not many studies that describe changes in TSH levels due to exposure to parabens and a possible relationship with neurological diseases. Despite this, there are positive correlations that endocrine disruptors, such as bisphenol A and chlorpyrifos, can lead to the development of behavioral diseases and that dysregulation in the HPT axis can be a target of the agents and explain the relationships described [90,104,105]. Age, duration, and the method of exposure may potentially have an impact on how parabens can affect TSH levels. Moreover, there might be several unreported mechanisms of action that also play a role in influencing hormonal levels. Despite the poor description of the effects of TSH levels when parabens are present in humans, studies seem to indicate that parabens can change TSH levels and that it could lead to problems in human health (Figure 1).

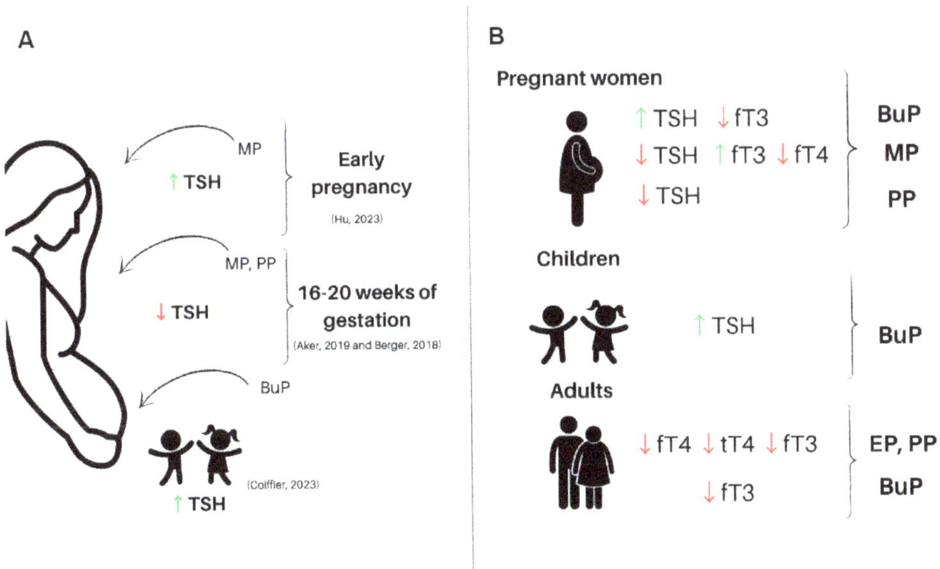

Figure 1. Main effects of exposure to parabens on human health [90,95,97,100]. (**A**) Effects of paraben exposureduring pregnancy on the HPT axis hormones and maternal and newborn health. (**B**) Effects of general human exposure to parabens on the HPT axis hormones. Legend: BuP—butylparaben; EP—ethylparaben; MP—methylparaben; PP—propylparaben; TSH—thyroid-stimulating hormone; THs—thyroid hormones; fT4—free thyroxine levels; fT3—free triiodothyronine levels; tT4—total thyroxine levels; tT3—total triiodothyronine levels. This figure was made using the Canva platform.

3.1.2. Parabens and TSH in Rodents

In murine models, some studies have shown how this exposure can affect the HPT axis. Gogoi, P. and Kalita, J.C. (2020) demonstrated that healthy adult female rats exposed to BuP for 7 and 21 days at low doses (1, 5, and 10 mg/kg BW/day) presented an increase in TSH levels, which caused an increase in the activity and gene expression of thyroid peroxidase and a decrease in the activity of type 1 iodothyronine deiodinase, both enzymes related to thyroid hormone biosynthesis [28]. Another study that used male rats treated for 28 days with a mixture of isopropylparaben (IPP) and isobutylparaben (IBP) (dermal exposure) at a dose of 600 mg/kg/day was able to induce a significant decrease in TSH levels in exposed animals compared to non-exposed animals. This finding shows that these compounds may have a synergistic action [94]. The dose used in this study was approximately three times greater than the known estimated level of human exposure to a single isolated paraben since it is difficult to estimate the total dose of parabens to which the human body is exposed daily [94]. Studies involving TSH and parabens are scarce, although there is some evidence that exposure to these agents leads to changes in TSH levels (Figure 2). The mechanisms involved in this pathway are not clear, and further studies are needed for a better understanding of the effects related to these agents, both in isolation and in combination, on the pituitary gland and on TSH levels. Little is discussed about the effects of exposure on the hypothalamus, and there are not many studies on how exposure to parabens affects TRH levels; perhaps exposure to parabens can affect the production and binding of TRH to its receptors as well as also affect feedback from thyroid hormones on the axis.

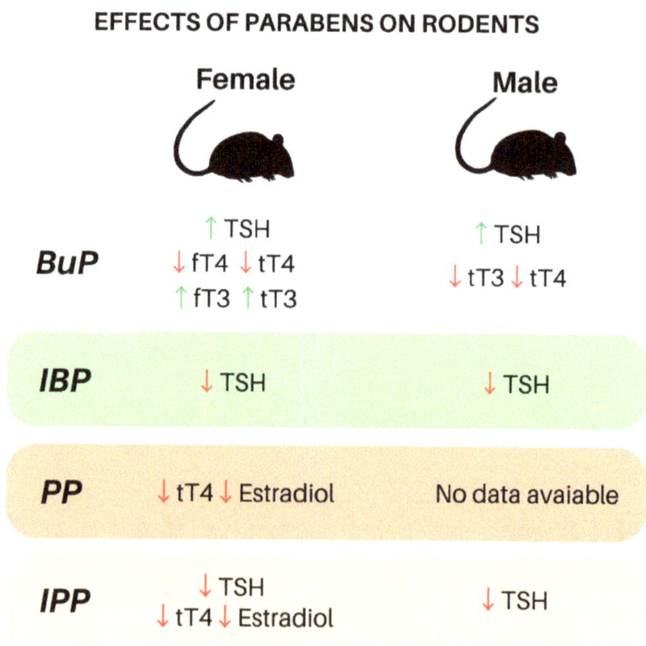

Figure 2. Main effects of exposure to parabens on rodents. Effects of parabens exposure in the HPT axis of female and male rodents. Legend: BuP—butylparaben; IBP—isobutylparaben; IPP—isopropylparaben; PP—propylparaben; TSH—thyroid-stimulating hormone; fT4—free thyroxine levels; fT3—free triiodothyronine levels; tT4—total thyroxine levels; tT3—total triiodothyronine levels. This figure was made using the Canva platform.

3.2. Parabens and Thyroid Hormones

THs are essential for development, growth, and metabolism, playing a key role in mammalian neurodevelopment. Therefore, alterations in the synthesis and secretion of these hormones can cause important disturbances in organisms. The literature demonstrates that exposure to parabens causes DNA damage [106], affects cell proliferative potency [106], and promotes tumorigenic processes [98], leading to a series of harmful effects on the health of individuals. Experimental studies have shown that exposure to parabens resulted in endocrine disturbances and adverse health effects. These compounds potentially bind to hormone receptors, interfering with the levels of synthesis and secretion of these hormones, relating to the increase or decrease in hormone action [107].

3.2.1. Parabens and Thyroid Hormones in Human

A prospective study conducted with pregnant women between 16 and 28 weeks of gestation in Puerto Rico (PROTECT) evaluated associations between parabens and reproductive hormones and urinary paraben concentrations for quantitative analysis. Serum samples were collected at three different time points of pregnancy for measurement of sex hormone-binding globulin (SHBG), thyroid-stimulating hormone (TSH), free thyroxine (fT3), free thyroxine (fT4), and progesterone/estradiol ratio. It was shown that progesterone and estradiol increased, and SHBG showed a tendency to increase throughout pregnancy, while fT4 and fT3 decreased without changes in TSH levels, raising some concerns regarding the correct fetal development, as maternal thyroid hormones are essential for the fetus throughout pregnancy [98]. Furthermore, another study conducted in Boston demonstrated that MP was associated with an increase in fT3 and fT4 levels at 15 weeks of gestation and BuP was associated with a decrease in fT3. However, after 20 gestational weeks, MP was associated with an increase in fT3 and a decrease in fT4, indicating that the impacts of exposure to parabens on the thyroid feedback and signaling system vary according to the moment of exposure throughout pregnancy [88].

In addition, epidemiological studies have shown the relationship between thyroid hormone disbalance and parabens exposure. According to data obtained from the 2007–2008 National Health and Nutrition Examination Survey (NHANES), increased levels of EP and PP in human urine samples were associated with reductions in tT4 levels in female and male serum samples, as well as fT4 in female serum. It was also shown that the serum level of fT3 was negatively associated with EP, PP, and BuP levels in adult females but not in serum samples from males [89]. In another study, high concentrations of parabens were found in urine samples from men compared to the levels found in women's urine samples, demonstrating that parabens can be found routinely in both men and women [103], suggesting that further association studies of various chemicals should be performed with potential common sources of exposure.

Several epidemiological studies have associated exposure to a variety of parabens found in human biological samples (ranging in concentrations between 0.1 and 38 μg/L), with disruption of thyroid function in humans, such as disruption of THs and TSH homeostasis in serum [88,89,97–99,108,109]. Furthermore, it has also been demonstrated that altered THs levels may be associated with an increased incidence of multiple tumors [110,111]. These data demonstrate that the effects of parabens on the THs are controversial, making it necessary to conduct more studies (experimental and epidemiological) to investigate the cellular and molecular mechanisms involved in exposure to parabens since these effects are not yet fully elucidated. In the same manner, good epidemiological studies have been conducted to establish a relationship between parabens exposure and thyroid dysfunctions, especially in pregnant women (Figure 1B).

3.2.2. Parabens and Thyroid Hormones in Rodents

A study carried out with male Wistar rats evaluated the effects of exposure to BuP (50 mg/kg/day) for 60 days on the HPT axis, demonstrating an increase in TSH levels and a decrease in fT4 levels as a response to oxidative stress resulting from conversion of parabens to glutathione hydroquinone conjugates by reaction with singlet oxygen and glutathione. Furthermore, decreased fT4 production may affect the antioxidant system and contribute to the generation of oxidative stress through catalase (peroxidase) modulation [112]. The effect of paraben on the HPT axis was reinforced by another study, which demonstrated that exposure to MP and BuP (1000 mg/kg/day) for 20 days decreased T4 levels and increased thyroid mass [93]. In contrast, another study carried out with pregnant rats exposed to EP (400 mg/kg/day) and BuP (200 and 400 mg/kg/day) for 14 days did not demonstrate significant changes in maternal or newborn THs serum levels [86].

3.2.3. Parabens and Thyroid Hormones in Vertebrates

A study with Zebrafish larvae showed that exposure to EP (0, 20, 50, and 100 μM), PP (0, 5, 10, and 20 μM), BuP (0, 2, 5, and 10 μM), and MP (0, 20, 100, and 200 μM) between 2 and 120 h post-fertilization (hpf) was able to decrease THs concentrations in most tested concentrations, and this exposure also showed a negative correlation with the survival rate of larvae. Furthermore, it has also been demonstrated that the toxicity of parabens increased according to the length of the alkyl carbon chain group, and the order of toxicity was BuP > PP > EP > MP (Figure 3A). These results reinforce the effects of exposure to these compounds on organisms in general, leading to a series of deleterious effects even on zebrafish larvae and affecting correct development by interfering with the correct synthesis and secretion of THs, as THs are fundamental for the processes of proliferation, migration, differentiation, and neuronal signaling, as well as brain myelination during neurodevelopment. Any interference in the levels of THs at this stage of development would have serious consequences for neurodevelopment, leading to several morphofunctional and physiological consequences, including the possibility of affecting the juvenile development process of these animals [91].

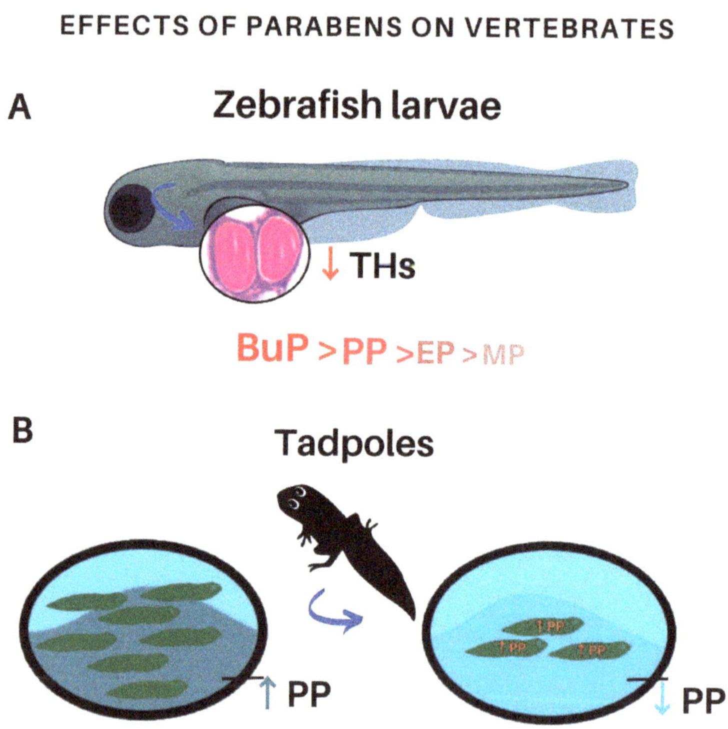

Figure 3. Main effects of exposure to parabens on vertebrates. (**A**) Consequences of paraben exposure in zebrafish larvae. The blue arrow indicates the effects of paraben exposure on the larvae's thyroid gland, resulting in a decrease in thyroid hormones (THs). The studied parabens are listed below the larva in order of toxicity: BuP > PP > EP > MP. (**B**) Impacts of paraben exposure on tadpoles. On the

left side of the image, we can see water with high levels of propylparaben (PP). Tadpoles exposed to parabens in the water experience a significant mortality rate, as shown on the right side of the image. The decrease in PP levels in the water indicates that tadpoles absorb PP from the water, accumulating the substance in their bodies. This figure was made using the Canva platform. Legend: THs—thyroid hormones; BuP—butylparaben; PP—propylparaben; EP—ethylparaben; MP—methylparaben.

In another experimental model, the authors evaluated the effects of exposure to environmental pollutants on tadpole metamorphosis. In this study, no significant differences were observed in the metamorphosis of tadpoles (between 12- and 14 days post fertilization) exposed to low concentrations of PP (5 mg/L) for 14 days and control animals. However, the authors observed a high mortality rate of tadpoles exposed to the highest concentration of PP (12.5 mg/L), revealing an acute toxic effect at increased concentrations of PP. In addition, a decrease in PP concentrations was also observed in the water after two days of exposure, indicating its rapid absorption (Figure 3B). This suggests that prolonged exposure to this compound may result in changes in the endocrine system of these individuals, even at low concentrations [87].

4. Conclusions

The studies presented in this review provide evidence, both from epidemiological and experimental models, of an association between paraben exposure and HPT axis dysfunctions. Previous studies have already correlated exposure to EDCs and disturbances in this axis. Although parabens are a class of EDCs widely used in industry, there are few studies that describe their effects on the HPT axis. We first pooled studies describing pituitary effects resulting from exposure to parabens. Although scarce, these studies described distinct effects. The data on humans are conflicting; however, there is a greater number of studies describing a decrease in TSH levels in humans compared to those that found increased TSH levels. In rodents, exposure appears to decrease TSH levels. Perhaps age, duration, dose, and mode of exposure may influence how parabens can affect TSH levels. Although the HPT axis functions similarly in rodents and humans, the sensitivity to paraben exposure can vary. Therefore, more investigation should be done to understand this discrepancy. We must also consider the possibility that the dose, timing, tissue response, and other variables may have altered these species-specific reactions.

Additionally, there could be various other yet undescribed mechanisms of action that influence the effects on hormone levels. Effects on THs levels are also poorly described. However, recent studies have shown different actions between different types of parabens on the HPT axis. While EP, PP, and BuP were associated with THs decrease, MP was associated with THs increase. This only allows us to speculate that the mechanisms of action of these parabens are different. Furthermore, studies of exposure to parabens are more evident in women and scarce in men. Women are more likely to have thyroid disease than men, and this difference may be due to the female thyroid gland having a higher amount of oxidative stress [59,113]. Additionally, a variety of thyroid-related processes, including iodide uptake, thyroperoxidase activity, and hydrogen peroxide production, which are necessary for the synthesis of thyroid hormones, seem to be impacted by gender and estrogen [114]. Compared to the male thyroid, the female thyroid is more susceptible to estrogen. In contrast to their male counterparts, adult female rats produce more reactive oxygen species (ROS) under the control of estrogen. Finally, disruption of estrogenic and androgenic receptors was caused by methylparaben, indicating that parabens may have estrogenic effects [93,115,116].

Therefore, future studies are needed to clarify the effects of exposure to different parabens and their mechanisms of action on the HPT axis since, between 1951 and 2023, few studies were published that show the effects of exposure to these compounds, and the mechanisms of action, as well as the physiological and/or molecular effects of these compounds in the organism, are still not fully known. To establish/characterize the effects of parabens exposure on human health in more detail, we will require more epidemiological research that makes use of large human cohorts from various world areas and life stages.

Finally, to prevent thyroid illnesses, we recommend that regulatory agencies and WHO work more effectively to limit human exposure to parabens.

Author Contributions: Conceptualization, D.B.C.A. and D.d.S.A.; methodology and investigation, D.B.C.A., D.d.S.A. and D.C.M.; writing—original draft preparation, D.B.C.A., D.d.S.A., J.B.G., P.S. and D.C.M.; writing—review and editing, all authors; visualization, L.M.-A., D.C.M., P.S. and J.B.G.; supervision, J.B.G., P.S., D.C.M. and L.M.-A.; project administration, L.M.-A.; funding acquisition, L.M-A. All authors have read and agreed to the published version of the manuscript.

Funding: This study is part of the projects supported by Fundação Carlos Chagas Filho de Amparo à Pesquisa do Estado do Rio de Janeiro (Cientista do Nosso Estado-FAPERJ E-26/200.798/2021; FAPERJ-Programa de Redes de Pesquisa em Saúde do Estado do Rio de Janeiro-2019 E-26/010.002429/2019; Programa de Apoio a Projetos Temáticos no Estado do Rio de Janeiro E-26/211.265/2021), Coordenação de Aperfeiçoamento de Pessoal de Nível Superior for students grants (D.B.C.A grant 88887.683091/2022-00 and D.S.A. grant 88887.683092/2022-00); and CNPq (L.M-A grant PQ2 310654/2021-1). The funders had no role in the study design, data collection and analysis, decision to publish, or preparation of the manuscript.

Institutional Review Board Statement: Not applicable.

Informed Consent Statement: Not applicable.

Data Availability Statement: The images in this study were generated using Canva for Education (www.canva.com).

Acknowledgments: The authors would like to thank Fundação de Amparo à Pesquisa do Estado do Rio de Janeiro (FAPERJ), Coordenação de Aperfeiçoamento de Pessoal de Nível Superior (CAPES) and Conselho Nacional de Desenvolvimento Científico e Tecnológico (CNPq) for financial support.

Conflicts of Interest: The authors declare no conflict of interest.

References

1. Mattiuzzi, C.; Lippi, G. Current Cancer Epidemiology. *J. Epidemiol. Glob. Health* **2019**, *9*, 217. [CrossRef]
2. Lauretta, R.; Sansone, A.; Sansone, M.; Romanelli, F.; Appetecchia, M. Endocrine Disrupting Chemicals: Effects on Endocrine Glands. *Front. Endocrinol.* **2019**, *10*, 178. [CrossRef] [PubMed]
3. Tran, T.-V.-T.; Kitahara, C.M.; de Vathaire, F.; Boutron-Ruault, M.-C.; Journy, N. Thyroid Dysfunction and Cancer Incidence: A Systematic Review and Meta-Analysis. *Endocr.-Relat. Cancer* **2020**, *27*, 245–259. [CrossRef] [PubMed]
4. Tran, T.-V.-T.; Kitahara, C.M.; Leenhardt, L.; de Vathaire, F.; Boutron-Ruault, M.-C.; Journy, N. The Effect of Thyroid Dysfunction on Breast Cancer Risk: An Updated Meta-Analysis. *Endocr.-Relat. Cancer* **2023**, *30*, e220155. [CrossRef] [PubMed]
5. Kogai, T.; Endo, T.; Saito, T.; Miyazaki, A.; Kawaguchi, A.; Onaya, T. Regulation by Thyroid-Stimulating Hormone of Sodium/Iodide Symporter Gene Expression and Protein Levels in FRTL-5 Cells. *Endocrinology* **1997**, *138*, 2227–2232. [CrossRef]
6. Dietrich, J.W.; Landgrafe, G.; Fotiadou, E.H. TSH and Thyrotropic Agonists: Key Actors in Thyroid Homeostasis. *J. Thyroid Res.* **2012**, *2012*, 351864. [CrossRef]
7. Haisenleder, D.J.; Ortolano, G.A.; Dalkin, A.C.; Yasin, M.; Marshall, J.C. Differential Actions of Thyrotropin (TSH)-Releasing Hormone Pulses in the Expression of Prolactin and TSH Subunit Messenger Ribonucleic Acid in Rat Pituitary Cells in Vitro. *Endocrinology* **1992**, *130*, 2917–2923. [CrossRef]
8. Zoeller, R.T.; Tan, S.W.; Tyl, R.W. General Background on the Hypothalamic-Pituitary-Thyroid (HPT) Axis. *Crit. Rev. Toxicol.* **2007**, *37*, 11–53. [CrossRef]
9. Fernandez, M.O.; Bourguignon, N.S.; Arocena, P.; Rosa, M.; Libertun, C.; Lux-Lantos, V. Neonatal Exposure to Bisphenol A Alters the Hypothalamic-Pituitary-Thyroid Axis in Female Rats. *Toxicol. Lett.* **2018**, *285*, 81–86. [CrossRef]
10. Rodrigues-Pereira, P.; Andrade, M.N.; Santos-Silva, A.P.; Teixeira, M.P.; Soares, P.; Graceli, J.B.; de Carvalho, D.P.; Dias, G.R.M.; Ferreira, A.C.F.; Miranda-Alves, L. Subacute and Low-Dose Tributyltin Exposure Disturbs the Mammalian Hypothalamus-Pituitary-Thyroid Axis in a Sex-Dependent Manner. *Comp. Biochem. Physiol. Part C Toxicol. Pharmacol.* **2022**, *254*, 109279. [CrossRef]
11. Graceli, J.B.; Dettogni, R.S.; Merlo, E.; Niño, O.; da Costa, C.S.; Zanol, J.F.; Ríos Morris, E.A.; Miranda-Alves, L.; Denicol, A.C. The Impact of Endocrine-Disrupting Chemical Exposure in the Mammalian Hypothalamic-Pituitary Axis. *Mol. Cell. Endocrinol.* **2020**, *518*, 110997. [CrossRef]
12. Santos-Silva, A.P.; Andrade, M.N.; Pereira-Rodrigues, P.; Paiva-Melo, F.D.; Soares, P.; Graceli, J.B.; Dias, G.R.M.; Ferreira, A.C.F.; de Carvalho, D.P.; Miranda-Alves, L. Frontiers in Endocrine Disruption: Impacts of Organotin on the Hypothalamus-Pituitary-Thyroid Axis. *Mol. Cell. Endocrinol.* **2018**, *460*, 246–257. [CrossRef]

13. Zhu, L.; Li, W.; Zha, J.; Wang, M.; Yuan, L.; Wang, Z. Butachlor Causes Disruption of HPG and HPT Axes in Adult Female Rare Minnow (*Gobiocypris rarus*). *Chem.-Biol. Interact.* **2014**, *221*, 119–126. [CrossRef] [PubMed]
14. Jia, P.-P.; Ma, Y.-B.; Lu, C.-J.; Mirza, Z.; Zhang, W.; Jia, Y.-F.; Li, W.-G.; Pei, D.-S. The Effects of Disturbance on Hypothalamus-Pituitary-Thyroid (HPT) Axis in Zebrafish Larvae after Exposure to DEHP. *PLoS ONE* **2016**, *11*, e0155762. [CrossRef] [PubMed]
15. Zhang, J.; Liu, H.; Li, J.; Lou, L.; Zhang, S.; Feng, D.; Feng, X. Exposure to Deltamethrin in Adolescent Mice Induced Thyroid Dysfunction and Behavioral Disorders. *Chemosphere* **2020**, *241*, 125118. [CrossRef] [PubMed]
16. McLanahan, E.D.; Campbell, J.L., Jr.; Ferguson, D.C.; Harmon, B.; Hedge, J.M.; Crofton, K.M.; Mattie, D.R.; Braverman, L.; Keys, D.A.; Mumtaz, M.; et al. Low-Dose Effects of Ammonium Perchlorate on the Hypothalamic-Pituitary-Thyroid Axis of Adult Male Rats Pretreated with PCB126. *Toxicol. Sci.* **2007**, *97*, 308–317. [CrossRef] [PubMed]
17. Taylor, P.N.; Albrecht, D.; Scholz, A.; Gutierrez-Buey, G.; Lazarus, J.H.; Dayan, C.M.; Okosieme, O.E. Global Epidemiology of Hyperthyroidism and Hypothyroidism. *Nat. Rev. Endocrinol.* **2018**, *14*, 301–316. [CrossRef]
18. Marotta, V.; Russo, G.; Gambardella, C.; Grasso, M.; La Sala, D.; Chiofalo, M.G.; D'Anna, R.; Puzziello, A.; Docimo, G.; Masone, S.; et al. Human Exposure to Bisphenol AF and Diethylhexylphthalate Increases Susceptibility to Develop Differentiated Thyroid Cancer in Patients with Thyroid Nodules. *Chemosphere* **2019**, *218*, 885–894. [CrossRef]
19. Sur, U.; Erkekoglu, P.; Bulus, A.D.; Andiran, N.; Kocer-Gumusel, B. Oxidative Stress Markers, Trace Elements, and Endocrine Disrupting Chemicals in Children with Hashimoto's Thyroiditis. *Toxicol. Mech. Methods* **2019**, *29*, 633–643. [CrossRef]
20. Coperchini, F.; Croce, L.; Ricci, G.; Magri, F.; Rotondi, M.; Imbriani, M.; Chiovato, L. Thyroid Disrupting Effects of Old and New Generation PFAS. *Front. Endocrinol.* **2021**, *11*, 612320. [CrossRef]
21. Sun, H.-J.; Li, H.-B.; Xiang, P.; Zhang, X.; Ma, L.Q. Short-Term Exposure of Arsenite Disrupted Thyroid Endocrine System and Altered Gene Transcription in the HPT Axis in Zebrafish. *Environ. Pollut.* **2015**, *205*, 145–152. [CrossRef] [PubMed]
22. de Souza, J.S.; Kizys, M.M.L.; da Conceição, R.R.; Glebocki, G.; Romano, R.M.; Ortiga-Carvalho, T.M.; Giannocco, G.; da Silva, I.D.C.G.; Dias da Silva, M.R.; Romano, M.A.; et al. Perinatal Exposure to Glyphosate-Based Herbicide Alters the Thyrotrophic Axis and Causes Thyroid Hormone Homeostasis Imbalance in Male Rats. *Toxicology* **2017**, *377*, 25–37. [CrossRef] [PubMed]
23. Phillips, L. EPA's Exposure Assessment Toolbox (EPA-Expo-Box). *J. Environ. Inform.* **2015**, *25*, 81–84. [CrossRef]
24. Damstra, T. Endocrine Disrupters: The Need for a Refocused Vision. *Toxicol. Sci.* **2003**, *74*, 231–232. [CrossRef] [PubMed]
25. La Merrill, M.A.; Vandenberg, L.N.; Smith, M.T.; Goodson, W.; Browne, P.; Patisaul, H.B.; Guyton, K.Z.; Kortenkamp, A.; Cogliano, V.J.; Woodruff, T.J.; et al. Consensus on the Key Characteristics of Endocrine-Disrupting Chemicals as a Basis for Hazard Identification. *Nat. Rev. Endocrinol.* **2019**, *16*, 45–57. [CrossRef] [PubMed]
26. Patisaul, H.B. Endocrine Disrupting Chemicals (EDCs) and the Neuroendocrine System: Beyond Estrogen, Androgen, and Thyroid. In *Endocrine-Disrupting Chemicals*; Elsevier: Amsterdam, The Netherlands, 2021; pp. 101–150.
27. Kabir, E.R.; Rahman, M.S.; Rahman, I. A Review on Endocrine Disruptors and Their Possible Impacts on Human Health. *Environ. Toxicol. Pharmacol.* **2015**, *40*, 241–258. [CrossRef]
28. Gogoi, P.; Kalita, J.C. Effects of Butylparaben Exposure on Thyroid Peroxidase (TPO) and Type 1 Iodothyronine Deiodinase (D1) in Female Wistar Rats. *Toxicology* **2020**, *443*, 152562. [CrossRef]
29. Macedo, S.; Teixeira, E.; Gaspar, T.B.; Boaventura, P.; Soares, M.A.; Miranda-Alves, L.; Soares, P. Endocrine-Disrupting Chemicals and Endocrine Neoplasia: A Forty-Year Systematic Review. *Environ. Res.* **2023**, *218*, 114869. [CrossRef]
30. Sonntag, J.; Vogel, M.; Geserick, M.; Eckelt, F.; Körner, A.; Raue, F.; Kiess, W.; Kratzsch, J. Age-Related Association of Calcitonin with Parameters of Anthropometry, Bone and Calcium Metabolism during Childhood. *Horm. Res. Paediatr.* **2020**, *93*, 361–370. [CrossRef]
31. Anast, C.; Arnaud, C.D.; Rasmussen, H.; Tenenhouse, A. Thyrocalcitonin and the Response to Parathyroid Hormone. *J. Clin. Investig.* **1967**, *46*, 57–64. [CrossRef]
32. Raue, F.; Scherübl, H. Extracellular Calcium Sensitivity and Voltage-Dependent Calcium Channels in C Cells. *Endocr. Rev.* **1995**, *16*, 752–764. [CrossRef] [PubMed]
33. Hall, J.E.; Hall, M.E. *Guyton and Hall Textbook of Medical Physiology E-Book: Guyton and Hall Textbook of Medical Physiology E-Book*; Elsevier Health Sciences: Amsterdam, The Netherlands, 2020.
34. Gerard, A.; Denef, J.; Colin, I.; van den Hove, M. Evidence for Processing of Compact Insoluble Thyroglobulin Globules in Relation with Follicular Cell Functional Activity in the Human and the Mouse Thyroid. *Eur. J. Endocrinol.* **2004**, *150*, 73–80. [CrossRef] [PubMed]
35. Citterio, C.E.; Targovnik, H.M.; Arvan, P. The Role of Thyroglobulin in Thyroid Hormonogenesis. *Nat. Rev. Endocrinol.* **2019**, *15*, 323–338. [CrossRef] [PubMed]
36. Rousset, B.; Dupuy, C.; Miot, F.; Dumont, J. *Thyroid Hormone Synthesis and Secretion*; MDText.com, Inc.: South Dartmouth, MA, USA, 2015; Chapter 2.
37. Deme, D.; Pommier, J.; Nunez, J. Kinetics of Thyroglobulin Iodination and of Hormone Synthesis Catalyzed by Thyroid Peroxidase. Role of Iodide in the Coupling Reaction. *Eur. J. Biochem.* **1976**, *70*, 435–440. [CrossRef]
38. Maurizis, J.-C.; Marriq, C.; Rolland, M.; Lissitzky, S. Thyroid Hormone Synthesis and Reactivity of Hormone-forming Tyrosine Residues of Thyroglobulin. *FEBS Lett.* **1981**, *132*, 29–32. [CrossRef]
39. Ma, Y.-A.; Sih, C.J. Mechanism of Thyroid Hormone Biosynthesis. Enzymatic Oxidative Coupling of 3,5-Diiodo-l-Tyrosine Derivatives. *Tetrahedron Lett.* **1999**, *40*, 9211–9214. [CrossRef]

40. Dunn, J.T.; Dunn, A.D. The Importance of Thyroglobulin Structure for Thyroid Hormone biosynthesis. *Biochimie* **1999**, *81*, 505–509. [CrossRef]
41. Carvalho, D.P.; Dupuy, C. Thyroid Hormone Biosynthesis and Release. *Mol. Cell. Endocrinol.* **2017**, *458*, 6–15. [CrossRef]
42. Coscia, F.; Taler-Verčič, A.; Chang, V.T.; Sinn, L.; O'Reilly, F.J.; Izoré, T.; Renko, M.; Berger, I.; Rappsilber, J.; Turk, D.; et al. The Structure of Human Thyroglobulin. *Nature* **2020**, *578*, 627–630. [CrossRef]
43. Carvalho, D.P.; Dupuy, C. Role of the NADPH Oxidases DUOX and NOX4 in Thyroid Oxidative Stress. *Eur. Thyroid J.* **2013**, *2*, 160–167. [CrossRef]
44. Deme, D.; Virion, A.; Hammou, N.A.; Pommier, J. NADPH-dependent Generation of H_2O_2 in a Thyroid Particulate Fraction Requires Ca^{2+}. *FEBS Lett.* **1985**, *186*, 107–110. [CrossRef] [PubMed]
45. Ameziane-El-Hassani, R.; Morand, S.; Boucher, J.-L.; Frapart, Y.-M.; Apostolou, D.; Agnandji, D.; Gnidehou, S.; Ohayon, R.; Noël-Hudson, M.-S.; Francon, J.; et al. Dual Oxidase-2 Has an Intrinsic Ca^{2+}-Dependent H_2O_2-Generating Activity. *J. Biol. Chem.* **2005**, *280*, 30046–30054. [CrossRef] [PubMed]
46. Song, Y.; Ruf, J.; Lothaire, P.; Dequanter, D.; Andry, G.; Willemse, E.; Dumont, J.E.; Van Sande, J.; De Deken, X. Association of Duoxes with Thyroid Peroxidase and Its Regulation in Thyrocytes. *J. Clin. Endocrinol. Metab.* **2010**, *95*, 375–382. [CrossRef] [PubMed]
47. Eskandari, S.; Loo, D.D.F.; Dai, G.; Levy, O.; Wright, E.M.; Carrasco, N. Thyroid Na+/I− Symporter. *J. Biol. Chem.* **1997**, *272*, 27230–27238. [CrossRef] [PubMed]
48. Schieber, M.; Chandel, N.S. ROS Function in Redox Signaling and Oxidative Stress. *Curr. Biol.* **2014**, *24*, R453–R462. [CrossRef] [PubMed]
49. Driessens, N.; Versteyhe, S.; Ghaddhab, C.; Burniat, A.; De Deken, X.; Van Sande, J.; Dumont, J.-E.; Miot, F.; Corvilain, B. Hydrogen Peroxide Induces DNA Single- and Double-Strand Breaks in Thyroid Cells and Is Therefore a Potential Mutagen for This Organ. *Endocr.-Relat. Cancer* **2009**, *16*, 845–856. [CrossRef]
50. Donkó, Á.; Morand, S.; Korzeniowska, A.; Boudreau, H.E.; Zana, M.; Hunyady, L.; Geiszt, M.; Leto, T.L. Hypothyroidism-Associated Missense Mutation Impairs NADPH Oxidase Activity and Intracellular Trafficking of Duox2. *Free Radic. Biol. Med.* **2014**, *73*, 190–200. [CrossRef]
51. Faria, C.C.; Peixoto, M.S.; Carvalho, D.P.; Fortunato, R.S. The Emerging Role of Estrogens in Thyroid Redox Homeostasis and Carcinogenesis. *Oxidative Med. Cell. Longev.* **2019**, *2019*, 2514312. [CrossRef]
52. Villanueva, I.; Alva-Sánchez, C.; Pacheco-Rosado, J. The Role of Thyroid Hormones as Inductors of Oxidative Stress and Neurodegeneration. *Oxidative Med. Cell. Longev.* **2013**, *2013*, 218145. [CrossRef]
53. Weyemi, U.; Caillou, B.; Talbot, M.; Ameziane-El-Hassani, R.; Lacroix, L.; Lagent-Chevallier, O.; Al Ghuzlan, A.; Roos, D.; Bidart, J.-M.; Virion, A.; et al. Intracellular Expression of Reactive Oxygen Species-Generating NADPH Oxidase NOX4 in Normal and Cancer Thyroid Tissues. *Endocr.-Relat. Cancer* **2010**, *17*, 27–37. [CrossRef]
54. Brieger, K.; Schiavone, S.; Miller, F.J., Jr.; Krause, K. Reactive Oxygen Species: From Health to Disease. *Swiss Med. Wkly.* **2012**, *142*, w13659. [CrossRef] [PubMed]
55. Sola, E.; Moyano, P.; Flores, A.; García, J.M.; García, J.; Anadon, M.J.; Frejo, M.T.; Pelayo, A.; de la Cabeza Fernandez, M.; del Pino, J. Cadmium-Promoted Thyroid Hormones Disruption Mediates ROS, Inflammation, Aβ and Tau Proteins Production, Gliosis, Spongiosis and Neurodegeneration in Rat Basal Forebrain. *Chem.-Biol. Interact.* **2023**, *375*, 110428. [CrossRef]
56. Macvanin, M.T.; Gluvic, Z.; Zafirovic, S.; Gao, X.; Essack, M.; Isenovic, E.R. The Protective Role of Nutritional Antioxidants against Oxidative Stress in Thyroid Disorders. *Front. Endocrinol.* **2023**, *13*, 1092837. [CrossRef] [PubMed]
57. Hybertson, B.M.; Gao, B.; Bose, S.K.; McCord, J.M. Oxidative Stress in Health and Disease: The Therapeutic Potential of Nrf2 Activation. *Mol. Asp. Med.* **2011**, *32*, 234–246. [CrossRef] [PubMed]
58. Lushchak, V.I. Free Radicals, Reactive Oxygen Species, Oxidative Stress and Its Classification. *Chem.-Biol. Interact.* **2014**, *224*, 164–175. [CrossRef]
59. Pop, A.; Kiss, B.; Vlase, L.; Popa, D.S.; Paltinean, R.; Iepure, R.; Loghin, F. Study of Oxidative Stress Induction after Exposure to Bisphenol a and Methylparaben in Rats. *Toxicol. Lett.* **2011**, *205*, S224. [CrossRef]
60. Soni, M.G.; Carabin, I.G.; Burdock, G.A. Safety Assessment of Esters of P-Hydroxybenzoic Acid (Parabens). *Food Chem. Toxicol.* **2005**, *43*, 985–1015. [CrossRef]
61. Andersen, F.A. Final Amended Report on the Safety Assessment of Methylparaben, Ethylparaben, Propylparaben, Isopropylparaben, Butylparaben, Isobutylparaben, and Benzylparaben as Used in Cosmetic Products. *Int. J. Toxicol.* **2008**, *27* (Suppl. S4), 1–82. [CrossRef]
62. Haman, C.; Dauchy, X.; Rosin, C.; Munoz, J.-F. Occurrence, Fate and Behavior of Parabens in Aquatic Environments: A Review. *Water Res.* **2015**, *68*, 1–11. [CrossRef]
63. Wang, L.; Liao, C.; Liu, F.; Wu, Q.; Guo, Y.; Moon, H.-B.; Nakata, H.; Kannan, K. Occurrence and Human Exposure of P-Hydroxybenzoic Acid Esters (Parabens), Bisphenol A Diglycidyl Ether (BADGE), and Their Hydrolysis Products in Indoor Dust from the United States and Three East Asian Countries. *Environ. Sci. Technol.* **2012**, *46*, 11584–11593. [CrossRef]
64. Liao, C.; Chen, L.; Kannan, K. Occurrence of Parabens in Foodstuffs from China and Its Implications for Human Dietary Exposure. *Environ. Int.* **2013**, *57–58*, 68–74. [CrossRef] [PubMed]
65. Guo, Y.; Kannan, K. A Survey of Phthalates and Parabens in Personal Care Products from the United States and Its Implications for Human Exposure. *Environ. Sci. Technol.* **2013**, *47*, 14442–14449. [CrossRef] [PubMed]

66. Ye, X.; Bishop, A.M.; Reidy, J.A.; Needham, L.L.; Calafat, A.M. Parabens as Urinary Biomarkers of Exposure in Humans. *Environ. Health Perspect.* **2006**, *114*, 1843–1846. [CrossRef] [PubMed]
67. Janjua, N.R.; Frederiksen, H.; Skakkebæk, N.E.; Wulf, H.C.; Andersson, A.-M. Urinary Excretion of Phthalates and Paraben after Repeated Whole-Body Topical Application in Humans. *Int. J. Androl.* **2008**, *31*, 118–130. [CrossRef]
68. Abbas, S.; Greige-Gerges, H.; Karam, N.; Piet, M.-H.; Netter, P.; Magdalou, J. Metabolism of Parabens (4-Hydroxybenzoic Acid Esters) by Hepatic Esterases and UDP-Glucuronosyltransferases in Man. *Drug Metab. Pharmacokinet.* **2010**, *25*, 568–577. [CrossRef]
69. Moos, R.K.; Angerer, J.; Dierkes, G.; Brüning, T.; Koch, H.M. Metabolism and Elimination of Methyl, Iso- and n-Butyl Paraben in Human Urine after Single Oral Dosage. *Arch. Toxicol.* **2015**, *90*, 2699–2709. [CrossRef]
70. Frederiksen, H.; Nielsen, J.K.S.; Mørck, T.A.; Hansen, P.W.; Jensen, J.F.; Nielsen, O.; Andersson, A.-M.; Knudsen, L.E. Urinary Excretion of Phthalate Metabolites, Phenols and Parabens in Rural and Urban Danish Mother–Child Pairs. *Int. J. Hyg. Environ. Health* **2013**, *216*, 772–783. [CrossRef]
71. Shirai, S.; Suzuki, Y.; Yoshinaga, J.; Shiraishi, H.; Mizumoto, Y. Urinary Excretion of Parabens in Pregnant Japanese Women. *Reprod. Toxicol.* **2013**, *35*, 96–101. [CrossRef]
72. Fransway, A.F.; Fransway, P.J.; Belsito, D.V.; Yiannias, J.A. Paraben Toxicology. *Dermatitis* **2019**, *30*, 32–45. [CrossRef]
73. Li, W.; Guo, J.; Wu, C.; Zhang, J.; Zhang, L.; Lv, S.; Lu, D.; Qi, X.; Feng, C.; Liang, W.; et al. Effects of Prenatal Exposure to Five Parabens on Neonatal Thyroid Function and Birth Weight: Evidence from SMBCS Study. *Environ. Res.* **2020**, *188*, 109710. [CrossRef]
74. Boberg, J.; Taxvig, C.; Christiansen, S.; Hass, U. Possible Endocrine Disrupting Effects of Parabens and Their Metabolites. *Reprod. Toxicol.* **2010**, *30*, 301–312. [CrossRef] [PubMed]
75. Nowak, K.; Ratajczak–Wrona, W.; Górska, M.; Jabłońska, E. Parabens and Their Effects on the Endocrine System. *Mol. Cell. Endocrinol.* **2018**, *474*, 238–251. [CrossRef] [PubMed]
76. 2021—Agência Nacional de Vigilância Sanitária. Anvisa. Available online: https://www.gov.br/anvisa/pt-br/assuntos/noticias-anvisa/2021/novas-normas-tratam-de-produtos-de-higiene-pessoal-cosmeticos-e-perfumes (accessed on 25 September 2023).
77. Scientific Committees. Public Health. Available online: http://ec.europa.eu/health/scientific_committees/consumer_safety/opinions/index_en.htm (accessed on 25 September 2023).
78. Opinion of the Scientific Panel on Food Additives, Flavourings, Processing Aids and Materials in Contact with Food (AFC) Related to Para Hydroxybenzoates (E 214-219). *EFSA J.* **2004**, *2*, 83. [CrossRef]
79. Hoberman, A.M.; Schreur, D.K.; Leazer, T.; Daston, G.P.; Carthew, P.; Re, T.; Loretz, L.; Mann, P. Lack of Effect of Butylparaben and Methylparaben on the Reproductive System in Male Rats. *Birth Defects Res. Part B Dev. Reprod. Toxicol.* **2008**, *83*, 123–133. [CrossRef] [PubMed]
80. Rivera-Núñez, Z.; Ashrap, P.; Barrett, E.S.; Llanos, A.A.M.; Watkins, D.J.; Cathey, A.L.; Vélez-Vega, C.M.; Rosario, Z.; Cordero, J.F.; Alshawabkeh, A.; et al. Personal Care Products: Demographic Characteristics and Maternal Hormones in Pregnant Women from Puerto Rico. *Environ. Res.* **2022**, *206*, 112376. [CrossRef] [PubMed]
81. Baker, B.H.; Wu, H.; Laue, H.E.; Boivin, A.; Gillet, V.; Langlois, M.-F.; Bellenger, J.-P.; Baccarelli, A.A.; Takser, L. Methylparaben in Meconium and Risk of Maternal Thyroid Dysfunction, Adverse Birth Outcomes, and Attention-Deficit Hyperactivity Disorder (ADHD). *Environ. Int.* **2020**, *139*, 105716. [CrossRef]
82. Park, N.-Y.; Cho, Y.H.; Choi, K.; Lee, E.-H.; Kim, Y.J.; Kim, J.H.; Kho, Y. Parabens in Breast Milk and Possible Sources of Exposure among Lactating Women in Korea. *Environ. Pollut.* **2019**, *255*, 113142. [CrossRef]
83. Valle-Sistac, J.; Molins-Delgado, D.; Díaz, M.; Ibáñez, L.; Barceló, D.; Silvia Díaz-Cruz, M. Determination of Parabens and Benzophenone-Type UV Filters in Human Placenta. First Description of the Existence of Benzyl Paraben and Benzophenone-4. *Environ. Int.* **2016**, *88*, 243–249. [CrossRef]
84. Philippat, C.; Wolff, M.S.; Calafat, A.M.; Ye, X.; Bausell, R.; Meadows, M.; Stone, J.; Slama, R.; Engel, S.M. Prenatal Exposure to Environmental Phenols: Concentrations in Amniotic Fluid and Variability in Urinary Concentrations during Pregnancy. *Environ. Health Perspect.* **2013**, *121*, 1225–1231. [CrossRef]
85. Geer, L.A.; Pycke, B.F.G.; Waxenbaum, J.; Sherer, D.M.; Abulafia, O.; Halden, R.U. Association of Birth Outcomes with Fetal Exposure to Parabens, Triclosan and Triclocarban in an Immigrant Population in Brooklyn, New York. *J. Hazard. Mater.* **2017**, *323*, 177–183. [CrossRef]
86. Taxvig, C.; Vinggaard, A.M.; Hass, U.; Axelstad, M.; Boberg, J.; Hansen, P.R.; Frederiksen, H.; Nellemann, C. Do Parabens Have the Ability to Interfere with Steroidogenesis? *Toxicol. Sci.* **2008**, *106*, 206–213. [CrossRef]
87. Carlsson, G.; Pohl, J.; Athanassiadis, I.; Norrgren, L.; Weiss, J. Thyroid Disruption Properties of Three Indoor Dust Chemicals Tested in Silurana Tropicalis Tadpoles. *J. Appl. Toxicol.* **2019**, *39*, 1248–1256. [CrossRef]
88. Aker, A.M.; Johns, L.; McElrath, T.F.; Cantonwine, D.E.; Mukherjee, B.; Meeker, J.D. Associations between Maternal Phenol and Paraben Urinary Biomarkers and Maternal Hormones during Pregnancy: A Repeated Measures Study. *Environ. Int.* **2018**, *113*, 341–349. [CrossRef]
89. Koeppe, E.S.; Ferguson, K.K.; Colacino, J.A.; Meeker, J.D. Relationship between Urinary Triclosan and Paraben Concentrations and Serum Thyroid Measures in NHANES 2007–2008. *Sci. Total Environ.* **2013**, *445–446*, 299–305. [CrossRef]
90. Aker, A.M.; Ferguson, K.K.; Rosario, Z.Y.; Mukherjee, B.; Alshawabkeh, A.N.; Calafat, A.M.; Cordero, J.F.; Meeker, J.D. A Repeated Measures Study of Phenol, Paraben and Triclocarban Urinary Biomarkers and Circulating Maternal Hormones during Gestation in the Puerto Rico PROTECT Cohort. *Environ. Health* **2019**, *18*, 28. [CrossRef]

91. Liang, J.; Yang, X.; Liu, Q.S.; Sun, Z.; Ren, Z.; Wang, X.; Zhang, Q.; Ren, X.; Liu, X.; Zhou, Q.; et al. Assessment of Thyroid Endocrine Disruption Effects of Parabens Using In Vivo, In Vitro, and In Silico Approaches. *Environ. Sci. Technol.* **2021**, *56*, 460–469. [CrossRef]
92. Wu, N.-X.; Deng, L.-J.; Xiong, F.; Xie, J.-Y.; Li, X.-J.; Zeng, Q.; Sun, J.-C.; Chen, D.; Yang, P. Risk of Thyroid Cancer and Benign Nodules Associated with Exposure to Parabens among Chinese Adults in Wuhan, China. *Environ. Sci. Pollut. Res.* **2022**, *29*, 70125–70134. [CrossRef]
93. Vo, T.T.B.; Yoo, Y.-M.; Choi, K.-C.; Jeung, E.-B. Potential Estrogenic Effect(s) of Parabens at the Prepubertal Stage of a Postnatal Female Rat Model. *Reprod. Toxicol.* **2010**, *29*, 306–316. [CrossRef]
94. Kim, M.J.; Kwack, S.J.; Lim, S.K.; Kim, Y.J.; Roh, T.H.; Choi, S.M.; Kim, H.S.; Lee, B.M. Toxicological Evaluation of Isopropylparaben and Isobutylparaben Mixture in Sprague–Dawley Rats Following 28 Days of Dermal Exposure. *Regul. Toxicol. Pharmacol.* **2015**, *73*, 544–551. [CrossRef]
95. Berger, K.; Gunier, R.B.; Chevrier, J.; Calafat, A.M.; Ye, X.; Eskenazi, B.; Harley, K.G. Associations of Maternal Exposure to Triclosan, Parabens, and Other Phenols with Prenatal Maternal and Neonatal Thyroid Hormone Levels. *Environ. Res.* **2018**, *165*, 379–386. [CrossRef]
96. Choi, S.; Kim, M.J.; Park, Y.J.; Kim, S.; Choi, K.; Cheon, G.J.; Cho, Y.H.; Jeon, H.L.; Yoo, J.; Park, J. Thyroxine-Binding Globulin, Peripheral Deiodinase Activity, and Thyroid Autoantibody Status in Association of Phthalates and Phenolic Compounds with Thyroid Hormones in Adult Population. *Environ. Int.* **2020**, *140*, 105783. [CrossRef]
97. Coiffier, O.; Nakiwala, D.; Rolland, M.; Malatesta, A.; Lyon-Caen, S.; Chovelon, B.; Faure, P.; Sophie Gauchez, A.; Guergour, D.; Sakhi, A.K.; et al. Exposure to a Mixture of Non-Persistent Environmental Chemicals and Neonatal Thyroid Function in a Cohort with Improved Exposure Assessment. *Environ. Int.* **2023**, *173*, 107840. [CrossRef]
98. Aker, A.M.; Watkins, D.J.; Johns, L.E.; Ferguson, K.K.; Soldin, O.P.; Anzalota Del Toro, L.V.; Alshawabkeh, A.N.; Cordero, J.F.; Meeker, J.D. Phenols and Parabens in Relation to Reproductive and Thyroid Hormones in Pregnant Women. *Environ. Res.* **2016**, *151*, 30–37. [CrossRef]
99. Genuis, S.J.; Birkholz, D.; Curtis, L.; Sandau, C. Paraben Levels in an Urban Community of Western Canada. *ISRN Toxicol.* **2013**, *2013*, 507897. [CrossRef]
100. Hu, L.; Mei, H.; Cai, X.; Hu, X.; Duan, Z.; Liu, J.; Tan, Y.; Yang, P.; Xiao, H.; Zhou, A. Maternal Paraben Exposure and Intra-Pair Thyroid-Stimulating Hormone Difference in Twin Neonates. *Ecotoxicol. Environ. Saf.* **2023**, *250*, 114502. [CrossRef]
101. Fortunato, R.S.; Braga, W.M.O.; Ortenzi, V.H.; Rodrigues, D.C.; Andrade, B.M.; Miranda-Alves, L.; Rondinelli, E.; Dupuy, C.; Ferreira, A.C.F.; Carvalho, D.P. Sexual Dimorphism of Thyroid Reactive Oxygen Species Production Due to Higher NADPH Oxidase 4 Expression in Female Thyroid Glands. *Thyroid* **2013**, *23*, 111–119. [CrossRef]
102. Bosch i Ara, L.; Katugampola, H.; Dattani, M.T. Congenital Hypopituitarism during the Neonatal Period: Epidemiology, Pathogenesis, Therapeutic Options, and Outcome. *Front. Pediatr.* **2021**, *8*, 600962. [CrossRef]
103. Egalini, F.; Marinelli, L.; Rossi, M.; Motta, G.; Prencipe, N.; Rossetto Giaccherino, R.; Pagano, L.; Grottoli, S.; Giordano, R. Endocrine Disrupting Chemicals: Effects on Pituitary, Thyroid and Adrenal Glands. *Endocrine* **2022**, *78*, 395–405. [CrossRef]
104. de Cock, M.; Maas, Y.G.H.; van de Bor, M. Does Perinatal Exposure to Endocrine Disruptors Induce Autism Spectrum and Attention Deficit Hyperactivity Disorders? Review. *Acta Paediatr.* **2012**, *101*, 811–818. [CrossRef]
105. Cowell, W.J.; Wright, R.J. Sex-Specific Effects of Combined Exposure to Chemical and Non-Chemical Stressors on Neuroendocrine Development: A Review of Recent Findings and Putative Mechanisms. *Curr. Environ. Health Rep.* **2017**, *4*, 415–425. [CrossRef]
106. Darbre, P.D.; Harvey, P.W. Paraben Esters: Review of Recent Studies of Endocrine Toxicity, Absorption, Esterase and Human Exposure, and Discussion of Potential Human Health Risks. *J. Appl. Toxicol.* **2008**, *28*, 561–578. [CrossRef] [PubMed]
107. Pitto, L.; Gorini, F.; Bianchi, F.; Guzzolino, E. New Insights into Mechanisms of Endocrine-Disrupting Chemicals in Thyroid Diseases: The Epigenetic Way. *Int. J. Environ. Res. Public Health* **2020**, *17*, 7787. [CrossRef] [PubMed]
108. Bernal, J. Thyroid Hormone Receptors in Brain Development and Function. *Nat. Clin. Pract. Endocrinol. Metab.* **2007**, *3*, 249–259. [CrossRef] [PubMed]
109. Meeker, J.D.; Yang, T.; Ye, X.; Calafat, A.M.; Hauser, R. Urinary Concentrations of Parabens and Serum Hormone Levels, Semen Quality Parameters, and Sperm DNA Damage. *Environ. Health Perspect.* **2011**, *119*, 252–257. [CrossRef] [PubMed]
110. Moeller, L.C.; Führer, D. Thyroid Hormone, Thyroid Hormone Receptors, and Cancer: A Clinical Perspective. *Endocr.-Relat. Cancer* **2013**, *20*, R19–R29. [CrossRef]
111. Lin, H.; Chin, Y.; Yang, Y.S.H.; Lai, H.; Whang-Peng, J.; Liu, L.F.; Tang, H.; Davis, P.J. Thyroid Hormone, Cancer, and Apoptosis. *Compr. Physiol.* **2016**, *6*, 1221–1237. [CrossRef]
112. Taha, M.; Marie, A.M.; Ahmed-Farid, O.A. Combined Approaches for Evaluation of Xenoestrogen Neural Toxicity and Thyroid Dysfunction: Screening of Oxido-nitrosative Markers, DNA Fragmentation, and Biogenic Amine Degradation. *J. Biochem. Mol. Toxicol.* **2020**, *34*, e22521. [CrossRef]
113. Miranda, R.A.; de Moura, E.G.; Soares, P.N.; Peixoto, T.C.; Lopes, B.P.; de Andrade, C.B.V.; de Oliveira, E.; Manhães, A.C.; de Faria, C.C.; Fortunato, R.S.; et al. Thyroid Redox Imbalance in Adult Wistar Rats That Were Exposed to Nicotine during Breastfeeding. *Sci. Rep.* **2020**, *10*, 15646. [CrossRef]
114. Terasaka, S.; Inoue, A.; Tanji, M.; Kiyama, R. Expression Profiling of Estrogen-Responsive Genes in Breast Cancer Cells Treated with Alkylphenols, Chlorinated Phenols, Parabens, or Bis- and Benzoylphenols for Evaluation of Estrogenic Activity. *Toxicol. Lett.* **2006**, *163*, 130–141. [CrossRef]

115. Costa, J.R.; Campos, M.S.; Lima, R.F.; Gomes, L.S.; Marques, M.R.; Taboga, S.R.; Biancardi, M.F.; Brito, P.V.A.; Santos, F.C.A. Endocrine-Disrupting Effects of Methylparaben on the Adult Gerbil Prostate. *Environ. Toxicol.* **2017**, *32*, 1801–1812. [CrossRef]
116. Nowak, K.; Jabłońska, E.; Garley, M.; Radziwon, P.; Ratajczak-Wrona, W. Methylparaben-Induced Regulation of Estrogenic Signaling in Human Neutrophils. *Mol. Cell. Endocrinol.* **2021**, *538*, 111470. [CrossRef] [PubMed]

toxics

MDPI

Article

Mixture Effects of Bisphenol A and Its Structural Analogs on Estrogen Receptor Transcriptional Activation

Handule Lee, Juyoung Park and Kwangsik Park *

College of Pharmacy, Dongduk Women's University, Seoul 02748, Republic of Korea
* Correspondence: kspark@dongduk.ac.kr

Abstract: Bisphenol A (BPA) exposure has been widely linked to endocrine-disrupting effects. Recently, many substitutes for BPA have been developed as safe structural analogs. However, they have still been reported to have similar adverse effects. The current study evaluated the effects of bisphenol A and eight structural analogs on the transcription of estrogen receptor alpha (ERα). The effects of binary and ternary mixtures prepared from different combinations of BPA analogs were also evaluated for transcription activity. The measured data of the mixtures were compared to the predicted data obtained by the full logistic model, and the model deviation ratio (MDR) was calculated to determine whether the effects were synergistic, antagonistic, or additive. Overall, the results suggest that the effect of bisphenol compound are additive in binary and ternary mixtures.

Keywords: endocrine disruption; mixture toxicity; estrogen receptor; bisphenol A; BPA analogs

Citation: Lee, H.; Park, J.; Park, K. Mixture Effects of Bisphenol A and Its Structural Analogs on Estrogen Receptor Transcriptional Activation. *Toxics* **2023**, *11*, 986. https://doi.org/10.3390/toxics11120986

Academic Editors: Miguel Machado Santos and Jodi Flaws

Received: 25 September 2023
Revised: 15 November 2023
Accepted: 30 November 2023
Published: 4 December 2023

1. Introduction

Bisphenol A [BPA, 4′4-(Propane-2,2-diyl)diphenol] is a crystalline solid substance used in the production of epoxy resins and polycarbonate plastics, which are found in many consumer and industrial products. Generally, humans are exposed to BPA through food packaging materials, water bottles, and lacquer coatings for cans. Furthermore, occupational workers are also exposed through direct contact with skin or inhalation. Due to its increased production and use worldwide (several million tons every year), it is now ubiquitously found in the environment including air, water, soil, house dust, and foodstuffs, as well as in human samples such as urine, blood, amniotic fluid, saliva, and breast milk [1–3]. BPA exposure has been widely linked to adverse health effects; it is known to be associated with the incidence of growth disruption, halting normal development, infertility, endocrine system disruption, immune system suppression, and carcinogenicity in animal models and epidemiological human studies [4–6]. BPA interacts with several biological receptors such as estrogen receptor (ER), androgen receptor (AR), and thyroid hormone receptor (THR), resulting in endocrine-disrupting effects. These effects lead to adverse outcomes such as reproductive and developmental issues in the offspring, reproductive dysfunction, neurotoxicity, mutagenesis, and even cancer development [7–11].

Due to the adverse health effects caused by BPA, the European Union (EU) and many countries including the United States (US), Canada, China, Japan, and Korea have successively enacted laws and regulations to restrict or prohibit the use of BPA. The US Food and Drug Administration (FDA) banned its use in food packaging in 2010; the EU banned its use in baby bottles in 2011; the European Food Safety Authority (EFSA) established a tolerable daily intake of 50 μg/kg/day in 2015; the European Chemicals Agency (ECHA) recognized it as an endocrine disruptor in humans in 2017. However, despite the regulations, BPA is still used in various fields including healthcare [5,12].

In recent years, the BPA structural analogs such as Bisphenol AF (BPAF), Bisphenol B (BPB), Bisphenol F (BPF), Bisphenol S (BPS), and Bisphenol Z (BPZ) have increasingly been adopted due to the hazard of BPA [13]. However, the reports on the endocrine toxicity of these analogs are increasing [14–16]. BPS, although it has been introduced as a safe, is known to have similar endocrine toxicity to BPA. Interactions between BPS and estrogen receptor alpha (ERα), subsequent disruption of the reproductive neuroendocrine system and immunotoxic potential via receptor binding have been evaluated in several studies. BPS was also found to alter the secretion of progesterone and estradiol in cultured ovine granulosa cells, which disrupts steroidogenesis [17,18]. Kitamura et al. reported the endocrine-disrupting activity of BPA and 19 related compounds. In the reports, several BPA analogs exhibited estrogenic activity in the human breast cancer cell line MCF-7 [19]. Testicular toxicity of BPS, BPAF, and tetrabromobisphenol A (TBBPA) was also investigated in mouse spermatogonial cells, and they exhibited higher toxicities including dose- and time-dependent alterations in nuclear morphology and cell cycle, DNA damage responses, and perturbation of the cytoskeleton [20]. To evaluate the toxicity potential of BPA substitutes, a toxicokinetic study was also performed after oral and intravenous administration in pigs [21]. Bioavailability and the internal dose of the substitutes are well-known key factors for the evaluation of human exposure risk. A test using *Diaphanosomacelebensis*, BPA, BPS, and BPF showed transcriptional modulation of genes which is related to the ecdysteroid pathway, indicating that these compounds can disrupt the normal endocrine system function of *D. celebensis* [22].

In human biomonitoring studies of BPA exposure, BPA and its substitutes were measured in the same blood and/or urine samples. This suggested that the co-deposition in target organs may increase the risk of bisphenol compounds [23,24]. It is therefore necessary to evaluate mixtures of bisphenol compounds to obtain a more realistic and reliable assessment of health risks compared to assessing individual chemicals [25]. Even with the increasing reports on the exposure of a mixture of bisphenol compounds, only a few of them have been published on the estrogenic activity of the mixtures [26,27]. Some tests on the mixture effects of BPA were performed with non-bisphenol compounds such as heavy metals, phthalates, and alkylphenols [28–30].

In this study, BPA and its eight analogs were selected and screened for estrogen receptor alpha (ERα) transactivation activity (agonistic/antagonistic) based on the OECD Test guideline 455 using HeLa 9903 cells [31]. Some of the compounds were found to be agonists, some were antagonists, and some showed no effects. After the evaluation of the single compounds, the transcriptional activity of binary and ternary mixtures (with combinations of agonist/agonist and antagonist/antagonist) was evaluated. In addition, calculated prediction values of the mixtures using the full logistic model (FLM) were also obtained and compared with the measured values. This study aimed to provide a concept for the prediction of endocrine-disrupting effects of bisphenol compounds that are simultaneously exposed to humans. The results of this study will help to develop and implement regulations for numerous BPA compound mixtures found in consumer products with potential endocrine-disrupting properties.

2. Materials and Methods

2.1. Test Chemicals

Test chemicals of BPA and its analogs were purchased from Sigma-Aldrich (St. Louis, MO, USA) and Tokyo Chemical Industry (TCI, Tokyo, Japan). They were first dissolved in 90% Dimethyl sulfoxide (DMSO) and then serially diluted to prepare test concentrations. The final concentration of the DMSO in media was 0.1%. The chemical name, abbreviation, structure, CAS number, test concentrations, and vendors are shown in Table 1. The reference chemicals for ERα transactivation assay, 17α-estradiol, and tamoxifen, were purchased from Sigma-Aldrich, and flutamide, 17β-estradiol (E2), corticosterone, and 17α-methyltestosterone were purchased from Tokyo Chemical Industry.

Table 1. Bisphenol A and its structural analogues for estrogen receptor transactivation assay.

No.	Chemical Name	CAS No.	M.W.	Test Con. (M)	Vendors	Structure
1	4,4′-Isopropylidenediphenol (BPA)	80-05-7	228.29	10^{-11}–10^{-5}	A	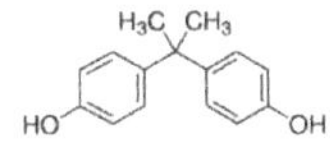
2	4,4′-(Hexafluoroisopropylidene)diphenol (BPAF)	1478-61-1	336.23	10^{-11}–10^{-5}	A	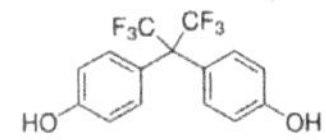
3	2,2-Bis(4-hydroxyphenyl)butane (BPB)	77-40-7	242.32	10^{-11}–10^{-5}	B	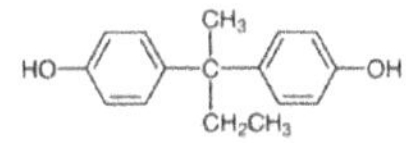
4	2,2-Bis(4-hydroxy-3-methylphenyl)propane (BPC)	79-97-0	256.35	10^{-11}–10^{-5}	B	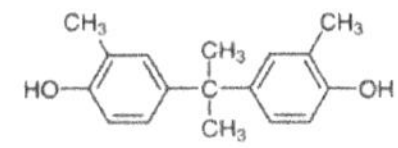
5	4,4′-Ethylidenediphenol (BPE)	2081-08-05	214.26	10^{-11}–10^{-5}	B	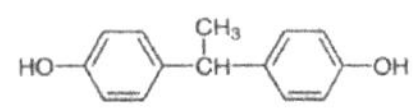
6	α,α′-Bis(4-hydroxyphenyl)-1,4-diisopropylbenzene (BPP)	2167-51-3	346.46	10^{-11}–10^{-6}	B	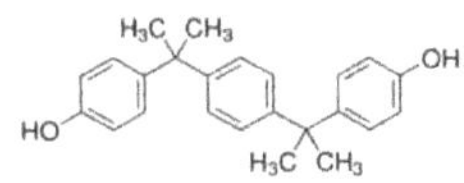
7	4-Benzyloxyphenyl 4-Hydroxyphenyl Sulfone (BPS-MPE)	63134-33-8	340.39	10^{-11}–10^{-6}	B	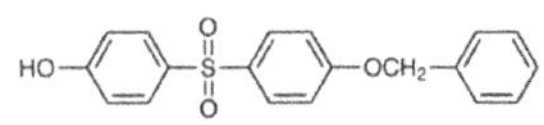
8	4,4′-Cyclohexylidenebisphenol (BPZ)	843-55-0	268.35	10^{-11}–10^{-5}	A	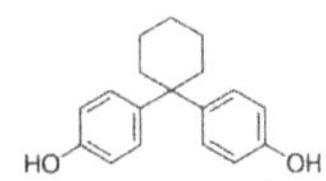
9	2,4′-Dihydroxydiphenyl Sulfone (2,4-BPS)	5397-34-2	250.27	10^{-11}–10^{-6}	B	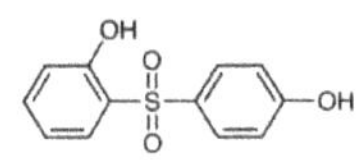

A: Sigma-Aldrich; B: Tokyo Chemical Industry.

2.2. Cell Culture

The hERα-HeLa-9903 (HeLa 9903) cell line (JCRB1318), which expresses human estrogen receptor α (hERα), was obtained from the Japanese Collection of Research Bioresources (JCRB) cell bank (Osaka, Japan), and cultured in phenol red-free Eagle's minimum essential medium (EMEM) supplemented with 10% dextran-coated charcoal-treated fetal bovine serum (DCC-FBS) and kanamycin (60 mg/L). Plastic wares with no estrogenic activity were used in this study according to OECD TG 455 [31].

2.3. Cytotoxicity Test

To achieve the appropriate test chemicals concentration for transactivation assay, the cytotoxicity tests were performed using 3-(4,5-Dimethylthiazol-2-yl)-2,5-Diphenyltetrazolium Bromide (MTT) assay. After 24 h stabilization in 96-well plates, HeLa 9903 cells were treated with test chemicals at wide ranges of concentrations (10^{-11} M~10^{-3} M) for 24 h. After exposure, 100 μL of the MTT solution (5 mg/mL) was added, and the cells were incubated for 2 h at 37 °C. The cells were treated with 100 μL of DMSO to solubilize the purple formazan, and the absorbance was quantified at 570 nm using the microplate spectrophotometer system (Spark®, TECAN, Männedorf, Switzerland). The concentration at which cell viability began to decrease was selected as the highest concentration for the ERα transactivation assay.

2.4. Preparation of Mixtures

Binary mixtures were prepared by combining two bisphenol compounds; agonist/agonist, and antagonist/antagonist. For the test concentrations of binary mixtures, the highest test concentration (master solution) was achieved by mixing individual chemicals where the final concentration was designed to be 100-fold EC_{50} of the individual compound for the agonist, and 100-fold IC_{50} of the individual compound for the antagonist (equi-effect basis). EC_{50} and IC_{50} are shown in Table 2. The assay was designed to obtain full concentration–response curve. The master solution of each binary mixture was then serially diluted for testing. Ternary mixtures were prepared by combining three bisphenol compounds—agonist/agonist/agonist, and antagonist/antagonist/antagonist—where the final concentration was designed to be 100-fold EC_{50} of the individual compound for the agonist, and 100-fold IC_{50} of the individual compound for the antagonist (equi-effect basis). In the case of BPB, the EC_{50} was too high to prepare 100-fold mixtures due to cytotoxicity at the master solution. So, the mixture containing BPB was designed to be 0.05-fold in binary mixture, and 0.25-fold in ternary mixture, as described in Table 3. Information on the master solution of the binary and ternary mixtures is presented in Table 3. No cytotoxicity was observed at the master solution.

Table 2. Chemicals for mixture preparations and their parameters obtained from concentration–response curves of agonistic activity or antagonistic activity.

Chemicals		EC_{50}		IC_{50}		MIN	MAX	Hill Slope
		(M)	(Log M)	(M)	(Log M)			
Agonist	BPA	2.31×10^{-7}	−6.64	-	-	2.66	95.14	0.66
	BPAF	2.40×10^{-7}	−6.62	-	-	5.00	93.34	1.11
	BPB	2.72×10^{-4}	−3.57	-	-	−6.41	424.1	0.32
	BPC	2.91×10^{-7}	−6.54	-	-	−3.95	180.3	0.45
	BPE	1.04×10^{-7}	−6.98	-	-	2.12	120.6	7.36
	BPZ	3.26×10^{-7}	−6.49	-	-	4.04	108.5	1.52
Antagonist	BPP	-	-	1.62×10^{-7}	−6.79	−5.64	92.12	−0.56
	BPS MPE	-	-	1.63×10^{-7}	−6.79	−4.52	96.26	−0.51
	2,4-BPS	-	-	1.50×10^{-8}	−7.83	19.75	93.86	−0.40

The parameters were calculated using GraphPad Prism. EC: effective concentration; MIN: minimum values of the curves; MAX: maximum values of the curves.

Table 3. Concentrations and ratios of component chemicals in the master solution of mixtures.

Chemicals			Ind. Conc. (M)	Total Conc.		Ratio
				(M)	(Log M)	
Binary	Ago + Ago	BPA + BPAF	$2.3 \times 10^{-5} + 2.4 \times 10^{-5}$	4.7×10^{-5}	−4.33	1.0:1.0
		BPA + BPB [(1)]	$1.2 \times 10^{-8} + 1.4 \times 10^{-5}$	1.4×10^{-5}	−4.87	1.0:1174.6
		BPA + BPC	$2.3 \times 10^{-5} + 2.9 \times 10^{-5}$	5.2×10^{-5}	−4.28	1.0:1.3.0
		BPA + BPE	$2.3 \times 10^{-5} + 1.1 \times 10^{-5}$	3.4×10^{-5}	−4.47	1.0:2.2
		BPA + BPZ	$2.3 \times 10^{-5} + 3.3 \times 10^{-5}$	5.6×10^{-5}	−4.25	1.0:1.4
	Anta + Anta	BPS-MPE + BPP	$1.6 \times 10^{-5} + 1.6 \times 10^{-5}$	3.3×10^{-5}	−4.49	1.0:1.0
		2,4-BPS + BPP	$1.5 \times 10^{-6} + 1.6 \times 10^{-5}$	1.8×10^{-5}	−4.75	1.0:10.8
		2,4-BPS + BPS-MPE	$1.5 \times 10^{-6} + 1.6 \times 10^{-5}$	1.8×10^{-5}	−4.75	1.0:10.9
Ternary	Ago + Ago + Ago	BPA + BPAF + BPZ	$2.3 \times 10^{-5} + 2.4 \times 10^{-5} + 3.3 \times 10^{-5}$	8.0×10^{-5}	−4.10	1.0:1.0:1.4
		BPAF + BPZ + BPB [(2)]	$6.0 \times 10^{-8} + 8.1 \times 10^{-8} + 6.8 \times 10^{-5}$	6.8×10^{-5}	−4.17	1.0:1.4:1131.6
		BPE + BPC + BPB [(2)]	$2.6 \times 10^{-8} + 7.3 \times 10^{-8} + 6.8 \times 10^{-5}$	6.8×10^{-5}	−4.17	1.0:2.8:2611.0
		BPE + BPA + BPC	$1.0 \times 10^{-5} + 2.3 \times 10^{-5} + 2.9 \times 10^{-5}$	6.3×10^{-5}	−4.20	1.0:2.2:2.8
		BPC + BPZ + BPB [(2)]	$7.3 \times 10^{-8} + 8.1 \times 10^{-8} + 6.8 \times 10^{-5}$	6.8×10^{-5}	−4.17	1.0:1.1:935.6
	Anta + Anta + Anta	2,4-BPS + BPP + BPS-MPE	$1.5 \times 10^{-6} + 1.6 \times 10^{-5} + 1.6 \times 10^{-5}$	3.4×10^{-5}	−4.47	1.0:10.8:10.9

The concentration of component chemicals in master solutions are 100-fold EC_{50} (agonists) or 100-fold IC50 (antagonists) except the mixture [(1)] and [(2)]. The master solution was diluted serially for the assay as designated in the figures. [(1)]: The concentration of component chemicals in master solutions are $0.05 \times EC_{50}$ (agonists). [(2)]: The concentration of component chemicals in master solutions are $0.25 \times$ EC50 (agonists).

2.5. Stably Transfected Transactivation (STTA) Assay for Estrogen Receptor Alpha

For the estrogen receptor agonist and antagonist tests, the ERa transactivation assay was performed based on the OECD TG 455 using cultured HeLa 9903 cells, which stably express the ERα gene containing a firefly luciferase gene as the reporter gene [31]. After subculturing the cells to a density of 75~90% in a 100 mm culture dish, the cells were transferred to a 96-well plate at 1×10^4 cells/100 μL/well, stabilized for 3 h, and then treated with the test chemicals for 24 h. After exposure, the luciferase activity was measured using a commercial luciferase assay reagent (E2510, Promega, Maidison, WI, USA). For the ERα transcription agonist assay, the normalized mean value of the luminescence signal (Luciferase activity) of the test wells was divided by that of the positive chemical (1 nM 17β-estradiol) to obtain relative transcriptional activity. The concentration–response curves of the test chemicals were processed to calculate the PC_{10} and PC_{50} values, which show 10% and 50% of the transcriptional activity of positive chemicals, respectively, using software provided by the OECD Test guideline. 17β-estradiol, 17α-estradiol, 17α-methyltestosterone, and corticosterone were used as the reference chemicals for the ERα transactivation assay to monitor the stability of the assay system. For the antagonist assay, test wells were spiked with 25 pM 17β-estradiol. Tamoxifen and flutamide were used as reference chemicals. The normalized mean value of the luminescence signal (Luciferase activity) of the test wells was divided by that of spiked 25 pM 17β-estradiol (100%) to obtain relative transcriptional activity. RPC_{Max} for the agonist and IC_{30} values for the antagonist were used for the positive and negative decision criteria, according to the OECD TG 455. The calculations of PC_{10}, PC_{50}, and RPC_{Max} in ERα agonist assay and IC_{30} and IC_{50} in ERα antagonist assay can be made by using the spreadsheet available together with the Test Guideline on the OECD public website [32,33].

2.6. The Prediction of ERα Transactivation of the Mixtures by the Full Logistic Model

The mixture's ERα transactivation assay was performed as described above and the measured data were obtained. Estimation of ERα transactivation by the chemical mixtures was performed with the full logistic model (FLM) with GraphPad Prism (V.9.3, Dan Diego, CA, USA) [34–36]. FLM used four parameters including Hill slope, the maximum and minimum response values, and EC_{50} of individual chemicals. The equation for FLM was as follows:

$$E_{mix} = \left[MAX_1 + \frac{MIN_1 - MAX_1}{1 + \left(\frac{C_1}{EC_{50,1}} + \frac{C_2}{EC_{50,2}}\right)^{P_1}}\right] \cdot \left[\frac{\frac{C_1}{EC_{50,1}}}{\left(\frac{C_1}{EC_{50,1}} + \frac{C_2}{EC_{50,2}}\right)}\right] + \left[MAX_2 + \frac{MIN_2 - MAX_2}{1 + \left(\frac{C_2}{EC_{50,2}} + \frac{C_1}{EC_{50,1}}\right)^{P_2}}\right] \cdot \left[\frac{\frac{C_2}{EC_{50,2}}}{\left(\frac{C_2}{EC_{50,2}} + \frac{C_1}{EC_{50,1}}\right)}\right]$$

MAX: the maximal value of the dose–response curve in the test of individual chemicals;
MIN: the minimal value of the dose–response curve in the test of individual chemicals;
EC_{50}: the half value of maximal effective concentration;
P: hill slope;
C: the concentration of individual chemicals in the test mixture;
The number indicates individual chemicals in the test mixture.

The model for the prediction of the mixture effect is based on the concentration–response curves of each chemical component obtained by the ERα transactivation assay. The parameters of the concentration–response curves for the individual chemicals were calculated using GraphPad Prism, as shown in Table 2. The parameter values were substituted into the FLM formulas. The measured values were compared to the predicted values obtained by the FLM.

2.7. The Assessment of Model Deviation Ratio (MDR)

To verify the difference between the measured and the predicted results, the Model Deviation Ratio (MDR) approach was applied. MDR is defined as the ratio obtained by dividing the expected value by the measured value at respective concentrations [37,38]. In this study, the ratio between the predicted effects calculated by FLM and the measured values at seven concentration points was obtained. MDR values lower than 0.50 are indicators of the synergistic behavior of compounds present in a mixture while values over 2.00 justify the statement of antagonistic action. MDR values within 0.50~2.00 indicate additive activity.

3. Results

3.1. Cell Viability and Test Concentration

To determine the appropriate test concentrations of the test chemicals, a cytotoxicity test was performed using the MTT assay, and the results are shown in Figure 1. The highest test concentration for transactivation assay, which was the concentration that showed the initial decrease in viability, was chosen from the cytotoxicity data, and most chemicals showed a decrease in viability at a concentration of 10^{-5}~10^{-4} M. The dotted lines in the figure indicate 80% viability compared with the media control group, which is the baseline for the determination of the highest concentration for the ERα transactivation assay. The concentration which showed a viability of less than 80% was not applied to the assay.

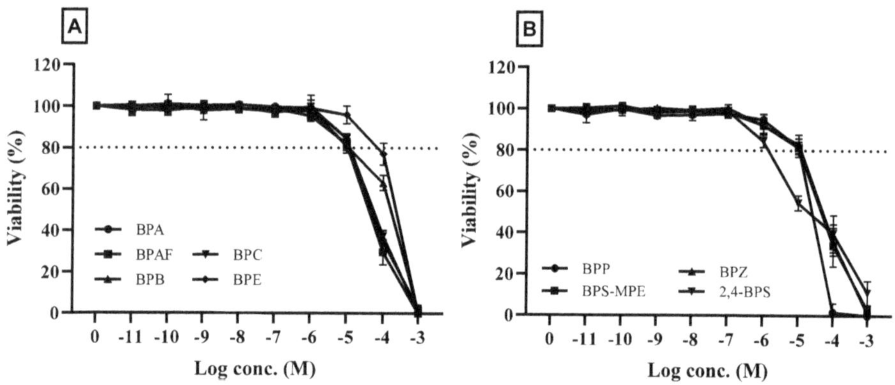

Figure 1. Effects of test chemicals on the viability of cultured HeLa 9903 cells. Cell viability was assessed by MTT assays, and the results are presented as a percentage of the control group viability. Cells were treated with indicated concentrations for 24 h. The results represent the means of three separate experiments, and the bars represent the standard error. The dotted line in the figure represents a viability of 80%. The figure shows the cytotoxicity of the chemicals that showed positive results in the agonistic or antagonistic tests of estrogen receptor transactivation (n = 3). (**A**): Bisphenol A (BPA), Bisphenol AF (BPAF), Bisphenol B (BPB), Bisphenol C (BPC), and Bisphenol E (BPE); (**B**): Bisphenol P (BPP), Bisphenol S(BPS)-MPE, 2,4-dihydroydiphenyl sulfone (2,4-BPS), and Bisphenol Z (BPZ).

3.2. Effects of Bisphenol A and Its Analogs on ERα Transactivation: Agonist Assay

The estrogenic activities of BPA and its eight analogs were evaluated based on OECD TG 455. For the agonist assay of ERα transactivation in cultured HeLa 9903 cells, the reference chemicals 17β-estradiol, 17α-estradiol, and 17α-methyltestosterone and corticosterone were first tested. As shown in Figure 2A, concentration–response transactivation was induced by 17β-estradiol, 17α-estradiol, and 17α-methyltestosterone, with different potencies and efficacies. The results for the reference chemicals were very similar to those provided in the test guideline. When the cultured cells were treated with corticosterone, no sign of transactivation was observed in all the test concentrations, 10^{-10}~10^{-4} M, as shown in TG 455. These results confirmed the validity of the performance criteria for the agonist assay in this study.

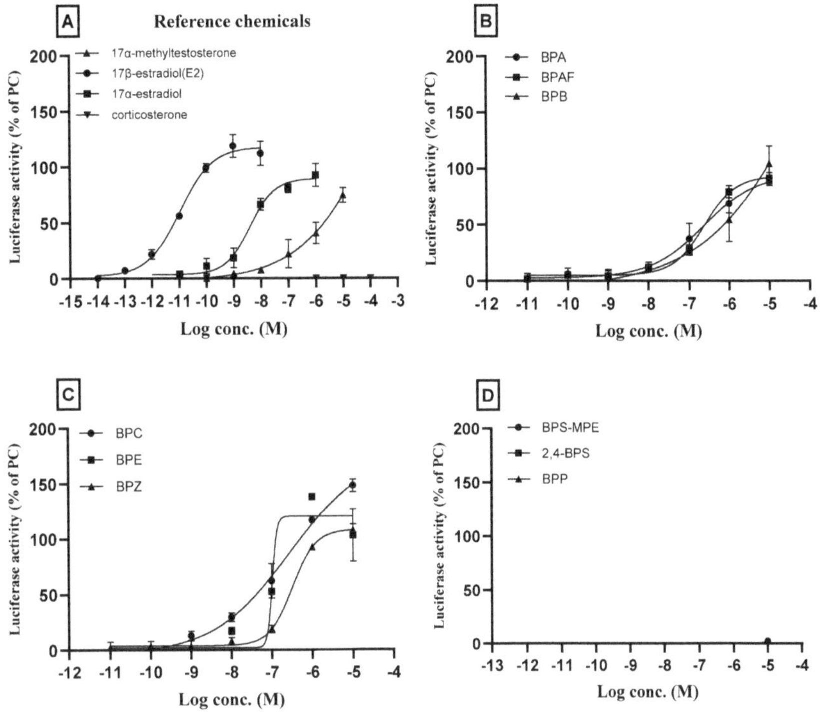

Figure 2. Agonistic assay of bisphenol compounds for estrogenic receptor alpha (ERα) transcriptional activity in cultured HeLa 9903 cell line. The cells were incubated with different concentrations of the respective bisphenol compounds based on the results of the cytotoxicity test, and a transactivation assay was performed. For the ERα transcription agonist assay, the normalized mean values of the luminescence signal (Luciferase activity) of the test wells were divided by that of the positive chemical (1 nM 17β-estradiol) to obtain the relative transcriptional activity. (**A**): The results of reference chemicals; 17β-estradiol, 17α-estradiol, and 17α-methyltestosterone induced ERα transactivation, while corticosterone showed no activity. (**B**): BPA, BPAF, BPB; (**C**): BPC, BPE, BPZ; (**D**): BPS-MPE, 2,4-BPS, BPP. Three separate experiments were performed in the transactivation assay and data were represented as mean ± standard error.

BPA and its analogs BPAF, BPB, BPC, BPE, and BPZ showed ERα transactivation agonistic activities toward ERα transcription. The concentration–response curves of the chemicals with agonistic effects are shown in Figure 2B,C. BPS-MPE, 2,4-BPS, and BPP did not exhibit agonistic activity (Figure 2D). RPC_{Max}, PC_{10}, and PC_{50} of individual chemicals which showed ERα agonistic effects were obtained from the concentration–response curve by using the spreadsheet provided by the OECD TG 455 and are presented in Table 4.

3.3. *Effects of Bisphenol A and Its Analogs on ERα Transactivation: Antagnoist Assay*

Tamoxifen and flutamide, as reference chemicals in the antagonist assay, were tested first, and the results are shown in Figure 3A. As shown in the figure, tamoxifen inhibited the ERα transactivation which was elevated by spiked 25 pM 17β-estradiol. The concentration of tamoxifen used in the test was 10^{-10}~10^{-5} M. The inhibitory activity was observed from a concentration of 10^{-8} M, and 50% inhibition was significant at 10^{-7} M. However, flutamide did not show any antagonistic effects in all the test concentrations. The results of the reference chemicals were very similar to those provided in the test guideline, confirming the validity of the performance criteria for the antagonist assay in this study.

Table 4. Parameters obtained from concentration–response curves of agonists or antagonists for the estrogen receptor transactivation.

Chemicals		RPC_{Max} (%)	PC_{10} (Log M)	PC_{50} (Log M)	IC_{30} (Log M)	IC_{50} (Log M)
Agonist	BPA	88.35	−8.36	−6.63	-	-
	BPAF	91.34	−8.35	−6.57	-	-
	BPB	104.53	−8.16	−6.00	-	-
	BPC	148.20	−9.24	−7.33	-	-
	BPE	138.06	−8.42	−7.07	-	-
	BPZ	107.87	−7.87	−6.58	-	-
Antagonist	BPP	-	-	-	−7.39	−6.68
	BPS-MPE	-	-	-	−7.57	−6.74
	2,4-BPS	-	-	-	−8.84	−8.00

PC: positive concentration; IC: inhibition concentration; RPC_{Max}: maximum level of response induced by a test chemical, expressed as a percentage of the response induced by 1 nM 17β-estradiol (positive chemical) on the same plate. The parameters were obtained using the spreadsheet provided by the OECD Test guideline.

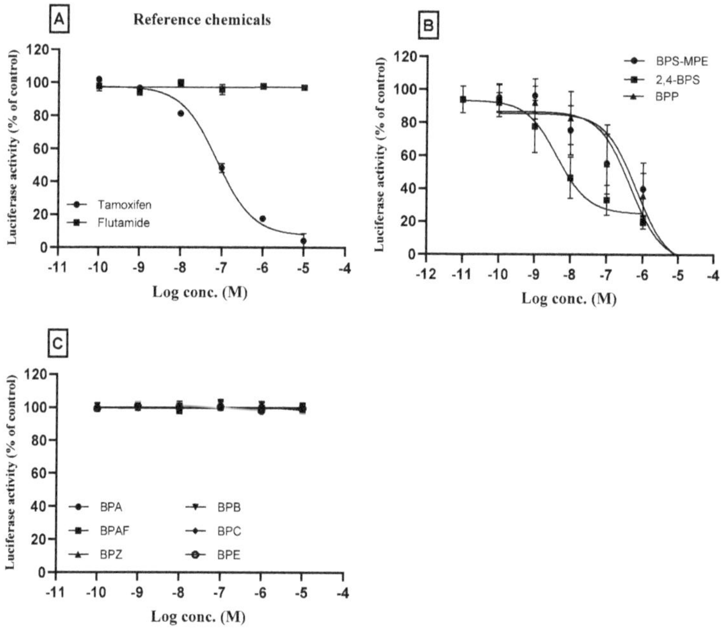

Figure 3. Antagonistic assay of bisphenol compounds for estrogenic receptor alpha (ERα) transcriptional activity in cultured HeLa 9903 cell line. The cells were incubated with different concentrations of bisphenol compounds based on the results of the cytotoxicity test, and an antagonistic assay was performed. For the antagonist assay, test wells were spiked with 25 pM 17β-estradiol. The normalized mean values of the luminescence signal (Luciferase activity) of the test wells were divided by that of the spiked-in control (100%) to obtain the relative transcriptional activity. (**A**): Tamoxifen (positive) and flutamide (negative) were used as the reference chemicals. (**B**): BPS-MPE, 2,4-BPS, and BPP showed positive antagonistic activity to the ERα transactivation. (**C**): BPA, BPAF, BPB, BPC, BPE, and BPZ did not show antagonistic activity to the 17β-estradiol-activated ERα transcription. Three separate experiments were performed in transactivation assay, and data were represented as mean ± standard error.

BPP, BPS-MPE, and 2,4-BPS did not show the agonistic activity of ERα transactivation but showed antagonistic activity as shown in Figure 3B. The six chemicals (BPA, BPB, BPC,

BPE, and BPZ) which showed agonistic effects on the ERα transactivation, did not inhibit ERα transcription activated by 25 pM 17β-estradiol in all the test concentrations (Figure 3C). IC_{30} and IC_{50} of the individual chemicals which showed ERα antagonistic effects were obtained from the concentration–response curve using the spreadsheet provided by the OECD TG 455 and are presented in Table 4.

3.4. *ERα Transactivation of Binary Mixtures*

3.4.1. Mixtures of Agonist/Agonist

The binary mixtures of BPA and other agonistic bisphenol compounds (BPAF, BPB, BPC, BPE, and BPZ), were prepared by mixing equi-effect concentrations as described in the methods section and agonistic effects of the binary mixtures were assayed. The results were compared to the predicted data obtained using FLM. The concentration–response curves of the binary mixtures (both the measured and the predicted values) are shown in Figure 4. As shown in the figures, the ERα transactivation increased as the mixture concentration increased both in the measured and predicted data. Furthermore, patterns of concentration–response curves of the measured and the predicted were very similar. The measured curves were very close to the predicted curves, respectively, which did not expect synergism or antagonism.

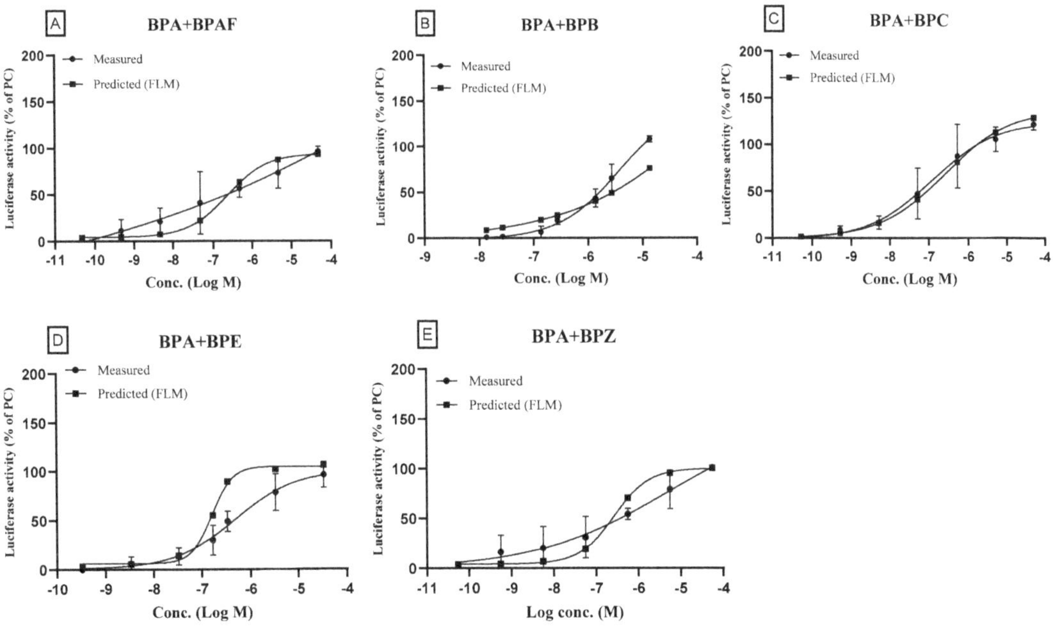

Figure 4. Effects of binary mixtures (Agonist + Agonist) on the estrogen receptor transcriptional activity: a comparison of the measured and the predicted values. For the test concentrations of binary mixtures, the highest concentration (master solution) was achieved by mixing the individual chemicals where the final concentration of each chemical was designed to be 100-fold EC_{50}, respectively, for the full concentration–response curve. In case of BPB, the EC_{50} was too high to prepare 100-fold mixtures due to cytotoxicity at the master solution. So, the mixture containing BPB was designed to be 0.05-fold in binary mixture. The solution of the highest concentration was then serially diluted to measure the ERα transactivation. The measured results were shown with the predicted ones obtained by the full logistic model. The compositions of binary mixtures were as follows: (**A**): BPA + BPAF; (**B**): BPA + BPB; (**C**): BPA + BPC; (**D**): BPA + BPE; (**E**): BPA + BPZ. Three separate experiments were performed in the transactivation assay and data were represented as mean ± standard error.

The RPC_{Max} (%), PC_{10}, and PC_{50} were also very similar as shown in Table 5. Due to the same mode of action of the individual compounds, the mixture effect of binary bisphenol compounds seemed to be additive. When we calculated the model deviation ratio (MDR), most of the ratios were more than 0.5 and less than 2, which means that they were not in the range of synergism or antagonism.

Table 5. Parameters obtained from concentration–response curves of binary mixtures for the estrogen receptor transactivation.

Chemicals		RPC_{Max} (%)		PC_{10} (Log M)		PC_{50} (Log M)		IC_{30} (Log M)		IC_{50} (Log M)	
		Meas	Pred	Meas	Pred	Meas	Pred	Meas	Pred	Meas	Pred
Ago + Ago	BPA + BPAF	96.7	92.8	−9.43	−8.16	−6.74	−6.63	-	-	-	-
	BPA + BPB	134.6	146.0	−6.79	<−7.56	−5.78	−5.54	-	-	-	-
	BPA + BPC	121.6	128.7	−9.12	−8.84	−7.22	−7.06	-	-	-	-
	BPA + BPE	96.8	106.5	−7.97	−8.00	−6.44	−6.86	-	-	-	-
	BPA + BPZ	100.8	100.4	−9.72	−8.00	−6.44	−6.65	-	-	-	-
Anta + Anta	BPS-MPE + BPP	101.2	90.7	-	-	-	-	−7.35	−7.77	−6.26	−6.98
	2,4-BPS + BPP	98.8	88.7	-	-	-	-	−8.01	−8.00	−7.16	−7.06
	BPS-MPE + 2,4-BPS	99.0	90.2	-	-	-	-	−8.10	−7.94	−7.26	−7.00

Ago: agonist; Anta: antagonist; Meas: measured values; Pred: predicted values; PC: positive concentration; IC: inhibition concentration; RPC_{Max}: maximum level of response. The parameters were obtained using the spreadsheet provided by the OECD TG 455.

3.4.2. Mixtures of Antagonist/Antagonist

The binary mixtures were prepared by mixing two compounds with an equi-effect ratio of IC_{50} as described in the methods section. Three mixtures were prepared: BPS-MPE/BPP, 2,4-BPS/BPP, and BPS-MPE/2,4-BPS. The highest test concentration of the binary mixture was based on the IC_{50} of each compound, and the concentration was serially diluted. Test wells were spiked with 25 pM 17β-estradiol, and the normalized mean values of the luminescence signal (luciferase activity) of the test wells were divided by that of spiked control (100%). Concentration–response curves of binary mixtures of the antagonist compounds, both the measured and the predicted, are shown in Figure 5. As shown in the figure, the measured curves were very close to the predicted curves, respectively, which did not expect the synergism or antagonism. When the parameters of the concentration–response curves were calculated, they were also very similar as shown in Table 5. The MDRs, calculated as the ratio of the expected values and the measured values in the test concentrations, were within 0.50~2.00 indicating additive activity.

3.5. ERα Transactivation of Ternary Mixtures

A mixture of three antagonists (BPP, BPS-MPE, and 2,4-BPS) was prepared and tested for the antagonistic effects on ERα transactivation activated by 25 pM 17β-estradiol. As described in the methods section, the highest test concentration of the ternary mixture for antagonist assay was based on the sum of IC_{50} of respective individual compounds. The mixture of three antagonists (BPS-MPE, 2,4-BPS, and BPP) also showed an antagonistic effect as shown in Figure 6A. Five ternary mixtures prepared by combining six agonists (BPA, BPAF, BPB, BPC, BPE, and BPZ) were tested for the agonistic effect on ERα transactivation. The highest test concentration of the ternary mixture for agonist assay was based on the EC_{50s} of the respective individual three compounds. The results are shown in Figure 6B–F. Similarly to the binary mixtures, the concentration–response parameters of ternary mixtures also showed similarity between the measured and the predicted results, as shown in Table 6.

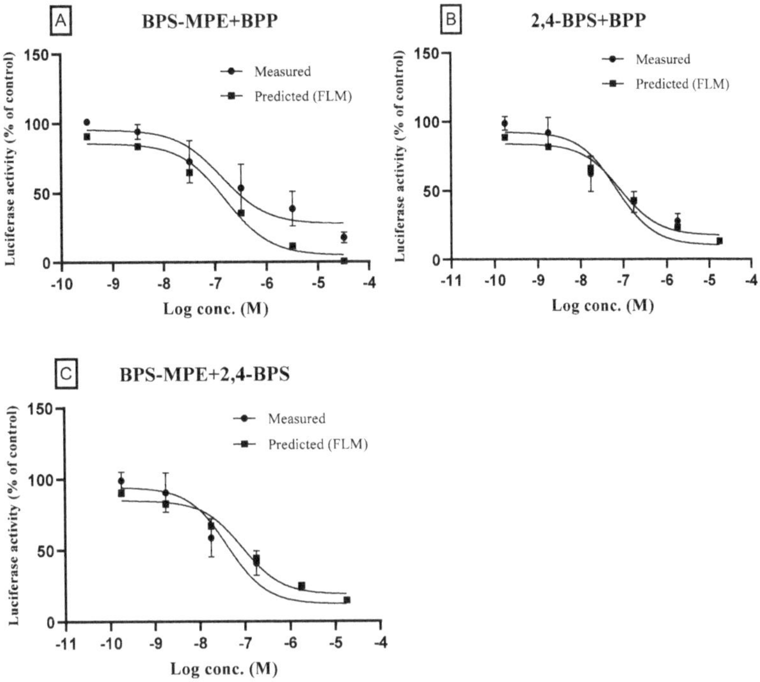

Figure 5. Effects of binary mixtures (Antagonist + Antagonist) on the estrogen receptor transcriptional activity: a comparison of the measured and the predicted values. For the test concentrations of binary mixtures, the highest concentration (master solution) was achieved by mixing the individual chemicals where the final concentration of each chemical was designed to be 100-fold IC_{50}, respectively, for full concentration–response curve. The solution of the highest concentration was then diluted for the measurement of antagonistic activity. Test wells were spiked with 25 pM 17β-estradiol. The normalized mean values of the luminescence signal (Luciferase activity) of the test wells were divided by that of the spiked-in control (100%) to obtain the relative transcriptional activity of the binary mixture. The measured results were shown with the predicted ones obtained by the full logistic model. The compositions of binary mixtures were as follows: (**A**): BPS-MPE + BPP; (**B**): 2,4-BPS + BPP; (**C**): BPS-MPE + 2,4-BPS. Three separate experiments were performed in the transactivation assay and data were represented as mean ± standard error.

Table 6. Parameters obtained from concentration–response curves of ternary mixtures for the estrogen receptor transactivation.

Chemicals	RPC_{Max} (%)		PC_{10} (Log M)		PC_{50} (Log M)		IC_{30} (Log M)		IC_{50} (Log M)	
	Meas	Pred	Meas	Pred	Meas	Pred	Meas	Pred	Meas	Pred
BPA + BPAF + BPZ	106.1	98.2	−8.50	−7.95	−6.33	−6.61	-	-	-	-
BPAF + BPZ + BPB	104.1	95.4	−8.00	−7.73	−5.60	−5.17	-	-	-	-
BPB + BPC + BPE	97.9	89.7	−8.00	−7.23	−5.03	−5.15	-	-	-	-
BPC + BPE + BPA	129.8	127.0	−8.53	−8.61	−6.40	−6.94	-	-	-	-
BPZ + BPB + BPC	105.8	108.7	−8.26	−7.51	−5.48	−5.40	-	-	-	-
BPS-MPE + 2,4-BPS + BPP	97.6	89.1	-	-	-	-	−7.65	−7.91	−6.71	−7.00

Meas: measured values; Pred: predicted values; PC: positive concentration; IC: inhibition concentration; RPC_{Max}: maximum level of response. The parameters were obtained using the spread sheet provided by the OECD Test guideline.

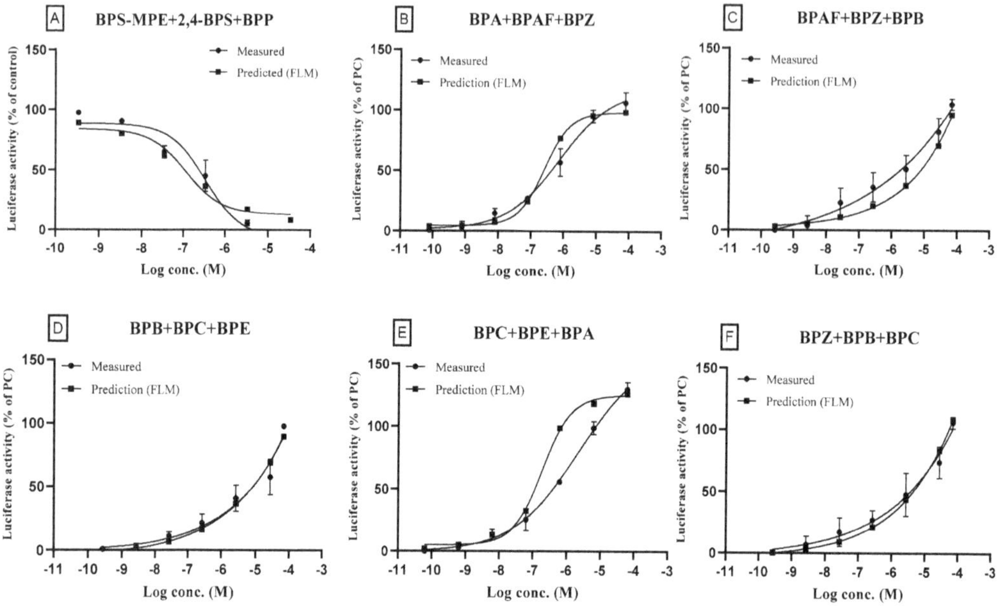

Figure 6. Effects of ternary mixtures on the estrogen receptor transcriptional activity: a comparison of the measured results and the predicted results. For the test concentrations of ternary mixtures, the highest test concentration was achieved by mixing the individual chemicals where the final concentration of each chemical was designed to be 100-fold EC_{50}, respectively, for the full concentration–response curve. In the case of BPB, the EC_{50} was too high to prepare 100-fold mixtures due to cytotoxicity at the master solution. So, the mixture containing BPB was designed to be 0.25-fold in ternary mixture. The concentration of component chemicals in master solutions was designed for the full concentration–response curve. The solution of the highest test concentration was then diluted for the measurement and prediction of ERα transactivation. The components of ternary mixtures for antagonists (**A**) and agonists (**B**–**F**) are described in the figures. The measured results of concentration–response curves are shown with the predicted ones obtained by the full logistic model. Three separate experiments were performed in the transcription assay and data were represented as mean ± standard error.

4. Discussion

Although the substitutes of BPA have been developed for safety or more useful products, these analogs also have adverse effects on the endocrine system. BPA and BPAF have been shown to activate the ERα and ERβ, while BPA (>100 nM) has a strong effect compared to BPAF (>1000 nM) [39]. On the ER transcriptional assay performed by Liu et al. in another study, BPAF, BPB, BPC, BPE, and BPZ were reported to function as full activators or agonists for ERα, but they completely inhibited the ERβ [40,41].

In our previous study, BPA, BPF, and BPS showed strong ERα transcriptional activity [29]. BPB has been demonstrated to be anti-androgenic as well as estrogenic and was also found to exert higher estrogenic activity than BPA [42,43]. In this study, as shown in Table 4, the RPC_{Max} of the BPC was higher and PC_{50} was lower than those of BPA, which supports the strong estrogenic activity of BPC. The estrogenic activity was in this order, BPC > BPE > BPZ > BPB > BPAF > BPA when they were compared based on the RPC_{Max}. It suggested that the BPA substitutes may also have strong endocrine-disrupting effects. Other BPA analogs including BPP, BPS-MPE, and 2,4-BPS did not show agonistic effects on the ERα transactivation but showed antagonistic effects (Figures 2D and 3).

Compared to BPS, only a few studies on 2,4-BPS and BPS-MPE have been reported. BPS is known to have strong agonistic activity in ERα transactivation, and the RPC_{Max} of ERα transactivation was shown to be 72% [32]. However, BPS-MPE and 2,4-BPS, which are the structural analogs of BPS, did not show agonistic activity but showed antagonistic activity in this study. The antagonistic activity of 2,4-BPS was stronger than that of BPS-MPE, compared based on $LogIC_{50}$, as shown in Table 4. Further investigations on the differences in the effects of BPS and 2,4-BPS are necessary because the two compounds are structurally similar and are only different in the position of the hydroxyl group. BPP also has antagonistic activity. Only a few reports for the BPP are available on the endocrine-disrupting effects [44].

Recently, an association study on the combined exposures to bisphenol compounds and endocrine disruption has been reported, where the concentration of BPA, BPF, and BPS in the urine of pregnant women was analyzed and the birth outcome was investigated [45]. Furthermore, it has been reported that the hazard index for the deterioration of semen quality might be increased by the combined exposure to a mixture of BPA, BPF, and BPS [46]. A study on the association of bisphenol compounds with odds of polycystic ovarian syndrome (PCOS) in women of childbearing age, showed that mixed exposure to seven bisphenol compounds (BPA, BPAP, BPAF, BPB, BPS, BPP, and BPZ) was found to be positively associated with the odds of PCOS (odds ratio = 1.26), while the odds ratio of single bisphenol compounds was less than 1 [47]. Although the independent action of bisphenol compound has been the main principle in mixture toxicity, co-exposure to these chemicals may induce hazardous health effects potentially associated with a complex body burden of different origins [25].

In our study, binary or ternary mixtures of bisphenol A and its structural analogs were evaluated for the interaction of single compounds on the ERα transactivation. When the mixtures were prepared and tested, they showed additive activity in both the agonists/agonist and antagonists/antagonist preparations in the test concentrations (Figures 4–6). There were no significant differences between the measured values and the predicted values. The MDR values were more than 0.5 and less than 2, which indicated the additive activity. Many chemicals with estrogenic activity could be contained in consumer products and may be exposed to humans simultaneously. Therefore, the mixture toxicity of estrogenic chemicals in consumer products needs to be assessed. However, the combinations of chemicals, the number of compounds, and the concentration of individual components in the mixture are too numerous to test efficiently. According to our study, we found that the measured ERα transcription activity results closely matched the calculated results obtained from FLM. This suggests that it may not be necessary to conduct a separate test for each combined mixture. Understanding the parameters of concentration–response curves for each chemical component, however, allows one to determine the transcription activity of multiple mixtures. Additionally, the in silico model used in this study may provide a good tool for the prediction of an in vitro transcription assay.

5. Conclusions

Bisphenol A and its eight substitutes, BPAF, BPB, BPC, BPE, BPP, BPS-MPE, BPZ, and 2,4-BPS, which are used in the manufacture of industrial and consumer products were selected and evaluated for the transactivation of estrogen receptor alpha (ER α). The test methods were based on the OECD TG 455. Among the chemicals, six compounds were found to be agonistic, and three compounds were antagonistic to the ERα transcription activity. Binary and ternary mixtures including different combinations of agonist/agonist and antagonist/antagonist were prepared on an equi-effect concentration basis and tested for ERα transcription activity. The activity of the mixtures showed concentration-dependent responses. The concentration–response curves were also predicted by using FLM. The measured and the predicted response curves showed close similarities. Overall results suggest that the effect of bisphenol compound is additive in binary and ternary mixtures.

Author Contributions: H.L.: test for the ERα/AR transactivation, data process, drawing figures and tables; J.P.: analysis of data, prediction using FLM, data process; K.P.: test design, reviewing the data and analysis, thesis writing, and revision. All authors have read and agreed to the published version of the manuscript.

Funding: This research was funded by the Korea Ministry of Environment (MOE) Project No. 2020002960006 and Project No. RS-2023-00215856.

Institutional Review Board Statement: Not applicable.

Informed Consent Statement: Not applicable.

Data Availability Statement: The data presented in this study are available on request from the corresponding author.

Conflicts of Interest: The authors declare no conflict of interest.

References

1. Chen, D.; Kannan, K.; Tan, H.; Zheng, Z.; Feng, Y.L.; Wu, Y.; Widelka, M. Bisphenol Analogues Other Than BPA: Environmental Occurrence, Human Exposure, and Toxicity-A Review. *Environ. Sci. Technol.* **2016**, *50*, 5438–5453. [CrossRef] [PubMed]
2. Abraham, A.; Chakraborty, P. A review on sources and health impacts of bisphenol A. *Rev. Environ. Health* **2020**, *35*, 201–210. [CrossRef] [PubMed]
3. Vandenberg, L.N.; Hauser, R.; Marcus, M.; Olea, N.; Welshons, W. Human exposure to bisphenol A (BPA). *Reprod. Toxicol.* **2007**, *24*, 139–177. [CrossRef] [PubMed]
4. Rochester, J.R. Bisphenol A and human health: A review of the literature. *Reprod. Toxicol.* **2013**, *42*, 132–155. [CrossRef] [PubMed]
5. vom Saal, F.S.; Vandenberg, L.N. Update on the Health Effects of Bisphenol A: Overwhelming Evidence of Harm. *Endocrinology* **2020**. [CrossRef] [PubMed]
6. Meenu Maniradhan, M.; Calivarathan, L. Bisphenol A-Induced Endocrine Dysfunction and its Associated Metabolic Disorders. *Endocr. Metab. Immune Disord. -Drug Targets* **2023**, *23*, 515–529. [CrossRef] [PubMed]
7. Costa, H.E.; Cairrao, E. Effect of bisphenol A on the neurological system: A review update. *Arch. Toxicol.* **2023**. [CrossRef]
8. Guo, J.; Liu, K.; Yang, J.; Su, Y. Prenatal exposure to bisphenol A and neonatal health outcomes: A systematic review. *Environ. Pollut.* **2023**, *335*, 122295. [CrossRef]
9. Fonseca, M.I.; Lorigo, M.; Cairrao, E. Endocrine-Disrupting Effects of Bisphenol A on the Cardiovascular System: A Review. *J. Xenobiot.* **2022**, *12*, 181–213. [CrossRef]
10. Rocca, Y.D.; Traini, E.M.; Diomede, F.; Fonticoli, L.; Trubiani, O.; Paganelli, A.; Pizzicannella, J.; Marconi, G.D. Current Evidence on Bisphenol A Exposure and the Molecular Mechanism Involved in Related Pathological Conditions. *Pharmaceutics* **2023**, *15*, 908. [CrossRef]
11. Ma, Y.; Liu, H.; Wu, J.; Yuan, L.; Wang, Y.; Du, X.; Wang, R.; Marwa, P.W.; Petlulu, P.; Chen, X.; et al. The adverse health effects of bisphenol A and related toxicity mechanisms. *Environ. Res.* **2019**, *176*, 108575. [CrossRef] [PubMed]
12. Sabour, A.; Helou, M.E.; Roger-Leroi, V.; Bauer, C. Release and toxicity of bisphenol-A (BPA) contained in orthodontic adhesives: A systematic review. *Int. Orthod.* **2021**, *19*, 1–14. [CrossRef] [PubMed]
13. Moon, M.K. Concern about the Safety of Bisphenol A Substitutes. *Diabetes Metab. J.* **2019**, *43*, 46–48. [CrossRef] [PubMed]
14. Keminer, O.; Teigeler, M.; Kohler, M.; Wenzel, A.; Arning, J.; Kaßner, F.; Windshügel, B.; Eilebrecht, E. A tiered high-throughput screening approach for evaluation of estrogen and androgen receptor modulation by environmentally relevant bisphenol A substitutes. *Sci. Total Environ.* **2020**, *717*, 134743. [CrossRef] [PubMed]
15. Delfosse, V.; Grimaldi, M.; Pons, J.; Boulahtouf, A.; Maire, A.; Vincent Cavailles, V.; Labesse, G.; Bourguet, W.; Balaguer, P. Structural and mechanistic insights into bisphenols action provide guidelines for risk assessment and discovery of bisphenol A substitutes. *PNAS* **2012**, *109*, 4930–14935. [CrossRef]
16. Rochester, J.R.; Bolden, A.L. Bisphenol S and F: A Systematic Review and Comparison of the Hormonal Activity of Bisphenol A Substitutes. *Environ. Health Perspect.* **2015**, *123*, 643–650. [CrossRef]
17. Qiu, W.; Zhan, H.; Hu, J.; Zhang, T.; Xu, H.; Wong, M.; Xu, B.; Zheng, C. The occurrence, potential toxicity, and toxicity mechanism of bisphenol S, a substitute of bisphenol A: A critical review of recent progress. *Ecotoxicol. Environ. Saf.* **2019**, *173*, 192–202. [CrossRef]
18. Téteau, O.; Vitorino Carvalho, A.; Papillier, P.; Mandon-Pépin, B.; Jouneau, L.; Jarrier-Gaillard, P.; Desmarchais, A.; Lebachelier de la Riviere, M.E.; Vignault, C.; Maillard, V.; et al. Bisphenol A and bisphenol S both disrupt ovine granulosa cell steroidogenesis but through different molecular pathways. *J. Ovarian Res.* **2023**, *16*, 30. [CrossRef]
19. Kitamura, S.; Suzuki, T.; Sanoh, S.; Kohta, R.; Jinno, N.; Sugihara, K.; Yoshihara, S.; Fujimoto, N.; Watanabe, H.; Ohta, S. Comparative study of the endocrine-disrupting activity of bisphenol A and 19 related compounds. *Toxicol. Sci.* **2005**, *84*, 249–259. [CrossRef]
20. Liang, S.; Yin, L.; Yu, S.K.; Hofmann, M.C.; Yu, X. High-Content Analysis Provides Mechanistic Insights into the Testicular Toxicity of Bisphenol A and Selected Analogues in Mouse Spermatogonial Cells. *Toxicol. Sci.* **2017**, *155*, 43–60. [CrossRef]

21. Gély, C.A.; Lacroix, M.Z.; Roques, B.B.; Toutain, P.L.; Gayrard, V.; Picard-Hagen, N. Comparison of toxicokinetic properties of eleven analogues of Bisphenol A in pig after intravenous and oral administrations. *Environ. Int.* **2023**, *171*, 107722. [CrossRef] [PubMed]
22. In, S.; Yoon, H.W.; Yoo, J.W.; Cho, H.; Kim, R.O.; Lee, Y.M. Acute toxicity of bisphenol A and its structural analogues and transcriptional modulation of the ecdysone-mediated pathway in the brackish water flea Diaphanosoma celebensis. *Ecotoxicol. Environ. Saf.* **2019**, *179*, 310–317. [CrossRef] [PubMed]
23. Hend, F.; Alharbi, H.F.; Algonaiman, R.; Alduwayghiri, R.; Aljutaily, T.; Algheshairy, R.M.; Almutairi, A.S.; Alhar, R.M. Exposure to Bisphenol A Substitutes, Bisphenol S and Bisphenol F, and Its Association with Developing Obesity and Diabetes Mellitus: A Narrative Review. *Int. J. Environ. Res. Public Health* **2022**, *19*, 15918. [CrossRef]
24. Mendy, A.; Salo, P.M.; Wilkerson, J.; Feinstein, L.; Ferguson, K.K.; Fessler, M.B.; Thorne, P.S.; Zeldin, D.C. Association of urinary levels of bisphenols F and S used as bisphenol A substitutes with asthma and hay fever outcomes. *Environ. Res.* **2020**, *183*, 108944. [CrossRef]
25. Ribeiro, E.; Ladeira, C.; Viegas, S. EDCs Mixtures: A Stealthy Hazard for Human Health? *Toxics* **2017**, *5*, 5. [CrossRef]
26. Park, C.; Song, H.; Choi, J.; Sim, S.; Kojima, H.; Park, J.; Iida, M.; Lee, Y. The mixture effects of bisphenol derivatives on estrogen receptor and androgen receptor. *Environ. Pollut.* **2020**, *260*, 114036. [CrossRef] [PubMed]
27. Skledar, D.G.; Mašič, L.P. In vitro estrogenic activity of binary and multicomponent mixtures with bisphenol A. *Sci. Total Environ.* **2020**, *707*, 135211. [CrossRef]
28. Mao, W.; Song, Y.; Sui, H.; Cao, P.; Liu, Z. Analysis of individual and combined estrogenic effects of bisphenol, nonylphenol and diethylstilbestrol in immature rats with mathematical models. *Environ. Health Prev. Med.* **2019**, *24*, 32. [CrossRef]
29. Baralić, K.; Buha Djordjevic, A.; Živančević, K.; Antonijević, E.; Anđelković, M.; Javorac, D.; Ćurčić, M.; Bulat, Z.; Antonijević, B.; Đukić-Ćosić, D. Toxic Effects of the Mixture of Phthalates and Bisphenol A-Subacute Oral Toxicity Study in Wistar Rats. *Int J. Environ. Res. Public Health* **2020**, *17*, 746. [CrossRef]
30. Yang, J.; Liao, A.; Hu, S.; Zheng, Y.; Liang, S.; Han, S.; Lin, Y. Acute and Chronic Toxicity of Binary Mixtures of Bisphenol A and Heavy Metals. *Toxics* **2022**, *10*, 255. [CrossRef]
31. OECD. *OECD TG 455. Performance-Based Test Guideline for Stably Transfected Transactivation In Vitro Assays to Detect Estrogen Receptor Agonists and Antagonists*; OECD: Paris, France, 2021.
32. Lee, H.; Park, J.; Park, K. Effects of consumer products chemicals ingredients and their mixtures on the estrogen receptor/androgen receptor transcriptional activation. *Chemosphere* **2022**, *302*, 134866. [CrossRef] [PubMed]
33. Yamasaki, K.; Takeyoshi, M.; Yakabe, Y.; Sawaki, M.; Imatanaka, N.; Takatsuki, M. Comparison of reporter gene assay and immature rat uterotrophic assay of twenty-three chemicals. *Toxicology* **2002**, *170*, 21–30. [CrossRef] [PubMed]
34. Ezechiáš, M.; Cajthaml, T. Novel full logistic model for estimation of the estrogenic activity of chemical mixtures. *Toxicology* **2016**, *359–360*, 58–70. [CrossRef] [PubMed]
35. Ezechiáš, M.; Cajthaml, T. Receptor partial agonism and method to express receptor partial activation concerning novel Full Logistic Model of mixture toxicology. *Toxicology* **2018**, *393*, 26–33. [CrossRef] [PubMed]
36. Michalíková, K.; Linhartová, L.; Ezechiáš, M.; Cajthaml, T. Assessment of agonistic and antagonistic properties of widely used oral care antimicrobial substances toward steroid estrogenic and androgenic receptors. *Chemosphere* **2018**, *217*, 534–541. [CrossRef] [PubMed]
37. Owczarek, K.; Kudłak, B.; Simeonov, V.; Mazerska, Z.; Namieśnik, J. Binary Mixtures of Selected Bisphenols in the Environment: Their Toxicity in Relationship to Individual Constituents. *Molecules* **2018**, *23*, 3226. [CrossRef] [PubMed]
38. Kudłak, B.; Wieczerzak, M.; Namieśnik, J. Bisphenols (A, S, and F) affect the basic hormonal activity determined for pharmaceuticals—Study of Saccharomyces cerevisiae. *Environ. Pollut.* **2019**, *246*, 914–920. [CrossRef]
39. Li, Y.; Burns, K.A.; Arao, Y.; Luh, C.J.; Korach, K.S. Differential estrogenic actions of endocrine-disrupting chemicals bisphenol A, bisphenol AF, and zearalenone through estrogen receptor α and β in vitro. *Environ. Health Perspect.* **2012**, *120*, 1029–1035. [CrossRef]
40. Liu, X.; Suyama, K.; Nose, T.; Shimohigashi, M.; Shimohigashi, Y. Bisphenol-C is the strongest bifunctional ERα-agonist and ERβ-antagonist due to magnified halogen bonding. *PLoS ONE* **2021**, *16*, e0246583. [CrossRef]
41. Liu, X.; Matsuyama, Y.; Shimohigashi, M.; Shimohigashi, Y. Toxicol ERα-agonist and ERβ-antagonist bifunctional next-generation bisphenols with no halogens: BPAP, BPB, and BPZ. *Toxicol. Lett.* **2021**, *345*, 24–33. [CrossRef]
42. Cao, L.Y.; Ren, X.M.; Li, C.H.; Zhang, J.; Qin, W.P.; Yang, Y.; Wan, B.; Guo, L.H. Bisphenol AF and Bisphenol B Exert Higher Estrogenic Effects than Bisphenol A via G Protein-Coupled Estrogen Receptor Pathway. *Environ. Sci. Technol.* **2017**, *51*, 11423–11430. [CrossRef] [PubMed]
43. Usman, A.; Ikhlas, S.; Ahmad, M. Occurrence, toxicity and endocrine disrupting potential of Bisphenol B and Bisphenol F: A mini-review. *Toxicol. Lett.* **2019**, *312*, 222–227. [CrossRef] [PubMed]
44. Wang, L.; Zhou, L.; Fan, D.; Wang, Z.; Gu, W.; Shi, L.; Liu, J.; Yang, J. Bisphenol P activates hormonal genes and introduces developmental outcomes in Chironomus tentans. *Ecotoxicol. Environ. Saf.* **2019**, *174*, 675–682. [CrossRef] [PubMed]
45. Kim, S.; Park, E.; Park, E.K.; Lee, S.; Kwon, J.A.; Shin, B.H.; Kang, S.; Park, E.Y.; Kim, B. Urinary Concentrations of Bisphenol Mixtures during Pregnancy and Birth Outcomes: The MAKE Study. *Int. J. Environ. Res. Public. Health.* **2021**, *18*, 10098. [CrossRef]

46. Kortenkamp, A.; Martin, O.; Ermler, S.; Baig, A.; Scholze, M. Bisphenol A and declining semen quality: A systematic review to support the derivation of a reference dose for mixture risk assessments. *Int. J. Hyg. Environ. Health* **2022**, *241*, 113942. [CrossRef]
47. Zhan, W.; Tang, W.; Shen, X.; Xu, H.; Zhang, J. Exposure to bisphenol A and its analogs and polycystic ovarian syndrome in women of childbearing age: A multicenter case-control study. *Chemosphere* **2023**, *313*, 137463. [CrossRef]

MDPI
St. Alban-Anlage 66
4052 Basel
Switzerland
www.mdpi.com

MDPI Books Editorial Office
E-mail: books@mdpi.com
www.mdpi.com/books

www.ingramcontent.com/pod-product-compliance
Lightning Source LLC
Chambersburg PA
CBHW061502220926
51850CB00004B/149
* 9 7 8 3 0 3 6 5 9 9 8 3 0 *